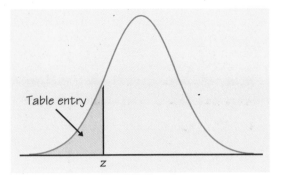

Table entry

Table entry for z is the area under the standard normal curve to the left of z.

TABLE A   Standard normal probabilities

z	.00	.01	.02	.03	.04	.05	.06	.07	.08	.09
−3.4	.0003	.0003	.0003	.0003	.0003	.0003	.0003	.0003	.0003	.0002
−3.3	.0005	.0005	.0005	.0004	.0004	.0004	.0004	.0004	.0004	.0003
−3.2	.0007	.0007	.0006	.0006	.0006	.0006	.0006	.0005	.0005	.0005
−3.1	.0010	.0009	.0009	.0009	.0008	.0008	.0008	.0008	.0007	.0007
−3.0	.0013	.0013	.0013	.0012	.0012	.0011	.0011	.0011	.0010	.0010
−2.9	.0019	.0018	.0018	.0017	.0016	.0016	.0015	.0015	.0014	.0014
−2.8	.0026	.0025	.0024	.0023	.0023	.0022	.0021	.0021	.0020	.0019
−2.7	.0035	.0034	.0033	.0032	.0031	.0030	.0029	.0028	.0027	.0026
−2.6	.0047	.0045	.0044	.0043	.0041	.0040	.0039	.0038	.0037	.0036
−2.5	.0062	.0060	.0059	.0057	.0055	.0054	.0052	.0051	.0049	.0048
−2.4	.0082	.0080	.0078	.0075	.0073	.0071	.0069	.0068	.0066	.0064
−2.3	.0107	.0104	.0102	.0099	.0096	.0094	.0091	.0089	.0087	.0084
−2.2	.0139	.0136	.0132	.0129	.0125	.0122	.0119	.0116	.0113	.0110
−2.1	.0179	.0174	.0170	.0166	.0162	.0158	.0154	.0150	.0146	.0143
−2.0	.0228	.0222	.0217	.0212	.0207	.0202	.0197	.0192	.0188	.0183
−1.9	.0287	.0281	.0274	.0268	.0262	.0256	.0250	.0244	.0239	.0233
−1.8	.0359	.0351	.0344	.0336	.0329	.0322	.0314	.0307	.0301	.0294
−1.7	.0446	.0436	.0427	.0418	.0409	.0401	.0392	.0384	.0375	.0367
−1.6	.0548	.0537	.0526	.0516	.0505	.0495	.0485	.0475	.0465	.0455
−1.5	.0668	.0655	.0643	.0630	.0618	.0606	.0594	.0582	.0571	.0559
−1.4	.0808	.0793	.0778	.0764	.0749	.0735	.0721	.0708	.0694	.0681
−1.3	.0968	.0951	.0934	.0918	.0901	.0885	.0869	.0853	.0838	.0823
−1.2	.1151	.1131	.1112	.1093	.1075	.1056	.1038	.1020	.1003	.0985
−1.1	.1357	.1335	.1314	.1292	.1271	.1251	.1230	.1210	.1190	.1170
−1.0	.1587	.1562	.1539	.1515	.1492	.1469	.1446	.1423	.1401	.1379
−0.9	.1841	.1814	.1788	.1762	.1736	.1711	.1685	.1660	.1635	.1611
−0.8	.2119	.2090	.2061	.2033	.2005	.1977	.1949	.1922	.1894	.1867
−0.7	.2420	.2389	.2358	.2327	.2296	.2266	.2236	.2206	.2177	.2148
−0.6	.2743	.2709	.2676	.2643	.2611	.2578	.2546	.2514	.2483	.2451
−0.5	.3085	.3050	.3015	.2981	.2946	.2912	.2877	.2843	.2810	.2776
−0.4	.3446	.3409	.3372	.3336	.3300	.3264	.3228	.3192	.3156	.3121
−0.3	.3821	.3783	.3745	.3707	.3669	.3632	.3594	.3557	.3520	.3483
−0.2	.4207	.4168	.4129	.4090	.4052	.4013	.3974	.3936	.3897	.3859
−0.1	.4602	.4562	.4522	.4483	.4443	.4404	.4364	.4325	.4286	.4247
−0.0	.5000	.4960	.4920	.4880	.4840	.4801	.4761	.4721	.4681	.4641

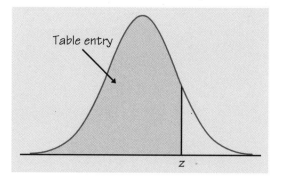

Table entry

Table entry for z is the area under the standard normal curve to the left of z.

Standard normal probabilities (*continued*)

z	.00	.01	.02	.03	.04	.05	.06	.07	.08	.09
0.0	.5000	.5040	.5080	.5120	.5160	.5199	.5239	.5279	.5319	.5359
0.1	.5398	.5438	.5478	.5517	.5557	.5596	.5636	.5675	.5714	.5753
0.2	.5793	.5832	.5871	.5910	.5948	.5987	.6026	.6064	.6103	.6141
0.3	.6179	.6217	.6255	.6293	.6331	.6368	.6406	.6443	.6480	.6517
0.4	.6554	.6591	.6628	.6664	.6700	.6736	.6772	.6808	.6844	.6879
0.5	.6915	.6950	.6985	.7019	.7054	.7088	.7123	.7157	.7190	.7224
0.6	.7257	.7291	.7324	.7357	.7389	.7422	.7454	.7486	.7517	.7549
0.7	.7580	.7611	.7642	.7673	.7704	.7734	.7764	.7794	.7823	.7852
0.8	.7881	.7910	.7939	.7967	.7995	.8023	.8051	.8078	.8106	.8133
0.9	.8159	.8186	.8212	.8238	.8264	.8289	.8315	.8340	.8365	.8389
1.0	.8413	.8438	.8461	.8485	.8508	.8531	.8554	.8577	.8599	.8621
1.1	.8643	.8665	.8686	.8708	.8729	.8749	.8770	.8790	.8810	.8830
1.2	.8849	.8869	.8888	.8907	.8925	.8944	.8962	.8980	.8997	.9015
1.3	.9032	.9049	.9066	.9082	.9099	.9115	.9131	.9147	.9162	.9177
1.4	.9192	.9207	.9222	.9236	.9251	.9265	.9279	.9292	.9306	.9319
1.5	.9332	.9345	.9357	.9370	.9382	.9394	.9406	.9418	.9429	.9441
1.6	.9452	.9463	.9474	.9484	.9495	.9505	.9515	.9525	.9535	.9545
1.7	.9554	.9564	.9573	.9582	.9591	.9599	.9608	.9616	.9625	.9633
1.8	.9641	.9649	.9656	.9664	.9671	.9678	.9686	.9693	.9699	.9706
1.9	.9713	.9719	.9726	.9732	.9738	.9744	.9750	.9756	.9761	.9767
2.0	.9772	.9778	.9783	.9788	.9793	.9798	.9803	.9808	.9812	.9817
2.1	.9821	.9826	.9830	.9834	.9838	.9842	.9846	.9850	.9854	.9857
2.2	.9861	.9864	.9868	.9871	.9875	.9878	.9881	.9884	.9887	.9890
2.3	.9893	.9896	.9898	.9901	.9904	.9906	.9909	.9911	.9913	.9916
2.4	.9918	.9920	.9922	.9925	.9927	.9929	.9931	.9932	.9934	.9936
2.5	.9938	.9940	.9941	.9943	.9945	.9946	.9948	.9949	.9951	.9952
2.6	.9953	.9955	.9956	.9957	.9959	.9960	.9961	.9962	.9963	.9964
2.7	.9965	.9966	.9967	.9968	.9969	.9970	.9971	.9972	.9973	.9974
2.8	.9974	.9975	.9976	.9977	.9977	.9978	.9979	.9979	.9980	.9981
2.9	.9981	.9982	.9982	.9983	.9984	.9984	.9985	.9985	.9986	.9986
3.0	.9987	.9987	.9987	.9988	.9988	.9989	.9989	.9989	.9990	.9990
3.1	.9990	.9991	.9991	.9991	.9992	.9992	.9992	.9992	.9993	.9993
3.2	.9993	.9993	.9994	.9994	.9994	.9994	.9994	.9995	.9995	.9995
3.3	.9995	.9995	.9995	.9996	.9996	.9996	.9996	.9996	.9996	.9997
3.4	.9997	.9997	.9997	.9997	.9997	.9997	.9997	.9997	.9997	.9998

The Practice of Statistics

The Practice of Statistics

TI-83 Graphing Calculator Enhanced

Daniel S. Yates
Virginia Satellite Educational Network and Henrico County Schools

David S. Moore
Purdue University

George P. McCabe
Purdue University

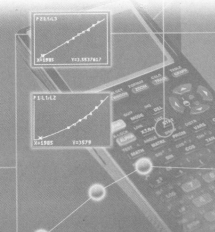

W. H. Freeman and Company
New York

Acquisitions editor: Patrick Farace
Publisher: Michelle Russel Julet
Project editor: Diane Cimino Maass
Marketing manager: Kimberly Manzi
Cover computer art: Salem Krieger
Text and cover designer: Blake Logan
Illustration coordinator: Bill Page
Illustrations: Publication Services
Production coordinator: Susan Wein
Composition: Publication Services
Manufacturing: R R Donnelly & Sons Company

Cover illustration: Line art of the TI-83 graphing calculator used by permission of Texas Instruments Incorporated.

TI-83 screens are used with permission of the publisher. Copyright © 1996, Texas Instruments, Incorporated.

Library of Congress Cataloging-in-Publication Data

Yates, Dan (Daniel S.)
    The practice of statistics : TI-83 Graphing Calculator Enhanced / Dan
Yates, David S. Moore, George P. McCabe.
        p.   cm.
    Includes bibliographical references and index.
    ISBN 0-7167-3370-6
    1. Mathematical statistics.   I. Moore, David S.   II. McCabe,
George P.   III. Title.
QA276.12.Y37   1998
519.5—dc21   98-24192
                                                                            CIP

Printed in the United States of America

Fourth printing, 2000

Contents

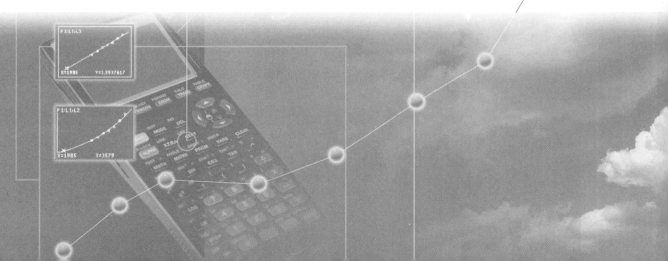

Preface

The Practice of Statistics: TI-83 Graphing Calculator Enhanced is intended for use by instructors whose preferred technology is the TI-83 calculator, both in colleges and universities and in Advanced Placement Statistics in secondary schools. It is based on the successful college texts *The Basic Practice of Statistics* (BPS) by David Moore and *Introduction to the Practice of Statistics* (IPS) by David Moore and George McCabe. This text also includes additional material drawn from several other W. H. Freeman and Company titles including *Statistics: Concepts and Controversies*, and *The Active Practice of Statistics*, both by David Moore.

The Advanced Placement Statistics syllabus as described by the College Board* resembles these texts in spirit and content, and is perhaps the best current expression of the general consensus among statisticians about the nature of a modern introductory statistics course. This text therefore follows the AP syllabus quite closely. In particular, the simulation capabilities of the TI-83 allow a more effective presentation of probability than is possible without this technology.

Guiding Principles

The American Statistical Association and the Mathematical Association of America recently formed a joint committee to study the teaching of introductory statistics. The following are their main recommendations:

- Emphasize statistical thinking
- Present more data and concepts with less theory and fewer recipes
- Foster active learning

The Practice of Statistics, like *BPS* and *IPS*, follows these principles. It takes advantage of the TI-83 to put more emphasis on active learning. Although the book is elementary in the level of mathematics required and in the statistical procedures presented, it aims to give students an understanding of the main ideas of statistics and useful skills for working with data. Examples and exercises, though intended for beginners, use real data and give enough background to allow students to consider the meaning of their calculations. Using the TI-83, for which full instruction is included in the text, allows students to focus on concepts and problem-solving rather than on calculations that are now automated.

Features

The following features of *The Practice of Statistics* indicate the similarities and differences between this text and the other Moore/McCabe books:

* **Additional topics** that appear on the 1998 AP Statistics Course Description but which are not included in *BPS* or *IPS* have been included. Some examples include transformations of non-linear data to include exponential and power regression; simulations; sufficient probability theory to provide a foundation for inference (as in *IPS*); geometric distributions; and goodness of fit. Several additional topics, such as normal probability plots, have been included because they are useful for checking test assumptions and can be easily done with the TI-83 graphing calculator. Other topics, such as Type I and Type II errors and power, have been added in anticipation of their inclusion on the AP syllabus.

* The **narrative, examples, and exercises** follow the acclaimed Moore/McCabe tradition. The text remains student-friendly and continues to promote student understanding and success. Most important, the **readability** of David Moore's books has been preserved, and the methodical sequencing of topics has not been disturbed.

* The **TI-83 graphing calculator** is the technology of choice and has been incorporated throughout the text. Step-by-step keyboarding instructions and many sample screens are provided.

* A special section has been added in Chapter 5, "Producing Data," to introduce the use of **simulations.** This section, which discusses the imitation of chance behavior, comes before we begin studying probability in Chapter 6. Simulations are accomplished using the random number table, spinners, and the graphing calculator. There are a plethora of examples and exercises involving simulations in this section and many more throughout the rest of the book to teach students statistical concepts with

a better level of understanding. This icon next to a problem indicates that it is a simulation exercise. There are approximately 60 of these in the text. A series of example TI-83 programs are included to carry out large numbers of repetitions. These simple, short programs can be easily modified by the student to simulate other phenomena in other problem settings.

* Some longer **chapters in *IPS* and *BPS* have been divided and topics realigned** to provide for shorter instructional blocks and more frequent student evaluation. For example, Chapter 1 in *IPS* and *BPS* has been split into the first two chapters of this text to present better the fundamental statistical ideas to high school students.

* Each chapter begins with a highly motivating active student investigation, and several exercises in the chapter development build on this **Activity,** which allows students to play an active role in learning statistical concepts.

* Each main idea is followed by a short section of exercises for **immediate reinforcement.** Each chapter review includes a list of specific skills so that students can check their learning.

Teacher Resource Binder

A very substantial teacher support package in the form of a Teacher Resource Binder (TRB) will be available to adopters. The TRB includes sample section quizzes, chapter tests, and semester exams, all intended to provide the teacher with starting points for student assessment. All these assessment items have been classroom tested. The TRB includes teaching recommendations and additional active student investigations by chapter, and a complete solutions guide. The TRB also contains suggestions by chapter for more in-depth investigations I call "Special Problems," which might culminate in a short (5–7 page) written report. These Special Problems, if assigned as a series of mini-term papers, could help students develop the written communication skills required on the AP examination. The Resources section of the TRB has an extensive list of references to book supplements and journal articles, video series, computer software, sources of data in both print and electronic formats, interesting statistics-related Web sites, and more.

Acknowledgments

First, I am extremely grateful to W. H. Freeman and Company for allowing me to undertake this project. They have been faithful and patient as I have worked feverishly to finish it on time. A particular thank you goes to: associate editor Patrick Farace; project editor Diane Cimino Maass; production coordinator Susan Wein; and copy editor Pamela Bruton.

I will always be grateful to Professors David Moore and George McCabe for their *Introduction to the Practice of Statistics,* which I have used as a text at both the college and high school levels. I have long regarded this book as the quintessential introductory statistics text, and it has made a major impact on my thinking about statistics and my own teaching of the subject. It is truly an honor for me to be allowed to have free reign with so much high quality text material to produce a text custom designed for this new Advanced Placement Statistics course.

I am especially grateful to my colleague Dr. Christopher Barat, Associate Professor at Virginia State University, who helped me get the manuscript ready for publication. To the colleagues from high schools, colleges, and universities across the country who took their time to review chapters of the manuscript and offer very helpful suggestions, I offer a sincere thank you.

Beverly Austin
Abraham Lincoln High School
San Francisco, California

Pat Bowler Johnson
New Trier High School
Winnetka, Illinois

Beth Chance
University of the Pacific
Stockton, California

John Chappelle
Brookstone School
Columbus, Georgia

Gretchen Davis
Santa Monica High School
Santa Monica, California

Harlan Goldberg
Highland Park High School
Northbrook, Illinois

Bill Harrington
State College Area HS
State College, Pennsylvania

Robert Keefer
Wichita Collegiate
Wichita, Kansas

Bruce King, retired
Connecticut State University
New Milford, Connecticut

Paul L. Myers
Woodward Academy
College Park, Georgia

Laura Jean Niland
MacArthur High School
San Antonio, Texas

Diann C. Resnick
Bellaire High School
Bellaire, Texas

Larry J. Peterson
Bonneville High School
Ogden, Utah

Murray Siegal
JL Mann Academy
Greer, South Carolina

I also wish to thank those new AP Statistics teachers across the country who have participated in workshops and classes with me, for sharing their ideas and lending their support. Thank you to those students who have participated in my AP Statistics class via satellite TV and to the Varina High School students in the studio classroom. A special thank you to the folks I work with in the Electronic Classroom at Varina High School for all their encouragement and support: David, Judy, Carolyn, Leslie, Patrick, Gary, and Greg.

And last, I am eternally grateful to my lovely bride, Betty Jo, and to my two children, Carla and Joey, for all their support and encouragement. Without Betty Jo's daily assistance and careful attention to detail, this project might still be in the vision stage. Thank you all.

Daniel S. Yates

Organizing Data: Looking for Patterns and Departures from Patterns

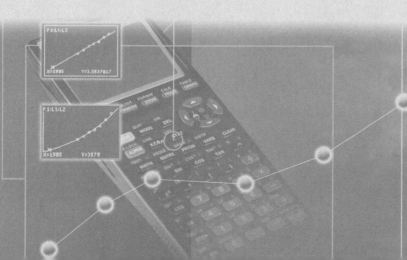

Exploring Data

■ **1.1 Displaying Distributions with Graphs**

■ **1.2 Describing Distributions with Numbers**

■ **Chapter Review**

1.1   DISPLAYING DISTRIBUTIONS WITH GRAPHS

Introduction

Statistics is the science of data. We therefore begin our study of statistics by mastering the art of examining data. Any set of data contains information about some group of *individuals*. The information is organized in *variables*.

INDIVIDUALS AND VARIABLES

Individuals are the objects described by a set of data. Individuals may be people, but they may also be animals or things.

A **variable** is any characteristic of an individual. A variable can take different values for different individuals.

Data for a study of a company's pay policies, for example, might contain data about every employee. These are the individuals described by the data set. For each individual, the data contain the values of variables such as the age in years, gender (female or male), job category, and annual salary in dollars. In practice, any set of data is accompanied by background information that helps us understand the data. We say that data are numbers collected in a particular context. When you meet a new set of data, ask yourself the following questions:

1. What **individuals** do the data describe? **How many** individuals appear in the data?

units

2. How many variables are there? What are the exact **definitions** of these variables? In what *units* is each variable recorded? Weights, for example, might be recorded in pounds, in thousands of pounds, or in kilograms. Is there any reason to mistrust the values of any variable?

3. What is the reason the data were gathered? Do we hope to answer some specific questions? Do we want to draw conclusions about individuals other than the ones we actually have data for?

The third set of questions is very important, so important that Chapter 5 addresses them in detail. For now, however, we are content to describe the individuals and the variables in a data set.

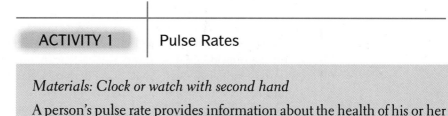

ACTIVITY 1 | Pulse Rates

Materials: Clock or watch with second hand

A person's pulse rate provides information about the health of his or her heart. This activity will have each student measure his or her resting pulse rate to see if there is a difference between males and females. Would you expect to find a difference? This activity works best if there are at least a dozen male subjects and a dozen female subjects, but it can be done with fewer.

1. To determine your pulse rate, hold the *fingers* of one hand on the artery in your neck or on the inside of the wrist. (The thumb should not be used, because there is a pulse in the thumb.) Count the number of pulse beats in one minute. Do this three times, and calculate your *average* individual pulse rate (add your three pulse rates and divide by 3.) Why is doing this three times better than doing it once?

2. Record the pulse rates for the class in a table, with one column for males and a second column for females. Are there any unusual pulse rates?

3. For now, simply calculate the average pulse rate for the males and the average pulse rate for the females, and compare.

Note: Keep these data; we will use them in later activities.

EXAMPLE 1.1 | Here is a small part of a data set that describes public education in the United States.

State	Region	Population (1,000)	SAT verbal	SAT math	Percent taking	Dollars per pupil	Teachers' pay ($1,000)
⋮							
CA	PAC	29,760	419	484	45	4,826	39.6
CO	MTN	3,294	456	513	28	4,809	31.8
CT	NE	3,287	430	471	74	7,914	43.8
⋮							

The *individuals* described are the states. There are 51 of them, the 50 states and the District of Columbia, but we give data for only 3. Each row in the table describes one individual. Each column contains the values of one variable for all the individuals. This is the usual arrangement in data tables. The first column identifies the state by its two-letter post office code. We give data for California, Colorado, and Connecticut.

The second column says which region of the country the state is in. The census bureau divides the nation into nine regions. These three are Pacific, Mountain, and New England. The third column contains state populations, in thousands of people. Be sure to notice that the *units* are thousands of people. California's 29,760 stands for 29,760,000 people. The population data come from the 1990 census. They are therefore quite accurate as of April 1, 1990, but don't show later changes in population.

The remaining five variables are the average scores of the states' high school seniors on the SAT verbal and mathematics exams, the percent of seniors who take the SAT, public school spending per pupil in dollars, and average teachers' salaries in thousands of dollars. Each of these variables needs more explanation before we can fully understand the data. For example, spending per pupil is total spending on public education divided by the number of pupils. Because the number of pupils in school varies from day to day, the data use average attendance. Attendance is reported by each school system. Attendance often determines how much state aid a school system gets, so some schools may report numbers that are too high. If this happens, the spending per pupil in the table will be too low.

exploratory data analysis

Statistical tools and ideas can help you examine data in order to describe their main features. This examination is called ***exploratory data analysis.*** Like an explorer crossing unknown lands, we first simply describe what we see. Each example we meet will have some background information to help us, but our emphasis is on examining the data. Here are two basic strategies that help us organize our exploration of a set of data:

- Begin by examining each variable by itself. Then move on to study relationships among the variables.
- Begin with a graph or graphs. Then add numerical summaries of specific aspects of the data.

We will organize our learning the same way. Chapters 1 and 2 examine single-variable data, and Chapters 3 and 4 look at relationships among variables. In both settings, we begin with graphs and then move on to numerical summaries.

Variables: categorical and quantitative

Some variables, like gender and job title, simply place individuals into categories. Others, like height and annual income, take numerical values for which we can do arithmetic. It makes sense to give an average income for a company's employees, but it does not make sense to give an "average" gender. We can, however, count the numbers of female and male employees and do arithmetic with these counts.

CATEGORICAL AND QUANTITATIVE VARIABLES

A **categorical variable** records which of several groups or categories an individual belongs to.

A **quantitative variable** takes numerical values for which it makes sense to do arithmetic operations like adding and averaging.

The **distribution** of a variable tells us what values the variable takes and how often it takes these values.

A variable generally takes values that vary. One variable may take values that are very close together while another variable takes values that are quite spread out. We say that the *pattern of variation* of a variable is its distribution.

The values of a categorical variable are just labels for the categories, like "male" and "female." The distribution of a categorical variable lists the categories and gives either the count or the percent of individuals who fall in each category. For example, here is the distribution of marital status for all Americans age 18 and over.

Marital status	Count (millions)	Percent
Single	41.8	22.6
Married	113.3	61.1
Widowed	13.9	7.5
Divorced	16.3	8.8

bar chart

To present such data to an audience, you may wish to use graphs like those in Figure 1.1. The *bar chart* in Figure 1.1(a) helps us compare the sizes of the four marital status groups. The heights of the four bars show the counts

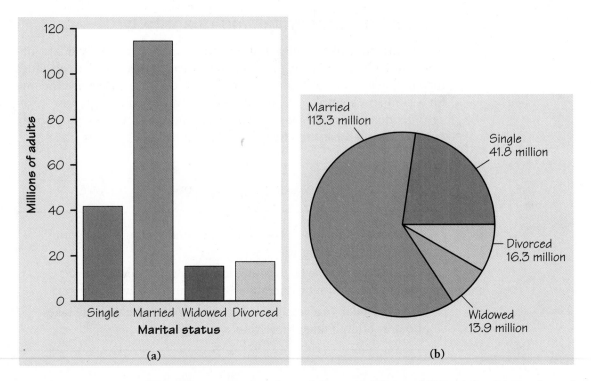

FIGURE 1.1   (*a*) Bar chart of the marital status of U.S. adults. (*b*) Pie chart of the same data.

pie chart

in the four categories. The **pie chart** in Figure 1.1(b) helps us see what part of the whole each group forms. For example, the "married" slice makes up 61% of the pie because 61% of adults are married. Bar charts and pie charts help an audience grasp the distribution quickly.

EXERCISES

1.1   In Activity 1, you collected some pulse rate data.

   (a) What individuals do the data describe?
   (b) How many variables are there? What are they?
   (c) In what units is each variable recorded?
   (d) Are the variables quantitative or categorical?

1.2   Example 1.1 presents data on the states. The first column identifies the states. Each of the remaining seven columns contains values of a variable. Which of these variables are categorical and which are quantitative?

1.3   Data from a medical study contain values of many variables for each of the people who were the subjects of the study. Which of the following variables are categorical and which are quantitative?

(a) Gender (female or male)

(b) Age (years)

(c) Race (Asian, black, white, or other)

(d) Smoker (yes or no)

(e) Systolic blood pressure (millimeters of mercury)

(f) Level of calcium in the blood (micrograms per milliliter)

Dotplots and histograms

EXAMPLE 1.2

dotplot

Table 1.1 presents the percent of residents aged 65 years and over in each of the 50 states. One way to quickly visualize a data set is to construct a ***dotplot.*** To make a dotplot, draw a horizontal line to represent the variable and impose a number scale for the values of the variable. Then mark a dot at the appropriate place for each observation. If you enter the data from Table 1.1 into a Minitab worksheet and ask for a dotplot, you would see the following:

```
MTB > Dotplot 'Pop%'.

                                              .
                                     :    .
                           .:  ..: .:: :
                 .        . :. .::.:::.:::.:        .
          +---------+---------+---------+---------+---------+-------Pop%
         3.0       6.0       9.0      12.0      15.0      18.0
MTB >
```

range

Notice that you need not begin with zero on the left; simply cover the ***range*** of the data. If you sort the data—the keystrokes on the TI-83 are STATS / 2:SortA(L_1)—you see that the smallest percent is 4.9 (Alaska) and the largest percent is 18.6 (Florida). We say that the range of the data is from 4.9 to 18.6. If you are constructing a dotplot by hand, you would mark a horizontal scale that extends from about 4 to about 19.

histogram

     Sometimes quantitiative variables take so many values that a graph of the distribution is clearer if nearby values are grouped together. A ***histogram*** is the most common graph of distributions with one quantitative variable. To illustrate the histogram, let's use the data from Table 1.1 to explore histograms on the TI-83. Enter the data into a list, say L_1, in your TI-83. In STAT PLOT, specify a histogram using the data from L_1; see Figure 1.2(a).

| TABLE 1.1 | Percent of population 65 years old and over by state (1995) | | |

State	Percent	State	Percent
Alabama	13.0	Montana	13.1
Alaska	4.9	Nebraska	13.9
Arizona	13.3	Nevada	11.4
Arkansas	14.5	New Hampshire	11.9
California	11.0	New Jersey	13.7
Colorado	10.0	New Mexico	10.9
Connecticut	14.3	New York	13.4
Delaware	12.6	North Carolina	12.5
Florida	18.6	North Dakota	14.5
Georgia	10.0	Ohio	13.4
Hawaii	12.6	Oklahoma	13.5
Idaho	11.4	Oregon	13.6
Illinois	12.5	Pennsylvania	15.9
Indiana	12.6	Rhode Island	15.7
Iowa	15.2	South Carolina	12.0
Kansas	13.7	South Dakota	14.4
Kentucky	12.6	Tennessee	12.5
Louisiana	11.4	Texas	10.2
Maine	13.9	Utah	8.8
Maryland	11.3	Vermont	12.0
Massachusetts	14.2	Virginia	11.1
Michigan	12.4	Washington	11.6
Minnesota	12.4	West Virginia	15.3
Mississippi	12.3	Wisconsin	13.3
Missouri	13.9	Wyoming	11.1

Source: *Statistical Abstract of the United States,* 1996.

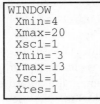

FIGURE 1.2(a)

FIGURE 1.2(b)

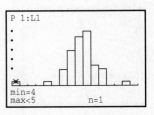

FIGURE 1.2(c)

Next we need to specify the widths of the bars and dimensions for the viewing window. Let's specify the width of the bars to be 1, beginning at $x = 4$ and ending at $x = 20$. This will cover the range of the data. To design our histogram, press WINDOW and enter the numbers given in Figure 1.2(b).

Press TRACE and hit the right cursor key several times. You should see the name of the list that contains the data, the left and right endpoints of the interval (class), and the *frequency*, that is, the number of observations in that class.

The algorithm for constructing a histogram by hand for this data set would go like this:

1. Divide the range of the data into classes of equal width. The data in Table 1.1 range from 4.9 to 18.6, so we choose as our classes

$$4.0 \le \text{percent over } 65 < 5.0$$
$$5.0 \le \text{percent over } 65 < 6.0$$
$$\vdots$$
$$18.0 \le \text{percent over } 65 < 19.0$$

Be sure to specify the classes precisely so that each observation falls into exactly one class. A state with 4.9% of its residents aged 65 or older would fall into the first class, but 5.0% falls into the second.

2. Count the number of observations in each class. Here are the counts:

Class	Count	Class	Count	Class	Count
4.0 to 4.9	1	9.0 to 9.9	0	14.0 to 14.9	5
5.0 to 5.9	0	10.0 to 10.9	4	15.0 to 15.9	4
6.0 to 6.9	0	11.0 to 11.9	9	16.0 to 16.9	0
7.0 to 7.9	0	12.0 to 12.9	12	17.0 to 17.9	0
8.0 to 8.9	1	13.0 to 13.9	13	18.0 to 18.9	1

3. Draw the histogram. First mark the scale for the variable whose distribution you are displaying on the horizontal axis. That's "percent of residents 65 or over" in this example. The scale runs from 4 to 19 because that is the span of the classes we chose. Then mark the scale of counts on the vertical axis. Each bar represents a class. The base of the bar covers the class, and the bar height is the class count. Draw the graph with no horizontal space between the bars (unless a class is empty, so that its bar has height zero).

The bars of a histogram should cover the entire range of values of a variable. When the possible values of a variable have gaps between them, extend the bases of the bars to meet halfway between two adjacent possible values. For example, in a histogram of the ages in years of university faculty, the bars representing 25 to 29 years and 30 to 34 years would meet at 29.5. There is no one right choice of the classes in a histogram. Too few classes will give a "skyscraper" graph, with all values in a few classes with tall bars. Too many will produce a "pancake" graph, with most classes having one or no observations. Neither choice will give a good picture of the shape of the distribution. You must use your judgment in choosing classes to display the shape. Our eyes respond to the *area* of the bars in a histogram, so be sure to choose classes that are all the same width. Then area is determined by height and all classes are fairly represented. If you use a computer or a graphing calculator, the software will choose the classes for you. The computer's choice is usually a good one, but you can change it if you want.

Interpreting histograms

Making a statistical graph is not an end in itself. After all, a computer or graphing calculator can make graphs faster than we can. The purpose of the graph is to help us understand the data. After you (or your computer) make a graph, always ask, "What do I see?" Here is a general tactic for looking at graphs:

• Look for an overall pattern and also for striking deviations from that pattern.

In the case of a histogram, the overall pattern is the overall shape of the distribution. *Outliers* are an important kind of deviation from the overall pattern.

OUTLIERS

An **outlier** in any graph of data is an individual observation that falls outside the overall pattern of the graph.

Three states stand out in the histogram of Figure 1.2(c). You can find them in the table once the histogram has called attention to them. Florida has 18.6% of its residents age 65 or over, and Alaska has only 4.9%. These states are clear outliers. You might also call Utah, with 8.8%, an outlier, though it is not as far from the overall pattern as Florida and Alaska. Whether an observation is an outlier is to some extent a matter of judgment. It is much easier to spot outliers in the histogram than in the data table.

Once you have spotted outliers, look for an explanation. Many outliers are due to mistakes, such as typing 4.0 as 40. Other outliers point to the special nature of some observations. Explaining outliers usually requires some background information. It is not surprising to an American that Florida, with its many retired people, has many residents over 65 and that Alaska, the northern frontier, has few.

What about the *overall pattern* of the distribution in Figure 1.2(c)?

OVERALL PATTERN OF A DISTRIBUTION

To describe the overall pattern of a distribution:

- Give the **center** and the **spread.**
- See if the distribution has a simple **shape** that you can describe in a few words.

Section 1.2 tells in detail how to measure center and spread. For now, describe the center by finding a value that divides the observations so that about half take larger values and about half have smaller values. The center in Figure 1.2(c) is about 13%. That is, about 13% of the residents of a typical state are at least 65 years old. You can describe the spread by the extent of the data from smallest to largest value. The spread in Figure 1.2(c) is about 10% to 16% if we ignore the outliers. The histogram in Figure 1.2(c) has an irregular shape that isn't easy to describe. Some distributions, however, do have simple shapes. Here are examples that illustrate some shapes to look for.

EXAMPLE 1.3

Look at the histograms in Figures 1.3 and 1.4. Figure 1.3 comes from a study of lightning storms in Colorado. It shows the distribution of the hour of the day during which the first lightning flash for that day occurred. The distribution has a single peak at noon and falls off on either side of this peak. The two sides of the histogram are roughly the same shape, so we call the distribution *symmetric.*

Figure 1.4 shows the distribution of lengths of words used in Shakespeare's plays.[1] This distribution is *skewed to the right.* That is, there are many short words (3 and 4 letters) and few very long words (10, 11, or 12 letters), so that the right tail of the histogram extends out much farther than the left tail.

Notice that the vertical scale in Figure 1.4 is not the count of words but the *percent* of all of Shakespeare's words that have each length. A histogram of percents rather than counts is convenient when the counts are very large or when we want to compare several distributions. Different kinds of writing have different distributions of word lengths, but all are right-skewed because short words are common and very long words are rare.

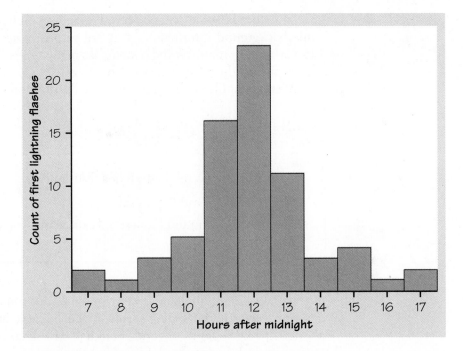

FIGURE 1.3   The distribution of the time of the first lightning flash each day at a site in Colorado.

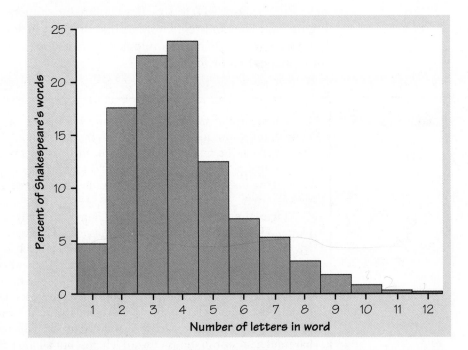

FIGURE 1.4   The distribution of lengths of words used in Shakespeare's plays.

SYMMETRIC AND SKEWED DISTRIBUTIONS

A distribution is **symmetric** if the right and left sides of the histogram are approximately mirror images of each other.

A distribution is **skewed to the right** if the right side of the histogram (containing the upper half of the observations) extends much farther out than the left side (containing the lower half of the observations). It is **skewed to the left** if the left side of the histogram extends much farther out than the right side.

In mathematics, symmetry means that the two sides of a figure like a histogram are exact mirror images of each other. Data are almost never exactly symmetric, so we are willing to call histograms like that in Figure 1.3 approximately symmetric as an overall description.

The overall shape of a distribution is important information about a variable. Some types of data regularly produce distributions that are symmetric or skewed. For example, the sizes of many living things of the same species (like lengths of cockroaches) tend to be symmetric. Data on incomes (whether of individuals, companies, or nations) are usually strongly skewed to the right. There are many moderate incomes, some large incomes, and a few very large incomes. Do remember that many histograms, like Figure 1.2(c), can't reasonably be called either symmetric or skewed. Some data show other patterns. Scores on an exam, for example, may have a cluster near the top of the scale if many students did well. Or they may show two distinct peaks if a tough problem divided the class into those who did and didn't solve it. Use your eyes and say what you see.

EXERCISES

1.4   There are many ways to measure the reading ability of children. One frequently used test is the Degree of Reading Power (DRP). In a research study on third-grade students, the DRP was administered to 44 students. Their scores were

40	26	39	14	42	18	25	43	46	27	19
47	19	26	35	34	15	44	40	38	31	46
52	25	35	35	33	29	34	41	49	28	52
47	35	48	22	33	41	51	27	14	54	45

Make a dotplot of these data. Then make a histogram. Which display do you prefer, and why? Describe the main features of the distribution.

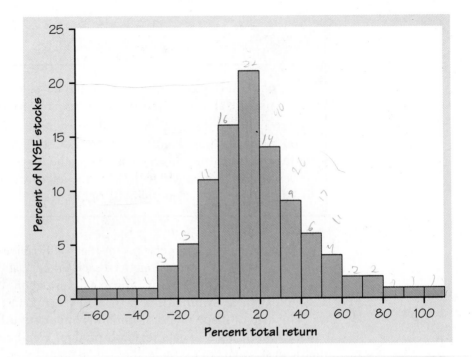

FIGURE 1.5   The distribution of percent total return for all New York Stock Exchange common stocks in one year, for Exercise 1.5.

1.5   The total return on a stock is the change in its market price plus any dividend payments made. Total return is usually expressed as a percent of the beginning price. Figure 1.5 is a histogram of the distribution of total returns for all 1528 stocks listed on the New York Stock Exchange in one year.[2] Like Figure 1.4, it is a histogram of the percents in each class rather than a histogram of counts.

(a)   Describe the overall shape of the distribution of total returns.

(b)   What is the approximate center of this distribution? (For now, take the center to be the value with roughly half the stocks having lower returns and half having higher returns.)

(c)   Approximately what were the smallest and largest total returns? (This describes the spread of the distribution.)

(d)   A return less than zero means that an owner of the stock lost money. About what percent of all stocks lost money?

1.6   Figure 1.6 is a histogram of the number of days in the month of April on which the temperature fell below freezing at Greenwich, England.[3] The data cover a period of 65 years.

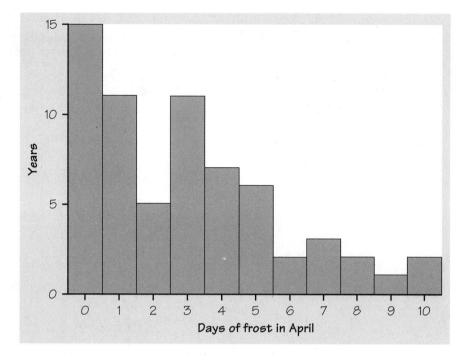

FIGURE 1.6   The distribution of the number of frost days during April at
Greenwich, England, over a 65-year period, for Exercise 1.6.

(a) Describe the shape, center, and spread of this distribution. Are there
any outliers?

(b) In what percent of these 65 years did the temperature never fall below
freezing in April?

1.7   How would you describe the center and spread of the distribution of first
lightning flash times in Figure 1.3? Of the distribution of Shakespeare's
word lengths in Figure 1.4?

1.8   The distribution of the ages of a nation's population has a strong influence
on economic and social conditions. The following table shows the age
distribution of U.S. residents in 1950 and 2075, in millions of persons.
The 1950 data come from that year's census. The 2075 data are projections
made by the Census Bureau.

(a) Because the total population in 2075 is much larger than the 1950
population, comparing percents in each age group is clearer than
comparing counts. Make a table of the percent of the total population in each age group for both 1950 and 2075.

(b) Make a histogram of the 1950 age distribution (in percents). Then describe the main features of the distribution. In particular, look at the percent of children relative to the rest of the population.

(c) Make a histogram of the projected age distribution for the year 2075. Use the same scales as in (b) for easy comparison. What are the most important changes in the U.S. age distribution projected for the 125-year period between 1950 and 2075?

Age group	1950	2075
Under 10 years	29.3	34.9
10 to 19 years	21.8	35.7
20 to 29 years	24.0	36.8
30 to 39 years	22.8	38.1
40 to 49 years	19.3	37.8
50 to 59 years	15.5	37.5
60 to 69 years	11.0	34.5
70 to 79 years	5.5	27.2
80 to 89 years	1.6	18.8
90 to 99 years	0.1	7.7
100 to 109 years	—	1.7
Total	151.1	310.6

Stemplots

stemplot

Dotplots and histograms are two of many graphical displays of distributions. For small data sets, a ***stemplot*** can be quicker to make and presents more detailed information.

EXAMPLE 1.4

stem, leaf

To make a stemplot, separate each observation into a ***stem*** consisting of all but the (rightmost) digit and a ***leaf***, the final digit. For the "65 or over" percents in Table 1.1, the whole-number part of the observation is the stem and the final digit (tenths) is the leaf. The Alabama entry, 13.0, had stem 13 and leaf 0. Stems can have as many digits as needed, but each leaf must consist of only a single digit.

- Write the stems vertically in increasing order from top to bottom, and draw a vertical line to the right of the stems.

- Go through the data, writing each leaf to the right of its stem.
- Write the stems again, and rearrange the leaves in increasing order out from each stem.

Here is the completed stemplot for the data in Table 1.1:

```
 4 | 9
 5 |
 6 |
 7 |
 8 | 8
 9 |
10 | 0029
11 | 011344469
12 | 003445556666
13 | 0133445677999
14 | 23455
15 | 2379
16 |
17 |
18 | 6
```

A stemplot looks like a dotplot or histogram turned on end. The stemplot in Example 1.4 is exactly like the histogram in Figure 1.2(c) because each stem is a class in the histogram. But the stemplot, unlike the dotplot or histogram, preserves the actual value of each observation. We interpret stemplots as we do dotplots and histograms, looking for the overall pattern and for any outliers.

You can choose the classes in a histogram. The classes (the stem) of a stemplot are given to you. There are two variations of a stemplot that give us more flexibility in graphing a distribution. First, you can **round** the data so that the final digit after rounding is suitable as a leaf. Do this when the data have too many digits. For example, data like

rounding

$$3.468 \quad 2.567 \quad 2.981 \quad 1.095 \ldots$$

would have too many stems if we took the first three digits as the stem and the final digit as the leaf. You would probably round these data to

$$3.5 \quad 2.6 \quad 3.0 \quad 1.1 \ldots$$

before making a stemplot.

splitting stems

You can also **split stems** to double the number of stems when all the leaves would otherwise fall on just a few stems. Each stem then appears twice. Leaves 0 to 4 go on the upper stem and leaves 5 to 9 go on the lower stem. If you split the stems in the stemplot of Example 1.4, for example, the 11 and 12 stems become

```
11 │ 0113444
11 │ 69
12 │ 00344
12 │ 5556666
```

Rounding and splitting stems are matters for judgment, like choosing the classes in a histogram. The stemplot in Example 1.4 does not need either change. Stemplots work well for small sets of data. When there are more than 100 observations, a histogram is almost always a better choice.

EXERCISES

1.9   The Survey of Study Habits and Attitudes (SSHA) is a psychological test that evaluates college students' motivation, study habits, and attitudes toward school. A private college gives the SSHA to a sample of 18 of its incoming first-year women students. Their scores are

| 154 | 109 | 137 | 115 | 152 | 140 | 154 | 178 | 101 |
| 103 | 126 | 126 | 137 | 165 | 165 | 129 | 200 | 148 |

Make a stemplot of these data. The overall shape of the distribution is irregular, as often happens when only a few observations are available. Are there any outliers? About where is the center of the distribution (the score with half the scores above it and half below)? What is the spread of the scores (ignoring any outliers)?

1.10   The Modern Language Association provides listening tests that measure understanding of spoken French. The range of scores is 0 to 36. Here are the scores of 20 high school French teachers at the beginning of an intensive summer course in French:[4]

32 31 29 10 30 33 22 25 32 20 30 20 24 24 31 30 15 32 23 23

(a)   Make a stemplot of these scores. Use split stems.

(b)   Describe in words the most important features of the distribution. Are there any outliers?

(c)   About what score would place a teacher in the center of this distribution, with roughly half the scores lower and half higher?

back-to-back 1.11   A *back-to-back stemplot* helps us compare two distributions. In this exercise
stemplot   you will construct a back-to-back stemplot of the pulse rate data you collected in Activity 1. Begin by making a stemplot of the male pulse rates. Then draw a vertical line to the left of the stems. Add the female pulse rates

as leaves on the same stems, but going out to the left rather than the right. Be sure to arrange the leaves on each stem in increasing order out from the stem. Here is an example:

Females		Males
0	10	
75431	9	0002
8864200	8	04688
88620	7	024578
742	6	00234679
5	5	488
	4	8

If you have enough data values, you should be able to visually compare the two distributions. Write a statement that compares the two distributions of pulse rates for your class.

Time plots

Many variables are measured at intervals over time. We might, for example, measure the height of a growing child or the price of a stock at the end of each month. In these examples, our main interest is change over time. To display change over time, make a *time plot*.

TIME PLOT

A **time plot** of a variable plots each observation against the time at which it was measured. Always mark the time scale on the horizontal axis and the variable of interest on the vertical axis. If there are not too many points, connecting the points by lines helps show the pattern of changes over time.

EXAMPLE 1.5

Here are data on the rate of deaths from cancer (deaths per 100,000 people) in the United States over the 50-year period 1940 to 1990:

Year	1940	1945	1950	1955	1960	1965	1970	1975	1980	1985	1990
Deaths	120.3	134.0	139.8	146.5	149.2	153.5	162.8	169.7	183.9	193.3	201.7

Figure 1.7 is a time plot of these data. It shows the steady increase in the cancer death rate during the past half-century. This increase does not mean that we have made no progress in treating cancer. Because cancer is primarily a disease of old age, the death rate from cancer increases when people live longer even if treatment improves.

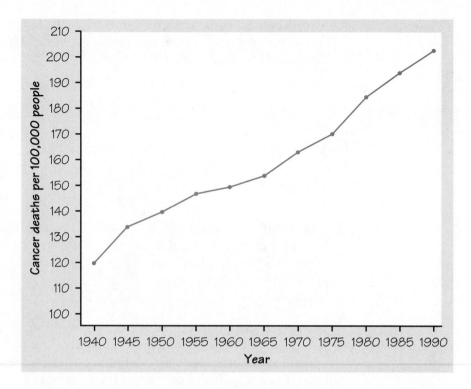

FIGURE 1.7   Time plot of the death rate from cancer (deaths per 100,000 people) from 1940 to 1990.

When you examine a time plot, look once again for an overall pattern and for strong deviations from the pattern. One common overall pattern is a *trend*, a long-term upward or downward movement over time. Figure 1.7 shows an upward trend in the cancer death rate, with no striking deviations such as short-term drops.

trend

EXERCISES

1.12   The buying power of a dollar changes over time. The Bureau of Labor Statistics measures the cost of a "market basket" of goods and services to compile its Consumer Price Index (CPI). If the CPI is 120, goods and services that cost $100 in the base period now cost $120. Here are the yearly average values of the CPI for the 1970s and 1980s. The base period is the years 1982 to 1984.

Year	CPI	Year	CPI
1970	38.8	1980	82.4
1971	40.5	1981	90.9
1972	41.8	1982	96.5
1973	44.4	1983	99.6
1974	49.3	1984	103.9
1975	53.8	1985	107.6
1976	56.9	1986	109.6
1977	60.6	1987	113.6
1978	65.2	1988	118.3
1979	72.6	1989	124.0

(a) With pencil and grid paper, make a time plot of the CPI for these two decades.

(b) Check your graph by doing the plot on your TI-83. Enter the years (the last two digits will suffice) into list L_1 and enter the CPI into list L_2. Then select STAT PLOT and specify the second graph icon, which is called the xyLine. In this graph, the data points are plotted and connected in order of appearance in Xlist and Ylist. Press ZOOM / 9:ZoomStat to see the graph.

```
Plot1 Plot2 Plot3
On  Off
Type: ⌐··  ⌐  dⅢb
       ₀-··  ·ⅢⅠ  ⌐
Xlist:L1
Ylist:L2
Mark: ▣  ◆  ·
```

(c) What was the overall trend in prices during this period? Were there any years in which this trend was reversed?

(d) In what period during these decades were prices rising fastest? In what period were they rising slowest?

1.13 The Virginia Department of Motor Vehicles publishes data each year on the number of road fatalities, pedestrian fatalities, and alcohol-related fatalities in the state. This information is used as a stimulus for public safety awareness programs, legislation on speed limits and the use of seat belts, highway engineering projects, and similar purposes. Here are the results for an eleven-year period:[5]

Year	Total	Pedestrian	Alcohol-related
1986	1118	141	492
1987	1022	117	418
1988	1069	131	522
1989	999	141	480
1990	1071	116	535
1991	938	112	429
1992	839	93	379
1993	875	112	397
1994	925	101	376
1995	900	93	360
1996	869	114	346

(a) Make a time plot for the total number of Virginia road fatalities. Does there appear to be a trend? If so, describe it. Can you give some possible reasons for what has happened?

(b) Make a time plot for the number of alcohol-related fatalities. Answer the same question as in (a).

SUMMARY

A data set contains information on a number of **individuals.** Individuals may be people, animals, or things. For each individual, the data give values for one or more **variables.** A variable describes some characteristic of an individual, such as a person's height, gender, or salary.

Exploratory data analysis uses graphs and numerical summaries to describe the variables in a data set and the relations among them.

Some variables are **categorical** and others are **quantitative.** A categorical variable places each individual into a category, like male or female. A quantitative variable has numerical values that measure some characteristic of each individual, like height in centimeters or annual salary in dollars.

The **distribution** of a variable describes what values the variable takes and how often it takes these values.

To describe a distribution, begin with a graph. **Dotplots, histograms,** and **stemplots** graph the distributions of quantitative variables.

When examining any graph, look for an **overall pattern** and for notable **deviations** from the pattern.

The **center, spread,** and **shape** describe the overall pattern of a distribution. Some distributions have simple shapes, such as **symmetric** and **skewed.** Not all distributions have a simple overall shape, especially when there are few observations.

Outliers are observations that lie outside the overall pattern of a distribution. Always look for outliers and try to explain them.

When observations on a variable are taken over time, make a **time plot** that graphs time horizontally and the values of the variable vertically. A time plot can reveal **trends** or other changes over time.

SECTION 1.1 EXERCISES

1.14 Here is a small part of the data set that a company keeps to record information about its employees:

Name	Age	Gender	Race	Salary	Job type
⋮					
Fleetwood, Delores	39	Female	White	52,100	Management
Fleming, LaVerne	27	Male	Black	37,500	Technical
Foo, Ruoh-Lin	22	Female	Asian	15,200	Clerical
⋮					

(a) What individuals does the complete data set describe?

(b) The data set records five variables in addition to the name. Which of these are categorical variables?

(c) Which of the variables are quantitative? Based on the data in the table, what do you think are the units of measurement for each of the quantitative variables?

1.15 The histogram in Figure 1.8 shows the number of hurricanes reaching the east coast of the United States each year over a 70-year period.[6] Give a brief description of the overall shape of this distribution. About where does the center of the distribution lie? (For now, take the center to be the value with roughly half the observations on either side of it.)

1.16 Figure 1.9 displays the distribution of batting averages for all 167 American League baseball players who batted at least 200 times in the 1980 season. (The outlier is the .390 batting average of George Brett, the highest batting average in the major leagues since Ted Williams hit .406 in 1941.)

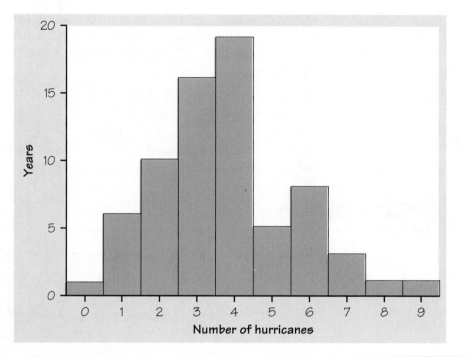

FIGURE 1.8   The distribution of the annual number of hurricanes on the U.S. east coast over a 70-year period, for Exercise 1.15.

(a) Is the overall shape (ignoring the outlier) roughly symmetric or clearly skewed or neither?

(b) What was the approximate batting average of a typical American League player? About what were the highest and lowest batting averages, leaving out George Brett?

1.17   Here are the numbers of home runs that Babe Ruth hit in his 15 years with the New York Yankees, 1920 to 1934:[7]

54   59   35   41   46   25   47   60   54   46   49   46   41   34   22

(a) Make a stemplot for these data. Is the distribution roughly symmetric, clearly skewed, or neither? About how many home runs did Ruth hit in a typical year? Is his famous 60 home runs in 1927 an outlier?

(b) Babe Ruth's home run record for a single year was broken by another Yankee, Roger Maris, who hit 61 home runs in 1961. Here are Maris's home run totals for his 10 years in the American League:

13   23   26   16   33   61   28   39   14   8

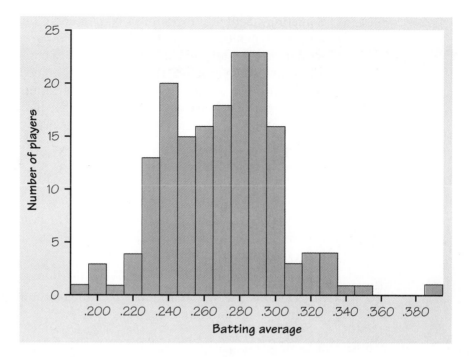

FIGURE 1.9 The distribution of batting averages of American League players in 1980, for Exercise 1.16.

Make a back-to-back stemplot with Ruth's data on the right and Maris's data on the left. (See Exercise 1.11, page 20, about back-to-back stemplots.) Is Maris's record 61 an outlier in his distribution of home runs hit? How does your plot show Ruth's superiority as a home run hitter?

1.18 Sometimes both a histogram or stemplot and a time plot give useful information about a variable. The following data are the *greatest* amounts of rain (in inches) that fell on one day in South Bend, Indiana, for each of 30 consecutive years. Successive years follow each other across the rows in the table.

1.88	2.23	2.58	2.07	2.94	2.29	3.14	2.15	1.95	2.51
2.86	1.48	1.12	2.76	3.10	2.05	2.23	1.70	1.57	2.81
1.24	3.29	1.87	1.50	2.99	3.48	2.12	4.69	2.29	2.12

(a) Make a stemplot for these data. Describe the general shape of the distribution and any prominent deviations from the overall pattern. (You are expected to round and split stems as needed in making stemplots.)

(b) Make a time plot of the data. Is there any suggestion of a long-term change in maximum rainfall at South Bend?

1.19   Sometimes both a histogram or stemplot and a time plot give useful information about a set of data. The data below are measurements of the tension on the wire grid behind the screen in a computer display. The display won't work properly if the tension is either too high or too low, so the manufacturer measures the tension of a small group of displays each hour. Here are the hourly averages for 20 consecutive hours of production, from left to right:

269.5   297.0   269.6   283.3   304.8   280.4   283.5   257.4   317.5   327.4
264.7   307.7   310.0   343.3   328.1   342.6   338.8   340.1   374.6   336.1

(a)   Make a stemplot of these data.

(b)   Describe the shape of the distribution. The distribution shows considerable variation in the tension, but nothing to alarm the manufacturer except for one outlier.

(c)   Make a time plot of the data. (Mark the time scale in hours, 1 to 20.)

(d)   Describe the overall pattern of the time plot and explain why the manufacturer should investigate right away.

1.20   The impression that a time plot gives depends on the scales you use on the two axes. If you stretch the vertical axis and compress the time axis, change appears to be more rapid. Compressing the vertical axis and stretching the time axis make change appear slower. Make two more time plots of the data in Example 1.5, one that makes cancer death rates appear to increase very rapidly and one that shows only a gentle increase. The moral of this exercise is: pay close attention to the scales when you look at a time plot.

Table 1.2 presents data about the individual states that relate to education. The data for these three states in Example 1.1 come from this table. Study of a data set with many variables begins by examining each variable by itself. Exercises 1.21 to 1.23 concern the data in Table 1.2.

1.21   Make a graph of the distribution of the percent of high school seniors who take the SAT in the various states. Briefly describe the overall pattern of the distribution and any outliers.

1.22   Make a graph to display the distribution of average teachers' salaries for the states. Is there a clear overall pattern? Are there any outliers or other notable deviations from the pattern?

1.23   Make a graph of the distribution of dollars per pupil spent on education for the states. Is there a clear overall pattern? Are there any outliers or other notable deviations from the pattern?

| TABLE 1.2 | Education and related data for the United States |

State	Region*	Population (1,000)	SAT verbal	SAT math	Percent taking	Dollars per pupil	Teachers' pay ($1,000)
AL	ESC	4,041	470	514	8	3,648	27.3
AK	PAC	550	438	476	42	7,887	43.4
AZ	MTN	3,665	445	497	25	4,231	30.8
AR	WSC	2,351	470	511	6	3,334	23.0
CA	PAC	29,760	419	484	45	4,826	39.6
CO	MTN	3,294	456	513	28	4,809	31.8
CT	NE	3,287	430	471	74	7,914	43.8
DE	SA	666	433	470	58	6,016	35.2
DC	SA	607	409	441	68	8,210	39.6
FL	SA	12,938	418	466	44	5,154	30.6
GA	SA	6,478	401	443	57	4,860	29.2
HI	PAC	1,108	404	481	52	5,008	32.5
ID	MTN	1,007	466	502	17	3,200	25.5
IL	ENC	11,431	466	528	16	5,062	34.6
IN	ENC	5,544	408	459	54	5,051	32.0
IA	WNC	2,777	511	577	5	4,839	28.0
KS	WNC	2,478	492	548	10	5,009	29.8
KY	ESC	3,865	473	521	10	4,390	29.1
LA	WSC	4,220	476	517	9	4,012	26.2
ME	NE	1,228	423	463	60	5,894	28.5
MD	SA	4,781	430	478	59	6,184	38.4
MA	NE	6,016	427	473	72	6,351	36.1
MI	ENC	9,295	454	514	12	5,257	38.3
MN	WNC	4,375	477	542	14	5,260	33.1
MS	ESC	2,573	477	519	4	3,322	24.4
MO	WNC	5,117	473	522	12	4,415	28.5
MT	MTN	799	464	523	20	5,184	26.7
NE	WNC	1,578	484	546	10	4,381	26.6
NV	MTN	1,202	434	487	24	4,564	32.2
NH	NE	1,109	442	486	67	5,504	31.3
NJ	MA	7,730	418	473	69	9,159	38.4
NM	MTN	1,515	480	527	12	4,446	26.2
NY	MA	17,990	412	470	70	8,500	42.1
NC	SA	6,629	401	440	55	4,802	29.2
ND	WNC	639	505	564	6	3,685	23.6
OH	ENC	10,487	450	499	22	5,639	32.6

| TABLE 1.2 | Education and related data for the United States (*continued*) | | | | | | |

State	Region*	Population (1,000)	SAT verbal	SAT math	Percent taking	Dollars per pupil	Teachers' pay ($1,000)
OK	WSC	3,146	478	523	9	3,742	24.3
OR	PAC	2,842	439	484	49	5,291	32.3
PA	MA	11,882	420	463	64	6,534	36.1
RI	NE	1,003	422	461	62	6,989	37.7
SC	SA	3,487	397	437	54	4,327	28.3
SD	WNC	696	506	555	5	3,730	22.4
TN	ESC	4,877	483	525	12	3,707	28.2
TX	WSC	16,987	413	461	42	4,238	28.3
UT	MTN	1,723	492	539	5	2,993	25.0
VT	NE	563	431	466	62	5,740	31.0
VA	SA	6,187	425	470	58	5,360	32.4
WA	PAC	4,867	437	486	44	5,045	33.1
WV	SA	1,793	443	490	15	5,046	26.0
WI	ENC	4,892	476	543	11	5,946	33.1
WY	MTN	454	458	519	13	5,255	29.0

SOURCE: *Statistical Abstract of the United States,* 1992.

*The census regions are East North Central, East South Central, Mountain, New England, Pacific, Middle Atlantic, South Atlantic, West North Central, and West South Central.

1.2　DESCRIBING DISTRIBUTIONS WITH NUMBERS

How old are presidents at their inauguration? Was Bill Clinton, at age 46, unusually young? Table 1.3 gives the data, the ages of all U.S presidents when they took office. On your TI-83, define a new list and name it PREZ. There are several ways to do this; here is one way. Get into STAT / EDIT mode, and move the cursor to highlight list L_1, press 2nd / INS to insert a new list. An empty list will be placed to the left of list L_1, "Name =" appears at the bottom of the screen, and the cursor flashes "A" for alphabetic mode. Notice the letters (in green) on the faceplate. Press the keys just below the letters P, R, E, Z to spell PREZ. When you finish, press ENTER, and then cursor down to the place where the first data item will go. Enter the 42 ages of the presidents in this list. We will use these data later. As the histogram in Figure 1.10 shows, there is a good deal of variation in the ages

TABLE 1.3		Ages of the presidents at inauguration			
President	Age	President	Age	President	Age
Washington	57	Buchanan	65	Harding	55
J. Adams	61	Lincoln	52	Coolidge	51
Jefferson	57	A. Johnson	56	Hoover	54
Madison	57	Grant	46	F. D. Roosevelt	51
Monroe	58	Hayes	54	Truman	60
J. Q. Adams	57	Garfield	49	Eisenhower	61
Jackson	61	Arthur	51	Kennedy	43
Van Buren	54	Cleveland	47	L. Johnson	55
W. H. Harrison	68	B. Harrison	55	Nixon	56
Tyler	51	Cleveland	55	Ford	61
Polk	49	McKinley	54	Carter	52
Taylor	64	T. Roosevelt	42	Reagan	69
Fillmore	50	Taft	51	Bush	64
Pierce	48	Wilson	56	Clinton	46

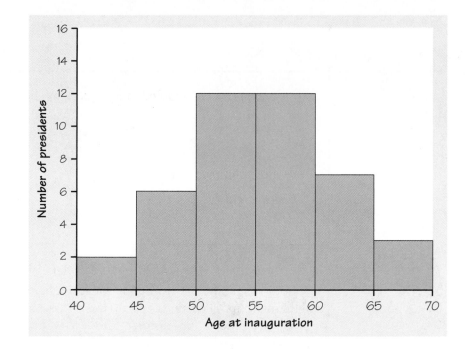

FIGURE 1.10 The distribution of the ages of presidents at their inaugurations, from Table 1.3.

at which people become president. Teddy Roosevelt was the youngest, at age 42, and Ronald Reagan, at age 69, was the oldest. The distribution is roughly symmetric. It appears that the age of a typical new president is about 55 years, because 55 is near the center of the histogram.

We just gave a brief description of a distribution that included its *shape* (fairly symmetric), a number describing its *center* (about 55), and numbers describing its *spread* (42 to 69). Shape, center, and spread provide a good description of the overall pattern of any distribution for a quantitative variable. Now we will learn specific ways to use numbers to measure the center and spread of a distribution. We can calculate these numerical measures for any quantitative variable. But to interpret measures of center and spread, and to choose among the several measures we will learn, you must think about the shape of the distribution and the meaning of the data. The numbers, like graphs, are aids to understanding, not "the answer" in themselves.

Measuring center: the mean

A description of a distribution almost always includes a measure of its center or average. The most common measure of center is the ordinary arithmetic average, or *mean*.

THE MEAN $\bar{x}$

If n observations are denoted by $x_1, x_2, \ldots, x_n$, their **mean** is

$$\bar{x} = \frac{x_1 + x_2 + \cdots + x_n}{n}$$

or in more compact notation

$$\bar{x} = \frac{1}{n} \sum x_i$$

The $\sum$ (capital Greek sigma) in the formula for the mean is short for "add them all up." The subscripts on the observations x_i are just a way of keeping the n observations distinct. They do not necessarily indicate order or any other special facts about the data. The bar over the x indicates the mean of all the x-values. Pronounce the mean $\bar{x}$ as "x-bar." This notation is very common. When writers who are discussing data use $\bar{x}$ or $\bar{y}$, they are talking about a mean.

EXAMPLE 1.6

Table 1.3 gives the ages of all 42 presidents when they were inaugurated. The mean age of a new president is

$$\bar{x} = \frac{x_1 + x_2 + \cdots + x_n}{n}$$

$$= \frac{57 + 61 + 57 + \cdots + 46}{42}$$

$$= \frac{2303}{42} = 54.8 \text{ years}$$

If you have previously entered the presidential ages into your TI-83, then use your calculator to calculate the mean. Here are the TI-83 keystrokes: 2nd / LIST / MATH / 3:mean(/ 2nd / LIST / $_L$PREZ / ENTER. Verify that the mean $\bar{x} = 54.8$ years.

EXAMPLE 1.7

A study in Switzerland examined the number of hysterectomies (removal of the uterus) performed in a year by doctors. Here are the data for a sample of 15 male doctors:

27   50   33   25   86   25   85   31   37   44   20   36   59   34   28

Enter the data into a new list named MHYST. A stemplot shows that the distribution is skewed to the right and that there are two outliers on the high side.

```
2 | 05578
3 | 13467
4 | 4
5 | 09
6 |
7 |
8 | 56
```

Use your calculator to verify that the mean number of operations that these 15 doctors performed is $\bar{x} = 41.3$. Then notice that only five of the 15 doctors performed more than the mean number of hysterectomies. That's because the two outliers (85 and 86) pull up the mean. Check that the mean of the other 13 observations is 34.5. The mean is the arithmetic average, but it is *not* the number of operations performed by a typical doctor.

nonresistant

Example 1.7 illustrates an important fact about the mean as a measure of center. We say that the mean is *nonresistant* because it is sensitive to the influence of extreme observations. These extreme observations may or may not be outliers. A skewed distribution that has no outliers will still pull the mean toward its long tail.

EXERCISES

1.24   Joey's first fourteen quiz grades in a marking period were

86   84   91   75   78   80   74
87   76   96   82   90   98   93

(a) Use the formula to calculate the mean. Use the calculator to check your work.

(b) Suppose Joey has an unexcused absence for the fifteenth quiz and he receives a score of zero. Determine his final quiz average. What property of the mean does this situation illustrate? Write a sentence about the effect of the zero on Joey's quiz average that mentions this property.

(c) What kind of plot would best show Joey's distribution of grades? Assume an 8-point grading scale (A:93–100, B: 85–92, etc.). Make an appropriate plot, and be prepared to justify your choice.

1.25   Here are the scores of 18 first-year college women on the Survey of Study Habits and Attitudes (SSHA):

154   109   137   115   152   140   154   178   101
103   126   126   137   165   165   129   200   148

(a) Find the mean score from the formula for the mean. Then enter the data into your calculator and specify 2nd / LIST / MATH / 3:mean to obtain the mean. Verify that you get the same result.

(b) A stemplot (Exercise 1.9) suggests that the score 200 is an outlier. Use your calculator to find the mean for the 17 observations that remain when you drop the outlier. Briefly describe how the outlier changes the mean.

Measuring center: the median

The mean is not the only way to describe the center of a distribution. Another natural idea is to use the "middle value" in a histogram or stemplot.

THE MEDIAN *M*

To find the **median** of a distribution:

1. Arrange all observations in order of size, from smallest to largest.
2. If the number of the observations n is odd, the median M is the center observation in the ordered list.
3. If the number of observations n is even, the median M is the mean of the two center observations in the ordered list.

That is, find the number such that half the observations are smaller and the other half are larger. This is the *median* of the distribution. We will call the median M for short. Although the idea of the median as the midpoint of a distribution is simple, we need precise rules for putting the idea into practice. The rules appear in the box above.

When working by hand, be sure to write down each individual observation in the data set, even if several observations repeat the same value. And be sure to arrange the observations in order of size before locating the median. The middle observation in the haphazard order in which the observations first come have no importance. Here is an example that shows how the rules for the median work for odd and even numbers of the observations.

EXAMPLE 1.8

To find the median number of hysterectomies performed by the fifteen doctors in Example 1.7, first arrange the observations in increasing order:

20 25 25 27 28 31 33 **34** 36 37 44 50 59 85 86

There is an odd number of observations, so there is one center observation. This is the median. The bold 34 in the list is the center observation, because there are seven observations to its left and seven to its right. So the median is $M = 34$.

The Swiss study also looked at a sample of 10 female doctors. The numbers of hysterectomies performed by these doctors (arranged in order) were

5 7 10 14 **18** **19** 25 29 31 33

Here $n = 10$ is even. There is no center observation, but there is a center pair. These are the bold 18 and 19 in the list, which have four observations to their left in the list and four to their right. The median is midway between these two observations:

$$M = \frac{18 + 19}{2} = 18.5$$

The typical female doctor performed many fewer hysterectomies than the typical male doctor. This was one of the important conclusions of the study.

Comparing the mean and the median

Examples 1.7 and 1.8 illustrate an important difference between the mean and the median. Example 1.7 shows how the two outliers pull the mean up. But they have no effect at all on the median. The outliers are just two observations in the upper half of the data. The median would stay the same even if one doctor had performed 1000 operations. We say that the *resistant* median is **resistant** to extreme observations. Don't conclude, however, that because the median is resistant, it is always preferred to the mean. The mean and median measure center in different ways, and both are useful. Use the median if you want the number of hysterectomies performed by a typical doctor. Use the mean if you are also interested in the total number of operations performed by all the doctors. The total is the mean times the number of doctors, but it has no connection with the median.

The mean and median of a symmetric distribution are close together. If the distribution is *exactly* symmetric, the mean and median are exactly the same. In a skewed distribution, the mean is farther out in the long tail than is the median. For example, the distribution of house prices is strongly skewed to the right. There are many moderately priced houses and a few very expensive mansions. The few expensive houses pull the mean up but do not affect the median. For example, the mean price of all houses sold in 1993 was $139,400, but the median price for these same houses was only $117,000. Reports about house prices, incomes and other strongly skewed distributions usually give the median ("typical value") rather than the mean ("arithmetic average value").

EXERCISES

1.26   Suppose a major league baseball team's mean yearly salary for a player is $1.2 million, and that the team has 25 players on its active roster. What is the team's annual payroll for players? If you knew only the median salary, would you be able to answer the question?

1.27   Here are the number of home runs Babe Ruth hit in each of his 15 years with the New York Yankees:

54   59   35   41   46   25   47   60   54   46   49   46   41   34   22

Roger Maris, who broke Ruth's single-year record, had these home run totals in his 10 years in the American League.

13   23   26   16   33   61   28   39   14   8

Find the median number of home runs hit in a season by each player.

1.28   In Exercise 1.25 you found the mean of the SSHA scores of 18 first-year college students. Now find the median of these scores. Is the median smaller or larger than the mean? Explain why this is so.

1.29   Last year a small accounting firm paid each of its five clerks $22,000, two junior accountants $50,000 each, and the firm's owner $270,000. What is the mean salary paid at this firm? How many of the employees earn less than the mean? What is the median salary? Write a sentence to describe how an unethical recruiter could use statistics to mislead prospective employees.

1.30   The mean and median salaries paid to major league baseball players in 1993 were $490,000 and 1,160,000. Which of these numbers is the mean, and which is the median? Explain your answer.

Measuring spread: the quartiles

The mean and median provide two different measures of the center of a distribution. A measure of center alone can be misleading. Two nations with the same median family income are very different if one has extremes of wealth and poverty and the other has little variation among families. A drug with the correct mean concentration of active ingredient is dangerous if some batches are much too high and others much too low. We are interested in the *spread* or *variability* of incomes and drug potencies as well as their centers. The simplest useful numerical description of a distribution consists of both a measure of center and a measure of spread.

range

One way to measure spread is to calculate the **range**, which is the difference between the largest and smallest observations.

EXAMPLE 1.9

Example 1.7 reports data on the number of hysterectomies performed in a year by 15 male doctors. The smallest number was 20 and the largest was 86, so the range is

$$\text{range} = 86 - 20 = 66$$

The range shows the full spread of the data. But it depends only on the smallest observation and the largest observation, which may be outliers. We can improve our description of spread by also looking at the spread of the middle half of the data. The **quartiles** mark out the middle

quartiles

half. Count up the ordered list of observations, starting from the smallest. The *first quartile* lies one-quarter of the way up the list. The *third quartile* lies three-quarters of the way up the list. In other words, the first quartile is larger than 25% of the observations, and the third quartile is larger than 75% of the observations. The second quartile is the median, which is larger than 50% of the observations. That is the idea of quartiles. We need a rule to make the idea exact. The rule for calculating the quartiles uses the rule for the median.

THE QUARTILES Q_1 AND Q_3

To calculate the **quartiles:**

1. Arrange the observations in increasing order and locate the median M in the ordered list of observations.

2. The first quartile Q_1 is the median of the observations whose position in the ordered list is to the left of the location of the overall median.

3. The third quartile Q_3 it the median of the observations whose position in the ordered list is to the right of the location of the overall median.

Here is an example that shows how the rules for the quartiles work for both odd and even numbers of observations.

EXAMPLE 1.10

The numbers of hysterectomies performed by our sample of 15 male doctors are (arranged in order)

20   25   25   27   28   31   33   **34**   36   37   44   50   59   85   86

There is an odd number of observations, so the median is the middle one, the bold 34 in the list. The first quartile is the median of the 7 observations to the left of $M = 34$. So $Q_1 = 27$. The third quartile is the median of the 7 observations to the right of $M = 34$. $Q_3 = 50$. Note that we don't include the median M when we're determining the quartiles.

For the 10 female doctors, the data are (again arranged in increasing order)

$$
\begin{array}{ccccc|ccccc}
 & & & & & 18.5 & & & & \\
5 & 7 & 10 & 14 & 18 & | & 19 & 25 & 29 & 31 & 33 \\
\text{Min} & & Q_1 & & & M & & & Q_3 & & \text{Max}
\end{array}
$$

There is an even number of observations, so the median lies midway between the middle pair. Its location is between the 5th and 6th values, marked by | in the list. The first quartile is the median of the first 5 observations, because these are the observations to the left of the location of the median. Check that $Q_1 = 10$ and $Q_3 = 29$. When the number of observations is even, all the observations enter into the calculation of the quartiles.

Be careful when several observations take the same numerical value. Write down all of the observations and apply the rules just as if they all had distinct values. For example, the median of

$$4 \quad 7 \quad 7 \quad \mathbf{7} \quad 8 \quad 9 \quad 9$$

is $M = 7$, because the bold 7 is the center observation in the list. The first quartile is the median of the three observations to the left of the bold 7, which are 4, 7, 7. So the first quartile is also 7. The third quartile is 9.

Some computer packages use a slightly different rule to find the quartiles, so computer results may be a bit different than your own work. Don't worry about this. The differences will always be too small to be important.

interquartile range   The *interquartile range*, abbreviated IQR, is defined as the distance between the first and third quartiles. The interquartile range measures the spread of the middle half of the data. If an observation falls between Q_1 and Q_3, then you know it's neither unusually high (upper 25%) nor unusually low (lower 25%). The IQR is useful for testing for outliers. We will *outlier* call an observation an *outlier* if it falls more than $1.5 \times IQR$ below Q_1 or above Q_3.

EXAMPLE 1.11

We might suspect that the 85 and the 86 hysterectomies performed by male doctors might be outliers on the high side. Let's test.

$$IQR = Q_3 - Q_1 = 50 - 27 = 23$$

$$1.5 \times IQR = 34.5$$

$$Q_3 + (1.5 \times IQR) = 50 + 34.5 = 84.5$$

Since 85 and 86 are both more extreme (farther away from the median) than 84.5, we identify them as outliers.

The five-number summary and boxplots

A convenient way to describe both the center and spread of a data set is to give the median to measure center and the quartiles and the smallest and largest individual observations to show the spread.

THE FIVE-NUMBER SUMMARY

The **five-number summary** of a data set consists of the smallest observation, the first quartile, the median, the third quartile, and the largest observation, written in order from smallest to largest. In symbols, the five-number summary is

$$\text{Minimum} \quad Q_1 \quad M \quad Q_3 \quad \text{Maximum}$$

These five numbers offer a reasonably complete description of center and spread. The five-number summaries from Example 1.10 are

$$20 \quad 27 \quad 34 \quad 50 \quad 86$$

for the male doctors and

$$5 \quad 10 \quad 18.5 \quad 29 \quad 33$$

for the female doctors.

boxplot

The five-number summary of a distribution leads to a new graph, the **boxplot.** Figure 1.11 shows boxplots for the Swiss doctors. The central box in a boxplot has its ends at the quartiles and therefore spans the middle half of the data. The line within the box marks the median. The "whiskers" at either end extend to the smallest and largest observations. We can see at once that the female doctors perform far fewer hysterectomies than male doctors, and also that there is less variation among female doctors.

You can draw boxplots either horizontally or vertically. Do be sure to include a numerical scale in the graph. When you look at a boxplot, first locate the median, which marks the center of the distribution. Then look at the spread. The quartiles show the spread of the middle half of the data, and the extremes (the smallest and largest observations) show the spread of the entire data set. The spacing of the quartiles and the extremes about the median gives an indication of the symmetry or skewness of the distribution. In a symmetric distribution, the first and third quartiles are equally distant from the median. In most distributions that are skewed to the right, on the other hand, the third quartile will be farther above the median than the first quartile is below it. The extremes behave the same way, but remember that they are just single observations and may say little about the distribution as a whole. Because boxplots show less detail than histograms or stemplots, they are best used for side-by-side comparison of more than one distribution, as in Figure 1.11. The TI-83 can plot up to three side-by-side boxplots, and it plots them horizontally.

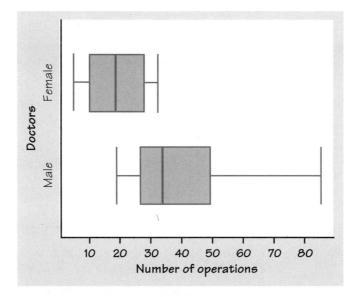

FIGURE 1.11   Side-by-side boxplots comparing the number of
hysterectomies performed by male and female doctors.

modified boxplot

Outliers usually deserve special attention. Because the regular boxplot conceals outliers, we will adopt the ***modified boxplot***, which plots
outliers as isolated points. Figures 1.12(a) and (b) show regular and
modified boxplots for the number of hysterectomies performed by male
doctors. The regular boxplot suggests a strongly right-skewed distribution. The modified boxplot shows that if not for the two outliers, the
distribution would still be right-skewed, but only slightly. (The two
outliers may not appear distinct in the modified boxplot figure because
they are so close in value.) Because the modified boxplot shows more
detail, when we say "boxplot" from now on, we will mean "modified
boxplot." The TI-82 can plot only regular boxplots, but the TI-83 gives
you a choice of regular or modified boxplot. When you construct a
(modified) boxplot by hand, extend the whisker out to the largest (or
smallest) data point that is not an outlier. Then plot outliers as isolated
points.

The TI-83 can calculate the mean, median, quartiles, and other one-
variable statistics for data stored in lists. To determine the mean and five-
number summary for the MHYST data, do the following: STAT / CALC
/ 1:1-Var Stats / 2nd / LIST and then select the MHYST list, and press
ENTER. Verify that $x = 41.3$, $n = 15$, minX $= 20$, $Q_1 = 27$, Med $=
34$, $Q_3 = 50$, and Max $= 86$.

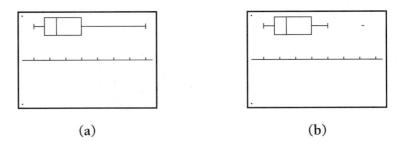

(a)   (b)

FIGURE 1.12   Regular (a) and modified (b) boxplots from the TI-83.

EXERCISES

1.31   Is the interquartile range a resistant measure of spread? Give an example of a small data set to illustrate your answer.

1.32   Exercise 1.27 (page 36) gives the number of home runs hit each year by Babe Ruth and Roger Maris.

   (a)   Find the five-number summary for each player by hand and by graphing calculator.

   (b)   Make side-by-side boxplots to compare Ruth and Maris as home run hitters. What do you conclude?

1.33   Return to the data on presidential ages in Table 1.3 (page 31). In Example 1.6 we saw that the mean age is 54.8 years.

   (a)   From the shape of the histogram (Figure 1.10, on page 31), do you expect the median to be much less than the mean, about the same as the mean, or much greater than the mean? Explain.

   (b)   Find the five-number summary and verify your expectation about the median.

   (c)   What is the IQR (the range of the middle half of the ages) of new presidents?

   (d)   Here's an interesting technique you can do with the TI-83. Define Plot1 to be the histogram, using the list named PREZ (did you enter the data earlier?). Define Plot2 to be a boxplot also using the list PREZ. Specify a point for the Mark instead of the box (□) or +. Then plot both (ZOOM / 9:ZoomStat). To adjust for the overlap: WINDOW / Ymin = −6 and Ymax = 22. Then press TRACE to inspect values. Press the up and down cursor keys to toggle between plots. Is there an outlier? If so, who was the outlier? What's the largest data point that is not an outlier? Who was that president?

1.34 Table 1.2 (page 29) contains data on education in the states. We want to compare the distributions of average SAT math and verbal scores. We enter these data into a computer with the names SATV for verbal scores and SATM for math scores. Here is output from the statistical software package Minitab that gives the five-number summary along with other information. (Other packages produce similar output.)

```
SATV
  N     MEAN    MEDIAN   STDEV     MIN      MAX       Q1       Q3
 51    448.16   443.00   30.82    397.00   511.00   422.00   476.00


SATM
  N     MEAN    MEDIAN   STDEV     MIN      MAX       Q1       Q3
 51    497.39   490.00   34.57    437.00   577.00   470.00   523.00
```

Use the output to make side-by-side boxplots of SAT math and verbal scores for the states. Briefly compare the two distributions in words.

Measuring spread: the standard deviation

The five-number summary is the most widely useful numerical description of a distribution, but it is not the most common. That distinction belongs to the combination of the mean to measure center with the *standard deviation* as a measure of spread. The standard deviation measures spread by looking at how far the observations are from their mean.

THE STANDARD DEVIATION s

The **variance** s^2 of a set of observations is the average of the squares of the deviations of the observations from their mean. In symbols, the variance of n observations $x_1, x_2, \ldots, x_n$ is

$$s^2 = \frac{(x_1 - \bar{x})^2 + (x_2 - \bar{x})^2 + \cdots + (x_n - \bar{x})^2}{n - 1}$$

or, more compactly,

$$s^2 = \frac{1}{n - 1} \sum (x_i - \bar{x})^2$$

The **standard deviation** s is the square root of the variance s^2:

$$s = \sqrt{\frac{1}{n - 1} \sum (x_i - \bar{x})^2}$$

In practice, use software or your calculator to obtain the standard deviation from keyed-in data. Doing a few examples step-by-step will help you understand how the variance and standard deviation work, however. Here is such an example.

EXAMPLE 1.12

A person's metabolic rate is the rate at which the body consumes energy. Metabolic rate is important in studies of weight gain, dieting, and exercise. Here are the metabolic rates of 7 men who took part in a study of dieting. (The units are calories per 24 hours. These are the same calories used to describe the energy content of foods.)

$$1792 \quad 1666 \quad 1362 \quad 1614 \quad 1460 \quad 1867 \quad 1439$$

The researchers reported $\bar{x}$ and s for these men.
   First find the mean:

$$\bar{x} = \frac{1792 + 1666 + 1362 + 1614 + 1460 + 1867 + 1439}{7} = \frac{11,200}{7}$$

$$= 1600$$

To see clearly the nature of the variance, start with a table of the deviations of the observations from this mean.

Observations x_i	Deviations $x_i - \bar{x}$		Squared deviations $(x_i - \bar{x})^2$	
1792	$1792 - 1600 =$	192	$192^2 =$	36,864
1666	$1666 - 1600 =$	66	$66^2 =$	4,356
1362	$1362 - 1600 =$	-238	$(-238)^2 =$	56,644
1614	$1614 - 1600 =$	14	$14^2 =$	196
1460	$1460 - 1600 =$	-140	$(-140)^2 =$	19,600
1867	$1867 - 1600 =$	267	$267^2 =$	71,289
1439	$1439 - 1600 =$	-161	$(-161)^2 =$	25,921
	sum $=$	0	sum $=$	214,870

The variance is the sum of the squared deviations by one less than the number of observations:

$$s^2 = \frac{1}{n-1}\sum(x_i - \bar{x})^2 = \frac{1}{6}(214,870) = 35,811.67$$

The standard deviation is the square root of the variance:

$$s = \sqrt{35,811.67} = 189.24$$

Compare these results for s^2 and s with those generated by your calculator or computer.

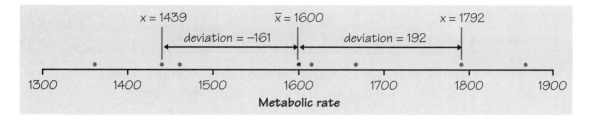

FIGURE 1.13   Metabolic rates for seven men, with their mean (*) and the deviations of two observations from the mean.

Figure 1.13 displays the data of Example 1.12 as points above the number line, with their mean marked as an asterisk (*). The arrows show two of the deviations from the mean. These deviations show how spread out the data are about their mean. Some of the deviations will be positive and some negative because some of the observations fall on each side of the mean. In fact, *the sum of the deviations of the observations from their mean will always be zero.* Check that this is true in Example 1.12. So we cannot simply add the deviations to get an overall measure of spread. Squaring the deviations makes them all positive, so that observations far from the mean in either direction will have large positive squared deviations. The variance s^2 is the average squared deviation. The variance is large if the observations are widely spread about their mean; it is small if the observations are all close to the mean.

Because the variance involves squaring the deviations, it does not have the same unit of measurement as the original observations. Lengths measured in centimeters, for example, have a variance measured in squared centimeters. Taking the square root remedies this. The standard deviation s measures spread about the mean in the original scale.

The idea of the variance is the average of the squares of the deviations of the observations from their mean. Why do we average by dividing by $n - 1$ rather than n? Because the sum of the deviations is always zero, the last deviation can be found once we know the other $n - 1$. So we are not averaging n unrelated numbers. Only $n - 1$ of the squared deviations can vary freely, and we average by dividing the total by $n - 1$. The number $n - 1$ is called the ***degrees of freedom*** of the variance or standard deviation. Many calculators offer a choice between dividing by n and dividing by $n - 1$, so be sure to use $n - 1$.

degrees
of freedom

Leaving the arithmetic to a calculator allows us to concentrate on what we are doing and why. What we are doing is measuring spread. Here are the basic properties of the standard deviation s as a measure of spread.

PROPERTIES OF THE STANDARD DEVIATION

- s measures spread about the mean and should be used only when the mean is chosen as the measure of center.
- $s = 0$ only when there is *no spread*. This happens only when all observations have the same value. Otherwise $s > 0$. As the observations become more spread out about their mean, s gets larger.
- s, like the mean $\bar{x}$, is strongly influenced by extreme observations. A few outliers can make s very large.

A skewed distribution with a few observations in the single long tail will have a large standard deviation. The number s does not give much helpful information in such a case. Because the two sides of a strongly skewed distribution have different spreads, no single number describes the spread well. The five-number summary, with its two quartiles and two extremes, does a better job. *The five-number summary is usually better than the mean and standard deviation for describing a skewed distribution. Use $\bar{x}$ and s only for reasonably symmetric distributions.*

You may rightly feel that the importance of the standard deviation is not yet clear. It is a bit complicated and is not a good description for skewed distributions. We will see in the next chapter that the standard deviation is the natural measure of spread for an important class of symmetric distributions, the normal distributions. The usefulness of many statistical procedures is tied to distributions of particular shapes. This is certainly true of the standard deviation.

Do remember that a graph gives the best overall picture of a distribution. Numerical measures of center and spread report specific facts about a distribution, but they do not describe its entire shape. Numerical summaries do not disclose the presence of multiple peaks or gaps, for example. Exercise 1.37 gives an example of a distribution for which numerical summaries alone are misleading. ALWAYS PLOT YOUR DATA.

EXERCISES

1.35   The level of various substances in the blood influences our health. Here are measurements of the level of phosphate in the blood of a patient, in milligrams of phosphate per deciliter of blood, made on 6 consecutive visits to a clinic.

$$5.6 \quad 5.2 \quad 4.6 \quad 4.9 \quad 5.7 \quad 6.4$$

A graph of only 6 observations gives little information, so we proceed to compute the mean and standard deviation.

(a) Find the mean from its definition. That is, find the sum of the 6 observations and divide by 6.

(b) Find the standard deviation from its definition. That is, find the deviations of each observation from the mean, square the deviations, then obtain the variance and the standard deviations. Example 1.11 shows the method.

(c) Now enter the data into your calculator to obtain $\bar{x}$ and s. Do the results agree with your hand calculations?

1.36 The number of hysterectomies performed in a year by the male doctors in the study described in Example 1.7 (page 33) were

$$27 \quad 50 \quad 33 \quad 25 \quad 86 \quad 25 \quad 85 \quad 31 \quad 37 \quad 44 \quad 20 \quad 36 \quad 59 \quad 34 \quad 28$$

We have previously shown that the highest two observations are outliers.

(a) The mean number of observations is $\bar{x} = 41.3$. Find the standard deviation s from its definition. That is, find the deviation of each observation from 41.3, square these deviations, then obtain the variance s^2 and the standard deviation s. Follow the model of Example 1.12.

(b) Use your calculator to verify your results. Then use your calculator to find $\bar{x}$ and s for the 13 observations that remain when you leave out the two outliers. How do the outliers affect the values of $\bar{x}$ and s? Is s a resistant measure of spread?

1.37 Exercise 1.34 (page 43) gives numerical summaries for the average SAT scores for the states. These numerical summaries (and the boxplots based on them) do *not* show one of the most important features of the distributions. Make a stemplot of the SAT math scores from Table 1.2 (page 29). What is the overall shape of the distribution? Remember to always start with a graph of your data—numerical summaries are not a complete description.

SUMMARY

A numerical summary of distribution should report its **center** and its **spread** or **variability**.

The **mean** $\bar{x}$ and the **median** M describe the center of a distribution in different ways. The mean is the arithmetic average of the observations, and the median is the midpoint of the values.

When you use the median to indicate the center of the distribution, describe its spread by giving the **quartiles**. The **first quartile** Q_1 has one-fourth of the observations below it, and the **third quartile** Q_3 has three-fourths of the observations below it. An extreme observation is an **outlier** if it is smaller than $Q_1 - (1.5 \times IQR)$ or larger than $Q_3 + (1.5 \times IQR)$.

The **five-number summary** consisting of the median, the quartiles, and the high and low extremes provides a quick overall description of a distribution. The median describes the center, and the quartiles and extremes show the spread.

Boxplots based on the five-number summary are useful for comparing several distributions. The box spans the quartiles and shows the spread of the central half of the distribution. The median is marked within the box. The whiskers extend to the smallest (or largest) observation that is not an outlier. Outliers are plotted as isolated points.

The **variance** s^2 and especially its square root, the **standard deviation** s, are common measures of spread about the mean as center. The standard deviation s is zero when there is no spread and gets larger as the spread increases.

The mean and standard deviation are strongly influenced by outliers or skewness in a distribution. They are good descriptions for symmetric distributions and are most useful for the normal distributions introduced in the next section.

The median and quartiles are not affected by outliers, and the two quartiles and two extremes describe the two sides of a distribution separately.

The five-number summary is the preferred numerical summary for skewed distributions.

SECTION 1.2 EXERCISES

1.38   Here are the percents of the popular vote won by the successful candidate in each of the presidential elections from 1948 to 1992:

Year	1948	1952	1956	1960	1964	1968	1972	1976	1980	1984	1988	1992
Percent	49.6	55.1	57.4	49.7	61.1	43.4	60.7	50.1	50.7	58.8	53.9	43.2

(a) Make a stemplot of the winners' percents. (Round to whole numbers and use split stems.)

(b) What is the median percent of the vote won by the successful candidates in presidential elections? (Work with the unrounded data.)

(c) Call an election a landslide if the winner's percent falls at or above the third quartile. Find the third quartile. Which elections were landslides?

1.39 Some people worry about how many calories they consume. *Consumer Reports* magazine, in a story on hot dogs, measured the calories in 20 brands of beef hot dogs, 17 brands of meat hot dogs, and 17 brands of poultry hot dogs.[8] The data are given in Table 1.4. Use this information to make side-by-side boxplots of the calorie counts for the three types of hot

TABLE 1.4		Calories and sodium in three types of hot dogs			
Beef hot dogs		*Meat hot dogs*		*Poultry hot dogs*	
Calories	Sodium	Calories	Sodium	Calories	Sodium
186	495	173	458	129	430
181	477	191	506	132	375
176	425	182	473	102	396
149	322	190	545	106	383
184	482	172	496	94	387
190	587	147	360	102	542
158	370	146	387	87	359
139	322	139	386	99	357
175	479	175	507	170	528
148	375	136	393	113	513
152	330	179	405	135	426
111	300	153	372	142	513
141	386	107	144	86	358
153	401	195	511	143	581
190	645	135	405	152	588
157	440	140	428	146	522
131	317	138	339	144	545
149	319				
135	298				
132	253				

SOURCE: *Consumer Reports*, June 1986, pp. 366–367.

dogs. Write a brief comparison of the distributions. Will eating poultry hot dogs usually lower your calorie consumption compared with eating beef or meat hot dogs?

1.40 We want to compare spending on public education in the northeastern and the southern states. Because states vary in population, we should compare dollars spent per pupil rather than total dollars spent. Table 1.2 (page 29) gives the dollars spent per pupil in each state. Take the northeastern states to be those in the MA (Middle Atlantic) and NE (New England) regions. The southern states are those in SA (South Atlantic) and ESC (East South Central) regions. Leave out the District of Columbia, which is a city rather than a state.

(a) List the dollars-per-pupil data for the northeastern and for the southern states from Table 1.2. These are the two data sets we want to compare.

(b) Make numerical summaries and graphs to compare the two distributions. Write a brief statement of what you find.

1.41 In 1798 the English scientist Henry Cavendish measured the density of the earth with great care. It is common practice to repeat careful measurements several times and use the mean as the final result. Cavendish repeated his work 29 times. Here are his results (the data give the density of the earth as a multiple of the density of water):[9]

5.50	5.61	4.88	5.07	5.26	5.55	5.36	5.29	5.58	5.65
5.57	5.53	5.62	5.29	5.44	5.34	5.79	5.10	5.27	5.39
5.42	5.47	5.63	5.34	5.46	5.30	5.75	5.68	5.85	

Present these measurements with a graph of your choice. Does the shape of the distribution allow the use of $\bar{x}$ and s to describe it? Find $\bar{x}$ and s. What is your estimate of the density of the earth based on these measurements?

1.42 Table 1.1 (page 10) records the percent of people age 65 and over living in each of the states. Figure 1.2(c) (page 10) is a histogram of these data. Do you prefer the five-number summary or $\bar{x}$ and s as a brief numerical description? Why? Calculate your preferred description.

1.43 The 1993 New York Mets had the worst won-lost record in major league baseball. They were paid well for playing poorly. Here are the Mets' salaries, in thousands of dollars. (For example, 6200 stands for Bobby Bonilla's salary of $6,200,000.)[10]

6200	5917	4000	3375	3000	2312	2300	2150	2100
1500	1012	850	650	635	500	475	220	205
195	195	158	145	109	109	109	109	109

Describe this salary distribution both with a graph and with an appropriate numerical summary. Then write a brief description of the important features of the distribution.

1.44 A study of the size of jury awards in civil cases (such as injury, product liability, and medical malpractice) in Chicago showed that the median award was about $8000. But the mean award was about $69,000. Explain how this great difference between the two measures of center can occur.

1.45 Which measure of center, the mean or the median, should you use in each of the following situations?

(a) Middletown is considering imposing an income tax on citizens. The city government wants to know the average income of citizens so that it can estimate the total tax base.

(b) In a study of the standard of living of typical families in Middletown, a sociologist estimates the average family income in that city.

1.46 You want to measure the average speed of vehicles on the interstate highway on which you are driving. You adjust your speed until the number of vehicles passing you equals the number you are passing. Have you found the mean speed or the median speed of vehicles on the highway?

1.47 This is a standard deviation contest. You must choose four numbers from the whole numbers 0 to 10, with repeats allowed.

(a) Choose four numbers that have the smallest possible standard deviation.

(b) Choose four numbers that have the largest possible standard deviation.

(c) Is more than one choice possible in either (a) or (b)? Explain.

CHAPTER REVIEW

Data analysis is the art of describing data using graphs and numerical summaries. The purpose of data analysis is to describe the most important features of a set of data. This chapter introduces data analysis by presenting statistical ideas and tools for describing the distribution of a single variable. To describe a distribution, we begin with a graph and add numerical descriptions of specific aspects of the distributions to describe the overall pattern. Here is a review list of the most important skills you should have acquired from your study of this chapter.

A. DATA

1. Identify the individuals and variables in a set of data.
2. Identify each variable as categorical or quantitative. Identify the units in which each quantitative variable is measured.

B. DOTPLOTS AND HISTOGRAMS

1. Make a dotplot that records dots for individual observations.
2. Make a histogram of the distribution of a quantitative variable when you are given counts for classes of equal width.
3. Make a histogram from a set of observations by choosing classes of equal widths, finding the class counts, and drawing the histogram.

C. STEMPLOTS

1. Make a stemplot of the distribution of a small set of observations.
2. Round leaves or split stems as needed to make an effective stemplot.

D. INSPECTING DISTRIBUTIONS

1. Look for the overall pattern and for major deviations from the pattern.
2. Describe the overall pattern by giving numerical measures of center and spread and a verbal description of shape.
3. Assess from a dotplot, stemplot, or histogram whether the shape of a distribution is roughly symmetric, distinctly skewed, or neither.
4. Decide which measures of center and spread are more appropriate: the mean and standard deviation (especially for symmetric distributions) or the five-number summary (especially for skewed distributions).
5. Recognize outliers.

E. TIME PLOTS

1. Make a time plot of data, with the time of each observation on the horizontal axis and the value of the observed variable on the vertical axis.
2. Recognize strong trends or other patterns in a time plot.

F. MEASURING CENTER

1. Calculate the mean $\bar{x}$ of a set of observations using a calculator.
2. Determine the median M of a set of observations.

3. Understand that the median is less affected by extreme observations than the mean. Recognize that skewness in a distribution moves the mean away from the median toward the long tail.

G. MEASURING SPREAD

1. Calculate the quartiles Q_1 and Q_3 and the *IQR* for a set of observations.
2. Give the five-number summary and draw a boxplot; assess symmetry and skewness from a boxplot.
3. Calculate the standard deviation s for a set of observations using a calculator.
4. Know the basic properties of s: $s \geq 0$ always; $s = 0$ only when all observations are identical and increases as the spread increases; s has the same units as the original measurements; s is pulled strongly up by outliers or skewness.

CHAPTER 1 REVIEW EXERCISES

1.48 The Province of Ontario carries out statistical studies of the working of Canada's national health care system in the province. The bar charts in Figure 1.14 come from a study of admissions and discharges from community hospitals in Ontario.[11] They show the number of heart attack patients admitted and discharged on each day of the week during 1992 and 1993.

(a) Explain why you expect the number of patients admitted with heart attacks to be roughly the same for all days of the week. Do the data show that this is true?

(b) Describe how the distribution of the day on which patients are discharged from the hospital differs from that of the day on which they are admitted. What do you think explains the difference?

1.49 Professor Moore, who lives a few miles outside a college town, records the time he takes to drive to the college each morning. Here are the times (in minutes) for 42 consecutive weekdays, with the dates in order along the rows:

8.25	7.83	8.30	8.42	8.50	8.67
8.17	9.00	9.00	8.17	7.92	9.00
8.50	9.00	7.75	7.92	8.00	8.08
8.42	8.75	8.08	9.75	8.33	7.83
7.92	8.58	7.83	8.42	7.75	7.42
6.75	7.42	8.50	8.67	10.17	8.75
8.58	8.67	9.17	9.08	8.83	8.67

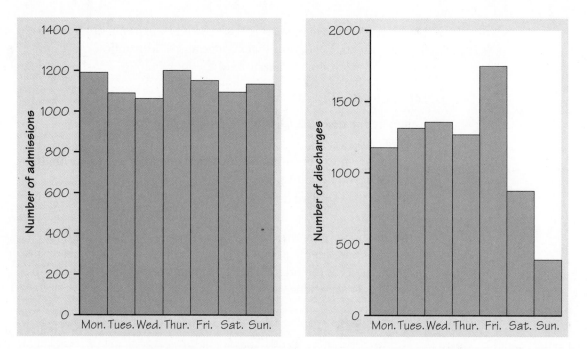

FIGURE 1.14   Bar charts of the number of heart attack victims admitted and discharged on each day of the week by hospitals in Ontario, Canada (Exercise 1.48).

(a) Make a stemplot of these drive times. (Round to the nearest tenth of a minute and use split stems.) Is the distribution roughly symmetric, clearly skewed, or neither? Are there any clear outliers?

(b) Make a time plot of the drive times. (Label the horizontal axis in days, 1 to 42.) The plot shows no clear trend, but it does show one unusually low drive time and two unusually high drive times. Circle these observations on your plot.

(c) All three unusual observations can be explained. The low time is the day after Thanksgiving (no traffic on campus). The two high times reflect delays due to an accident and icy roads. Remove these three observations, and find the mean $\bar{x}$ and standard deviation s of the remaining 39 drive times.

1.50   Corn is an important animal food. Normal corn lacks certain amino acids, which are building blocks for protein. Plant scientists have developed new corn varieties that have more of these amino acids. To test a new corn as an animal food, a group of 20 one-day-old male chicks was fed a ration containing the new corn. A control group of another 20 chicks was fed a ration that was identical expect that it contained normal corn. Here are the weight gains (in grams) after 21 days.[12]

Normal corn		New corn	
380	366	361	401
283	402	434	393
356	329	406	467
350	316	427	477
345	360	430	410
321	356	447	375
349	462	403	426
410	399	318	407
384	272	420	392
455	431	339	326

(a) Compute five-number summaries for the weight gains of the two groups of chicks. Then make boxplots to compare the two distributions. What do the data show about the effect of the new corn?

(b) The researchers actually reported means and standard deviations for the two groups of chicks. What are they? How much larger is the mean weight gain of chicks fed the new corn?

1.51 Joe DiMaggio played center field for the Yankees for 13 years. He was succeeded by Mickey Mantle, who played for 18 years. Here is the number of home runs hit each year by DiMaggio:

29   46   32   30   31   30   21   25   20   39   14   32   12

Here are Mantle's home run counts:

13   23   21   27   37   52   34   42   31   40   54   30   15   35   19   23   22   18

Compute the five-number summary for each player, and make side-by-side boxplots of the home run distributions. What does your comparison show about DiMaggio and Mantle as home run hitters?

1.52 Environmental Protection Agency regulations require automakers to give the city and highway gas mileages for each model of car. The highway mileages (miles per gallon) for 30 midsize and large 1994 car models are given below.[13]

Model	MPG	Model	MPG
BMW 740i	23	Hyundai Sonata	27
Buick Century	31	Infiniti Q45	22
Buick LeSabre	28	Lexus LS400	23
Buick Park Avenue	27	Lincoln Continental	26
Buick Regal	29	Lincoln Mark VIII	25
Buick Roadmaster	25	Mazda 626	31
Cadillac DeVille	25	Mazda 929	24
Chevrolet Caprice	26	Mercedes-Benz S320	24
Chevrolet Lumina	29	Mercedes-Benz S420	20
Chrysler Concorde	28	Nissan Maxima	26
Chrysler New Yorker	26	Rolls-Royce Silver Spur	15
Dodge Spirit	27	Saab 900	26
Ford LTD	25	Saab 9000	27
Ford Taurus	29	Toyota Camry	28
Ford Thunderbird	26	Volvo 850	26

(a) Make a graph that displays the distribution. Describe its main features (overall patterns and any outliers) in words.

(b) The government imposes a "gas guzzler" tax on cars with low gas mileage. Which of these cars do you think are certainly subject to the gas guzzler tax? (In fact, four of these cars are subject to the tax, but highway mileage alone doesn't point to all four.)

(c) Find the five-number summary for these data. What is the highway gas mileage of a "typical" 1994 midsize or large car? How many miles per gallon must a car achieve to be in the top quarter of these models?

(d) Find the mean highway gas mileage for these cars. What feature of the distribution explains the difference between the mean and the median?

1.53 The data below are survival times in days of 72 guinea pigs after they were injected with tubercle bacilli in a medical experiment.[14] Survival times, whether of machines under stress or cancer patients after treatment, usually have distributions that are skewed to the right.

43	45	53	56	56	57	58	66	67	73	74	79
80	80	81	81	81	82	83	83	84	88	89	91
91	92	92	97	99	99	100	100	101	102	102	102
103	104	107	108	109	113	114	118	121	123	126	128
137	138	139	144	145	147	156	162	174	178	179	184
191	198	211	214	243	249	329	380	403	511	522	598

(a) Graph the distribution and describe its main features. Does it show the expected right skew?

(b) Here is the output from the computer statistical package Data Desk for these data:

```
Summary statistics for days

Mean 141.84722
Median 102.50000
Cases 72
StdDev 109.20863
Min 43
Max 598
25th%ile 82.250000
75th%ile 153.75000
```

(Data Desk uses "Cases" for the number of observations, and "25th%ile" for the first quartile, which is also called the 25th percentile because 25% of the data lie below it. Similarly, "75th%ile" is the third quartile.) Explain how the relationship between the mean and the median reflects the skewness of the data.

(c) Give the five-number summary and explain briefly how it reflects the skewness of the data.

1.54 The rate of return on a stock is its change in price plus any dividends paid, usually measured in percent of the starting value. We have data on the monthly rate of return for the stock of Wal-Mart stores for the years 1973 to 1991, the first 19 years Wal-Mart was listed on the New York Stock Exchange. There are 228 observations. Here is output from the computer statistical package SPLUS that describes the distribution of these data:

```
Mean      =    3.064
Standard deviation   =    11.49

N = 228   Median = 3.4691
Quartiles = -2.950258, 8.4511

Decimal point is 1 place to the right of the colon

Low:   -34.04255 -31.25000 -27.06271 -26.61290

-1 : 985
-1 : 444443322222110000
-0 : 99998877766666665555
-0 : 444444443333333222222222221111111100
```

```
0 :  000001111111111122222233333333344444444
0 :  5555555555555555555556666666666667777777888888888899999
1 :  0000000001111111122233334444
1 :  55566667889
2 :  011334
```

```
High:   32.01923 41.80531 42.05607 57.89474 58.67769
```

SPLUS gives high and low outliers separately from the stemplot rather than spreading out the stemplot to include them. Notice that the stems in the plot are the tens digits of the percent returns. The leaves are the ones digits.

(a) Give the five-number summary for monthly returns on Wal-Mart stock.

(b) If you had $1000 worth of Wal-Mart stock at the beginning of the best month during these 19 years, how much would your stock be worth at the end of the month? If you had $1000 worth of stock at the beginning of the worst month, how much would your stock be worth at the end of the month?

(c) Find the interquartile range IQR for the Wal-Mart data. Are there any outliers according to the $1.5 \times IQR$ criterion? Does it appear to you that SPLUS uses this criterion in choosing which observations to report separately as outliers?

(d) Describe in words the main features of the distribution.

1.55 Has the behavior of Wal-Mart stock changed over the 19 years 1973 to 1991? In Exercise 1.54 we saw the distribution of all 228 monthly returns. That display can't answer questions about change over time. Figure 1.15 is a variation of a time plot. Rather than plotting all 228 observations, we give side-by-side boxplots that compare the 19 years with each other. There are 12 monthly returns in each year.

(a) Is there a long term trend in the typical monthly return over time?

(b) Is there a trend in the spread of the monthly returns?

(c) The stemplot in Exercise 1.54 reports several outliers. Which of these can you spot in the boxplots? In what years did they occur? Does this reinforce your conclusions from part (b)? Are there any outliers that are especially striking after taking your result from (b) into account?

1.56 Julie says, "People are living longer now, so presidents are likely to be older than in the past." John replies, "No—modern voters like youth and don't respect age, so presidents are likely to be younger now than earlier in our history."

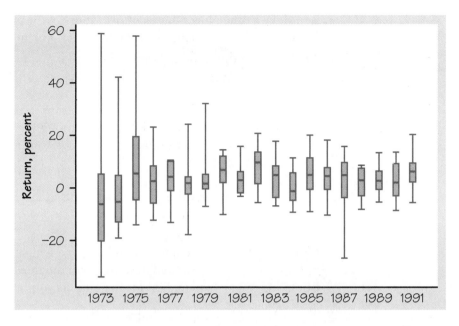

FIGURE 1.15   Side-by-side boxplots comparing the distributions of monthly rates of return on Wal-Mart stock for 19 years (Exercise 1.55).

Make a time plot of the presidential ages in Table 1.3 (page 31). Take the horizontal axis to be the order of the presidents, from 1 for Washington to 42 for Clinton. Are there any clear trends over time? Do the data support either Julie or John?

1.57   The years around 1970 brought unrest to many U.S. cities. Here are government data on the number of civil disturbances in each 3-month period during the years 1968 to 1972:

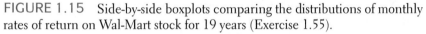

Period	Count	Period	Count
1968, Jan.–Mar.	6	1970, July–Sept.	20
Apr.–June	46	Oct.–Dec.	6
July–Sept.	25	1971, Jan.–Mar.	12
Oct.–Dec.	3	Apr.–June	21
1969, Jan.–Mar.	5	July–Sept.	5
Apr.–June	27	Oct.–Dec.	1
July–Sept.	19	1972, Jan.–Mar.	3
Oct.–Dec.	6	Apr.–June	8
1970, Jan.–Mar.	26	July–Sept.	5
Apr.–June	24	Oct.–Dec.	5

(a) Make a time plot of these counts. Connect the points in your plot with straight line segments to make the pattern clearer.

(b) Describe the trend and any other striking pattern in your time plot. Can you suggest an explanation for the pattern in civil disorders?

1.58   Table 1.2 reports data on the states. Much more information is available. Do your own exploration of differences among the states. Find in the library a recent edition of the annual *Statistical Abstract of the United States.* Look up data on either

(a)   marriage rates per thousand inhabitants or

(b)   death rates per thousand inhabitants

for the 50 states. Make a graph and a numerical summary to display the distribution and write a brief description of the most important characteristics. Suggest an explanation for the outliers you see.

1.59   The NASDAQ Composite Index describes the average price of common stock traded over the counter, that is, not on one of the stock exchanges. In 1991, the mean capitalization of the companies in the NASDAQ index was $80 million and the median capitalization was $20 million. (A company's capitalization is the total market value of its stock.) Explain why the mean capitalization is much higher than the median.

NOTES AND DATA SOURCES

1. The Shakespeare data appear in C. B. Williams, *Style and Vocabulary: Numerological Studies,* Griffin, London, 1970.

2. Data from John K. Ford, "Diversification: how many stocks will suffice?" *American Association of Individual Investors Journal,* January 1990, pp. 14–16.

3. Data on frosts from C. E. Brooks and N. Carruthers, *Handbook of Statistical Methods in Meteorology,* H. M. Stationery Office, 1953.

4. Data provided by Joseph A Wipf, Department of Foreign Languages and Literatures, Purdue University.

5. The Virginia highway fatalities data provided by the Virginia Department of Motor Vehicles.

6. Hurricane data from H. C. S. Thom, *Some Methods of Climatological Analysis,* World Meteorological Organization, Geneva, Switzerland, 1966.

7. Data from *The Baseball Encyclopedia,* 3rd ed., Macmillan, New York, 1976. Maris's home run data are from the same source.

8. *Consumer Reports*, June 1986, pp. 366–367. A more recent study of hot dogs appears in *Consumer Reports*, July 1993, pp. 415–419. The newer data cover few brands of poultry hot dogs and take calorie counts mainly from the package labels, resulting in suspiciously round numbers.

9. Cavendish's data and the background information about his work appear in S. M. Stigler, "Do robust estimators work with real data?" *Annals of Statistics*, 5 (1977), pp. 1055–1078.

10. The Mets' salaries were reported in the *New York Times*, April 11, 1993.

11. Based on Antoni Basinski, "Almost never on Sunday: implications of the patterns of admission and discharge for common conditions," Institute for Clinical Evaluative Sciences in Ontario, October 18, 1993.

12. Based on data summaries in G. L. Cromwell, et al., "A comparison of the nutritive value of *opaque-2, floury-2*, and normal corn for the chick," *Poultry Science*, 57 (1968), pp. 840–847.

13. The gas mileage data are from the U.S. Department of Energy's *1994 Gas Mileage Guide*, October 1993. In the table, the data are for the basic engine/transmission combination for each model, and models that are essentially identical (such as the Ford Taurus and the Mercury Sable) appear only once.

14. Data from T. Bjerkedal, "Acquisition of resistance in guinea pigs infected with different doses of virulent tubercle bacilli." *American Journal of Hygiene*, 72 (1960), pp. 130–148.

The Normal Distributions

2.1 DENSITY CURVES AND THE NORMAL DISTRIBUTIONS

We now have a kit of graphical and numerical tools for describing distributions. What is more, we have a clear strategy for exploring data on a single quantitative variable:

• Start with a graph, usually a stemplot or histogram.
• Look for the overall pattern and for striking deviations such as outliers.
• Choose a numerical summary to briefly describe center and spread.

Here is one more step to add to this strategy:

• Sometimes the overall pattern of a large number of observations is so regular that we can describe it by a smooth curve.

| ACTIVITY 2A | A Fine-Grained Distribution |

Materials: Sheet of grid paper; salt; can of spray paint; paint easel; newspapers

1. Place the grid paper on the easel with a horizontal fold as shown, at about a 45° angle to the horizontal. Provide a "lip" at the bottom to catch the salt. Place newspaper behind the grid and extending out on all sides so you will not get paint on the easel.

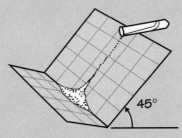

2. Pour a stream of salt slowly from a point near the middle of the top edge of the grid. The grains of salt will hop and skip their way down the grid as they collide with one another and bounce left and right. They will accumulate at the bottom, piled against the grid, with the smooth profile of a bell-shaped curve, known as a normal distribution. We will learn about the normal distribution in this chapter.

(continued)

3. Now carefully spray the grid—salt and all—with paint. Then discard the salt. You should be able to easily measure the height of the curve at different places by simply counting lines on the grid, or you could approximate areas by counting small squares or portions of squares in the grid.

How could you get a tall, narrow curve? How could you get a short, broad curve? What factors might affect the height and breadth of the curve? From the members of the class, collect a set of normal curves that differ from one another.

ACTIVITY 2B | Roll a Normal Distribution

Materials: Several marbles, all the same size; two metersticks for a "ramp"; a ruled sheet of paper; a flat table about 4 feet long; carbon paper; Scotch Tape or masking tape

1. At one end of the table prop up the two metersticks in a "V" shape to provide a ramp for the marbles to roll down. The marble will roll down the chute, continue across the table, and fall off the table to the floor below. Make sure that the ramp is secure and that the tabletop does not have any grooves or obstructions.

2. Roll the marble down the ramp several times to get a good idea of the area of the floor where the marble will fall.

3. Center the ruled sheet of paper (see Figure 2.1) over this area, face up, with the bottom edge toward the table and parallel to the edge of the table. The ruled lines should go in the same direction as the marble's path. Tape the sheet securely to the floor. Place the sheet of carbon paper, carbon side down, over the ruled sheet.

4. Roll the marble for a class total of 200 times. The spots where it hits the floor will be recorded on the ruled paper as black dots. When the marble hits the floor, it will probably bounce, so try to catch it in midair after the impact so that you don't get any extra marks. After the first 100 rolls, replace the sheet of paper. This will make it easier for you to count the spots. Make sure that the second sheet is in exactly the same position as the first one.

(continued)

5. When the marble has been rolled 200 times, make a histogram of the distribution of the points as follows. First, count the number of dots in each column. Then graph this number by drawing horizontal lines in the columns at the appropriate level. Use the scale on the left-hand side of the sheet.

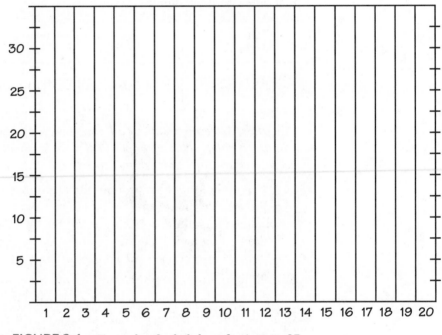

FIGURE 2.1   Example of ruled sheet for Activity 2B.

Density curves

Look at Figure 2.2. The histogram in that figure displays the distribution of the scores of all 947 seventh-grade students in Gary, Indiana, on the vocabulary part of the Iowa Test of Basic Skills.[1] Scores of many students on this national test have a quite regular distribution. The histogram is symmetric, and both tails fall off quite smoothly from a single center peak. There are no large gaps or obvious outliers. The smooth curve drawn through the tops of the histogram bars is a good description of the overall pattern of the data.

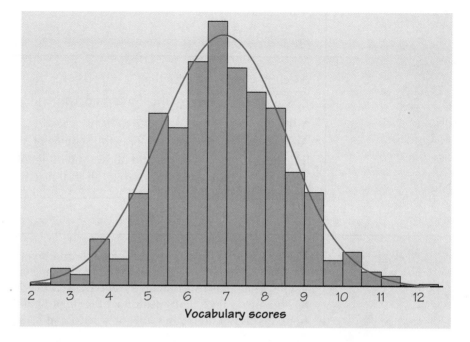

FIGURE 2.2 Histogram of the vocabulary scores of all seventh-grade students in Gary, Indiana. The smooth curve shows the overall shape of the distribution.

We will see that it is easier to work with the smooth curve in Figure 2.2 than with the histogram. The reason is that the histogram depends on our choice of classes, while with a little care we can use a curve that does not depend on any choices we make. Here's how we do it.

- We will use a smooth curve to describe what *proportions* of the observations fall in each range of values, not the counts of observations.

- Our eyes respond to the areas of the bars in a histogram. The bar areas represent the proportions of observations falling into each class. The same is true of the smooth curve: *areas under the curve represent proportions of the observations.*

- Adjust the scale of the graph so that *the total area under the curve is exactly 1.* This area represents the proportion 1, that is, all the observations. The area under the curve and above any range of values on the horizontal axis is equal to the proportion of observations falling in this range, thought of as a fraction of the whole. The curve is a *density curve* for the distribution.

DENSITY CURVE

A **density curve** is a curve that
- is always on or above the horizontal axis, and
- has an area exactly 1 underneath it.

A density curve describes the overall pattern of a distribution. The area under the curve and above any range of values is the proportion of all observations that fall in that range.

EXAMPLE 2.1

Figure 2.3 shows the density curve for a distribution that is slightly skewed to the left. The smooth curve makes the overall shape of the distribution clearly visible. The shaded area under the curve covers the range of values from 7 to 8. This area is 0.12. This means that the proportion 0.12 of all observations from this distribution have values between 7 and 8.

   Density curves, like distributions, come in many shapes. Figure 2.4 shows two density curves: a symmetric density curve and a right-skewed

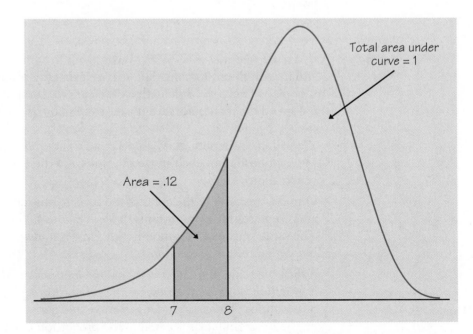

FIGURE 2.3   The shaded area under this density curve is the proportion of observations taking values between 7 and 8 (Example 2.1).

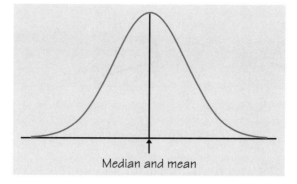

FIGURE 2.4(a)   The median and mean of a symmetric density curve.

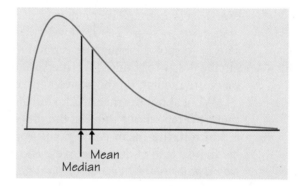

FIGURE 2.4(b)   The median and mean of a right-skewed density curve.

curve. A density curve is often an adequate description of the overall pattern of a distribution. The curve doesn't describe outliers, which are deviations from the overall pattern. Of course, no set of real data is exactly described by a density curve. The curve is an approximation that is easy to use and accurate enough for practical use.

The median and mean of a density curve

Our measures of center and spread apply to density curves as well as to actual sets of observations. The median and quartiles are easy. Areas under a density curve represent proportions of the total number of observations. The median is the point with half the observations on either side. So *the median of a density curve is the equal-areas point*, the point with half the area under the curve to its left and the remaining half of the area to its right. The quartiles divide the area under the curve into quarters. One-fourth of the area under the curve is to the left of the first quartile, and

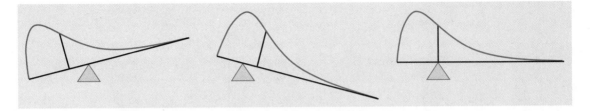

FIGURE 2.5   The mean is the balance point of a density curve.

three-fourths of the area is to the left of the third quartile. You can roughly locate the median and quartiles of any density curve by eye by dividing the area under the curve into four equal parts.

A symmetric density curve, unlike most distributions of real data, is exactly symmetric. The median of a symmetric density curve is therefore at its center. Figure 2.4(a) shows the median of a symmetric curve. It isn't so easy to spot the equal-areas point on a skewed curve. There are mathematical ways of finding the median for any density curve. We did that to mark the median on the skewed curve in Figure 2.4(b).

What about the mean? The mean of a set of observations is their arithmetic average. If we think of the observations as weights strung out along a thin rod, the mean is the point at which the rod would balance. This fact is also true of density curves. *The mean is the point at which the curve would balance if made of solid material.* Figure 2.5 illustrates this fact about the mean. A symmetric curve balances at its center because the two sides are identical. So *the mean and median of a symmetric density curve are equal,* as in Figure 2.4(a). We know that the mean of a skewed distribution is pulled toward the long tail. Figure 2.4 (b) shows how the mean of a skewed density curve is pulled toward the long tail more than is the median. It's hard to locate the balance point by eye on a skewed curve. There are mathematical ways of calculating the mean for any density curve, so we are able to mark the mean as well as the median in Figure 2.4(b).

MEDIAN AND MEAN OF A DENSITY CURVE

The **median** of a density curve is the equal-areas point, the point that divides the area under the curve in half.

The **mean** of a density curve is the balance point, at which the curve would balance if made of solid material.

The median and mean are the same for a symmetric density curve. They both lie at the center of the curve.

We can roughly locate the mean, median, and quartiles of any density curve by eye. This is not true of the standard deviation. Remember the earlier remark that the standard deviation is not a natural measure for most distributions. When necessary, we can once again call on more advanced mathematics to learn the value of the standard deviation. The study of mathematical methods for doing calculations with density curves is part of theoretical statistics. Though we are concentrating on statistical practice, we often make use of the results of mathematical study.

mean μ

standard deviation σ

Because a density curve is an idealized description of the distribution of data, we need to distinguish between the mean and standard deviation of the density curve and the mean $\bar{x}$ and standard deviation s computed from the actual observations. The usual notation for the mean of an idealized distribution is μ (the Greek letter mu). We write the standard deviation of a density curve as σ (the Greek letter sigma).

EXERCISES

2.1   (a)   Sketch a density curve that is symmetric but has a shape different from that of a curve in Figure 2.4(a).

(b)   Sketch a density curve that is strongly skewed to the left.

2.2   Figure 2.6 displays the density curve of a *uniform distribution*. The curve takes the constant value 1 over the interval from 0 to 1 and is zero outside the range of values. This means that data described by this distribution take values that are uniformly spread between 0 and 1. Use areas under this density curve to answer the following questions.

(a)   What percent of the observations lie above 0.8?

(b)   What percent of the observations lie below 0.6?

(c)   What percent of the observations lie between 0.25 and 0.75?

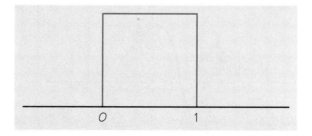

FIGURE 2.6   The density curve of a uniform distribution (Exercise 2.2).

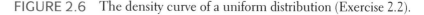

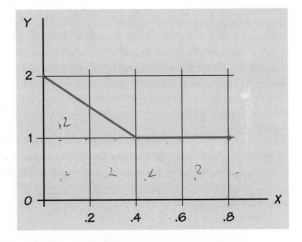

FIGURE 2.7   An unusual "broken line" density curve.

2.3   Figure 2.7 shows an unusual density curve. Use areas under this density curve to find the proportion of observations within the following intervals:

(a) $0.6 \leq X \leq 0.8$
(b) $0 \leq X \leq 0.4$
(c) $0 \leq X \leq 0.8$
(d) $0 \leq X \leq 0.2$

2.4   Figure 2.8 displays three density curves, each with three points indicated. At which of these points on each curve do the mean and the median fall?

2.5   In this exercise we will pretend to roll a regular, 6-sided die 120 times. Each time we roll the die, we will record the number on the up face. We say that the numbers 1, 2, 3, 4, 5, and 6 are the **outcomes** of this experiment.

outcomes

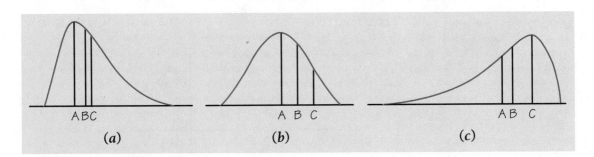

(a)   *(b)*   *(c)*

FIGURE 2.8   Three density curves (Exercise 2.4).

simulation

In 120 rolls, how many of each number would you expect to roll? The TI-83 is a useful device for conducting experiments and is particularly handy for performing many repetitions of an experiment like this. Because we are only pretending to roll the die repeatedly, we call this experiment a *simulation.* There will be a more formal treatment of simulations in a later chapter. Begin by clearing the list L_2 on your TI-83. The command `randInt(1,6,120)→L2` uses the calculator's random number generator to generate 120 random whole numbers between 1 and 6 and then to store these numbers in list L_2. Enter the command `randInt`, which is in the MATH / PROB menu. Then set the viewing window parameters as shown.

```
WINDOW
 Xmin=1
 Xmax=7
 Xscl=1
 Ymin=-5
 Ymax=25
 Yscl=5
 Xres=1
```

Specify a histogram using the data in list L_2. Finally, press GRAPH to plot the results. Are you surprised? This is called a **frequency histogram** because it plots the **frequency** of each outcome (number of times each outcome occurred).

```
Plot1  Plot2  Plot3
On  Off
Type: ⌐⌐  ⌐⌐  ⊞⊞
       ⊹⊹  ⊞⊞  ⌐⌐
Xlist:L2
Freq:1
```

Repeat the simulation several times. You can recall and reuse the previous command by pressing 2nd / ENTRY. It's a good habit to clear list L_2 before you roll the die again.

In theory, of course, each number should come up 20 times. But in practice, there is chance variation, so that the bars in the histogram will probably be different heights. Theoretically, what should the distribution look like?

Normal Distributions

One particularly important class of density curves has already appeared in Figures 2.2 and 2.4(a) and the "fine-grained distribution" of Activity 2A. These density curves are symmetric, single-peaked, and bell-shaped. They

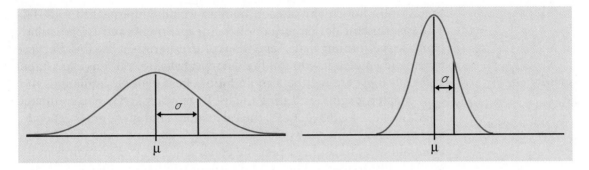

FIGURE 2.9   Two normal curves, showing the mean μ and standard deviation σ.

normal distributions

are called *normal curves,* and they describe **normal distributions.** All normal distributions have the same overall shape. The exact density curve for a particular normal distribution is described by giving its mean μ and its standard deviation σ. The mean is located at the center of the symmetric curve, and is the same as the median. Changing μ without changing σ moves the normal curve along the horizontal axis without changing its spread. The standard deviation σ controls the spread of a normal curve. Figure 2.9 shows two normal curves with different values of σ. The curve with the larger standard deviation is more spread out.

The standard deviation σ is the natural measure of spread for normal distributions. Not only do μ and σ completely determine the shape of a normal curve, but we can locate σ by eye on the curve. Here's how. As we move out in either direction from the center μ, the curve changes from falling ever more steeply

to falling ever less steeply.

inflection points

*The points at which this change of curvature takes place are called **inflection points** and are located at distance σ on either side of the mean μ.* Figure 2.9 shows σ for two different normal curves.

Although there are many normal curves, they all have common properties. In particular, all normal distributions have the properties described by the following rule.

THE 68–95–99.7 RULE

In the normal distribution with mean μ and standard deviation σ:
- 68% of the observations fall within σ of the mean μ.
- 95% of the observations fall within 2σ of μ.
- 99.7% of the observations fall within 3σ of μ.

Figure 2.10 illustrates the 68–95–99.7 rule. By remembering these three numbers, you can think about normal distributions without constantly making detailed calculations.

EXAMPLE 2.2

The distribution of heights of young women aged 18 to 24 is approximately normal with mean $\mu = 64.5$ inches and the standard deviation $\sigma = 2.5$ inches. Figure 2.11 shows the application of the 68–95–99.7 rule in this example.

Two standard deviations is 5 inches for this distribution. The 95 part of the 68–95–99.7 rule says that the middle 95% of young women are between $64.5 - 5$ and $64.5 + 5$ inches tall, that is, between 59.5 and 69.5 inches. This fact is exactly true for an exactly normal distribution. It is approximately true for the heights of young women because the distribution of heights is approximately normal.

The other 5% of young women have heights outside the range from 59.5 to 69.5 inches. Because the normal distributions are symmetric, half of these women are on the tall side. So the tallest 2.5% of young women are taller than 69.5 inches.

The 99.7 part of the 68–95–99.7 rule says that almost all young women (99.7% of them) have heights between $\mu - 3\sigma$ and $\mu + 3\sigma$. This range of heights is 57 to 72 inches.

percentile   The term **percentile** is probably familiar to most students because national test scores are frequently reported in terms of percentiles, rather than raw scores. Suppose your score on the math portion of such a test was reported as the 90th percentile. That means that 90% of the students who took the math test scored *lower than or equal to* your score. Percentiles are used when we are most interested in seeing where an individual observation stands relative to the other individuals in the distribution. Typically, in practice, the number of observations is quite large so that it makes sense to talk about the distribution as a density curve. The median score would

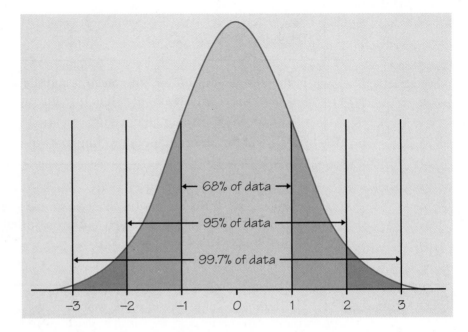

FIGURE 2.10   The 68–95–99.7 rule for normal distributions.

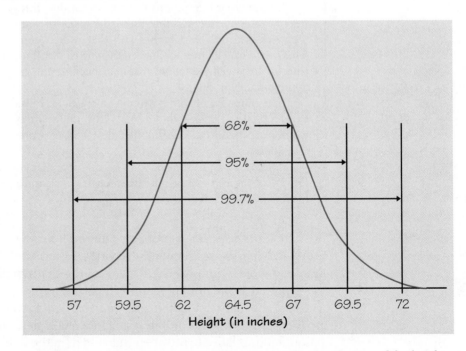

FIGURE 2.11   The 68–95–99.7 rule applied to the distribution of the heights of young women (Example 2.2).

be the 50th percentile because half the scores are to the left of (i.e., lower than) the median. The first quartile is the 25th percentile and the third quartile is the 75th percentile.

Because we will mention normal distributions often, a short notation is helpful. We abbreviate the normal distribution with mean μ and standard deviation σ as $N(\mu, \sigma)$. For example, the distribution of young women's heights is $N(64.5, 2.5)$.

Why are the normal distributions important in statistics? Here are three reasons. First, normal distributions are good descriptions for some distributions of *real data*. Distributions that are often close to normal include scores on tests taken by many people (such as SAT exams and many psychological tests), repeated careful measurements of the same quantity, and characteristics of biological populations (such as lengths of cockroaches and yields of corn). Second, normal distributions are good approximations to the results of many kinds of *chance outcomes*, such as tossing a coin many times. Third, and most important, we will see that many *statistical inference* procedures based on normal distributions work well for other roughly symmetric distributions. HOWEVER... even though many sets of data follow a normal distribution, many do not. Most income distributions, for example, are skewed to the right and so are not normal. Nonnormal data, like nonnormal people, are not only common, but are sometimes more interesting than their normal counterparts.

EXERCISES

2.6 The distribution of heights of adult American men is approximately normal with mean 69 inches and standard deviation 2.5 inches. Draw a normal curve on which this mean and standard deviation are correctly located. (Hint: Draw the curve first, locate the points where the curvature changes, then mark the horizontal axis.)

2.7 The distribution of heights of adult American men is approximately normal with mean 69 inches and standard deviation 2.5 inches. Use the 68–95–99.7 rule to answer the following questions.

(a) What percent of men are taller than 74 inches?

(b) Between what heights do the middle 95% of men fall?

(c) What percent of men are shorter than 66.5 inches?

(d) A height of 71.5 inches corresponds to what percentile of adult male American heights?

2.8   Scores on the Wechsler Adult Intelligence Scale (WAIS, a standard "IQ test") for the 20 to 34 age group are approximately normally distributed with $\mu = 110$ and $\sigma = 25$. Use the 68–95–99.7 rule to answer these questions.

(a) About what percent of people in this age group have scores above 110?

(b) About what percent have scores above 160?

(c) In what range do the middle 95% of all IQ scores lie?

2.9   The distribution of heights of young women aged 18 to 24 is discussed in Example 2.2. Find the percentiles for the following heights.

(a) 64.5 inches

(b) 59.5 inches

(c) 67 inches

(d) 72 inches

2.10   You can do this exercise if you spray-painted a normal distribution in Activity 2A. On your "fine-grained distribution," first count the number of whole squares and parts of squares under the curve. Approximate as best you can. This represents the total area under the curve.

(a) Mark vertical lines at $\mu - 1\sigma$ and $\mu + 1\sigma$. Count the number of squares or parts of squares between these two vertical lines. Now divide the number of squares within one standard deviation of μ by the total number of squares under the curve and express your answer as a percent. How does this compare with 68%? Why would you expect your answer to differ somewhat from 68%?

(b) Count squares to determine the percent of area within 2σ of μ. How does your answer compare with 95%?

(c) Count squares to determine the percent of area within 3σ of μ. How does your answer compare with 99.7%?

SUMMARY

We can sometimes describe the overall pattern of a distribution by a **density curve**. A density curve always remains on or above the horizontal axis and has total area 1 underneath it. An area under a density curve gives the proportion of observations that fall in a range of values.

A density curve is an idealized description of the overall pattern of a distribution that smooths out the irregularities in the actual data. Write the mean of a density curve as μ and the standard deviation of a density curve as σ to distinguish them from the mean $\bar{x}$ and the standard deviation s of the actual data.

The **mean,** the **median,** and the **quartiles** of a density curve can be located by eye. The mean μ is the balance point of the curve. The median divides the area under the curve in half. The quartiles with the median divide the area under the curve into quarters. The **standard deviation** σ cannot be located by eye on most density curves.

The mean and median are equal for symmetric density curves. The mean of a skewed curve is located farther toward the long tail than is the median.

The **normal distributions** are described by a special family of bell-shaped symmetric density curves, called **normal curves.** The mean μ and standard deviation σ completely specify a normal distribution $N(\mu, \sigma)$. The mean is the center of the curve and σ is the distance from μ to the inflection points on either side.

In particular, all normal distributions satisfy the **68–95–99.7 rule,** which describes what percent of observations lie within one, two, and three standard deviations of the mean.

An observation's **percentile** is the percent of the distribution that is at or to the left of the observation.

SECTION 2.1 EXERCISES

2.11 Figure 2.12 shows two normal curves, both with mean 0. Approximately what is the standard deviation of each of these curves?

2.12 The army reports that the distribution of head circumference among male soldiers is approximately normal with mean 22.8 inches and standard deviation 1.1 inches. Use the 68–95–99.7 rule to answer the following questions.

(a) What percent of soldiers have head circumference greater than 23.9 inches?

(b) A head circumference of 23.9 inches would be what percentile?

(c) What percent of soldiers have head circumference between 21.7 inches and 23.9 inches?

2.13 The length of human pregnancies from conception to birth varies according to a distribution that is approximately normal with mean 266 days and

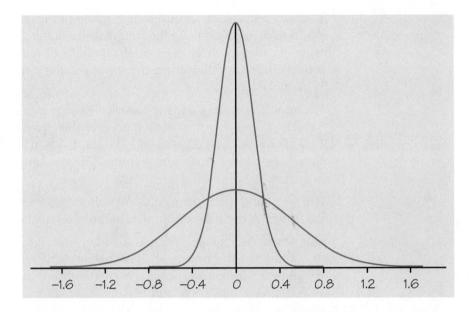

FIGURE 2.12   Two normal curves with the same mean but different standard deviations (Exercise 2.11).

standard deviation 16 days. Use the 68–95–99.7 rule to answer the following questions.

(a) Between what values do the lengths of the middle 95% of all pregnancies fall?

(b) How short are the shortest 2.5% of all pregnancies?

(c) How long are the longest 2.5% of all pregnancies?

2.14   The TI-83 can be used to confirm your answers to exercises in this section. Consider Exercise 2.8. Specify a viewing window as follows: X[35, 185]$_{25}$ and Y[−.01, .02]$_{.01}$. Here are the keystrokes:

```
WINDOW
 Xmin=35
 Xmax=185
 Xscl=25
 Ymin=⁻.01
 Ymax=.02
 Yscl=.01
 Xres=1
```

(a) 2nd / DISTR / DRAW / 1:ShadeNorm(110,1000,110,25)
Note that we are finding the area under the curve to the right of X = 110. We take X = 1000 to be a sufficiently large number that, for all

practical purposes, there is no area beyond this point. The last two numbers in the parentheses are the mean and standard deviation, in that order. When you finish, you need to clear the graphic: 2nd / DRAW / 1:ClrDraw. Describe your findings in a sentence.

(b) 2nd / DISTR / DRAW / 1:ShadeNorm(160,1000,110,25)
Don't forget that 95% is an approximation, correct to two decimal places. Describe your findings in a sentence.

(c) 2nd / DISTR / DRAW / 1:ShadeNorm(60,160,110,25)
Describe your findings in a sentence.

2.15 Wechsler Adult Intelligence Scale (WAIS) scores for young adults are N(110, 25). Use your TI-83 and the method described in the previous exercise to show that the area under the entire curve is equal to 1. Note that you can't specify the interval $(-\infty, +\infty)$, so you'll have to decide on some endpoints that are far enough from the center (110) of the distribution to give at least four-decimal-place accuracy. Record the interval that you use and the area that the TI-83 reports. Will it suffice to go out four standard deviations on either side of the center? Five standard deviations?

2.16 Wechsler Adult Intelligence Scale (WAIS) scores for young adults are N(110, 25).

(a) If someone's score were reported as the 16th percentile, exactly what score would that individual have? Use your TI-83 to confirm your answer.

(b) Answer the same question for the 84th percentile and the 97.5 percentile.

2.17 A study of elite distance runners found a mean body weight of 63.1 kilograms(kg), with a standard deviation of 4.8 kg.

(a) Assuming that the distribution of weights is normal, sketch the density curve of the weight distribution with the horizontal axis marked in kilograms.

(b) Use the 68–95–99.7 rule to find intervals centered at the mean that will include 68%, 95%, and 99.7% of the weights of the runners.

2.18 Like Minitab and similar computer utilities, the TI-83 has a "random number generator" that produces decimal numbers that are uniformly distributed between 0 and 1. Copy the command rand to the Home screen (MATH / PRB / 1:rand) and press the ENTER key several times to see the results. The command 2rand produces a random number between 0 and 2. The density curve of the outcomes has constant height between 0 and 2, and height 0 elsewhere.

(a) What is the height of the density curve between 0 and 2? Draw a graph of the density curve.

(b) Use your graph from (a) and the fact that areas under the curve are relative frequencies of outcomes to find the proportion of outcomes that are less than 1.

(c) What is the median of the distribution? What are the quartiles?

(d) Find the proportion of outcomes that lie between 0.5 and 1.3.

2.19   The program FLIP50 simulates flipping a fair coin 50 times and counts the number of times the coin comes up heads. It prints the number of heads on the screen. Then it repeats the experiment for a total of 100 times, each time displaying the number of heads in 50 flips. When it finishes, it draws a histogram of the 100 results.

(a) What outcomes are likely? What outcome(s) are the most likely? If you made a histogram of the results of the 100 replications, what shape distribution would you expect?

(b) The program is listed below. Enter the program carefully, or link it from a classmate or your teacher. Run the program and observe the variations in the results of the 100 applications.

(c) When the histogram appears, TRACE to see the classes and frequencies. Record the results in a frequency table.

(d) Describe the distribution: symmetric versus non-symmetric; center; spread; number of peaks; gaps; suspected outliers. What shape density curve would best fit your distribution?

```
prgm:FLIP50
100→DIM(L₁)
For(I,1,100)
0→H
For(J,1,50)
randInt(0,1)→N
If N=1:H+1→H
End
Disp H
H→L₁(I)
End
```

```
PlotsOff
10→Xmin
40→Xmax
2→Xscl
-6→Ymin
25→Ymax
5→Yscl
Plot1(Histogram,L₁)
DispGraph
```

2.2 STANDARD NORMAL CALCULATIONS

The standard normal distribution

As the 68–95–99.7 rule suggests, all normal distributions share many common properties. In fact, all normal distributions are the same if we measure in units of size σ about the mean μ as center. Changing to these units is called *standardizing*. To standardize a value, subtract the mean of the distribution and then divide by the standard deviation.

STANDARDIZED OBSERVATIONS

If x is an observation from a distribution that has mean μ and standard deviation σ, the **standardized value** of x is

$$z = \frac{x - \mu}{\sigma}$$

---

z-scores

The letter z is commonly used for standardized observations. In fact, standardized observations are sometimes called **z-scores**. We often standardize observations from symmetric distributions to express them in a common scale. A standardized observation tells us how many standard deviations the original observation falls away from the mean, and in which direction. Observations larger than the mean are positive when standardized, and observations smaller than the mean are negative.

EXAMPLE 2.3

The heights of young women are approximately normal with $\mu = 64.5$ inches and $\sigma = 2.5$ inches. The standardized height is

$$z = \frac{\text{height} - 64.5}{2.5}$$

A woman's standardized height is the number of standard deviations by which her height differs from the mean height of all young women. A woman 68 inches tall, for example, has standardized height

$$z = \frac{68 - 64.5}{2.5} = 1.4$$

or 1.4 standard deviations above the mean. Similarly, a woman 5 feet (60 inches) tall has standardized height

$$z = \frac{60 - 64.5}{2.5} = -1.8$$

or 1.8 standard deviations less than the mean height.

If the variable we standardize has a normal distribution, standardizing does more than give a common scale. It makes all normal distributions into a single distribution, and this distribution is still normal. Standardizing a variable that has any normal distribution produces a new variable that has the *standard normal distribution*.

STANDARD NORMAL DISTRIBUTION

The **standard normal distribution** is the normal distribution $N(0, 1)$ with mean 0 and standard deviation 1.

If a variable x has any normal distribution $N(\mu, \sigma)$ with mean μ and standard deviation σ, then the standardized variable

$$z = \frac{x - \mu}{\sigma}$$

has the standard normal distribution.

EXERCISES

2.20   Eleanor scores 680 on the mathematics part of the SAT. The distribution of SAT scores in a reference population is normal, with mean 500 and standard deviation 100. Gerald takes the American College Testing (ACT) mathematics test and scores 27. ACT scores are normally distributed with mean 18 and standard deviation 6. Find the standardized scores for both students. Assuming that both tests measure the same kind of ability, who has the higher score?

2.21   The normal density curves are defined by a particular equation:

$$y = \frac{1}{\sigma \sqrt{2\pi}} e^{-\frac{1}{2}\left(\frac{x-\mu}{\sigma}\right)^2}$$

We can obtain individual members of this family of curves by specifying particular values for the mean μ and the standard deviation σ. If we specify the values $\mu = 0$ and $\sigma = 1$, then we have the equation for the *standard* normal distribution. This exercise will explore two functions.

   Step 1. Enter as Y_1 the following equation for the standard normal distribution:

$$Y_1 = (1/(\sqrt{(2\pi)})(e^\wedge(-.5x^2)))$$

As Y_2 enter `normalpdf(X)`. Position the cursor after $Y_2 =$ and then enter

2nd / DIST / 1:normalpdf(

Finish defining Y_2.

   Highlight the slash \ to the left of Y_2 and press ENTER once to make a heavy graphing line. When you graph these two functions (sequentially), the first will plot with the usual line, and the second function will plot with the heavier line. From the Y= screen, if you notice any stat plots active, move the cursor to them and press ENTER to toggle them off. Delete or turn off all other defined functions.

   Step 2. Specify a viewing window X$[-3, 3]_1$ and Y$[-.1, .5]_{.1}$.

   Step 3. Press GRAPH.

   Write a sentence that describes the connection between these two functions. *Note:* `normalpdf` stands for "normal probability density function." We'll learn more about pdf's in Chapter 8.

Normal distribution calculations

An area under a density curve is a proportion of the observations in a distribution. Any question about what proportion of observations lie in some range of values can be answered by finding an area under the curve. Because all normal distributions are the same when we standardize, we can find areas under any normal curve from a single table, a table that gives areas under the curve for the standard normal distribution.

EXAMPLE 2.4

What proportion of all young women are less than 68 inches tall? This proportion is the area under the $N(64.5, 2.5)$ curve to the left of the point 68. Because the standardized height corresponding to 68 inches is

$$z = \frac{x - \mu}{\sigma} = \frac{68 - 64.5}{2.5} = 1.4$$

this area is the same as the area under the *standard normal* curve to the left of the point $z = 1.4$. Figure 2.13(a) shows this area.

Table A in the back of the book gives areas under the standard normal curve. Table A also appears on the inside front cover.

THE STANDARD NORMAL TABLE

Table A is a table of areas under the standard normal curve. The table entry for each value z is the area under the curve to the left of z.

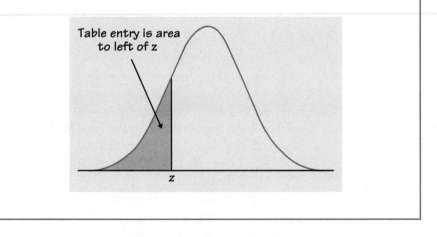

EXAMPLE 2.5

Problem: Find the proportion of observations from the standard normal distribution that are less than 1.4.

Solution: To find the area to the left of 1.40, locate 1.4 in the left-hand column of Table A, then locate the remaining digit 0 as .00 in the top row. The entry opposite 1.4 and under .00 is 0.9192. This is the area we seek. Figure 2.13(a) illustrates the relationship between the value $z = 1.40$ and the area 0.9192. Because $z = 1.40$ is the standardized value of height 68 inches, the proportion of young women who are less than 68 inches tall is 0.9192 (about 92%).

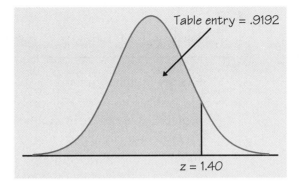

FIGURE 2.13(a)   The area under a standard normal curve to the left of the point $z = 1.4$ is 0.9192. Table A gives areas under the standard normal curve.

Problem: Find the proportion of observations from the standard normal distribution that are greater than -2.15.

Solution: Enter Table A under $z = -2.15$. That is, find -2.1 in the left-hand column and .05 in the top row. The table entry is 0.0158. This is the area to the *left* of -2.15. Because the total area under the curve is 1, the area lying to the *right* of -2.15 is $1 - .0158 = .9842$. Figure 2.13(b) illustrates these areas.

We can answer any question about proportions of observations in a normal distribution by standardizing and then using the standard normal table. Here is an outline of the method for finding the proportion of the distribution in any region.

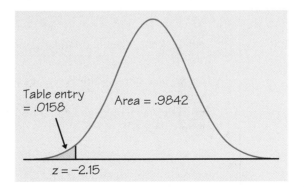

FIGURE 2.13(b)   Areas under the standard normal curve to the right and left of $z = -2.15$. Table A gives only areas to the left.

FINDING NORMAL PROPORTIONS

- State the problem in terms of the observed variable x.
- Standardize x to restate the problem in terms of a standard normal variable z. Draw a picture to show the area under the standard normal curve.
- Find the required area under the standard normal curve, using Table A and the fact that the total area under the curve is 1.

EXAMPLE 2.6

The level of cholesterol in the blood is important because high cholesterol levels may increase the risk of heart disease. The distribution of blood cholesterol levels in a large population of people of the same age and sex is roughly normal. For 14-year-old boys, the mean is $\mu = 170$ milligrams of cholesterol per deciliter of blood (mg/dl) and the standard deviation is $\sigma = 30$ mg/dl.[2] Levels above 240 mg/dl may require medical attention. What percent of 14-year-old boys have more than 240 mg/dl of cholesterol?

- *State the problem.* Call the level of cholesterol in the blood x. The variable x has the $N(170, 30)$ distribution. We want the proportion of boys with $x > 240$.
- *Standardize.* Subtract the mean, then divide by the standard deviation, to turn x into a standard normal z:

$$x > 240$$

$$\frac{x - 170}{30} > \frac{240 - 170}{30}$$

$$z > 2.33$$

- *Draw a picture.* Sketch a standard normal curve, and shade the area of interest. See Figure 2.14.
- *Use the table.* From Table A, we see that the proportion of observations less than 2.33 is 0.9901. About 99% of boys have cholesterol levels less than 240. The area to the right of 2.33 is therefore $1 - .9901 = .0099$. This is about 0.01, or 1%. Only about 1% of boys have high cholesterol.

In a normal distribution, the proportion of observations with $x > 240$ is the same as the proportion with $x \geq 240$. There is no area under the curve and exactly over 240, so the areas under the curve with $x > 240$ and $x \geq 240$ are the same. This isn't true of the actual data. There may be a boy with exactly 240 mg/dl of blood cholesterol. The normal distribution is just an easy-to-use approximation, not a description of every detail in the actual data.

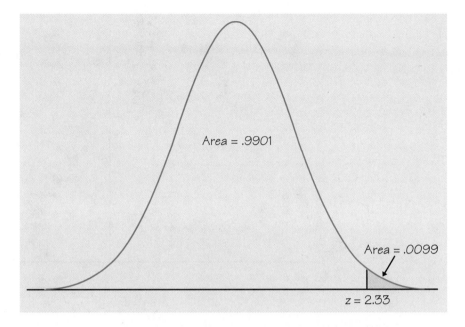

FIGURE 2.14   Areas under the standard normal curve (Example 2.6).

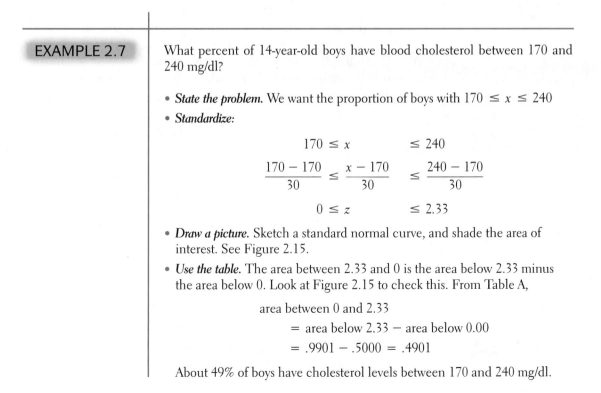

EXAMPLE 2.7

What percent of 14-year-old boys have blood cholesterol between 170 and 240 mg/dl?

- *State the problem.* We want the proportion of boys with $170 \leq x \leq 240$
- *Standardize:*

$$170 \leq x \qquad \leq 240$$

$$\frac{170 - 170}{30} \leq \frac{x - 170}{30} \leq \frac{240 - 170}{30}$$

$$0 \leq z \qquad \leq 2.33$$

- *Draw a picture.* Sketch a standard normal curve, and shade the area of interest. See Figure 2.15.
- *Use the table.* The area between 2.33 and 0 is the area below 2.33 minus the area below 0. Look at Figure 2.15 to check this. From Table A,

area between 0 and 2.33

= area below 2.33 − area below 0.00

= .9901 − .5000 = .4901

About 49% of boys have cholesterol levels between 170 and 240 mg/dl.

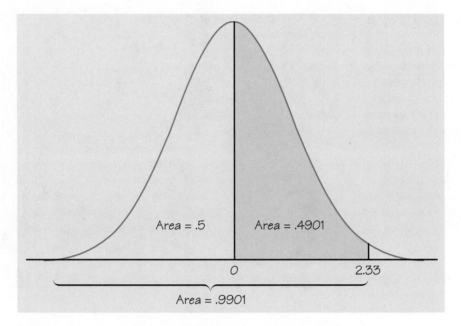

Area = .5   Area = .4901

0   2.33

Area = .9901

FIGURE 2.15   Areas under the standard normal curve (Example 2.7).

What if we meet a z that falls outside the range covered by Table A? For example, the area to the left of $z = -4$ does not appear in the table. But since -4 is less than -3.4, this area is smaller than the entry for $z = -3.40$, which is 0.0003. There is very little area under the standard normal curve outside the range covered by Table A. You can take this area to be zero with little loss of accuracy.

Finding a value given a proportion

Examples 2.6 and 2.7 illustrate the use of Table A to find what proportion of the observations satisfies some condition, such as "blood cholesterol between 170 mg/dl and 240 mg/dl." We may instead want to find the observed value with a given proportion of the observations above or below it. To do this, use Table A backward. Find the given proportion in the body of the table, read the corresponding z from the left column and top row, then "unstandardize" to get the observed value. Here is an example.

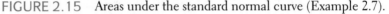

EXAMPLE 2.8   Scores on the SAT verbal test in recent years follow approximately the $N(505, 110)$ distribution. How high must a student score in order to place in the top 10% of all students taking the SAT?

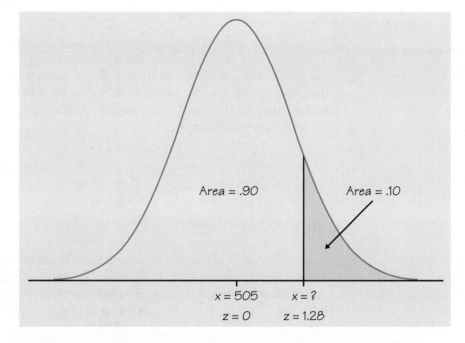

FIGURE 2.16   Locating the point on a normal curve with area 0.10 to its right (Example 2.8).

- *State the problem.* We want to find the SAT score x with area 0.1 to its *right* under the normal curve with mean $\mu = 505$ and standard deviation $\sigma = 110$. That's the same as finding the SAT score x with area 0.9 to its *left*. Figure 2.16 poses the question in graphical form. Because Table A gives the areas to the left of z-values, always state the problem in terms of the area to the left of x.

- *Use the table.* Look in the body of Table A for the entry closest to 0.9. It is 0.8997. This is the entry corresponding to $z = 1.28$. So $z = 1.28$ is the standardized value with area 0.9 to its left.

- *Unstandardize* to transform the solution from the z back to the original x scale. We know that the standardized value of the unknown x is $z = 1.28$. So x itself satisfies

$$\frac{x - 505}{110} = 1.28$$

Solving this equation for x gives

$$x = 505 + (1.28)(110) = 645.8$$

This equation should make sense: it finds the x that lies 1.28 standard deviations above the mean on this particular normal curve. That is the "unstandardized" meaning of $z = 1.28$. We see that a student must score at least 646 to place in the highest 10%.

2.22   Use Table A to find the proportion of observations from a standard normal distribution that satisfies each of the following statements. In each case, sketch a standard normal curve and shade the area under the curve that is the answer to the question.

    (a) $z < 2.85$

    (b) $z > 2.85$

    (c) $z > -1.66$

    (d) $-1.66 < z < 2.85$

2.23   Use Table A to find the value z of a standard normal variable that satisfies each of the following conditions. (Use the value of z from Table A that comes closest to satisfying the condition.) In each case, sketch a standard normal curve with your value of z marked on the axis.

    (a)   The point z with 25% of the observations falling below it.

    (b)   The point z with 40% of the observations falling above it.

2.24   The distribution of heights of adult American men is approximately normal with mean 69 inches and standard deviation 2.5 inches.

    (a)   What percent of men are at least 6 feet (72 inches) tall?

    (b)   What percent of men are between 5 feet (60 inches) and 6 feet tall?

    (c)   How tall must a man be to be in the tallest 10% of all adult men?

2.25   Scores on the Wechsler Adult Intelligence Scale (a standard "IQ test") for the 20 to 34 age group are approximately normally distributed with $\mu = 110$ and $\sigma = 25$.

    (a)   What percent of people age 20 to 34 have IQ scores above 100?

    (b)   What percent have scores above 150?

    (c)   How high an IQ score is needed to be in the highest 25%?

Assessing normality

In the latter part of this course we will want to invoke various tests of significance to try to answer questions that are important to us. These tests involve sampling people or objects and inspecting them carefully to gain insights into the populations from which they come. Many of these procedures are based on the assumption that the host population is approximately normally distributed. Consequently, we will need to be able to develop methods for assessing normality.

Method 1 Construct a frequency histogram or a stemplot. See if the shape of the graph is approximately bell-shaped and symmetric about the mean.

A histogram or stemplot can reveal distinctly nonnormal features of a distribution, such as outliers, pronounced skewness, or gaps and clusters. You can improve the effectiveness of these plots for assessing whether a distribution is normal by marking the points $\bar{x}$, $\bar{x} \pm s$, and $\bar{x} \pm 2s$ on the x axis. This gives the scale natural to normal distributions. Then compare the count of observations in each interval with the 68–95–99.7 rule.

EXAMPLE 2.9

The histogram in Figure 2.2 suggests that the distribution of the 947 Gary vocabulary scores is close to normal. It is hard to assess by eye how close to normal a histogram is. Let's use the 68–95–99.7 rule to check more closely. We enter the scores into a statistical computing system and ask for the mean and standard deviation. The computer replies,

$$\text{MEAN} \;=\; 6.8585$$
$$\text{STDEV} \;=\; 1.592$$

Now that we know that $\bar{x} = 6.8585$ and $s = 1.5952$, we check the 68–95–99.7 rule by finding the actual counts of Gary vocabulary scores in intervals of length s about the mean $\bar{x}$. The computer will also do this for us. Here are the counts:

1	21	129	331	318	125	21	1

2.07	3.67	5.26	6.86	8.45	10.05	11.64
$\bar{x} - 3s$	$\bar{x} - 2s$	$\bar{x} - s$	$\bar{x}$	$\bar{x} + s$	$\bar{x} + 2s$	$\bar{x} + 3s$

The distribution is very close to symmetric. It also follows the 68–95–99.7 rule closely: there are 68.5% of the scores (649 out of 947) within one standard deviation of the mean, 95.4% (903 of 947) within two standard deviations, and 99.8% (945 of the 947 scores) within three. These counts confirm that the normal distribution with $\mu = 6.86$ and $\sigma = 1.595$ fits these data well.

Smaller data sets rarely fit the 68–95–99.7 rule as well as the Gary vocabulary scores. This is true even of observations taken from a larger population that really has a normal distribution. There is more chance variation in small data sets.

normal
probability plot

Method 2   Construct a *normal probability plot*. A normal probability plot provides a good assessment of the adequacy of the normal model for a set of data. Most statistics utilities, including Minitab and Data Desk, can construct normal probability plots from entered data. The TI-83 will also do normal probability plots. You will need to be able to produce a normal probability plot (either with a TI-83 or with computer software) and interpret it. We will do this part first, and then we will describe the steps the calculator goes through to produce the plot.

EXAMPLE 2.10

If you ran the program FLIP50 in Exercise 2.19, and you still have the data (100 numbers mostly in the 20s) in list L_1, then use these data. If you have not entered the program and run it, take a few minutes to do that now. Duplicate this example with *your* data. Here is the histogram that was generated at the end of one run of this simulation:

```
P1:L1


min=24
max<26          n=28
```

Asking for one-variable statistics (STAT / CALC / 1:1 Var Stats / L_1) gives us the following:

```
1-Var Stats              1-Var Stats
x̄=25.16161616           ↑n=99
Σx=2491                    minX=16
Σx²=63895                  Q1=23
Sx=3.524569943            Med=25
σx=3.506723904           Q3=27
↓n=99                     maxX=32
```

Comparing the mean $\bar{x} = 25.16$ with the median $M = 25$ suggests that the distribution is fairly symmetric. A boxplot confirms the symmetric shape (and shows an outlier).

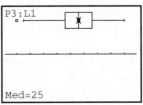

```
P3:L1




Med=25
```

To construct a normal probability plot of the data, disable other STATPLOTs and define Plot1 like this:

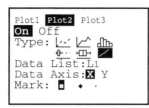

The last plot-type icon represents a normal probability plot. Make sure you select this icon. Then ZOOM / 9:ZoomStat to see the following graph.

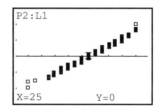

Interpretation: If the data distribution is close to a normal distribution, the plotted points will lie close to a straight line. Conversely, nonnormal data will show a nonlinear trend. Outliers appear as points that are far away from the overall pattern of the plot. Since the above plot is quite linear, our conclusion is that it is reasonable to believe that the data are from a normal distribution.

For those who like to know how things work, the TI-83 uses an algorithm similar to the following to construct the normal probability plot.

1. Arrange the data values x_i from smallest to largest. Calculate the percentile for each data value. For example, the smallest observation in a set of 20 is at the 5% point, the second smallest is at the 10% point, and so on.

2. Determine the z-values for each of these same percentiles. For example, the 5% point of the standard normal distribution corresponds to $z = -1.645$, and the 10% point corresponds to $z = -1.282$.

3. Plot the observed data values x_i on the x axis, and the z-values of the percentiles on the y (vertical) axis.

Any normal distribution produces a straight line on the plot because standardizing is a transformation that can change the slope and intercept of the line in our plot but cannot change a line into a curved pattern.

EXAMPLE 2.11

The stemplot for the number of hysterectomies performed by male doctors (Example 1.7) is clearly right-skewed. The two large outliers (85, 86) pull the mean up and away from the median, $M = 34$. The mean is 41.33. The histogram and boxplot also show that the data are right-skewed.

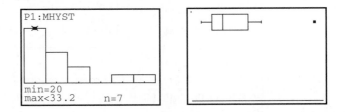

Thus we would expect that a normal probability plot would not show a linear trend, and it does not.

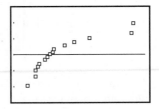

In this case it is reasonable to believe that the data do *not* come from a normal distribution.

EXERCISES

2.26   Repeated careful measurements of the same physical quantity often have a distribution that is close to normal. Here are Henry Cavendish's 29 measurements of the density of the earth, made in 1798. (The data give the density of the earth as a multiple of the density of water.)

5.50	5.61	4.88	5.07	5.26	5.55	5.36	5.29	5.58	5.65
5.57	5.53	5.62	5.29	5.44	5.34	5.79	5.10	5.27	5.39
5.42	5.47	5.63	5.34	5.46	5.30	5.75	5.68	5.85	

(a)   A stemplot (Exercise 1.35, page 46) shows that the data are reasonably symmetric. Now check how closely they follow the 68–95–99.7 rule. Find $\bar{x}$ and s, then count the number of observations that fall between $\bar{x} - s$ and $\bar{x} + s$, between $\bar{x} - 2s$ and $\bar{x} + 2s$, and between

$\bar{x} - 3s$ and $\bar{x} + 3s$. Compare the percents of the 29 observations in each of these intervals with the 68–95–99.7 rule.

(b) Use your TI-83 to construct a normal probability plot for Cavendish's density of the earth data, and write a brief statement about the normality of the data. Does the normal probability plot reinforce your findings in (a)?

We expect that when we have only a few observations from a normal distribution, the percents will show some deviation from 68, 95, and 99.7. Cavendish's measurements are in fact close to normal.

SUMMARY

All normal distributions are the same when measurements are made in units of σ about the mean. These are called **standardized observations.** The standardized value z of an observation x is

$$z = \frac{x - \mu}{\sigma}$$

If x has the $N(\mu, \sigma)$ distribution, then the **standardized variable** $z = (x - \mu)/\sigma$ has the **standard normal distribution** $N(0, 1)$ with mean 0 and standard deviation 1. Table A gives the proportions of standard normal observations that are less than z for many values of z. By standardizing, we can use Table A for any normal distribution.

In order to perform certain tests of significance in later chapters, we will need to know that the data come from populations that are approximately normally distributed. To assess normality, one can observe the shape of histograms, stemplots, and boxplots and see how well the data fit the 68–95–99.7 rule for normal distributions. A good method for assessing normality is to construct a **normal probability plot.**

SECTION 2.2 EXERCISES

2.27   Three landmarks of baseball achievement are Ty Cobb's batting average of .420 in 1911, Ted Williams's .406 in 1941, and George Brett's .390 in 1980. These batting averages cannot be compared directly because the distribution of major league batting averages has changed over the years. The distributions are quite symmetric and (except for outliers such as Cobb, Williams, and Brett) reasonably normal. While the mean batting average has been held roughly constant by rule changes and the balance between hitting and pitching, the standard deviation has dropped over time. Here are the facts:

Decade	Mean	Std. dev.	
1910s	.266	.0371	
1940s	.267	.0326	
1970s	.261	.0317	

Compute the standardized batting averages for Cobb, Williams, and Brett to compare how far each stood above his peers.[3]

2.28   Use Table A to find the proportion of observations from a standard normal distribution that falls in each of the following regions. In each case, sketch a standard normal curve and shade the area representing the region.

(a) $z \leq -2.25$

(b) $z \geq -2.25$

(c) $z > 1.77$

(d) $-2.25 < z < 1.77$

2.29   (a) Find the number z such that the proportion of observations that are less than z in a standard normal distribution is 0.8.

(b) Find the number z such that 35% of all observations from a standard normal distribution are greater than z.

2.30   The rate of return on stock indexes (which combine many individual stocks) is approximately normal. Since 1945, the Standard & Poor's 500 index has had a mean yearly return of 11.8%, with a standard deviation of 16.6%. Take this normal distribution to be the distribution of yearly returns over a long period.

(a) In what range do the middle 95% of all yearly returns lie?

(b) The market is down for the year if the return on the index is less than zero. In what percent of years is the market down?

(c) In what percent of years does the index gain 25% or more?

2.31   The length of human pregnancies from conception to birth varies according to a distribution that is approximately normal with mean 266 days and standard deviation 16 days.

(a) What percent of pregnancies last less than 240 days (that's about 8 months)?

(b) What percent of pregnancies last between 240 and 270 days (roughly between 8 months and 9 months)?

(c) How long do the longest 20% of pregnancies last?

2.32 The quartiles of any density curve are the points with area 0.25 and 0.75 to their left under the curve.

(a) What are the quartiles of a standard normal distribution?

(b) How many standard deviations away from the mean do the quartiles lie in any normal distribution? What are the quartiles for the lengths of human pregnancies? (Use the distribution in the previous exercise.)

2.33 The *deciles* of any distribution are the points that mark off the lowest 10% and the highest 10%. The deciles of a density curve are therefore the points with area 0.1 and 0.9 to their left under the curve.

(a) What are the deciles of the standard normal distribution?

(b) The heights of young women are approximately normal with mean 64.5 inches and standard deviation 2.5 inches. What are the deciles of this distribution?

2.34 The TI-83's normalcdf function approximates the area under the normal distribution and above a specified interval. The keystrokes are

2nd / DIST / 2:normalcdf(

If you are using standard normal values, you need only specify the left and right endpoints of the interval within the parentheses. For example, normalcdf(-1, 1) returns .6827, meaning that the area above the interval from $z = -1$ to $z = 1$ is approximately 0.6827, correct to four decimal places (recall the 68–95–99.7 rule). If you want to find the area to the left of $z = 1$, you could specify the interval $(-10, 1)$, or to be absolutely certain, you could use -1E99 (which is scientific notation for -1×10^{99}) as a negative number very large in absolute value, the largest, in fact, the TI-83 can handle. If you're using raw data, simply include the mean and standard deviation as the third and fourth numbers in the parentheses, separated by commas. For example, normalcdf(150, 1E99, 100, 15) would find the area to the right of $x = 150$ for a distribution whose mean is 100 and whose standard deviation is 15.

Use the calculator's normalcdf function to verify your answers to Exercise 2.28.

2.35   The TI-83's `invNorm` function calculates the standardized z-value corresponding to a known area or relative frequency. The keystrokes are

2nd / DIST / 3:invNorm(

If you are using standard normal values, you need only specify the area in the parentheses. Use this feature to verify your answers to Exercise 2.29.

2.36   This exercise uses the TI-83 to calculate standardized values for a familiar data set and then calculates the mean and standard deviation for these transformed values. Without knowing the data set, can you guess the mean and standard deviation? Use the SetUp Editor on your TI-83 to make the list PREZ (the presidents' ages, in Table 1.3, on page 31) the first list: STAT / 5:SetUpEditor / 2nd / LIST / PREZ / ENTER. Move the cursor to the name of the next list in the Edit screen, and insert a new list named STDSC (for standard scores). With the name of this list, STDSC, highlighted, define the list by carefully entering

($_L$PREZ − mean($_L$PREZ))/stdev($_L$PREZ)

(The mean and stdev commands are found under 2nd / LIST / MATH.) Then scroll through the list STDSC to verify that the values range from about −3 to 3. Construct a histogram of STDSC, and calculate one-variable statistics for STDSC. What are the mean and standard deviation?

2.37   The histogram for the ages of the 42 presidents was very symmetric (see Figure 1.10, page 31). Use the list that we named PREZ to construct a normal probability plot for this data set, and confirm the linear trend. Write a statement about your assessment of normality of the presidents' ages.

CHAPTER REVIEW

Here is a review list of the most important skills you should have acquired from your study of this chapter.

A. DENSITY CURVES

1. Know that areas under a density curve represent proportions of all observations and that the total area under a density curve is 1.

2. Approximately locate the median (equal-areas point) and the mean (balance point) on a density curve.

3. Know that the mean and median both lie at the center of a symmetric density curve and that the mean moves farther toward the long tail of a skewed curve.

B. NORMAL DISTRIBUTIONS

1. Recognize the shape of normal curves and be able to estimate both the mean and standard deviation from such a curve.

2. Use the 68–95–99.7 rule and symmetry to state what percent of the observations from a normal distribution fall between two points when both points lie at the mean or one, two, or three standard deviations on either side of the mean.

3. Given that a variable has the normal distribution with a stated mean μ and standard deviation σ, calculate the proportion of values above a stated number, below a stated number, or between two stated numbers.

4. Given that a variable has the normal distribution with a stated mean μ and standard deviation σ, calculate the point having a stated proportion of all values above it. Also calculate the point having a stated proportion of all values below it.

C. ASSESSING NORMALITY

1. Plot a histogram, stemplot, and/or boxplot to determine if a distribution is bell-shaped.

2. Determine the proportion of observations within one, two, and three standard deviations of the mean, and compare with the 68–95–99.7 rule for normal distributions.

3. Construct and interpret normal probability plots.

CHAPTER 2 REVIEW EXERCISES

2.38   A density curve consists of a straight-line segment that begins at the origin, $(0, 0)$, and has slope 1.

(a) Sketch the density curve. What are the coordinates of the right endpoint of the segment? (*Note:* The right endpoint should be fixed so that the total area under the curve is 1. This is required for a valid density curve.)

(b) Determine the median, the first quartile (Q_1), and the third quartile (Q_3).

(c) Relative to the median, where would you expect the mean of the distribution?

(d) What percent of the observations lie below 0.5? Above 1.5?

2.39   A density curve looks like an inverted letter "V." The first segment goes from the point $(0, 0.6)$ to the point $(0.5, 1.4)$. The second segment goes from $(0.5, 1.4)$ to $(1, 0.6)$.

   (a) Sketch the curve. Verify that the area under the curve is 1, so that it is a valid density curve.

   (b) Determine the median. Mark the median and the approximate locations of the quartiles Q_1 and Q_3 on your sketch.

   (c) What percent of the observations lie below 0.3?

   (d) What percent of the observations lie between 0.3 and 0.7?

2.40   The Chapin Social Insight Test evaluates how accurately the subject appraises other people. In the reference population used to develop the test, scores are approximately normally distributed with mean 25 and standard deviation 5. The range of possible scores is 0 to 41.

   (a) What proportion of the population has scores below 20 on the Chapin test?

   (b) What proportion has scores below 10?

   (c) What proportion has scores above 35?

   (d) How high a score must you have in order to be in the top quarter of the population in social insight?

2.41   The scores of a reference population on the Wechsler Intelligence Scale for Children (WISC) are normally distributed with $\mu = 100$ and $\sigma = 15$. A school district classified children as "gifted" if their WISC score exceeds 135. There are 1300 sixth-graders in the school district. About how many of them are gifted?

2.42   The Acculturation Rating Scale for Mexican Americans (ARSMA) is a psychological test that measures the degree to which Mexican Americans are adapted to Mexican/Spanish versus Anglo/English culture. The range of possible scores is 1.0 to 5.0, with higher scores showing more Anglo/English acculturation. The distribution of ARSMA scores in a population used to develop the test is approximately normal with mean 3.0 and standard deviation 0.8. A researcher believes that Mexicans will have an average score near 1.7 and that first-generation Mexican Americans will average about 2.1 on the ARSMA scale. What proportion of the population used to develop the test has scores below 1.7? Between 1.7 and 2.1?

2.43   The army reports that the distribution of head circumference among soldiers is approximately normal with mean 22.8 inches and standard deviation 1.1 inches. Helmets are mass-produced for all except the smallest 5% and the largest 5% of head sizes. Soldiers in the smallest or largest 5% get custom-made helmets. What head sizes get custom-made helmets?

2.44 The ARSMA test is described in Exercise 2.42. How high a score on this test must a Mexican American obtain to be among the 30% of the population used to develop the test who are most Anglo/English in cultural orientation? What scores make up the 30% who are most Mexican/Spanish in their acculturation?

2.45 Exercise 1.49 (page 53) shows driving times between home and college for Professor Moore.

(a) Make a histogram of these drive times. Is the distribution roughly symmetric, clearly skewed, or neither? Are there any clear outliers?

(b) The data show three unusual situations: the day after Thanksgiving (no traffic on campus); a delay due to an accident; and a day with icy roads. Identify and remove these three observations. Are the remaining observations reasonably close to having a normal distribution? Write a short statement that describes your analyses and your conclusions.

2.46 Exercise 1.50 (page 54) presents data on the weight gains of chicks fed two types of corn. The researchers use $\bar{x}$ and s to summarize each of the two distributions. Make a normal probability plot for each group and report your findings. Is the use of $\bar{x}$ and s justified?

NOTES AND DATA SOURCES

1. Data from Gary Community School Corporation, courtesy of Celeste Foster. Department of Education, Purdue University.

2. Detailed data appear in P. S. Levy et al., "Total serum cholesterol values for youths 12–17 years," *Vital and Health Statistics Series 11*, No. 150 (1975), U.S. National Center for Health Statistics.

3. Data from Stephen Jay Gould, "Entropic homogeneity isn't why no one hits 400 anymore," *Discover*, August 1986, pp. 60–66. Gould does not standardize but gives a speculative discussion instead.

Examining Relationships

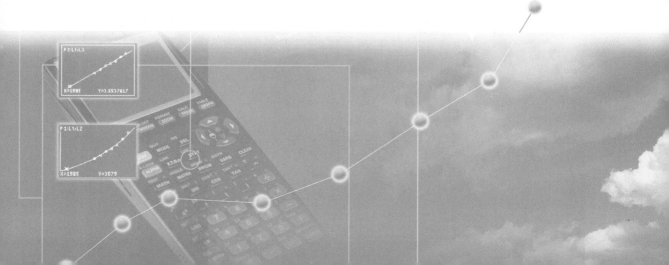

| ACTIVITY 3 | SAT/ACT Scores |

Materials: Pencil, grid paper

Is there an association between SAT math scores and SAT verbal scores? If a student performs well on the math part of the SAT exam, will he or she do well on the verbal part, too? If a student performs well on one part, does that suggest that the student will not do as well on the other? Is it rare or fairly common for students to score about the same on both parts of the SAT? In this activity you will collect, anonymously of course, the SAT math and SAT verbal scores for each member of the class who has taken the SAT exam. You will then plot these data and inspect the graph to see if a pattern is evident. If your school is in a state where the ACT exam is the principal college placement test, then use ACT scores.

1. Begin by writing your math score and verbal score on an index card or similar uniform "ballot." Label your math score M, and your verbal score V. A selected student should collect the folded index cards in a box or other container. When all of the index cards have been placed in the box, they should be mixed without looking, so that the privacy of each student is protected.

   If the size of your class is "small," then you may need to supplement your data with the scores of students in other classes. Perhaps your teacher can request that scores from other AP classes be provided to make a larger data set. Try to obtain data from at least 25 or 30 students.

2. The scores should be called out by the student who collects the data and recorded on the blackboard as ordered pairs in the form (math, verbal).

3. Each student should construct a plot of the data with pencil and paper. Since the math scores appear first in the ordered pairs, label your horizontal axis "Math" and label the vertical axis "Verbal." Determine the range of the math scores and the range of the verbal scores, and then construct scales for both axes. Note that axes don't have to intersect at the point (0,0), but the scales on both axes should be uniform.

(continued)

4. When you finish constructing your graph, look to see if there is any discernible pattern. If so, can you describe the pattern? Does the graph provide any insight into a possible association between SAT math and SAT verbal scores?

We will return to analyze these data in more detail after we develop some methodology.

INTRODUCTION

Most statistical studies involve more than one variable. Sometimes we want to compare the distributions of the same variable for several groups. For example, we might compare the distributions of SAT scores among students at several colleges. Side-by-side boxplots, stemplots, or histograms make the comparison visible. In this chapter, however, we concentrate on relationships among several variables for the same group of individuals. For example, Table 3.1 records seven variables that describe education in the United States. We have already examined some of these variables one at a time. Now we might ask how SAT mathematics scores are related to SAT verbal scores or to the percent of a state's high school seniors who take the SAT or to what region a state is in.

When you examine the relationship between two or more variables, first ask the preliminary questions that are familiar from Chapters 1 and 2.

- What *individuals* do the data describe?
- What exactly are the *variables*? How are they measured?
- Are all the variables *quantitative* or is at least one a *categorical* variable?

We have concentrated on quantitative variables until now. When we have data on several variables, however, categorical variables are often present and help organize the data. Categorical variables will play a larger role in this chapter. There is one more question you should ask when you are interested in relations among several variables:

- Do you want simply to explore the nature of the relationship, or do you think that some of the variables explain or even cause changes in others? That is, are some of the variables *response variables* and others *explanatory variables*?

| TABLE 3.1 | | Education and related data for the United States | | | | | |

State	Region*	Population (1,000)	SAT verbal	SAT math	Percent taking	Dollars per pupil	Teachers' pay ($1,000)
AL	ESC	4,041	470	514	8	3,648	27.3
AK	PAC	550	438	476	42	7,887	43.4
AZ	MTN	3,665	445	497	25	4,231	30.8
AR	WSC	2,351	470	511	6	3,334	23.0
CA	PAC	29,760	419	484	45	4,826	39.8
CO	MTN	3,294	456	513	28	4,809	31.8
CT	NE	3,287	430	471	74	7,914	43.8
DE	SA	666	433	470	58	6,016	35.2
DC	SA	607	409	441	68	8,210	39.6
FL	SA	12,938	418	466	44	5,154	30.6
GA	SA	6,478	401	443	57	4,860	29.2
HI	PAC	1,108	404	481	52	5,008	32.5
ID	MTN	1,007	466	502	17	3,200	25.5
IL	ENC	11,431	466	528	16	5,062	34.6
IN	ENC	5,544	408	459	54	5,051	32.0
IA	WNC	2,777	511	577	5	4,839	28.0
KS	WNC	2,478	492	548	10	5,009	29.8
KY	ESC	3,685	473	521	10	4,390	29.1
LA	WSC	4,220	476	517	9	4,012	26.2
ME	NE	1,228	423	463	60	5,894	28.5
MD	SA	4,781	430	478	59	6,184	38.4
MA	NE	6,016	427	473	72	6,351	36.1
MI	ENC	9,295	454	514	12	5,257	38.3
MN	WNC	4,375	477	542	14	2,260	33.1
MS	ESC	2,573	477	519	4	3,322	24.4
MO	WNC	5,117	473	522	12	4,415	28.5
MT	MTN	799	464	523	20	5,184	26.7
NE	WNC	1,578	484	546	10	4,381	26.6
NV	MTN	1,202	434	487	24	4,564	32.2
NH	NE	1,109	442	486	67	5,504	31.3
NJ	MA	7.730	418	473	69	9,159	38.4
NM	MTN	1,515	480	527	12	4,446	26.2
NY	MA	17,990	412	470	70	8,500	42.1

		Population	SAT	SAT	Percent	Dollars	Teachers'
State	Region*	(1,000)	verbal	math	taking	per pupil	pay ($1,000)
NC	SA	6,629	401	440	55	4,802	29.2
ND	WNC	639	505	564	6	3,685	23.6
OH	ENC	10,487	450	499	22	5,639	32.6
OK	WSC	3,146	478	523	9	3,742	24.3
OR	PAC	2,842	439	484	49	5,291	32.3
PA	MA	11,882	420	463	64	6,534	36.1
RI	NE	1,003	422	461	62	6,989	37.7
SC	SA	3,487	397	437	54	4,327	28.3
SD	WNC	696	506	555	5	3,730	22.4
TN	ESC	4,877	483	525	12	3,707	28.2
TX	WSC	16,987	413	461	42	4,238	28.3
UT	MTN	1,723	492	539	5	2,993	25.0
VT	NE	563	431	466	62	5,740	31.0
VA	SA	6,187	425	470	58	5,360	32.4
WA	PAC	4,867	437	486	44	5,045	33.1
WV	SA	1,793	443	490	15	5,046	26.0
WI	ENC	4,892	476	543	11	5,946	33.1
WY	MTN	454	458	519	13	5,255	29.0

TABLE 3.1 Education and related data for the United States (*continued*)

SOURCE: *Statistical Abstract of the United States*, 1992.

*The census regions are East North Central, East South Central, Mountain, New England, Pacific, Middle Atlantic, South Atlantic, West North Central, and West South Central.

RESPONSE VARIABLE, EXPLANATORY VARIABLE

A **response variable** measures an outcome of a study. An **explanatory variable** attempts to explain the observed outcomes.

It is easiest to identify explanatory and response variables when we actually set values of one variable in order to see how it affects another variable.

EXAMPLE 3.1

Alcohol has many effects on the body. One effect is a drop in body temperature. To study this effect, researchers give several different amounts of alcohol to mice, then measure the change in each mouse's body temperature in the 15 minutes after taking the alcohol. Amount of alcohol is the explanatory variable, and change in body temperature is the response variable.

When you don't set the values of either variable but just observe both variables, there may or may not be explanatory and response variables. Whether there are depends on how you plan to use the data.

EXAMPLE 3.2

Jim wants to know how the median SAT math and verbal scores in the 51 states (including the District of Columbia) are related to each other. He doesn't think that either score explains or causes the other. Jim has two related variables, and neither is an explanatory variable.

Julie looks at some data. She asks, "Can I predict a state's SAT math score if I know its SAT verbal score?" Julie is treating the verbal score as the explanatory variable and the math score as the response variable.

In Example 3.1 alcohol actually *causes* a change in body temperature. There is no cause-and-effect relationship between SAT math and verbal scores in Example 3.2. Because the scores are closely related, we can nonetheless use a state's SAT verbal score to predict its math score. We will learn how to do the prediction in Section 3.3. Prediction requires that we identify an explanatory variable and a response variable. Other statistical techniques ignore this distinction. Do remember that calling one variable explanatory and the other response doesn't necessarily mean that changes in one *cause* changes in the other.

You will often find explanatory variables called *independent variables*, and response variables called *dependent variables*. The idea behind this language is that the response variable depends on the explanatory variable. Because the words "independent" and "dependent" have other, unrelated meanings in statistics, we won't use them here.

The statistical techniques used to study relations among variables are more complex than the one-variable methods in Chapters 1 and 2. Fortunately, analysis of several-variable data builds on the tools used for examining individual variables. The principles that guide examination of data are also the same:

1. Start with a graph.
2. Look for an overall pattern and deviations from the pattern.
3. Add numerical descriptions of specific aspects of the data.
4. Sometimes there is a way to describe that overall pattern very briefly.

EXERCISES

3.1   How well does a child's height at age 6 predict height at age 16? To find
        out, measure the heights of a large group of children at age 6, wait until
        they reach age 16, then measure their heights again. What are the ex-
        planatory and response variables here? Are these variables categorical or
        quantitative?

3.2   There may be a "gender gap" in political party preference in the United
        States, with women more likely than men to prefer Democratic candi-
        dates. A political scientist selects a large sample of registered voters, both
        men and women. She asks each voter whether they voted for the Demo-
        cratic or for the Republican candidate in the last congressional election.
        What are the explanatory and response variables in this study? Are they
        categorical or quantitative variables?

3.3   The most common treatment for breast cancer was once removal of the
        breast. It is now usual to remove only the tumor and nearby lymph nodes,
        followed by radiation. The change in policy was due to a large medical
        experiment that compared the two treatments. Some breast cancer pa-
        tients, chosen at random, were given each treatment. The patients were
        closely followed to see how long they lived following surgery. What are
        the explanatory and response variables? Are they categorical or quantita-
        tive?

3.4   What are the variables in Activity 3 (page 106)? Is there an explanatory/
        response relationship? If so, which is the explanatory variable and which
        is the response variable? Are the variables quantitative or categorical?

3.1 SCATTERPLOTS

The most effective way to display the relation between two quantitative
variables is a *scatterplot*. Here is an example of a scatterplot.

EXAMPLE 3.3

Some people use median SAT scores to rank state or local school systems.
This is not proper, because the percent of high school seniors who take the
SAT varies from place to place. Let us examine the relationship between the
percent in a state who take the exam and the state median SAT mathematics
score, using data from Table 3.1.

 We think that "percent taking" will help explain "median score." Therefore, "percent taking" is the explanatory variable and "median score" is the
response variable. We want to see how median score changes when percent
taking changes, so we put percent taking (the explanatory variable) on the

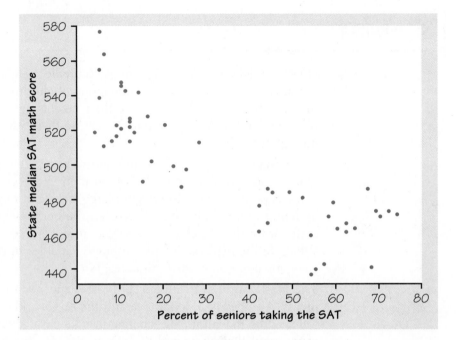

FIGURE 3.1   Scatterplot of the median SAT math score in each state against the percent of that state's high school seniors who take the SAT, from Table 3.1.

horizontal axis. Figure 3.1 is the scatterplot. Each point represents a single state. In Alabama, for example, 8% take the SAT, and the median SAT math score is 514. Find 8 on the x (horizontal) axis and 514 on the y (vertical) axis. Alabama appears as the point (8,514) above 8 and to the right of 514. Alaska appears as the point (42,476), and so on.

SCATTERPLOT

A **scatterplot** shows the relationship between two quantitative variables measured on the same individuals. The values of one variable appear on the horizontal axis, and the values of the other variable appear on the vertical axis. Each individual in the data appears as the point in the plot fixed by the values of both variables for that individual.

Always plot the explanatory variable, if there is one, on the horizontal axis (the x axis) of a scatterplot. As a reminder, we usually call the explanatory variable x and the response variable y. If there is no explanatory-response distinction, either variable can go on the horizontal axis.

EXERCISES

3.5 Propelled by a stream of pressurized water, jet skis and other so-called wet bikes carry from one to three people, retail for an average price of $5,700, and have become one of the most popular types of recreational vehicle sold today. But critics say that they're noisy, dangerous, and damaging to the environment. An article in the August 1997 issue of the *Journal of the American Medical Association* reported on a survey that tracked emergency room visits at randomly selected hospitals nationwide. Here are data on the number of jet skis in use, the number of accidents, and the number of fatalities for the years 1987 to 1996:[1]

Year	Number in use	Accidents	Fatalities
1987	92,756	376	5
1988	126,881	650	20
1989	178,510	844	20
1990	241,376	1,162	28
1991	305,915	1,513	26
1992	372,283	1,650	34
1993	454,545	2,236	35
1994	600,000	3,002	56
1995	760,000	4,028	68
1996	900,000	4,010	55

(a) We want to examine the relationship between the number of jet skis in use and the number of accidents. Which is the explanatory variable?

(b) Make a scatterplot of these data. (Be sure to label the axes with the variable names, not just x and y.) What does the scatterplot show about the relationship between these variables?

3.6 Make a scatterplot of the (math SAT/ACT score, verbal SAT/ACT score) data from Activity 3, if you haven't done so already. Does the scatterplot describe a strong association, a moderate association, a weak association, or no association between these variables?

Interpreting scatterplots

To interpret a scatterplot, look first for an overall pattern. This pattern should reveal the *direction, form,* and *strength* of the relationship between the two variables.

EXAMPLE 3.4

clusters

The *form* of the relationship in Figure 3.1 strikes us at once: there are two distinct **clusters** of states. In one cluster, more than 40% of high school seniors take the SAT, and the state median score is low. Less than 30% of seniors in states in the other cluster take the SAT, and these states have higher median scores. What explains the clusters? There are two widely used college entrance exams: the SAT and the ACT exam. Each state favors one or the other. The left cluster in Figure 3.1 contains the ACT states, and the SAT states make up the right cluster. In ACT states, most students who take the SAT are applying to a selective college that requires SAT scores. This selected group of students has a higher median score than the much larger group of students who take the SAT in SAT states.

The *direction* of the overall pattern in Figure 3.1 is also clear. Even within the clusters, states where a higher percent of seniors take the SAT tend to have lower median scores. We say that percent taking and median score are *negatively associated*.

POSITIVE ASSOCIATION, NEGATIVE ASSOCIATION

> Two variables are **positively associated** when above-average values of one tend to accompany above-average values of the other and below-average values also tend to occur together. Two variables are **negatively associated** when above-average values of one accompany below-average values of the other, and vice versa.

The relationship in Figure 3.1 has a clear *direction*, the negative association. The *form* is dominated by the clusters and may have a curved shape overall. We say "may have" because, aside from the clusters, the *strength* of the relationship is weak. There is a wide spread in SAT scores among states with about the same percent of students taking the exam. Here is an example of a stronger relationship with a clearer form.

EXAMPLE 3.5

The Sanchez household is about to install solar panels to reduce the cost of heating their house. In order to know how much the solar panels help, they record their consumption of natural gas before the panels are installed. Gas consumption is higher in cold weather, so the relationship between outside temperature and gas consumption is important.

Table 3.2 gives data for 16 months.[2] The response variable y is the average amount of natural gas consumed each day during the month, in hundreds of cubic feet. The explanatory variable x is the average number of heating degree-

TABLE 3.2		Average degree-days and natural gas consumption			
Month	Degree-days	Gas (100 cu ft)	Month	Degree-days	Gas (100 cu ft)
Nov.	24	6.3	July	0	1.2
Dec.	51	10.9	Aug.	1	1.2
Jan.	43	8.9	Sep.	6	2.1
Feb.	33	7.5	Oct.	12	3.1
Mar.	26	5.3	Nov.	30	6.4
Apr.	13	4.0	Dec.	32	7.2
May	4	1.7	Jan.	52	11.0
June	0	1.2	Feb.	30	6.9

days each day during the month. (Heating degree-days are the usual measure of demand for heating. One degree-day is accumulated for each degree a day's average temperature falls below 65° F. An average temperature of 20° F, for example, corresponds to 45 degree-days.)

linear relationship

The scatterplot in Figure 3.2 shows a strong positive association. More degree-days means colder weather and so more gas consumed. The form of the relationship is **linear**. That is, the points lie in a straight-line pattern. It is a strong relationship because the points lie close to a line, with little scatter. If we know how cold a month is, we can predict gas consumption quite accurately from the scatterplot. That strong relationships make accurate predictions possible is an important point that we will soon discuss in more detail.

Of course, not all relationships are linear in form. What is more, not all relationships have a clear direction that we can describe as positive association or negative association. Exercise 3.8 gives an example that is not linear and has no clear direction.

EXERCISES

3.7 In Exercise 3.5 you made a scatterplot of jet skis in use and number of accidents.

(a) Describe the direction of the relationship. Are the variables positively or negatively associated?

(b) Describe the form of the association. Is it linear?

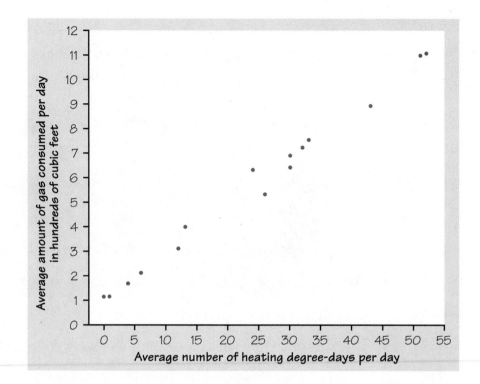

FIGURE 3.2   Scatterplot of the average amount of natural gas used per day by the Sanchez household in 16 months against the average number of heating degree-days per day in those months, from Table 3.2.

(c) Describe the strength of the association. Can the number of accidents be predicted accurately from the number of jet skis in use? If there were 1 million jet skis in use, about how many accidents would occur?

3.8   How does the fuel consumption of a car change as its speed increases? Here are data for a British Ford Escort. Speed is measured in kilometers per hour, and fuel consumption is measured in liters of gasoline used per 100 kilometers traveled.[3]

Speed (km/h)	Mileage (liters/100 km)	Speed (km/h)	Mileage (liters/100 km)	Speed (km/h)	Mileage (liters/100 km)
10	21.00	60	5.90	110	9.03
20	13.00	70	6.30	120	9.87
30	10.00	80	6.95	130	10.79
40	8.00	90	7.57	140	11.77
50	7.00	100	8.27	150	12.83

(a) Make a scatterplot. Here are steps for the TI-83. The command `seq(10X,X,1,15)→SPEED` will create a list named SPEED and assign the numbers 10, 20, ..., 150 to the list. (Note that `seq` is found under 2nd / LIST / OPS.) Next get into the STAT / EDIT mode and insert a second list named FUEL, and enter the number of liters/100 km into this list.

SPEED	FUEL	L1 2
10	21	- - - -
20	13	
30	10	
40	8	
50	7	
60	5.9	
70	**6.3**	

FUEL(7)=6.3

Under STAT PLOT, specify a scatterplot using these two lists.

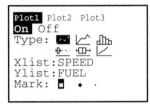

Then ZOOM / 9 to graph the scatterplot. (The calculator looks at the data in the two lists and automatically sets the window dimensions.)

(b) Describe the form of the relationship. Why is it not linear? Explain why the form of the relationship makes sense.

(c) It does not make sense to describe the variables as either positively associated or negatively associated. Why?

(d) Is the relationship reasonably strong or quite weak? Explain your answer.

Adding categorical variables to scatterplots

After examining the overall pattern of a scatterplot, look for deviations from the patterns such as *outliers*.

OUTLIERS

An **outlier** in any graph of data is an individual observation that falls outside the overall pattern of the graph.

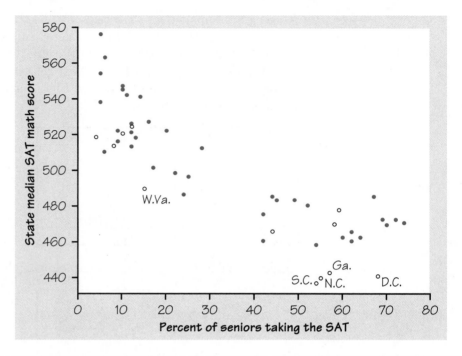

FIGURE 3.3   Median SAT math score and percent of high school seniors who took the test, by state, with the southern states differentiated.

Neither of the scatterplots in Figures 3.1 and 3.2 has strong outliers. Figures 3.1 does show a group of four points at the bottom right that have unusually low SAT math scores. Let's investigate.

EXAMPLE 3.6

One of the four low points is the District of Columbia, which is a city rather than a state. The other three are Georgia, North Carolina, and South Carolina. This finding suggests that perhaps the South as a whole has low SAT scores even after we take into account the effect of the percent taking the test.

Figure 3.3 explores that guess by plotting the southern states with a different plot symbol. (We took the South to be the states in the East South Central and South Atlantic regions.) The states we already noticed, and perhaps West Virginia, do have low SAT scores. The other southern states blend in with the rest of the country. The data defeat our guess that the South as a whole has low scores.

In dividing the states into "southern" and "nonsouthern," we introduced a third variable into the scatterplot. This is a categorical variable that has only two values. The two values are displayed by the two different plotting symbols. *Use different colors or symbols to plot points when you want to add a categorical variable to a scatterplot.*[4]

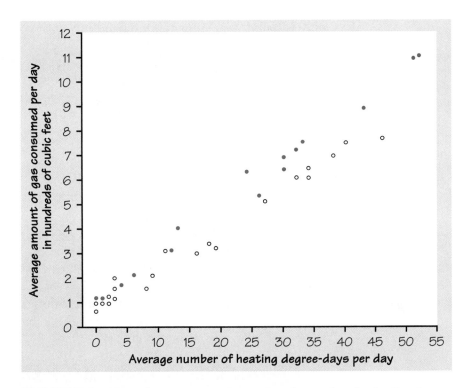

FIGURE 3.4   Natural gas consumption against degree-days for the Sanchez household. The solid blue observations are for 16 months before installing solar panels. The open black observations are for 23 months with the panels in use.

EXAMPLE 3.7

After the Sanchez household gathered the information recorded in Table 3.2 and Figure 3.2, they added solar panels to their house. Then they measured their natural gas consumption for 23 more months. To see how the solar panels affected gas consumption, add the new data (including degree-days for these months) to the scatterplot. Figure 3.4 is the result. We use different symbols to distinguish before from after. The "after" data form a linear pattern that is close to the "before" pattern in warm months (few degree-days). In colder months, with more degree-days, gas consumption after installing the solar panels is less than similar months before the panels were added. The scatterplot shows the energy savings from the panels.

Both of our examples suffer from a common problem in drawing scat-terplots that you may not notice when a computer does the work. When several individuals have exactly the same data, they occupy the same point on the scatterplot. Look at Delaware and Virginia in Table 3.1, or at June and July in Table 3.2. Table 3.2 contains data for 16 months, but there are only 15 points in Figure 3.2. June and July both occupy the same point. You can use a different plotting symbol to call attention to points that stand for more than one individual. Some computer software, such as Minitab and Data Desk, does this automatically, but some—including the software used for our scatterplots—does not. We recommend that you do use a different symbol for repeated observations when you plot a small number of observations by hand.

EXERCISE

3.9   Metabolic rate, the rate at which the body consumes energy, is important in studies of weight gain, dieting, and exercise. The table below gives data on the lean body mass and resting metabolic rate for 12 women and 7 men who are subjects in a study of dieting. Lean body mass, given in kilograms, is a person's weight leaving out all fat. Metabolic rate is measured in calories burned per 24 hours, the same calories used to describe the energy content of foods. The researchers believe that lean body mass is an important influence on metabolic rate.

Subject	Sex	Mass	Rate	Subject	Sex	Mass	Rate
1	M	62.0	1792	11	F	40.3	1189
2	M	62.9	1666	12	F	33.1	913
3	F	36.1	995	13	M	51.9	1460
4	F	54.6	1425	14	F	42.4	1124
5	F	48.5	1396	15	F	34.5	1052
6	F	42.0	1418	16	F	51.1	1347
7	M	47.4	1362	17	F	41.2	1204
8	F	50.6	1502	18	M	51.9	1867
9	F	42.0	1256	19	M	46.9	1439
10	M	48.7	1614				

(a) Make a scatterplot of the data for the female subjects. If you do this on your TI-83, define two lists, MASSF for female mass, and METF for female metabolic rate. Enter the data into these two lists, and then define Plot1 as a scatterplot.

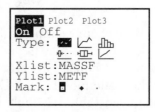

(b) Is the association between these variables positive or negative? What is the form of the relationship? How strong is the relationship?

(c) Now add the data for the male subjects to your graph, using a different color or a different plotting symbol. For TI-83 users, define lists MASSM and METM and enter the male data. Then define Plot2. Notice that the plotting symbol is different for the male data so that we will be able to see the difference when we plot male and female data together. Make sure that Plot1 and Plot2 are both turned on. Then ZOOM / 9 to see all of the subjects in one scatterplot and still be able to distinguish males from females. Does the pattern of relationship that you observed in (b) hold for men also? How do the male subjects as a group differ from the female subjects as a group?

SUMMARY

When we think that changes in a variable x explain or even cause changes in a second variable y, we call x an **explanatory variable** and y a **response variable**.

A **scatterplot** displays the relationship between two quantitative variables measured on the same individuals. Mark values of one variable on the horizontal axis (x axis) and values of the other variable on the vertical axis (y axis). Plot each individual's data as a point on the graph.

Values of the explanatory variable, if there is one, always go on the horizontal axis of a scatterplot.

In examining a scatterplot, look for an overall pattern showing the **direction, form,** and **strength** of the relationship, and then for **outliers** or other deviations from this pattern.

If the relationship has a clear direction, we speak of either **positive association** (high values of the two variables tend to occur together) or **negative association** (high values of one variable tend to occur with low values of the other variable).

Linear relationships, where the points show a straight-line pattern, are an important form of relationship between two variables. Curved relationships and clusters are other forms to watch for.

The **strength** of a relationship is determined by how close the points in the scatterplot lie to a simple form such as a line.

You can show the effect of a **categorical variable** by plotting points on a scatterplot with different colors or symbols.

SECTION 3.1 EXERCISES

3.10   Figure 3.5 is a scatterplot that displays the heights of 53 pairs of parents. The mother's height is plotted on the vertical axis and the father's height on the horizontal axis.[5]

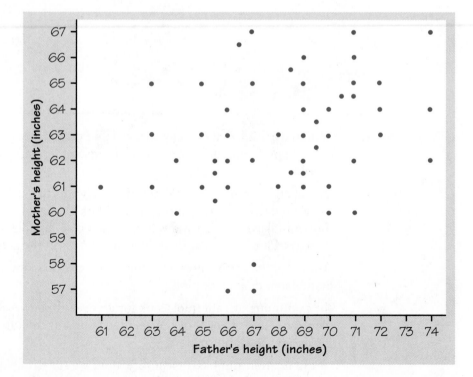

FIGURE 3.5   Scatterplot of the heights of the mother and father in 53 pairs of parents.

(a) What is the smallest height of any mother in the group? How many mothers have that height? What are the heights of the fathers in these pairs?

(b) What is the greatest height of any father in the group? How many fathers have that height? How tall are the mothers in these pairs?

(c) Are there clear explanatory and response variables, or could we freely choose which variable to plot horizontally?

(d) Say in words what a positive association between these variables means. The scatterplot shows a weak positive association. Why do we say the association is weak?

3.11 Are hot dogs that are high in calories also high in salt? Figure 3.6 is a scatterplot of the calories and salt content (measured as milligrams of sodium) in 17 brands of meat hot dogs.[6]

(a) Roughly what are the lowest and highest calorie counts among these brands? Roughly what is the sodium level in the brands with the fewest and with the most calories?

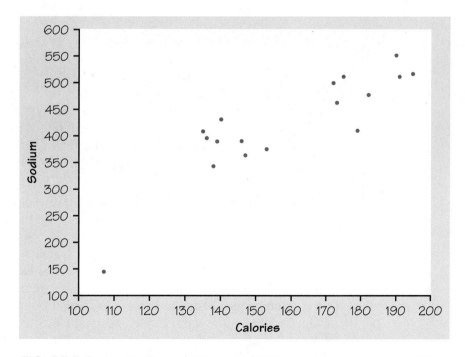

FIGURE 3.6   Scatterplot of milligrams of sodium and calories in each of 17 brands of meat hot dogs.

(b) Does the scatterplot show a clear positive or negative association? Say in words what this association means about calories and salt in hot dogs.

(c) Are there any outliers? Is the relationship (ignoring any outliers) roughly linear in form? Still ignoring outliers, how strong would you say the relationship between calories and sodium is?

3.12   A food industry group asked 3368 people to guess the number of calories in each of several common foods. Here is a table of the average of their guesses and the correct number of calories:[7]

Food	Guessed calories	Correct calories
8 oz. whole milk	196	159
5 oz. spaghetti with tomato sauce	394	163
5 oz. macaroni with cheese	350	269
One slice wheat bread	117	61
One slice white bread	136	76
2-oz. candy bar	364	260
Saltine cracker	74	12
Medium-size apple	107	80
Medium-size potato	160	88
Cream-filled snack cake	419	160

(a) We think that how many calories a food actually has helps explain people's guesses of how many calories it has. With this in mind, make a scatterplot of these data. (Because both variables are measured in calories, you should use the same scale on both axes. Your plot will be square.)

(b) Describe the relationship. Is there a positive or negative association? Is the relationship approximately linear? Are there any outliers?

3.13   *Archaeopteryx* is an extinct beast having feathers like a bird but teeth and a long bony tail like a reptile. Only six fossil specimens are known. Because these specimens differ greatly in size, some scientists think they are different species rather than individuals from the same species. We will examine some data. If the specimens belong to the same species and differ in size because some are younger than others, there should be a positive linear relationship between the lengths of a pair of bones from all individuals. An outlier from this relationship would suggest a different species. Here are data on the lengths in centimeters of the femur (a leg bone) and the humerus (a bone in the upper arm) for the five specimens that preserve both bones:[8]

Femur	38	56	59	64	74
Humerus	41	63	70	72	84

Make a scatterplot. Do you think that all five specimens come from the same species?

3.14 Table 3.3 shows the Indianapolis 500 automobile race winners for the years 1967 to 1997. Enter the data into your calculator or computer package and then plot a scatterplot. Is there an apparent trend? During this period, the race was less than 500 miles in 1973 (332.5 miles), in 1975 (435 miles), and in 1976 (255 miles). Are these three shortened races apparent in the scatterplot? If so, in what way? Can you suggest an explanation?

3.15 How much corn per acre should a farmer plant to obtain the highest yield? Too few plants will give a low yield. On the other hand, if there are too many plants, they will compete with each other for moisture and nutrients, and yields will fall. To find the best planting rate, plant at different rates on several plots of ground and measure the harvest. (Be sure to treat all the plots the same except for the planting rate.) Here are the data from such an experiment:[9]

Plants per acre	Yield (bushels per acre)			
12,000	150.1	113.0	118.4	142.6
16,000	166.9	120.7	135.2	149.8
20,000	165.3	130.1	139.6	149.9
24,000	134.7	138.4	156.1	
28,000	119.0	150.5		

(a) Is yield or planting rate the explanatory variable?

(b) Make a scatterplot of yield and planting rate.

(c) Describe the overall pattern of the relationship. Is it linear? Is there a positive or negative association, or neither?

(d) Find the mean yield for each of the five planting rates. Plot each mean yield against its planting rate on your scatterplot and connect these five points with lines. This combination of numerical description and graphing makes the relationship clearer. What planting rate would you recommend to a farmer whose conditions were similar to those in the experiment?

TABLE 3.3	Indianapolis 500 automobile race winners

Year	Driver	Average speed (MPH)
1967	A. J. Foyt	151.207
1968	Bobby Unser	152.882
1969	Mario Andretti	156.867
1970	Al Unser	155.749
1971	Al Unser	157.735
1972	Mark Donohue	162.962
1973	Gordon Johncock	159.036
1974	Johnny Rutherford	158.589
1975	Bobby Unser	149.213
1976	Johnny Rutherford	148.725
1977	A. J. Foyt	161.331
1978	Al Unser	161.363
1979	Rick Mears	158.899
1980	Johnny Rutherford	142.862
1981	Bobby Unser	139.084
1982	Gordon Johncock	162.026
1983	Tom Sneva	162.117
1984	Rick Mears	163.612
1985	Danny Sullivan	152.982
1986	Bobby Rahal	170.722
1987	Al Unser	162.175
1988	Rick Mears	144.809
1989	Emerson Fittipaldi	167.581
1990	Arie Luyendyk	185.981
1991	Rick Mears	176.457
1992	Al Unser, Jr.	134.477
1993	Emerson Fittipaldi	157.207
1994	Al Unser, Jr.	160.872
1995	Jacques Villeneuve	153.616
1996	Buddy Lazier	147.956
1997	Arie Luyendyk	145.827

SOURCE: *The World Almanac and Book of Facts, 1997.*

3.16   Table 3.1 gives educational data for the states. We are interested in the relationship between how much states spend on education (dollars per pupil) and how much they pay their teachers (median teacher salaries, in thousands of dollars.)

   (a) Explain why you expect a positive association between these variables.
   (b) Make a scatterplot, with education spending (dollars per pupil) as the explanatory variable.
   (c) Describe the relationship. Is there a positive association? Is the relationship approximately linear?
   (d) On the plot, identify a state where teacher salaries are unusually high relative to the state's education spending. (This state is an outlier, though not an extreme outlier.) What state is this?
   (e) How do the Mountain states compare with the rest of the country in education spending and teacher salaries? Mark the points for states in the MTN region with a different color on your scatterplot. Based on the plot, briefly answer the question.

3.17   A scatterplot shows the relationship between two quantitative variables. Here is a similar plot to study the relationship between a categorical explanatory variable and a quantitative response variable.
        The presence of harmful insects in farm fields is detected by putting up boards covered with a sticky material and then examining the insects trapped on the board. Which colors attract insects best? Experimenters placed six boards of each of four colors in a field of oats and measured the number of cereal leaf beetles trapped.[10]

Board color	Insects trapped						
Lemon yellow	45	59	48	46	38	47	
White	21	12	14	17	13	17	
Green	37	32	15	25	39	41	
Blue	16	11	20	21	14	7	

   (a) Make a plot of the counts of insects trapped against board color (space the four colors equally on the horizontal axis). Compute the mean count for each color, add the means to your plot, and connect the means with line segments.
   (b) Based on the data, what do you conclude about the attractiveness of these colors to the beetles?
   (c) Does it make sense to speak of a positive or negative association between board color and insect count?

3.2 CORRELATION

A scatterplot displays the direction, form, and strength of the relationship between two quantitative variables. Linear relations are particularly important because a straight line is a simple pattern that is quite common. We say a linear relation is strong if the points lie close to a straight line, and weak if they are widely scattered about a line. Our eyes are not good judges of how strong a linear relationship is. The two scatterplots in Figure 3.7

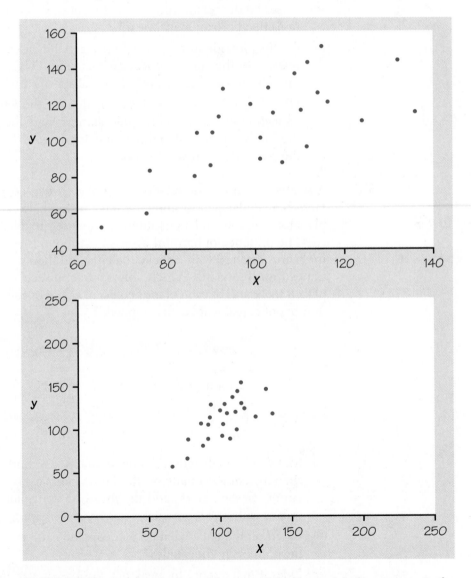

FIGURE 3.7 Two scatterplots of the same data; the straight-line pattern in the lower plot appears stronger because of the surrounding white space.

depict exactly the same data, but the lower plot is drawn smaller in a large field. The lower plot seems to show a stronger linear relationship. Our eyes can be fooled by changing the plotting scales or the amount of white space around the cloud of points in a scatterplot.[11] We need to follow our strategy for data analysis by using a numerical measure to supplement the graph. *Correlation* is the measure we use.

The correlation r

CORRELATION

The **correlation** measures the strength and direction of the linear relationship between two quantitative variables. Correlation is usually written as r.

Suppose that we have data on variables x and y for n individuals. The values for the first individual are x_1 and y_1, the values for the second individual are x_2 and y_2, and so on. The means and standard deviations of the two variables are $\bar{x}$ and s_x for the x-values, and $\bar{y}$ and s_y for the y-values. The correlation r between x and y is

$$r = \frac{1}{n-1} \sum \left(\frac{x_i - \bar{x}}{s_x} \right) \left(\frac{y_i - \bar{y}}{s_y} \right)$$

As always, the summation sign $\sum$ means "add these terms for all the individuals." The formula for r begins by standardizing the observations. Suppose, for example, that x is height in centimeters and y is weight in kilograms and that we have height and weight measurements for n people. Then $\bar{x}$ and s_x are the mean and standard deviation of the n heights, both in centimeters. The value

$$\frac{x_i - \bar{x}}{s_x}$$

is the standardized height of the ith person, familiar from Chapter 2. The standardized height says how many standard deviations above or below the mean a person's height lies. Standardized values have no units—in this example, they are no longer measured in centimeters. Standardize the weights also. The correlation r is an average of the products of the standardized height and standardized weight for the n people. Since the standardized value is positive when that value is above average, then r is likely to be large and positive when there are many pairs in which both x and y are above average or in which both are below average. When will r be negative, with a "large" absolute value? It helps to see how the correlation r is calculated, but in practice you should use your calculator or computer software to find r from keyed-in values of x and y.

Correlation makes no use of the distinction between explanatory and response variables. It makes no difference which variable you call x and which you call y in calculating the correlation. Correlation does require that both variables be quantitative, so that it makes sense to do the arithmetic required by the formula for r. You can't calculate a correlation between the incomes of a group of people and what city they live in, because city is a categorical variable.

EXERCISES

3.18   We will calculate the correlation r step-by-step in some very simple examples to see how the formula works.

(a) Enter the following data into lists L_1 and L_2 in your calculator. Then draw a scatterplot.

x	4	4	−4	−4
y	−4	4	4	−4

- Briefly describe in words the association between x and y.
- Use your calculator to find the mean and standard deviation for both x and y. (Use STAT / CALC / 2-Var Stats. Unless you specify otherwise, the calculator thinks the x-values are in L_1 and the y-values are in L_2.)
- Find the standardized values of x and y, then use the formula to find the correlation r.

(b) Do the same as in (a) for this set of data.

x	4	3	0	−3	−4
y	−4	−2	0	2	4

(c) Enter the following data into lists L_1 and L_2 in your calculator. Then draw a scatterplot on your calculator.

x	4	2	−2	−4
y	4	−2	2	−4

Briefly describe in words the association between x and y.

In this part we will instruct the TI-83 to perform all of the calculations. In STAT / EDIT mode, clear list L_3 and highlight L_3. Define $L_3 = L_1 - \bar{x}$. (Note that $\bar{x}$ is found under VARS / 5:Statistics....) List L_3 will consist of differences of the form $x - \bar{x}$. Then clear list L_4 and define $L_4 = L_2 - \bar{y}$. Clear list L_5 and define $L_5 = L_3 \times L_4$. Then in the Home screen, enter the formula for the correlation r

$$\frac{1}{n-1} \sum \frac{(x - \bar{x})(y - \bar{y})}{s_x s_y}$$

as follows: `(1/(n-1))*sum(L5/(Sx*Sy))`. Note that `n`, `Sx`, and `Sy` are found under the VARS / 5:Statistics... menu, and `sum` is found under the 2nd / LIST / MATH menu. What is the value of r?

3.19   Exercise 3.13 (page 124) gives the lengths of two bones in five fossil specimens of the extinct beast *Archaeopteryx*.

(a)   Find the correlation r step-by-step. That is, find the mean and standard deviation of the femur lengths and of the humerus lengths. Then find the five standardized values for each variable and use the formula for r.

(b)   Enter these data into your calculator in lists L_1 and L_2 (explanatory in L_1 and response in L_2). Then key this sequence: 2nd / CATALOG / x^{-1} (for the letter D) / (arrow down to select DiagnosticOn) / ENTER / ENTER. The calculator should say "Done." (Selecting DiagnosticOn is a one-time action.) Then key this sequence: STAT / CALC / 8:LinReg(a+bx) / ENTER. The calculator will print several lines, the last of which shows the correlation coefficient r. (We will explain the LinReg function in the next section.) Check that your calculator produces the same r-value as in (a).

Facts about correlation

The formula for correlation helps us see that r is positive when there is a positive association between the variables. Height and weight, for example, have a positive association. People who are above average in height tend to also be above average in weight. Both the standardized height and the standardized weight for such a person are positive. People who are below average in height tend to also have below-average weight. Then both standardized height and standardized weight are negative. In both cases, the products in the formula for r are mostly positive and so r is positive. In the same way, we can see that r is negative when the association between

x and y is negative. More detailed study of the formula gives more detailed properties of r. Here is what you need to know in order to interpret correlation.

1. Positive r indicates positive association between the variables, and negative r indicates negative association.

2. The correlation r always falls between -1 and 1. Values of r near 0 indicate a very weak linear relationship. The strength of the linear relationship increases as r moves away from 0 toward either -1 or 1. Values of r close to -1 or 1 indicate that the points lie close to a straight line. The extreme values $r = -1$ and $r = 1$ occur only in the case of a perfect linear relationship, when the points in a scatterplot lie exactly along a straight line.

3. Because r uses the standardized values of the observations, r does not change when we change the units of measurement of x, y, or both. Changing from centimeters to inches and from kilograms to pounds does not change the correlation between height and weight. The correlation r itself has no unit of measurement; it is just a number between -1 and 1.

4. Correlation measures the strength of only a linear relationship between two variables. Correlation does not describe curved relationships between variables, no matter how strong they are.

5. Like the mean and standard deviation, the correlation is strongly affected by a few outlying observations. Use r with caution when outliers appear in the scatterplot.

The scatterplots in Figure 3.8 illustrate how values of r closer to 1 or -1 correspond to stronger linear relationships. To make the essential meaning of r clear, the standard deviations of both variables in these plots are equal and the horizontal and vertical scales are the same. In general, it is not so easy to guess the value of r from the appearance of a scatterplot. Remember that changing the plotting scales in a scatterplot may mislead our eyes, but it does not change the correlation.

The real data we have examined also illustrate how correlation measures the strength and direction of linear relationships. Figure 3.2 shows a very strong positive linear relationship between degree-days and natural gas consumption. The correlation is $r = .9953$. Check this on your calculator using the data in Table 3.2. Figure 3.1 shows a clear but weaker negative association between percent of students taking the SAT and the median SAT math score in a state. The correlation is $r = -.8581$.

Do remember that *correlation is not a complete description of two-variable data*, even when the relationship between the variables is linear. You should give the means and standard deviations of both x and y along

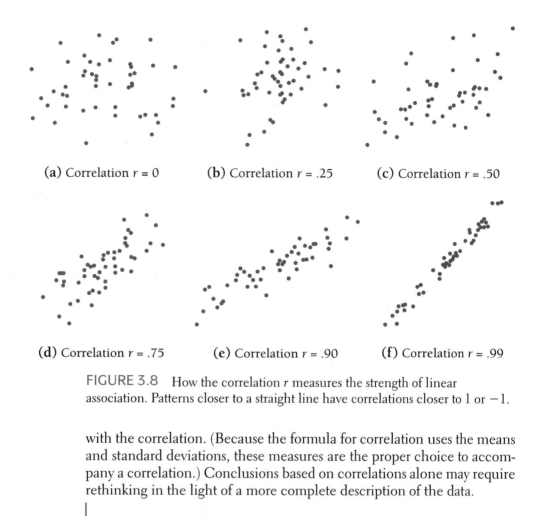

(a) Correlation $r = 0$ (b) Correlation $r = .25$ (c) Correlation $r = .50$

(d) Correlation $r = .75$ (e) Correlation $r = .90$ (f) Correlation $r = .99$

FIGURE 3.8   How the correlation r measures the strength of linear association. Patterns closer to a straight line have correlations closer to 1 or -1.

with the correlation. (Because the formula for correlation uses the means and standard deviations, these measures are the proper choice to accompany a correlation.) Conclusions based on correlations alone may require rethinking in the light of a more complete description of the data.

EXAMPLE 3.8

Competitive divers are scored on their form by a panel of judges who use a scale from 1 to 10. The subjective nature of the scoring often results in controversy. We have the scores awarded by two judges, Ivan and George, on a large number of dives. How well do they agree? We do some calculation and find that the correlation between their scores is $r = .9$. But the mean of Ivan's scores is 3 points lower than George's mean.

These facts do not contradict each other. They are simply different kinds of information. The mean scores show that Ivan awards much lower scores than George. But because Ivan gives *every* dive a score about 3 points lower than George, the correlation remains high. Adding or subtracting the same number to all values of either x or y does not change the correlation. If Ivan and George both rate several divers, the contest is fairly scored because Ivan and George agree on which dives are better than others. The high r shows their agreement. But if Ivan scores one diver and George another, we must add 3 points to Ivan's scores to arrive at a fair comparison.

3.20   Figure 3.5 (page 122) is a scatterplot that displays the heights of 53 pairs of parents. Do you think the correlation *r* for these data is near −1, clearly negative but not near −1, near 0, clearly positive but not near 1, or near 1? Explain your answer.

3.21   Figure 3.6 (page 123) is a scatterplot of the calories and sodium content in 17 brands of meat hot dogs. Do you think the correlation *r* for these data is near −1, clearly negative but not near −1, near 0, clearly positive but not near 1, or near 1? Explain your answer.

3.22   If women always married men who were 2 years older than themselves, what would be the correlation between the ages of husband and wife?

3.23   Exercise 3.13 (page 124) gives the lengths of two bones in five fossil specimens of the extinct beast *Archaeopteryx*. You found the correlation *r* in Exercise 3.19.

   (a)   Make a scatterplot if you did not do so earlier. Explain why the value of *r* matches the scatterplot.

   (b)   The lengths were measured in centimeters. If we changed to inches, how would *r* change? (There are 2.54 centimeters in an inch.)

3.24   The gas mileage of an automobile first increases and then decreases as the speed increases. Suppose that this relationship is very regular, as shown by the following data on speed (miles per hour) and mileage (miles per gallon):

Speed	20	30	40	50	60
MPG	24	28	30	28	24

Make a scatterplot of mileage versus speed. Show that the correlation between speed and mileage is *r* = 0. Explain why the correlation is 0 even though there is a strong relationship between speed and mileage.

SUMMARY

The **correlation** *r* measures the strength and direction of the linear association between two quantitative variables *x* and *y*. Although you can calculate a correlation for any scatterplot, *r* measures only straight-line relationships.

Correlation indicates the direction of a linear relationship by its sign: $r > 0$ for a positive association and $r < 0$ for a negative association.

Correlation always satisfies $-1 \leq r \leq 1$ and indicates the strength of a relationship by how close it is to -1 or 1. Perfect correlation, $r = \pm 1$, occurs only when the points lie exactly on a straight line.

Correlation ignores the distinction between explanatory and response variables. The value of r is not affected by changes in the unit of measurement of either variable. But r can be strongly affected by outlying observations.

SECTION 3.2 EXERCISES

3.25 Exercise 3.9 (page 120) gives data on the lean body mass and metabolic rate for 12 women and 7 men.

(a) Make a scatterplot if you did not do so in Exercise 3.9. Use different symbols or colors for women and men. Do you think the correlation will be about the same for men and women or quite different for the two groups? Why?

(b) Calculate r for women alone and also for men alone. (Use your calculator.)

(c) Calculate the mean body mass for the women and for the men. Does the fact that the men are heavier than the women on the average influence the correlations? If so, in what way?

(d) Lean body mass was measured in kilograms. How would the correlations change if we measured body mass in pounds? (There are about 2.2 pounds in a kilogram.)

3.26 Exercise 3.12 (page 124) gives data on the true calorie counts in ten foods and the average guesses made by a large group of people.

(a) Make a scatterplot if you did not do so in Exercise 3.12. Then calculate the correlation r (use your calculator). Explain why your r is reasonable based on the scatterplot.

(b) The guesses are all higher than the true calorie counts. Does this fact influence the correlation in any way? How would r change if every guess were 100 calories higher?

(c) The guesses are much too high for spaghetti and snack cake. Circle these points on your scatterplot. Calculate r for the other eight foods, leaving out these two points. Explain why r changed in the direction that it did.

3.27   Changing the units of measurement can dramatically alter the appearance
        of a scatterplot. Consider the following data:

$$
\begin{array}{ccccccc}
x & -4 & -4 & -3 & 3 & 4 & 4 \\
y & .5 & -.6 & -.5 & .5 & .5 & -.6
\end{array}
$$

(a) Enter the data into lists L_1 and L_2. Then use Plot1 to plot the scatter-
    plot. Use the plotting symbol □.

(b) Define new variables $x^* = x/10$ and $y^* = 10y$, and enter these into
    lists L_3 and L_4 as follows: $L_3 = L_1/10$ and $L_4 = 10L_2$. Define Plot2
    to be a scatterplot with Xlist: L_3 and Ylist: L_4, and Mark: +. Plot both
    scatterplots at the same time, and on the same axes, using ZOOM /
    9:ZoomStat. The two plots are very different in appearance.

(c) Use your calculator to find the correlation between x and y. Then
    find the correlation between x^* and y^*. How are the two correlations
    related? Explain why this isn't surprising.

3.28   A college newspaper interviews a psychologist about student ratings of the
        teaching of faculty members. The psychologist says, "The evidence indi-
        cates that the correlation between the research productivity and teaching
        rating of faculty members is close to zero." The paper reports this as "Pro-
        fessor McDaniel said that good researchers tend to be poor teachers, and
        vice versa." Explain why the paper's report is wrong. Write a statement in
        plain language (don't use the word "correlation") to explain the psychol-
        ogist's meaning.

3.29   The data in Exercise 3.24 were made up to create an example of a strong
        curved relationship for which, nonetheless, $r = 0$. Exercise 3.8 (page 116)
        gives actual data on gas used versus speed for a British Ford Escort. Make
        a scatterplot if you did not do so in Exercise 3.8. Calculate the correlation,
        and explain why r is close to 0 despite a strong relationship between speed
        and gas used.

3.30   Each of the following statements contains a blunder. Explain in each case
        what is wrong.

(a) "There is a high correlation between the sex of American workers and
    their income."

(b) "We found a high correlation ($r = 1.09$) between students' ratings of
    faculty teaching and ratings made by other faculty members."

(c) "The correlation between planting rate and yield of corn was found
    to be $r = .23$ bushel."

3.3   LEAST-SQUARES REGRESSION

Correlation measures the strength and direction of the linear relationship between any two quantitative variables. If a scatterplot shows a linear relationship, we would like to summarize this overall pattern by drawing a line through the scatterplot. *Least-squares regression* is a method for finding a line that summarizes the relationship between two variables, but only in a specific setting.

REGRESSION LINE

A **regression line** is a straight line that describes how a response variable y changes as an explanatory variable x changes. We often use a regression line to predict the value of y for a given value of x. Regression, unlike correlation, requires that we have an explanatory variable and a response variable.

---

model

The least-squares regression line, which we will occasionally abbreviate LSRL, is a *model*—or more formally, a ***mathematical model***—for the data. If we believe that the data show a linear trend, then it would be appropriate to try to fit an LSRL to the data. In the next chapter, we will explore data that are not linear and for which a curve is a more appropriate model. At the beginning, though, we will focus our discussion on linear trends.

EXAMPLE 3.9

A scatterplot shows that there is a strong linear relationship between the average outside temperature (measured by heating degree-days) in a month and the average amount of natural gas that the Sanchez household uses per day during the month. The Sanchez household wants to use this relationship to predict their natural gas consumption. "If a month averages 20 degree-days per day (that's 45° F), how much gas will we use?"

prediction

In Figure 3.9 we have drawn a regression line on the scatterplot. To use this line to ***predict*** gas consumption at 20 degree-days, first locate 20 on the x axis. Then go "up and over" as in the figure to find the gas consumption y that corresponds to $x = 20$. We predict that the Sanchez household will use about 4.9 hundreds of cubic feet of gas each day in such a month.

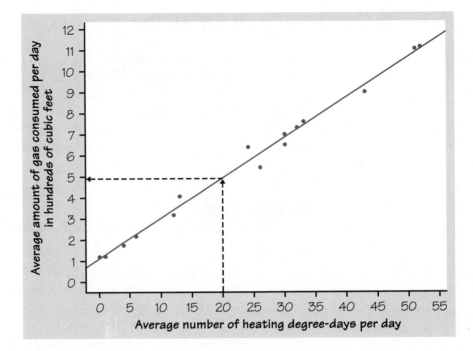

FIGURE 3.9   The Sanchez household gas consumption data, with a regression line for predicting gas consumption from degree-days. The dashed lines illustrate how to use the regression line to predict gas consumption for a month averaging 20 degree-days per day.

The least-squares regression line

Different people might draw different lines by eye on a scatterplot. This is especially true when the points are more widely scattered than those in Figure 3.9. We need a way to draw a regression line that doesn't depend on our guess as to where the line should go. No line will pass exactly through all the points, so we want one that is as close as possible. We will use the line to predict y from x, so we want a line that is as close as possible to the points in the *vertical* direction. That's because the prediction errors we make are errors in y, which is the vertical direction in the scatterplot. If we predict 4.9 hundreds of cubic feet for a month with 20 degree-days and the actual usage turns out to be 5.1 hundreds of cubic feet, our error is

$$\text{error} = \text{observed} - \text{predicted}$$
$$= 5.1 - 4.9 = .2$$

We want a regression line that makes the vertical distances of the points in a scatterplot from the line as small as possible. Figure 3.10(a) illustrates the idea. For clarity, the plot shows only three of the points from

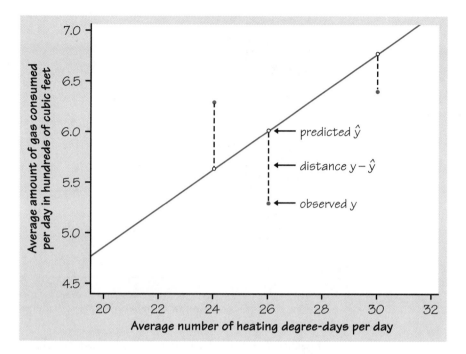

FIGURE 3.10(a)   The least-squares idea. For each observation, find the vertical distance of each point on the scatterplot from a regression line. The least-squares regression line makes the sum of the squares of these distances as small as possible.

Figure 3.9, along with the line, on an expanded scale. The line passes above two of the points and below one of them. The vertical distances of the data points from the line appear as vertical line segments. A "good" regression line makes these distances as small as possible. There are many ways to make "as small as possible" precise. The most common is the *least-squares* idea.

LEAST-SQUARES REGRESSION LINE

The **least-squares regression line** of y on x is the line that makes the sum of the squares of the vertical distances of the data points from the line as small as possible.

Figure 3.10(b) gives a geometric interpretation to the phrase "sum of the squares of the vertical distances of the data points from the line."

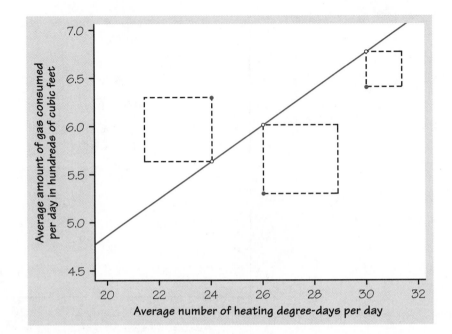

FIGURE 3.10(b)   Equivalently, the least-squares regression line is that line which minimizes the total *area* in the squares.

One reason for the popularity of the least-squares regression line is that the problem of finding the line has a simple answer. We can give the recipe for the least-squares line in terms of the means and standard deviations of the two variables and their correlation.

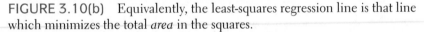

EQUATION OF THE LEAST-SQUARES REGRESSION LINE

We have data on an explanatory variable x and a response variable y for n individuals. From the data, calculate the means $\bar{x}$ and $\bar{y}$ and the standard deviations s_x and s_y of the two variables, and their correlation r. The least-squares regression line is the line

$$\hat{y} = a + bx$$

with **slope**

$$b = r\frac{s_y}{s_x}$$

and **intercept**

$$a = \bar{y} - b\bar{x}$$

We write $\hat{y}$ (read "y hat") in the equation of the regression line to emphasize that the line gives a *predicted* response $\hat{y}$ for any x. The predicted response will usually not be exactly the same as the actually *observed* response y. We will see that the equation gives insight into the behavior of least-squares regression. But in practice, you don't need to calculate the means, standard deviations, and correlation first. Statistical software or your calculator will give the slope b and intercept a of the least-squares line from keyed-in values of the variables x and y. You can then concentrate on understanding and using the regression line.

EXAMPLE 3.10

The line in Figure 3.9 is in fact the least-squares regression line of gas consumption on degree-days. Enter the degree-days into list L_1 in your TI-83, and enter the gas consumption into list L_2. Under STAT PLOT define a scatterplot using L_1 and L_2 and then ZOOM / 9 to plot the scatterplot.

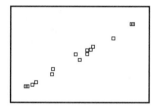

To determine the LSRL, press STAT / CALC / 8:LinReg(a+bx) and then finish the command to read: `LinReg(a+bx) L₁,L₂,Y₁`. (Y_1 is found under VARS / Y-VARS / 1:Function.) This says to use x-values in L_1, y-values in L_2, calculate the LSRL equation, and store the equation as Y_1.

```
LinReg
  y=a+bx
  a=1.089210843
  b=.1889989538
  r2=.9905504416
  r=.995264006
```

Deselect any other equation in the Y= screen, and then press GRAPH to overlay the LSRL on the scatterplot.

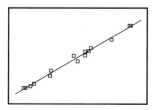

Although the calculator will report the values for a and b to nine decimal places, we usually round off to four decimal places. You would write the LSRL equation as

$$\hat{y} = 1.0892 + .1890x$$

When you write the equation, don't forget the hat symbol over the y; this means *predicted value*.

slope

The **slope** of a regression line is usually important for the interpretation of the data. The slope is the rate of change, the amount of change in $\hat{y}$ when x increases by 1. The slope $b = .1890$ in this example says that, on the average, each additional degree-day predicts consumption of 0.1890 more hundreds of cubic feet of natural gas per day.

intercept

The **intercept** of the regression line is the value of $\hat{y}$ when $x = 0$. Although we need the value of the intercept to draw the line, it is statistically meaningful only when x can actually take values close to zero. In our example, $x = 0$ occurs when the average outdoor temperature is at least 65° F. We predict that the Sanchez household will use an average of $a = 1.0892$ hundreds of cubic feet of gas per day when there are no degree-days. They use this gas for cooking and heating water, which continue in warm weather.

The equation of the regression line makes prediction easy. Just substitute an x-value into the equation. To predict gas consumption at 20 degree-days, substitute $x = 20$.

$$\hat{y} = 1.0892 + (.1890)(20)$$
$$= 1.0892 + 3.78 = 4.869$$

plotting a line

To **plot the line** on the scatterplot by hand, use the equation to find $\hat{y}$ for two values of x, one near each end of the range of x in the data. Plot each $\hat{y}$ above its x and draw the line through the two points.

EXERCISES

3.31   Example 3.10 gives the equation of the regression line of gas consumption y on degree-days x for the data in Table 3.2 as

$$\hat{y} = 1.0892 + .1890x$$

Enter the data from Table 3.2 into your calculator. Use your calculator to find the mean and standard deviation of both x and y and their correlation r. Find the slope b and the intercept a of the regression line from these, using the facts in the box *Equation of the least-squares regression line*. Verify that you get the equation in Example 3.10. (Results may differ slightly because of rounding off.)

3.32   If you previously plotted a scatterplot for the ordered-pairs (math SAT scores, verbal SAT scores) data collected by the class in Activity 3, then ask yourself, "Do these data describe a linear trend?" If so, then use your calculator to determine the LSRL equation and correlation coefficient. Overlay this regression line on your scatterplot. Considering the appearance of the scatterplot, the regression line, and the correlation, write a brief statement about the appropriateness of this regression line to model the data. Is the line useful?

3.33　Researchers studying acid rain measured the acidity of precipitation in a Colorado wilderness area for 150 consecutive weeks. Acidity is measured by pH. Lower pH values show higher acidity. The acid rain researchers observed a linear pattern over time. They reported that the least-squares regression line

$$pH = 5.43 - (.0053 \times weeks)$$

fit the data well.[12]

(a) Draw a graph of this line. Is the association positive or negative? Explain in plain language what this association means.

(b) According to the regression line, what was the pH at the beginning of the study (weeks $= 1$)? At the end (weeks $= 150$)?

(c) What is the slope of the regression line? Explain clearly what this slope says about the change in the pH of the precipitation in this wilderness area.

3.34　Concrete road pavement gains strength over time as it cures. Highway builders use regression lines to predict the strength after 28 days (when curing is complete) from measurements made after 7 days. Let x be strength after 7 days (in pounds per square inch) and y the strength after 28 days. One set of data gives this least-squares regression line:

$$\hat{y} = 1389 + .96x$$

(a) Draw a graph of this line, with x running from 3000 to 4000 pounds per square inch.

(b) Explain what the slope $b = .96$ in this equation says about how concrete gains strength as it cures.

(c) A test of some new pavement after 7 days shows that its strength is 3300 pounds per square inch. Use the equation of the regression line to predict the strength of this pavement after 28 days. Also draw the "up and over" lines from $x = 3300$ on your graph (as in Figure 3.9).

3.35　Verify that every regression line passes through the point $(\bar{x}, \bar{y})$.

Facts about least-squares regression

Least-squares regression looks at the distances of the data points from the line only in the y direction. So the two variables x and y play different roles in regression.

EXAMPLE 3.11

Figure 3.11 is a scatterplot of data that played a central role in the discovery that the universe is expanding. They are the distances from earth of 24 spiral galaxies and the speed at which these galaxies are moving away from us, reported by the astronomer Edwin Hubble in 1929.[13] There is a positive linear relationship, $r = .7842$, so that more distant galaxies are moving away more rapidly. Astronomers believe that there is in fact a perfect linear relationship, and that the scatter is caused by imperfect measurements.

The two lines on the plot are the two least-squares regression lines. The regression line of velocity on distance is solid. The regression line of distance on velocity is dashed. *Regression of velocity on distance and regression of distance on velocity give different lines.* In the regression setting you must know clearly which variable is explanatory.

Although the correlation r ignores the distinction between explanatory and response variables, there is a close connection between correlation and regression.

SLOPE OF THE LEAST-SQUARES REGRESSION LINE

The slope of the least-squares regression line is

$$b = r\frac{s_y}{s_x}$$

This equation says that along the regression line, *a change of one standard deviation in x corresponds to a change of r standard deviations in y.*

When the variables are perfectly correlated ($r = 1$ or $r = -1$), the change in the predicted response $\hat{y}$ is the same (in standard deviation units) as the change in x. Otherwise, because $-1 \leq r \leq 1$, the change in $\hat{y}$ is less than the change in x. As the correlation grows less strong, the prediction $\hat{y}$ moves less in response to changes in x.

Another connection between correlation and regression is even more important. In fact, the numerical value of r as a measure of the strength of a linear relationship is best interpreted by thinking about regression.

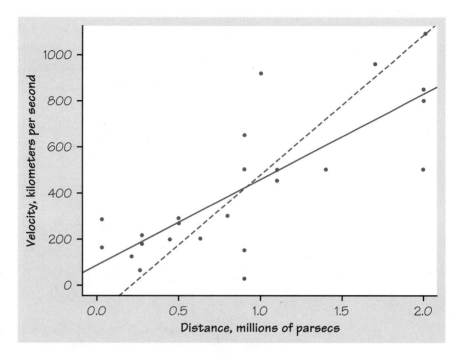

FIGURE 3.11 Scatterplot of Hubble's data on the distance from earth of 24 galaxies and the velocity at which they are moving away from us. The two lines are the two least-squares regression lines: of velocity on distance (solid) and of distance on velocity (dashed).

EXAMPLE 3.12

One way to determine the usefulness of the least-squares regression model is to measure the contribution of x in predicting y. A simple example will help clarify the reasoning. Consider data set A

$$\begin{array}{cccc} x & 0 & 3 & 6 \\ y & 0 & 10 & 2 \end{array}$$

and its scatterplot in Figure 3.12(a). The association between x and y appears to be positive but weak. The sample means are easily calculated to be $\bar{x} = 3$ and $\bar{y} = 4$. Knowing that x is 0 or 3 or 6 gives us very little information to predict y, and so we have to fall back to $\bar{y}$ as a predictor of y. The deviations of the three points about the mean $\bar{y}$ are shown in Figure 3.12(b). The horizontal line in Figure 3.12(b) is at height $\bar{y} = 4$. The sum of the squares of the deviations for the prediction equation $\hat{y} = \bar{y}$ is

$$\text{SSM} = \sum (y - \bar{y})^2$$

We call this quantity SSM for "Sum of Squares about the Mean, $\bar{y}$". Geometric squares have been constructed on the graph with the deviations from the mean as one side. The total area of these three squares is a measure of the total sample variability.

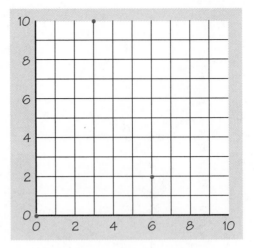

FIGURE 3.12(a)   Scatterplot for data set A.

The LSRL has equation $\hat{y} = 3 + (1/3)x$; see Figure 3.12(c). It has y-intercept 3 and passes through the point $(\bar{x}, \bar{y}) = (3, 4)$. Now we want to consider the sum of the squares of the deviations of the points about this regression line. We call this SSE for "sum of squares for error."

$$\text{SSE} = \sum (y - \hat{y})^2$$

Figure 3.12(c) also shows geometric squares with deviations from the regression line as one side. The calculations can be summarized in a table:

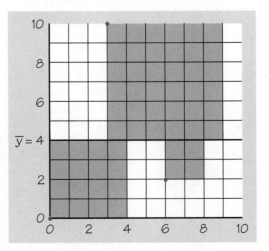

FIGURE 3.12(b)   Squares of deviations about $\bar{y}$.

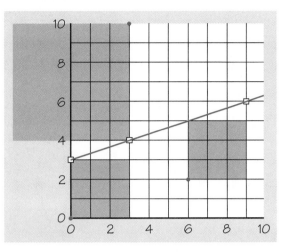

FIGURE 3.12(c)   Squares of deviations about $\hat{y}$.

x	y	$(y - \bar{y})^2$	$(y - \hat{y})^2$
0	0	16	9
3	10	36	36
6	2	4	9
		$\overline{56}$	$\overline{54}$
		SSM	SSE

If x is a poor predictor of y, then the sum of squares of deviations about the mean $\bar{y}$ and the sum of squares of deviations about the regression line $\hat{y}$ would be approximately the same. This is the case in our example. If SSM = 56 measures the total sample variation of the observations about the mean $\bar{y}$, then SSE = 54 is the remaining "unexplained sample variability" after fitting the regression line. The difference, SSM − SSE, measures the amount of variation of y that can be explained by the regression line of y on x. The ratio of these two quantities

$$\frac{\text{SSM} - \text{SSE}}{\text{SSM}}$$

is interpreted as the proportion of the total sample variability that is explained by the least-squares regression of y on x. It can be shown algebraically that this fraction is equal to the square of the correlation coefficient. For this reason, we call this fraction r^2 and refer to it as the *coefficient of determination*. For data set A,

coefficient of determination

$$r^2 = \frac{\text{SSM} - \text{SSE}}{\text{SSM}} = \frac{56 - 54}{56} = .0357$$

We say that 3.57% of the variation in y is explained by least-squares regression of y on x.

EXAMPLE 3.13

Consider data set B and its accompanying scatterplot in Figure 3.13(a):

$$\begin{array}{cccc} x & 0 & 5 & 10 \\ y & 0 & 7 & 8 \end{array}$$

The association between x and y appears to be positive and strong. The sample means are $\bar{x} = 5$ and $\bar{y} = 5$. The squares of the deviations about the mean $\bar{y}$ are shown in Figure 3.13(b), and the squares of the deviations about the regression line $\hat{y}$ are shown in Figure 3.13(c). The LSRL has equation $\hat{y} = 1 + 0.8x$. It has y-intercept 1 and passes through the points $(\bar{x}, \bar{y}) = (5, 5)$ and $(10, 9)$. Here are the calculations:

x	y	$(y - \bar{y})^2$	$(y - \hat{y})^2$
0	0	25	1
5	7	4	4
10	8	9	1
		38	6
		SSM	SSE

If x is a good predictor of y, then the deviations and hence the SSE would be small; in fact, if all of the points fell exactly on the regression line, SSE would be 0. For data set B, we have

$$r^2 = \frac{\text{SSM} - \text{SSE}}{\text{SSM}} = \frac{38 - 6}{38} = .842$$

We say that 84% of the variation in y is explained by least-squares regression of y on x.

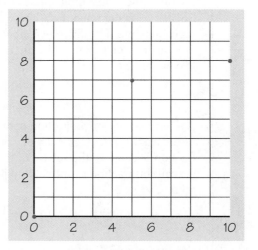

FIGURE 3.13(a)   Scatterplot for data set B.

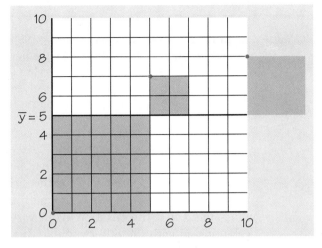

FIGURE 3.13(b)   Squares of deviations about $\bar{y}$.

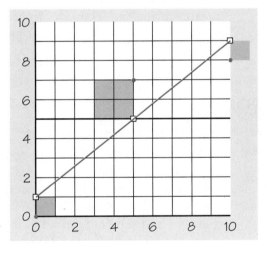

FIGURE 3.13(c)   Squares of deviations about $\hat{y}$.

The last two examples illustrate the role of r^2 in regression.

r^2 IN REGRESSION

The **coefficient of determination**, r^2, is the fraction of the variation in the values of y that is explained by least-squares regression of y on x.

EXAMPLE 3.14

In Figure 3.9, $r = .9953$ and $r^2 = .9906$. Over 99% of the variation in gas consumption is accounted for by the linear relationship with degree-days. In Figure 3.11, $r = .7842$ and $r^2 = .6150$. The linear relationship between distance and velocity explains 61.5% of the variation *in either variable*. There are two regression lines, but just one correlation, and r^2 helps interpret both regressions.

When you report a regression, give r^2 as a measure of how successful the regression was in explaining the response. When you see a correlation, square it to get a better feel for the strength of the association. Perfect correlation ($r = -1$ or $r = 1$) means the points lie exactly on a line. Then $r^2 = 1$ and all of the variation in one variable is accounted for by the linear relationship with the other variable. If $r = -.7$ or $r = .7$, $r^2 = .49$ and about half the variation is accounted for by the linear relationship. In the r^2 scale, correlation $\pm.7$ is about halfway between 0 and ±1.

These connections with correlation are special properties of least-squares regression. They are not true for other methods of fitting a line to data. Another reason that least-squares is the most common method for fitting a regression line to data is that it has many convenient special properties.

EXERCISES

3.36   In Professor Smith's economics course the correlation between the students' total scores prior to the final examination and their final examination scores is $r = .6$. The pre-exam totals for all students in the course have mean 280 and standard deviation 30. The final exam scores have mean 75 and standard deviation 8. Professor Smith has lost Julie's final exam but knows that her total before the exam was 300. He decides to predict her final exam score from her pre-exam total.

(a)   What is the slope of the least-squares regression line of final exam scores on pre-exam total scores in this course? What is the intercept?

(b)   Use the regression line to predict Julie's final exam score.

(c)   Julie doesn't think this method accurately predicts how well she did on the final exam. Calculate r^2 and use the value you get to argue that her actual score should have been much higher (or much lower) than the predicted value.

3.37   Good runners take more steps per second as they speed up. Here are the average numbers of steps per second for a group of top female runners at different speeds. The speeds are in feet per second.[14]

Speed (ft/s)	15.86	16.88	17.50	18.62	19.97	21.06	22.11
Steps per second	3.05	3.12	3.17	3.25	3.36	3.46	3.55

(a) You want to predict steps per second from running speed. Make a scatterplot of the data with this goal in mind.

(b) Describe the pattern of the data and find the correlation.

(c) Find the least-squares regression line of steps per second on running speed. Draw this line on your scatterplot.

(d) Does running speed explain most of the variation in the number of steps a runner takes per second? Calculate r^2 and use it to answer this question.

(e) If you wanted to predict running speed from a runner's steps per second, would you use the same line? Explain your answer. Would r^2 stay the same?

3.38   A study of class attendance and grades among first-year students at a state university showed that in general students who attended a higher percent of their classes earned higher grades. Class attendance explained 16% of the variation in grade index among the students. What is the numerical value of the correlation between percent of classes attended and grade index?

Residuals

A regression line is a simple way to describe the overall pattern of a linear relationship between an explanatory variable and a response variable. Deviations from the overall pattern are also important. In the regression setting, we see deviations by looking at the scatter of the data points about the regression line. The vertical distances from the points to the least-squares regression line are as small as possible, in the sense that they have the smallest possible sum of squares. So we give these distances a special name, *residuals*.

RESIDUALS

A **residual** is the difference between an observed value of the response variable and the value predicted by the regression line. That is,

$$\text{residual} = \text{observed } y - \text{predicted } y$$
$$= y - \hat{y}$$

EXAMPLE 3.15

Does the age at which a child begins to talk predict later score on a test of mental ability? A study of the development of young children recorded the age in months at which each of the 21 children spoke their first word and Gesell Adaptive Score, the result of an aptitude test taken much later. The data appear in Table 3.4.[15]

Figure 3.14 is a scatterplot, with age at first word as the explanatory variable x and Gesell score as the response variable y. Children 3 and 13, and also Children 16 and 21, have identical values of both variables. We use a different plotting symbol to show that one point stands for two individuals. The plot shows a negative association. That is, children who begin to speak later tend to have lower test scores than early talkers. The overall pattern is moderately linear. The correlation describes both the direction and strength of the linear relationship. It is $r = -0.640$.

The line on the plot is the least-squares regression line of Gesell score on age at first word. Its equation is

$$\hat{y} = 109.8738 - 1.1270x$$

For Child 1, who first spoke at 15 months, we predict the score

$$\hat{y} = 109.8738 - (1.1270)(15) = 92.97$$

The child's actual score was 95. The residual is

$$\text{residual} = \text{observed } y - \text{predicted } y$$
$$= 95 - 92.97 = 2.03$$

The residual is positive because the data point lies above the line.

There is residual for each data point. Here are the 21 residuals for the Gesell data, as output by a statistical software package:

```
residuals:
    2.0310   -9.5721  -15.6040   -8.7309    9.0310   -0.3341    3.4120
    2.5230    3.1421    6.6659   11.0151   -3.7309  -15.6040  -13.4770
    4.5230    1.3960    8.6500   -5.5403   30.2850  -11.4770    1.3960
```

Examine the residuals carefully to check how well the regression line fits the data. You can do this by looking at the vertical deviations of the points from the line in a scatterplot of the original data like Figure 3.14. A

residual plot

residual plot plots the residuals on the vertical axis against the explanatory variable on the horizontal axis. Such a plot magnifies the residuals and makes patterns easier to see. Figure 3.15 is a residual plot for the data in Figure 3.14. Residuals from least-squares regression have a special property: *the mean of the residuals is always zero.* You can check that the sum of the residuals above is -0.0002. The sum is not exactly 0 because the

roundoff error

software rounded the residuals to four decimal places. This is *roundoff error.*

Child	Age	Score	Child	Age	Score
1	15	95	12	9	96
2	26	71	13	10	83
3	10	83	14	11	84
4	9	91	15	11	102
5	15	102	16	10	100
6	20	87	17	12	105
7	18	93	18	42	57
8	11	100	19	17	121
9	8	104	20	11	86
10	20	94	21	10	100
11	7	113			

TABLE 3.4 — Age (in months) at first word and Gesell score

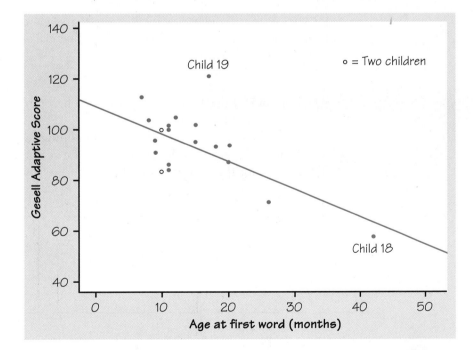

FIGURE 3.14 Scatterplot of Gesell Adaptive Scores versus the age at first word for 21 children from Table 3.4. The line is the least-squares regression line for predicting Gesell score from age at first word.

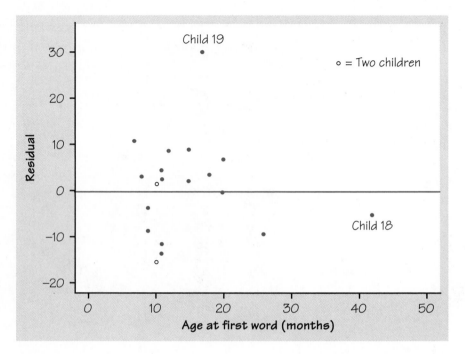

FIGURE 3.15 Residual plot for the regression of Gesell score on age at first word, Child 19 is an outlier, Child 18 is an influential observation that does not have a large residual.

The horizontal line at $y = 0$ in Figure 3.15 helps orient us. It corresponds to the regression line in Figure 3.14.

The following is a procedure for calculating residuals on your TI-83 and then displaying a residual plot. Enter the ages and Gesell scores from Table 3.4. Plot the scatterplot and perform the linear regression. Install the regression equation as Y_1.

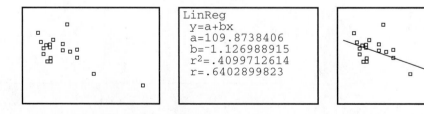

This sets the stage. What follows are instructions to graph the residual plot.

Do STATS / SetUpEditor / ENTER to restore the six default lists. Insert a new list named RES between lists L_2 and L_3. Highlight RES and define this list as the observed value minus the predicted value, $L_2 - Y_1(L_1)$.

```
 L1       L2       RES      3
  15       95       2.031
  26       71      -9.572
  10       83      -15.6
   9       91      -8.731
  15      102       9.031
  20       87      -.3341
  18       93       3.412
RES(1)=2.03099313…
```

Next, in the Y= screen, deselect Y_1, the regression equation, and define $Y_2 = 0$. Y_2 will serve as a reference line, with points above this line corresponding to positive residuals and points below the line corresponding to negative residuals. Specify Plot2 with L_1 as the *x* variable and RES as the *y* variable. ZOOM / 9 to see the residual plot. We TRACEd to see the regression outlier at $x = 17$.

```
P2:L1,RES

        □
     □  □
    □ □  □  □
   □ □   □ □
    □
    □            □              □
     □
    □
```
```
X=17            Y=30.284971
```

You should be aware that some computer utilities, such as Data Desk, prefer to plot the residuals against the fitted values $\hat{y}_i$ instead of against the values x_i of the explanatory variable. The information in the two plots is the same because $\hat{y}$ is linearly related to *x*.

Finally, we have previously noted that an important property of residuals is that their sum is zero. Calculate one-variable statistics on the RES list to verify that $\sum(\text{residuals}) = 0$ and that, consequently, the mean of the residuals is also 0.

```
1-Var Stats
 x̄=3.857143E-12
 Σx=8.1E-11
 Σx²=2308.58578
 Sx=10.74380235
 σx=10.4848775
↓n=21
```

Note that the calculator is showing some roundoff error; you should recognize these peculiar looking numbers as equivalent to zero.

We would like a residual plot to look something like the simplified pattern in Figure 3.16(a). That plot shows a uniform scatter of the points above and below the fitted line, with no unusual individual observations. Here are some things to look for when you examine the residuals, using either a scatterplot of the data or a residual plot.

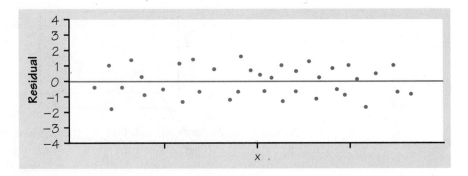

FIGURE 3.16(a)   Idealized patterns in plots of least-squares residuals. Plot (a) indicates that the regression line fits the data well, so the line is a good model.

- A curved pattern, which shows that the overall pattern is not linear. Figure 3.16(b) is a simplified example. A straight line is not a good model for such data.
- Increasing or decreasing spread about the line as *x* increases. Figure 3.16(c) is a simplified example. Prediction of *y* will be less accurate for larger *x* in that example.
- Individual points with large residuals, like Child 19 in Figures 3.14 and 3.15. These points are *outliers* because they lie outside the straight-line pattern.
- Individual points that are extreme in the *x* direction, like Child 18 in Figures 3.14 and 3.15. Such points may not have large residuals, but they can be very important.

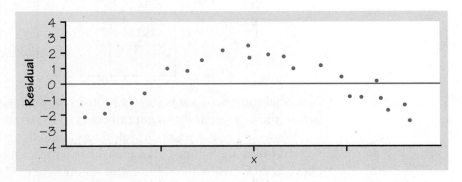
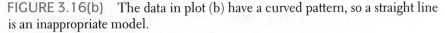

FIGURE 3.16(b)   The data in plot (b) have a curved pattern, so a straight line is an inappropriate model.

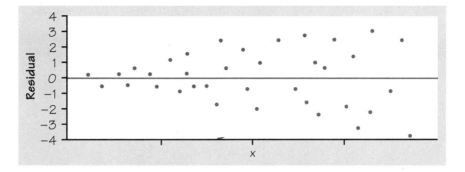

FIGURE 3.16(c) The response variable y in plot (c) has more spread for larger values of the explanatory variable x, so prediction will be less accurate when x is large.

Influential observations

Children 18 and 19 are both unusual in the Gesell example. They are unusual in different ways. Child 19 lies far from the regression line. Child 18 is close to the line but far out in the x direction. Child 19 is an *outlier*, with a Gesell score so high that we should check for a mistake in recording it. In fact, the score is correct.

Child 18 began to speak much later than any of the other children. *Because of its extreme position on the age scale, this point has a strong influence on the position of the regression line.* Figure 3.17 adds a second regression line, calculated after leaving out Child 18. You can see that this one point moves the line quite a bit. Least-squares lines make the sum of squares of the vertical distances to the points as small as possible. A point that is extreme in the x direction with no other points near it pulls the line toward itself. We call such points *influential*.

OUTLIERS AND INFLUENTIAL OBSERVATIONS IN REGRESSION

An **outlier** is an observation that lies outside the overall pattern of the other observations in a scatterplot. An observation can be an outlier in the x direction, in the y direction, or in both directions.

An observation is **influential** if removing it would markedly change the position of the regression line. Points that are outliers in the x direction are often influential.

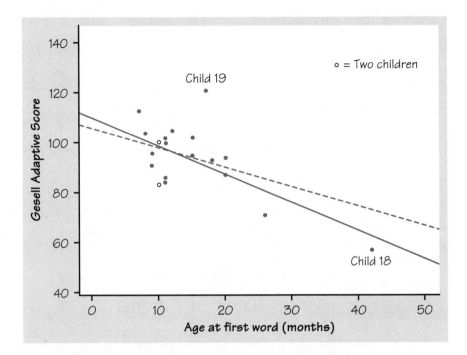

FIGURE 3.17   Two least-squares regression lines of Gesell score on age at first word. The solid line is calculated from all the data. The dashed line is calculated leaving out Child 18. Child 18 is an influential observation because leaving out this point moves the regression line quite a bit.

Child 18 is an outlier in the *x* direction and is influential. Child 19 is an outlier in the *y* direction. It has less influence on the regression line because the many other points with similar values of *x* anchor the line well below the outlying point. Influential points often have small residuals, because they pull the regression line toward themselves. If you just look at residuals, you will miss influential points. Influential observations can greatly change the interpretation of data.

EXAMPLE 3.16   Correlation and least-squares regression are strongly influenced by extreme observations. In the Gesell example, the original data have $r^2 = 0.41$. That is, the age at which a child begins to talk explains 41% of the variation on a later test of the mental ability. This relationship is strong enough to be interesting to parents. But if we leave out child 18, r^2 drops to only 11%.

What should the child development researcher do? She must decide whether Child 18 is so slow to speak that this individual should not be allowed to influence the analysis. If she excludes Child 18, much of the evidence for the connection between the age at which a child begins to talk and later ability score vanishes. If she keeps Child 18, she needs data on other children who were also slow to begin talking, so that the analysis no longer depends so heavily on just one child.

EXERCISES

3.39   Exercise 3.8 (page 116) gives data on the fuel consumption y of a car at various speeds x. Fuel consumption is measured in liters of gasoline per 100 kilometers driven and speed is measured in kilometers per hour. A statistical software package gives the least-squares regression line and also the residuals. The regression line is

$$\hat{y} = 11.058 - .01466x$$

The residuals, in the same order as the observations, are

10.09	2.24	−0.62	−2.47	−3.33	−4.28	−3.73	−2.94
−2.17	−1.32	−0.42	0.57	1.64	2.76	3.97	

(a) Make a scatterplot of the observations and draw the regression line on your plot.

(b) Would you use the regression line to predict y from x? Explain your answer.

(c) Check that the residuals have sum zero (up to roundoff error).

(d) Make a residual plot. Notice that the residuals show the same pattern about the reference line $y = 0$ as the data points show about the regression line in the scatterplot in (a). What do you conclude from the residual plot?

3.40   Exercise 3.12 (page 124) gives data on the true calories in ten foods and the average guesses made by a large group of people. Exercise 3.26 (page 135) explored the influence of two outlying observations on the correlation.

(a) Make a scatterplot suitable for predicting guessed calories from true calories. Circle the points for spaghetti and snack cake on your plot. These points lie outside the linear pattern of the other eight points.

(b) Use your calculator to find the least-squares regression line of guessed calories on true calories. Do this twice, first for all ten data points and then leaving out spaghetti and snack cake.

(c) Plot both lines on your graph. (Make one dashed so that you can tell them apart.) Are spaghetti and snack cake, taken together, influential observations? Explain your answer.

3.41   The discussion of Example 3.15 shows that Child 18 in the Gesell data in Table 3.4 is an influential observation. Now we will examine the effect of Child 19, who is also an outlier in Figure 3.14.

(a) Find the least-squares regression line of Gesell score on age at first word, leaving out Child 19. Example 3.15 gives the regression line from all the children. Plot both lines on the same graph. (You do not have to make a scatterplot of all the points—just plot the two lines.) Would you call Child 19 very influential? Why?

(b) How does removing Child 19 change the r^2 for this regression? Explain why r^2 changes in this direction when you drop Child 19.

SUMMARY

A **regression line** is a straight line that describes how a response variable y changes as an explanatory variable x changes.

The **least-squares regression** line is the straight line $\hat{y} = a + bx$ that minimizes the sum of the squares of the vertical distances of the observed y-values from the line.

You can use a regression line to **predict** the value of y for any value of x by substituting this x into the equation of the line.

The **slope** b of a regression line $\hat{y} = a + bx$ is the rate at which the predicted response $\hat{y}$ changes along the line as the explanatory variable x changes. Specifically, b is the change in $\hat{y}$ when x increases by 1.

The **intercept** a of a regression line $\hat{y} = a + bx$ is the predicted response $\hat{y}$ when the explanatory variable $x = 0$. This prediction is of no statistical use unless x can actually take values near 0.

Correlation and **regression** are closely connected. The correlation r is the slope of the least-squares regression line when we measure both x and y in standardized units. The square of the correlation r^2 is the fraction of the variation of one variable that is explained by least-squares regression on the other variable.

You can examine the fit of a regression line by studying the **residuals**, which are the differences between the observed and predicted values of y. Be on the lookout for outlying points with unusually large residuals and also for nonlinear patterns and uneven variation about the line.

Also look for **influential observations**, individual points that substantially change the regression line. Influential observations are often outliers in the x direction, but they need not have large residuals.

SECTION 3.3 EXERCISES

3.42   (Review of straight lines) Fred keeps his savings in his mattress. He began with $500 from his mother and adds $100 each year. His total savings y after x years are given by the equation

$$y = 500 + 100x$$

(a) Draw a graph of this equation. (Choose two values of x, such as 0 and 10. Compute the corresponding values of y from the equation. Plot these two points on graph paper and draw the straight line joining them.)

(b) After 20 years, how much will Fred have in his mattress?

(c) If Fred had added $200 instead of $100 each year to his initial $500, what is the equation that describes his savings after x years?

3.43   (Review of straight lines) During the period after birth, a male white rat gains exactly 40 grams (g) per week. (This rat is unusually regular in his growth, but 40 g per week is a realistic rate.)

(a) If the rat weighed 100 g at birth, give an equation for his weight after x weeks. What is the slope of this line?

(b) Draw a graph of this line between birth and 10 weeks of age.

(c) Would you be willing to use this line to predict the rat's weight at age 2 years? Do the prediction and think about the reasonableness of the result. (There are 454 grams in a pound. To help you assess the result, note that a large cat weighs about 10 pounds.)

3.44   The solid line in Figure 3.11 (page ~~123~~ 145) is the least-squares regression line of a galaxy's speed on its distance. Use this graph to predict the speed of a galaxy that is 1.5 million parsecs away. Do you expect your prediction to be very accurate? Why?

3.45   Sarah's parents are concerned that she seems short for her age. Their doctor has the following record of Sarah's height:

Age (months)	36	48	51	54	57	60
Height (cm)	86	90	91	93	94	95

(a) Make a scatterplot of these data. Note the strong linear pattern.

(b) Using your calculator, find the equation of the least-squares regression line of height on age.

(c) Predict Sarah's height at 40 months and at 60 months. Use your results to draw the regression line on your scatterplot.

(d)   What is Sarah's rate of growth, in centimeters per month? Normally growing girls gain about 6 cm in height between ages 4 (48 months) and 5 (60 months). What rate of growth is this in centimeters per month? Is Sarah growing more slowly than normal?

3.46   Investors ask about the relationship between returns on investments in the United States and on investments overseas. Here are the data on the total returns on U.S. and overseas common stocks over a 22-year period. (The total return is change in price plus any dividends paid, converted into U.S. dollars. Both returns are averages over many individual stocks.)[16]

Year	Overseas % return	U.S. % return	Year	Overseas % return	U.S. % return
1971	29.6	14.6	1984	7.4	6.1
1972	36.3	18.9	1985	56.2	31.6
1973	−14.9	−14.8	1986	69.4	18.6
1974	−23.2	−26.4	1987	24.6	5.1
1975	35.4	37.2	1988	28.5	16.8
1976	2.5	23.6	1989	10.6	31.5
1977	18.1	−7.4	1990	−23.0	−3.1
1978	32.6	6.4	1991	12.8	30.4
1979	4.8	18.2	1992	−12.1	7.6
1980	22.6	32.3	1993	32.9	10.1
1981	−2.3	−5.0	1994	6.2	1.3
1982	−1.9	21.5	1995	11.2	37.6
1983	23.7	22.4			

(a)   Make a scatterplot suitable for predicting overseas returns from U.S. returns.

(b)   Find the correlation and r^2. Describe the relationship between U.S. and overseas returns in words, using r and r^2 to make your description more precise.

(c)   Find the least-squares regression line of overseas returns on U.S. returns. Draw the line on the scatterplot.

(d)   In 1996, the return on U.S. stocks was 23.0%. Use the regression line to predict the return on overseas stocks. The actual overseas return was 6.4%. Are you confident that predictions using the regression line will be quite accurate? Why?

(e)   Circle the point that has the largest residual (either positive or negative). What year is this? Are there any points that seem likely to be very influential?

3.47   Use the least-squares regression line for data in Exercise 3.45 to predict Sarah's height at age 40 years (480 months). Your prediction is in centimeters. Convert it to inches using the fact that a centimeter is 0.3937 inch.

   The prediction is impossibly large. It is not reasonable to use data for 36 to 60 months to predict height at 480 months. Don't use regression lines for prediction for values of x far outside the range used to find the line.

3.48   Exercise 3.46 examined the relationship between returns on U.S. and overseas stocks. Investors also want to know what typical returns are and how much year-to-year variability (called *volatility* in finance) there is. Regression and correlation don't answer questions about center and spread.

   (a) Find the five-number summaries for both U.S. and overseas returns, and make side-by-side boxplots to compare the two distributions.

   (b) Were returns generally higher in the United States or overseas during this period? Explain your answer.

   (c) Were returns more volatile (more variable) in the United States or overseas during this period? Explain your answer.

3.49   The mathematics department of a large state university would like to use the number of freshmen entering the university x to predict the number of students y who will sign up for freshman-level math courses in the fall semester. Here are the data for the years 1991 to 1998:

Year	1991	1992	1993	1994	1995	1996	1997	1998
x	4595	4827	4427	4258	3995	4330	4265	4351
y	7364	7547	7099	6894	6572	7156	7232	7450

Computer software gives the correlation $r = .8333$ and the least-squares regression line

$$\hat{y} = 2492.69 + 1.0663x$$

The software also gives a table of the residuals:

Year	1991	1992	1993	1994	1995	1996	1997	1998
Residual	−28.44	−92.83	−114.30	−139.09	−180.65	46.13	191.44	317.74

   (a) Make a scatterplot and draw the regression line on it. The regression line does not predict very accurately. What percent of variation in class enrollment is explained by the linear relationship with the count of freshmen?

   (b) Check that the residuals have sum zero (at least up to roundoff error).

   (c) Plots of the residuals against other variables are often revealing. Plot the residuals against year. One of the schools in the university recently changed its program to require that entering students take another mathematics course. How does the residual plot show the change? In what year was the change effective?

CHAPTER REVIEW

Chapters 1 and 2 dealt with data analysis for a single variable. In this chapter, we have studied analysis of data for two or more variables. The proper analysis depends on whether the variables are categorical or quantitative and on whether one is an explanatory variable and the other a response variable.

Data analysis begins with graphs and then adds numerical summaries of specific aspects of the data.

This chapter concentrates on relations between two quantitative variables. Scatterplots show the relationship, whether or not there is an explanatory-response distinction. Correlation describes the strength of a linear relationship, and least-squares regression fits a line to data that have an explanatory-response relation.

Here is a review list of the most important skills you should have gained from studying this chapter.

A. DATA

1. Recognize whether each variable is quantitative or categorical.
2. Identify the explanatory and response variables in situations where one variable explains or influences another.

B. SCATTERPLOTS

1. Make a scatterplot for two quantitative variables, placing the explanatory variable (if any) on the horizontal scale.
2. Add a categorical variable to the scatterplot by using a different plotting symbol.
3. Recognize positive or negative association, a linear pattern, and outliers in a scatterplot.

C. CORRELATION

1. Compute the correlation coefficient r for small sets of observations, using a calculator.
2. Know the basic properties of correlation: r measures the strength and direction of linear relations only; $-1 \leq r \leq 1$ always; $r = \pm 1$ only for perfect straight-line relations; r moves away from 0 toward ± 1 as the linear relation gets stronger.

D. STRAIGHT LINES

1. Explain what the slope b and the intercept a mean in the equation $y = a + bx$ of a straight line.
2. Draw a graph of the straight line when you are given its equation.

E. REGRESSION

1. Calculate the least-squares regression line of a response variable y on an explanatory variable x from data, using a calculator.
2. Find the slope and the intercept of the least-squares regression line from the means and standard deviations of x and y and their correlation.
3. Use the regression line to predict y for a given x. Recognize extrapolation and be aware of its dangers.
4. Use r^2 to describe how much of the variation in one variable can be accounted for by a straight-line relationship with another variable.
5. Recognize outliers and potentially influential observations from a scatterplot with the regression line drawn on it.
6. Calculate the residuals and plot them against the explanatory variable x or against other variables. Recognize unusual patterns.

CHAPTER 3 REVIEW EXERCISES

3.50 Manatees are large, gentle sea creatures that live along the Florida coast. Many manatees are killed or injured by powerboats. Here are data on powerboat registrations (in thousands) and the number of manatees killed by boats in Florida in the years 1977 to 1994:

Year	Powerboat registrations (1000)	Manatees killed	Year	Powerboat registrations (1000)	Manatees killed
1977	447	13	1986	614	33
1978	460	21	1987	645	39
1979	481	24	1988	675	43
1980	498	16	1989	711	50
1981	513	24	1990	719	47
1982	512	20	1991	716	53
1983	526	15	1992	716	38
1984	559	34	1993	716	35
1985	585	33	1994	735	49

(a) We want to examine the relationship between number of powerboats and number of manatees killed by boats. Which is the explanatory variable?

(b) Make a scatterplot of these data. Be sure to label the axes with the variable names, not just x and y. What does the scatterplot show about the relationship between these variables?

Here is part of the output from the regression command in the Minitab statistical software:

```
The regression equation is
Killed = - 35.2 + 0.113 Boats

Unusual Observations
Obs.    Boats    Killed       Fit Stdev.Fit   Residual    St.Resid
  17      716     35.00     45.51      1.92     -10.51       -2.08R

R denotes an obs. with a large st. resid.
```

(c) From your scatterplot, does it appear that there is a strong straight-line pattern? What is r^2 for these data?

(d) Draw the regression line given by Minitab on your scatterplot. Predict how many manatees would be killed each year if Florida decided to freeze the number of boats at 700,000. (Use Minitab's work.)

(e) Minitab checks for large residuals and influential observations. It calls attention to one observation that has a somewhat large residual. Circle this observation on your plot. We have no reason to remove it.

(f) Residuals from least-squares regression often have a distribution that is roughly normal. So Minitab reports the *standardized* residuals— that's what St.Resid means. Use the 68–95–99.7 rule for normal distributions to say how surprising a residual with standardized value −2.08 is.

3.51   Table 3.5 shows world records in the men's mile run between 1868 and 1993.[17] The times have been converted to seconds to facilitate plotting. Enter the data into your calculator or computer package and then plot a scatterplot. Is there an apparent trend? Perform least-squares regression and calculate the correlation. Comment on the suitability of the LSRL as a model for the data and interpret the correlation. Are there any regression outliers? Influential observations? On average, how many seconds are lopped off this record each year? Would you feel comfortable predicting the world record for this event in the year 2000? In the year 2005?

TABLE 3.5	Mile Run World Records (Men)

Year	Record holder	Record (min.)	Record (sec.)
1868	W. C. Gibbs (G.B.)	4:38.8	278.8
1874	Walter Slade (G.B.)	4:26	266.0
1875	Walter Slade (G.B.)	4:24.5	264.5
1880	Walter George (G.B.)	4:23.2	263.2
1882	Walter George (G.B.)	4:19.4	259.4
1884	Walter George (G.B.)	4:18.4	258.4
1894	Fred Bacon (Scotland)	4:18.2	258.2
1895	Fred Bacon (Scotland)	4:17.0	257.0
1911	J. P. Jones (U.S.)	4:15.4	255.4
1913	J. P. Jones (U.S.)	4:14.6	254.6
1915	Norman Taber (U.S.)	4:12.6	252.6
1923	Paavo Nurmi (Finland)	4:10.4	250.4
1931	J. Ladoumegue (France)	4:09.2	249.2
1933	Jack Lovelock (N.Z.)	4:07.6	247.6
1934	G. Cunningham (U.S.)	4:06.8	246.8
1937	S. Wooderson (G.B.)	4:06.4	246.4
1942	Gunder Haegg (Sweden)	4:04.6	244.6
1943	A. Andersson (Sweden)	4:02.6	242.6
1944	A. Andersson (Sweden)	4:01.6	241.6
1945	Gunder Haegg (Sweden)	4:01.4	241.4
1954	John Landry (Australia)	3:58.0	238.0
1957	Derek Ibbotson (G.B.)	3:57.2	237.2
1958	Herb Elliot (Australia)	3:54.5	234.5
1962	Peter Snell (N.Z.)	3:54.4	234.4
1964	Peter Snell (N.Z.)	3:54.1	234.1
1965	Michel Jazy (France)	3:43.6	233.6
1966	Jim Ryun (U.S.)	3:51.3	231.3
1967	Jim Ryun (U.S.)	3:51.1	231.1
1975	John Walker (N.Z.)	3:49.4	229.4
1979	Sebastian Coe (G.B.)	3:49.0	229.0
1980	Steve Ovett (G.B.)	348.8	228.8
1981	Sebastian Coe (G.B.)	3:47.3	227.3
1985	Steve Cram (G.B.)	3:46.3	226.3
1993	Noureddin Morcelli (Algeria)	3:44.39	224.39

Note: Where more than one record was set in a year, only the final record is shown.

3.52   Match the following scatterplots with their regression equations and correlations.

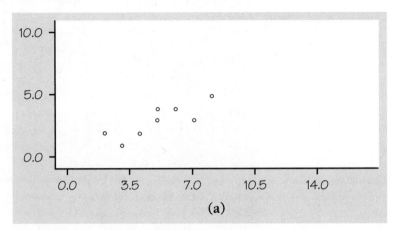

(a)

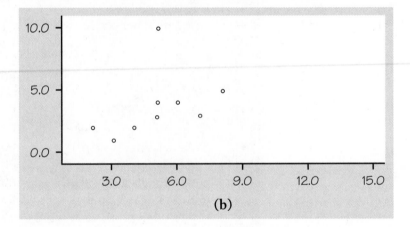

(b)

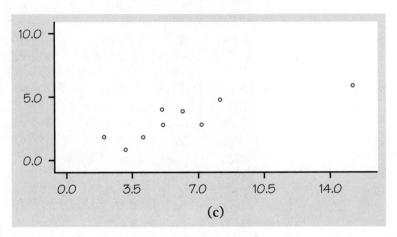

(c)

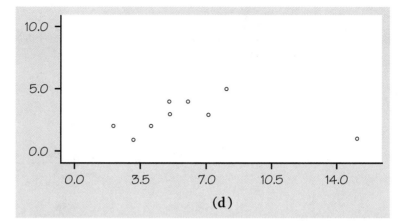

(d)

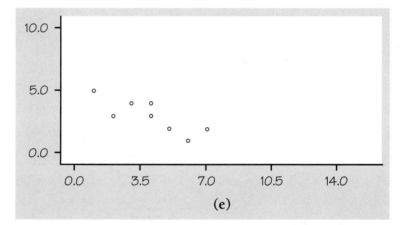

(e)

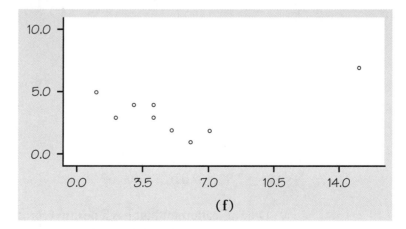

(f)

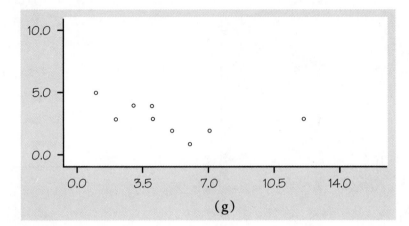

(g)

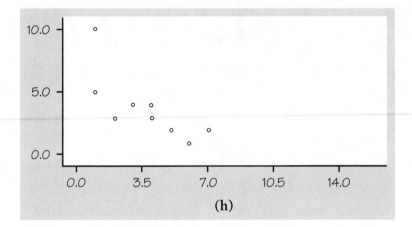

(h)

1. $\hat{y} = 7.21 - 0.94x, r = -.75$
2. $\hat{y} = 1.10 + 0.54x, r = .38$
3. $\hat{y} = 5.14 - 0.54x, r = -.82$
4. $\hat{y} = 3.86 - 0.18x, r = -.47$
5. $\hat{y} = 1.15 + 0.36x, r = .86$
6. $\hat{y} = 0.32 + 0.54x, r = .82$
7. $\hat{y} = 2.52 + 0.18x, r = .40$
8. $\hat{y} = 2.92 - 0.02x, r = -.07$

3.53   (a) Draw a scatterplot that has a positive correlation such that when one point is added, the correlation becomes negative. Circle the influential point.

(b) Draw a scatterplot that has a correlation close to 0 (say less than 0.1) such that when one point is added, the correlation is close to 1 (say greater than 0.9). Circle the influential point.

3.54   In Exercise 3.14 you constructed a scatterplot for Indianapolis 500 automobile race winners for the years 1967 to 1997. Perform least-squares regression on the data and calculate the correlation, r. Comment on the suitability of the LSRL as a model for the data and interpret the correlation. Are there any regression outliers? If so, can you find explanations? Are there influential observations? If so, identify them. Would you feel comfortable predicting the average speed for the Indy 500 winner in the year 2000? In 2010? What is fundamentally different about this problem compared with the mile run world record problem (Exercise 3.51)?

3.55   The Franklin National Bank failed in 1974. Franklin was one of the 20 largest banks in the nation, and the largest ever to fail. Could Franklin's weakened condition have been detected in advance by simple data analysis? The table below gives the total assets (in billions of dollars) and net income (in millions of dollars) for the 20 largest banks in 1973, the year before Franklin failed.[18] Franklin is bank number 19.

	Bank									
	1	2	3	4	5	6	7	8	9	10
Assets	49.0	42.3	36.3	16.4	14.9	14.2	13.5	13.4	13.2	11.8
Income	218.8	265.6	170.9	85.9	88.1	63.6	96.9	60.9	144.2	53.6

	Bank									
	11	12	13	14	15	16	17	18	19	20
Assets	11.6	9.5	9.4	7.5	7.2	6.7	6.0	4.6	3.8	3.4
Income	42.9	32.4	68.3	48.6	32.2	42.7	28.9	40.7	13.8	22.2

(a) We expect banks with more assets to earn higher income. Make a scatterplot of these data that displays the relation between assets and income. Mark Franklin (Bank 19) with a separate symbol.

(b) Describe the overall pattern of your plot. Are there any banks with unusually high or low income relative to their assets? Does Franklin stand out from other banks in your plot?

(c) Find the least-squares regression line for predicting a bank's income from its assets. Draw the regression line on your scatterplot.

TABLE 3.6	Men's and women's world records in the 800-meter run

Year	Men's record	Women's record
1905	113.4	—
1915	111.9	—
1925	111.9	144.0
1935	109.7	135.6
1945	106.6	132.0
1955	105.7	125.0
1965	104.3	118.0
1975	104.1	117.5
1985	101.73	113.28
1995	101.73	113.28

(d) Use the regression line to predict Franklin's income. Was the actual income higher or lower than predicted? What is the residual?

3.56   Table 3.6 shows the men's and women's world records in the 800-meter run.[19]

(a) For each gender separately, do the following: Enter the data into your calculator or computer package and then plot a scatterplot. (Use the box plotting symbol for the men, and use the + plotting symbol for the women.) Describe the trend, if there is one. Perform least-squares regression and calculate the correlation. Comment on the suitability of the LSRL as a model for the data and interpret the correlation. Identify any regression outliers and influential observations.

(b) Brian Whipp and Susan Ward wrote an article based on the 800-meter run data entitled "Will Women Soon Outrun Men?" which appeared in the British journal *Nature* in 1992. They suggested in the article that women have made more progress in track events over the last half-century than men, hence the title of the article. Extend your calculator viewing window so that you can see both data sets and best-fit lines, and determine the intersection of the two LSRLs. Then comment on the premise of the *Nature* article.

3.57 Nematodes are microscopic worms. Here are data from an experiment to study the effect of nematodes in the soil on plant growth. The experimenter prepared 16 planting pots and introduced different numbers of nematodes. Then he placed a tomato seedling in each pot and measured its growth (in centimeters) after 16 days.[20]

Nematodes	Seedling growth (cm)			
0	10.8	9.1	13.5	9.2
1,000	11.1	11.1	8.2	11.3
5,000	5.4	4.6	7.4	5.0
10,000	5.8	5.3	3.2	7.5

Analyze these data and give your conclusions about the effects of nematodes on plant growth.

3.58 The mean height of American women in their early twenties is about 64.5 inches and the standard deviation is about 2.5 inches. The mean height of men the same age is about 68.5 inches, with standard deviation about 2.7 inches. If the correlation between the heights of husbands and wives is about $r = .5$, what is the slope of the regression line of the husband's height on the wife's height in young couples? Draw a graph of this regression line. Predict the height of the husband of a woman who is 67 inches tall.

3.59 (a) Is correlation a resistant measure? Give an example to support your answer.

(b) Is the least-squares regression line resistant? Give an example to support your answer.

3.60 There are several different procedures one could use to find a "best-fit" line for a set of data. For example, you could measure the perpendicular distance from each point to the line, add these distances, and minimize the sum. Write at least one other way to find a best-fit line for a data set. Then for the method given and your method, give one or more possible drawbacks.

NOTES AND DATA SOURCES

1. Data from Personal Watercraft Industry Association, U.S. Coast Guard.

2. Data provided by Robert Dale, Purdue University.

3. Based on T. N. Lam, "Estimating fuel consumption from engine size," *Journal of Transportation Engineering*, 111 (1985), pp. 339–357. The data for 10 to 50 km/h

are measured; those for 60 and higher are calculated from a model given in the paper and are therefore smoothed.

4. A sophisticated treatment of improvements and additions to scatterplots is W. S. Cleveland and R. McGill, "The many faces of a scatterplot," *Journal of the American Statistical Association,* 79 (1984), pp. 807–822.

5. The data are a random sample of 53 from the 1079 pairs recorded by K. Pearson and A. Lee, "On the laws of inheritance in man," *Biometrika,* November 1903, p. 408.

6. Data from *Consumer Reports,* June 1986, pp. 366–367.

7. From a survey by the Wheat Industry Council reported in *USA Today,* October 20, 1983.

8. The data are from M. A. Houck et al., "Allometric scaling in the earliest fossil bird, *Archaeopteryx lithographica,*" *Science,* 247 (1990), pp. 195–198. The authors conclude from a variety of evidence that all specimens represent the same species.

9. The data are from W. L. Colville and D. P. McGill, "Effect of rate and method of planting on several plant characters and yield of irrigated corn," *Agronomy Journal,* 54 (1962), pp. 235–238.

10. Modified from M. C. Wilson and R. E. Shade, "Relative attractiveness of various luminescent colors to the cereal leaf beetle and the meadow spittlebug," *Journal of Economic Entomology,* 60 (1967), pp. 578–580.

11. A careful study of this phenomenon is W. S. Cleveland, P. Diaconis, and R. McGill, "Variables on scatterplots look more highly correlated when the scales are increased," *Science,* 216 (1982), pp. 1138–1141.

12. From W. M. Lewis and M. C. Grant, "Acid precipitation in the western United States," *Science,* 207 (1980), pp. 176–177.

13. Data from E. P. Hubble, "A relation between distance and radial velocity among extra-galactic nebulae," *Proceeding of the National Academy of Sciences,* 15 (1929), pp. 168–173.

14. Data from R. C. Nelson, C. M. Brooks, and N. L. Pike, "Biomechanical comparison of male and female distance runners," in P. Milvy (ed.), *The Marathon: Physiological, Medical, Epidemiological, and Psychological Studies,* New York Academy of Sciences, 1977, pp. 793–807.

15. These data were originally collected by L. M. Linde of UCLA but were first published by M. R. Mickey, O. J. Dunn, and V. Clark, "Note on the use of stepwise regression in detecting outliers," *Computers and Biomedical Research,* 1 (1967), pp. 105–111. The data have been used by several authors. We found them in N. R. Draper and J. A. John, "Influential observations and outliers in regression," *Technometrics,* 23 (1981), pp. 21–26.

16. The U.S. returns are for the Standard & Poor's 500 stock index. The overseas returns are for the Morgan Stanley Europe, Australia, Far East (EAFE) index.

17. This exercise was suggested in an article by Robert Plummer et al. in *Mathematics Teacher,* 86, no. 8 (November 1993), pp. 636–641.

18. Data from D. E. Booth, *Regression Methods and Problem Banks,* COMAP, Inc., 1986.

19. This exercise was suggested in an article by Edward Wallace in *Mathematics Teacher,* 86, no. 9 (December 1993), p. 741.

20. Data provided by Matthew Moore.

More on Two–Variable Data

4.1   MODELING NONLINEAR DATA

As we have seen in the previous chapter, many scatterplots that we encounter are linear in nature and can be modeled with a straight line. If
the data are linear, then the differences in y-values, Δy, are constant for
uniform differences in x. For example, consider the line whose equation
is $y = 0.5x + 1$:

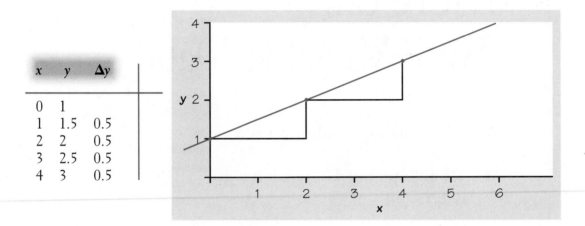

x	y	Δy
0	1	
1	1.5	0.5
2	2	0.5
3	2.5	0.5
4	3	0.5

The difference between y-values *for consecutive-integer x-values* is a constant, 0.5, and you recognize this number as the slope of the line. If you
have (x, y) data and the x-values are at equal-increment intervals, then
an approximately constant difference in the y-values is a useful test for a
linear trend.

Many situations in the real world exhibit growth that is not linear.
In this section we will investigate two specific types of nonlinear growth,
namely, growth that can be modeled by an ***exponential function*** of the form
$y = ab^x$ and growth that can be modeled by a ***power function*** of the form
$y = ax^b$. Equations in both of these forms can be transformed into linear forms. Then we can use the linear regression methods of Section
3.3 to find the equation of a best-fitting line for the transformed data.
The last step is to perform an inverse transformation to obtain a curve
that models the original data. In performing these transformations, we
will make extensive use of the rules of logarithms and the rules for exponents. Recall that $\log_2 8 = 3$ because 3 is the exponent to which 2 must
be raised to yield 8. Here is a quick summary of algebraic properties of
logarithms.

*exponential
function*
power function

$$\log_b x = y \quad \text{if and only if} \quad b^y = x$$

The rules for logarithms are

1. $\log(AB) = \log A + \log B$
2. $\log(A/B) = \log A - \log B$
3. $\log X^p = p \log X$

ACTIVITY 4 | Designing a Game of Chance

Materials: 6 to 8 disks of various sizes: coins, plastic lids, etc.

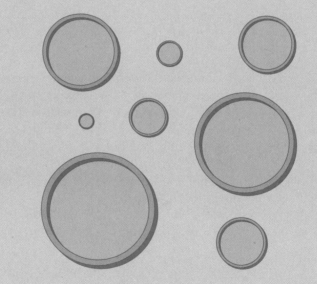

The PTA has asked your class to design a game of chance as part of its annual fund-raising activities.[1] Contestants will pay \$1 to toss a disk onto the cafeteria floor, which is covered with 9 inch × 9 inch tiles. If the disk comes to rest entirely inside a tile, they win \$2; if the disk comes to rest on an edge, they lose. To give players a relatively good chance of

(*continued*)

winning, the PTA wants you to determine the diameter of the disk that would enable players to win 40% of the time.

First locate a tiled floor that can be used for the activity: classroom, hall, or cafeteria. Although 9-inch square tiles are assumed, other sizes could be used with similar analyses. Students should work in pairs with a disk whose diameter is different from other teams' disks. Each team will toss their disk onto the grid a total of 100 times and record the wins and losses. Students can take turns tossing the disk while the partner records the results. Make a chart like this:

Win

Lose

and use tally marks to record a win (completely inside a tile, not touching an edge) or a lose (overlapping an edge). There are only two rules:

1. If the disk hits anything or anyone before coming to rest, disregard that toss, and toss again.
2. If the disk is too close to call as a winner or loser, ignore that toss, and toss again.

Once you understand what you need to do, carry out the activity at your teacher's signal. When you finish, calculate the percent of times you won with your particular disk. Report the diameter of your disk, your win/lose data, and the percent of wins to your teacher. We will return to this activity later in the chapter after we develop some new ideas.

After you have done the activity and reported your results, answer the following questions:

(a) Would you expect the percents to be different from team to team? Explain.

(b) What diameter disk would have a 0% chance of winning? What *theoretical* diameter disk would have a 100% chance of winning?

(c) Visualize a scatterplot that plots the diameters of the disks on the horizontal axis and the winning percents on the vertical axis. Draw a sketch of what you think the scatterplot would look like.

(d) Using the percent of wins that your team calculated, how much money would the PTA earn (or lose) from this game if you had 100 players? Based on your answer, should the disk be larger or smaller?

Exponential growth and decay

Bankers and financial planners frequently talk about the power of compounding. If you deposit money in an interest-bearing account, the account balance will grow slowly at first, but in each successive period it will grow more than it did in the previous period, because you are receiving interest on a larger amount of money. To illustrate, suppose you deposit $2 in a savings account that pays 6% annually, compounded monthly. At the end of each month you receive interest in the amount $.06/12 = .005$ times the amount in your account at the beginning of the month. Table 4.1 shows how your money grows, month by month.

  The amount in the account after 1 month is the principal $2 plus the interest, $2(0.005)$ dollars. Do you see why the table shows this amount as $2(1.005)$ dollars?

  We now have a formula for the amount of money in the account after n months. It's an exponential equation of the form $y = ab^x$. Also note that the amount in the account is always the amount from the previous month times 1.005. This latter number is called the ***common ratio***. Whenever there is a common ratio y_n/y_{n-1} for equal-interval values of x, you have exponential growth.

common ratio

LINEAR GROWTH AND EXPONENTIAL GROWTH

A variable grows ***linearly*** over time if it *adds* a fixed increment in each equal time period.

A variable grows ***exponentially*** if it is *multiplied* by a fixed number greater than 1 in each equal time period. Exponential decay occurs when the factor is less than 1.

EXAMPLE 4.1

We'll use the graphing calculator to explore the growth of a $2 deposit over 30 years. In your TI-83, clear list L_1 and use the command `seq(X,X,0,360, 12)`→L_1 to enter the months 0, 12, 24, . . . , 360 into the list L_1. Clear list L_2 and define $L_2 = 2(1.005)\wedge L_1$. Plot a scatterplot where Xlist = L_1, Ylist = L_2, and the plotting symbol is the dot.

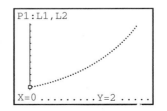

	TABLE 4.1		Growth of money in a savings account

After n months	Principal	Interest	Amount in account
0	2		\$2
1	2	2(.005)	2(1.005)
2	2(1.005)	2(1.005)(.005)	2(1.005)^2
3	2(1.005)^2	2(1.005)^2(.005)	2(1.005)^3
$\vdots$	$\vdots$	$\vdots$	$\vdots$
n	2(1.005)$^{n-1}$	2(1.005)$^{n-1}$(.005)	2(1.005)^n

We know that the money is growing exponentially because we defined L_2 as an exponential function. Notice also that the x-increment is constant (12 months) and the y-values (entries in L_2) show a common ratio (1.061677812).

Now we want to transform the points in our scatterplot into a linear set. You may recall that the exponential and logarithm functions are inverses of each other. That is, if

we start with a number	x
and raise 10 to that number	10^x
then take the logarithm of the result	$\log 10^x$
we get the original number	x

We know that numbers in L_2 are powers, but in problems of this type we generally won't know the base. There are two handy logarithms with which you're familiar:

$$\log(\text{base } 10) \qquad \text{and} \qquad \ln(\text{base } e \ = \ 2.71828\ldots)$$

We'll use $\log_{10}$. Insert a new list after L_2 and name it LOGY. Then define LOGY = $\log(L_2)$.

L1	L2	LOGY 3
0	2	.30103
12	2.1234	.32702
24	2.2543	.35302
36	2.3934	.37901
48	2.541	.405
60	2.6977	.43099
72	2.8641	.45699

LOGY(1) = .30102999 …

Turn off Plot1 and define Plot2 to be a scatterplot with Xlist = L_1, Ylist = LOGY, and specify the dot as the plotting symbol. Graph Plot2.

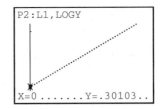

Because L_2 was a perfect exponential, this plot should be a perfectly straight line. Regress LOGY on L_1 to obtain the LSRL and confirm that the correlation for the transformed points is $r = 1$.

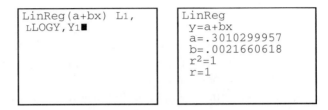

So far we have used mathematical methods to transform an exponential pattern of points into a linear pattern so that we could perform least-squares regression on the transformed points. One task remains, and that is to perform an inverse transformation on the linear equation to find the curve that will model our original exponential data. We begin with the LSRL, and we want to solve for $\hat{y}$.

$$\log \hat{y} = .3010 + .0022x$$

If we raise 10 to both sides of the equation (that is, perform the inverse of log), we have

$$10^{\log \hat{y}} = 10^{0.3010 + 0.0022x}$$

$$\hat{y} = (10^{0.3010})(10^{0.0022x})$$

Notice that this curve is now in the exponential form $\hat{y} = ab^x$ where $b = 10^{0.0022}$. This exponential curve should fit the original data. To see this, deselect Y_1 and define $Y_2 = 10\verb|^|(\text{RegEQ})$. Here are the keystrokes for RegEQ: VARS / 5:Statistics / EQ / 1:RegEQ. Make sure that you put parentheses around RegEQ when you define Y_2.

Plot the original data (Xlist=L_1 and Ylist=L_2 and overlay the exponential curve Y_2.

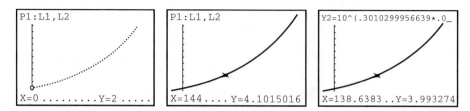

Tracing over Y_2 lets you see that it will take you 11 years and 7 months to double your \$2 investment. How many months will it take to double again, from \$4 to \$8? What will your \$2 be worth at the end of 30 years?

Petroleum has become the most important single source of energy for the developed nations in this century. In recent years it has been the cause of economic dislocation and even war. Table 4.2 shows the growth of annual world crude oil production, measured in millions of barrels per year.[2] A scatterplot (Figure 4.1) shows that there is an increasing trend; the pattern of growth of oil production appears to be exponential until 1973, when a Mideast war touched off a dramatic price increase and a change in the previous pattern of production. You want to investigate whether you can describe the growth of oil production from 1880 to 1973 by an exponential curve. Minitab was used to do the analyses and produce the

TABLE 4.2		Annual world crude oil production, 1880 to 1994 (millions of barrels)			
Year	Mbbl	Year	Mbbl	Year	Mbbl
1880	30	1945	2,595	1976	20,188
1890	77	1950	3,803	1978	21,922
1900	149	1955	5,626	1980	21,722
1905	215	1960	7,674	1982	19,411
1910	328	1962	8,882	1984	19,837
1915	432	1964	10,310	1986	20,246
1920	689	1966	12,016	1988	21,338
1925	1,069	1968	14,104	1990	22,100
1930	1,412	1970	16,690	1992	22,028
1935	1,655	1972	18,584	1994	22,234
1940	2,150	1974	20,389		

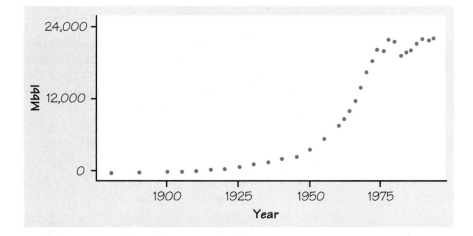

FIGURE 4.1   Annual world crude oil production, 1880 to 1994, from Table 4.2.

plots, but you should use your TI-83 to work through Example 4.2 and compare your results with the results shown below.

EXAMPLE 4.2

A preliminary test for exponential growth is to calculate ratios of the form y_n/y_{n-1} to see whether a common ratio exists. Because the year increments are not uniform in the table, the data for the even-decade years are used in the test (Table 4.3). This test reveals that although there is some variation in the ratios due to extraordinary events such as World War I, the Depression, and World War II, these ratios are approximately constant and equal to 2. We will accept this as tentative confirmation of exponential growth. The next step is to transform the points into a linear pattern. Because the logarithm is the inverse of the exponential, we take the log(base 10) of the y-values (the Mbbl. column). See Table 4.4. If you were to calculate the differences between consecutive logs in Table 4.4, you would find that the common difference is approximately constant. This would be good evidence that the scatterplot of $\log y$ on x is linear.

Next we have Minitab do a *linear* regression of $\log y$ versus year and report the least-squares regression equation. Here is partial output:

```
The regression equation is
Logy = - 52.7 + 0.0289 Year

Unusual Observations
Obs.      Year       Logy        Fit Stdev.Fit     Residual      St.Resid
   1      1880     1.4771     1.6079     0.0279      -0.1307       -2.66R
   8      1925     3.0290     2.9074     0.0133       0.1215        2.21R

R denotes an obs. with a large st. resid.
```

TABLE 4.3	Annual world crude oil production, 1880 to 1970 (millions of barrels)

Year (x)	Mbbl (y)	Ratio (y_n/y_{n-1})
1880	30	
1890	77	2.6
1900	149	1.9
1910	328	2.2
1920	689	2.1
1930	1,412	2.1
1940	2,150	1.5
1950	3,803	1.8
1960	7,674	2.0
1970	16,690	2.2

TABLE 4.4	Annual world crude oil production, 1880 to 1972 (millions of barrels)

Year	Mbbl	Log y	Year	Mbbl	Log y
1880	30	1.47712	1945	2,595	3.41414
1890	77	1.88649	1950	3,803	3.58013
1900	149	2.17319	1955	5,626	3.75020
1905	215	2.33244	1960	7,674	3.88502
1910	328	2.51587	1962	8,882	3.94851
1915	432	2.63548	1964	10,310	4.01326
1920	689	2.83822	1966	12,016	4.07976
1925	1,069	3.02898	1968	14,104	4.14934
1930	1,412	3.14983	1970	16,690	4.22246
1935	1,655	3.21880	1972	18,584	4.26914
1940	2,150	3.33244			

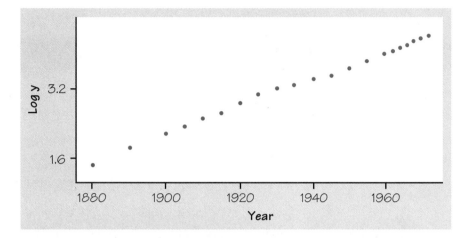

FIGURE 4.2 Scatterplot of the logarithm of oil production data for the years 1880 to 1972.

Figure 4.2 shows the scatterplot of log(Mbbl) on year. Figure 4.3 shows the scatterplot with the LSRL superimposed.

The last step in formulating the exponential function that models this growth is to perform the inverse transformation and solve for y:

$$\log \hat{y} = -52.7 + .0289x$$
$$10^{\log \hat{y}} = 10^{(-52.7 + 0.0289x)}$$
$$\hat{y} = (10^{-52.7})(10^{0.0289x})$$

This exponential function, as shown in Figure 4.4, appears to be an excellent model for the oil production data from 1880 to 1972.

Residuals again

Just as in the case of linear growth, calculating and plotting residuals allow a closer examination of the fit of the exponential growth model to data. When we performed a least-squares regression of log y on x, we used data for years up to 1972 only, because we know that the pattern of exponential growth ended in 1973. To produce a residual plot with Minitab, you specify a scatterplot with response variable RESID and explanatory (Minitab calls it Predictor) variable YEAR. Figure 4.5 shows the residual plot for the data in Figure 4.3.

Recall that the residuals of the data from the fitted line are computed

$$\text{residual} = \text{observed value} - \text{predicted value}$$

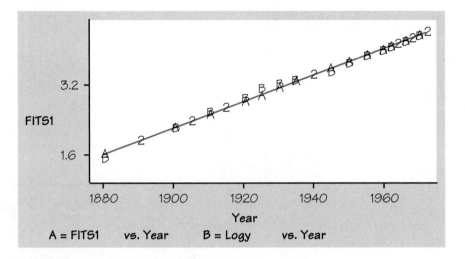

FIGURE 4.3   Scatterplot of the logarithm of oil production data for the years 1880 to 1972 with least-squares line overlayed.

For the year 1945, the value of the logarithm *predicted* by the regression line is

$$\log(\text{production}) = -52.7 + (.0289 \times 1945)$$
$$= 3.5105$$

The *observed* value for 1945 is the logarithm of the 2595 million barrels of oil produced in that year, so the residual corresponding to $x = 1945$ is

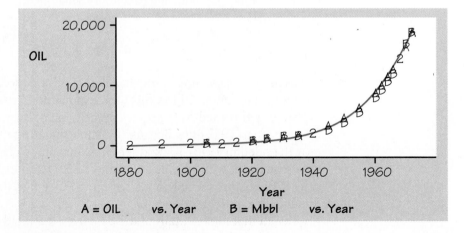

FIGURE 4.4   An exponential model for oil production data for the years 1880 to 1972.

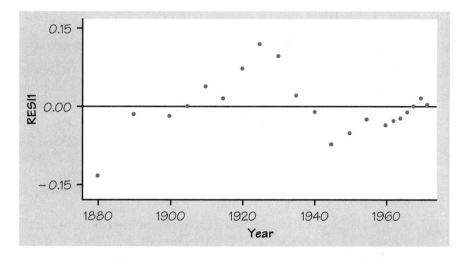

FIGURE 4.5 Residual plot for the least-squares line for world oil production.

$$\text{residual} = \text{observed value} - \text{predicted value}$$
$$= \log 2595 - 3.5105$$
$$= 3.4141 - 3.5105$$
$$= -0.0964$$

The residual plot can give us a great deal of information about the fit of the exponential growth model.

The deviations from the overall pattern of exponential growth in oil production are easier to see in the residual plot (Figure 4.5) than in Figure 4.4. The pattern of residuals about the horizontal line at 0 shows systematic departures from the overall fit. Because we are plotting logarithms, straight-line patterns of residuals have a specific interpretation. The line of fit (zero residual) represents exponential growth at the average rate observed in the entire period 1880 to 1972. A straight-line rising pattern of residuals shows a period of growth at a faster rate, while a declining pattern shows a slower rate of growth.

Oil production was increasing more rapidly than the long-term rate in the years between 1900 and the beginning of the Depression in 1929. Production increased more slowly not only during the Depression but also during World War II. That is, the turning points in the residual plot occur in 1925 (the last point before 1929) and 1945 (the end of World War II). Only after the war did oil production return to an above-average growth rate, which lasted until 1973. (The production of oil was of course increasing during the entire period up to 1973—the runs of declining residuals show periods of slower growth, not an actual drop in oil production.)

Our tools for data analysis have given us a quite detailed understanding of the history of world oil production. The long-term picture is one of exponential growth from the beginnings of commercial petroleum use to the OPEC boycott of 1973. But the long-term rate of growth is an average over several periods of somewhat slower or somewhat faster growth, periods that coincide with major events in the economic history of the twentieth century.

Could we have avoided logarithms by fitting an exponential growth curve to the actual production data displayed in Figure 4.1? Yes. But there are several reasons for transforming before fitting the model and calculating residuals. First, straight lines are simple, while the explicit mathematical form of the exponential growth model is more involved. A second reason for working with the logarithms is that it is easy to calculate the least-squares line. Fitting an exponential curve to a set of data is more difficult. Many statistical computing systems (not all) will carry out such nonlinear fits, but the calculations are not feasible without a computer.

EXERCISES

4.1   Biological populations can grow exponentially if not restrained by predators or lack of food. The gypsy moth outbreaks that occasionally devastate the forests of the Northeast illustrate approximate exponential growth. It is easier to count the number of acres defoliated by the moths than to count the moths themselves. Here are data on an outbreak in Massachusetts.[3]

Year	Acres
1978	63,042
1979	226,260
1980	907,075
1981	2,826,095

(a) Plot the number of acres defoliated y against the year x. The pattern of growth appears exponential.

(b) Verify that y is being multiplied by about 4 each year by calculating the ratio of acres defoliated each year to the previous year. (Start with 1979 to 1978, when the ratio is $226,260/63,042 = 3.6$.)

(c) Take the logarithm of each number y and plot the logarithms against the year x. The linear pattern confirms that the growth is exponential.

(d) Verify that the least-squares line fitted to the four points is

$$\log \hat{y} = -1094.51 + .5558 \times \text{year}$$

(e) Construct and interpret a residual plot for $\log \hat{y}$ on year.

(f) Perform the inverse transformation to express $\hat{y}$ as an exponential equation. Display a scatterplot of the original data with the exponential curve model superimposed. Is your exponential function a satisfactory model for the data?

(g) Use your model to predict the number of acres defoliated in 1982.

(*Postscript:* A viral disease reduced the gypsy moth population between the readings in 1981 and 1982. The actual count of defoliated acres in 1982 was 1,383,265.)

4.2 In recent years there has been growing concern about the federal debt (officially the Public Debt of the United States) because the government has been spending more each year than it takes in through taxes and other revenues. The following table shows the federal debt from 1981 through 1996.[4] (Note that $1 trillion = $1,000,000,000,000.)

Fiscal year	Coded year	Debt (trillions of $)	Fiscal year	Coded year	Debt (trillions of $)
1981	1	0.998	1989	9	2.857
1982	2	1.142	1990	10	3.233
1983	3	1.377	1991	11	3.665
1984	4	1.572	1992	12	4.065
1985	5	1.823	1993	13	4.412
1986	6	2.125	1994	14	4.693
1987	7	2.350	1995	15	4.974
1988	8	2.602	1996	16	5.225

(a) Make a scatterplot of the data. Let $x =$ the number of years after 1980 (to simplify the computations).

(b) Test for exponential growth by calculating y_n/y_{n-1} ratios of consecutive debt figures. Although there is variation in these ratios, verify that they center at 1.12. We will accept this as evidence of exponential growth.

(c) Perform a logarithmic transformation of the data. Show the steps. Then construct a scatterplot of $\log y$ versus year code.

(d) Your colleagues suggest that the pattern of points resembles two linked segments rather than one single linear pattern. Since you are mostly interested in predicting the debt for the immediate future, you decide to discard the first 6 data points and retain the last 10. Do this, and construct a scatterplot for the remaining points (years 1987 to 1996).

(e) Perform a least-squares regression for the reduced set of transformed points. Record the correlation coefficient and interpret the strength of association.

(f) A residual plot reveals a weakness of the model, but you decide to continue. Perform the inverse transformation to obtain the exponential function model, and plot the original data for 1987 to 1996 with your exponential model superimposed. Show the steps for the inverse transformation and the final exponential equation.

(g) The last several points in the residual plot show that the debt is rapidly falling away from the regression line. Offer a possible explanation for this phenomenon. Use your model to predict the federal debt for the year 1997.

(h) The public debt for 1997 was $5.413 trillion. Compare your prediction for 1997 to the actual figure and give an explanation for any discrepancy.

(i) Considering the economy as well as political initiatives and legislation in the mid-1990s, would you be comfortable using your model to predict the federal debt for the year 2000? Explain briefly.

(j) Consult the latest *World Almanac* or similar resource to obtain the most recent actual data. Or try these World Wide Web sites: opd.opdpenny.htm or www.PublicDebt.treas.gov. Based on this new information, how would you proceed to construct a more reliable model?

Power regression

Consider the power function $y = ax^b$. This form represents a large family of familiar functions such as the quadratic functions $y = x^2$ and $y = -.5x^2$ and the cubic functions $y = x^3$ and $y = -2x^3$. Many other, more complicated functions fit this form, but these will suffice for illustrative purposes. As with our development of exponential regression, the form of the power function suggests that we might again be able to use logarithms to transform the equation into a linear form. Then we can employ least-squares regression to obtain an equation for the model. Finally we can perform an inverse transformation to be back to the original form, $y = ax^b$. Here are the steps:

1. Begin with the form for the power function:

$$y = ax^b$$

2. Take the logarithm of both sides:

$$\log y = \log(ax^b)$$
$$= \log a + (b \log x)$$

3. Recognize that this is the *form* of a linear equation:

$$Y = A + bX$$

Note: Don't forget that $A = \log a$ is a constant.

4. Perform least-squares regression on the linear equation, using $\log x$ as the horizontal-axis variable and $\log y$ as the vertical-axis variable. Note that we can also find the correlation r for the transformed data in the process.

5. Use the rules of logarithms to write b as an exponent:

$$\log y = \log a + (b \log x)$$
$$= \log a + \log x^b$$

6. Now perform the inverse transformation. That is, raise 10 to the left side of the equation, and set this equal to 10 raised to the right side:

$$10^{\log y} = 10^{\log a + (b \log x)}$$
$$= 10^{\log a + \log x^b}$$

7. Use the rules of exponents and the fact that $10^{\log x} = x$ to simplify the right side:

$$y = (10^{\log a})(10^{\log x^b})$$
$$= (10^{\log a})(10^{\log x})^b$$
$$= (10^{\log a})(x^b)$$
$$= ax^b$$

In applications there are many situations where consideration of dimension can lead us to a promising model. Many pizza restaurants display different-sized pans on the wall with the diameters clearly marked. One might wonder if the cost of a plain pizza (no additional toppings) is related to the area of the pizza. Area is a two-dimensional attribute and we have a handy formula that relates circular area to diameter:

$$\text{area} = \pi r^2 = \pi(d/2)^2 = \pi d^2/4 = (\pi/4)d^2$$

where r = radius and d = diameter. So it seems reasonable to conjecture that the cost of a plain pizza might be proportional to the square of the diameter, that is,

$$\text{cost} = a \times \text{diameter}^2$$

where a = constant of proportionality.

As another example, consider the business of buying a house. People who wish to buy a new home pay particular attention to the area in square feet of the living portion of the house (excluding garages, porches, and decks). Could it be that there is a relationship between the purchase price of a house and the living area? A reasonable guess would be to try the model

$$\text{cost} = a \times \text{area}$$

Perhaps you have seen charts at doctors' offices or fitness centers that list various male and female heights and the corresponding "ideal" weights. Since height is a one-dimensional attribute and weight is a three-dimensional attribute, a possible model for the relationship between height and weight would be

$$\text{weight} = a \times \text{height}^3$$

power regression

All of the preceding examples are good candidates for *power regression* (one quantity is proportional to a second quantity raised to a power). You should also notice that when these power functions are plotted, all of them will pass through the origin $(0, 0)$ since the pair $x = 0, y = 0$ will satisfy the equations. We will focus on these special cases of mostly quadratic and cubic equations for two important reasons. First, many pairs of variables, like height and weight, are such that when one variable is zero, the other is also zero. So this special class of applications is quite extensive. Second, the mathematics required for fitting a curve of the more general forms $y = ax^2 + bx + c$ and $y = ax^3 + bx^2 + cx + d$ is considerably more complicated.

EXAMPLE 4.3

Imagine that you have been put in charge of organizing a fishing tournament in which prizes will be given for the heaviest fish caught. You know that many of the fish caught during the tournament will be measured and released. You are also aware that trying to weigh a fish that is flipping around, in a boat that is rolling with the swells, using delicate scales will probably not yield very reliable results.

It would be much easier to measure the *length* of the fish on the boat. What you need is a way to convert the length of the fish to its weight. You reason that since length is one-dimensional and weight is three-dimensional, and since a fish 0 units long would weigh 0 pounds, the weight of a fish should be proportional to the cube of its length. Thus, a model of the form weight $= a \times \text{length}^3$ should work. You contact the nearby marine research laboratory and they provide the average length and weight catch data for the Atlantic Ocean rockfish *Sebastes mentella* (Table 4.5).[5] The lab also advises you that the model relationship between body length and weight has been found to be accurate for most fish species growing under normal feeding conditions.

TABLE 4.5	Average length and weight at different ages for Atlantic Ocean rockfish, *Sebastes mentella*				
Age (yr)	Length (cm)	Weight (g)	Age (yr)	Length (cm)	Weight (g)
1	5.2	2	11	28.2	318
2	8.5	8	12	29.6	371
3	11.5	21	13	30.8	455
4	14.3	38	14	32.0	504
5	16.8	69	15	33.0	518
6	19.2	117	16	34.0	537
7	21.3	148	17	34.9	651
8	23.3	190	18	36.4	719
9	25.0	264	19	37.1	726
10	26.7	293	20	37.7	810

Entering the data into lists LEN and WT in your calculator, you plot the scatterplot.

Although the growth might appear to be exponential, we know that it is frequently misleading to trust too much to the eye. Moreover, we have already decided on a model that makes sense in this context: weight $= a \times \text{length}^3$.

If we take the $\log_{10}$ of both sides, we obtain

$$\log(\text{weight}) = \log a + [3 \times \log(\text{length})]$$

This equation looks like a linear equation

$$Y = A + BX$$

so we plot log(weight) against log(length).

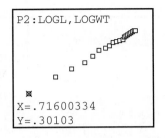

We visually confirm that the relationship appears very linear. We perform a least-squares regression on the transformed points [log(length), log(weight)].

```
LinReg
 y=a+bx
 a=-1.899397207
 b=3.049417866
 r²=.9985228463
 r=.9992611502
■
```

We see that the correlation r of the logarithms of length and weight is virtually 1. (Remember, however, that correlation was defined only for linear fits.) Despite the very high r-value, it's still important to look at a residual plot.

```
P3:LOGL,RESID
```

```
X=.71600334
Y=.01703381
```

The random scatter of the points tells us that the line is a good model for the logs of length and weight. The last step is to perform an inverse transformation on the linear regression equation:

$$\log(\text{weight}) = -1.8994 + [3.0494\,\log(\text{length})]$$
$$= -1.8994 + \log(\text{length})^{3.0494}$$

This is the critical step: to remember to use a property of logarithms to write the multiplicative constant 3.0494 as an exponent. Let's continue. Raise 10 to the left side of the equation and set this equal to 10 raised to the right side:

$$10^{\log(\text{weight})} = 10^{-1.8994+\log(\text{length})^{3.0494}}$$

$$\text{weight} = 10^{-1.8994} \times \text{length}^{3.0494}$$

This is the final power equation for the original data. Since we entered the LSRL equation as Y_1, we enter the power function as Y_2.

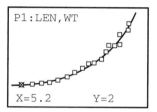

```
Plot1  Plot2  Plot3
\Y1=-1.899397207
3453+3.049417866
2105X
\Y2■(10^(-1.8994
))(X^3.0494)■
\Y3=
\Y4=
```

The scatterplot of the original data along with the power function model appears as shown.

```
P1:LEN,WT



X=5.2          Y=2
```

The fit of this model has visual appeal. We will leave it as an exercise to calculate the sum of the squares of the deviations. It should be noted that the power of x that we obtained for the model, 3.0494, is very close to the value 3 that we conjectured when we proposed the form for our model.

The original purpose for developing this model was to approximate the weight of a fish given its length. Suppose your catch measured 36 centimeters. Our model predicts a weight of $Y_2(36) = 702.0836281$, or about 702 grams. If you entered a fishing contest, would you be comfortable with this procedure for determining the weights of the fish caught, and hence for determining the winner of the contest?

EXERCISES

4.3   (a) Use the model we derived for approximating the weight of *Sebastes mentella*, $\hat{y} = 10^{-1.8994}x^{3.0494}$, to determine the sum of the squares of the deviations between the observed weights (in grams) and the predicted values. Did we minimize this quantity in the process of constructing our model? If not, what quantity was minimized?

(b) When we performed least-squares regression of log(weight) on log(length) on the TI-83, residuals were calculated and stored in a list named RESID. Use this list and the 1-Var Stats command to calculate the sum of the squares of the residuals. Compare this sum of squares with the sum of squares you calculated in (a).

(c) Would you expect the answers in (a) and (b) to be the same or different? Explain.

4.4   The U.S. Department of Health and Human Services characterizes adults as "seriously overweight" if they meet a certain weight criterion for their height as shown in the table below (only a portion of the chart is reproduced here).

Height (ft, in)	Height (in)	Severely overweight (lb)	Height (ft, in)	Height (in)	Severely overweight (lb)
4'10"	58	138	5'8"	68	190
5'0"	60	148	6'0"	72	213
5'2"	62	158	6'2"	74	225
5'4"	64	169	6'4"	76	238
5'6"	66	179	6'6"	78	250

Weights are given in pounds, without clothes. Height is measured without shoes. There is no distinction between men and women; a note accompanying the table states, "The higher weights apply to people with more muscle and bone, such as many men." Despite any reservations you may have about the department's common standard for both genders, do the following:

(a) Without looking at the data, hypothesize a relationship between height and weight of U.S. adults. That is, write a general form of an equation that you believe will model the relationship.

(b) Which variable would you select as explanatory and which would be the response? Plot the data from the table.

(c) Perform a transformation to linearize the data. Do a least-squares regression on the transformed data and check the correlation coefficient.

(d) Construct a residual plot of the transformed data. Interpret the residual plot.

(e) Perform the inverse transformation and write the equation for your model. Use your model to predict how many pounds a 5'10" adult would have to weigh in order to be classified by the department as "seriously overweight." Do the same for a 7-foot individual.

SUMMARY

Nonlinear data may exhibit *exponential growth or decay* or may be modeled by a *power function* that passes through the origin. Since both forms, $y = ab^x$ and $y = ax^b$, involve exponents, our techniques for constructing the models begin in a similar fashion. We first take the *logarithm* of both sides. For exponential data, we plot $\log y$ on x, and if that produces a linear pattern, we perform least-squares regression on the transformed points. Then we do the inverse transformation and see if the resulting exponential function captures the trend of the data. For power functions, we again take the logarithm of both sides but plot $\log y$ versus $\log x$. If the transformed points are linear, then we find the least-squares regression line for $\log y$ versus $\log x$ and do the inverse transformation to obtain the power function.

This section makes heavy use of the properties of logarithms, rules of exponents, linear regression, and data-handling techniques with the graphing calculator. If you're a little rusty in any of these areas, make sure you work through enough practice exercises that the steps become second nature to you.

SECTION 4.1 EXERCISES

4.5   The cost of sending a 1-ounce letter and a postcard through the U.S. mail has risen over the years. The table below shows the rates and the dates new rates became effective.[6]

Date of rate change		Letter, first oz	Postcard	Date of rate change		Letter, first oz	Postcard
Aug. 1,	1958	.04	.03	May 29,	1978	.15	.10
Jan. 7,	1963	.05	.04	Mar. 22,	1981	.18	.12
Jan. 7,	1968	.06	.05	Nov. 1,	1981	.20	.13
May 16,	1971	.08	.06	Feb. 17,	1985	.22	.14
Mar. 2,	1974	.10	.08	Apr. 3,	1988	.25	.15
Sept. 14,	1975	.10	.07	Feb. 3,	1991	.29	.19
Dec. 31,	1975	.13	.09	Jan. 1,	1995	.32	.20

Plot the data and use the methods of this chapter to construct an appropriate model for the cost of a 1-ounce letter. Then use your model to predict the cost of mailing a 1-ounce letter in the year 2005. How many years will it take for the cost to rise to 50¢ per 1-ounce letter?

4.6   The following table gives the major league baseball average salaries as compiled by the Major League Baseball Players Association:[7]

Year	Average salary
1989	$512,804
1990	$578,930
1991	$891,188
1992	$1,084,408
1993	$1,120,254
1994	$1,188,679
1995	$1,071,029
1996	$1,176,967
1997	$1,383,578

(a) Plot the data. The points display neither a linear nor an exponential trend. Nevertheless, extend the table by recording the ratio of salaries from year to year. Which year produced the salary that most disrupts the trend?

(b) Transform the data as if the salaries were increasing exponentially. In the plot of log(salary) versus year, observe that five of the points seem to line up nicely in a straight-line pattern, while the remaining four points appear "out of line." Which four points are not in the linear pattern?

(c) Delete the four years referenced in (b), the corresponding salaries, and their logarithms. Then perform least-squares regression on the remaining five transformed points. How well does the LSRL fit these five points [year, log(salary)]? What is the correlation? Find the exponential equation and overlay this equation on the five original data points.

(d) Look at your original scatterplot of the data. In 1991, the average salary increased a whopping 54% over the previous year, fueled in part by several relatively large contracts to high-visibility players. There followed several years of some dissatisfaction among the players, culminating in a general strike in 1995. That year, the average salary decreased (the only such year in major league history). What does your analysis in (c) suggest about the net results after the strike on the growth of player salaries?

(e) The *average* salaries for baseball players for the last several years have been significantly higher than the *median* salaries. How could you explain this?

4.7 Electronic Funds Transfer (EFT) includes automated teller machine (ATM) transactions and transactions at point-of-sale (POS) terminals. POS terminals are electronic terminals in retail stores that allow a customer to pay for goods through a direct debit to that customer's bank account. EFT machines are ubiquitous now, but they first appeared as recently as the early 1980s and their use has increased dramatically since then. The following table shows the total number of EFT transactions (in millions) per year from 1985 to 1996:[8]

Year	Number of transactions (millions)
1985	3,579
1990	5,942
1991	6,642
1992	7,537
1993	8,135
1994	8,958
1995	10,464
1996	11,830

(a) Plot the data, and verify that the number of EFT transactions is growing exponentially.

(b) Transform the data to achieve a linear scatterplot. Perform least-squares regression on the transformed data, and report the correlation for the transformed data. Construct a residual plot to assess the linearity.

(c) Perform the inverse transformation to obtain an exponential equation. Plot the original EFT volume data and overlay the exponential model. Use this model to predict the number of EFT transactions for a later year, such as 2000. If you can find the actual data, evaluate your prediction.

4.8 Stamp collectors know that the United States and other countries have issued a very large number of different stamps over the years. They may not realize, however, that postage stamp production in the United States has been increasing at an exponential rate. The following table shows the cumulative number of regular and commemorative U.S. postage stamp issues by 10-year intervals from 1848 to 1988:[9]

Year	Number of U.S. stamps issued (cumulative)	Year	Number of U.S. stamps issued (cumulative)	Year	Number of U.S. stamps issued (cumulative)
1848	2	1898	293	1948	980
1858	30	1908	341	1958	1123
1868	88	1918	529	1968	1364
1878	181	1928	647	1978	1769
1888	218	1938	838	1988	2400

(a) Use a logarithmic transformation to produce a linear scatterplot of $\log y$ on x = year.

(b) Notice that the first three points do not fit the linear pattern of the rest of the transformed points. Delete the first three data points, and complete the model with the remaining data points. Record the exponential equation for your model.

(c) Use your model to predict the cumulative number of U.S. postage stamp issues in the year 2000.

4.9 The data in the following table show the population per square mile in the United States from 1790 to 1990:[10]

Year	People per square mile	Year	People per square mile	Year	People per square mile
1790	4.5	1860	10.6	1930	34.7
1800	6.1	1870	10.9	1940	37.2
1810	4.3	1880	14.2	1950	42.6
1820	5.5	1890	17.8	1960	50.6
1830	7.4	1900	21.5	1970	57.5
1840	9.8	1910	26.0	1980	64.0
1850	7.9	1920	29.9	1990	70.3

(a) Find a mathematical model that can be used to predict the population density in the year 2000.

(b) Comment on any unusual features, such as the decline in population density in the 1810 figure. What important historical event happened between 1800 and 1810, during the administration of President Thomas Jefferson, that would affect the population density in the United States?

4.10 The health care industry in the United States has grown dramatically in recent years due to advances in knowledge and technology. As new medicines are developed and new methods are refined for treating ailments, the public's expectations and demands increase accordingly. The following table shows the total national health expenditures per person from 1960 to 1994:[11]

Year	Expenditure per capita ($)	Year	Expenditure per capita ($)	Year	Expenditure per capita ($)
1960	141	1976	662	1986	1849
1965	202	1978	827	1988	2201
1970	341	1980	1052	1990	2688
1972	415	1982	1346	1992	3144
1974	513	1984	1594	1994	3510

(a) Plot the data, and use the methods of this chapter to develop an appropriate model for the data.

(b) Use your model to predict the national per capita health expenditure in the year 2000.

4.11 Each year the Federal Bureau of Investigation (FBI) issues a report that provides information about crimes in the United States. The following table gives the total number of violent crimes in the United States for the years 1984 to 1994:[12]

Year	Number of violent crimes (thousands)	Year	Number of violent crimes (thousands)
1984	1273	1990	1820
1985	1329	1991	1912
1986	1489	1992	1932
1987	1484	1993	1926
1988	1566	1994	1864
1989	1646		

(a) Plot the data. Observe that there is a pattern but that several points don't fit the pattern. Which points don't fit?

(b) Are violent crimes increasing linearly or exponentially? Calculate the ratios of current number of violent crimes to previous number of violent crimes. Are the ratios approximately constant and greater than 1? What is the average ratio for the first eight data points?

(c) You decide to discard the last three points and develop an exponential model for the years 1984 to 1991. Delete the last three points and transform the remaining data to achieve a linear scatterplot. Then perform a least-squares regression on the transformed points and record the correlation r.

(d) Perform the inverse transformation and record the equation that models the data for the years 1984 to 1991.

(e) Use your exponential model from (c) to predict the number of violent crimes in 1986. This year produces the largest residual. What is the residual for 1986?

4.12   Use the methods discussed in this section to analyze the following data on the hearts of various mammals.[13] Write your findings and conclusions in a short narrative.

Mammal	Heart weight (grams)	Length of cavity of left ventricle (centimeters)
Mouse	.13	.55
Rat	.64	1.0
Rabbit	5.8	2.2
Dog	102	4.0
Sheep	210	6.5
Ox	2030	12.0
Horse	3900	16.0

4.13   The new manager of a pizza restaurant wants to add variety to the pizza offerings at the restaurant. She also wants to determine if the prices for existing sizes of pizzas are consistent. Prices for plain (cheese only) pizzas are shown below:[14]

Size	Diameter (inches)	Cost
Small	10	$4.00
Medium	12	$6.00
Large	14	$8.00
Giant	18	$10.00

(a) Construct an appropriate model for these data. Comment on your choice of model.

(b) Based on your analysis, would you advise the manager to adjust the price on any of the pizza sizes? If so, explain briefly.

(c) Use your model to suggest a price for a new "personal pizza," with a 6-inch diameter.

(d) Use your model to suggest a price for a new "soccer team" size, with a 24-inch diameter (assuming the oven is large enough to hold it).

4.14 Return to the 800-meter world record times for men and women of Exercise 3.56 (page 172). Suppose you are uncomfortable with a linear model for the decline in winning times that will eventually intersect the horizontal axis.

(a) Construct exponential and power regression models for the *men's* record times. Which do you consider to be a better model?

(b) Based on your answer to (a), construct a similar model for the *women's* record times.

(c) Will either of these curves eventually reach zero? Will the curves intersect each other? If so, in what year will the curves intersect?

(d) Is this a satisfactory model, or is there a better model for these data?

4.15 Return to Activity 4, where your assignment was to design a disk toss game that would give players a 40% chance of winning (and the PTA a 60% chance of taking the players' money).

(a) Make a table of the results for different-sized disks, and record the results for all of the class teams:

Type of disk	Diameter (mm)	No. of winners/ no. of tosses	Winning percent
Dime	18		
Quarter	24		
Coffee can lid			

(b) Plot the scatterplot of winning percents on disk diameters. If there is a pattern to the points, describe the pattern.

(c) With the assumption that the winning percent is related to the *areas* of the tile and the disk, use the TI-83's quadratic regression function to construct a model for the data. Does the model appear to fit the data well?

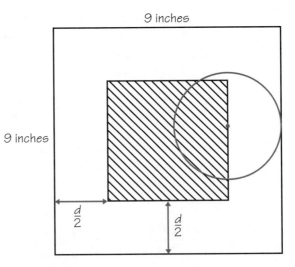

FIGURE 4.6   Geometric solution to the Activity 4 problem.

(d) What should $y(0)$ equal? What diameter should produce a y-value
of zero? Evaluate your model at these two values and compare these
predicted values with the theoretical values.

4.16   This is a continuation of the previous exercise. In this exercise you will
derive an analytical solution to the game of chance problem in Activity 4.

(a) Determine the quadratic equation that expresses y (the winning percentage) as a function of the diameter of the disk. If the center of the
disk lands anywhere in the shaded square shown in Figure 4.6, the
toss is a win. Do you see why? Begin with the following equation:

$$y = \text{winning percent} = \frac{\text{area of shaded square}}{\text{area of tile}}$$

where winning percent is written as a decimal; use Figure 4.6 to help
you rewrite the numerator and denominator. Let d represent the
diameter of the disk. Write the quadratic equation in the form
$y = ad^2 + bd + c$. What is $y(0)$? What diameter d gives a y-value of
0.40?

(b) Compare the equation you obtained in (a) with the quadratic regression equation you obtained in Exercise 4.15. In particular, compare
the theoretical solution to the disk-tossing problem with the experimental results.

(c) Suppose that the PTA decides that having a 40% chance of winning will not make a big enough profit and that they decide to drop to 35%. Use both equations to determine the size of the disk (to the nearest millimeter) needed for a 35% chance of winning.

4.17 Weight lifting in the Summer Olympics is divided into classes according to how much the participants weigh. Winners at the 1996 Summer Olympics in Atlanta and the amounts lifted are shown below (the weight limits of the participants have been converted from kilograms to pounds):[15]

Weight limit (lb)	Winner	Country	Amount lifted (lb)
119	Halil Mutlu	Turkey	633
130	Tang Ningsheng	China	678
141	Naim Sulemanoglu	Turkey	739
154	Zhan Xugang	China	787
167.5	Pablo Lara	Cuba	809
183	Pyrros Dimas	Greece	864
200.5	Alexi Petrov	Russia	886
218	Kakhi Kakhiasvili	Greece	926
238	Timur Taimazov	Russia	948

(a) Plot a scatterplot and see if there is a relationship between the weight limit for the athlete and the number of pounds lifted.

(b) Use the TI-83's quadratic regression option to find a model for these data. Comment on how well the curve appears to fit the data.

(c) Is your model helpful outside the range of weight limits given? In particular, what amount lifted would your model predict for a weight limit of 0 pounds? For a weight limit of 300 pounds? Comment on the implications.

(d) What parts of the problem would change if you used the original kilograms instead of pounds?

4.18 The following data were collected to investigate the relationship between the amount of electricity used (in kilowatt-hours) in a given month in all-electric houses and the size of the houses (in square feet):

Size of house (ft^2)	Electrical usage (kWh)
1420	1222
1510	1327
1650	1535
1790	1680
1980	1812
2220	1830
2400	2005
2800	2025

In trying to formulate a model, the investigator conjectured that the amount of electricity used was proportional to the number of square feet in the living area of the house. A scatterplot of electrical usage on house size showed a pattern that appeared to fit the shape of a parabola. But a plot of log y on log x revealed a pattern that was not linear, so the proportionality idea was abandoned. Confirm the preceding, and then use the TI-83's built-in quadratic regression function to try to fit a generalized parabola to the data. Can you use this model to extrapolate outside the range of the data? Explain briefly.

4.2   INTERPRETING CORRELATION AND REGRESSION

Correlation and regression are powerful tools for describing the relationship between two variables. When you use these tools, you must be aware of their limitations, beginning with the fact that they describe *only linear* relationships. Also remember that *both r and the least-squares regression line can be strongly influenced by a few extreme observations*. One influential observation or incorrectly entered data point can greatly change these measures. Always plot your data before interpreting regression or correlation. Here are some other cautions to keep in mind when you apply correlation and regression or read accounts of their use.

Extrapolation

Suppose that you have data on a child's growth between 3 and 8 years of age. You find a strong linear relationship between age x and height y. If you fit a regression line to these data and use it to predict height at age 25 years, you will predict that the child will be 8 feet tall. Growth slows down and stops at maturity, so extending the straight line to adult ages is foolish.

Few relationships are linear for all values of x. So don't stray far from the domain of x that actually appears in your data.

EXTRAPOLATION

Extrapolation is the use of a regression line or curve for prediction outside the domain of values of the explanatory variable x that you used to obtain the line or curve. Such predictions cannot be trusted.

EXERCISE

4.19 The number of people living on American farms has declined steadily during this century. Here are data on the farm population (millions of persons) from 1935 to 1980.

Year	1935	1940	1945	1950	1955	1960	1965	1970	1975	1980
Population	32.1	30.5	24.4	23.0	19.1	15.6	12.4	9.7	8.9	7.2

(a) Make a scatterplot of these data and find the least-squares regression line of farm population on year.

(b) According to the regression line, how much did the farm population decline each year on the average during this period? What percent of the observed variation in farm population is accounted for by linear change over time?

(c) Use the regression equation to predict the number of people living on farms in 1990. Is this result reasonable? Why?

Lurking variables

In our study of correlation and regression we looked at just two variables at a time. Often the relationship between two variables is strongly influenced by other variables. More advanced statistical methods allow the study of many variables together, so that we can take other variables into account. But sometimes the relationship between two variables is influenced by other variables that we did not measure or even think about. Because these variables are lurking in the background, we call them *lurking variables*.

LURKING VARIABLE

A **lurking variable** is a variable that has an important effect on the relationship among the variables in a study but is not included among the variables studied.

A lurking variable can falsely suggest a strong relationship between x and y, or it can hide a relationship that is really there. Here are examples of each of these effects.

EXAMPLE 4.4

The National Halothane Study was a major investigation of the safety of the anesthetics used in surgery. Records of over 850,000 operations performed in 34 major hospitals showed the following death rates for four common anesthetics:[16]

Anesthetic	A	B	C	D
Death rate	1.7%	1.7%	3.4%	1.9%

There is a clear association between the anesthetic used and the death rate of patients. Anesthetic C appears dangerous. But there are obvious lurking variables: the age and condition of the patient and the seriousness of the surgery. In fact, Anesthetic C was more often used in serious operations on older patients in poor condition. The death rate would be higher among these patients no matter what anesthetic they received. After measuring the lurking variables and adjusting for their effect, the apparent relationship between anesthetic and death rate is very much weaker.

EXAMPLE 4.5

A study of housing conditions in the city of Hull, England, measured a large number of variables for each of the wards in the city. Two of the variables were a measure x of overcrowding and a measure y of the lack of indoor toilets. Because x and y are both measures of inadequate housing, we expect a high correlation. In fact the correlation was only $r = .08$. How can this be?

Investigation found that some poor wards had a lot of public housing. These wards had high values of x but low values of y because public housing always includes indoor toilets. Other poor wards lacked public housing, and these wards had high values of both x and y. Within wards of each type, there

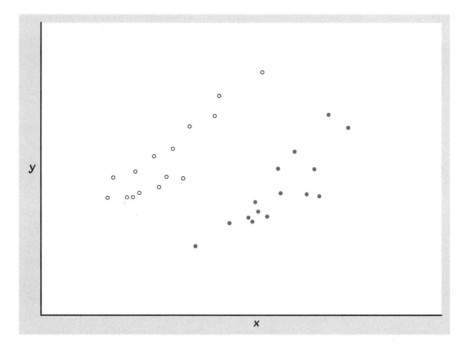

FIGURE 4.7   The variables in this scatterplot have a small correlation even though there is a strong correlation within each of the two clusters.

was a strong positive association between x and y. Analyzing all wards together ignored the lurking variable—amount of public housing—and hid the nature of the relationship between x and y.[17]

Figure 4.7 shows in simplified form how groups formed by a lurking variable can make correlation and regression misleading. The groups appear as clusters of points in the scatterplot. There is a strong relationship between x and y within each of the clusters. In fact, $r = .85$ and $r = .91$ in the two clusters. However, because similar values of x correspond to quite different values of y in the two clusters, x alone is of little value for predicting y. The correlation for all the points together is only $r = .14$.

EXERCISE

4.20   A group of college students believes that herbal tea has remarkable pow-ers. To test this belief, they make weekly visits to a local nursing home, where they visit with the residents and serve them herbal tea. The nurs-ing home staff reports that after several months many of the residents are more cheerful and healthy. A skeptical sociologist commends the stu-dents for their good deeds but scoffs at the idea that herbal tea helped the residents. Identify the explanatory and response variables in this informal study. Then explain what lurking variables account for the observed association.

Using averaged data

Many regression or correlation studies work with averages or other measures that combine information from many individuals. You should note this carefully and resist the temptation to apply the results of such studies to individuals. We have seen, starting with Figure 3.2 (page 116), a strong relationship between outside temperature and the Sanchez household's natural gas consumption. Each point on the scatterplot represents a month. Both degree-days and gas consumed are averages over all the days in the month. Data for individual days would show more scatter about the regression line and lower correlation. Averaging over an entire month smoothes out the day-to-day variation due to doors left open, houseguests using more gas to heat water, and so on. *Correlations based on averages are usually too high when applied to individuals.* This is another reminder that it is important to note exactly what variables were measured in a statistical study.

EXERCISE

4.21   The data in Exercise 3.37 (page 150) give the average steps per second for a group of top female runners at each of several running speeds. There is a high positive correlation between steps per second and speed. Suppose that you had the full data, which record steps per second for each runner separately at each speed. If you plotted each individual observation and computed the correlation, would you expect the correlation to be lower than, about the same as, or higher than the correlation for the published data? Why?

Association is not causation

When we study the relationship between two variables, we often want to show that changes in the explanatory variable *cause* changes in the response variable. But a strong association between two variables is not enough to draw conclusions about cause and effect. Sometimes an observed association really does reflect cause and effect. The Sanchez household uses more natural gas in colder months because cold weather requires burning more gas to stay warm. In other cases, an association is explained by lurking variables, and the conclusion that x causes y is either wrong or not proved.

EXAMPLE 4.6   An article in a women's magazine reported that mothers who nurse their babies feel more receptive toward their infants than mothers who bottle-feed. The author concluded that breast-feeding (x) leads to a more positive attitude (y) toward the child. But women choose whether to nurse or bottle-feed, and this choice may reflect already existing attitudes toward their infants. Perhaps mothers who already feel more positive about the child choose to nurse, while those to whom the baby is a nuisance more often choose the bottle. The mother's already established attitude is a lurking variable that prevents conclusions about whether breast-feeding itself changes mothers' attitudes.

A strong association between two variables x and y can reflect any of several underlying relationships.

- *Causation:* Changes in x cause changes in y—for example, a drop in outdoor temperature causes an increase in natural gas consumption for heating. If we can change x, we can bring about a change in y. Quitting smoking will reduce a person's chance of getting lung cancer if causation holds.

- *Common response:* Both x and y respond to changes in some unobserved variable or variables—for example, both GRE scores and graduate GPA respond to a student's ability and level of knowledge. In this case y can sometimes be predicted from x, but intervening to change x would not bring about a change in y. The genetic hypothesis claims that smoking behavior and lung cancer are both responses to a genetic predisposition; quitting smoking does not change heredity and will have no effect on future lung cancer.

- *Confounding:* The effect on y of the explanatory variable x is hopelessly mixed up with the effects on y of other variables. Minority students, for example, have lower average scores on college entrance exams such as the SAT than do whites; but minorities (again on the average) grew up in poorer households and attended poorer schools than did whites. The effects of social and economic conditions on test scores are mixed together in a way that makes any cause-and-effect conclusion suspect. Figure 4.8 illustrates these three kinds of relationship in schematic form. Both common response and confounding involve the influence of a lurking variable (or variables) z on the response variable y that explains the observed association at least in part.

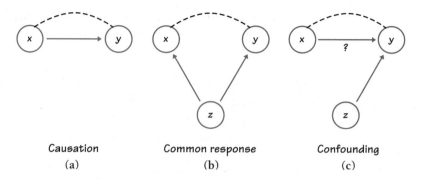

FIGURE 4.8   Variables *x* and *y* show a strong association (*dashed line*). This association may be the result of any of several causal relationships (*solid arrows*). (a) Causation: Changes in *x* cause changes in *y*. (b) Common response: Changes in both *x* and *y* are caused by changes in a lurking variable *z*. (c) Confounding: The effect (if any) of *x* on *y* is confounded with the effect of a lurking variable *z*.

experiment

    The best way to get good evidence that *x* causes *y* is to do an **experiment** in which *x* is changed and lurking variables are kept under control. We might, for example, flip a coin to decide which new mothers would nurse their infants and which would use a bottle. That eliminates the influence of the mothers' attitudes and helps us see the effects of the feeding method. Of course, this experiment is morally and legally impossible.

    When experiments cannot be done, finding the explanation for an observed association is often difficult and controversial. Data show that a company pays women less than men. The issue goes to court, where the plaintiffs claim that the pay differences are the result of sex discrimination. The company replies that pay depends on other variables, including education, experience, and seniority in the company. These are lurking variables if we only look at sex and pay. Both sides hire statisticians, who try to show that the difference in pay between men and women is or is not explained by the other variables. Which is true depends on the facts in the particular case.

ASSOCIATION DOESN'T IMPLY CAUSATION

An association between an explanatory variable *x* and a response variable *y*, even if it is very strong, is not by itself good evidence that changes in *x* actually cause changes in *y*.

EXAMPLE 4.7

Sometimes claims that x causes y are clearly wrong. For example, there is a strong negative correlation between teachers' salaries and students' scores on the SAT over time. The columnist William F. Buckley pointed this out and seemed to claim that paying teachers more causes lower SAT scores. "Teachers' salaries in 1945 averaged \$14,770, and by 1992 had risen to \$35,334. The results? In 1950, SAT math scores were 493; in 1992, down to 476."[18] But the 1992 SAT was taken by a much broader group of students because many more students plan to go to college than in 1950. And think of the social changes between 1950 and 1992 that probably reduce SAT scores: television, more divorce, more births to unmarried mothers, and so on. Society changes over time, so that a correlation between variables that are both changing over time is often meaningless. (Buckley did restate 1950 salaries in terms of 1992 buying power. Otherwise the effect of inflation in driving up salaries would make the negative correlation even stronger.)

EXERCISES

For Exercises 4.22 through 4.24, answer the question. State whether the relationship between the two variables involves causation, common response, or confounding. Identify possible lurking variable(s). Draw a diagram of the relationship in which each circle represents a variable. Write a brief description of the variable by each circle.

4.22   Someone says, "There is a strong positive correlation between the number of firefighters at a fire and the amount of damage the fire does. So sending lots of firefighters just causes more damage." Why is this reasoning wrong?

4.23   A study of elementary school children, ages 6 to 11, finds a high positive correlation between shoe size x and score y on a test of reading comprehension. What explains this correlation?

4.24   A study shows that there is a positive correlation between the size of a hospital (measured by its number of beds x) and the median number of days y that patients remain in the hospital. Does this mean that you can shorten a hospital stay by choosing a small hospital?

SUMMARY

Correlation and regression should be **interpreted with caution.** Remember to **plot the data.** Watch out for the effects of **extreme observations** and remember that correlation and regression describe **only linear** relations.

Avoid **extrapolation**, which is use of a regression line or curve for prediction for values of the explanatory variable outside the domain of the data from which the line or curve was calculated.

Lurking variables that you did not measure can explain the relations between the variables you did measure. Correlation and regression can be misleading if you ignore important lurking variables.

Remember that **correlations based on averages** are usually too high when applied to individuals.

Most of all, be careful not to conclude that there is a cause-and-effect relationship between two variables just because they are strongly associated. The relationship could involve common response or confounding. **High correlation does not imply causation.** The best evidence that an association is due to causation comes from an **experiment** in which the explanatory variable is directly changed and other influences on the response are controlled.

SECTION 4.2 EXERCISES

4.25   Table 3.1 (page 108) gives education data for the states. The correlation between the median SAT math scores and the median SAT verbal scores for the states is $r = .9620$.

(a) Find r^2 and explain in simple language what this number tells us.

(b) If you calculated the correlation between the SAT math and verbal scores of a large number of individual students, would you expect the correlation to be about 0.96 or quite different? Explain your answer.

For Exercises 4.26 through 4.29, carry out the instructions. Then state whether the relationship between the two variables involves causation, common response, or confounding. Then identify possible lurking variable(s). Draw a diagram of the relationship in which each circle represents a variable. By each circle, write a brief description of the variable.

4.26   There is a strong positive correlation between years of education and income for economists employed by business firms. (In particular, economists with doctorates earn more than economists with only a bachelor's degree.) There is also a strong positive correlation between years of education and income for economists employed by colleges and universities. But when all economists are considered, there is a *negative* correlation between education and income. The explanation for this is that business pays high salaries and employs mostly economists with bachelor's degrees, while colleges pay lower salaries and employ mostly economists with doctorates. Sketch a scatterplot with two groups of cases (business and academic) illustrating how a strong positive correlation within each group and a negative overall correlation can occur together. (Hint: Begin by studying Figure 4.7.)

4.27   Members of a high school language club believe that study of a foreign language improves a student's command of English. From school records, they obtain the scores on an English achievement test given to all seniors. The mean score of seniors who studied a foreign language for at least two years is much higher than the mean score of seniors who studied no foreign language. The club's advisor says that these data are not good evidence that language study strengthens English skills. Identify the explanatory and response variables in this study. Then explain what lurking variable prevents the conclusion that language study improves students' English scores.

4.28   There is a strong positive correlation between years of schooling completed x and lifetime earnings y for American men. One possible reason for this association is causation: more education leads to higher-paying jobs. But lurking variables may explain some of the correlation. Suggest some lurking variables that would explain why men with more education earn more.

4.29   A study of London double-decker bus drivers and conductors found that drivers had twice the death rate from heart disease as conductors. Because drivers sit while conductors climb up and down stairs all day, it was at first thought that this association reflected the effect of physical activity on heart disease. Then a look at bus company records showed that drivers were issued consistently larger-size uniforms when hired than were conductors. This fact suggested an alternative explanation of the observed association between job type and deaths. What is it?

4.3   RELATIONS IN CATEGORICAL DATA

To this point we have concentrated on relationships in which at least the response variable was quantitative. Now we will shift to describing relationships between two or more categorical variables. Some variables—such as sex, race, and occupation—are inherently categorical. Other categorical variables are created by grouping values of a quantitative variable into classes. Published data are often reported in grouped form to save space. To analyze categorical data, we use the *counts* or *percents* of individuals that fall into various categories.

TABLE 4.6	Years of school completed, by age, 1995 (thousands of persons)

| | Age group | | | |
Education	25 to 34	35 to 54	55 and over	Total
Did not complete high school	5,325	9,152	16,035	30,512
Completed high school	14,061	24,070	18,320	56,451
1 to 3 years of college	11,659	19,926	9,662	41,247
4 or more years of college	10,342	19,878	8,005	38,225
Total	41,388	73,028	52,022	166,438

EXAMPLE 4.8

two-way table

row and column variables

Table 4.6 presents Census Bureau data on the years of school completed by Americans of different ages. Many people under 25 years of age have not completed their education, so they are left out of the table. Both variables, age and education, are grouped into categories. This is a *two-way table* because it describes two categorical variables. Education is the *row variable* because each row in the table describes people with one level of education. Age is the *column variable* because each column describes one age group. The entries in the table are the counts of persons in each age-by-education class. Although both age and education in this table are categorical variables, both have a natural order from least to most. The order of the rows and the columns in Table 4.6 reflects the order of the categories.

Marginal distributions

How can we best grasp the information contained in Table 4.6? First, *look at the distribution of each variable separately*. The distribution of a categorical variable just says how often each outcome occurred. The "Total" column at the right of the table contains the totals for each of the rows. These row totals give the distribution of education level (the row variable) among all people over 25 years of age: 30,512,000 did not complete high school, 56,451,000 finished high school but did not attend college, and so on. In the same way, the "Total" row at the bottom gives the age distribution. If the row and column totals are missing, the first thing to do in studying a two-way table is to calculate them. The distributions of education alone and age alone are often called *marginal distributions* because they appear at the right and bottom margins of the two-way table.

marginal distributions

roundoff error

If you check the column totals in Table 4.6, you will notice a few discrepancies. For example, the sum of the entries in the "25 to 34" column is 41,387. The entry in the "Total" row for that column is 41,388. The explanation is *roundoff error*. The table entries are in thousands of persons, and each is rounded to the nearest thousand. The Census Bureau obtained the "Total" entry by rounding the exact number of people age 25 to 34 to the nearest thousand. The result was 41,388,000. Adding the column entries, each of which is already rounded, gives a slightly different result.

Percents are often easier to grasp than counts. We can display the marginal distribution of education level in terms of percents by dividing each row total by the table total and converting to a percent.

EXAMPLE 4.9

The percent of people over 25 years of age who have at least 4 years of college is

$$\frac{\text{total with 4 years of college}}{\text{table total}} = \frac{38,225}{166,438} = .230 = 23.0\%$$

Do three more such calculations to obtain the marginal distribution of education level in percents. Here it is.

Education	Did not finish high school	Completed high school	1–3 years of college	≥4 years of college
Percent	18.3	33.9	24.8	23.0

The total is 100% because everyone is in one of the four education categories.

In working with two-way tables, you must calculate percents—lots of them. Here's a tip to help decide what fraction gives the percent you want. Ask, "What group represents the total that I want a percent of?" The count for that group is the denominator of the fraction that leads to the percent. In Example 4.9, we wanted a percent "of people over 25 years of age," so the count of people over 25 (the table total) is the denominator.

EXERCISES

4.30   Sum the counts in the "35 to 54" age column in Table 4.6. Then explain why the sum is not the same as the entry for this column in the "Total" row.

4.31   Give the marginal distribution of age among people over 25 years of age in percents, starting from the counts in Table 4.6.

4.32   Here are data from eight high schools on smoking among students and among their parents:[19]

	Student smokes	Student does not smoke
Both parents smoke	400	1380
One parent smokes	416	1823
Neither parent smokes	188	1168

(a) How many students do these data describe?

(b) What percent of these students smoke?

(c) Give the marginal distribution of parents' smoking behavior, both in counts and in percents.

Describing relationships

The marginal distributions of age and of education separately do not tell us how the two variables are related. That information is in the body of the table. How can we describe the relationship between age and years of school completed? No single graph (such as a scatterplot) portrays the form of the relationship between categorical variables, and no single numerical measure (such as the correlation) summarizes the strength of an association. *To describe relationships among categorical variables, calculate appropriate percents from the counts given.* We use percents because counts are often hard to compare. For example, 10,342,000 people age 25 to 34 have completed college, and only 8,005,000 people in the 55 and over age group have done so. But the older age group is larger, so we can't directly compare these counts.

EXAMPLE 4.10   What percent of people age 25 to 34 have completed 4 years of college? This is the count who are 25 to 34 and have 4 years of college as a percent of the age group total:

$$\frac{10,342}{41,388} = .250 = 25.0\%$$

"People age 25 to 34" is the group we want a percent of, so the count for that group is the denominator. In the same way, the percent of people in the 55 and over age group who completed college is

$$\frac{8,005}{52,022} = .154 = 15.4\%$$

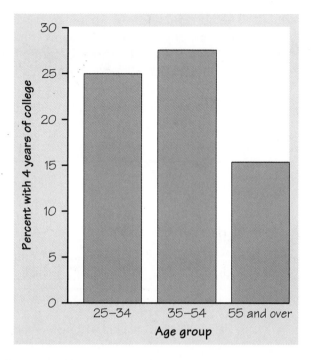

FIGURE 4.9 Bar chart comparing the percents of three age groups who have completed 4 or more years of college. The height of each bar is the percent of people in one age group who have 4 years of college.

Here are the results for all three age groups:

Age group	25 to 34	35 to 54	55 and over
Percent with 4 years of college	25.0	27.2	15.4

These percents help us see how the education of Americans varies with age. Older people are less likely to have completed college, but the data don't show a steady drop in college graduates as age increases. The percent of college graduates does not drop off sharply until we reach ages over 55.

bar chart

Although graphs are not as useful for describing categorical variables as they are for quantitative variables, a graph still helps an audience to grasp the data quickly. You can use a ***bar chart*** like Figure 4.9 to present the information in Example 4.10. Each bar represents one age group. The height of the bar is the percent of that age group with at least 4 years of college. Although bar charts look a bit like histograms, their details and uses are different. A histogram shows the distribution of the values of a quantitative variable. A bar chart compares the sizes of different items.

The horizontal axis of a bar chart need not have any measurement scale but may simply identify the items being compared. The items compared in Figure 4.9 are the three age groups. Because each bar in a bar chart describes a different item, we draw the bars with space between them.

Example 4.10 does not compare the complete distributions of years of schooling in the three age groups. It compares only the percents who finished college. Let's look at the complete picture.

EXAMPLE 4.11

Information about the 25 to 34 age group occupies the first column in Table 4.6. To find the complete distribution of education in this age group, look only at that column. Compute each count as a percent of the column total: 41,388. Here is the distribution:

Education	Did not finish high school	Completed high school	1–3 years of college	≥ 4 years of college
Percent	12.9	34.0	28.2	25.0

These percents add to 100% (with some roundoff error), because all 25- to 34-year-olds fall in one of the educational categories. The four percents together are the ***conditional distribution*** of education, given that a person is 25 to 34 years of age. We use the term "conditional" because the distribution refers only to people who satisfy the condition that they are 25 to 34 years old.

conditional distribution

For comparison, here is the conditional distribution of years of school completed among people age 55 and over. To find these percents, look only at the "55 and over" column in Table 4.6. The column total is the denominator for each percent calculation.

Education	Did not finish high school	Completed high school	1–3 years of college	≥ 4 years of college
Percent	30.8	35.2	18.6	15.4

The percent who did not finish high school is much higher in the older age group, and the percents with some college and who finished college are much lower. Comparing the conditional distributions of education in different age groups describes the association between age and education. There are three different conditional distributions of education given age, one for each of the three age groups. All of these conditional distributions differ from the marginal distribution of education found in Example 4.9.

In Example 4.11 we compared the education of different age groups. That is, we thought of age as the explanatory variable and education as the response variable. We might also be interested in the distribution of age among persons having a certain level of education. To do this, look only at one *row* in Table 4.6. Calculate each entry in that row as a percent of the row total, the total of that education group. The result is another conditional distribution, the conditional distribution of age given a certain level of education.

A two-way table contains a great deal of information in compact form. Making that information clear almost always requires finding percents. You must decide which percents you need. If you are studying trends in the training of the American workforce, comparing the distributions of education for different age groups reveals the more extensive education of younger people. If, on the other hand, you are planning a program to improve the skills of people who did not finish high school, the age distribution within this educational group is important information.

EXERCISES

4.33   Using the counts in Table 4.6, find the percent of people in each age group who did not complete high school. Draw a bar chart that compares these percents. State briefly what the data show.

4.34   Verify that the results for the conditional distribution of education level among people age 55 and over given in Example 4.11 are correct.

4.35   Example 4.11 gives the conditional distributions of education level among 25- to 34-year-olds and among people 55 and over. Find the conditional distribution of education level among 35- to 54-year-olds in percents. Is this distribution more like the distribution for 25- to 34-year-olds or the distribution for people 55 and over?

4.36   Find the conditional distribution of age among people with at least 4 years of college.

4.37   Here are the row and column totals for a two-way table with two rows and two columns:

	Col. 1	Col. 2	Total
Row 1			50
Row 2			50
Total	60	40	100

Find *two different* sets of four entries for the body of the table that give these same totals. This shows that the relationship between the two variables can't be described by their marginal distributions.

4.38   Is high blood pressure dangerous? Medical researchers classified each of a group of men as "high" or "low" blood pressure, then watched them for 5 years. (Men with systolic blood pressure 140 mm Hg or higher were "high"; the others, "low.") The following two-way table gives the results of the study:[20]

	Died	Survived
Low blood pressure	21	2655
High blood pressure	55	3283

(a) How many men took part in the study? What percent of these men died during the 5 years of the study?

(b) The two categorical variables in the table are blood pressure (high or low) and outcome (died or survived). Which is the explanatory variable?

(c) Is high blood pressure associated with a higher death rate? Calculate and compare percents to answer this question.

4.39   Do the smoking habits of parents help explain whether or not their children smoke? Exercise 4.32 gives these data from eight Arizona high schools:

	Student smokes	Student does not smoke
Both parents smoke	400	1380
One parent smokes	416	1823
Neither parent smokes	188	1168

(a) What percent of students smoke among those with two smoking parents, among those with one smoking parent, and among those with neither parent smoking?

(b) Draw a bar chart that compares the three percents of students who smoke that you found in (a).

(c) Briefly describe the relationship between parents' smoking and students' smoking.

Simpson's paradox

As is the case with quantitative variables, the effects of lurking variables can change or even reverse relationships between two categorical variables. Here is a hypothetical example that demonstrates the surprises that can await the unsuspecting user of data.

EXAMPLE 4.12

To help consumers make informed decisions about health care, the government releases data about patient outcomes in hospitals. You want to compare Hospital A and Hospital B, which serve your community. Here is a two-way table of data on the survival of patients after surgery in these two hospitals. All patients undergoing surgery in a recent time period are included. "Survived" means that the patient lived at least 6 weeks following surgery.

	Hospital A	Hospital B
Died	63	16
Survived	2037	784
Total	2100	800

The evidence seems clear: Hospital A loses 3% (63/2100) of its surgery patients, and Hospital B loses only 2% (16/800). It seems that you should choose Hospital B if you need surgery.

Not all surgery cases are equally serious, however. Later in the government report you find data on the outcome of surgery broken down by the condition of the patient before the operation. Patients are classified as being in either "poor" or "good" condition. Here are the more detailed data. Check that the entries in the original two-way table are just the sums of the "poor" and "good" entries in this pair of tables.

Good Condition			Poor Condition		
	Hospital A	Hospital B		Hospital A	Hospital B
Died	6	8	Died	57	8
Survived	594	592	Survived	1443	192
Total	600	600	Total	1500	200

Aha! Hospital A beats Hospital B for patients in good condition: only 1% (6/600) died in Hospital A, compared with 1.3% (8/600) in Hospital B. And Hospital A wins again for patients in poor condition, losing 3.8% (57/1500) to Hospital B's 4% (8/200). So Hospital A is safer for both patients in good condition and patients in poor condition. If you are facing surgery, you should choose Hospital A.

The patient's condition is a lurking variable when we compare the death rates at the two hospitals. When we ignore the lurking variable, Hospital B seems safer, even though Hospital A does better for both classes of patients. How can A do better in each group, yet do worse overall? Look at the data. Hospital A is a medical center that attracts seriously ill patients from a wide region. It had 1500 patients in poor condition. Hospital B had only 200 such cases. Because patients in poor condition are more likely to die, Hospital A has a higher death rate despite its superior performance for each class of patients. The original two-way table, which did not take account of the condition of the patients, was misleading. Example 4.12 illustrates *Simpson's paradox*.

SIMPSON'S PARADOX

Simpson's paradox refers to the reversal of the direction of a comparison or an association when data from several groups are combined to form a single group.

The lurking variables in Simpson's paradox are categorical. That is, they break the individuals into groups, as when surgery patients are classified as "good condition" or "poor condition." Simpson's paradox is just an extreme form of the fact that observed associations can be misleading when there are lurking variables. Although the hospital data in Example 4.12 were made up, the setting is real. The hospital death rates compiled and released by the Federal Health Care Financing Agency do allow for the patient's condition and diagnosis. Hospitals continue to complain that there are differences among patients that can make the government's mortality data misleading.

EXERCISES

4.40   Upper Wabash Tech has two professional schools, business and law. Here are two-way tables of applicants to both schools, categorized by sex and admission decision. (Although these data are made up, similar situations occur in reality.[21])

	Business			Law	
	Admit	Deny		Admit	Deny
Male	480	120	Male	10	90
Female	180	20	Female	100	200

(a) Make a two-way table of sex by admission decision for the two professional schools together by summing entries in this table.

(b) From the two-way table, calculate the percent of male applicants who are admitted and the percent of female applicants who are admitted. Wabash admits a higher percent of male applicants.

(c) Now compute separately the percents of male and female applicants admitted by the business school and by the law school. Each school admits a higher percent of female applicants.

(d) This is Simpson's paradox: both schools admit a higher percent of the women who apply, but overall Wabash admits a lower percent of female applicants than of male applicants. Explain carefully, as if speaking to a skeptical reporter, how it can happen that Wabash appears to favor males when each school individually favors females.

4.41   Whether a convicted murderer gets the death penalty seems to be influenced by the race of the victim. Here are data on 326 cases in which the defendant was convicted of murder:[22]

White Defendant			*Black Defendant*		
	Death penalty			*Death penalty*	
	Yes	No		Yes	No
White victim	19	132	White victim	11	52
Black victim	0	9	Black victim	6	97

(a) Use these data to make a two-way table of defendant's race (white or black) versus death penalty (yes or no).

(b) Show that Simpson's paradox holds: a higher percent of white defendants are sentenced to death overall, but for both black and white victims a higher percent of black defendants are sentenced to death.

(c) Use the data to explain why the paradox holds in language that a judge could understand.

SUMMARY

A **two-way table** of counts describes the relationship between two categorical variables. Values of the **row variable** label the rows of the table, and values of the **column variable** label the columns. Two-way tables are often used to summarize large amounts of data by grouping outcomes into categories.

You can present various aspects of the information in a two-way table by calculating and comparing **percents** from the counts in the table. Start each calculation by deciding what count represents the total that you want a percent of. That count is the denominator of the fraction you must calculate.

The **row totals** and **column totals** in a two-way table give the **marginal distributions** of the two variables separately. They do not give any information about the relationship between the variables.

To find the **conditional distribution** of the row variable for one specific value of the column variable, look only at that one column in the table. Find each entry in the column as a percent of the column total. There is a conditional distribution of the row variable for each column in the table. Comparing these conditional distributions is one way of showing the association between the row and the column variables.

Compare the conditional distributions of the row variable for the different categories of the column variable when you think of the column variable as the explanatory variable. If the row variable is your explanatory variable, look at each row separately and show the association by comparing the conditional distributions of the column variable in the rows.

A comparison between two variables that holds for each individual value of a third variable can be changed or even reversed when the data for all values of the third variable are combined. This is **Simpson's paradox.**

SECTION 4.3 EXERCISES

Exercises 4.42 to 4.45 are based on Table 4.7. This two-way table reports Census Bureau data on undergraduate students enrolled in U.S. colleges and universities in the fall of 1991.

4.42   (a) How many students aged 30 to 44 were enrolled part-time at 2-year colleges?

        (b) How many students of all ages were enrolled part-time at 2-year colleges?

        (c) The sum of the entries in the "2-year part-time" column is not equal to the "Total" entry for that column. How do you explain this difference?

4.43   (a) How many undergraduate students were enrolled in colleges and universities?

	2-year full-time	2-year part-time	4-year full-time	4-year part-time
Age				
15–17	44	4	79	0
18–21	1,345	456	3,869	159
22-29	489	690	1,358	494
30–44	287	704	289	627
≥ 45	49	209	62	160
Total	2,212	2,065	5,657	1,440

TABLE 4.7 Undergraduate college enrollment by age of students—autumn, 1991 (thousands of students)

SOURCE: Current Population Survey, October 1991.

(b) What percent of all undergraduate students were 18 to 21 years old in the fall of the academic year?

(c) Find the percent of the undergraduates enrolled in each of the four types of program who were 18 to 21 years old. Make a bar chart to compare these percents.

(d) The 18 to 21 group is the "traditional" age group for college students. Briefly summarize what you have learned from the data about the extent to which this group predominates in different kinds of college programs.

4.44 (a) What percent of students enrolled part-time at 2-year colleges are 30 to 44 years old?

(b) What percent of all 30- to 44-year-old students are enrolled part-time at 2-year colleges?

4.45 (a) Find the marginal distribution of age among all undergraduate students, first in counts and then in percents.

(b) Find the conditional distribution of age (in percents) among students enrolled part-time in 2-year colleges.

(c) Briefly describe the most important differences between the two age distributions.

4.46   Do child restraints and seat belts prevent injuries to young passengers in automobile accidents? Here are data on the 26,971 passengers under the age of 15 in accidents reported in North Carolina during two years before the law required restraints:[23]

	Restrained	Unrestrained
Injured	197	3,844
Uninjured	1,749	21,181

(a)   What percent of these young passengers were restrained?

(b)   Do the data provide evidence that young passengers are less likely to be injured in an accident if they wear restraints? Calculate and compare percents to answer this question.

In Exercises 4.47 and 4.48 determine whether each situation represents causation, confounding, or common response. Draw a diagram of the relationship in which each circle represents a variable. Next to each circle, write a brief description of the variable.

4.47   There is a negative correlation between the number of flu cases reported each week throughout the year and the amount of ice cream sold in that particular week. It's unlikely that ice cream prevents flu. What is a more plausible explanation for this observed correlation?

4.48   Over the past 30 years in the United States there has been a strong negative correlation between number of infant deaths at birth and number of people over age 65.

4.49   A business school conducted a survey of companies in its state. They mailed a questionnaire to 200 small companies, 200 medium-sized companies, and 200 large companies. The rate of nonresponse is important in deciding how reliable survey results are. Here are the data on response to this survey:

	Response	No response	Total
Small	125	75	200
Medium	81	119	200
Large	40	160	200

(a)   What was the overall percent of nonresponse?

(b)   Describe how nonresponse is related to the size of the business. (Use percents to make your statements precise.)

(c) Draw a bar chart to compare the nonresponse percents for the three size categories.

4.50 Most baseball hitters perform differently against right-handed and left-handed pitching. Consider two players, Joe and Moe, both of whom bat right-handed. The table below records their performance against right-handed and left-handed pitchers.

Player	Pitcher	Hits	At bats
Joe	Right	40	100
	Left	80	400
Moe	Right	120	400
	Left	10	100

(a) Make a two-way table of player (Joe or Moe) versus outcome (hit or no hit) by summing over both kinds of pitcher.

(b) Find the overall batting average (hits divided by total times at bat) for each player. Who has the higher batting average?

(c) Make a separate two-way table of player versus outcome for each kind of pitcher. From these tables, find the batting averages of Joe and Moe against right-handed pitching. Who does better? Do the same for left-handed pitching. Who does better?

(d) The manager doesn't believe that one player can hit better against both left-handers and right-handers yet have a lower overall batting average. Explain in simple language why this happens to Joe and Moe.

CHAPTER REVIEW

In Chapter 3, we learned how to analyze two-variable data that show a linear pattern. We learned about positive and negative associations and how to measure the strength of association between two variables. We also developed a procedure for constructing a model (the least-squares regression line) that captures the trend of the data. This LSRL is useful for prediction purposes. A recurring theme is that data analysis begins with graphs and then adds numerical summaries of specific aspects of the data.

In this chapter we learned how to construct mathematical models for data that fit an exponential function or a power function through the origin. We also learned that although correlation and regression are powerful tools for understanding two-variable data when both variables are quantitative, both correlation and regression have their limitations. In particular, we are cautioned that a strong observed association between two variables may exist without a cause-and-effect link between them. If both variables are categorical, there is no satisfactory graph for displaying the data, although bar charts can be helpful. We describe the relationship by comparing percents.

Here is a review list of the most important skills you should have gained from studying this chapter.

A. MODELING NONLINEAR DATA

1. Recognize that when a variable is multiplied by a fixed number greater than 1 in each equal time period, exponential growth results; when the ratio is a positive number less than 1, it's called exponential decay.

2. Recognize that when one variable is proportional to a power of a second variable, the result is a power function.

3. In the case of both exponential growth and power function, perform a logarithmic transformation and obtain points that lie in a linear pattern. Then use least-squares regression on the transformed points. An inverse transformation then produces a curve that is a model for the original points.

4. Know that deviations from the overall pattern are most easily examined by fitting a line to the transformed points and plotting the residuals from this line against the explanatory variable (or fitted values).

B. INTERPRETING CORRELATION AND REGRESSION

1. Understand that both r and the least-squares regression line can be strongly influenced by a few extreme observations.

2. Recognize possible lurking variables that may explain the observed association between two variables x and y.

3. Understand that even a strong correlation does not mean that there is a cause-and-effect relationship between x and y.

C. RELATIONS IN CATEGORICAL DATA

1. From a two-way table of counts, find the marginal distributions of both variables by obtaining the row sums and column sums.

2. Express any distribution in percents by dividing the category counts by their total.

3. Describe the relationship between two categorical variables by computing and comparing percents. Often this involves comparing the conditional distributions of one variable for the different categories of the other variable.

CHAPTER 4 REVIEW EXERCISES

4.51   (Exact exponential growth) The common intestinal bacterium *E. coli* is one of the fastest-growing bacteria. Under ideal conditions, the number of *E. coli* in a colony doubles about every 15 minutes until restrained by lack of resources. Starting from a single bacterium, how many *E. coli* will there be in 1 hour? In 5 hours?

4.52   (Exact exponential growth) A clever courtier, offered a reward by an ancient king of Persia, asked for a grain of rice on the first square of a chess board, 2 grains on the second square, then 4, 8, 16, and so on.

(a) Make a table of the number of grains on each of the first 10 squares of the board.

(b) Plot the number of grains on each square against the number of the square for squares 1 to 10, and connect the points with a smooth curve. This is an exponential curve.

(c) How many grains of rice should the king deliver for the 64th (and final) square?

(d) Take the logarithm of each of your numbers of grains from (a). Plot these logarithms against the number of squares from 1 to 10. You should get a straight line.

(e) From your graph in (d) find the approximate values of the slope b and the intercept a for the line. Use the equation $y = a + bx$ to predict the logarithm of the amount for the 64th square. Check your result by taking the logarithm of the amount you found in (c).

4.53   (Exact exponential growth) Alice is given a savings bond at birth. The bond is initially worth $500 and earns interest at 7.5% each year. This means that the value is multiplied by 1.075 each year.

(a) Find the value of the bond at the end of 1 year, 2 years, and so on up to 10 years.

(b) Plot the value y against years x on graph paper. Connect the points with a smooth curve. This is an exponential curve.

(c) Take the logarithm of each of the values y that you found in (a). Plot the logarithm $\log y$ against years x on graph paper. You should obtain a straight line.

4.54   Fred and Alice were born the same year, and each began life with $500. Fred added $100 each year, but earned no interest. Alice added nothing, but earned interest at 7.5% annually. After 25 years, Fred and Alice are getting married. Who has more money?

4.55   Federal expenditures on social insurance (chiefly social security and Medicare) increased rapidly after 1960. Here are the amounts spent, in millions of dollars:

Year	1960	1965	1970	1975	1980	1985	1990
Spending	14,307	21,807	45,246	99,715	191,162	310,175	422,257

(a) Plot social insurance expenditures against time. Does the pattern appear closer to linear growth or to exponential growth?

(b) Take the logarithm of the amounts spent. Plot these logarithms against time. Do you think that the exponential growth model fits well?

(c) After entering the data into the Minitab statistical system, with year as C1 and expenditures as C2, we obtain the least-squares line for the logarithms as follows:

```
MTB> LET C3 = LOGT(C2)
MTB> REGRESS C3 ON 1, C1

The regression equation is
C3 = -98.63833 + 0.05244 C1
```

That is, the least-squares line is

$$\log y = -98.63833 + (.05244 \times \text{year})$$

Draw this line on your graph from (b).

(d) Use this line to predict the logarithm of social insurance outlays for 1988. Then compute

$$y = 10^{\log y}$$

to predict the amount y spent in 1988.

(e) The actual amount (in millions) spent in 1988 was $358,412. Take the logarithm of this amount and add the 1988 point to your graph in (b). Does it fall close to the line? In 1981, President Reagan took office with a policy of slowing growth in spending on social programs. Did the trend of exponential growth in spending for social insurance change in a major way during the Reagan years 1981 to 1988?

4.56 The following table shows the growth of the population of Europe (millions of persons) between 400 B.C. and 1950:

Date	Population	Date	Population	Date	Population
400 B.C.	23	1200	61	1600	90
A.D. 1	37	1250	69	1650	103
200	67	1300	73	1700	115
700	27	1350	51	1750	125
1000	42	1400	45	1800	187
1050	46	1450	60	1850	274
1100	48	1500	69	1900	423
1150	50	1550	78	1950	594

(a) Plot population against time.

(b) The graph shows that the population of Europe dropped at the collapse of the Roman Empire (A.D. 200–500) and at the time of the Black Death (A.D. 1348). Growth has been uninterrupted since 1400. Compute the ratios of consecutive terms to test for exponential growth between 1400 and 1950.

(c) Transform the data to obtain a linear plot. Then perform least-squares linear regression, and report the correlation for the transformed points.

(d) Perform an inverse transformation to yield an appropriate model for the population data. Record the equation for your model and show the scatterplot and prediction curve plotted together. Describe the growth of the population.

(e) Why did we restrict the test for exponential growth to the period from 1400 to 1950?

4.57 The following table gives the resident population of the United States from 1790 to 1990, in millions of persons:

Date	Population	Date	Population	Date	Population
1790	3.9	1860	31.4	1930	122.8
1800	5.3	1870	39.8	1940	131.7
1810	7.2	1880	50.2	1950	151.3
1820	9.6	1890	62.9	1960	179.3
1830	12.9	1900	76.0	1970	203.3
1840	17.1	1910	92.0	1980	226.5
1850	23.2	1920	105.7	1990	248.7

(a) Plot population against time. The growth of the American population appears roughly exponential.

(b) Plot the logarithms of population against time. The pattern of growth is now clear. An expert says that "the population of the United States increased exponentially from 1790 to about 1880. After 1880 growth was still approximately exponential, but at a slower rate." Explain how this description is obtained from the graph.

(c) Use an appropriate subset of the data to construct a model for the purpose of predicting the U.S. population in the near future.

(d) Use your model to predict the population of the United States in the year 2000.

4.58   The productivity of American agriculture has grown rapidly due to improved technology (crop varieties, fertilizers, mechanization). Here are data on the output per hour of labor on American farms. The variable is an "index number" that gives productivity as a percent of the 1967 level.

Year	1940	1945	1950	1955	1960	1965	1970	1975	1980	1985
Productivity	21	27	35	47	67	91	113	137	166	217

Plot the data. Briefly describe the pattern of growth that your plot suggests. Then use the methods discussed in this chapter to construct an appropriate model that will capture the trend of the data. Show the algebraic steps and enough plots to document the model-building process.

4.59   Here are data on infant mortality (deaths per thousand live births) in the New England states:[24]

Year	1980	1981	1985	1988	1990	1991	1992	1993
Mortality	10.5	9.7	9.2	8.1	7.2	6.8	6.8	6.5

(a) Make a plot of infant mortality against time. Describe the pattern that you see.

(b) Use a transformation to construct a model for the data. Report the equation, and display the scatterplot with the curve (model) superimposed.

(c) Extend the table by recording the predicted value for each year from 1994 to 1999. Would you feel comfortable using this model to predict infant mortality for the year 2005?

4.60   An experiment was conducted with a pendulum of variable length. The *period*, or length of time to complete one complete oscillation, was recorded for several lengths. Here are the data:

Length (feet)	1	2	3	4	5	6	7
Period (seconds)	1.10	1.56	1.92	2.20	2.50	2.71	2.93

(a) Make a plot of period against length. Describe the pattern that you see.

(b) Propose a model form. Then use a transformation to construct a model for the data. Report the equation, and plot the original data with the model on the same axes.

(c) Describe the relationship between the length of a pendulum and its period.

4.61 In physics class, the intensity of a 100-watt light bulb was measured by a sensing device at various distances from the light source, and the following data were collected. Note that a *candela* (cd) is an international unit of luminous intensity.

Distance (meters)	Intensity (candelas)
1	.2965
1.1	.2522
1.2	.2055
1.3	.1746
1.4	.1534
1.5	.1352
1.6	.1145
1.7	.1024
1.8	.0923
1.9	.0832
2.0	.0734

(a) Plot the data. Based on the pattern of points, propose a model form for the data. Then use a transformation followed by linear regression and then an inverse transformation to construct a model.

(b) Report the equation, and plot the original data with the model on the same axes.

(c) Describe the relationship between the intensity and the distance from the light source.

(d) Consult the physics textbooks used in your school and find the formula for the intensity of light as a function of distance from the light source. How do your experimental results compare with the theoretical formula?

4.62   Does taking aspirin regularly help prevent heart attacks? The Physicians' Health Study tried to find out. The subjects were 22,071 healthy male doctors at least 40 years old. Half the subjects, chosen at random, took aspirin every other day. The other half took a placebo, a dummy pill that looked and tasted like aspirin. Here are the results.[25] (The row for "None of these" is left out of the two-way table.)

	Aspirin group	Placebo group
Fatal heart attacks	10	26
Other heart attacks	129	213
Strokes	119	98
Total	11,037	11,034

What do the data show about the association between taking aspirin and heart attacks and stroke? Use percents to make your statements precise. Do you think the study provides evidence that aspirin actually reduces heart attacks (cause and effect)?

4.63   Here is a two-way table of suicides committed in 1990, categorized by the sex of the victim and the method used. Based on these data, write a brief account of differences in suicide between men and women. Be sure to cite appropriate counts or percents to justify your statements.

	Male	Female
Firearms	16,285	2,600
Poison	3,221	2,203
Hanging	3,688	756
Other	1,530	623
Total	24,724	6,182

4.64   Cocaine addiction is hard to break. Addicts need cocaine to feel any pleasure, so perhaps giving them an antidepressant drug will help. A 3-year study with 72 chronic cocaine users compared an antidepressant drug called desipramine with lithium and a placebo. (Lithium is a standard drug to treat cocaine addiction. A placebo is a dummy drug, used so that the effect of being in the study but not taking any drug can be seen.) One-third of the subjects, chosen at random, received each drug. Here are the results.[26]

	Cocaine relapse?	
	Yes	No
Desipramine	10	14
Lithium	18	6
Placebo	20	4

(a) Compare the effectiveness of the three treatments in preventing relapse. Use percents and draw a bar chart.

(b) Do you think that this study gives good evidence that desipramine actually *causes* a reduction in relapses?

4.65 The following two-way table describes the age and marital status of American women in 1991. The table entries are in thousands of women.

	Marital status				
Age	Single	Married	Widowed	Divorced	Total
18–24	9,008	3,352	8	257	12,627
25–39	6,658	21,769	248	3,224	31,899
40–64	1,975	24,462	2,570	4,755	33,762
≥65	900	7,255	8,464	925	17,545
Total					95,833

(a) Find the sum of the entries in the 18–24 row. Why does this sum differ from the "Total" entry for that row?

(b) Give the marginal distribution of marital status for all adult women (use percents). Draw a bar chart to display this distribution.

(c) Compare the conditional distributions of marital status for women aged 18 to 24 and women aged 40 to 64. Briefly describe the most important differences between the two groups of women, and back up your description with percents.

(d) You are planning a magazine aimed at single women who have never been married. (That's what "single" means in government data.) Find the conditional distribution of ages among single women.

4.66 A study by the National Science Foundation[27] found that the median salary of newly graduated female engineers and scientists was only 73% of the median salary for males. When the new graduates were broken down by field, however, the picture changed. Women earned at least 84% as much as men in *every* field of engineering and science. The median salary for women was higher than that of men in many engineering disciplines. How can women earn nearly as much as men in every field yet fall far behind men when we look at all young engineers and scientists?

4.67   Recent studies have shown that earlier reports underestimated the health risks associated with being overweight. The error was due to overlooking some important variables. In particular, smoking tends both to reduce weight and to lead to earlier death. Illustrate Simpson's paradox by a simplified version of this situation. That is, make up a table of overweight (yes or no) by early death (yes or no) by smoker (yes or no) such that

- Overweight smokers and overweight nonsmokers both tend to die earlier than those not overweight.
- But when smokers and nonsmokers are combined into a two-way table of overweight by early death, persons who are not overweight tend to die earlier.

NOTES AND DATA SOURCES

1. This activity is described in Mako E. Haruta et al., "Coin tossing," *Mathematics Teacher*, 89, no. 8 (Nov. 1996), pp. 642–645.

2. Oil production data for 1880 to 1972 are from the Energy Information Administration, recorded in Robert H. Romer, *Energy: An Introduction to Physics*, W. H. Freeman, San Francisco, 1976; for more recent years, see *Statistical Abstracts of the United States*.

3. Gypsy moth data provided by Chuck Schwalbe, U.S. Department of Agriculture.

4. Public debt data from the Bureau of Public Debt, U.S. Department of the Treasury, as reported in *The World Almanac and Book of Facts, 1997*.

5. Fish data from Gordon L. Swartzman and Stephen P. Kaluzny, *Ecological Simulation Primer*, Macmillan, New York, 1987, p. 98.

6. Postal rates data from the *Statistical Abstract of the United States, 1997*.

7. Salary data provided by the Major League Baseball Players Association. Visit the Web site www.canoe.ca/Baseball/salaries.html.

8. EFT data from the *Statistical Abstracts of the United States, 1997*.

9. Data provided by *Scott Standard Postage Stamp Catalog*, vol. 1, Amos Press, Signey, Ohio, 1989. The exercise was suggested in an article by David Kullman in *Mathematics Teacher*, 85, no. 3 (March 1992), pp. 188–189.

10. Population density data from *The World Almanac and Book of Facts, 1997*.

11. Health care expenditure data from the *Statistical Abstracts of the United States, 1997*.

12. Violent crime data from the U.S. Federal Bureau of Investigation, *Crime in the United States* annual.

13. Data originally from A. J. Clark, *Comparative Physiology of the Heart*, Macmillan, New York, 1927, p. 84. Obtained from Frank R. Giordano and Maurice D. Weir, *A First Course in Mathematical Modeling*, Brooks/Cole, Belmont, Calif., 1985, p. 56.

14. Data from Little Caesar's Pizza in Ashland, Va.

15. Olympic weight-lifting results from the *Sports Illustrated Sports Almanac, 1997*.

16. L. E. Moses and F. Mosteller, "Safety of anesthetics," in J. Tanur et al. (eds.), *Statistics: A Guide to the Unknown*, 3rd ed., Wadsworth, 1989, pp. 15–24.

17. This example is drawn from M. Goldstein, "Preliminary inspection of multivariate data," *The American Statistician*, 36 (1982), pp. 358–362.

18. From his column of June 28, 1993.

19. From S. V. Zagona (ed.), *Studies and Issues in Smoking Behavior*, University of Arizona Press, 1967, pp. 157–180.

20. From J. Stamler, "The mass treatment of hypertensive disease: defining the problem," *Mild Hypertension: To Treat or Not to Treat*, New York Academy of Sciences, 1978, pp. 333–358.

21. See P. J. Bickel and J. W. O'Connell, "Is there a sex bias in graduate admissions?" *Science*, 187 (1975), pp. 398–404.

22. From M. Radelet, "Racial characteristics and imposition of the death penalty," *American Sociological Review*, 46 (1981), pp. 918–927.

23. Adapted from data of Williams and Zador in *Accident Analysis and Prevention*, 9 (1977), pp. 69–76.

24. From National Center for Health Statistics, *Monthly Vital Statistics Report*, December 1983.

25. Reported in the *New York Times*, July 20, 1989, from an article appearing that day in the *New England Journal of Medicine*.

26. From D. M. Barnes, "Breaking the cycle of addiction," *Science*, 241 (1988), pp. 1029–1030.

27. National Science Board, *Science and Engineering Indicators, 1991*, U.S. Government Printing Office, 1991. The detailed data appear in Appendix Table 3-5, p. 274.

Producing Data: Samples and Experiments

■ 5. Producing Data

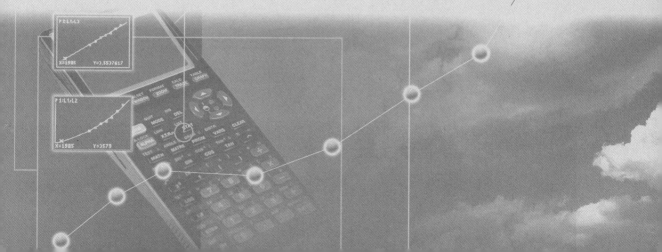

Producing Data

ACTIVITY 5A | A Class Survey

A class survey is a quick way to collect interesting data. Certainly there are things about the class as a group that you would like to know. Your task here is to construct a *draft* of a class survey, a questionnaire that would be used to gather data about the members of your class. Here are the steps to take:

1. As a class, discuss the questions you would like to include on the survey. In addition to *what* you want to ask, you should also consider *how many questions* you want to ask. Have one student serve as recorder and make a list on the blackboard or overhead projector of topics to include.

2. Once you have identified the topics, then work on the wording of the questions. Try to achieve as much consensus as possible. If there is a computer in the room, a student could use a word-processing program to enter the questions as they are developed.

3. Make one copy of the final draft of the survey for each student, but do not distribute the surveys at this time. The surveys are to be put aside for the time being. As you complete this chapter, you will return to take another look at the survey you have constructed, make final adjustments, and then administer the survey to all of the members of your class. This survey should provide some interesting data that can be analyzed during the remainder of the course.

As a starting point, here is a sample of a short survey:

CLASS SURVEY

Your answers to the questions below will help describe your class. DO NOT PUT YOUR NAME ON THIS PAPER. Your answers are completely private. They just help us describe the entire class.

1. Are you MALE or FEMALE? (Circle one.)
2. How many brothers and sisters do you have? _____
3. How tall are you in inches, to the nearest inch? _____
4. Estimate the number of pairs of shoes you own. _____
5. How much money in coins are you carrying right now? (Don't count any paper money, just coins.) _____
6. On a typical school night, how much time do you spend doing homework? (Answer in minutes. For example, 2 hours is 120 minutes.) _____
7. On a typical school night, how much time do you spend watching television? (Answer in minutes.) _____

INTRODUCTION

Exploratory data analysis seeks to discover and summarize what data say by using graphs and numerical summaries. The conclusions we draw from data analysis apply to the specific data that we examine. Often we want to extend those conclusions to some larger group of individuals. If our data don't fairly represent the larger group, our conclusions from the data don't apply to the larger group.

EXAMPLE 5.1

The advice columnist Ann Landers once asked her readers, "If you had it to do over again, would you have children?" A few weeks later, her column was headlined "70% OF PARENTS SAY KIDS NOT WORTH IT." Indeed, 70% of the nearly 10,000 parents who wrote in said they would not have children if they could make the choice again.

These data are worthless as indicators of opinion among all American parents. The people who responded felt strongly enough to take the trouble to write Ann Landers. Their letters showed that many of them were angry at their children. These people don't fairly represent all parents. It is not surprising that a statistically designed opinion poll on the same issue a few months later found that 91% of parents *would* have children again. Ann Landers announced a 70% "No" result when the truth about parents was close to 90% "Yes."

sampling   Both Ann Landers's write-in poll and the more careful opinion poll that contradicted her are examples of **sampling.** The idea is to study a part—a sample—in order to gain information about an entire group. Ann Landers used a *voluntary response sample.*

VOLUNTARY RESPONSE SAMPLE

A **voluntary response sample** consists of people who choose themselves by responding to a general appeal.

Voluntary response samples overrepresent people with strong opinions, most often negative opinions. If sample data are to give reliable information about a larger group of people or things, we must think carefully about how to choose the sample.

experiment   In other settings, we gather data from an **experiment.** In doing an experiment, we don't just observe individuals or ask them questions. We actively impose some treatment in order to observe the response. If the data we gather are to be trustworthy, we must design the experiment carefully.

EXAMPLE 5.2   In 1940, a psychologist conducted an experiment to study the effect of propaganda on attitude toward a foreign government. He made up a test of attitude toward the German government and administered it to a group of American students. After they read German propaganda for several months, he tested the students again to see if their attitudes had changed.

Unfortunately, Germany attacked and conquered France while the experiment was in progress. The students did change their attitudes toward the German government between the test and the retest—but we shall never know how much of the change was due to the explanatory variable (reading propaganda) and how much to the historical events of that time. The data give no information about the effect of reading propaganda.

In Example 5.2, the effects of the explanatory variable are hopelessly mixed up with the effects of the events of history. We say that the variables are *confounded.* Unless experiments are carefully designed, the effects of the explanatory variables can't be seen because of confounding with lurking variables.

CONFOUNDING

Two variables (explanatory variables or lurking variables) are
confounded when their effects on a response variable cannot be
distinguished from each other.

*statistical
inference*

In both Examples 5.1 and 5.2, the purpose of producing data was to
answer specific questions: What percent of parents would have children
again? Does reading propaganda change attitudes toward a foreign government? ***Statistical inference*** provides ways to answer specific questions from
data with some guarantee that the answers are good ones. Inference is the
topic of much of the rest of this book. Voluntary response and confounding can both make it impossible to answer the questions we had in mind.
These examples make it clear that when we do statistical inference we
must think about how to *produce* data as well as about how to *analyze*
data. If we design our production of data well, inference will be straightforward and the results will be convincing. Section 5.1 presents statistical
ideas for choosing samples that will yield trustworthy data. Section 5.2
discusses how to design experiments. Section 5.3 introduces methods of
simulating experiments. The ideas in these sections are simple, but they
are among the most important in statistics.

5.1   DESIGNING SAMPLES

An opinion poll wants to know what fraction of the public approves the
president's performance in office. A quality engineer must estimate what
fraction of the bearings rolling off an assembly line are defective. Government economists inquire about household income. In all these situations,
we want to gather information about a large group of people or things.
Time, cost, and inconvenience usually forbid inspecting every bearing or
contacting every household. In such cases, we gather information about
only part of the group in order to draw conclusions about the whole.

POPULATION, SAMPLE

The entire group of individuals that we want information about is
called the **population.**
A **sample** is a part of the population that we actually examine in
order to gather information.

Notice that the population is defined in terms of our desire for knowledge. If we wish to draw conclusions about all U.S. college students, that group is our population even if only local students are available for questioning. The sample is the part from which we draw conclusions about *sample design* the whole. The **design** of a sample refers to the method used to choose the sample from the population. Poor sample designs can produce misleading conclusions. Voluntary response (Example 5.1) is one common type *convenience* of bad sample design. Another is **convenience sampling**, which chooses the *sampling* individuals easiest to reach. Here is an example of convenience sampling.

EXAMPLE 5.3

Manufacturers and advertising agencies often use interviews at shopping malls to gather information about the habits of consumers and the effectiveness of ads. A sample of mall shoppers is fast and cheap. "Mall interviewing is being propelled primarily as a budget issue," one expert told the *New York Times*. But people contacted at shopping malls are not representative of the entire U.S. population. They are richer, for example, and more likely to be teenagers or retired. Moreover, mall interviewers tend to select neat, safe-looking individuals from the stream of customers. Decisions based on mall interviews may not reflect the preferences of all consumers.[1]

Both voluntary response samples and convenience samples choose a sample that is almost guaranteed not to represent the entire population. These sampling methods display *bias*, or systematic error, in favoring some parts of the population over others.

BIAS

The design of a study is **biased** if it systematically favors certain outcomes.

EXERCISES

5.1   A sociologist wants to know the opinions of employed adult women about government funding for day care. She obtains a list of the 520 members of a local business and professional women's club and mails a questionnaire to 100 of these women selected at random. Only 48 questionnaires are returned. What is the population in this study? What is the sample?

5.2   For each of the following sampling situations, identify the population as exactly as possible. That is, say what kind of individuals the population consists of and say exactly which individuals fall in the population. If the information given is not complete, complete the description of the population in a reasonable way.

(a) Each week, the Gallup Poll questions a sample of about 1500 adult U.S. residents to determine national opinion on a wide variety of issues.

(b) The 1990 census tried to gather basic information from every household in the United States. But a "long form" requesting much additional information was sent to a sample of about 17% of households.

(c) A machinery manufacturer purchases voltage regulators from a supplier. There are reports that variation in the output voltage of the regulators is affecting the performance of the finished products. To assess the quality of the supplier's production, the manufacturer sends a sample of 5 regulators from the last shipment to a laboratory for study.

5.3 A newspaper advertisement for *USA Today: The Television Show* once said:

> *Should handgun control be tougher? You call the shots in a special call-in poll tonight. If yes, call 1-900-720-6181. If no, call 1-900-720-6182. Charge is 50 cents for the first minute.*

Explain why this opinion poll is almost certainly biased.

5.4 You are on the staff of a member of Congress who is considering a bill that would provide government-sponsored insurance for nursing home care. You report that 1128 letters have been received on the issue, of which 871 oppose the legislation. "I'm surprised that most of my constituents oppose the bill. I thought it would be quite popular," says the congresswoman. Are you convinced that a majority of the voters oppose the bill? How would you explain the statistical issue to the congresswoman?

Simple random samples

In a voluntary response sample, people choose whether to respond. In a convenience sample, the interviewer makes the choice. In both cases, personal choice produces bias. The statistician's remedy is to allow impersonal chance to choose the sample. A sample chosen by chance allows neither favoritism by the sampler nor self-selection by respondents. Choosing a sample by chance attacks bias by giving all individuals an equal chance to be chosen. Rich and poor, young and old, black and white, all have the same chance to be in the sample.

The simplest way to use chance to select a sample is to place names in a hat (the population) and draw out a handful (the sample). This is the idea of *simple random sampling*.

SIMPLE RANDOM SAMPLE

A **simple random sample (SRS)** of size n consists of n individuals from the population chosen in such a way that every set of n individuals has an equal chance to be the sample actually selected.

An SRS not only gives each individual an equal chance to be chosen (thus avoiding bias in the choice) but also gives every possible sample an equal chance to be chosen. There are other random sampling designs that give each individual, but not each sample, an equal chance. Exercise 5.22 describes one such design, called systematic random sampling.

The idea of an SRS is to choose our sample by drawing names from a hat. In practice, computer software can choose an SRS almost instantly from a list of the individuals in the population. If you don't use software, you can randomize by using a *table of random digits*.

RANDOM DIGITS

A **table of random digits** is a long string of the digits 0, 1, 2, 3, 4, 5, 6, 7, 8, 9 with these two properties:

1. Each entry in the table is equally likely to be any of the 10 digits 0 through 9.
2. The entries are independent of each other. That is, knowledge of one part of the table gives no information about any other part.

Table B at the back of the book and inside the rear cover is a table of random digits. You can think of Table B as the result of asking an assistant (or a computer) to mix the digits 0 to 9 in a hat, draw one, then replace the digit drawn, mix again, draw a second digit, and so on. The assistant's mixing and drawing save us the work of mixing and drawing when we need to choose at random. The entries in Table B appear in groups of five, but this is only to make the table easier to read. The rows are numbered, but this just makes it easier to say where you started in the table. These conveniences do not affect the nature of the table as a long string of randomly chosen digits. Because the digits in Table B are random:

- Each entry is equally likely to be any of the 10 possibilities 0, 1, . . . , 9.
- Each pair of entries is equally likely to be any of the 100 possible pairs 00, 01, . . . , 99.
- Each triple of entries is equally likely to be any of the 1000 possibilities 000, 001, . . . , 999, and so on.

These "equally likely" facts make it easy to use Table B to choose an SRS. Here is an example that shows how.

EXAMPLE 5.4

Joan's small accounting firm serves 30 business clients. Joan wants to interview a sample of 5 clients in detail to find ways to improve client satisfaction. To avoid bias, she chooses an SRS of size 5.

Step 1: Label. Give each client a numerical label, using as few digits as possible. Two digits are needed to label 30 clients, so we use labels

$$01, 02, 03, \ldots, 29, 30$$

It is also correct to use labels 00 to 29 or even another choice of 30 two-digit labels. Here is the list of clients, with labels attached:

01	A-1 Plumbing	02	Accent Printing
03	Action Sport Shop	04	Anderson Construction
05	Bailey Trucking	06	Balloons Inc.
07	Bennett Hardware	08	Best's Camera Shop
09	Blue Print Specialties	10	Central Tree Service
11	Classic Flowers	12	Computer Answers
13	Darlene's Dolls	14	Fleisch Realty
15	Hernandez Electronics	16	Johnson Commodities
17	JL Records	18	Keiser Construction
19	Liu's Chinese Restaurant	20	MagicTan
21	Peerless Machine	22	Photo Arts
23	River City Books	24	Riverside Tavern
25	Rustic Boutique	26	Satellite Services
27	Scotch Wash	28	Sewer's Center
29	Tire Specialties	30	Von's Video Store

Step 2: Table. Enter Table B anywhere and read two-digit groups. Suppose we enter at line 130, which is

69051 64817 87174 09517 84534 06489 87201 97245

The first 10 two-digit groups in this line are

69 05 16 48 17 87 17 40 95 17

Each successive two-digit group is a label. The labels 00 and 31 to 99 are not used in this example, so we ignore them. The first 5 labels between 01 and 30 that we encounter in the table choose our sample. Of the first 10 labels in line 130, we ignore 5 because they are too high (over 30). The others are 05, 16, 17, 17, and 17. The clients labeled 05, 16, and 17 go into the sample. Ignore the second and third 17s because that client is already in the sample. Now run your finger across line 130 (and continue to line 131 if needed) until 5 clients are chosen.

The sample is the clients labeled 05, 16, 17, 20, 19. These are Bailey Trucking, Johnson Commodities, JL Records, MagicTan, and Liu's Chinese Restaurant.

CHOOSING AN SRS

Choose an SRS in two steps:

Step 1: Label. Assign a numerical label to every individual in the population.

Step 2: Table. Use Table B to select labels at random.

Don't try to scramble the labels as you assign them. Table B will do the required randomizing. You can assign labels in any convenient manner, such as alphabetical order for names of people. Be certain that all labels have the same number of digits. Only then will all individuals have the same chance to be chosen. Use the shortest possible labels: one digit for a population of up to 10 members, 2 digits for 11 to 100 members, three digits for 101 to 1000 members, and so on. As standard practice, we recommend that you begin with label 1 (or 01, or 001, as needed). You can read digits from Table B in any order—across a row, down a column, and so on—because the table has no order. As standard practice, we recommend reading across rows.

EXERCISES

5.5   A firm wants to understand the attitudes of its minority managers toward its system for assessing management performance. Below is a list of all the firm's managers who are members of minority groups. Use Table B at line 139 to choose 6 to be interviewed in detail about the performance appraisal system.

Agarwal	Gates	Peters
Anderson	Goel	Pliego
Baxter	Gomez	Puri
Bonds	Hernandez	Richards
Bowman	Huang	Rodriguez
Castillo	Kim	Santiago
Cross	Liao	Shen
Dewald	Mourning	Vega
Fernandez	Naber	Wang
Fleming		

5.6 Your class in ancient Ugaritic religion is poorly taught and wants to complain to the dean. The class decides to choose 4 of its members at random to carry the complaint. The class list appears below. Choose an SRS of 4 using the table of random digits, beginning at line 145.

Anderson	Gupta	Patnaik
Aspin	Gutierrez	Pirelli
Bennett	Harter	Rao
Bock	Henderson	Rider
Breiman	Hughes	Robertson
Castillo	Johnson	Rodriguez
Dixon	Kempthorne	Sosa
Edwards	Laskowsky	Tran
Gonzalez	Liang	Trevino
Green	Olds	Wang

5.7 You must choose an SRS of 10 of the 440 retail outlets in New York that sell your company's products. How would you label this population? Use Table B, starting at line 105, to choose your sample.

Other sampling designs

The general framework for sampling is a *probability sample*.

PROBABILITY SAMPLE

A **probability sample** gives each member of the population a known chance (greater than zero) to be selected.

Some probability sampling designs (such as an SRS) give each member of the population an equal chance to be selected. This may not be true in more elaborate sampling designs. In every case, however, the use of chance to select the sample is the essential principle of statistical sampling.

Designs for sampling from large populations spread out over a wide area are usually more complex than an SRS. For example, it is common to sample important groups within the population separately, then combine these samples. This is the idea of a *stratified sample*.

STRATIFIED RANDOM SAMPLE

To select a **stratified random sample**, first divide the population into groups of similar individuals, called **strata**. Then choose a separate SRS in each stratum and combine these SRSs to form the full sample.

Choose the strata based on facts known before the sample is taken. For example, a population of election districts might be divided into urban, suburban, and rural strata. A stratified design can produce more exact information than an SRS of the same size by taking advantage of the fact that individuals in the same stratum are similar to one another. If all individuals in each stratum are identical, for example, just one individual from each stratum is enough to completely describe the population.

EXAMPLE 5.5

A radio or television station that broadcasts a piece of music owes a royalty to the composer. ASCAP (the American Society of Composers, Authors, and Publishers) sells licenses that permit broadcast of works by any of its members. ASCAP must then pay the proper royalties to the composers whose music was played. Television networks keep program logs of all music played, but local radio and television stations do not. Because there are over a billion ASCAP-licensed performances each year, a detailed accounting is too expensive and cumbersome. Here is a case for sampling.

ASCAP divides its royalties among its members by taping a stratified sample of broadcasts. The sample of local commercial radio stations, for example, consists of 60,000 hours of broadcast time each year. Radio stations are stratified by type of community (metropolitan, rural), geographic location (New England, Pacific, etc.), and the size of the license fee paid to ASCAP, which reflects the size of the audience. In all, there are 432 strata. Tapes are made at random hours for randomly selected members of each stratum. The tapes are reviewed by experts who can recognize almost every piece of music ever written, and the composers are then paid according to their popularity.[2]

Another common means of restricting random selection is to choose the sample in stages. This is usual practice for national samples of households or people. For example, government data on employment and unemployment are gathered by the Current Population Survey, which conducts interviews in about 60,000 households each month. It is not practical to maintain a list of all U.S. households from which to select an SRS. Moreover, the cost of sending interviewers to the widely scattered households in an SRS would be too high. The Current Population Survey therefore uses a **multistage sample design.** The final sample consists of clusters of nearby households. Most opinion polls and other national samples are also multistage, though interviewing in most national samples today is done by telephone rather than in person, eliminating the economic need for clustering.

multistage sample

A national multistage sample proceeds somewhat as follows:

Stage 1. Take a sample from the 3000 counties in the United States.

Stage 2. Select a sample of townships within each of the counties chosen.

Stage 3. Select a sample of blocks within each chosen township.

Stage 4. Take a sample of households within each block.

Lists of counties and townships are easily available. From that point, the sampling can be based on a local map if necessary. We don't need a national list of households. Moreover, the sample households occur in clusters in the same block and so are easy for a single interviewer to contact. The sample at any stage of a multistage design may be an SRS. Stratified samples are also common—for example, we might group counties into rural, suburban, and urban strata before sampling.

Analysis of data from sampling designs more complex than an SRS takes us beyond basic statistics. But the SRS is the building block of more elaborate designs, and analysis of other designs differs more in complexity of detail than in fundamental concepts.

EXERCISES

5.8 A club has 30 student members and 10 faculty members. The students are

Abel	Fisher	Huber	Miranda	Reinmann
Carson	Ghosh	Jimenez	Moskowitz	Santos
Chen	Griswold	Jones	Neyman	Shaw
David	Hein	Kim	O'Brien	Thompson
Deming	Hernandez	Klotz	Pearl	Utts
Elashoff	Holland	Liu	Potter	Varga

The faculty members are

Andrews	Fernandez	Kim	Moore	West
Besicovitch	Gupta	Lightman	Phillips	Yang

The club can send 4 students and 2 faculty members to a convention. It decides to choose those who will go by random selection. Use Table B to choose a stratified random sample of 4 students and 2 faculty members.

5.9   Accountants often use stratified samples during audits to verify a company's records of such things as accounts receivable. The stratification is based on the dollar amount of the item and often includes 100% sampling of the largest items. One company reports 5000 accounts receivable. Of these, 100 are in amounts over $50,000; 500 are in amounts between $1000 and $50,000; and the remaining 4400 are in amounts under $1000. Using these groups as strata, you decide to verify all of the largest accounts and to sample 5% of the midsize accounts and 1% of the small accounts. How would you label the two strata from which you will sample? Use Table B, starting at line 115, to select *only the first 5* accounts from each of these strata.

Cautions about sample surveys

Random selection eliminates bias in the choice of a sample from a list of the population. When the population consists of human beings, however, accurate information from a sample requires much more than a good sampling design.[3] To begin, we need an accurate and complete list of the population. Because such a list is rarely available, most samples suffer from some degree of *undercoverage*. A sample survey of households, for example, will miss not only homeless people but prison inmates and students in dormitories. An opinion poll conducted by telephone will miss the 7% to 8% of American households without residential phones. The results of national sample surveys therefore have some bias if the people not covered—who most often are poor people—differ from the rest of the population.

A more serious source of bias in most sample surveys is *nonresponse*, which occurs when a selected individual cannot be contacted or refuses to cooperate. Nonresponse to sample surveys often reaches 30% or more, even with careful planning and several callbacks. Because nonresponse is higher in urban areas, most sample surveys substitute other people in the same area to avoid favoring rural areas in the final sample. If the people contacted differ from those who are rarely at home or who refuse to answer questions, some bias remains.

UNDERCOVERAGE AND NONRESPONSE

Undercoverage occurs when some groups in the population are left out of the process of choosing the sample.

Nonresponse occurs when an individual chosen for the sample can't be contacted or refuses to cooperate.

EXAMPLE 5.6

Even the 1990 census, backed by the resources and authority of the federal government, suffered from undercoverage and nonresponse. The census begins by mailing forms to every household in the country. The Census Bureau buys lists of addresses from private firms, then tries to fill in missing addresses. The final list is still incomplete, resulting in undercoverage. Despite special efforts to count homeless people (who can't be reached at any address), homelessness causes more undercoverage.

About 35% of the households who were mailed census forms did not mail them back. In New York City, 47% did not return the form. That's nonresponse. The Census Bureau sends interviewers to these households. In central cities, the interviewers could not contact about one in five of the nonresponders, even after six tries.

The Census Bureau estimates that the 1990 census missed about 1.6% of the total population due to undercoverage and nonresponse. Because the undercount was greater in the poorer sections of large cities, the Census Bureau estimates that it failed to count 4.6% of blacks and 5.0% of Hispanics.[4]

response bias

In addition, the behavior of the respondent or of the interviewer can cause *response bias* in sample results. Respondents may lie, especially if asked about illegal or unpopular behavior. The sample then underestimates the presence of such behavior in the population. An interviewer whose attitude suggests that some answers are more desirable than others will get these answers more often. The race or sex of the interviewer can influence responses to questions about race relations or attitudes toward feminism. Answers to questions that ask respondents to recall past events are often inaccurate because of faulty memory. For example, many people "telescope" events in the past, bringing them forward in memory to more recent time periods. "Have you visited a dentist in the last 6 months?" will often draw a "Yes" from someone who last visited a dentist 8 months ago.[5] Careful training of interviewers and careful supervision to avoid variation among the interviewers can greatly reduce response bias. Good interviewing technique is another aspect of a well-done sample survey.

wording effects

The *wording of questions* is the most important influence on the answers given to a sample survey. Confusing or leading questions can introduce strong bias, and even minor changes in wording can change a survey's outcome. Leading questions are common in surveys sponsored by companies and intended to persuade rather than inform. Here are two examples.[6]

EXAMPLE 5.7

When Levi Strauss & Co. asked college students to choose the most popular clothing item from a list, 90% chose Levi's 501 jeans—but they were the only jeans listed.

A survey paid for by makers of disposable diapers found that 84% of the sample opposed banning disposable diapers. Here is the actual question:

> *It is estimated that disposable diapers account for less than 2% of the trash in today's landfills. In contrast, beverage containers, third-class mail and yard wastes are estimated to account for about 21% of the trash in landfills. Given this, in your opinion, would it be fair to ban disposable diapers?*

This question gives information on only one side of an issue, then asks an opinion. That's a sure way to bias the responses. A different question that described how long disposable diapers take to decay and how many tons they contribute to landfills each year would draw a quite different response.

Never trust the results of a sample survey until you have read the exact questions posed. The sampling design, the amount of nonresponse, and the date of the survey are also important. Good statistical design is a part, but only a part, of a trustworthy survey.

EXERCISES

5.10

sampling frame

The list of individuals from which a sample is actually selected is called the **sampling frame.** Ideally, the frame should list every individual in the population, but in practice this is often difficult. A frame that leaves out part of the population is a common source of undercoverage.

(a) Suppose that a sample of households in a community is selected at random from the telephone directory. What households are omitted from this frame? What types of people do you think are likely to live in these households? These people will probably be underrepresented in the sample.

(b) It is more common in telephone surveys to use random digit dialing equipment that selects the last four digits of a telephone number at random after being given the exchange (the first three digits). Which of the households you mentioned in your answer to (a) will be included in the sampling frame by random digit dialing?

5.11   A common form of nonresponse in telephone surveys is "ring-no-answer." That is, a call is made to an active number but no one answers. The Italian National Statistical Institute looked at nonresponse to a government survey of households in Italy during the periods January 1 to Easter and July 1 to August 31. All calls were made between 7 and 10 p.m., but 21.4% gave "ring-no-answer" in one period versus 41.5% "ring-no-answer" in the other period.[7] Which period do you think had the higher rate of no answers? Why? Explain why a high rate of nonresponse makes sample results less reliable.

5.12   Here are two wordings for the same question:

A. Should laws be passed to eliminate all possibilities of special interests giving huge sums of money to candidates?

B. Should laws be passed to prohibit interest groups from contributing to campaigns, or do groups have a right to contribute to the candidates they support?

One of these questions drew 40% favoring banning contributions; the other drew 80% with this opinion.[8] Which question produced the 40% and which got 80%? Explain why the results were so different.

Inference about the population

Despite the many practical difficulties in carrying out a sample survey, using chance to choose a sample does eliminate bias in the actual selection of the sample from the list of available individuals. But it is unlikely that results from a sample are exactly the same as for the entire population. Sample results, like the official unemployment rate obtained from the monthly Current Population Survey, are only estimates of the truth about the population. If we select two samples at random from the same population, we will draw different individuals. So the sample results will almost certainly differ somewhat. Two runs of the Current Population Survey would produce somewhat different unemployment rates. Properly designed samples avoid systematic bias, but their results are rarely exactly correct and they vary from sample to sample.

How accurate is a sample result like the monthly unemployment rate? We can't say for sure, because the result would be different if we took another sample. But the results of random sampling don't change haphazardly from sample to sample. Because we deliberately use chance, the results obey the laws of *probability* that govern chance behavior. We can say how large an error we are likely to make in drawing conclusions about the population from a sample. Results from a sample survey usually come

probability

with a margin of error that sets bounds on the size of the likely error. How to do this is part of the business of statistical inference. We will describe the reasoning in Chapter 10.

One point is worth making now: *larger samples give more accurate results than smaller samples.* By taking a very large sample, you can be confident that the sample result is very close to the truth about the population. The Current Population Survey's sample of 60,000 households estimates the national unemployment rate very accurately. Of course, only probability samples carry this guarantee. Ann Landers's voluntary response sample is worthless even though 10,000 people wrote in. Using a probability sampling design and taking care to deal with practical difficulties reduce bias in a sample. The size of the sample then determines how close to the population truth the sample result is likely to fall.

EXERCISE

5.13   Just before a presidential election, a national opinion polling firm increases the size of its weekly sample from the usual 1500 people to 4000 people. Why do you think the firm does this?

SUMMARY

A sample survey selects a **sample** from the **population** of all individuals about which we desire information. We base conclusions about the population on data about the sample.

The **design** of a sample refers to the method used to select the sample from the population. **Probability sampling designs** use random selection to give each member of the population a known chance (greater than zero) to be selected for the sample.

The basic probability sample is a **simple random sample (SRS)**. An SRS gives every possible sample of a given size the same chance to be chosen.

Choose an SRS by labeling the members of the population and using a **table of random digits** to select the sample.

To choose a **stratified random sample**, divide the population into **strata**, groups of individuals that are similar in some way that is important to the response. Then choose a separate SRS from each stratum.

Multistage samples select successively smaller groups within the population in stages. Each stage may employ an SRS, a stratified sample, or another type of sample.

Failure to use probability sampling often results in **bias,** or systematic errors in the way the sample represents the population. **Voluntary response samples,** in which the respondents choose themselves, are particularly prone to large bias.

In human populations, even probability samples can suffer from bias due to **undercoverage** or **nonresponse,** from **response bias,** or from misleading results due to **poorly worded questions.** Sample surveys must deal expertly with these potential problems in addition to using a probability sampling design.

SECTION 5.1 EXERCISES

5.14 Different types of writing can sometimes be distinguished by the lengths of the words used. A student interested in this fact wants to study the lengths of words used by Tom Wolfe in his novels. She opens a Wolfe novel at random and records the lengths of each of the first 250 words on the page.

What is the population in this study? What is the sample? What is the variable measured?

5.15 For each of the following sampling situations, identify the population as exactly as possible. That is, say what kind of individuals the population consists of and say exactly which individuals fall in the population. If the information given is not complete, complete the description of the population in a reasonable way.

(a) A business school researcher wants to know what factors affect the survival and success of small businesses. She selects a sample of 150 eating-and-drinking establishments from those listed in the telephone directory Yellow Pages for a large city.

(b) A member of Congress wants to know whether his constituents support proposed legislation on health care. His staff reports that 228 letters have been received on the subject, of which 193 oppose the legislation.

(c) An insurance company wants to monitor the quality of its procedures for handling loss claims from its auto insurance policyholders. Each month the company selects an SRS of all auto insurance claims filed that month to examine them for accuracy and promptness.

5.16 The author Shere Hite undertook a study of women's attitudes toward sex and love by distributing 100,000 questionnaires through women's groups. Only 4.5% of the questionnaires were returned. Based on this sample of women, Hite wrote *Women and Love*, a best-selling book claiming that women are fed up with men. For example, 91% of the divorced women who responded said that they had initiated the divorce, and 70% of the married women said that they had committed adultery.

Explain briefly why Hite's sampling method is nearly certain to produce a strong bias. Are the sample results cited (91% and 70%) much higher or much lower than the truth about the population of all adult American women?

5.17   Some television stations take quick polls of public opinion by announcing a question on the air and asking viewers to call one of two telephone numbers to register their opinion as "Yes" or "No." Telephone companies make available "900" numbers for this purpose. Dialing a 900 number results in a small charge to your telephone bill. The first major use of call-in polling was by the ABC television network in October 1980. At the end of the first Reagan-Carter presidential election debate, ABC asked its viewers which candidate won. The call-in poll proclaimed that Reagan had won the debate by a 2 to 1 margin. But a random survey by CBS News showed only a 44% to 36% margin for Reagan, with the rest undecided. Why are call-in polls likely to be biased? Can you suggest why this bias might have favored the Republican Reagan over the Democrat Carter?

5.18   A flour company wants to know what fraction of Toronto households bake some or all of their own bread. The company selects an SRS of 500 residential addresses in Toronto and sends interviewers to these addresses. The interviewers visit households only during regular working hours on weekdays. Explain why this sampling method is biased. Is the percent of the sample who bake bread probably higher or lower than the percent of the population who bake bread?

5.19   A manufacturer of chemicals chooses 3 from each lot of 25 containers of a reagent to test for purity and potency. Below are the control numbers stamped on the bottles in the current lot. Use Table B at line 111 to choose an SRS of 3 of these bottles.

A1096	A1097	A1098	A1101	A1108
A1112	A1113	A1117	A2109	A2211
A2220	B0986	B1011	B1096	B1101
B1102	B1103	B1110	B1119	B1137
B1189	B1223	B1277	B1286	B1299

5.20   Figure 5.1 is map of a census tract in a fictitious town. Census tracts are small, homogeneous areas averaging 4000 in population. On the map, each block is marked with a Census Bureau identification number. An SRS of blocks from a census tract is often the next-to-last stage in a multistage sample. Use Table B beginning at line 125 to choose an SRS of 5 blocks from this census tract.

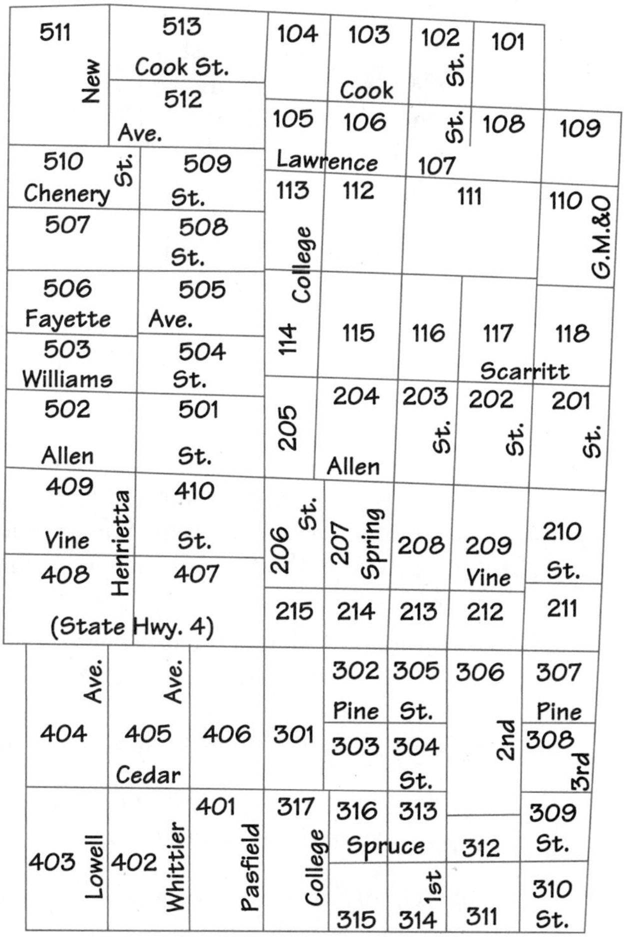

FIGURE 5.1   Map of a census tract (Exercise 5.20).

5.21   Which of the following statements are true of a table of random digits, and which are false? Briefly explain your answers.

(a)   There are exactly four 0s in each row of 40 digits.
(b)   Each pair of digits has chance 1/100 of being 00.
(c)   The digits 0000 can never appear as a group, because this pattern is not random.

systematic   5.22   Sample surveys often use a *systematic random sample* to choose a sample of
random sampling   apartments in a large building or dwelling units in a block at the last stage of a multistage sample. An example will illustrate the idea of a systematic sample.

Suppose that we must choose 4 addresses out of 100. Because 100/4 = 25, we can think of the list as four lists of 25 addresses. Choose 1 of the first 25 addresses at random using Table B. The sample contains this address and the addresses 25, 50, and 75 places down the list from it. If the table gives 13, for example, then the systematic random sample consists of the addresses numbered 13, 38, 63, and 88.

(a) Use Table B to choose a systematic random sample of 5 addresses from a list of 200. Enter the table at line 120.

(b) Like an SRS, a systematic random sample gives all individuals the same chance to be chosen. Explain why this is true. Then explain carefully why a systematic sample is nonetheless *not* an SRS.

5.23   A corporation employs 2000 male and 500 female engineers. A stratified random sample of 200 male and 50 female engineers gives each engineer 1 chance in 10 to be chosen. This sample design gives every individual in the population the same chance to be chosen for the sample. Is it an SRS? Explain your answer.

5.24   Comment on each of the following as a potential sample survey question. Is the question clear? Is it slanted toward a desired response?

(a) Which of the following best represents your opinion on gun control?

1. The government should confiscate our guns.
2. We have the right to keep and bear arms.

(b) A freeze in nuclear weapons should be favored because it would begin a much-needed process to stop everyone in the world from building nuclear weapons now and reduce the possibility of nuclear war in the future. Do you agree or disagree?

(c) In view of escalating environmental degradation and incipient resource depletion, would you favor economic incentives for recycling of resource-intensive consumer goods?

5.25   A *New York Times* opinion poll on women's issues contacted a sample of 1025 women and 472 men by randomly selecting telephone numbers. The *Times* publishes complete descriptions of its polling methods. Here is part of the description for this poll.[9]

> In theory, in 19 cases out of 20 the results based on the entire sample will differ by no more than three percentage points in either direction from what would have been obtained by seeking out all adult Americans.
>   The potential sampling error for smaller subgroups is larger. For example, for men it is plus or minus five percentage points.

Explain why the margin of error is larger for conclusions about men alone than for conclusions about all adults.

| ACTIVITY 5B | The Class Survey Revisited |

Each student should have a copy of the survey that the class constructed in Activity 5A at the beginning of the chapter. Now that you are experts on good and bad characteristics of survey questions, do the following:

1. Consider the questions in order. As you look at each item, see if the question contains bias. Does it advocate a position? Does the question contain any complicated words that might be misinterpreted? Will any questions evoke response bias?
2. Make any changes that the group feels are needed. Remember that the survey should be *anonymous* (no names on the papers) so that students are assured that the class *as a whole* rather than themselves as individuals will be described.
3. Print the final version of the survey. Make one copy for each member of the class and an extra copy on which to tally the results.
4. Each student should complete the survey.
5. Place the completed surveys, upside down, in a pile, in the middle of the room. The last student finished should shuffle the pile of surveys to ensure anonymity.
6. Designate someone (the teacher?) to tally the responses as homework and prepare a cumulative summary. Give a copy of the results to each student in the class for later analysis.

5.2 DESIGNING EXPERIMENTS

A sample survey collects information about a population by selecting and measuring a sample from the population. The goal is a picture of the population, disturbed as little as possible by the act of gathering information. Sample surveys are one kind of *observational study*.

OBSERVATION AND EXPERIMENT

An **observational study** observes individuals and measures variables of interest but does not attempt to influence the responses.

An **experiment,** on the other hand, deliberately imposes some treatment on individuals in order to observe their responses.

An observational study, even one based on a sound statistical sample, is a poor way to gauge the effect of an intervention. To see how nature responds to a change, we must actually impose the change. When our goal is to understand cause and effect, experiments are the only source of fully convincing data. Here is the basic vocabulary of experiments.

EXPERIMENTAL UNITS, SUBJECTS, TREATMENT

The individuals on which the experiment is done are the **experimental units.** When the units are human beings, they are called **subjects.** A specific experimental condition applied to the units is called a **treatment.**

factor

level

Because the purpose of an experiment is to reveal the response of one variable to changes in other variables, the distinction between explanatory and response variables is essential. The explanatory variables in an experiment are often called *factors.* Many experiments study the joint effects of several factors. In such an experiment, each treatment is formed by combining a specific value (often called a *level*) of each of the factors.

EXAMPLE 5.8   Researchers studying the absorption of a drug into the bloodstream inject the drug (the treatment) into 25 people (the subjects). The response variable is the concentration of the drug in a subject's blood, measured 30 minutes after the injection. This experiment has a single factor with only one level. If three different doses of the drug are injected, there is still a single factor (the dosage of the drug), now with three levels. The three levels of the single factor are the treatments that the experiment compares.

		Factor B Repetitions		
		1 time	3 times	5 times
Factor A Length	30 seconds	1	2	3
	90 seconds	4	5	6

FIGURE 5.2 The treatments in the experimental design of Example 5.9. Combinations of levels of the two factors form six treatments.

EXAMPLE 5.9

What are the effects of repeated exposure to an advertising message? The answer may depend both on the length of the ad and on how often it is repeated. An experiment investigated this question using undergraduate students as subjects. All subjects viewed a 40-minute television program that included ads for a 35 mm camera. Some subjects saw a 30-second commercial; others, a 90-second version. The same commercial was repeated either 1, 3, or 5 times during the program. After viewing, all of the subjects answered questions about their recall of the ad, their attitude toward the camera, and their intention to purchase it. These are the response variables.[10]

This experiment has two factors: length of the commercial, with 2 levels, and repetitions, with 3 levels. The 6 combinations of one level of each factor form 6 treatments. Figure 5.2 shows the layout of the treatments.

Examples 5.8 and 5.9 illustrate the advantages of experiments over observational studies. Experimentation allows us to study the effects of the specific treatments we are interested in. Moreover, we can control the environment of the experimental units to hold constant factors that are of no interest to us, such as the specific product advertised in Example 5.9. The ideal case is a laboratory experiment in which we control all outside factors. Like most ideals, such control is not always realized in practice. Nonetheless, a well-designed experiment makes it possible to draw conclusions about the effect of one variable on another.

Another advantage of experiments is that we can study the combined effects of several factors simultaneously. The interaction of several factors can produce effects that could not be predicted from looking at the effect of each factor alone. Perhaps longer commercials increase interest in a product, and more commercials also increase interest, but if we both make a commercial longer and show it more often, viewers get annoyed and their interest in the product drops. The two-factor experiment in Example 5.9 will help us find out.

5.26   There may be a "gender gap" in political party preference in the United States, with women more likely than men to prefer Democratic candidates. A political scientist selects a large sample of registered voters, both men and women. She asks each voter whether they voted for the Democratic or the Republican candidate in the last congressional election. Is this study an experiment? Why or why not? What are the explanatory and response variables?

5.27   A manufacturer of food products uses package liners that are sealed at the top by applying heated jaws after the package is filled. The customer peels the sealed pieces apart to open the package. What effect does the temperature of the jaws have on the force needed to peel the liner? To answer this question, engineers obtain 20 pairs of pieces of package liner. They seal five pairs of each at 250° F, 275° F, 300° F, and 325° F. Then they measure the force needed to peel each seal.

(a)   What are the experimental units?

(b)   There is one factor (explanatory variable). What is it, and what are its levels?

(c)   What is the response variable?

5.28   An educator wants to compare the effectiveness of computer software that teaches reading with that of a standard reading curriculum. She tests the reading ability of each student in a class of fourth graders, then divides them into two groups. One group uses the computer regularly, while the other studies a standard curriculum. At the end of the year, she retests all the students and compares the increase in reading ability in the two groups.

(a)   Is this an experiment? Why or why not?

(b)   What are the explanatory and response variables?

5.29   A chemical engineer is designing the production process for a new product. The chemical reaction that produces the product may have higher or lower yield, depending on the temperature and the stirring rate in the vessel in which the reaction takes place. The engineer decides to investigate the effects of combinations of two temperatures (50° C and 60° C) and three stirring rates (60 rpm, 90 rpm, and 120 rpm) on the yield of the process. She will process two batches of the product at each combination of temperature and stirring rate.

(a) What are the experimental units and the response variable in this experiment?

(b) How many factors are there? How many treatments? Use a diagram like that in Figure 5.2 to lay out the treatments.

(c) How many experimental units are required for the experiment?

Comparative experiments

The design of an experiment first describes the response variable or variables, the factors (explanatory variables), and the specific treatments. Figure 5.2 illustrates this aspect of the design of a marketing experiment. Laboratory experiments in the sciences and engineering often have a simple design with only one treatment, which is applied to all of the units. The design of such an experiment looks like this

$$\text{Treatment} \longrightarrow \text{Observation}$$

or, if before-and-after measurements are made, like this

$$\text{Observation 1} \longrightarrow \text{Treatment} \longrightarrow \text{Observation 2}$$

For example, we may subject a beam to a load (treatment) and measure how far it bends (observation). Alas, when experiments are conducted in the field or with living subjects, these simple designs often yield worthless data because of confounding. Example 5.2 (page 246) shows what can go wrong. Here is another example.

EXAMPLE 5.10

Ulcers in the upper intestine are unfortunately common in modern society. "Gastric freezing" is a clever treatment for ulcers. The patient swallows a deflated balloon with tubes attached, then a refrigerated solution is pumped through the balloon for an hour. The idea is that cooling the stomach will reduce its production of acid and so relieve ulcers. An experiment reported in the *Journal of the American Medical Association* showed that gastric freezing did reduce acid production and relieve ulcer pain. The treatment was safe and easy and was widely used for several years.

placebo effect

The gastric freezing experiment was poorly designed. The patients' response may have been due to the *placebo effect*. A placebo is a dummy treatment that can have no physical effect. Many patients respond favorably to *any* treatment, even a placebo, presumably because of trust in the doctor and expectations of a cure. This response to a dummy treatment is the placebo effect.

A second experiment, done several years later, divided ulcer patients into two groups. One group was treated by gastric freezing as before. The other group received a placebo treatment in which the solution in the balloon was at body temperature rather than freezing. The results: 34% of the 82 patients in the treatment group improved, but so did 38% of the 78 patients in the placebo group. This and other properly designed experiments showed that gastric freezing was no better than a placebo, and its use was abandoned.[11]

The original gastric freezing experiment had the simple design

$$\text{Observe pain} \longrightarrow \text{Gastric freezing} \longrightarrow \text{Observe pain}$$

The experimental data were misleading because of the placebo effect. The data also reflect any special features of that particular study, such as a physician with a soothing manner. The effect of gastric freezing is confounded with these lurking variables.

Fortunately, the remedy is simple. Experiments should *compare* treatments rather than attempt to assess a single treatment in isolation. When we compare the two groups of patients in the second gastric freezing experiment, the placebo effect and other lurking variables operate on both groups. The only difference between the groups is the actual effect of gastric freezing. The group of patients who received a sham treatment is called a *control group*, because it enables us to control the effects of lurking variables on the outcome. *Control of the effects of lurking variables is the first principle of statistical design of experiments.* Comparison of several treatments is the simplest form of control.

control group

Without comparison of treatments, experimental results in such areas as consumer behavior and medicine can be dominated by the details of the experimental arrangement, the selection of subjects, and the placebo effect. The result is often *bias*, systematic favoritism toward one outcome. An uncontrolled study of a new ulcer treatment, for example, is biased in favor of finding the treatment effective because of the placebo effect. It should not surprise you to learn that uncontrolled studies in medicine give new therapies a much higher success rate than proper comparative experiments.

bias

Completely randomized experiments

The first step in an experimental design is the choice of treatments, with comparison as the leading principle. Next, we must say how we will assign the experimental units to the treatments. Comparison of the effects of several treatments is valid only if we apply all treatments to similar groups of experimental units. If one corn variety is planted on more fertile ground, or if one TV commercial is shown to more receptive subjects, comparisons among several corn varieties or among several commercials are meaningless. Systematic differences among the groups of experimental units in a comparative experiment are a possible source of bias. How should we allocate the available units or subjects among the treatments?

matching

Experimenters often try to **match** the treatment groups in a systematic way. A comparison of two TV ads, for example, may try to match the subjects in the experimental group that will see the new ad with the subjects in the control group that will see the current ad by age, sex, race, and so on. If there is a 27-year-old black female accountant in one group, there should be a matching young black female professional in the other group. Attempts at matching are helpful but not adequate—there are too many lurking variables that might affect the outcome. The experimenter can't measure some of these variables and will not think of others until after the experiment.

The remedy is to use impersonal chance to make the assignment. Then the groups don't depend on any characteristic of the experimental units or on the judgment of the experimenter. The use of chance can be combined with matching, but the simplest design creates groups by chance alone. Here is an example.

EXAMPLE 5.11

A food company assesses the nutritional quality of a new "instant breakfast" product by feeding it to newly weaned male white rats. The response variable is a rat's weight gain over a 28-day period. A control group of rats eats a standard diet but otherwise receives exactly the same treatment as the experimental group.

This experiment has one factor (the diet) with two levels. The researchers use 30 rats for the experiment and so must divide them into two groups of 15. To do this in an unbiased fashion, put the cage numbers of the 30 rats in a hat, mix them up, and draw 15. These rats form the experimental group and the remaining 15 make up the control group. That is, *each group is an SRS of the available rats.*

In practice, we use the table of random digits to randomize. Label the rats 01 to 30. Enter Table B at (say) line 130. Run your finger along this line (and continue to lines 131 and 132 as needed) until 15 rats are chosen. They are the rats labeled

05, 16, 17, 20, 19, 04, 25, 29, 18, 07, 13, 02, 23, 27, 21

These rats form the experimental group; the remaining 15 are the control group.

---

randomization

The use of chance to divide experimental units into groups is called **randomization.** *Randomization is the second major principle of statistical design of experiments.* Combining comparison and randomization, we arrive at the simplest randomized comparative design:

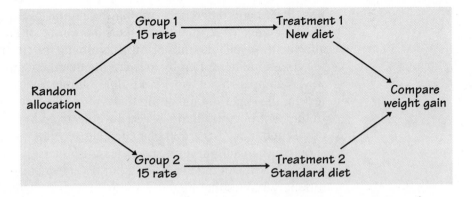

EXAMPLE 5.12

Many utility companies have introduced programs to encourage energy conservation among their customers. An electric company considers placing electronic indicators in households to show what the cost would be if the electricity use at that moment continued for a month. Will indicators reduce electricity use? Would cheaper methods work almost as well? The company decides to design an experiment.

One cheaper approach is to give customers a chart and information about monitoring their electricity use. The experiment compares these two approaches (indicator, chart) and also a control. The control group of customers receives information about energy conservation but no help in monitoring electricity use. The response variable is total electricity used in a year. The company finds 60 single-family residences in the same city willing to participate, so it assigns 20 residences at random to each of the 3 treatments. The outline of the design is

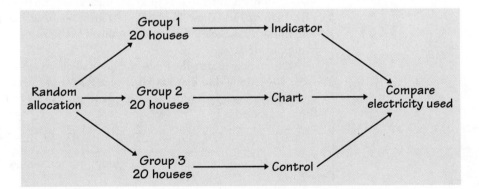

To carry out the random assignment, label the 60 households 01 to 60. Enter Table B and select an SRS of 20 to receive the indicators. Continue in Table B, selecting 20 more to receive charts. The remaining 20 form the control group.

The designs used for the experiments in Examples 5.11 and 5.12 are *completely randomized*.

COMPLETELY RANDOMIZED DESIGN

> In a **completely randomized** experimental design, all the experimental units are allocated at random among all the treatments.

Completely randomized designs can compare any number of treatments. To do this, assign the experimental units at random to as many groups as there are treatments, as in Example 5.12. The diagrams of the experiments in Examples 5.11 and 5.12 display the randomization and comparison used in the design. They also display other essential information: the treatments, the response variable, and the number of units in each group. We will see later that for statistical inference there are advantages to having equal numbers of units in all groups. You should choose groups of equal size unless there is some compelling reason not to do so.

Examples 5.11 and 5.12 involve completely randomized experimental designs to compare levels of a single factor. In Example 5.11, the factor is the diet fed to the rats. In Example 5.12, it is the method used to encourage energy conservation. Completely randomized designs can have more than one factor. The advertising experiment of Example 5.9 has two factors: the length and the number of repetitions of a television commercial. Their combinations form the six treatments outlined in Figure 5.2 (page 267). A completely randomized design assigns subjects at random to these six treatments. Once the layout of treatments is set, the randomization needed for a completely randomized design is tedious but straightforward.

EXERCISES

5.30   A large study used records from Canada's national health care system to compare the effectiveness of two ways to treat prostate disease. The two treatments are traditional surgery and a new method that does not require surgery. The records described many patients whose doctors had chosen each method. The study found that patients treated by the new method were significantly more likely to die within 8 years.[12]

(a) Further study of the data showed that this conclusion was wrong. The extra deaths among patients who got the new method could be explained by lurking variables. What lurking variables might be confounded with a doctor's choice of surgical or nonsurgical treatment?

(b) You have 300 prostate patients who are willing to serve as subjects in an experiment to compare the two methods. Use a diagram to outline the design of a randomized comparative experiment. (When using a diagram to outline the design of an experiment, be sure to indicate the size of the treatment groups and the response variable. The diagrams in Examples 5.11 and 5.12 are models.)

5.31   Use a diagram to describe a completely randomized experimental design for the package liner experiment of Exercise 5.27. (When using a diagram to outline the design of an experiment, be sure to indicate the size of the treatment groups and the response variable. The diagrams in Examples 5.11 and 5.12 are models.)

5.32   Will providing child care for employees make a company more attractive to women, even those who are unmarried? You are designing an experiment to answer this question. You prepare recruiting material for two fictitious companies, both in similar businesses in the same location. Company A's brochure does not mention child care. There are two versions of Company B's material, identical except that one describes the company's on-site child-care facility. Your subjects are 40 unmarried women who are college seniors seeking employment. Each subject will read recruiting material for both companies and choose the one she would prefer to work for. You will give each version of Company B's brochure to half the women. You expect that a higher percentage of those who read the description that includes child care will choose Company B.

(a) Outline an appropriate design for the experiment.

(b) The names of the subjects appear below. Use Table B, beginning at line 131, to do the randomization required by your design. List the subjects who will read the version that mentions child care.

Abrams	Danielson	Gutierrez	Lippman	Rosen
Adamson	Durr	Howard	Martinez	Sugiwara
Afifi	Edwards	Hwang	McNeill	Thompson
Brown	Fluharty	Iselin	Morse	Travers
Cansico	Garcia	Janle	Ng	Turing
Chen	Gerson	Kaplan	Quinones	Ullmann
Cortez	Green	Kim	Rivera	Williams
Curzakis	Gupta	Lattimore	Roberts	Wong

5.33   Use Table B, starting at line 120, to do the randomization required by your design for the package liner experiment in Exercise 5.31.

5.34   You decide to use a completely randomized design in the two-factor experiment on response to advertising described in Example 5.9 (page 267). You have 36 students who will serve as subjects. Outline the design. Then use Table B at line 130 to randomly assign the subjects to the 6 treatments.

The logic of experimental design

The logic behind a randomized comparative design is as follows.

- Randomization produces groups of experimental units that should be similar in all respects before the treatments are applied.
- Comparative design ensures that influences other than the experimental treatments operate equally on all groups.
- Therefore, differences in the response variable must be due to the effects of the treatments. That is, the treatments not only are associated with the observed differences in the response but must actually cause them.

The great advantage of randomized comparative experiments is that they can produce data that give good evidence for a cause-and-effect relationship between the explanatory and response variables. We know that in general a strong association does not imply causation. A strong association in data from a well-designed experiment does imply causation.

Experimenters were slow to accept randomized comparative designs when they were first put forward in the 1920s. One reason is that the outcome of a randomized experiment depends on chance. If the researchers in Example 5.11 drew cage numbers out of the hat a second time, they would get a different 15 rats in the experimental group and no doubt a somewhat different result when they compare the average weight gains of the two groups.

The outcomes of a randomized experiment, like the result of a random sample, do depend on chance. But in both settings, the laws of probability describe how much chance variation is present. *"Random" in statistics does not mean "haphazard."* Feeding the new breakfast food to the first 15 rats that the technician could catch would be haphazard, and perhaps biased if these rats were smaller, slower, or friendlier than the others. Randomization, on the other hand, gives each rat an equal chance to be chosen.

The presence of chance variation does require us to look more closely at the logic of randomized comparative experiments. We cannot say that *any* difference in average weight gain between the two groups of rats, however small, must be due to the diets. Some differences will appear even if the diets are identical, because the rats are not exactly alike. Some rats will grow faster than others, and by chance more of the faster-growing rats may end up in one group than in the other. Even though there are no systematic differences between the groups, there will still be chance differences. It is the business of statistical inference, using the laws of probability, to say whether the observed difference is too large to occur plausibly as a result of chance alone.

STATISTICAL SIGNIFICANCE

An observed effect too large to attribute plausibly to chance is called **statistically significant.**

If we observe statistically significant differences among the groups after a comparative randomized experiment, we have good evidence that the treatments actually caused these differences. We will explore statistical significance in detail in Chapter 10. One important point should be made immediately, however: *experiments with many subjects are better able to detect differences among the effects of the treatments than similar experiments with fewer subjects.* You would not trust the results of an experiment that fed each diet to only one rat. The role of chance is too large if we use two rats and toss a coin to decide which is fed the new diet. The more rats we use, the more likely it is that randomization will create groups that are alike on the average. When differences among the rats are averaged out, only the effects of the different treatments remain. Here is *replication*   a third principle of statistical design of experiments, called *replication*: repeat each treatment on a large enough number of experimental units or subjects to allow the systematic effects of the treatments to be seen.

PRINCIPLES OF EXPERIMENTAL DESIGN

The basic principles of statistical design of experiments are

1. **Control** of the effects of lurking variables on the response, most simply by comparing several treatments.
2. **Randomization,** the use of impersonal chance to assign subjects to treatments.
3. **Replication** of the experiment on many subjects to reduce chance variation in the results.

5.35   Example 5.12 describes an experiment to learn whether providing house-holds with electronic indicators or charts will reduce their electricity consumption. An executive of the electric company objects to including a control group. He says, "It would be simpler to just compare electricity use last year (before the indicator or chart was provided) with consumption in the same period this year. If households use less electricity this year, the indicator or chart must be working." Explain clearly why this design is inferior to that in Example 5.12.

5.36   Does regular exercise reduce the risk of a heart attack? Here are two ways to study this question. Explain clearly why the second design will produce more trustworthy data.

1. A researcher finds 2000 men over 40 who exercise regularly and have not had heart attacks. She matches each with a similar man who does not exercise regularly, and she follows both groups for 5 years.

2. Another researcher finds 4000 men over 40 who have not had heart attacks and are willing to participate in a study. She assigns 2000 of the men to a regular program of supervised exercise. The other 2000 continue their usual habits. The researcher follows both groups for 5 years.

5.37   The financial aid office of a university asks a sample of students about their employment and earnings. The report says that "for academic year earnings, a significant difference was found between the sexes, with men earning more on the average. No significant difference was found between the earnings of black and white students." Explain the meaning of "a significant difference" and "no significant difference" in plain language.

Cautions about experimentation

Randomized comparative experiments are the best means of gaining knowledge about the effects of explanatory variables on a response. You should nonetheless examine even experimental evidence with a critical eye.

hidden bias   First, the way the experiment is conducted may produce **hidden bias** despite the use of comparison and randomization. Experimenters must take great care to deal with all experimental units or subjects in exactly the same way, so that the treatments are the only systematic differences present. Unequal conditions introduce bias.

EXAMPLE 5.13

A study of the roasting of meat in large commercial ovens found mysteriously large weight losses in the roasts cooked at the right front corner of the oven. The reason: in every run of the experiment, a meat thermometer was thrust into the right front roast, allowing juice to escape.

Less obvious violations of equal treatment can occur in medical and behavioral experiments. Suppose that in the second gastric freezing experiment of Example 5.10 the person interviewing the subjects knew which patients received the real freezing and which received "only" the placebo. This knowledge may subconsciously affect the interviewer's attitude toward the patient and recording of the patient's degree of pain relief. The experiment should therefore be *double-blind*.

DOUBLE-BLIND EXPERIMENT

In a double-blind experiment, neither the subjects nor the people who have contact with them know which treatment a subject received.

The gastric freezing experiment was double-blind, with attention to such details as ensuring that the tube in the mouth of each subject was cold, whether or not the fluid in the balloon was refrigerated. Careful planning and attention to detail are the keys to avoiding hidden bias.

lack of realism in experiments

The most serious potential weakness of experiments is **lack of realism.** The subjects or treatments or setting of an experiment may not realistically duplicate the conditions we really want to study. Here are some examples.

EXAMPLE 5.14

The study of television advertising in Example 5.9 showed a 40-minute videotape to students who knew an experiment was going on. We can't be sure that the results apply to everyday television viewers. Many behavioral science experiments use as subjects students who know they are subjects in an experiment. That's not a realistic setting.

An industrial experiment uses a small-scale pilot production process to find the choices of catalyst concentration and temperature that maximize yield. These may not be the best choices for the operation of a full-scale plant.

Lack of realism can limit our ability to apply the conclusions of an experiment to the settings of greatest interest. Most experimenters want to generalize their conclusions to some setting wider than that of the actual experiment. Statistical analysis of the original experiment cannot tell us how far the results will generalize. Rather, the experimenters in Example 5.14 must argue based on an understanding of psychology or chemical engineering that the experimental results do describe the wider world. Other psychologists or engineers may disagree. This is one reason why a single experiment is rarely completely convincing, despite the compelling logic of experimental design. The true scope of a new finding must usually be explored by a number of experiments in various settings.

A convincing case that an experiment is sufficiently realistic to produce useful information is based not on statistics but on the experimenter's knowledge of the subject matter of the experiment. The attention to detail required to avoid hidden bias also rests on subject matter knowledge. Good experiments combine statistical principles with understanding of a specific field of study.

EXERCISES

5.38 An experiment that claimed to show that meditation lowers anxiety proceeded as follows. The experimenter interviewed the subjects and rated their level of anxiety. Then the subjects were randomly assigned to two groups. The experimenter taught one group how to meditate and they meditated daily for a month. The other group was simply told to relax more. At the end of the month, the experimenter interviewed all the subjects again and rated their anxiety level. The meditation group now had less anxiety. Psychologists said that the results were suspect because the ratings were not blind. Explain what this means and how lack of blindness could bias the reported results.

5.39 Fizz Laboratories, a pharmaceutical company, has developed a new pain-relief medication. Sixty patients suffering from arthritis and needing pain relief are available. Each patient will be treated and asked an hour later, "About what percentage of pain relief did you experience?"

(a) Why should Fizz not simply administer the new drug and record the patients' responses?

(b) Outline the design of an experiment to compare the drug's effectiveness with that of aspirin and of a placebo.

(c) Should patients be told which drug they are receiving? How would this knowledge probably affect their reactions?

(d) If patients are not told which treatment they are receiving, the experiment is single-blind. Should this experiment be double-blind also? Explain.

Other experimental designs

Completely randomized designs are the simplest statistical designs for experiments. They are the analog of simple random samples. In fact, each treatment group is an SRS drawn from the available subjects. Completely randomized designs illustrate clearly the principles of control, randomization, and replication. However, just as in sampling, more elaborate statistical designs are often superior. In particular, matching the subjects in various ways can produce more precise results than simple randomization.

EXAMPLE 5.15

Women and men respond differently to advertising. An experiment to compare the effectiveness of three television commercials for the same product will want to look separately at the reactions of men and women, as well as assess the overall response to the ads.

A completely randomized design considers all subjects, both men and women, as a single pool. The randomization assigns subjects to three treatment groups without regard to their sex. This ignores the differences between men and women. A better design considers women and men separately. Randomly assign the women to three groups, one to view each commercial. Then separately assign the men at random to three groups. Figure 5.3 outlines this improved design.

The design of Figure 5.3 uses the principles of comparison, randomization, and replication. However, the randomization is not complete (all subjects randomly assigned to treatment groups) but restricted to operate

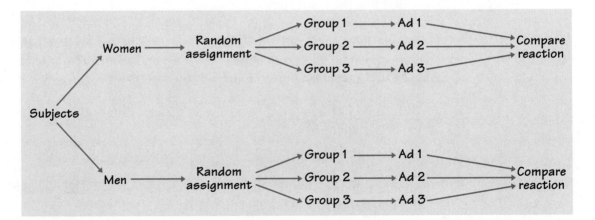

FIGURE 5.3   The outline of a block design to compare the effectiveness of three TV advertisements (Example 5.15). Male and female subjects form two blocks.

only within groups of similar subjects. The groups are called *blocks*, and the design is called a *block design*.

BLOCK DESIGN

A **block** is a group of experimental units or subjects that are similar in ways that are expected to affect the response to the treatments. In a **block design,** the random assignment of units to treatments is carried out separately within each block.

Blocks are another form of *control*. They control the effects of some lurking variables by bringing those variables into the experiment to form the blocks. Block designs are similar to stratified designs for sampling. Blocks and strata both group similar units. We use two different names only because the idea developed separately for sampling and experiments. Blocks allow us to draw separate conclusions about each block, for example, about men and women in Example 5.15. Blocking also allows more precise overall conclusions, because the systematic differences between men and women can be removed when we study the overall effects of the three commercials.

matched pairs design

A simple and common special type of block design is the *matched pairs design.* Matched pairs designs compare just two treatments. Each block consists of just two units, as closely matched as possible. These units are assigned at random to the treatments, by tossing a coin or reading odd and even digits from Table B. Alternatively, each block in a matched pairs design may consist of just one subject, who gets both treatments one after the other. Each subject serves as his or her own control. The *order* of the treatments can influence the subject's response, so the order is randomized for each subject, again by a coin toss.

EXAMPLE 5.16

Pepsi once wanted to demonstrate that Coke drinkers in fact prefer Pepsi when they taste both colas blind. The subjects, all people who said they were Coke drinkers, tasted both colas from glasses without brand markings and said which they liked better. This is a matched pairs design in which each subject compares the two colas. Because responses may depend on which cola is tasted first, the order of tasting should be chosen at random for each subject.

When more than half the Coke drinkers chose Pepsi, Coke claimed that the experiment was biased. The Pepsi glasses were marked *M* and Coke glasses were marked *Q*. Aha, said Coke, this just shows that people like the letter *M* better than the letter *Q*. A careful experiment would in fact take care to avoid any distinction other than the actual treatments.[13]

EXERCISES

5.40   Twenty overweight females have agreed to participate in a study of the effectiveness of four weight-loss treatments: A, B, C, and D. The researcher first calculates how overweight each subject is by comparing the subject's actual weight with her "ideal" weight. The subjects and their excess weights in pounds are

Birnbaum	35	Hernandez	25	Moses	25	Smith	29
Brown	34	Jackson	33	Nevesky	39	Stall	33
Brunk	30	Kendall	28	Obrach	30	Tran	35
Cruz	34	Loren	32	Rodriguez	30	Wilansky	42
Deng	24	Mann	28	Santiago	27	Williams	22

The response variable is the weight lost after 8 weeks of treatment. Because a subject's excess weight will influence the response, a block design is appropriate.

(a)   Arrange the subjects in order of increasing excess weight. Form 5 blocks of 4 subjects each by grouping the 4 least overweight, then the next 4, and so on.

(b)   Use Table B to randomly assign the 4 subjects in each block to the 4 weight-loss treatments. Be sure to explain exactly how you used the table.

5.41   Return to the advertising experiment of Example 5.9 (page 267). You have 36 subjects: 24 women and 12 men. Men and women often react differently to advertising. You therefore decide to use a block design with the two genders as blocks. You must assign the 6 treatments at random within each block separately.

(a)   Outline the design with a diagram.

(b)   Use Table B, beginning at line 140, to do the randomization. Report your result in a table that lists the 24 women and 12 men and the treatment you assigned to each.

5.42   Is the right hand generally stronger than the left in right-handed people? You can crudely measure hand strength by placing a bathroom scale on a shelf with the end protruding, then squeezing the scale between the thumb below and the four fingers above it. The reading of the scale shows the force exerted. Describe the design of a matched pairs experiment to compare the strength of the right and left hands, using 10 right-handed people as subjects. (You need not actually do the randomization.)

5.43   Some investment advisors believe that charts of past trends in the prices of securities can help predict future prices. Most economists disagree. In an experiment to examine the effects of using charts, business students trade (hypothetically) a foreign currency at computer screens. There are 20 student subjects available, named for convenience A, B, C, . . . , T. Their goal is to make as much money as possible, and the best performances are rewarded with small prizes. The student traders have the price history of the foreign currency in dollars in their computers. They may or may not also have software that highlights trends. Describe *two* designs for this experiment, a completely randomized design and a matched pairs design in which each student serves as his or her own control. In both cases, carry out the randomization required by the design.

5.44   An expert on worker performance is interested in the effect of room temperature on the performance of tasks requiring manual dexterity. She chooses temperatures of 70° F and 90° F as treatments. The response variable is the number of correct insertions, during a 30-minute period, in a peg-and-hole apparatus that requires the use of both hands simultaneously. Each subject is trained on the apparatus and then asked to make as many insertions as possible in 30 minutes of continuous effort.

(a)   Outline a completely randomized design to compare dexterity at 70° and 90°. Twenty subjects are available.

(b)   Because individuals differ greatly in dexterity, the wide variation in individual scores may hide the systematic effect of temperature unless there are many subjects in each group. Describe in detail the design of a matched pairs experiment in which each subject serves as his or her own control.

SUMMARY

An **observational study** gathers data without trying to influence the responses. An **experiment** imposes some treatment in order to observe the response.

In an experiment, we impose one or more **treatments** on the experimental **units** or **subjects**. Each treatment is a combination of levels of the explanatory variables, which we call **factors**.

The **design** of an experiment refers to the choice of treatments and the manner in which the experimental units or subjects are assigned to the treatments.

The basic principles of statistical design of experiments are **control, randomization**, and **replication**.

The simplest form of control is **comparison.** Experiments should compare two or more treatments in order to avoid **confounding** of the the effect of a treatment with other influences, such as lurking variables.

Randomization uses chance to assign subjects to the treatments. Randomization creates treatment groups that are similar (except for chance variation) before the treatments are applied. Randomization and comparison together prevent **bias,** or systematic favoritism, in experiments.

Randomization can be carried out by giving numerical labels to the experimental units and using a **table of random digits** to choose treatment groups.

Replication of the treatments on many units reduces the role of chance variation and makes the experiment more sensitive to differences among the treatments.

The validity of experimental results may be threatened by **hidden bias** or **lack of realism.** Statistics alone does not protect against these threats.

In addition to comparison, a second form of control is to restrict randomization by forming **blocks** of experimental units that are similar in some way that is important to the response. Randomization is then carried out separately within each block.

Matched pairs are a common form of blocking for comparing just two treatments. In some matched pairs designs, each subject receives both treatments in a random order. In others, the subjects are matched in pairs as closely as possible, and one subject in each pair receives each treatment.

SECTION 5.2 EXERCISES

5.45 (a) Exercise 4.20 (page 209) describes a study of the effects of herbal tea on the attitude of nursing home residents. Is this study an experiment? Why?

(b) Exercise 4.27 (page 215) describes a study of the effect of learning a foreign language on scores on an English test. Is this study an experiment? Why?

5.46 If children are given more choices within a class of products, will they tend to prefer that product to a competing product that offers fewer choices? Marketers want to know. An experiment prepared three "choice sets" of beverages. The first contained two milk drinks and two fruit drinks. The second had the same two fruit drinks but four milk drinks. The third contained four fruit drinks but only the original two milk drinks. The researchers divided 210 children aged 4 to 12 years into 3 groups at random. They offered each group one of the choice sets. As each child chose a beverage to drink from the choice set presented, the researchers noted whether the choice was a milk drink or a fruit drink.

(a) What are the experimental units or subjects?

(b) What is the factor, and what are its levels?

(c) What is the response variable?

5.47 Can aspirin help prevent heart attacks? The Physicians' Health Study, a large medical experiment involving 22,000 male physicians, attempted to answer this question. One group of about 11,000 physicians took an aspirin every second day, while the rest took a placebo. After several years the study found that subjects in the aspirin group had significantly fewer heart attacks than subjects in the placebo group.

(a) Identify the experimental subjects, the factor and its levels, and the response variable in the Physicians' Health Study.

(b) Use a diagram to outline a completely randomized design for the Physicians' Health Study.

5.48 Use a diagram to outline a completely randomized design for the children's choice study of Exercise 5.46.

5.49 Some medical researchers suspect that added calcium in the diet reduces blood pressure. You have available 40 men with high blood pressure who are willing to serve as subjects.

(a) Outline an appropriate design for the experiment, taking the placebo effect into account.

(b) The names of the subjects appear below. Use Table B, beginning at line 119, to do the randomization required by your design, and list the subjects to whom you will give the drug.

Alomar	Denman	Han	Liang	Rosen
Asihiro	Durr	Howard	Maldonado	Solomon
Bennett	Edwards	Hruska	Marsden	Tompkins
Bikalis	Farouk	Imrani	Moore	Townsend
Chen	Fratianna	James	O'Brian	Tullock
Clemente	George	Kaplan	Ogle	Underwood
Cranston	Green	Krushchev	Plochman	Willis
Curtis	Guillen	Lawless	Rodriguez	Zhang

5.50 The children's choice experiment in Exercise 5.46 has 210 subjects. Explain how you would assign labels to the 210 children in the actual experiment. Then use Table B at line 125 to choose *only the first 5* children assigned to the first treatment.

5.51   A survey of physicians found that some doctors give a placebo to a patient who complains of pain for which the physician can find no cause. If the patient's pain improves, these doctors conclude that it had no physical basis. The medical school researchers who conducted the survey claimed that these doctors do not understand the placebo effect. Why?

5.52   There are several psychological tests that measure the extent to which Mexican Americans are oriented toward Mexican/Spanish or Anglo/English culture. Two such tests are the Bicultural Inventory (BI) and the Acculturation Rating Scale for Mexican Americans (ARSMA). To study the correlation between the scores on these two tests, researchers will give both tests to a group of 22 Mexican Americans.

(a)   Briefly describe a matched pairs design for this study. In particular, how will you use randomization in your design?

(b)   You have an alphabetized list of the subjects (numbered 1 to 22). Carry out the randomization required by your design and report the result.

5.53   You are participating in the design of a medical experiment to investigate whether a calcium supplement in the diet will reduce the blood pressure of middle-aged men. Preliminary work suggests that calcium may be effective and that the effect may be greater for black men than for white men.

(a)   Outline in graphical form the design of an appropriate experiment.

(b)   Choosing the sizes of the treatment groups requires more statistical expertise. We will learn more about this aspect of design in later chapters. Explain in plain language the advantage of using larger groups of subjects.

5.3   SIMULATING EXPERIMENTS

Toss a coin 10 times. What is the likelihood of a run of 3 or more consecutive heads or tails? A couple plans to have children until they have a girl or until they have four children, whichever comes first. What are the chances that they will have a girl among their children? An airline knows from past experience that a certain percentage of customers who have purchased tickets will not show up to board the airplane. If the airline "overbooks" a particular flight (i.e., sells more tickets than they have seats), what are the chances that the airline will encounter more ticketed passengers than they have seats for? There are three methods we can use to answer questions involving chance like these:

1. Try to estimate the likelihood of a result of interest by actually carrying out the experiment many times and calculating the result's relative frequency. That's slow, sometimes costly, and often impractical or logistically difficult.

probability model

2. Develop a ***probability model*** and use it to calculate a theoretical answer. This requires that we know something about the rules of probability and therefore may not be feasible. (We will develop a probability model in the next chapter.)

3. Start with a model that, in some fashion, reflects the truth about the experiment, and then develop a procedure for imitating—or simulating—a number of repetitions of the experiment. This is quicker than repeating the real experiment, especially if we can use the TI-83 or a computer, and it allows us to do problems that are hard when done with formal mathematical analysis.

Here is an example of a simulation.

EXAMPLE 5.17

Suppose we are interested in estimating the likelihood of a couple's having a girl among their first four children. Let a flip of a fair coin represent a birth, with heads corresponding to a girl and tails a boy. Since girls and boys are equally likely to occur on any birth, the coin flip is an accurate imitation of the situation. Flip the coin until a head appears or until the coin has been flipped 4 times, whichever comes first. The appearance of a head within the first 4 flips corresponds to the couple's having a girl among their first four children.

If this coin-flipping procedure is repeated many times, to represent the births in a large number of families, then the proportion of times that a head appears within the first 4 flips should be a good estimate of the true likelihood of the couple's having a girl.

A single die (one of a pair of dice) could also be used to simulate the birth of a son or daughter. Let an even number of spots (called pips) represent a girl, and let an odd number of spots represent a boy.

SIMULATION

The imitation of chance behavior, based on a model that accurately reflects the experiment under consideration, is called a **simulation**.

SIMULATION BASICS

Simulation is an effective tool for finding likelihoods of complex results once we have a trustworthy model. In particular, we can use random digits from a table, graphing calculator, or computer software to simulate many repetitions quickly. The proportion of repetitions on which a result occurs will eventually be close to its true likelihood, so simulation can give good estimates of probabilities. The art of random-digit simulation can be illustrated by a series of examples.

EXAMPLE 5.18

Step 1: State the problem or describe the experiment. Toss a coin 10 times. What is the likelihood of a run of at least 3 consecutive heads or 3 consecutive tails?

Step 2: State the assumptions. There are two:

- A head or a tail is equally likely to occur on each toss.
- Tosses are independent of each other (i.e., what happens on one toss will not influence the next toss).

Step 3: Assign digits to represent outcomes. In a random number table, such as Table B in the back of the book, the digits 0, 1, 2, 3, 4, 5, 6, 7, 8, and 9 occur with the same long-term relative frequency (1/10). We also know that the successive digits in the table are independent. It follows that even digits and odd digits occur with the same long-term relative frequency, 50%. Here is one assignment of digits for coin tossing:

- One digit simulates one toss of the coin.
- Odd digits represent heads; even digits represent tails.

Successive digits in the table simulate independent tosses.

Step 4: Simulate many repetitions. Looking at 10 consecutive digits in Table B simulates one repetition. Read many groups of 10 digits from the table to simulate many repetitions. Be sure to keep track of whether or not the event we want (a run of 3 heads or 3 tails) occurs on each repetition.

Here are the first three repetitions, starting at line 101 in Table B. Runs of 3 or more heads or tails have been underlined.

Digits	1 9 2 2 3	9 5 0 3 4	0 5 7 5 6	2 8 7 1 3	9 6 4 0 9	1 2 5 3 1
Heads/tails	H H T T H	H H T H T	T H H H T	T T H H H	H T T T H	H T H H H
Run of 3		YES		YES		YES

Twenty-two additional repetitions were done for a total of 25 repetitions; 23 of them did have a run of 3 or more heads or tails.

Step 5: State your conclusions. We estimate the probability of a run by the proportion

$$\text{estimated probability} = \frac{23}{25} = .92$$

Of course, 25 repetitions are not enough to be confident that our estimate is accurate. Now that we understand how to do the simulation, we can tell a computer to do many thousands of repetitions. A long simulation (or mathematical analysis) finds that the true probability is about 0.826.

Once you have gained some experience in simulation, establishing a correspondence between random numbers and outcomes in the experiment is usually the hardest part, and must be done carefully. Although coin tossing may not fascinate you, the model in Example 5.18 is typical of many probability problems because it consists of independent trials (the tosses) all having the same possible outcomes and probabilities. Shooting 10 free throws and observing the sexes of 10 children have similar models and are simulated in much the same way.

The idea is to state the basic structure of the random phenomenon and then use simulation to move from this model to the probabilities of more complicated events. The model is based on opinion and past experience. If it does not correctly describe the random phenomenon, the probabilities derived from it by simulation will also be incorrect.

Step 3 (assigning digits) can usually be done in several different ways, but some assignments are more efficient than others. Here are some examples of this step.

EXAMPLE 5.19

(a) Choose a person at random from a group of which 70% are employed. One digit simulates one person:

$$0, 1, 2, 3, 4, 5, 6 = \text{employed}$$

$$7, 8, 9 = \text{not employed}$$

The following correspondence is also satisfactory:

$$00, 01, \ldots, 69 = \text{employed}$$

$$70, 71, \ldots, 99 = \text{not employed}$$

This assignment is less efficient, however, because it requires twice as many digits and ten times as many numbers.

(b) Choose one person at random from a group of which 73% are employed. Now *two* digits simulate one person:

$$00, 01, 02, \ldots, 72 = \text{employed}$$

$$73, 74, 75, \ldots, 99 = \text{not employed}$$

We assigned 73 of the 100 two-digit pairs to "employed" to get probability 0.73. Representing "employed" by $01, 02, \ldots, 73$ would also be correct.

(c) Choose one person at random from a group of which 50% are employed, 20% are unemployed, and 30% are not in the labor force. There are now three possible outcomes, but the principle is the same. One digit simulates one person:

$$0, 1, 2, 3, 4 = \text{employed}$$

$$5, 6 = \text{unemployed}$$

$$7, 8, 9 = \text{not in the labor force}$$

Another valid assignment of digits might be

$$0, 1 = \text{unemployed}$$

$$2, 3, 4 = \text{not in the labor force}$$

$$5, 6, 7, 8, 9 = \text{employed}$$

What is important is the number of digits assigned to each outcome, not the order of the digits.

As the last example shows, simulation methods work just as easily when outcomes are not equally likely. Consider the following slightly more complicated example.

EXAMPLE 5.20

Orders of frozen yogurt flavors (based on sales) have the following relative frequencies: 38% chocolate, 42% vanilla, and 20% strawberry. The experiment consists of customers entering the store and ordering yogurt. The task is to simulate 10 frozen yogurt sales based on this recent history. Instead of considering the random number table to be made up of single digits, we now consider it to be made up of pairs of digits. This is because the relative frequencies of interest have a maximum of *two* significant digits. The range of the pairs of digits is 00 to 99, and since all the pairs are equally likely to occur, the pairs 00, 01, 02, ..., 99 all have relative frequency 0.01.

Thus we may assign the numbers in the random number table as follows:

- 00 to 37 to correspond to the outcome chocolate (C)
- 38 to 79 to correspond to the outcome vanilla (V)
- 80 to 99 to correspond to the outcome strawberry (S)

The sequence of random numbers (starting at the 21st column of row 112 in Table B) is as follows:

$$19352 \quad 73089 \quad 84898 \quad 45785$$

This yields the following two-digit numbers:

$$19 \quad 35 \quad 27 \quad 30 \quad 89 \quad 84 \quad 89 \quad 84 \quad 57 \quad 85$$

which correspond to the outcomes

$$C \quad C \quad C \quad C \quad S \quad S \quad S \quad S \quad V \quad S$$

EXAMPLE 5.21

A couple plans to have children until they have a girl or until they have four children, whichever comes first. We will show how to use random digits to estimate the likelihood that they will have a girl.

The model is the same as for coin tossing.

- Each child has probability 0.5 of being a girl and 0.5 of being a boy.
- The sexes of successive children are independent.

Assigning digits is also easy. One digit simulates the sex of one child:

$$0, 1, 2, 3, 4 = \text{girl}$$
$$5, 6, 7, 8, 9 = \text{boy}$$

To simulate one repetition of this child-bearing strategy, read digits from Table B until the couple has either a girl or four children. Notice that the number of digits needed to simulate one repetition depends on how quickly the couple gets a girl. Here is the simulation, using line 130 of Table B. To interpret the digits, G for girl and B for boy are written under them, space separates repetitions, and under each repetition "+" indicates if a girl was born and "−" indicates one was not.

690	51	64	81	7871	74	0
BBG	BG	BG	BG	BBBG	BG	G
+	+	+	+	+	+	+
951	784	53	4	0	64	8987
BBG	BBG	BG	G	G	BG	BBBB
+	+	+	+	+	+	−

In these 14 repetitions, a girl was born 13 times. Our estimate of the probability that this strategy will produce a girl is therefore

$$\text{estimated probability} = \frac{13}{14} = .93$$

Some mathematics shows that if our probability model is correct, the true likelihood of having a girl is 0.938. Our simulated answer came quite close. Unless the couple is unlucky, they will succeed in having a girl.

EXERCISES

5.54   An opinion poll selects adult Americans at random and asks them, "Which political party, Democratic or Republican, do you think is better able to manage the economy?" Explain carefully how you would assign digits from Table B to simulate the response of one person in each of the following situations.

(a) Of all adult Americans, 50% would choose the Democrats and 50% the Republicans.

(b) Of all adult Americans, 60% would choose the Democrats and 40% the Republicans.

(c) Of all adult Americans, 40% would choose the Democrats, 40% would choose the Republicans, and 20% would be undecided.

(d) Of all adult Americans, 53% would choose the Democrats and 47% the Republicans.

5.55   Use Table B to simulate the responses of 10 independently chosen adults in each of the four situations of Exercise 5.54.

(a) For situation (a), use line 110.

(b) For situation (b), use line 111.

(c) For situation (c), use line 112.

(d) For situation (d), use line 113.

5.56   Suppose that 80% of a university's students favor abolishing evening exams. You ask 10 students chosen at random. What is the likelihood that all 10 favor abolishing evening exams?

(a) Describe how you would pose this question to 10 students independently of each other. How would you model the procedure?

(b) Assign digits to represent the answers "Yes" and "No."

(c) Simulate 25 repetitions, starting at line 129 of Table B. What is your estimate of the likelihood of the desired result?

5.57 A basketball player makes 70% of her free throws in a long season. In a tournament game she shoots 5 free throws late in the game and misses 3 of them. The fans think she was nervous, but the misses may simply be chance. You will shed some light by estimating a probability.

(a) Describe how to simulate a single shot if the probability of making each shot is 0.7. Then describe how to simulate 5 independent shots.

(b) Simulate 50 repetitions of the 5 shots and record the number missed on each repetition. Use Table B starting at line 125. What is the approximate likelihood that the player will miss 3 or more of the 5 shots?

SIMULATIONS WITH THE CALCULATOR OR COMPUTER

The TI-83 can be extremely useful in conducting simulations because it can be easily programmed to quickly perform a large number of repetitions. Study the reasoning and the steps involved in the following example so that you may become adept at using the capabilities of the TI-83 to design and carry out simulations.

EXAMPLE 5.22

With the TI-83, the command `randInt` (found under MATH/PRB/ 5:randInt) can be used to generate random digits between any two specified values. Here are three applications.

The command `randInt(0,9,5)` generates 5 random integers between 0 and 9. This could serve as a block of 5 random digits in the random number table. The command `randInt(1,6,7)` could be used to simulate rolling a die 7 times. Generating 10 two-digit numbers between 00 and 99 from Example 5.20 could be done with the command `randInt(0,99,10)`.

```
randInt(0,9,5)
        {5 6 5 7 1}
randInt(1,6,7)
      {5 6 5 5 3 4 1}
randInt(0,99,10)

{81 23 86 2 40...
■
```

Using the statistical software package Minitab, the following set of commands will generate a set of 10 random numbers in the range 00 to 99 and store these numbers in column C1.

```
MTB > random 10 c1;
SUBC> integer 0 99.
MTB > Print C1.
```

```
C1
    38      93      14      30      50      92      16      18      84      20
```

When you combine the power and simplicity of simulations with the power of technology, you have formidable tools for answering questions involving chance behavior. To illustrate, consider the following two applications.

EXAMPLE 5.23

Exercise 5.57 states, "A basketball player makes 70% of her free throws in a long season. In a tournament game she shoots 5 free throws late in the game and misses 3 of them. The fans think she was nervous, but the misses may simply be chance." The task is to simulate 50 repetitions of the 5 shots and record the number missed on each repetition, then calculate the relative frequency that the player will miss 3 or more of the 5 shots.

First assign the digits 0 to 6 (70%) to the outcome "hit," and 7 to 9 (30%) to the outcome "miss." Next clear lists L_1 and L_2 (ClrList L_1, L_2) so that the results of our simulation can be stored there, and set a counter to 1 (1→C). The variable C is a count of the number of repetitions (in programming, such a variable is known as a counter). The command randInt $(0,9,5)$ →L_1 simulates 5 free throws and stores the results in L_1. The command sum($L_1 \geq 7$ and $L_1 \leq 9$)→L_2(C) uses Boolean logic to count the number of misses (7, 8, or 9) in the 5 simulated free throws and stores this count as the C-th item in list L_2 (C is currently 1). Increase the count to 2 (1+C→C) and repeat the process. Continue executing these three commands until you have performed several repetitions. When you stop, you should have several numbers in L_2, one for each repetition.

The command sum($L_2 \geq 3$ and $L_2 \leq 5$)/N calculates the relative frequency of misses and approximates the probability that the player will miss 3 or more of the 5 shots. Rather than continue accumulating results this way, do the following. First enter these two lines:

```
ClrList L₁,L₂
1→C
```

Next, combine the remaining commands, separated by a colon, as follows:

randInt $(0,9,5)$ →L_1 : sum($L_1 \geq 7$ and $L_1 \leq 9$) →L_2(C) :1+C→C

After you execute the third line, go to the STAT/EDIT mode and look at the two lists L_1 and L_2. There should be 5 numbers in L_1 simulating the first repetition of 5 free throws, and there should be one number in L_2, which is the count of misses in this first repetition. Quit to the Home screen, press 2nd/ENTRY to recall the last command, and press ENTER. Continue with the keystrokes 2nd/ENTRY/ENTER until you have 25 repetitions. Then you can either scroll through list L_2 and count the number of 3s, 4s, and 5s, or you can have the calculator determine the relative frequency with the command $\text{sum}(L_2 \geq 3 \text{ and } L_2 \leq 5)/25$. Compare your results with those of your classmates. If you worked Exercise 5.57 using the random number table, compare those results with your calculator results.

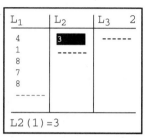

```
randInt(0,9,5)→L₁
:sum(L1≥7 and L₁
≤9)→L₂(C):1+C→C₂
```

EXERCISES

5.58 The program FREETHRO implements the algorithm described in Example 5.23. Enter the program or link it from a classmate or your teacher.

```
PROGRAM:FREETHRO
:ClrHome
:ClrList L₁,L₂
:Disp "HOW MANY
TRIALS"
:Prompt N
:1→C
:While C≤N
:randInt(0,9,5)→
L₁
:sum(L₁≥7 and L₁
≤9)→L₂(C)
:1+C→C
:End
:Disp "REL FREQ=
"
:Disp sum(L₂≥3 a
nd L₂≤5)/N
:
```

(a) Compare the steps in the program with the steps in the algorithm described in Example 5.23 and make sure they accomplish the same tasks. If the problem stated that the player made 78% of her free throws, how would you change the program?

(b) Execute the program three times, specifying 25 trials, 50 trials, and 100 trials. Record the results in each case. Based on these results, what is your best guess as to the probability that the player will miss 3 or more out of 5 attempts?

5.59 Use your TI-83 to simulate a couple's having children until they have a girl or until they have four children, whichever comes first. (See Example 5.21.) Use the simulation to estimate the probability that they will have a girl among their children. Compare your calculator results with those of Example 5.21.

SUMMARY

There are times when actually carrying out an experiment is too costly, too slow, or simply impractical. In situations like these, a carefully designed **simulation** can provide approximate answers to our questions.

A simulation is an imitation of chance behavior, most often carried out with random numbers. The **steps of a simulation** are:

1. State the problem or describe the experiment.
2. State the assumptions.
3. Assign digits to represent outcomes.
4. Simulate many repetitions.
5. State your conclusions.

Programmable calculators, like the TI-83, and computers are particularly useful for conducting simulations because they can perform many repetitions quickly.

SECTION 5.3 EXERCISES

5.60 Suppose a major league baseball player has a current batting average of .320. Note that the batting average = (number of hits)/(number of at-bats).

(a) Describe an assignment of random numbers to possible results in order to simulate the player's next 20 at-bats.

(b) Carry out the simulation for 20 repetitions, and report your results. What is the relative frequency of at-bats in which the player gets a hit?

(c) Compare your simulated experimental results with the player's actual batting average of .320.

5.61 Use your TI-83 and the simulation method to show that in a class of 23 unrelated students, the chances of at least 2 students with the same birthday are about 50%. Show that in a room of 41 people, the chances of at least 2 people having the same birthday are about 90%. What assumptions are you using in your simulations?

5.62 On a small island there are 25 inhabitants. One of these inhabitants, named Jack, starts a rumor which spreads around the isle. Any person who hears the rumor continues spreading it until he or she meets someone who has heard the story before. At that point, the person stops spreading it, since nobody likes to spread stale news.

(a) Do you think that all 25 inhabitants will eventually hear the rumor or will the rumor die out before that happens? Estimate the proportion of inhabitants who will hear the rumor.

(b) In the first time increment, Jack randomly selects one of the other inhabitants, named Jill, to tell the rumor to. In the second time increment, both Jack and Jill each randomly select one of the remaining 24 inhabitants to tell the rumor to. (*Note:* They could conceivably pick each other again.) In the next time increment, there are 4 rumor spreaders, and so on. If a randomly selected person has already heard the rumor, that rumor teller stops spreading the rumor. Design a record-keeping chart, and simulate this procedure. Use your TI-83 to help with the random selection. Continue until all 25 inhabitants hear the rumor or the rumor dies out. How many inhabitants out of 25 eventually heard the rumor?

(c) Combine your results with those of other students in the class. What is the mean number of inhabitants who hear the rumor?

5.63 Amarillo Slim is a cardsharp who likes to play the following game. Draw two cards from the deck of 52 cards. If at least one of the cards is a heart, then you win $1. If neither card is a heart, then you lose $1.

(a) Describe a correspondence between random numbers and possible outcomes in this game.

(b) Simulate playing the game for 25 rounds. Shuffle the cards after each round. See if you can beat Amarillo Slim at his own game. Remember to write down the results of each game. When you finish, combine your results with those of 3 other students to obtain a total of 100 trials. Report your cumulative proportion of wins. Do you think this is a "fair" game? That is, do both you and Slim have an equal chance of winning?

5.64   A certain game of chance is based on randomly selecting three numbers from 00 to 99, inclusive (allowing repetitions), and adding the numbers. A person wins the game if the resulting sum is a multiple of 5.

(a) Describe your scheme for assigning random numbers to outcomes in this game.

(b) Use simulation to estimate the proportion of times a person wins the game.

5.65   A nuclear reactor is equipped with two independent automatic shutdown systems to shut down the reactor when the core temperature reaches the danger level. Neither system is perfect. System A shuts down the reactor 90% of the time when the danger level is reached. System B does so 80% of the time. The reactor is shut down if *either* system works.

(a) Explain how to simulate the response of System A to a dangerous temperature level.

(b) Explain how to simulate the response of System B to a dangerous temperature level.

(c) Both systems are in operation simultaneously. Combine your answers to (a) and (b) to simulate the response of both systems to a dangerous temperature level. Explain why you cannot use the same entry in Table B to simulate both responses.

(d) Now simulate 100 trials of the reactor's response to an emergency of this kind. Estimate the probability that it will shut down. This probability is higher than the probability that either system working alone will shut down the reactor.

CHAPTER REVIEW

Designs for producing data are essential parts of statistics in practice. Random sampling and randomized comparative experiments are perhaps the most important statistical inventions in this century. Both were slow to gain acceptance, and you will still see many voluntary response samples and uncontrolled experiments. This chapter has explained good techniques for producing data and has also explained why bad techniques often produce worthless data.

The deliberate use of chance in producing data is a central idea in statistics. It allows use of the laws of probability to analyze data, as we will see in the following chapters. Here are the major skills you should have now that you have studied this chapter.

A. SAMPLING

1. Identify the population in a sampling situation.
2. Recognize bias due to voluntary response samples and other inferior sampling methods.
3. Use Table B of random digits to select a simple random sample (SRS) from a population.
4. Recognize the presence of undercoverage and nonresponse as sources of error in a sample survey. Recognize the effect of the wording of questions on the responses.
5. Use random digits to select a stratified random sample from a population when the strata are identified.

B. EXPERIMENTS

1. Recognize whether a study is an observational study or an experiment.
2. Recognize bias due to confounding of explanatory variables with lurking variables in either an observational study or an experiment.
3. Identify the factors (explanatory variables), treatments, response variables, and experimental units or subjects in an experiment.
4. Outline the design of a completely randomized experiment using a diagram like those in Examples 5.11 and 5.12. The diagram in a specific case should show the sizes of the groups, the specific treatments, and the response variable.
5. Use Table B of random digits to carry out the random assignment of subjects to groups in a completely randomized experiment.
6. Recognize the placebo effect. Recognize when the double-blind technique should be used.
7. Explain why a randomized comparative experiment can give good evidence for cause-and-effect relationships.

C. SIMULATIONS

1. Recognize that many random phenomena can be investigated by means of a carefully designed simulation.

2. Use the following steps to construct and run a simulation:

   a. State the problem or describe the experiment.

   b. State the assumptions.

   c. Assign digits to represent outcomes.

   d. Simulate many repetitions.

   e. Calculate relative frequencies and state your conclusions.

3. Use a random number table, the TI-83, or a computer utility such as Minitab, Data Desk, or a spreadsheet to conduct simulations.

CHAPTER 5 REVIEW EXERCISES

5.66   The Ministry of Health in the Canadian Province of Ontario wants to know whether the national health care system is achieving its goals in the province. Much information about health care comes from patient records, but that source doesn't allow us to compare people who use health services with those who don't. So the Ministry of Health conducted the Ontario Health Survey, which interviewed a random sample of 61,239 people who live in the Province of Ontario.[14]

   (a) What is the population for this sample survey? What is the sample?

   (b) The survey found that 76% of males and 86% of females in the sample had visited a general practitioner at least once in the past year. Do you think these estimates are close to the truth about the entire population? Why?

5.67   What is the preferred treatment for breast cancer that is detected in its early stages? The most common treatment was once removal of the breast. It is now usual to remove only the tumor and nearby lymph nodes, followed by radiation. To study whether these treatments differ in their effectiveness, a medical team examines the records of 25 large hospitals and compares the survival times after surgery of all women who have had either treatment.

   (a) What are the explanatory and response variables?

   (b) Explain carefully why this study is not an experiment.

   (c) Explain why confounding will prevent this study from discovering which treatment is more effective. (The current treatment was in fact recommended after a large randomized comparative experiment.)

5.68   A study of the relationship between physical fitness and leadership uses as subjects middle-aged executives who have volunteered for an exercise program. The executives are divided into a low-fitness group and a high-fitness group on the basis of a physical examination. All subjects then take a psychological test designed to measure leadership, and the results for the two groups are compared. Is this study an experiment? Explain your answer.

5.69   A university's financial aid office wants to know how much it can expect students to earn from summer employment. This information will be used to set the level of financial aid. The population contains 3478 students who have completed at least one year of study but have not yet graduated. The university will send a questionnaire to an SRS of 100 of these students, drawn from an alphabetized list.

   (a) Describe how you will label the students in order to select the sample.

   (b) Use Table B, beginning at line 105, to select the *first 5* students in the sample.

5.70   A labor organization wants to study the attitudes of college faculty members toward collective bargaining. These attitudes appear to be different depending on the type of college. The American Association of University Professors classifies colleges as follows:

   Class I. Offer doctorate degrees and award at least 15 per year.
   Class IIA. Award degrees above the bachelor's but are not in Class I.
   Class IIB. Award no degrees beyond the bachelor's.
   Class III. Two-year colleges.

   Discuss the design of a sample of faculty from colleges in your state, with total sample size about 200.

5.71   You want to investigate the attitudes of students at your school about the school's policy on sexual harassment. You have a grant that will pay the costs of contacting about 500 students.

   (a) Specify the exact population for your study. For example, will you include part-time students?

   (b) Describe your sample design. Will you use a stratified sample?

   (c) Briefly discuss the practical difficulties that you anticipate. For example, how will you contact the students in your sample?

5.72   New varieties of corn with altered amino acid content may have higher nutritional value than standard corn, which is low in the amino acid lysine.

An experiment compares two new varieties, called opaque-2 and floury-2, with normal corn. The researchers mix corn-soybean meal diets using each type of corn at each of three protein levels, 12% protein, 16% protein, and 20% protein. They feed each diet to 10 one-day-old male chicks and record their weight gains after 21 days. The weight gain of the chicks is a measure of the nutritional value of their diet.

(a) What are the experimental units and the response variable in this experiment?

(b) How many factors are there? How many treatments? Use a diagram like Figure 5.2 to describe the treatments. How many experimental units does the experiment require?

(c) Use a diagram to describe a completely randomized design for this experiment. (You do not need to actually do the randomization.)

5.73   A chemical engineer is designing the production process for a new product. The chemical reaction that produces the product may have higher or lower yield, depending on the temperature and the stirring rate in the vessel in which the reaction takes place. The engineer decides to investigate the effects of all combinations of two temperatures (50° C and 60° C) and three stirring rates (60 rpm, 90 rpm, and 120 rpm) on the yield of the process. She will process two batches of the product at each combination of temperature and stirring rate. In Exercise 5.29 you identified the treatments.

(a) Outline in graphic form the design of an appropriate experiment.

(b) The randomization in this experiment determines the order in which batches of the product will be processed according to each treatment. Use Table B, starting at line 128, to carry out the randomization and state the result.

5.74   Is the number of days a letter takes to reach another city affected by the time of day it is mailed and whether or not the zip code is used? Describe briefly the design of a two-factor experiment to investigate this question. Be sure to specify the treatments exactly and to tell how you will handle lurking variables such as the day of the week on which the letter is mailed.

5.75   Do consumers prefer the taste of a cheeseburger from McDonald's or from Wendy's in a blind test in which neither burger is identified? Describe briefly the design of a matched pairs experiment to investigate this question.

5.76   The previous two exercises illustrate the use of statistically designed experiments to answer questions that arise in everyday life. Select a question of interest to you that an experiment might answer and briefly discuss the design of an appropriate experiment.

5.77   A study on predicting job performance reports that there is a statistically significant correlation between the score on a screening test given to potential employees and a new employee's score on an evaluation made after a year on the job. What does "statistically significant" mean here?

5.78   A game of chance is based on spinning a 1–10 spinner like the one shown in the illustration two times in succession. The player wins if the larger of the two numbers is greater than 5.

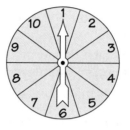

(a) What constitutes a single run of this experiment? What are the possible outcomes resulting in win or lose?

(b) Describe a correspondence between random digits from a random number table and outcomes in the game.

(c) Describe a technique using the `randInt` command on the TI-83 to simulate the result of a single run of the experiment.

(d) Use either the random number table or your calculator to simulate 20 trials. Report the proportion of times you win the game. Then combine your results with those of other students to obtain results for a large number of trials.

5.79   The owner of a bakery knows that the daily demand for a highly perishable cheesecake is as follows:

Number/day	0	1	2	3	4	5
Relative frequency	.05	.15	.25	.25	.20	.10

(a) Use simulation to find the demand for the cheesecake on 30 consecutive business days.

(b) Suppose that it cost the baker $5 to produce a cheesecake, and that the unused cheesecakes must be discarded at the end of the business day.

Suppose also that the selling price of a cheesecake is $13. Use simulation to estimate the number of cheesecakes that he should produce each day in order to maximize his profit.

5.80   Joey is interested in investigating so-called hot streaks in foul shooting among basketball players. He's a fan of Carla, who has been making approximately 80% of her free throws. Specifically, Joey wants to use simulation methods to determine Carla's longest *run* of baskets on average, for 20 consecutive free throws.

   (a) Describe a correspondence between random numbers and outcomes.

   (b) What will constitute one repetition in this simulation? Carry out 20 repetitions and record the longest run for each repetition. Combine your results with those of 4 other students to obtain at least 100 replications.

   (c) What is the mean run length? Are you surprised? Determine the five-number summary for the data.

   (d) Construct a histogram of the results.

5.81   Elaine is enrolled in a self-paced course that allows three attempts to pass an examination on the material. She does not study and has 2 out of 10 chances of passing on any one attempt by luck. What is Elaine's likelihood of passing on at least one of the three attempts? (Assume the attempts are independent because she takes a different examination on each attempt.)

   (a) Explain how you would use random digits to simulate one attempt at the exam. Elaine will of course stop taking the exam as soon as she passes.

   (b) Simulate 50 repetitions. What is your estimate of Elaine's likelihood of passing the course?

   (c) Do you think the assumption that Elaine's likelihood of passing the exam is the same on each trial is realistic? Why?

5.82   A more realistic model for Elaine's attempts to pass an exam in the previous exercise is as follows: On the first try she has probability 0.2 of passing. If she fails on the first try, her probability on the second try increases to 0.3 because she learned something from her first attempt. If she fails on two attempts, the probability of passing on a third attempt is 0.4. She will stop as soon as she passes. The course rules force her to stop after three attempts in any case.

   (a) Explain how to simulate one repetition of Elaine's tries at the exam. Notice that she has different probabilities of passing on each successive try.

(b) Simulate 50 repetitions and estimate the probability that Elaine eventually passes the exam.

NOTES AND DATA SOURCES

1. Based in part on Randall Rothenberger, "The trouble with mall interviewing," *New York Times*, August 16, 1989.

2. The information in this example is taken from *The ASCAP Survey and Your Royalties*, ASCAP, New York, undated.

3. For more detail on the material of this section, along with references, see P. E. Converse and M. W. Traugott, "Assessing the accuracy of polls and surveys," *Science*, 234 (1986), pp. 1094–1098.

4. The estimates of the census undercount come from Howard Hogan, "The 1990 post-enumeration survey: operations and results," *Journal of the American Statistical Association*, 88 (1993), pp. 1047–1060. The information about nonresponse appears in Eugene P. Eriksen and Teresa K. DeFonso, "Beyond the net undercount: how to measure census error," *Chance*, 6, no. 4 (1993), pp. 38–43 and 14.

5. For more detail on the limits of memory in surveys, see N. M. Bradburn, L. J. Rips, and S. K. Shevell, "Answering autobiographical questions: the impact of memory and inference on surveys," *Science*, 236 (1987), pp. 157–161.

6. The Levi jeans and disposable diaper examples are taken from Cynthia Crossen, "Margin of error: studies galore support products and positions, but are they reliable?" *Wall Street Journal*, November 14, 1991, and reprinted in her book *Tainted Truth: The Manipulation of Fact in America*, Simon & Schuster, New York, 1994.

7. Giuliana Coccia, "An overview of non-response in Italian telephone surveys," *Proceedings of the 99th Session of the International Statistical Institute, 1993*, Book 3, pp. 271–272.

8. The first question was asked by Ross Perot, and the second by a Time/CNN poll, both in March 1993. The example comes from W. Mitofsky, "Mr. Perot, you're no pollster," *New York Times*, March 27, 1993.

9. From the *New York Times* of August 21, 1989.

10. Simplified from Arno J. Rethans, John L. Swasy, and Lawrence J. Marks, "Effects of television commercial repetition, receiver knowledge, and commercial length: a test of the two-factor model," *Journal of Marketing Research*, 23 (February 1986), pp. 50–61.

11. L. L. Miao, "Gastric freezing: an example of the evaluation of medical therapy by randomized clinical trials," in J. P. Bunker, B. A. Barnes, and F. Mosteller (eds.), *Costs, Risks and Benefits of Surgery*, Oxford University Press, New York, 1977, pp. 198–211.

12. Based on Christopher Anderson, "Measuring what works in health care," *Science*, 263 (1994), pp. 1080–1082.

13. Taken from "Advertising: the cola war," *Newsweek*, August 30, 1976, p. 67.

14. Information from Warren McIsaac and Vivek Goel, "Is access to physician services in Ontario equitable?" Institute for Clinical Evaluative Sciences in Ontario, October 18, 1993.

Probability: Foundations for Inference

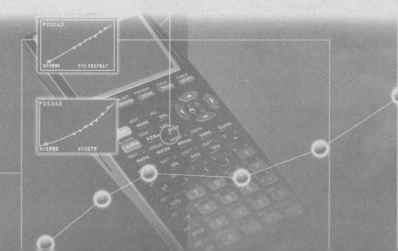

Probability: The Study of Randomness

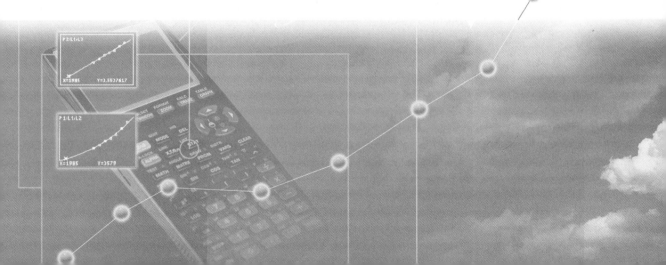

6.1 RANDOMNESS

Chance is all around us. Sometimes chance results from human design, as in the casino's games of chance and the statistician's random samples. Sometimes nature uses chance, as in choosing the sex of a child. Sometimes the reasons for chance behavior are mysterious, as when the number of deaths each year in a large population is as regular as the number of heads in many tosses of a coin. Probability is the branch of mathematics that describes the pattern of chance outcomes.

The reasoning of statistical inference rests on asking, "How often would this method give a correct answer if I used it very many times?" When we produce data by random sampling or randomized comparative experiments, the laws of probability answer the question "What would happen if we did this many times?" This chapter presents the fundamental concepts of probability. Probability calculations are the basis for inference. The tools you acquire in this chapter will help you describe the behavior of statistics from random samples and randomized comparative experiments in later chapters. Even our brief acquaintance with probability will enable us to answer questions like these:

- If we know the blood types of a man and a woman, what can we say about the blood types of their future children?
- Give a test for the AIDS virus to the employees of a small company. What is the chance of at least one positive test if all the people tested are free of the virus?
- An opinion poll asks a sample of 1500 adults what they consider the most serious problem facing our schools. How often will the poll percent who answer "drugs" come within two percentage points of the truth about the entire population?
- If you buy a ticket to Vermont's Green Mountain Numbers state lottery game every day for many years, how much will each ticket win on the average?

ACTIVITY 6 | The Spinning Wheel

Materials: Margarine tub spinner or graphing calculator or table of random numbers

Imagine a spinner with three sectors, all the same size, marked 1, 2, and 3 as shown.

(continued)

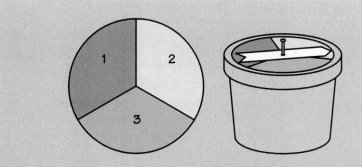

The experiment consists of spinning the spinner three times and recording the numbers as they occur (e.g., 123). We want to determine the proportion of times that *at least one digit occurs in its correct position*. For example, in the numbers 123, all of the digits are in their proper positions, but in the number 331, none are. For this activity, use a spinner like the one in the illustration, a table of random digits, or your TI-83.

1. Guess the proportion of times at least one digit will occur in its proper place.
2. To use your calculator to randomly generate the three-digit number, enter the command `randInt(1,3,3)`. Continue to press ENTER to generate more three-digit numbers. Use a tally mark to record the results in a table like the one below. Do 20 trials and then calculate the relative frequency for the event "at least one digit in the correct position."

At least one digit in the correct position	
Not	

To use a random number table, select a row, and discarding digits 4 to 9 and 0, record digits in the 1 to 3 range in groups of three.

3. Combine your results with those of your classmates to obtain as many trials as possible (at least 100 randomly generated three-digit numbers; 200 would be better).
4. Count the number of times at least one digit occurred in its correct position, and calculate the proportion.
5. The program SPIN123 implements the experiment for the TI-83. It is modeled after the program FREETHRO (see Exercise 5.58).

(continued)

As in FREETHRO, the key step uses the calculator's Boolean logic to count the number of "hits." Enter the program or link it from a classmate or your teacher.

```
PROGRAM:SPIN123
:ClrHome
:ClrList L₁,L₂
:Disp "HOW MANY
TRIALS"
:Prompt N
:1→C
:While C≤N
:randInt(1,3,3)→
L₁
:(L₁(1)=1 or L₁(
2)=2 or L₁(3)=3)
→L₂(C)
:1+C→C
:End
:Disp "REL FREQ=
"
:Disp sum(L₂=1)/
N
:
```

Execute the program for 25, 50, and 100 repetitions. Compare the calculator results with the results you obtained in steps 2 to 4.

Later in the chapter we will calculate the theoretical probability of this event happening, so keep your data at hand so that you can compare the theoretical probability with your experimental results.

The idea of probability

The mathematics of probability begins with the observed fact that some phenomena are random—that is, the relative frequencies of their outcomes seem to settle down to fixed values in the long run. Consider tossing a single coin. The relative frequency of heads is quite erratic in 2 or 5 or 10 tosses. But after several thousand tosses it remains stable, changing very little over further thousands of tosses.

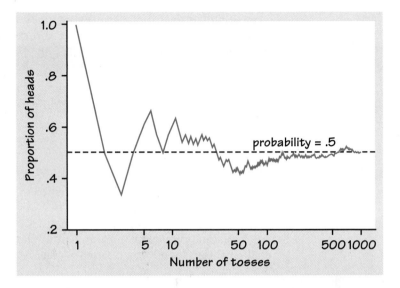

FIGURE 6.1 The behavior of the proportion of coin tosses that give a head, from 1 to 1000 tosses of a coin (Example 6.1). In the long run, the proportion of heads approaches 0.5, the probability of a head.

EXAMPLE 6.1

Figure 6.1 shows the results of tossing a coin 1000 times. For each number of tosses from 1 to 1000, we have plotted the proportion of those tosses that gave a head. The first toss was a head, so the proportion of heads starts at 1. The second toss was a tail, reducing the proportion of heads to 0.5 after two tosses. The next three tosses gave a tail followed by two heads, so the proportion of heads after five tosses is 3/5, or 0.6.

The proportion of tosses that produce heads is quite variable at first, but settles down as we make more and more tosses. Eventually this proportion gets close to 0.5 and stays there. We say that 0.5 is the *probability* of a head. The probability 0.5 appears as a horizontal line on the graph.

The language of probability

"Random" in statistics is not a synonym for "haphazard" but a description of a kind of order that emerges only in the long run. We often encounter the unpredictable side of randomness in our everyday experience, but we rarely see enough repetitions of the same random phenomenon to observe the long-term regularity that probability describes. You can see that regularity emerging in Figure 6.1. In the very long run, the proportion of tosses that give a head is 0.5. This is the intuitive idea of probability. Probability 0.5 means "occurs half the time in a very large number of trials."

We might suspect that a coin has probability 0.5 of coming up heads just because the coin has two sides. As Exercises 6.1 and 6.2 will illustrate, such suspicions are not always correct. The idea of probability is empirical (based on observation). Probability describes what happens in very many trials, and we must actually observe many trials to pin down a probability. In the case of tossing a coin, some diligent people have in fact made thousands of tosses.

EXAMPLE 6.2

The French naturalist Count Buffon (1707–1788) tossed a coin 4040 times. Result: 2048 heads, or proportion 2048/4040 = .5069 for heads.

Around 1900, the English statistician Karl Pearson heroically tossed a coin 24,000 times. Result: 12,012 heads, a proportion of .5005.

While imprisoned by the Germans during World War II, the Australian mathematician John Kerrich tossed a coin 10,000 times. Result: 5067 heads, proportion of heads .5067.

RANDOMNESS AND PROBABILITY

We call a phenomenon **random** if individual outcomes are uncertain but there is nonetheless a regular distribution of outcomes in a large number of repetitions.

The **probability** of any outcome of a random phenomenon is the proportion of times the outcome would occur in a very long series of repetitions.

Thinking about randomness

That some things are random is an observed fact about the world. The outcome of a coin toss, the time between emissions of particles by a radioactive source, and the sexes of the next litter of lab rats are all random. So is the outcome of a random sample or a randomized experiment. Probability theory is the branch of mathematics that describes random behavior. Of course, we can never observe a probability exactly. We could always continue tossing the coin, for example. Mathematical probability is an idealization based on imagining what would happen in an indefinitely long series of trials.

The best way to understand randomness is to observe random behavior—not only the long-run regularity but the unpredictable results of short runs. You can do this with physical devices, as in Exercises

6.1 to 6.4, but computer simulations (imitations) of random behavior allow faster exploration. Exercises 6.6 and 6.7 suggest some simulations of random behavior. As you explore randomness, remember:

independence

- You must have a long series of **independent** trials. That is, the outcome of one trial must not influence the outcome of any other. Imagine a crooked gambling house where the operator of a roulette wheel can stop it where she chooses—she can prevent the proportion of "red" from settling down to a fixed number. These trials are not independent.

- The idea of probability is empirical. Computer simulations start with given probabilities and imitate random behavior, but we can estimate a real-world probability only by actually observing many trials.

- Nonetheless, computer simulations are very useful because we need long runs of trials. In situations such as coin tossing, the proportion of an outcome often requires several hundred trials to settle down to the probability of that outcome. The kinds of physical random devices suggested in the exercises are too slow for this. Short runs give only rough estimates of probability.

The uses of probability

Probability theory originated in the study of games of chance. Tossing dice, dealing shuffled cards, and spinning a roulette wheel are examples of deliberate randomization that are similar to random sampling. Although games of chance are ancient, they were not studied by mathematicians until the sixteenth and seventeenth centuries. It is only a mild simplification to say that probability as a branch of mathematics arose when seventeenth-century French gamblers asked the mathematicians Blaise Pascal and Pierre de Fermat for help. Gambling is still with us, in casinos and state lotteries. We will make use of games of chance as simple examples that illustrate the principles of probability. Careful measurements in astronomy and surveying led to further advances in probability in the eighteenth and nineteenth centuries because the results of repeated measurements are random and can be described by distributions much like those arising from random sampling. Similar distributions appear in data on human life span (mortality tables) and in data on lengths or weights in a population of skulls, leaves, or cockroaches.[1] In the twentieth century, we employ the mathematics of probability to describe the flow of traffic through a highway system, a telephone interchange, or a computer processor; the genetic makeup of individuals or populations; the energy states of subatomic particles; the spread of epidemics or rumors; and the rate of return on risky investments. Although we are interested in probability because of its usefulness in statistics, the mathematics of chance is important in many fields of study.

SUMMARY

A **random phenomenon** has outcomes that we cannot predict but that nonetheless have a regular distribution in very many repetitions.

The **probability** of an event is the proportion of times the event occurs in many repeated trials of a random phenomenon.

SECTION 6.1 EXERCISES

6.1 Hold a penny upright on its edge under your forefinger on a hard surface, then snap it with your other forefinger so that it spins for some time before falling. Based on 50 spins, what is the probability of heads?

6.2 Bend one end of a paper clip so that it is perpendicular to the rest of the paper clip (see sketch). Then drop the clip onto a hard surface. When it comes to rest it points up (U) or down (D). Do you think that one of these outcomes is more likely to occur than the other? If so, which one? Guess the proportion of times the paper clip will land pointing up. Drop the paper clip 50 times and calculate the proportion of times the clip points up. Is this proportion the probability of a paper clip coming to rest in the U position, or an estimate of this probability?

6.3 Toss a thumbtack on a hard surface 100 times. How many times did it land with the point up? What is the approximate probability of landing point up?

6.4 Obtain 10 identical thumbtacks and toss all 10 at once. Do this 50 times and record the number that land point up on each trial. (a) What is the approximate probability that at least one lands point up? (b) What is the approximate probability that more than one lands point up?

6.5 You read in a book on poker that the probability of being dealt three of a kind in a five-card poker hand is 1/50. Explain in simple language what this means.

6.6 In the game of Heads or Tails, Betty and Bob toss a coin four times. Betty wins a dollar from Bob for each head and pays Bob a dollar for each tail—that is, she wins or loses the difference between the number of heads and the number of tails. For example, if there are one head and three tails, Betty loses $2. You can check that Betty's possible outcomes are

$$\{-4, -2, 0, 2, 4\}$$

Assign probabilities to these outcomes by playing the game 20 times and using the proportions of the outcomes as estimates of the probabilities. If possible, combine your trials with those of other students to obtain long-run proportions that are closer to the probabilities.

6.7   The basketball player Shaquille O'Neal makes about half of his free throws over an entire season. We will use the TI-83 to simulate 100 free throws shot independently by a player who has probability 0.5 of making each shot. We let the number 1 represent the outcome "Hit" and 0 represent a "Miss."

(a)   Define a new list named SHAK. Then enter the command `randInt` `(0,1,100)→`L`SHAK`. This tells the calculator to randomly select a hit (1) or a miss (0), do this 100 times in succession, and store the results in the list named SHAK.

(b)   What percent of the 100 shots are hits?

(c)   Examine the sequence of hits and misses. How long was the longest run of shots made? Of shots missed? (Sequences of random outcomes often show runs longer than our intuition thinks likely.)

6.8   Probability is a measure of how likely an event is to occur. Match one of the probabilities that follow with each statement about an event. (The probability is usually a much more exact measure of likelihood than is the verbal statement.)

$$0, .01, .3, .6, .99, 1$$

(a)   This event is impossible. It can never occur.

(b)   This event is certain. It will occur on every trial of the random phenomenon.

(c)   This event is very unlikely, but it will occur once in a while in a long sequence of trials.

(d)   This event will occur more often than not.

6.2   PROBABILITY MODELS

Earlier chapters gave mathematical models for linear relationships (in the form of the equation of a line) and for some distributions of data (in the form of normal density curves). Now we must give a mathematical description or model for randomness. To see how to proceed, think first about a very simple random phenomenon, tossing a coin once. When we toss a coin, we cannot know the outcome in advance. What do we know? We are willing to say that the outcome will be either heads or tails. Because the coin appears to be balanced, we believe that each of these outcomes has probability 1/2. This description of coin tossing has two parts:

• A list of possible outcomes
• A probability for each outcome

Such a description is the basis for all probability models. We will begin by describing the outcomes of a random phenomenon and then learn how to assign probabilities to the outcomes.

Sample spaces

A probability model first tells us what outcomes are possible.

SAMPLE SPACE

The **sample space** S of a random phenomenon is the set of all possible outcomes.

The name "sample space" is natural in random sampling, where each possible outcome is a sample and the sample space contains all possible samples.

To specify S, we must state what constitutes an individual outcome and then state which outcomes can occur. We often have some freedom in defining the sample space, so the choice of S is a matter of convenience as well as correctness. The idea of a sample space, and the freedom we may have in specifying it, are best illustrated by examples.

EXAMPLE 6.3

Toss a coin. There are only two possible outcomes, and the sample space is

$$S = \{\text{heads, tails}\}$$

or more briefly, $S = \{H, T\}$.

EXAMPLE 6.4

Let your pencil point fall blindly into Table B of random digits; record the value of the digit it lands on. The possible outcomes are

$$S = \{0, 1, 2, 3, 4, 5, 6, 7, 8, 9\}$$

EXAMPLE 6.5

An experiment consists of flipping a coin and rolling a die. Possible outcomes are a head (H) followed by any of the digits 1 to 6, or a tail (T) followed by any of the digits 1 to 6. The sample space contains 12 outcomes:

$$S = \{H1, H2, H3, H4, H5, H6, T1, T2, T3, T4, T5, T6\}$$

Being able to properly enumerate the outcomes in a sample space will be critical to determining probabilities. Two techniques are very helpful in making sure you don't accidentally overlook any outcomes. The first is called a ***tree diagram*** because it resembles the branches of a tree. The first action in Example 6.5 is to toss a coin. To construct the tree diagram, begin with a point and draw a line from the point to H and a second line from the point to T. The second action is to roll a die; there are six possible faces that can come up on the die. So draw a line from each of H and T to these six outcomes. See Figure 6.2.

tree diagram

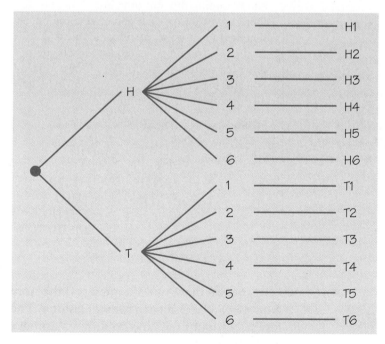

FIGURE 6.2   Tree diagram for Example 6.5.

The second technique is to make use of the following rule.

MULTIPLICATION PRINCIPLE

If you can do one task in *a* number of ways and a second task in *b* number of ways, then both tasks can be done in $a \times b$ number of ways.

To determine the number of outcomes in the sample space for Example 6.5, there are 2 ways the coin can come up, and there are 6 ways the die can come up, so there are 2×6 possible outcomes in the sample space. To see why this is true, just sketch a tree diagram.

EXAMPLE 6.6

An experiment consists of flipping four coins. You can think of either tossing four coins onto the table all at once or flipping a coin four times in succession and recording the four outcomes. One possible outcome is HHTH. Because there are two ways each coin can come up, the multiplication principle says that the total number of outcomes is $2 \times 2 \times 2 \times 2 = 16$. This is the easy part. Listing all 16 outcomes requires a scheme or systematic method so that you don't leave out any possibilities. One way is to list all the ways you can obtain 0 heads, then list all the ways you can get 1 head, 2 heads, 3 heads, and finally all 4 heads. Here is an enumeration:

0 heads	1 heads	2 heads	3 heads	4 heads
TTTT	HTTT	HHTT	HHHT	HHHH
	THTT	HTHT	HHTH	
	TTHT	HTTH	HTHH	
	TTTH	THHT	THHH	
		THTH		
		TTHH		

Suppose that our only interest is the number of heads in four tosses. Now we can be exact in a simpler fashion. The random phenomenon is to toss a coin four times and count the number of heads. The sample space contains only five outcomes:

$$S = \{0, 1, 2, 3, 4\}$$

This example also illustrates the importance of carefully specifying what constitutes an individual outcome.

Although these examples seem remote from the practice of statistics, the connection is surprisingly close. Suppose that in the course of conducting an opinion poll you select four people at random from a large population and ask each if he or she favors reducing federal spending on low-interest student loans. The answers are "Yes" or "No." The possible outcomes—the sample space—are exactly as in Example 6.3 if we replace heads by "Yes" and tails by "No." Similarly, the possible outcomes of an SRS of 1500 people are the same in principle as the possible outcomes of tossing a coin 1500 times. One of the great advantages of mathematics is that the essential features of quite different phenomena can be described by the same mathematical model.

Of course, some sample spaces are simply too large to allow all of the possible outcomes to be listed, as the next example shows.

EXAMPLE 6.7

Many computing systems have a function that will generate a random number between 0 and 1. The sample space is

$$S = \{\text{all numbers between 0 and 1}\}$$

This S is a mathematical idealization. Any specific random number generator produces numbers with some limited number of decimal places so that, strictly speaking, not all numbers between 0 and 1 are possible outcomes. The entire interval from 0 to 1 is easier to think about. It also has the advantage of being a suitable sample space for different computers that produce random numbers with different numbers of significant digits.

---

replacement

If you are selecting objects from a collection of distinct choices, such as drawing playing cards from a standard deck of 52 cards, then much depends on whether each choice is exactly like the previous choice. If you are selecting random digits by drawing numbered slips of paper from a hat, and you want all ten digits to be equally likely to be selected each draw, then after you draw a digit and record it, you must put it back into the hat. Then the second draw will be exactly like the first. This is referred to as sampling **with replacement**. If you do not replace the slips you draw, however, there are only nine choices for the second slip picked, and eight for the third. This is called sampling *without replacement*. So if the question is "How many three-digit numbers can you make?" the answer is, by the multiplication principle, $10 \times 10 \times 10 = 1000$, providing all ten numbers are eligible for each of the three positions in the number. On the other had, there are $10 \times 9 \times 8 = 720$ different ways to construct a three-digit number *without replacement*. You should be able to determine from the context of the problem whether the selection is with or without replacement, and this will help you properly identify the sample space.

EXERCISES

6.9   In each of the following situations, describe a sample space S for the indicated random phenomenon. In some cases, you have some freedom in your choice of S.

(a) A seed is planted in the ground. It either germinates or fails to grow.

(b) A patient with a usually fatal form of cancer is given a new treatment. The response variable is the length of time that the patient lives after treatment.

(c) A student enrolls in a statistics course and at the end of the semester receives a letter grade.

(d) A basketball player shoots two free throws.

(e) A year after knee surgery, a patient is asked to rate the amount of pain in the knee. A seven-point scale is used, with 1 corresponding to no pain and 7 corresponding to extreme discomfort.

6.10   In each of the following situations, describe a sample space S for the indicated random phenomenon. In some cases you have some freedom in specifying S, especially in setting the largest and smallest value in S.

(a) Choose a student in your class at random. Ask how much time that student spent studying during the past 24 hours.

(b) The Physicians' Health Study asked 11,000 physicians to take an aspirin every other day and observed how many of them had a heart attack in a five-year period.

(c) In a test of a new package design, you drop a carton of a dozen eggs from a height of 1 foot and count the number of broken eggs.

(d) Choose a student in your class at random. Ask how much cash that student is carrying.

(e) A nutrition researcher feeds a new diet to a young male white rat. The response variable is the weight (in grams) that the rat gains in 8 weeks.

6.11   Give a reasonable sample space for the number of calories in a hot dog. (Table 1.4 on page 49 contains some typical values to guide you.)

6.12   For each of the following, use a tree diagram or the multiplication principle to determine the number of outcomes in the sample space. Then write the sample space using set notation.

(a) Toss 2 coins.

(b) Toss 3 coins.

(c) Toss 4 coins.

6.13   For each of the following, use a tree diagram or the multiplication principle to determine the number of outcomes in the sample space.

(a) Suppose a county license tag has a four-digit number for identification. If any digit can occupy any of the four positions, how many county license tags can you have?

(b) If the county license tags described in (a) do not allow duplicate digits, how many county license tags can you have?

(c) Suppose the county license tags described in (a) can have *up to* four digits. How many county license tags will this scheme allow?

6.14   Refer to the experiment described in Activity 6.

(a) Determine the number of outcomes in the sample space.

(b) List the outcomes in the sample space.

6.15   Suppose a state's license plates have three digits, then the state seal, and then three more digits.

(a) If any digit can go in any position, how many license plates are possible?

(b) Suppose the state used up all of the six-digit combinations and decided to replace the first three digits with letters of the alphabet. How many license plates are now possible?

(c) The state had to expand once more and added a space for a fourth digit at the end. Example: ABC 1234. How many license plates are possible?

6.16   (a) If a telephone company adds a new exchange (the first three digits in a seven-digit number), how many additional telephone numbers are possible?

(b) If the company adds a new area code, how many additional telephone numbers are possible? (Note: The first digit can't be 0 or 1.)

Intuitive probability

A sample space S lists the possible outcomes of a random phenomenon. To complete a mathematical model for the random phenomenon, we must also give the probabilities with which these outcomes occur.

The true long-term proportion of any outcome—say, "exactly 2 heads in four tosses of a coin"—can be found only empirically, and then only approximately. How then can we describe probability mathematically? Rather than immediately attempt to give "correct" probabilities, let's confront the easier task of laying down rules that any assignment of probabilities must satisfy. We need to assign probabilities not only to single outcomes but also to sets of outcomes.

EVENT

An **event** is an outcome or a set of outcomes of a random phenomenon. That is, an event is a subset of the sample space.

EXAMPLE 6.8

Take the sample space S for four tosses of a coin to be the list of all possible outcomes in the form HTHH. Then "exactly 2 heads" is an event. Call this event A. The event A expressed as a set of outcomes is

$$A = \{HHTT, HTHT, HTTH, THHT, THTH, TTHH\}$$

In a probability model, events have probabilities. What properties must any assignment of probabilities to events have? Here are some basic facts about any probability model. These facts follow from the idea of probability as "the long-run proportion of repetitions in which an event occurs."

1. **Any probability is a number between 0 and 1.** Any proportion is a number between 0 and 1, so any probability is also a number between 0 and 1. An event with probability 0 never occurs, and an event with probability 1 occurs on every trial. An event with probability 0.5 occurs in half the trials in the long run.

2. **All possible outcomes together must have probability 1.** Because some outcome must occur on every trial, the sum of the probabilities for all possible outcomes must be exactly 1.

3. **The probability that an event does not occur is 1 minus the probability that the event does occur.** If an event occurs in (say) 70% of all trials,

it fails to occur in the other 30%. The probability that an event occurs and the probability that it does not occur always add to 100%, or 1.

4. **If two events have no outcomes in common, the probability that one or the other occurs is the sum of their individual probabilities.** If one event occurs in 40% of all trials, a different event occurs in 25% of all trials, and the two can never occur together, then one or the other occurs on 65% of all trials because 40% + 25% = 65%.

Probability rules

Formal probability uses mathematical notation to state Facts 1 to 4 more concisely. We use capital letters near the beginning of the alphabet to denote events. If A is any event, we write its probability as $P(A)$. Here are our probability facts in formal language. As you apply these rules, remember that they are just another form of intuitively true facts about long-run proportions.

PROBABILITY RULES

Rule 1. The probability $P(A)$ of any event A satisfies $0 \leq P(A) \leq 1$.

Rule 2. If S is the sample space in a probability model, then $P(S) = 1$.

Rule 3. The **complement** of any event A is the event that A does not occur, written as A^c. The **complement rule** states that

$$P(A^c) = 1 - P(A)$$

Rule 4. Two events A and B are **disjoint** if they have no outcomes in common and so can never occur simultaneously. If A and B are disjoint,

$$P(A \text{ or } B) = P(A) + P(B)$$

This is the **addition rule** for disjoint events.

Venn diagram

You may find it helpful to draw a picture to remind yourself of the meaning of complements and disjoint events. A picture like Figure 6.3 that shows the sample space S as a rectangular area and events as areas within S is called a **Venn diagram**. The events A and B in Figure 6.3 are disjoint because they do not overlap. As Figure 6.4 shows, the complement A^c contains exactly the outcomes that are not in A.

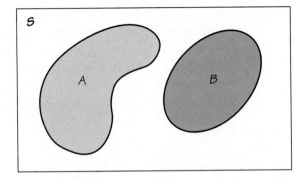

FIGURE 6.3   Venn diagram showing disjoint events A and B.

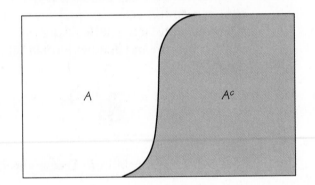

FIGURE 6.4   Venn diagram showing the complement A^c of an event A.

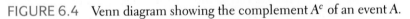

EXAMPLE 6.9

Select a woman aged 25 to 29 years old at random and record her marital status. "At random" means that we give every such woman the same chance to be the one we choose. That is, we choose an SRS of size 1. The probability of any marital status is just the proportion of all women aged 25 to 29 who have that status—if we selected many women, this is the proportion we would get. Here is the probability model:

Marital status	Never married	Married	Widowed	Divorced
Probability	.353	.574	.002	.071

Each probability is between 0 and 1. The probabilities add to 1 because these outcomes together make up the sample space S.

The probability that the woman we draw is not married is, by the complement rule,

$$P(\text{not married}) = 1 - P(\text{married})$$
$$= 1 - .574 = .426$$

That is, if 57.4% are married, then the remaining 42.6% are not married.

"Never married" and "Divorced" are disjoint events, because no woman can be both never married and divorced. So the addition rule says that

$$P(\text{never married or divorced}) = P(\text{never married}) + P(\text{divorced})$$
$$= .353 + .071 = .424$$

That is, 42.4% of women in this age group are either never married or divorced.

Assigning probabilities: finite number of outcomes

The events "never married" and "divorced" in Example 6.9 contain just one outcome each. The addition rule applies to individual outcomes and provides a way to assign probabilities to events: start with probabilities for individual outcomes and add to get probabilities for events. This idea works well when there are only a finite (fixed and limited) number of outcomes.

PROBABILITIES IN A FINITE SAMPLE SPACE

Assign a probability to each individual outcome. These probabilities must be numbers between 0 and 1 and must have sum 1.

The probability of any event is the sum of the probabilities of the outcomes making up the event.

EXAMPLE 6.10

A sociologist studies social mobility in England by recording the social class of a large sample of fathers and their sons. Social class is determined by such factors as education and occupation. The social classes are ordered from Class 1 (lowest) to Class 5 (highest). Here are the probabilities that the son of a lower-class (Class 1) father will end up in each social class:

Son's class	1	2	3	4	5
Probability	.48	.38	.08	.05	.01

Consider the events

$$A = \{\text{son remains in Class 1}\}$$
$$B = \{\text{son reaches one of the two highest classes}\}$$

From the table of probabilities,

$$P(A) = .48$$
$$P(B) = .05 + .01 = .06$$

What is the probability that the son of a Class 1 father does *not* remain in Class 1? This is

$$P(A^c) = 1 - P(A)$$
$$= 1 - .48 = .52$$

The events A and B are disjoint, so the probability that the son of a lower-class father either remains in the lower class or reaches one of the two top classes is

$$P(A \text{ or } B) = P(A) + P(B)$$
$$= .48 + .06 = .54$$

Assigning probabilities: equally likely outcomes

Assigning correct probabilities to individual outcomes often requires long observation of the random phenomenon. In some special circumstances, however, we are willing to assume that individual outcomes are equally likely because of some balance in the phenomenon. Ordinary coins have a physical balance that should make heads and tails equally likely, for example, and the table of random digits comes from a deliberate randomization.

EXAMPLE 6.11

The successive digits in Table B were produced by a careful randomization that makes each entry equally likely to be any of the 10 candidates. Because the total probability must be 1, the probability of each of the 10 outcomes must be 1/10. That is, the assignment of probabilities to outcomes is

Outcome	0	1	2	3	4	5	6	7	8	9
Probability	.1	.1	.1	.1	.1	.1	.1	.1	.1	.1

The probability of any event is the sum of the probabilities of the outcomes making up the event. For example, the probability that an odd digit is chosen is

$$P \text{ (odd outcome)} = P(1) + P(3) + P(5) + P(7) + P(9) = .5$$

Here are two events, with the outcomes that they contain:

$$A = \{\text{odd outcome}\} = \{1, 3, 5, 7, 9\}$$
$$B = \{\text{outcome less than or equal to 3}\} = \{0, 1, 2, 3\}$$

We saw that $P(A) = .5$. Because the event B contains 4 outcomes, $P(B) = .4$. The event $\{A \text{ or } B\}$ contains 7 outcomes,

$$\{A \text{ or } B\} = \{0, 1, 2, 3, 5, 7, 9\}$$

so it has probability 0.7. This is *not* the sum of $P(A)$ and $P(B)$, because A and B are *not* disjoint events. Outcomes 1 and 3 belong to both A and B.

In Example 6.11 all outcomes have the same probability. Because there are 10 equally likely outcomes, each must have probability 1/10. Because exactly 5 of the 10 equally likely outcomes are odd, the probability of an odd outcome is 5/10, or 0.5. In the special situation where all outcomes are equally likely, we have a simpler rule for assigning probabilities to events.

EQUALLY LIKELY OUTCOMES

If a random phenomenon has k possible outcomes, all equally likely, then each individual outcome has probability $1/k$. The probability of any event A is

$$P(A) = \frac{\text{count of outcomes in } A}{\text{count of outcomes in } S}$$

$$= \frac{\text{count of outcomes in } A}{k}$$

Most random phenomena do not have equally likely outcomes, so the general rule for finite sample spaces is more important than the special rule for equally likely outcomes.

EXERCISES

6.17   Refer to the experiment described in Activity 6 and Exercise 6.14.

(a) Determine the theoretical probability that at least one digit will occur in its correct place.

(b) Compare the theoretical probability with your experimental (empirical) results.

6.18   All human blood can be typed as one of O, A, B, or AB, but the distribution of the types varies a bit with race. Here is the distribution of the blood type of a randomly chosen black American:

Blood type	O	A	B	AB
Probability	.49	.27	.20	? .04

(a) What is the probability of type AB blood? Why?

(b) Maria has type B blood. She can safely receive blood transfusions from people with blood types O and B. What is the probability that a randomly chosen black American can donate blood to Maria?

6.19   If you draw an M&M candy at random from a bag of the candies, the candy you draw will have one of six colors. The probability of drawing each color depends on the proportion of each color among all candies made.

(a) The table below gives the probability of each color for a randomly chosen plain M&M:

Color	Brown	Red	Yellow	Green	Orange	Blue
Probability	.3	.2	.2	.1	.1	?

What must be the probability of drawing a blue candy?

(b) The probabilities for peanut M&Ms are a bit different. Here they are:

Color	Brown	Red	Yellow	Green	Orange	Blue
Probability	.2	.1	.2	.1	.1	?

What is the probability that a peanut M&M chosen at random is blue?

(c) What is the probability that a plain M&M is any of red, yellow, or orange? What is the probability that a peanut M&M has one of these colors?

6.20   A sociologist studying social mobility in Denmark finds that the probability that the son of a lower-class father remains in the lower class is 0.46. What is the probability that the son moves to one of the higher classes?

6.21   Government data assign a single cause for each death that occurs in the United States. The data show that the probability is 0.45 that a randomly chosen death was due to cardiovascular (mainly heart) disease, and 0.22 that it was due to cancer. What is the probability that a death was due either to cardiovascular disease or to cancer? What is the probability that the death was due to some other cause?

6.22   Choose an acre of land in Canada at random. The probability is 0.35 that it is forest and 0.03 that it is pasture.

(a) What is the probability that the acre chosen is not forested?
(b) What is the probability that it is either forest or pasture?
(c) What is the probability that a randomly chosen acre in Canada is something other than forest or pasture?

6.23   Select a first-year college student at random and ask what his or her academic rank was in high school. Here are the probabilities, based on proportions from a large sample survey of first-year students:

Rank	Top 20%	Second 20%	Third 20%	Fourth 20%	Lowest 20%
Probability	.41	.23	.29	.06	.01

(a) What is the sum of these probabilities? Why do you expect the sum to have this value?
(b) What is the probability that a randomly chosen first-year college student was not in the top 20% of his or her high school class?
(c) What is the probability that a first-year student was in the top 40% in high school?

Independence and the multiplication rule

Rule 4, the addition rule for disjoint events, describes the probability that *one or the other* of two events A and B will occur in the special situation when A and B cannot occur together because they are disjoint. Our final rule describes the probability that *both* events A and B occur, again only in a special situation. More general rules appear in Section 6.3, but in our study of statistics we will need only the rules that apply to special situations.

Suppose that you toss a balanced coin twice. You are counting heads, so two events of interest are

$$A = \text{first toss is a head}$$

$$B = \text{second toss is a head}$$

The events A and B are not disjoint. They occur together whenever both tosses give heads. We want to compute the probability of the event $\{A \text{ and } B\}$ that *both* tosses are heads. The Venn diagram in Figure 6.5

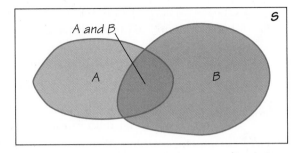

FIGURE 6.5   Venn diagram showing the event {A and B}.

illustrates the event {A and B} as the overlapping area that is common to both A and B.

The coin tossing of Buffon, Pearson, and Kerrich described at the beginning of this chapter makes us willing to assign probability 1/2 to a head when we toss a coin. So

$$P(A) = .5$$
$$P(B) = .5$$

What is $P(A \text{ and } B)$? Our common sense says that it is 1/4. The first toss will give a head half the time and then the second will give a head on half of those trials, so both tosses will give heads on $1/2 \times 1/2 = 1/4$ of all trials in the long run. This reasoning assumes that the second toss still has probability 1/2 of a head after the first has given a head. This is true—we can verify it by performing many trials of two tosses and observing the proportion of heads on the second toss after the first toss has produced a head. We say that the events "head on the first toss" and "head on the second toss" are *independent*. Here is our final probability rule.

THE MULTIPLICATION RULE FOR INDEPENDENT EVENTS

Rule 5. Two events A and B are **independent** if knowing that one occurs does not change the probability that the other occurs. If A and B are independent,

$$P(A \text{ and } B) = P(A)P(B)$$

This is the **multiplication rule** for independent events.

Our definition of independence is rather informal. A more precise definition appears in Section 6.3. In practice, though, we rarely need a precise definition of independence, because independence is usually *assumed* as part of a probability model when we want to describe random phenomena that seem to be physically unrelated to each other.

EXAMPLE 6.12

Because a coin has no memory and most coin tossers cannot influence the fall of the coin, it is safe to assume that successive coin tosses are independent. For a balanced coin this means that after we see the outcome of the first toss, we still assign probability 1/2 to heads on the second toss.

On the other had, the colors of successive cards dealt from the same deck are not independent. A standard 52-card deck contains 26 red and 26 black cards. For the first card dealt from a shuffled deck, the probability of a red card is 26/52 = .50 (equally likely outcomes). Once we see that the first card is red, we know that there are only 25 reds among the remaining 51 cards. The probability that the second card is red is therefore only 25/51 = .49. Knowing the outcome of the first deal changes the probability for the second.

If a doctor measures your blood pressure twice, it is reasonable to assume that the two results are independent because the first result does not influence the instrument that makes the second reading. But if you take an IQ test or other mental test twice in succession, the two test scores are not independent. The learning that occurs on the first attempt influences your second attempt.

When independence is part of a probability model, the multiplication rule applies. Here is an example.

EXAMPLE 6.13

Gregor Mendel used garden peas in some of the experiments that revealed that inheritance operates randomly. The seed color of Mendel's peas can be either green or yellow. Two parent plants are "crossed" (one pollinates the other) to produce seeds. Each parent plant carries two genes for seed color, and each of these genes has probability 1/2 of being passed to a seed. The two genes that the seed receives, one from each parent, determine its color. The parents contribute their genes independently of each other.

Suppose that both parents carry the G and the Y genes. The seed will be green if both parents contribute a G gene; otherwise it will be yellow. If M is the event that the male contributes a G gene and F is the event that the female contributes a G gene, then the probability of a green seed is

$$P(M \text{ and } F) = P(M)P(F)$$
$$= (.5)(.5) = .25$$

In the long run, 1/4 of all seeds produced by crossing these plants will be green.

The multiplication rule $P(A \text{ and } B) = P(A)P(B)$ holds if A and B are independent but not otherwise. The addition rule $P(A \text{ or } B) = P(A) + P(B)$ holds if A and B are disjoint but not otherwise. Resist the temptation to use these simple formulas when the circumstances that justify them are not present. You must also be certain not to confuse disjointness and independence. If A and B are disjoint, then the fact that A occurs tells us that B cannot occur—look again at Figure 6.3. So disjoint events are not independent. Unlike disjointness or complements, independence cannot be pictured by a Venn diagram, because it involves the probabilities of the events rather than just the outcomes that make up the events.

Applying the probability rules

If two events A and B are independent, then their complements A^c and B^c are also independent and A^c is independent of B. Suppose, for example, that 75% of all registered voters in a suburban district are Republicans. If an opinion poll interviews two voters chosen independently, the probability that the first is a Republican and the second is not a Republican is $(.75)(.25) = .1875$. The multiplication rule also extends to collections of more than two events, provided that all are independent. Independence of events A, B, and C means that no information about any one or any two can change the probability of the remaining events. The formal definition is a bit messy. Fortunately, independence is usually assumed in setting up a probability model. We can then use the multiplication rule freely, as in this example.

EXAMPLE 6.14

A transatlantic telephone cable contains repeaters at regular intervals to amplify the signal. If a repeater fails, it must be replaced by fishing the cable to the surface at great expense. Each repeater has probability 0.999 of functioning without failure for 10 years. Repeaters fail independently of each other. (This assumption means that there are no "common causes" such as earthquakes that would affect several repeaters at once.) Denote by A_i the event that the ith repeater operates successfully for 10 years.

The probability that two repeaters both last 10 years is

$$P(A_1 \text{ and } A_2) = P(A_1)P(A_2)$$
$$= .999 \times .999 = .998$$

For a cable with 10 repeaters the probability of no failures in 10 years is

$$P(A_1 \text{ and } A_2 \text{ and } \dots \text{ and } A_{10}) = P(A_1)P(A_2)\dots P(A_{10})$$
$$= .999 \times .999 \times \dots \times .999$$
$$= .999^{10} = .990$$

Cables with 2 or 10 repeaters are quite reliable. Unfortunately, a transatlantic cable has 300 repeaters. The probability that all 300 work for 10 years is

$$P(A_1 \text{ and } A_2 \text{ and} \ldots \text{ and } A_{300} = .999^{300} = .741$$

There is therefore about one chance in four that the cable will have to be fished up for replacement of a repeater sometime during the next 10 years. Repeaters are in fact designed to be much more reliable than 0.999 in 10 years. Some transatlantic cables have served for more than 20 years with no failures.

By combining the rules we have learned, we can compute probabilities for rather complex events. Here is an example.

EXAMPLE 6.15

A diagnostic test for the presence of the AIDS virus has probability 0.005 of producing a false positive. That is, when a person free of the AIDS virus is tested, the test has probability 0.005 of falsely indicating that the virus is present. If the 140 employees of a medical clinic are tested and all 140 are free of AIDS, what is the probability that at least one false positive will occur?

It is reasonable to assume as part of the probability model that the test results for different individuals are independent. The probability that the test is positive for a single person is 0.005, so the probability of a negative result is $1 - .005 = .995$ by the complement rule. The probability of at least one false positive among the 140 people tested is therefore

$$
\begin{aligned}
P(\text{at least one positive}) &= 1 - P(\text{no positives}) \\
&= 1 - P(140 \text{ negatives}) \\
&= 1 - .995^{140} \\
&= 1 - .496 = .504
\end{aligned}
$$

The probability is greater than 1/2 that at least one of the 140 people will test positive for AIDS, even though no one has the virus.

EXERCISES

6.24   A general can plan a campaign to fight one major battle or three small battles. He believes that he has probability 0.6 of winning the large battle and probability 0.8 of winning each of the small battles. Victories or defeats in the small battles are independent. The general must win either the large battle or all three small battles to win the campaign. Which strategy should he choose?

6.25   An automobile manufacturer buys computer chips from a supplier. The supplier sends a shipment containing 5% defective chips. Each chip chosen from this shipment has probability 0.05 of being defective, and each automobile uses 12 chips selected independently. What is the probability that all 12 chips in a car will work properly?

6.26   Government data show that 27% of the civilian labor force have at least 4 years of college and that 16% of the labor force work as laborers or operators of machines or vehicles. Can you conclude that because $(.27)(.16) = .043$, about 4% of the labor force are college-educated laborers or operators? Explain your answer.

6.27   Choose at random a U.S. resident at least 25 years of age. We are interested in the events

$$A = \{\text{The person chosen completed 4 years of college}\}$$
$$B = \{\text{The person chosen is 55 years old or older}\}$$

Government data recorded in Table 4.6 on page 216 allow us to assign probabilities to these events.

(a) Explain why $P(A) = .230$.
(b) Find $P(B)$.
(c) Find the probability that the person chosen is at least 55 years old *and* has 4 years of college education, $P(A \text{ and } B)$. Are the events A and B independent?

SUMMARY

A **probability model** for a random phenomenon consists of a sample space S and an assignment of probabilities P.

The **sample space** S is the set of all possible outcomes of the random phenomenon. Sets of outcomes are called **events.** P assigns a number $P(A)$ to an event A as its probability.

The **complement** A^c of an event A consists of exactly the outcomes that are not in A. Events A and B are **disjoint** if they have no outcomes in common. Events A and B are **independent** if knowing that one event occurs does not change the probability we would assign to the other event.

Any assignment of probability must obey the rules that state the basic properties of probability:

1. $0 \leq P(A) \leq 1$ for any event A.
2. $P(S) = 1$.
3. **Complement rule:** For any event A, $P(A^c) = 1 - P(A)$.

4. **Addition rule:** If events A and B are **disjoint,** then $P(A \text{ or } B) = P(A) + P(B)$.

5. **Multiplication rule:** If events A and B are **independent,** then $P(A \text{ and } B) = P(A)P(B)$.

SECTION 6.2 EXERCISES

6.28   Figure 6.6 displays several assignments of probabilities to the six faces of a die. We can learn which assignment is actually *accurate* for a particular die only by rolling the die many times. However, some of the assignments are not *legitimate* assignments of probability. That is, they do not obey the rules. Which are legitimate and which are not? In the case of the illegitimate models, explain what is wrong.

6.29   In each of the following situations, state whether or not the given assignment of probabilities to individual outcomes is legitimate, that is, satisfies the rules of probability. If not, give specific reasons for your answer.

(a)   When a coin is spun, $P(H) = .55$ and $P(T) = .45$.

(b)   When two coins are tossed, $P(HH) = .4, P(HT) = .4$, $P(TH) = .4$, and $P(TT) = .4$.

(c)   Plain M&Ms have not always had the mixture of colors given in Exercise 6.19. In the past there were no red candies and no blue candies.

Outcome	Model 1	Model 2	Model 3	Model 4
⚀	$\frac{1}{3}$	$\frac{1}{6}$	$\frac{1}{7}$	$\frac{1}{3}$
⚁	0	$\frac{1}{6}$	$\frac{1}{7}$	$\frac{1}{3}$
⚂	$\frac{1}{6}$	$\frac{1}{6}$	$\frac{1}{7}$	$-\frac{1}{6}$
⚃	0	$\frac{1}{6}$	$\frac{1}{7}$	$-\frac{1}{6}$
⚄	$\frac{1}{6}$	$\frac{1}{6}$	$\frac{1}{7}$	$\frac{1}{3}$
⚅	$\frac{1}{3}$	$\frac{1}{6}$	$\frac{1}{7}$	$\frac{1}{3}$

FIGURE 6.6   Four assignments of probabilities to the six faces of a die (Exercise 6.28).

Tan had probability 0.10 and the other four colors had the same probabilities that are given in Exercise 6.19.

6.30   Choose an American farm at random and measure its size in acres. Here are the probabilities that the farm chosen falls in several acreage categories:

Acres	< 10	10–49	50–99	100–179	180–499	500–999	1000–1999	≥ 2000
Probability	0.09	0.20	0.15	0.16	0.22	0.09	0.05	0.04

Let A be the event that the farm is less than 50 acres in size, and let B be the event that it is 500 acres or more.

(a) Find $P(A)$ and $P(B)$.

(b) Describe A^c in words and find $P(A^c)$ by the complement rule.

(c) Describe {A or B} in words and find its probability by the addition rule.

6.31   Choose an American worker at random and classify his or her occupation into one of the following classes. These classes are used in government employment data.

    A   Managerial and professional
    B   Technical, sales, administrative support
    C   Service occupations
    D   Precision production, craft, and repair
    E   Operators, fabricators, and laborers
    F   Farming, forestry, and fishing

The table below gives the probabilities that a randomly chosen worker falls into each of 12 sex-by-occupation classes:

	A	B	C	D	E	F
Male	.14	.11	.06	.11	.12	.03
Female	.09	.20	.08	.01	.04	.01

(a) Verify that this is a legitimate assignment of probabilities to these outcomes.

(b) What is the probability that the worker is female?

    (c) What is the probability that the worker is not engaged in farming, forestry, or fishing?

    (d) Classes D and E include most mechanical and factory jobs. What is the probability that the worker holds a job in one of these classes?

    (e) What is the probability that the worker does not hold a job in Classes D or E?

6.32   A roulette wheel has 38 slots, numbered 0, 00, and 1 to 36. The slots 0 and 00 are colored green, 18 of the others are red, and 18 are black. The dealer spins the wheel and at the same time rolls a small ball along the wheel in the opposite direction. The wheel is carefully balanced so that the ball is equally likely to land in any slot when the wheel slows. Gamblers can bet on various combinations of numbers and colors.

    (a) What is the probability that the ball will land in any one slot?

    (b) If you bet on "red," you win if the ball lands in a red slot. What is the probability of winning?

    (c) The slot numbers are laid out on a board on which gamblers place their bets. One column of numbers on the board contains all multiples of 3, that is, 3, 6, 9, ..., 36. You place a "column bet" that wins if any of these numbers comes up. What is your probability of winning?

6.33   A six-sided die has four green and two red faces and is balanced so that each face is equally likely to come up. The die will be rolled several times. You must choose one of the following three sequences of colors; you will win \$25 if the first rolls of the die give the sequence that you have chosen.

RGRRR

RGRRRG

GRRRRR

Which sequence do you choose? Explain your choice. (In a psychological experiment, 63% of 260 students who had not studied probability chose the second sequence. This is evidence that our intuitive understanding of probability is not very accurate.)[2]

6.34   The gene for albinism in humans is recessive. That is, carriers of this gene have probability 1/2 of passing it to a child, and the child is albino only if both parents pass the albinism gene. Parents pass their genes independently of each other. If both parents carry the albinism gene, what is the probability that their first child is albino? If they have two children (who inherit independently of each other), what is the probability that both are albino? That neither is albino?

6.35   The "random walk" theory of securities prices holds that price movements in disjoint time periods are independent of each other. Suppose that we record only whether the price is up or down each year, and that the probability that our portfolio rises in price in any one year is 0.65. (This probability is approximately correct for a portfolio containing equal dollar amounts of all common stocks listed on the New York Stock Exchange.)

(a) What is the probability that our portfolio goes up for 3 consecutive years?

(b) If you know that the portfolio has risen in price 2 years in a row, what probability do you assign to the event that it will go down next year?

(c) What is the probability that the portfolio's value moves in the same direction in both of the next 2 years?

6.36   The type of medical care a patient receives may vary with the age of the patient. A large study of women who had a breast lump investigated whether or not each woman received a mammogram and a biopsy when the lump was discovered. Here are some probabilities estimated by the study. The entries in the table are the probabilities that *both* of two events occur; for example, 0.321 is the probability that a patient is under 65 years of age *and* the tests were done. The four probabilities in the table have sum 1 because the table lists all possible outcomes.

| | Tests done? | |
	Yes	No
Age under 65	.321	.124
Age 65 or over	.365	.190

(a) What is the probability that a patient in this study is under 65? That a patient is 65 or over?

(b) What is the probability that the tests were done for a patient? That they were not done?

(c) Are the events A = {patient was 65 or older} and B = {the tests were done} independent? Were the tests omitted on older patients more or less frequently than would be the case if testing were independent of age?

6.3 MORE ABOUT PROBABILITY

Now we return to the laws that govern any assignment of probabilities. The purpose of learning more laws of probability is to be able to give probability models for more complex random phenomena. We have already met and used five rules.

RULES OF PROBABILITY

Rule 1. $0 \leq P(A) \leq 1$ for any event A.

Rule 2. $P(S) = 1$.

Rule 3. Complement rule: For any event A,

$$P(A^c) = 1 - P(A)$$

Rule 4. Addition rule: If A and B are disjoint events, then

$$P(A \text{ or } B) = P(A) + P(B)$$

Rule 5. Multiplication rule: If A and B are independent events, then

$$P(A \text{ and } B) = P(A)P(B)$$

General addition rules

Probability has the property that if A and B are disjoint events, then $P(A$ or $B) = P(A) + P(B)$. What if there are more than two events, or if the events are not disjoint? These circumstances are covered by more general addition rules for probability.

UNION

The **union** of any collection of events is the event that at least one of the collection occurs.

For two events A and B, the union is the event $\{A$ or $B\}$ that A or B or both occur. From the addition rule for two disjoint events, we can obtain rules for more general unions. Suppose first that we have several

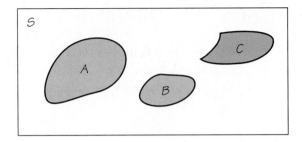

FIGURE 6.7   The addition rule for disjoint events: $P(A \text{ or } B \text{ or } C) = P(A) +$ $P(B) + P(C)$ when events A, B, and C are disjoint.

events—say A, B, and C—that are disjoint in pairs. That is, no two can occur simultaneously. The Venn diagram in Figure 6.7 illustrates three disjoint events. The addition rule for two disjoint events extends to the following law:

ADDITION RULE FOR DISJOINT EVENTS

If events A, B, and C are disjoint in the sense that no two have any outcomes in common, then

$$P(\text{one or more of } A, B, C) = P(A) + P(B) + P(C)$$

This rule extends to any number of disjoint events.

EXAMPLE 6.16

Generate a random number X between 0 and 1. What is the probability that the first digit will be odd? We will learn in Chapter 7 that the variable X has the density curve of a uniform distribution (see Exercise 2.2). This density curve has constant height 1 between 0 and 1 and is 0 elsewhere. The event that the first digit of X is odd is the union of five disjoint events. These events are

$$.10 \leq X < .20$$
$$.30 \leq X < .40$$
$$.50 \leq X < .60$$
$$.70 \leq X < .80$$
$$.90 \leq X < 1.00$$

Figure 6.8 illustrates the probabilities of these events as areas under the density curve. Each has probability 0.1 equal to its length. The union of the five therefore has probability equal to the sum, or 0.5. As we should expect, a random number is equally likely to begin with an odd or an even digit.

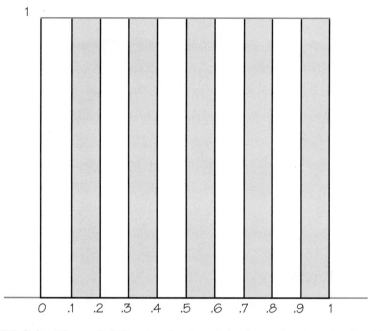

FIGURE 6.8   The probability that the first digit of a random number is odd is the sum of the probabilities of the 5 disjoint events shown (Example 6.16).

If events A and B are not disjoint, they can occur simultaneously. The probability of their union is then *less* than the sum of their probabilities. As Figure 6.9 suggests, the outcomes common to both are counted twice when we add probabilities, so we must subtract this probability once. Here is the addition rule for the union of any two events, disjoint or not.

GENERAL ADDITION RULE FOR UNIONS OF TWO EVENTS

For any two events A and B,

$$P(A \text{ or } B) = P(A) + P(B) - P(A \text{ and } B)$$

If A and B are disjoint, the event $\{A$ and $B\}$ that both occur has no outcomes in it. This *empty event* is the complement of the sample space S and must have probability 0. So the general addition rule includes Rule 4, the addition rule for disjoint events.

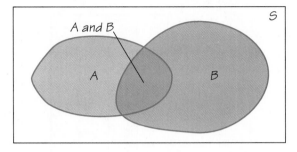

FIGURE 6.9   The general addition rule for the union of two events: $P(A$ or $B) = P(A) + P(B) - P(A$ and $B)$ for any events A and B.

EXAMPLE 6.17

Deborah and Matthew are anxiously awaiting word on whether they have been made partners of their law firm. Deborah guesses that her probability of making partner is 0.7 and that Matthew's is 0.5. (These are personal probabilities reflecting Deborah's assessment of chance.) This assignment of probabilities does not give us enough information to compute the probability that at least one of the two is promoted. In particular, adding the individual probabilities of promotion gives the impossible result 1.2. If Deborah also guesses that the probability that *both* she and Matthew are made partners is 0.3, then by the addition rule for unions

$$P(\text{at least one is promoted}) = .7 + .5 - .3 = .9$$

The probability that *neither* is promoted is then 0.1 by the complement rule.

Venn diagrams are a great help in finding probabilities for unions, because you can just think of adding and subtracting areas. Figure 6.10 shows some events and their probabilities for Example 6.17. What is the probability that Deborah is promoted and Matthew is not? The Venn diagram shows that this is the probability that Deborah is promoted minus the probability that both are promoted, $.7 - .3 = .4$. Similarly, the probability that Matthew is promoted and Deborah is not is $.5 - .3 = .2$. The four probabilities that appear in the figure add to 1 because they refer to four disjoint events whose union is the entire sample space.

joint event

The simultaneous occurrence of two events, such as $A =$ Deborah is promoted *and* $B =$ Matthew is promoted, is called a **joint event**. The probability of a joint event, such as P(Deborah is promoted *and* Matthew is

joint probability

promoted$) = P(A$ and $B)$, is called a **joint probability**. Determining joint probabilities when you have equally likely outcomes can be as easy as counting outcomes. For most situations, however, we will need more powerful methods, which will be developed later in this section.

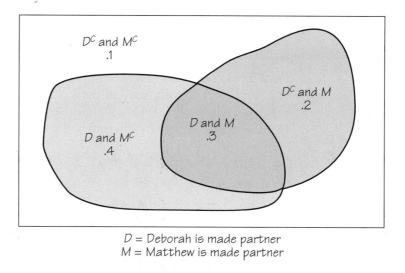

D = Deborah is made partner
M = Matthew is made partner

FIGURE 6.10   Venn diagram and probabilities for Example 6.17.

EXERCISES

6.37   Call a household prosperous if its income exceeds $75,000. Call the household educated if the householder completed college. Select an American household at random, and let A be the event that the selected household is prosperous and B the event that it is educated. According to the Census Bureau, $P(A) = .125$, $P(B) = .237$, and the joint probability that a household is both prosperous and educated is $P(A \text{ and } B) = .077$. What is the probability $P(A \text{ or } B)$ that the household selected is either prosperous or educated?

6.38   Consolidated Builders has bid on two large construction projects. The company president believes that the probability of winning the first contract (event A) is 0.6, that the probability of winning the second (event B) is 0.4, and that the joint probability of winning both jobs (event $\{A \text{ and } B\}$) is 0.2. What is the probability of the event $\{A \text{ or } B\}$ that Consolidated will win at least one of the jobs?

6.39   Draw a Venn diagram that shows the relation between the events A and B in Exercise 6.37. Indicate each of the following events on your diagram and use the information in Exercise 6.37 to calculate the probability of each joint event. Finally, describe in words what each event is.

(a) $\{A \text{ and } B\}$
(b) $\{A \text{ and } B^c\}$

(c) $\{A^c \text{ and } B\}$

(d) $\{A^c \text{ and } B^c\}$

6.40   Draw a Venn diagram that illustrates the relation between events A and B in Exercise 6.38. Write each of the following events in terms of A, B, A^c, and B^c. Indicate the events on your diagram and use the information in Exercise 6.38 to calculate the probability of each.

(a) Consolidated wins both jobs.

(b) Consolidated wins the first job but not the second.

(c) Consolidated does not win the first job but does win the second.

(d) Consolidated does not win either job.

Conditional probability

In Section 4.3 we met the idea of a *conditional distribution*, the distribution of a variable given that a condition is satisfied. Now we will introduce the probability language for this idea. Table 6.1 gives data on the age and marital status of adult American women. Choose an adult American woman at random. What is the probability that the woman you choose is married? Table 6.1 shows that there are 99,585,000 adult women in the United States, and that 58,929,000 of them are married. Because "choose at random" gives an equal chance to all women, the probability we want is just the proportion of all women in the population who are married. That is,

$$P(\text{married}) = \frac{58,929}{99,585} = .592$$

Suppose we now learn that the woman chosen is between 18 and 24 years old. Does this information change the probability that she is married? We know that young women are less likely to be married, so we are quite sure that the probability for young women is lower than the probability for all women. In fact, there are 12,614,000 women aged 18 to 24, and 3,046,000 of these are married. So the probability that a woman is married *given the information that she is between 18 and 24 years old* is

$$P(\text{married} \mid \text{age 18 to 24}) = \frac{3,046}{12,614} = .241$$

conditional probability   This is a *conditional probability*. That is, it gives the probability of one event (the woman chosen is married) under the condition that we know another event (she is between ages 18 and 24). You can read the bar $\mid$ as

TABLE 6.1	Age and marital status of women (thousands of women)

| | Age | | | |
	18 to 24	25 to 64	65 and over	Total
Married	3,046	48,116	7,767	58,929
Never married	9,289	9,252	768	19,309
Widowed	19	2,425	8,636	11,080
Divorced	260	8,916	1,091	10,267
Total	12,614	68,709	18,262	99,585

"given the information that." It is common sense that if we know a woman is young, we lower our guess of the probability that she is married. If we have the data, we can compute the actual conditional probability. Our task is to turn this common sense into something more general.

The counts in Table 6.1 show that if we choose a woman at random, the joint probability that we draw a married woman between 18 and 24 years old is

$$P(\text{age 18 to 24 and married}) = \frac{3,046}{99,585} = .031$$

This calculation used a count from the body of the table: there are 3,046,000 women who are both age 18 to 24 and married.

If we are given no information about marital status, we find probabilities about a woman's age from the "Total" row at the bottom of the table:

18 to 24	25 to 64	65 and over	Total
12,614	68,709	18,262	99,585

For example, the probability that the woman we choose is between 18 and 24 years old is

$$P(\text{age 18 to 24}) = \frac{12,614}{99,585} = .127$$

If we are given some information, we look only at the women described by the information given. For example, if we are told that the woman is between 18 and 24 years old, we are confined to the "18 to 24" column in Table 6.1:

Married	3,046
Never married	9,289
Widowed	19
Divorced	260
Total	12,614

Thus the *conditional* probability that a woman is married, given the information that she is between 18 and 24 years of age, is

$$P(\text{married} \mid \text{age 18 to 24}) = \frac{3,046}{12,614} = .241$$

There is a relationship among these three probabilities. The joint probability that a woman is both age 18 to 24 and married is the product of the probabilities that she is age 18 to 24 and that she is married *given* that she is age 18 to 24. That is,

$$P(\text{age 18 to 24 and married}) = P(\text{age 18 to 24}) \times P(\text{married} \mid \text{age 18 to 24})$$

$$= \frac{12,614}{99,585} \times \frac{3,046}{12,614}$$

$$= \frac{3,046}{99,585} = .031 \quad \text{(as before)}$$

Try to think your way through this in words before looking at the formal notation. We have just discovered the general multiplication rule of probability.

GENERAL MULTIPLICATION RULE

The joint probability that both of two events A and B happen together can be found by

$$P(A \text{ and } B) = P(A)P(B \mid A)$$

Here $P(B \mid A)$ is the conditional probability that B occurs given the information that A occurs.

In words, this rule says that for both of two events to occur, first one must occur and then, given that the first event has occurred, the second must occur. In our example, the joint probability that a randomly chosen woman is both age 18 to 24 (event A) and married (event B) is

$$P(A \text{ and } B) = P(A)P(B \mid A)$$
$$= (.127)(.241) = .031$$

If $P(A)$ and $P(A \text{ and } B)$ are given, we can rearrange the general multiplication rule to produce a *definition* of the conditional probability $P(B \mid A)$ in terms of unconditional probabilities.

DEFINITION OF CONDITIONAL PROBABILITY

When $P(A) > 0$, the conditional probability of B given A is

$$P(B \mid A) = \frac{P(A \text{ and } B)}{P(A)}$$

Be sure to keep in mind the distinct roles in $P(B \mid A)$ of the event B whose probability we are computing and the event A that represents the information we are given. The conditional probability $P(B \mid A)$ makes no sense if the event A can never occur, so we require that $P(A) > 0$ whenever we talk about $P(B \mid A)$.

EXAMPLE 6.18

What is the conditional probability that a woman is a widow, given that she is at least 65 years old? We see from Table 6.1 that

$$P(\text{at least } 65) = \frac{18,262}{99,585} = .183$$

$$P(\text{widowed } and \text{ at least } 65) = \frac{8,636}{99,585} = .087$$

The conditional probability is therefore

$$P(\text{widowed} \mid \text{at least } 65) = \frac{P(\text{widowed } and \text{ at least } 65)}{P(\text{at least } 65)}$$

$$= \frac{.087}{.183} = .475$$

Check that this agrees (up to roundoff error) with the result obtained from the "65 and over" column of Table 6.1.

EXERCISES

6.41　Choose an adult American woman at random. Table 6.1 describes the population from which we draw. Use the information in that table to answer the following questions.

(a) What is the probability that the woman chosen is 65 years old or older?

(b) What is the conditional probability that the woman chosen is married, given that she is 65 or over?

(c) How many women are *both* married and in the over-65 age group? What is the joint probability that the woman we choose is a married woman at least 65 years old?

(d) Verify that the three probabilities you found in (a), (b), and (c) satisfy the general multiplication rule.

6.42　Choose an adult American woman at random. Table 6.1 describes the population from which we draw. Use the information in that table to find the following probabilities.

(a) The probability that the woman chosen is a widow.

(b) The conditional probability that the woman chosen is a widow, given that she is at least 65 years old.

(c) The conditional probability that the woman chosen is a widow, given that she is between 25 and 64 years old.

(d) Are the events "widow" and "at least 65 years old" independent? How do you know?

6.43　Choose an adult American woman at random. Table 6.1 describes the population from which we draw.

(a) What is the conditional probability that the woman chosen is 18 to 24 years old, given that she is married?

(b) Earlier, we found the $P(\text{married} \mid \text{age 18 to 21}) = .241$. Complete this sentence: 0.241 is the proportion of women who are _____ among those women who are _____.

(c) In (a), you found $P(\text{age 18 to 24} \mid \text{married})$. Write a sentence of the form given in (b) that describes the meaning of this result. The two conditional probabilities give us very different information.

General multiplication rules

The definition of conditional probability reminds us that in principle all probabilities, including conditional probabilities, can be found from the assignment of probabilities to events that describe a random phenomenon. More often, however, conditional probabilities are part of the information given to us in a probability model, and the multiplication rule is used to compute $P(A \text{ and } B)$.

EXAMPLE 6.19	Slim is a professional poker player. At the moment, he wants very much to draw two diamonds in a row. As he sits at the table looking at his hand and at the upturned cards on the table, Slim sees 11 cards. Of these, 4 are diamonds. The full deck contains 13 diamonds among its 52 cards, so 9 of the 41 unseen cards are diamonds. Because the deck was carefully shuffled, each card that Slim draws is equally likely to be any of the cards that he has not seen.

To find Slim's probability of drawing two diamonds, first calculate

$$P(\text{first card diamond}) = \frac{9}{41}$$

$$P(\text{second card diamond} \mid \text{first card diamond}) = \frac{8}{40}$$

Slim finds both probabilities by counting cards. The probability that the first card drawn is a diamond is 9/41 because 9 of the 41 unseen cards are diamonds. If the first card is a diamond, that leaves 8 diamonds among the 40 remaining cards. So the *conditional* probability of another diamond is 8/40. The multiplication rule now says that

$$P(\text{both cards diamonds}) = \frac{9}{41} \times \frac{8}{40} = .044$$

Slim will need luck to draw his diamonds.

The union of a collection of events is the event that *any* of them occur. Here is the corresponding term for the event that *all* of them occur.

INTERSECTION

> The **intersection** of any collection of events is the event that *all* of the events occur.

The multiplication rule extends to the probability that all of several events occur. The key is to condition each event on the occurrence of *all* of the preceding events. For example, the intersection of three events A, B, and C has probability

$$P(A \text{ and } B \text{ and } C) = P(A)P(B \mid A)P(C \mid A \text{ and } B)$$

EXAMPLE 6.20

Only 5% of male high school basketball, baseball, and football players go on to play at the college level. Of these, only 1.7% enter major league professional sports. About 40% of the athletes who compete in college and then reach the pros have a career of more than 3 years.[3] Define these events:

$$A = \{\text{competes in college}\}$$
$$B = \{\text{competes professionally}\}$$
$$C = \{\text{pro career longer than 3 years}\}$$

What is the probability that a high school athlete competes in college and then goes on to have a pro career of more than 3 years? We know that

$$P(A) = .05$$
$$P(B \mid A) = .017$$
$$P(C \mid A \text{ and } B) = .4$$

The probability we want is therefore

$$P(A \text{ and } B \text{ and } C) = P(A)P(B \mid A)P(C \mid A \text{ and } B)$$
$$= .05 \times .017 \times .40 = .00034$$

Only about 3 of every 10,000 high school athletes can expect to compete in college and have a professional career of more than 3 years. High school students would be wise to concentrate on studies rather than on unrealistic hopes of fortune from pro sports.

Tree diagrams revisited

Probability problems often require us to combine several of the basic rules into a more elaborate calculation. Here is an example that illustrates how to solve problems that have several stages.

EXAMPLE 6.21

What is the probability that a high school athlete will go on to professional sports? In the notation of Example 6.20, this is $P(B)$. To find $P(B)$ from the information in Example 6.20, use the tree diagram in Figure 6.11 to organize your thinking.

Each segment in the tree is one stage of the problem. Each complete branch shows a path that an athlete can take. The probability written on each segment is the conditional probability that an athlete follows that segment

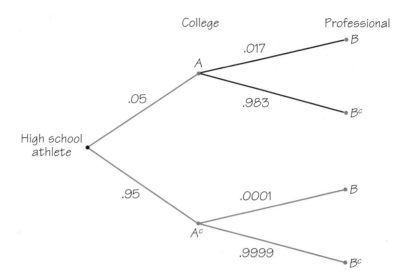

FIGURE 6.11   Tree diagram for Example 6.21. The probability $P(B)$ is the sum of the probabilities of the two branches ending at B.

given that he has reached the point from which it branches. Starting at the left, high school athletes either do or do not compete in college. We know that the probability of competing in college is $P(A) = .05$, so the probability of not competing is $P(A^c) = .95$. These probabilities mark the leftmost branches in the tree.

Conditional on competing in college, the probability of playing professionally is $P(B \mid A) = .017$. So the conditional probability of *not* playing professionally is

$$P(B^c \mid A) = 1 - P(B \mid A) = 1 - .017 = .983$$

These conditional probabilities mark the paths branching out from A in Figure 6.11.

The lower half of the tree diagram describes athletes who do not compete in college (A^c). It is unusual for these athletes to play professionally, but a few go straight from high school to professional leagues. Suppose that the conditional probability that a high school athlete reaches professional play given that he does not compete in college is $P(B \mid A^c) = .0001$. We can now mark the two paths branching from A^c in Figure 6.11.

There are two disjoint paths to B (professional play). By the addition rule, $P(B)$ is the sum of their probabilities. The probability of reaching B through college (top half of the tree) is

$$P(B \text{ and } A) = P(A)P(B \mid A)$$
$$= .05 \times .017 = .00085$$

The probability of reaching B without college is

$$P(B \text{ and } A^c) = P(A^c)P(B \mid A^c)$$
$$= .95 \times .0001 = .000095$$

The final result is

$$P(B) = .00085 + .000095 = .000945$$

About 9 high school athletes out of 10,000 will play professional sports.

Tree diagrams combine the addition and multiplication rules. The multiplication rule says that the probability of reaching the end of any complete branch is the product of the probabilities written on its segments. The probability of any outcome, such as the event B that an athlete reaches professional sports, is then found by adding the probabilities of all branches that are part of that event.

Independence again

The conditional probability $P(B \mid A)$ is generally not equal to the unconditional probability $P(B)$. That is because the occurrence of event A generally gives us some additional information about whether or not event B occurs. If knowing that A occurs gives no additional information about B, then A and B are independent events. The precise definition of independence is expressed in terms of conditional probability.

INDEPENDENT EVENTS

Two events A and B that both have positive probability are **independent** if

$$P(B \mid A) = P(B)$$

This is a more exact description of independence than that given in Section 6.1. We now see that the multiplication rule for independent events, $P(A \text{ and } B) = P(A)P(B)$, is a special case of the general multiplication rule, $P(A \text{ and } B) = P(A)P(B \mid A)$, just as the addition rule for disjoint events is a special case of the general addition rule.

SUMMARY

The **complement** A^c of an event A contains all outcomes that are not in A. The **union** $\{A \text{ or } B\}$ of events A and B contains all outcomes in A, in B, or in both A and B. The **intersection** $\{A \text{ and } B\}$ contains all outcomes that are in both A and B, but not outcomes in A alone or B alone.

The **conditional probability** $P(B \mid A)$ of an event B given an event A is defined by

$$P(B \mid A) = \frac{P(A \text{ and } B)}{P(A)}$$

when $P(A) > 0$ but in practice is most often found from directly available information.

The essential general rules of elementary probability are

Legitimate values: $0 \le P(A) \le 1$ for any event A

Total probability 1: $P(S) = 1$

Complement rule: $P(A^c) = 1 - P(A)$

Addition rule: $P(A \text{ or } B) = P(A) + P(B) - P(A \text{ and } B)$

Multiplication rule: $P(A \text{ and } B) = P(A)P(B \mid A)$

If A and B are **disjoint**, then $P(A \text{ and } B) = 0$. The general addition rule for unions then becomes the special addition rule, $P(A \text{ or } B) = P(A) + P(B)$.

A and B are **independent** when $P(B \mid A) = P(B)$. The multiplication rule for intersections then becomes $P(A \text{ and } B) = P(A)P(B)$.

In problems with several stages, draw a **tree diagram** to organize use of the multiplication and addition rules.

SECTION 6.3 EXERCISES

6.44   Here are the counts (in thousands) of earned degrees in the United States in a recent year, classified by level and by the sex of the degree recipient:

	Bachelor's	Master's	Professional	Doctorate	Total
Female	616	194	30	16	856
Male	529	171	44	26	770
Total	1145	365	74	42	1626

(a) If you choose a degree recipient at random, what is the probability that the person you choose is a woman?

(b) What is the conditional probability that you choose a woman, given that the person chosen received a professional degree?

(c) Are the events "choose a woman" and "choose a professional degree recipient" independent? How do you know?

6.45   The previous exercise gives the counts (in thousands) of earned degrees in the United States in a recent year. Use these data to answer the following questions.

(a) What is the probability that a randomly chosen degree recipient is a man?

(b) What is the conditional probability that the person chosen received a bachelor's degree, given that he is a man?

(c) Use the multiplication rule to find the joint probability of choosing a male bachelor's degree recipient. Check your result by finding this probability directly from the table of counts.

6.46   Here is a two-way table of all suicides committed in a recent year by sex of the victim and method used:

	Male	Female
Firearms	15,802	2,367
Poison	3,262	2,233
Hanging	3,822	856
Other	1,571	571
Total	24,457	6,027	30484

(a) What is the probability that a randomly selected suicide victim is male?

(b) What is the probability that the suicide victim used a firearm?

(c) What is the conditional probability that a suicide used a firearm, given that it was a man? Given that it was a woman?

(d) Describe in simple language (don't use the word "probability") what your results in (a) tell you about the difference between men and women with respect to suicide.

6.47   Common sources of caffeine in the diet are coffee, tea, and cola drinks. Suppose that

> 55% of adults drink coffee
>
> 25% of adults drink tea
>
> 45% of adults drink cola

and also that

> 15% drink both coffee and tea
>
> 5% drink all three beverages
>
> 25% drink both coffee and cola
>
> 5% drink only tea

Draw a Venn diagram marked with this information. Use it along with the addition rules to answer the following questions.

(a) What percent of adults drink only cola?

(b) What percent drink none of these beverages?

6.48   Choose an employed person at random. Let A be the event that the person chosen is a woman, and B the event that the person holds a managerial or professional job. Government data tells us that $P(A) = .46$ and the probability of managerial and professional jobs among women is $P(B \mid A) = .32$. Find the probability that a randomly chosen employed person is a woman holding a managerial or professional position.

6.49   Functional Robotics Corporation buys electrical controllers from a Japanese supplier. The company's treasurer feels that there is probability 0.4 that the dollar will fall in value against the Japanese yen in the next month. The treasurer also believes that *if* the dollar falls, there is probability 0.8 that the supplier will demand renegotiation of the contract. What probability has the treasurer assigned to the event that the dollar falls and the suppler demands renegotiation?

6.50   In the language of government statistics, the "labor force" includes all civilians over 16 years of age who are working or looking for work. Select a member of the U.S. labor force at random. Let A be the event that the person selected is white and B the event that he or she is employed. In 1995, 84.6% of the labor force was white. Of the whites in the labor force, 95.1% were employed. Among nonwhite members of the labor force, 91.9% were employed.

(a) Express each of the percents given as a probability involving the events A and B; for example, $P(A) = .846$.

(b) Draw a tree diagram for the outcomes of recording first the race (white or nonwhite) of a randomly chosen member of the labor force and then whether or not the person is employed.

(c) Find the probability that the person chosen is an employed white. Also find the probability that an employed nonwhite is chosen. What is the probability $P(B)$ that the person chosen is employed?

6.51 Suppose that in Exercise 6.49 the treasurer also feels that if the dollar does not fall, there is probability 0.2 that the Japanese supplier will demand that the contract be renegotiated. What is the probability that the supplier will demand renegotiation?

6.52 Use your results from Exercise 6.50 and the definition of conditional probability to find the probability $P(A \mid B)$ that a randomly selected member of the labor force is white, given that he or she is employed.

6.53 An examination consists of multiple-choice questions, each having five possible answers. Linda estimates that she has probability 0.75 of knowing the answer to any question that may be asked. If she does not know the answer, she will guess, with conditional probability 1/5 of being correct. What is the probability that Linda gives the correct answer to a question? (Draw a tree diagram to guide the calculation.)

6.54 The voters in a large city are 40% white, 40% black, and 20% Hispanic. (Hispanics may be of any race in official statistics, but in this case we are speaking of political blocks.) A black mayoral candidate anticipates attracting 30% of the white vote, 90% of the black vote, and 50% of the Hispanic vote. Draw a tree diagram with probabilities for the race (white, black, or Hispanic) and vote (for or against the candidate) of a randomly chosen voter. What percent of the overall vote does the candidate expect to get?

6.55 In the setting of Exercise 6.53, find the conditional probability that Linda knows the answer, given that she supplies the correct answer. (Hint: Use the result of Exercise 6.53 and the definition of conditional probability.)

6.56 Choose a point at random in the square $\square$ with sides $0 \leq x \leq 1$ and $0 \leq y \leq 1$. This means that the probability that the point falls in any region within the square is the area of that region. Let X be the x coordinate and Y the y coordinate of the point chosen. Find the conditional probability $P(Y < 1/2 \mid Y > X)$. (Hint: Draw a diagram of the square and the events $Y < 1/2$ and $Y > X$.)

CHAPTER REVIEW

Probability describes the pattern of chance outcomes. Probability calculations provide the basis for inference. When data are produced by random sampling or randomized comparitive experiments, the laws of probability answer the question, "What would happen if we did this very many times?" Probability is used to describe the long-term regularity that results from many repetitions of the same random phenomenon. The reasoning of statistical inference rests on asking "How often would this method give a correct answer if I used it very many times?" This chapter developed a probability model, including rules and tools that will help you describe the behavior of statistics from random samples in later chapters. Here are the most important things you should be able to do after studying this chapter.

THE PROBABILITY MODEL

1. Describe the sample space of a random phenomenon. For a finite number of outcomes, use the multiplication principle to determine the number of outcomes, and use counting techniques, Venn diagrams, and tree diagrams to determine simple probabilities. For the continuous case, use geometric areas to find probabilities (areas under simple density curves) of events (intervals on the horizontal axis).

2. Know the probability rules and be able to apply them to determine probabilities of defined events. In particular, determine if a given assignment of probabilities is valid.

3. Determine if two events are disjoint, complementary, or independent. Find unions and intersections of two or more events.

4. Know the general addition rule for the union of two events, and define joint probability. Apply these characterizations to solve problems.

5. Define conditional probability and use the definition to find conditional probabilities of events.

6. Use the multiplication rule to find the joint probability of two events.

7. Construct tree diagrams to organize the use of the multiplication and addition rules to solve problems with several stages.

CHAPTER 6 REVIEW EXERCISES

6.57   Consider the quarter circle of radius $r = 1$ in the first quadrant and the enclosing unit square.

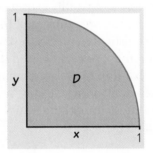

Suppose we select points in the square at random and then calculate the proportion of points that also fall within the quarter disk. The ratio of points in the quarter disk to the total number of points in the square will be approximately equal to the ratio of their respective areas.

(a) Equate these two ratios and then solve to obtain an *approximation* for π.

(b) Use your TI-83 to generate a point in the square and then test to see if it is in the quarter disk: `rand→X:rand→Y:X`^{2}`+Y`^{2}` ≤1`. If the condition is satisfied (i.e., the point is inside or on the quarter disk), the calculator will report a 1, meaning the statement is true. This is Boolean logic. If the inequality is false, the calculator will return a 0. Press ENTER 20 times and use tally marks to record hits or misses. When you finish, multiply the fraction: hits/20 by 4 to obtain an approximation of π.

(c) The program PI listed below automates the process. Read through the steps of the program and make sure you understand what each step does. Then enter the program into your calculator (or link from your instructor or another student). Execute the program for different numbers of trials, such as 50, 100, 200. Does a larger number of trials generally result in greater accuracy? How do your experimental results compare with the true value of $\pi = 3.14159265\ldots$?

```
PROGRAM:PI
:ClrHome
:Disp "HOW MANY
TRIALS"
:Prompt N
:0→C:0→J
:While J≤N
:rand→X
:rand→Y
:If X²+Y²≤1
:C+1→C
:J+1→J
:End
:ClrHome
:Disp "PI="
:Disp 4*C/N
```

6.58   At a carnival game at the state fair, contestants toss a disk of diameter 4 centimeters (cm) onto a grid composed of squares 6 cm on a side. The contestant wins a stuffed animal if the tossed disk lies entirely within a square. Assuming that the disk's center lands at random on the grid, find the theoretical probability of winning a prize. (*Hint:* Refer to the commentary for Activity 4.)

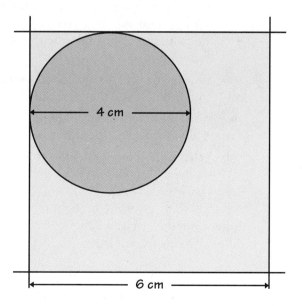

6.59   Abby, Deborah, Julie, Sam, and Roberto work in a firm's public relations office. Their employer must choose two of them to attend a conference in

Paris. To avoid unfairness, the choice will be made by drawing two names from a hat. (This is an SRS of size 2.)

(a) Write down all possible choices of two of the five names. This is the sample space.

(b) The random drawing makes all choices equally likely. What is the probability of each choice?

(c) What is the probability that Julie is chosen?

(d) What is the probability that neither of the two men (Sam and Roberto) is chosen?

6.60 A string of Christmas lights contains 20 lights. The lights are wired in series, so that if any light fails the whole string will go dark. Each light has probability 0.02 of failing during a 3-year period. The lights fail independently of each other. What is the probability that the string of lights will remain bright for 3 years?

6.61 The distribution of blood types among white Americans is approximately as follows: 37% type A, 13% type B, 44% type O, and 6% type AB. Suppose that the blood types of married couples are independent and that both the husband and wife follow this distribution.

(a) An individual with type B blood can safely receive transfusions only from persons with type B or type O blood. What is the probability that the husband of a woman with type B blood is an acceptable blood donor for her?

(b) What is the probability that in a randomly chosen couple the wife has type B blood and the husband has type A?

(c) What is the probability that one of a randomly chosen couple has type A blood and the other has type B?

(d) What is the probability that at least one of a randomly chosen couple has type O blood?

6.62 Exercise 6.31 gives the probability distribution of the sex and occupation of a randomly chosen American worker. Use this distribution to answer the following questions:

(a) Given that the worker chosen holds a managerial (Class A) job, what is the conditional probability that the worker is female?

(b) Classes D and E include most mechanical and factory jobs. What is the conditional probability that a worker is female, given that he or she holds a job in one of these classes?

6.63   Here is the distribution of the adjusted gross income X (in thousands of dollars) reported on individual federal income tax returns in 1993:

Income	< 10	10–24	25–49	50–99	≥ 100
Probability	.29	.27	.25	.14	.05

(a) What is the probability that a randomly chosen return shows an adjusted gross income of $50,000 or more?

(b) Given that a return shows an income of at least $50,000, what is the conditional probability that the income is at least $100,000?

6.64   You have torn a tendon and are facing surgery to repair it. The orthopedic surgeon explains the risks to you. Infection occurs in 3% of such operations, the repair fails in 14%, and both infection and failure occur together in 1%. What percent of these operations succeed and are free from infection?

6.65   It is difficult to conduct sample surveys on sensitive issues because many people will not answer questions if the answers might embarrass them. "Randomized response" is an effective way to guarantee anonymity while collecting information on topics such as student cheating or sexual behavior. Here is the idea. To ask a sample of students whether they have plagiarized a term paper while in college, have each student toss a coin in private. If the coin lands "heads" *and* they have not plagiarized, they are to answer "No." Otherwise they are to give "Yes" as their answer. Only the student knows whether the answer reflects the truth or just the coin toss, but the researchers can use a proper random sample with follow-up for nonresponse and other good sampling practices.

    Suppose that in fact the probability is 0.3 that a randomly chosen student has plagiarized a paper. Draw a tree diagram in which the first stage is tossing the coin and the second is the truth about plagiarism. The outcome at the end of each branch is the answer given to the randomized-response question. What is the probability of a "No" answer in the randomized-response poll? If the probability of plagiarism were 0.2, what would be the probability of a "No" response on the poll? Now suppose that you get 39% "No" answers in a randomized-response poll of a large sample of students at your college. What do you estimate to be the percent of the population who have plagiarized a paper?

6.66   ELISA tests are used to screen donated blood for the presence of the AIDS virus. The test actually detects antibodies, substances that the body produces when the virus is present. When antibodies are present, ELISA is positive with probability about 0.997 and negative with probability 0.003. When the blood tested is not contaminated with AIDS antibodies, ELISA gives a positive result with probability about 0.015 and a negative result with probability 0.985.[4] (Because ELISA is designed to keep the AIDS virus out of blood supplies, the higher probability 0.015 of a false positive is acceptable in exchange for the low probability 0.003 of failing to detect contaminated blood. These probabilities depend on the expertise of the particular laboratory doing the test.) Suppose that 1% of a large population carries the AIDS antibody in their blood.

(a) Draw a tree diagram for selecting a person from this population (outcomes: the person does or does not carry the AIDS antibody) and for testing his or her blood (outcomes: positive or negative).

(b) What is the probability that the ELISA test for AIDS is positive for a randomly chosen person from this population?

(c) What is the probability that a person has the antibody given that the ELISA test is positive? (This exercise illustrates a fact that is important when considering proposals for widespread testing for AIDS or illegal drugs: if the condition being tested is uncommon in the population, most positives will be false positives.)

NOTES AND DATA SOURCES

1. An informative and entertaining account of the origins of probability theory is Florence N. David, *Games, Gods and Gambling*, Charles Griffin, London, 1962.

2. This and similar psychology experiments are reported by A. Tversky and D. Kahneman, "Extensional versus intuitive reasoning: the conjunction fallacy in probability judgement," *Psychological Review*, 90 (1983), pp. 293–315.

3. These probabilities come from studies by the sociologist Harry Edwards, reported in the *New York Times*, February 25, 1986.

4. These probabilities are estimated from a large national study reported in E. M. Sloand et al., "HIV testing: state of the art," *Journal of the American Medical Association*, 266 (1991), pp. 2861–2866.

Random Variables

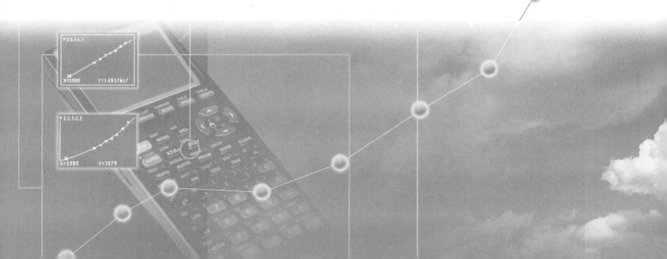

ACTIVITY 7 | The Game of Craps

Materials: Pair of dice for each pair of students

The game of craps is one of the most famous (or notorious) of all gambling games played with dice. In this game, the player rolls a pair of six-sided dice, and the *sum* of the numbers that turn up on the two faces is noted. If the sum is 7 or 11, then the player wins immediately. If the sum is 2, 3, or 12, then the player loses immediately. If any other sum is obtained, then the player continues to throw the dice until he either wins by repeating the first sum he obtained or loses by rolling a 7. Your mission in this activity is to estimate the probability of a player winning at craps. But first, let's get a feel for the game. For this activity, your class will be divided into groups of two. Your instructor will provide a pair of dice for each group of two students.

1. In your group of two students, play a total of 20 games of craps. One person will roll the dice; the other will keep track of the sums and record the end result (win or lose). If you like, you can switch jobs after 10 games have been completed. How many times out of 20 does the player win? What is the relative frequency (i.e., percentage, written as a decimal) of wins?

2. Combine your results with those of the other two-student groups in the class. What is the relative frequency of wins for the entire class?

3. Use simulation techniques to represent 25 games of craps, using either the table of random numbers or the random number generating feature of your TI-83. (To use the random number table, round the relative frequency of wins from the data you collected for the class to two decimal places, then use appropriate groups of two digits to represent wins and losses.) What is the relative frequency of wins based on the 25 simulations? How does this number compare to the relative frequency you found in step 2?

(continued)

4. One of the ways you can win at craps is to roll a sum of 7 or 11 on your first roll. Using your results and those of your fellow students, determine the number of times a player won by rolling a sum of 7 of the first roll. What is the relative frequency of rolling a sum of 7? Repeat these calculations for a sum of 11. Which of these sums appears more likely to occur than the other, based on the class results?

5. One of the ways you can lose at craps is to roll a sum of 2, 3, or 12 on your first roll. Using your results and those of your fellow students, determine the number of times a player lost by rolling a sum of 2 on the first roll. What is the relative frequency of rolling a sum of 2? Repeat these calculations for a sum of 3 and a sum of 12. Which of these sums appears more likely to occur than the others, based on the class results?

6. Clearly, the key quantity of interest in craps is the *sum* of the numbers on the two dice. Let's try to get a better idea of how this sum behaves in general by conducting a simulation. First, determine how you would simulate the roll of a single fair die. (*Hint:* Just use digits 1 to 6 and ignore the others.) Then determine how you would simulate a roll of two fair dice. Using this model, simulate 36 rolls of a pair of dice and determine the relative frequency of each of the possible sums.

7. Construct a relative frequency histogram of the relative frequency results in step 6. What is the approximate shape of the distribution? What sum appears most likely to occur? Which appears least likely to occur?

8. From the relative frequency data in step 6, compute the relative frequency of winning and the relative frequency of losing on your first roll in craps. How do these simulated results compare with what the class obtained?

7.1 DISCRETE AND CONTINUOUS RANDOM VARIABLES

Introduction

Sample spaces need not consist of numbers. When we toss four coins, we can record the outcome as a string of heads and tails, such as HTTH.

In statistics, however, we are most often interested in numerical outcomes such as the count of heads in the four tosses. It is convenient to use a short-hand notation: Let X be the number of heads. If our outcome is HTTH, then $X = 2$. If the next outcome is TTTH, the value of X changes to $X = 1$. The possible values of X are 0, 1, 2, 3, and 4. Tossing a coin four times will give X one of these possible values. Tossing four more times will give X another and probably different value. We call X a *random variable* because its values vary when the coin tossing is repeated.

RANDOM VARIABLE

A **random variable** is a variable whose value is a numerical outcome of a random phenomenon.

We usually denote random variables by capital letters near the end of the alphabet, such as X or Y. Of course, the random variables of greatest interest to us are outcomes such as the mean $\bar{x}$ of a random sample, for which we will keep the familiar notation.[1] As we progress from general rules of probability toward statistical inference, we will concentrate on random variables. When a random variable X describes a random phenomenon, the sample space S just lists the possible values of the random variable. We usually do not mention S separately. There remains the second part of any probability model, the assignment of probabilities to events. In this section, we will learn two ways of assigning probabilities to the values of a random variable. The two types of probability models that result will dominate our application of probability to statistical inference.

Discrete random variables

We have learned several rules of probability but only one method of assigning probabilities: state the probabilities of the individual outcomes and assign probabilities to events by summing over the outcomes. The outcome probabilities must be between 0 and 1 and have sum 1. When the outcomes are numerical, they are values of a random variable. We will now attach a name to random variables having probability assigned in this way.[2]

DISCRETE RANDOM VARIABLE

A **discrete random variable** X has a countable number of possible values. The **probability distribution** of X lists the values and their probabilities:

Value of X	x_1	x_2	x_3	$\cdots$	x_k
Probability	p_1	p_2	p_3	$\cdots$	p_k

The probabilities p_i must satisfy two requirements:

1. Every probability p_i is a number between 0 and 1.
2. $p_1 + p_2 + \cdots + p_k = 1$.

Find the probability of any event by adding the probabilities p_i of the particular values x_i that make up the event.

EXAMPLE 7.1

The instructor of a large class gives 15% each of A's and D's, 30% each of B's and C's, and 10% F's. Choose a student at random from this class. To "choose at random" means to give every student the same chance to be chosen. The student's grade on a four-point scale (A = 4) is a random variable X.

The value of X changes when we repeatedly choose students at random, but it is always one of 0, 1, 2, 3, or 4. Here is the distribution of X:

Grade	0	1	2	3	4
Probability	.10	.15	.30	.30	.15

The probability that the student got a B or better is the sum of the probabilities of an A and a B:

$$P(\text{grade is 3 or 4}) = P(X = 3) + P(X = 4)$$
$$= .30 + .15 = .45$$

probability histogram

We can use a **probability histogram** to picture the probability distribution of a discrete random variable. Figure 7.1 displays two probability histograms. Figure 7.1(a) is the distribution of an entry in the table of random digits, with probability distributed uniformly over all 10 digits. Part (b) is

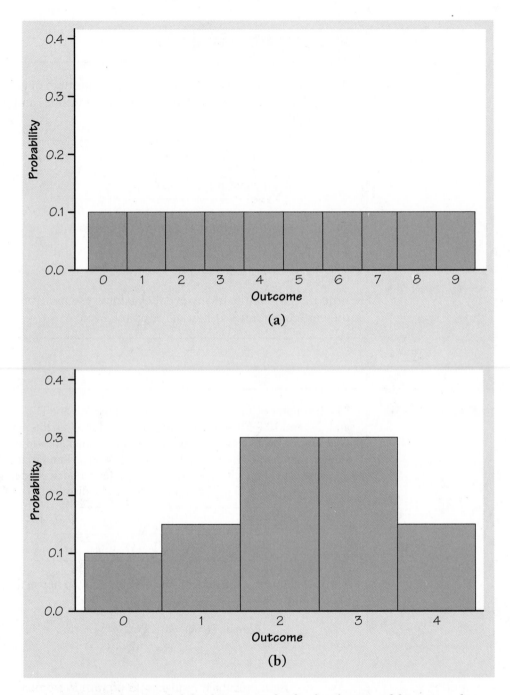

FIGURE 7.1   Probability histograms for the distributions of discrete random variables: (a) generating a random digit and (b) grades in a large class (see Example 7.1).

```
                                     HTTH
                                     HTHT
                        HTTT         THTH         HHHT
                        THTT         HHTT         HHTH
                        TTHT         THHT         HTHH
            TTTT         TTTH         TTHH         THHH         HHHH

            X= O         X= 1         X= 2         X= 3         X= 4
```

FIGURE 7.2   Possible outcomes in four tosses of a coin. The random variable X is the number of heads.

the distribution of grades in Example 7.1. The horizontal scale shows the possible values of X, and the height of each bar is the probability for the value at its base. A probability histogram is in effect a relative frequency histogram for a very large number of trials.

EXAMPLE 7.2

What is the probability distribution of the discrete random variable X that counts the number of heads in four tosses of a coin? We can derive this distribution if we make two reasonable assumptions:

1. The coin is balanced, so each toss is equally likely to give H or T.
2. The coin has no memory, so tosses are independent.

The outcome of four tosses is a sequence of heads and tails such as HTTH. There are 16 possible outcomes in all. Figure 7.2 lists these outcomes along with the value of X for each outcome. The multiplication rule for independent events tells us that, for example,

$$P(\text{HTTH}) = \frac{1}{2} \times \frac{1}{2} \times \frac{1}{2} \times \frac{1}{2} = \frac{1}{16}$$

Each of the 16 possible outcomes similarly has probability 1/16. That is, these outcomes are equally likely.

The number of heads X has possible values 0, 1, 2, 3, and 4. These values are *not* equally likely. As Figure 7.2 shows, there is only one way that $X = 0$ can occur: namely when the outcome is TTTT. So $P(X = 0) = 1/16$. But the event $\{X = 2\}$ can occur in six different ways, so that

$$P(X = 2) = \frac{\text{count of ways } X = 2 \text{ can occur}}{16}$$

$$= \frac{6}{16}$$

We can find the probability of each value of X from Figure 7.2 in the same way. Here is the result:

$$P(X = 0) = \frac{1}{16} = .0625$$

$$P(X = 1) = \frac{4}{16} = .25$$

$$P(X = 2) = \frac{6}{16} = .375$$

$$P(X = 3) = \frac{4}{16} = .25$$

$$P(X = 4) = \frac{1}{16} = .0625$$

These probabilities have sum 1, so this is a legitimate probability distribution. In table form the distribution is

Number of heads	0	1	2	3	4
Probability	.0625	.25	.375	.25	.0625

Figure 7.3 is a probability histogram for this distribution. The probability distribution is exactly symmetric. It is an idealization of the relative frequency distribution of the number of heads after many tosses of four coins, which would be nearly symmetric but is unlikely to be exactly symmetric.

Any event involving the number of heads observed can be expressed in terms of X, and its probability can be found from the distribution of X. For example, the probability of tossing at least two heads is

$$P(X \geq 2) = .375 + .25 + .0625 = .6875$$

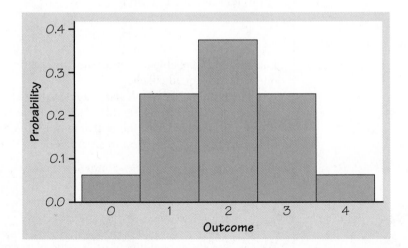

FIGURE 7.3   Probability histogram for the number of heads in four tosses of a coin.

The probability of at least one head is most simply found by use of the complement rule:

$$P(X \geq 1) = 1 - P(X = 0)$$
$$= 1 - .0625 = .9375$$

Recall that tossing a coin n times is similar to choosing an SRS of size n from a large population and asking a yes-or-no question. We will extend the results of Example 7.2 when we return to sampling distributions in the next two chapters.

EXERCISES

7.1 If a carefully made die is rolled once, it is reasonable to assign probability 1/6 to each of the six faces.

(a) What is the probability of rolling a number less than 3?

(b) Use your TI-83 to simulate rolling a die 100 times, and assign the values to list L_1. Sort the list in ascending order, and then count the outcomes that are either 1s or 2s. Record the relative frequency.

(c) Repeat part (b) four more times, and then average the five relative frequencies. Is this number close to your result in (a)?

7.2 A couple plans to have three children. There are 8 possible arrangements of girls and boys. For example, GGB means the first two children are girls and the third child is a boy. All 8 arrangements are (approximately) equally likely.

(a) Write down all 8 arrangements of the sexes of three children. What is the probability of any one of these arrangements?

(b) Let X be the number of girls the couple has. What is the probability that $X = 2$?

(c) Starting from your work in (a), find the distribution of X. That is, what values can X take, and what are the probabilities for each value?

7.3 A study of social mobility in England looked at the social class reached by the sons of lower-class fathers. Social classes are numbered from 1 (low) to 5 (high). Take the random variable X to be the class of a randomly chosen son of a father in Class 1. The study found that the distribution of X is

Son's class	1	2	3	4	5
Probability	.48	.38	.08	.05	.01

(a) What percent of the sons of lower-class fathers reach the highest class, Class 5?

(b) Check that this distribution satisfies the two requirements for a discrete probability distribution.

(c) What is $P(X \leq 3)$?

(d) What is $P(X < 3)$?

(e) Write the event "a son of a lower-class father reaches one of the two highest classes" in terms of values of X. What is the probability of this event?

(f) Briefly describe how you would use simulation to answer the question in (c).

Continuous random variables

When we use the table of random digits to select a digit between 0 and 9, the result is a discrete random variable. The probability model assigns probability 1/10 to each of the 10 possible outcomes, as Figure 7.1(a) shows. Suppose that we want to choose a number at random between 0 and 1, allowing *any* number between 0 and 1 as the outcome. Software random number generators will do this. You can visualize such a random number by thinking of a spinner (Figure 7.4) that turns freely on its axis and slowly comes to a stop. The pointer can come to rest anywhere on a circle that is marked from 0 to 1. The sample space is now an entire interval of numbers:

$$S = \{\text{all numbers } x \text{ such that } 0 \leq x \leq 1\}$$

How can we assign probabilities to such events as $.3 \leq x \leq .7$? As in the case of selecting a random digit, we would like all possible outcomes to be equally likely. But we cannot assign probabilities to each individual value of x and then sum, because there are infinitely many possible values. Instead we use a new way of assigning probabilities directly to events—as *areas under a density curve*. Any density curve has area exactly 1 underneath it, corresponding to total probability 1.

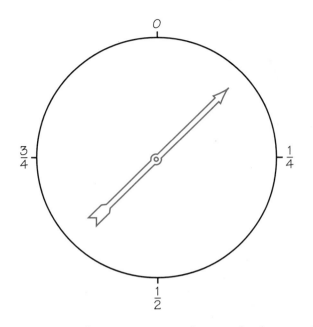

FIGURE 7.4   A spinner that generates a random number between 0 and 1.

EXAMPLE 7.3	The random number generator will spread its output uniformly across the entire interval from 0 to 1 as we allow it to generate a long sequence of numbers. The results of many trials are represented by the density curve of a **uniform distribution** (Figure 7.5). This density curve has height 1 over the interval from 0 to 1. The area under the density curve is 1, and the probability of any event is the area under the density curve and above the event in question.

uniform distribution

As Figure 7.5(a) illustrates, the probability that the random number generator produces a number X between 0.3 and 0.7 is

$$P(.3 \leq X \leq .7) = .4$$

because the area under the density curve and above the interval from 0.3 to 0.7 is 0.4. The height of the density curve is 1 and the area of a rectangle is the product of height and length, so the probability of any interval of outcomes is just the length of the interval.
Similarly,

$$P(X \leq .5) = .5$$
$$P(X > .8) = .2$$
$$P(X \leq .5 \text{ or } X > .8) = .7$$

Notice that the last event consists of two nonoverlapping intervals, so the total area above the event is found by adding two areas, as illustrated by Figure 7.5(b). This assignment of probabilities obeys all of our rules for probability.

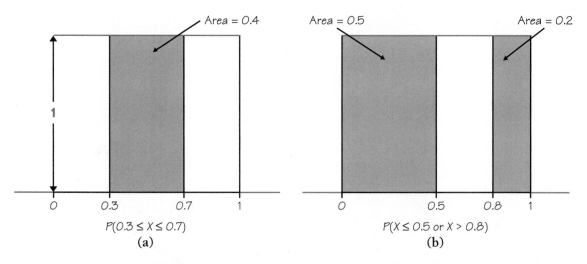

FIGURE 7.5 Assigning probability for generating a random number between 0 and 1. The probability of any interval of numbers is the area above the interval and under the curve.

Probability as area under a density curve is a second important way of assigning probabilities to events. Figure 7.6 illustrates this idea in general form. We call X in Example 7.3 a *continuous random variable* because its values are not isolated numbers but an entire interval of numbers.

CONTINUOUS RANDOM VARIABLE

A **continuous random variable** X takes all values in an interval of numbers. The **probability distribution** of X is described by a density curve. The probability of any event is the area under the density curve and above the values of X that make up the event.

The probability model for a continuous random variable assigns probabilities to intervals of outcomes rather than to individual outcomes. In fact, *all continuous probability distributions assign probability 0 to every individual outcome*. Only intervals of values have positive probability. To see that this is true, consider a specific outcome such as $P(X = .8)$ in Example 7.3. The probability of any interval is the same as its length. The

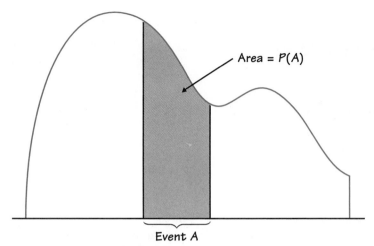

FIGURE 7.6   The probability distribution of a continuous random variable assigns probabilities as areas under a density curve.

point 0.8 has no length, so its probability is 0. Although this fact may seem odd at first glance, it does make intuitive, as well as mathematical, sense. The random number generator produces a number between 0.79 and 0.81 with probability 0.02. An outcome between 0.799 and 0.801 has probability 0.002, and a result between 0.7999 and 0.8001 has probability 0.0002. Continuing to home in on 0.8, we can see why an outcome *exactly* equal to 0.8 should have probability 0. Because there is no probability exactly at $X = .8$, the two events $\{X > .8\}$ and $\{X \geq .8\}$ have the same probability. We can ignore the distinction between $>$ and $\geq$ when finding probabilities for continuous (but not discrete) random variables.

Normal distributions as probability distributions

The density curves that are most familiar to us are the normal curves. (We discussed normal curves in Section 2.1.) Because any density curve describes an assignment of probabilities, *normal distributions are probability distributions*. Recall that $N(\mu, \sigma)$ is our shorthand notation for the normal distribution having mean μ and standard deviation σ. In the language of random variables, if X has the $N(\mu, \sigma)$ distribution, then the standardized variable

$$Z = \frac{X - \mu}{\sigma}$$

is a standard normal random variable having the distribution $N(0, 1)$.

EXAMPLE 7.4

An opinion poll asks an SRS of 1500 American adults what they consider to be the most serious problem facing our schools. Suppose that if we could ask all adults this question, 30% would say "drugs." We will learn in Chapter 9 that the proportion $p = .3$ is a population parameter and that the proportion $\hat{p}$ of the sample who answer "drugs" is a statistic used to estimate p. We will see in Chapter 9 that $\hat{p}$ is a random variable that has approximately the $N(0.3, 0.0118)$ distribution. The mean 0.3 of this distribution is the same as the population parameter. The standard deviation is controlled mainly by the sample size, which is 1500 in this case.

What is the probability that the poll result differs from the truth about the population by more than two percentage points? Figure 7.7 shows this probability as an area under a normal density curve. By the addition rule for disjoint events, the desired probability is

$$P(\hat{p} < .28 \text{ or } \hat{p} > .32) = P(\hat{p} < .28) + P(\hat{p} > .32)$$

You can find the two individual probabilities from software or by standardizing and using Table A.

$$P(\hat{p} < .28) = P\left(Z < \frac{.28 - .3}{.0118}\right)$$

$$= P(Z < -1.69) = .0455$$

$$P(\hat{p} > .32) = P\left(Z > \frac{.32 - .3}{.0118}\right)$$

$$= P(Z > 1.69) = .0455$$

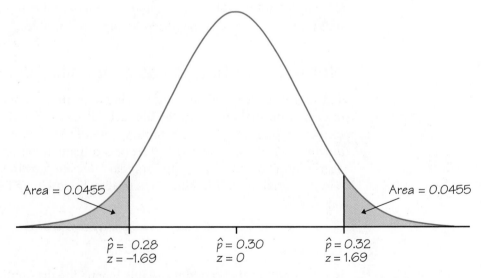

Area = 0.0455   Area = 0.0455

$\hat{p} = 0.28$ $\hat{p} = 0.30$ $\hat{p} = 0.32$
$z = -1.69$ $z = 0$ $z = 1.69$

FIGURE 7.7   Probability in Example 7.4 as area under a normal density curve.

Therefore,

$$P(\hat{p} < .28 \text{ or } \hat{p} > .32) = .0455 + .0455 = .0910$$

The probability that the sample result will miss the truth by more than two percentage points is 0.091. The arrangement of this calculation is familiar from our earlier work with normal distributions. Only the language of probability is new.

We could also do the calculation by first finding the probability of the complement:

$$P(.28 \le \hat{p} \le .32) = P\left(\frac{.28 - .3}{.0118} \le Z \le \frac{.32 - .3}{.0118}\right)$$
$$= P(-1.69 \le Z \le 1.69)$$
$$= .9545 - .0455 = .9090$$

Then by the complement rule,

$$P(\hat{p} < .28 \text{ or } \hat{p} > .32) = 1 - P(.28 \le \hat{p} \le .32)$$
$$= 1 - .9090 = .0910$$

There is often more than one correct way to use the rules of probability to answer a question.

EXERCISES

7.4 Let X be a random number between 0 and 1 produced by the idealized uniform random number generator described in Example 7.3 and Figure 7.5. Find the following probabilities:

(a) $P(0 \le X \le .4)$

(b) $P(.4 \le X \le 1)$

(c) $P(.3 \le X \le .5)$

(d) $P(.3 < X < .5)$

(e) $P(.226 \le X \le .713)$

7.5 Let the random variable X be a random number with the uniform density curve in Figure 7.5, as in the previous exercise. Find the following probabilities:

(a) $P(X \le .49)$

(b) $P(X \ge .27)$

(c) $P(.27 < X < 1.27)$

(d) $P(.1 \leq X \leq .2 \text{ or } .8 \leq X \leq .9)$

(e) The probability that X is not in the interval 0.3 to 0.8.

(f) $P(X = 0.5)$

SUMMARY

The previous chapter included a general discussion of the idea of probability and the properties of probability models. Two very useful specific types of probability models are distributions of discrete and continuous random variables. In our study of statistics we will employ only these two types of probability models.

A **random variable** is a variable taking numerical values determined by the outcome of a random phenomenon. The **probability distribution** of a random variable X tells us what the possible values of X are and how probabilities are assigned to those values. A random variable X and its distribution can be discrete or continuous.

A **discrete random variable** has a countable number of possible values. The probability distribution assigns each of these values a probability between 0 and 1 such that the sum of all the probabilities is exactly 1. The probability of any event is the sum of the probabilities of all the values that make up the event.

A **continuous random variable** takes all values in some interval of numbers. A **density curve** describes the probability distribution of a continuous random variable. The probability of any event is the area under the curve above the values that make up the event.

Normal distributions are one type of continuous probability distribution.

You can picture a probability distribution by drawing a **probability histogram** in the discrete case or by graphing the density curve in the continuous case.

SECTION 7.1 EXERCISES

7.6   Some games of chance rely on tossing two dice. Each die has six faces, marked with 1, 2, ..., 6 spots called pips. The dice used in casinos are carefully balanced so that each face is equally likely to come up. When two dice are tossed, each of the 36 possible pairs of faces is equally likely to come up. The outcome of interest to a gambler is the sum of the pips on the two up faces. Call this random variable X.

(a) Write down all 36 possible pairs of faces.

(b) If all pairs have the same probability, what must be the probability of each pair?

(c) Write the value of X next to each pair of faces and use this information with the result of (b) to give the probability distribution of X. Draw a probability histogram to display the distribution.

(d) One bet available in craps wins if a 7 or 11 comes up on the next roll of two dice. What is the probability of rolling a 7 or 11 on the next roll? Compare your answer with your experimental results (relative frequency) in Activity 7, part 4.

(e) After the dice are rolled the first time, several bets lose if a 7 is then rolled. If any outcome other than a 7 occurs, these bets either win or continue to the next roll. What is the probability that anything other than a 7 is rolled?

7.7   Choose an American household at random and let the random variable X be the number of persons living in the household. If we ignore the few households with more than seven inhabitants, the probability distribution of X is as follows:

Inhabitants	1	2	3	4	5	6	7
Probability	.25	.32	.17	.15	.07	.03	.01

(a) Verify that this is a legitimate discrete probability distribution and draw a probability histogram to display it.

(b) What is $P(X \geq 5)$?

(c) What is $P(X > 5)$?

(d) What is $P(2 < X \leq 4)$?

(e) What is $P(X \neq 1)$?

(f) Write the event that a randomly chosen household contains more than two persons in terms of the random variable X. What is the probability of this event?

7.8   A study of education followed a large group of fifth-grade children to see how many years of school they eventually completed. Let X be the highest year of school that a randomly chosen fifth grader completes. (Students who go on to college are included in the outcome $X = 12$.) The study found this probability distribution for X:

Years	4	5	6	7	8	9	10	11	12
Probability	.010	.007	.007	.013	.032	.068	.070	.041	.752

(a) What percent of fifth graders eventually finished twelfth grade?

(b) Check that this is a legitimate discrete probability distribution.

(c) Find $P(X \geq 6)$.

(d) Find $P(X > 6)$.

(e) What values of X make up the event "the student completed at least one year of high school"? (High school begins with the ninth grade.) What is the probability of this event?

7.9   Weary of the low turnout in student elections, a college administration decides to choose an SRS of three students to form an advisory board that represents student opinion. Suppose that 40% of all students oppose the use of student fees to fund student interest groups and that the opinions of the three students on the board are independent. Then the probability is 0.4 that each opposes the funding of interest groups.

(a) Call the three students A, B, and C. What is the probability that A and B support funding and C opposes it?

(b) List all possible combinations of opinions that can be held by students A, B, and C.. (*Hint:* There are eight possibilities.) Then give the probability of each of these outcomes. Note that they are not equally likely.

(c) Let the random variable X be the number of student representatives who oppose the funding of interest groups. Give the probability distribution of X.

(d) Express the event "a majority of the advisory board opposes funding" in terms of X and find its probability.

7.10   Many random number generators allow users to specify the range of the random numbers to be produced. Suppose that you specify that the range is to be $0 \leq y \leq 2$. Then the density curve of the outcomes has constant height between 0 and 2, and height 0 elsewhere.

(a) What is the height of the density curve between 0 and 2? Draw a graph of the density curve.

(b) Use your graph from (a) and the fact that probability is area under the curve to find $P(y \leq 1)$.

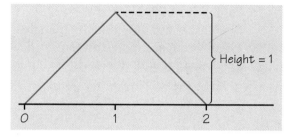

FIGURE 7.8   The density curve for the sum of two random numbers. This continuous random variable takes values between 0 and 2 (Exercise 7.11).

(c) Find $P(.5 < y < 1.3)$.

(d) Find $P(y \geq .8)$.

7.11   Generate *two* random numbers between 0 and 1 and take Y to be their sum. Then Y is a continuous random variable that can take any value between 0 and 2. The density curve of Y is the triangle shown in Figure 7.8.

(a) Verify that the area under this curve is 1.

(b) What is the probability that Y is less than 1? (Sketch the density curve, shade the area that represents the probability, then find that area. Do this for (c) also.)

(c) What is the probability that Y is less than 0.5?

(d) Use simulation methods to answer the questions in (b) and (c). Here's one way using the TI-83: Clear lists L_1, L_2, and L_3 and enter these commands:

`rand(200)→L₁`	Generates 200 random numbers and stores them in L_1
`rand(200)→L₂`	Generates 200 random numbers and stores them in L_2
`L₁+L₂→L₃`	Adds the first number in L_1 and the first number in L_2 and stores the sum in L_3, and so forth, from $i = 1$ to $i = 200$
`SortA(L₃)`	Sorts the sums in L_3 in ascending order

Now simply scroll through L_3 and count the number of sums that satisfy the conditions stated in (b) and (c), and determine the relative frequency.

7.12   (Optional) This is a continuation of the previous exercise. If you carried out the simulation in 7.11(d), you can picture the distribution as follows: Deselect any active functions in the Y= screen, and turn off all STAT PLOTs. Define PLOT1 to be a histogram using list L_3. Set WINDOW dimensions as follows: X[0, 2]$_1$ and Y[−6, 25]$_5$. Then press GRAPH. Does the result-ing histogram resemble the triangle in Figure 7.8? Can you imagine the triangle superimposed on top of the histogram? Of course, some bars will be too short and others will be too long, but this is due to chance varia-tion. To overlay the triangle, define Y$_1$ to be Y$_1$ = (25X)(X ≥ 0 and X ≤ 1) + (-25X+50)(X ≥ 1 and X ≤ 2) and then press GRAPH again. How well does this "curve" fit your histogram?

7.13   An SRS of 400 American adults is asked, "What do you think is the most serious problem facing our schools?" Suppose that in fact 30% of all adults would answer "drugs" if asked this question. The proportion $\hat{p}$ of the sam-ple who answer "drugs" will vary in repeated sampling. In fact, we can assign probabilities to values of $\hat{p}$ using the normal density curve with mean 0.3 and standard deviation 0.023. Use this density curve to find the probabilities of the following events:

(a) At least half of the sample believes that drugs are the schools' most serious problem.

(b) Less than 25% of the sample believes that drugs are the most serious problem.

(c) The sample proportion is between 0.25 and 0.35.

7.14   This is a continuation of the previous exercise. How could you design a simulation to answer part (b) of Exercise 7.13? What we need to do is simulate 400 observations from the N(0.3, 0.023) distribution. This is easily done on the TI-83. Here's one way: Clear list L$_1$ and enter the commands

```
randNorm(0.3,.023,400)→L₁    Selects 400 random observations from
                             the N(0.3, 0.023) distribution
SortA(L₁)                    Sorts the 400 observations in L₁
                             in ascending order
```

(Note that randNorm is found under the MATH / PRB menu.) Then scroll through list L$_1$. How many entries (observations) are less than 0.25? What is the relative frequency of this event? Compare the results of your simu-lation with your answer to Exercise 7.13(b).

7.15   An opinion poll asks an SRS of 1500 adults, "Do you happen to jog?" Suppose that the population proportion who jog is p = .15. To esti-mate p, we use the proportion $\hat{p}$ in the sample who answer "Yes." The statistic $\hat{p}$ is a random variable that is approximately normally distributed with mean μ = .15 and standard deviation σ = .0092. Find the following probabilities:

(a) $P(\hat{p} \geq .16)$

(b) $P(.14 \leq \hat{p} \leq .16)$

7.16   In continuation of the previous exercise, describe the details of a simulation you could carry out to approximate an answer to Exercise 7.15(a). Then carry out the simulation. About how many repetitions do you need to get a result close to your answer to Exercise 7.15(a)?

7.2   MEANS AND VARIANCES OF RANDOM VARIABLES

Probability is the mathematical language that describes the long-run regular behavior of random phenomena. The probability distribution of a random variable is an idealized relative frequency distribution. The probability histograms and density curves that picture probability distributions resemble our earlier pictures of distributions of data. In describing data, we moved from graphs to numerical measures such as means and standard deviations. Now we will make the same move to expand our descriptions of the distributions of random variables. We can speak of the mean winnings in a game of chance or the standard deviation of the randomly varying number of calls a travel agency receives in an hour. In this section we will learn more about how to compute these descriptive measures and about the laws they obey.

The mean of a random variable

The mean $\bar{x}$ of a set of observations is their ordinary average. The mean of a random variable X is also an average of the possible values of X, but with an essential change to take into account the fact that not all outcomes need be equally likely. An example will show what we must do.

EXAMPLE 7.5

Thirty-seven states, five Canadian provinces, and most European countries have government-sponsored lotteries. New Hampshire started the fad in the United States, in 1964. Here is a simple lottery wager, the Tri-State Pick 3 game that New Hampshire shares with Maine and Vermont. You choose a three-digit number; the state chooses a three-digit winning number at random and pays you $500 if your number is chosen. Because there are 1000 three-digit numbers, you have probability 1/1000 of winning. Taking X to be the amount you win, the probability distribution of X is

Outcome	$0	$500
Probability	.999	.001

What are your average winnings? The ordinary average of the two possible outcomes $0 and $500 is $250, but that makes no sense as the average winnings because $500 is much less likely than $0. In the long run you win $500 once in every 1000 tickets and $0 on the remaining 999 of 1000 tickets. Your long-run average winnings from a ticket are

$$\$500\frac{1}{1000} + \$0\frac{999}{1000} = \$.50$$

or fifty cents. That number is the mean of the random variable X. (Tickets cost $1, so in the long run the state keeps half the money you wager.)

If you play Tri-State Pick 3 several times, we would as usual call the mean of the actual amounts you win $\bar{x}$. The mean in Example 7.5 is a different quantity—it is the long-run average winnings you expect if you play a very large number of times. Just as probabilities are an idealized description of long-run proportions, the mean of a probability distribution describes the long-run average outcome. We can't call this mean $\bar{x}$, so we need a different symbol. The common symbol for the mean of a proba-

mean μ

bility distribution is μ, the Greek letter mu. We used μ in Chapter 2 for the mean of a normal distribution, so this is not a new notation. We will often be interested in several random variables, each having a different probability distribution with a different mean. To remind ourselves that we are talking about the mean of X we often write μ_X rather than simply μ. In Example 7.5, $\mu_X = \$0.50$. Notice that, as often happens, the mean is not a possible value of X. You will often find the mean of a random vari-

expected value

able X called the *expected value* of X. This term can be misleading, for we don't necessarily expect one observation on X to be close to its expected value.

The mean of any discrete random variable is found just as in Example 7.5. It is an average of the possible outcomes, but a weighted average in which each outcome is weighted by its probability. Because the probabilities add to 1, we have total weight 1 to distribute among the outcomes. An outcome that occurs half the time has probability one-half and so gets one-half the weight in calculating the mean. Here is the general definition.

MEAN OF A DISCRETE RANDOM VARIABLE

Suppose that X is a discrete random variable whose distribution is

Value of X	x_1	x_2	x_3	$\cdots$	x_k
Probability	p_1	p_2	p_3	$\cdots$	p_k

To find the **mean** of X, multiply each possible value by its probability, then add all the products:

$$\mu_X = x_1 p_1 + x_2 p_2 + \ldots + x_k p_k$$
$$= \Sigma x_i p_i$$

EXAMPLE 7.6

The distribution of the count X of heads in four tosses of a balanced coin was found in Example 7.2 to be

Number of heads x_i	0	1	2	3	4
Probability p_i	.0625	.25	.375	.25	.0625

The mean of X is therefore

$$\mu_X = (0)(.0625) + (1)(.25) + (2)(.375) + (3)(.25) + (4)(.0625)$$
$$= 2$$

This discrete distribution is symmetric, as the probability histogram in Figure 7.9(a) reminds us. The mean therefore falls at the center of symmetry.

EXAMPLE 7.7

What is the mean size of an American household? Here is the distribution of the size of households according to Census Bureau studies:

Inhabitants	1	2	3	4	5	6	7
Proportion of households	.25	.32	.17	.15	.07	.03	.01

If we imagine selecting a single household at random, the size of the household chosen is a random variable X with probability distribution given by the

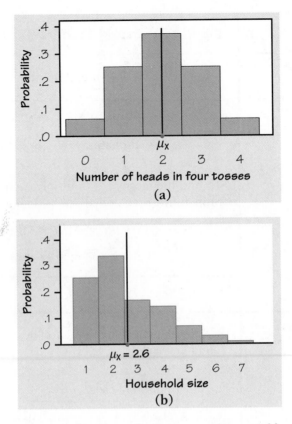

FIGURE 7.9   Locating the mean of a discrete random variable on the probability histogram for (a) the number of heads in four tosses of a coin (Example 7.6) and (b) the number of persons in a family (Example 7.7).

table. The mean μ_X is the mean household size in the population. This mean is

$$\mu_X = (1)(.25) + (2)(.32) + (3)(.17) + (4)(.15) + (5)(.07) + (6)(.03) + (7)(.01)$$
$$= 2.6$$

Figure 7.9(b) locates the mean on a probability histogram. (In this example, we have ignored the few households with 8 or more members. The Census Bureau reports that the actual mean size of American households is 2.65 people when these few large households are included.)

What about continuous random variables? The probability distribution of a continuous random variable X is described by a density curve. Chapter 2 showed how to find the mean of the distribution: it is the point at which the area under the density curve would balance if it were made out of solid material. The mean lies at the center of symmetric density curves such as the normal curves. Exact calculation of the mean of a distribution with a skewed density curve requires advanced mathematics.[3]

The idea that the mean is the balance point of the distribution applies to discrete random variables as well, but in the discrete case we have a formula that gives us this point.

EXERCISES

7.17 Example 7.1 gives the distribution of grades (A = 4, B = 3, and so on) in a large class as

Grade	0	1	2	3	4
Probability	.10	.15	.30	.30	.15

Find the average (that is, the mean) grade in this course.

7.18 Keno is a favorite game in casinos, and similar games are popular with the states that operate lotteries. Balls numbered 1 to 80 are tumbled in a machine as the bets are placed, then 20 of the balls are chosen at random. Players select numbers by marking a card. The simplest of the many wagers available is "Mark 1 Number." Your payoff is $3 on a $1 bet if the number you select is one of those chosen. Because 20 of 80 numbers are chosen, your probability of winning is 20/80, or 0.25.

(a) What is the probability distribution (the outcomes and their probabilities) of the payoff X on a single play?

(b) What is the mean payoff μ_X?

(c) In the long run, how much does the casino keep from each dollar bet?

7.19 The Tri-State Pick 3 lottery game offers a choice of several bets. You choose a three-digit number. The lottery commission announces the winning three-digit number, chosen at random, at the end of each day. The "box" pays $83.33 if the number you choose has the same digits as the winning number, in any order. Find the expected payoff for a $1 bet on the box. (Assume that you chose a number having three different digits.)

Statistical estimation and the law of large numbers

We would like to estimate the mean height μ of the population of all American women between the ages of 18 and 24 years. This μ is the mean

μ_X of the random variable X obtained by choosing a young woman at random and measuring her height. To estimate μ, we choose an SRS of young women and use the sample mean $\bar{x}$ to estimate the unknown population mean μ. Statistics obtained from probability samples are random variables, because their values would vary in repeated sampling. The sampling distributions of statistics are just the probability distributions of these random variables. We will study sampling distributions in Chapter 9.

It seems reasonable to use $\bar{x}$ to estimate μ. An SRS should fairly represent the population, so the mean $\bar{x}$ of the sample should be somewhere near the mean μ of the population. Of course, we don't expect $\bar{x}$ to be exactly equal to μ, and we realize that if we choose another SRS, the luck of the draw will probably produce a different $\bar{x}$.

If $\bar{x}$ is rarely exactly right and varies from sample to sample, why is it nonetheless a reasonable estimate of the population mean μ? Here is one answer: if we keep on adding observations to our random sample, the statistic $\bar{x}$ is *guaranteed* to get as close as we wish to the parameter μ and then stay that close. We have the comfort of knowing that if we can afford to keep on measuring more young women, eventually we will estimate the mean height of all young women very accurately. This remarkable fact is called the *law of large numbers*. It is remarkable because it holds for *any* population, not just for some special class such as normal distributions.

LAW OF LARGE NUMBERS

Draw independent observations at random from any population with finite mean μ. Decide how accurately you would like to estimate μ. As the number of observations drawn increases, the mean $\bar{x}$ of the observed values eventually approaches the mean μ of the population as closely as you specified and then stays that close.

The behavior of $\bar{x}$ is similar to the idea of probability. In the long run, the proportion of outcomes taking any value gets close to the probability of that value, and the average outcome gets close to the distribution mean. Figure 6.1 shows how proportions approach probability in one example. Here is an example of how sample means approach the distribution mean.

EXAMPLE 7.8

The distribution of the heights of all young women is close to the normal distribution with mean 64.5 inches and standard deviation 2.5 inches. Suppose that $\mu = 64.5$ were exactly true. Figure 7.10 shows the behavior of the mean height $\bar{x}$ of n women chosen at random from a population whose heights follow the $N(64.5, 2.5)$ distribution. The graph plots the values of $\bar{x}$ as we add women to our sample. The first woman drawn had height 64.21 inches, so the line starts there. The second had height 64.35 inches, so for $n = 2$ the mean is

$$\bar{x} = \frac{64.21 + 64.35}{2} = 64.28$$

This is the second point on the line in the graph.

At first, the graph shows that the mean of the sample changes as we take more observations. Eventually, however, the mean of the observations gets close to the population mean $\mu = 64.5$ and settles down at that value. The law of large numbers says that this *always* happens.

The mean μ of a random variable is the average value of the variable in two senses. By its definition, μ is the average of the possible values,

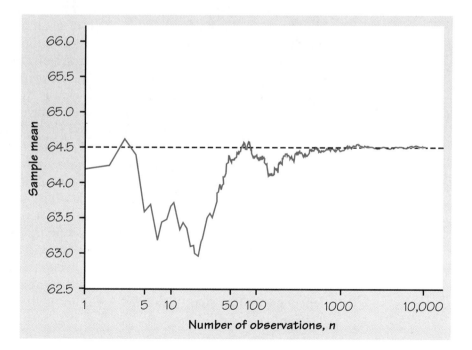

FIGURE 7.10   The law of large numbers in action. As we increase the size of our sample, the sample mean $\bar{x}$ always approaches the mean μ of the population.

weighted by their probability of occurring. The law of large numbers says that μ is also the long-run average of many independent observations on the variable. The law of large numbers can be proved mathematically starting from the basic laws of probability.*

Thinking about the law of large numbers

The law of large numbers says broadly that the average results of many independent observations are stable and predictable. Casinos are not the only businesses that base forecasts on this fact. A grocery store deciding how many gallons of milk to stock and a fast-food restaurant deciding how many beef patties to prepare can predict demand even though their many customers make independent decisions. The law of large numbers says that these many individual decisions will produce a stable result. It is worth the effort to think a bit more closely about so important a fact.

The "law of small numbers"   Both the rules of probability and the law of large numbers describe the regular behavior of chance phenomena *in the long run*. Psychologists have discovered that the popular understanding of randomness is quite different from the true laws of chance.[4] Most people believe in an incorrect "law of small numbers." That is, we expect even short sequences of random events to show the kind of average behavior that in fact appears only in the long run.

Try this experiment: Write down a sequence of heads and tails that you think imitates 10 tosses of a balanced coin. How long was the longest string (called a *run*) of consecutive heads or consecutive tails in your tosses? Most people will write a sequence with no runs of more than two consecutive heads or tails. Longer runs don't seem "random" to us. In fact, the probability of a run of three or more consecutive heads or tails in 10 tosses is greater than 0.8, and the probability of *both* a run of three or more heads and a run of three or more tails is almost 0.2.[5] This and other probability calculations suggest that a short sequence of coin tosses will often not appear random to us. The runs of consecutive heads or consecutive tails that appear in real coin tossing (and that are predicted by the mathematics of probability) seem surprising to us. Because we don't expect to see long runs, we may conclude that the coin tosses are not independent or that some influence is disturbing the random behavior of the coin.

---

*The earliest version of the law of large numbers was proved by the Swiss mathematician Jacob Bernoulli (1654–1705). The Bernoullis were a remarkable mathematical family; five of them contributed to the early study of probability.

Belief in the law of small numbers influences behavior. If a basketball player makes several consecutive shots, both the fans and his teammates believe that he has a "hot hand" and is more likely to make the next shot. This is doubtful. Careful study suggests that runs of baskets made or missed are no more frequent in basketball than would be expected if each shot were independent of the player's previous shots. Players perform consistently, not in streaks. (Of course, some players make a higher percent of their shots in the long run than others.) Our perception of hot or cold streaks simply shows that we don't perceive random behavior very well.[6]

Gamblers often follow the hot-hand theory, betting that a run will continue. At other times, however, they draw the opposite conclusion when confronted with a run of outcomes. If a coin gives 10 straight heads, some gamblers feel that it must now produce some extra tails to get back to the average of half heads and half tails. Not so. If the next 10,000 tosses give about 50% tails, those 10 straight heads will be swamped by the later thousands of heads and tails. No compensation is needed to get back to the average in the long run. Remember that it is *only* in the long run that the regularity described by probability and the law of large numbers takes over.

Our inability to accurately distinguish random behavior from systematic influences points out once more the need for statistical inference to supplement exploratory analysis of data. Probability calculations can help verify that what we see in the data is more than a random pattern.

How large is a large number?

The law of large numbers says that the actual mean outcome of many trials gets close to the distribution mean μ as more trials are made. It doesn't say how many trials are needed to guarantee a mean outcome close to μ. That depends on the *variability* of the random outcomes. The more variable the outcomes, the more trials are needed to ensure that the mean outcome $\bar{x}$ is close to the distribution mean μ.

The law of large numbers is the foundation of such business enterprises as gambling casinos and insurance companies. Games of chance must be quite variable if they are to hold the interest of gamblers. Even a long evening in a casino has an unpredictable outcome. Gambles with extremely variable outcomes, like state lottos with their very large but very improbable jackpots, require impossibly large numbers of trials to ensure that the average outcome is close to the expected value. Though most forms of gambling are less variable than lotto, the layman's answer to the applicability of the law of large numbers is usually that the house plays often enough to rely on it, but you don't. Much of the psychological allure of gambling is its unpredictability for the player. The business of gambling rests on the fact that the result is not unpredictable for the house. The average winnings of the house on tens of thousands of bets will be very close to the mean of the distribution of winnings. Needless to say, this mean guarantees the house a profit.

EXERCISES

7.20 This exercise is based on Example 7.8 and uses the TI-83 to simulate the law of large numbers and the sampling process. Begin by clearing lists L_1, L_2, L_3, and L_4. Then enter the following commands:

`seq(X,X,1,200)→L`$_1$	Enters the positive integers 1 to 200 into list L_1 (for seq, look under 2nd / LIST/ OPS)
`randNorm(64.5,2.5,200)→L`$_2$	Generates 200 random heights (in inches) from the $N(64.5, 2.5)$ distribution and stores these values in list L_2 (for randNorm, look under MATH / PRB)
`cumSum(L`$_2$`)→L`$_3$	Provides a cumulative sum of the observations and stores these values in list L_3 (for cumSum, look under 2nd / LIST / OPS)
`L`$_3$`/L`$_1$`→L`$_4$	Calculates the average heights of the women and stores these values in list L_4

Specify PLOT1 as follows: xyLine (2nd Type icon); Xlist: L_1; Ylist: L_4; Mark: . Set the viewing WINDOW as follows: $X[1,10]_{10}$. To set the Y dimensions, scan the values in L_4. Or start with $Y[60,69]_1$ and adjust as necessary. Press GRAPH. In the WINDOW screen, change Xmax to 100, and press GRAPH again. In your own words, write a short description of the principle that this exercise demonstrates.

7.21 One consequence of the law of large numbers is that once we have a probability distribution for a random variable, we can find its mean by simulating many outcomes and averaging them. The law of large numbers says that if we take enough outcomes, their average value is sure to approach the mean of the distribution.

I have a little bet to offer you. Toss a coin ten times. If there is no run of three or more straight heads or tails in the ten outcomes, I'll pay you $2. If there is a run of three or more, you pay me just $1. Surely you will want to take advantage of me and play this game?

Simulate enough plays of this game (the outcomes are +$2 if you win and -$1 if you lose) to estimate the mean outcome. Is it to your advantage to play?

7.22 (a) A gambler knows that red and black are equally likely to occur on each spin of a roulette wheel. He observes five consecutive reds and bets heavily on red at the next spin. Asked why, he says that "red is hot" and that the run of reds is likely to continue. Explain to the gambler what is wrong with this reasoning.

(b) After hearing you explain why red and black remain equally probable after five reds on the roulette wheel, the gambler moves to a poker game. He is dealt five straight red cards. He remembers what you said and assumes that the next card dealt in the same hand is equally likely to be red or black. Is the gambler right or wrong? Why?

7.23   The baseball player Tony Gwynn gets a hit about 35% of the time over an entire season. After he has failed to hit safely in six straight at-bats, the TV commentator says, "Tony is due for a hit by the law of averages." Is that right? Why?

Rules for means

You are studying flaws in the painted finish of refrigerators made by your firm. Dimples and paint sags are two kinds of surface flaw. Not all refrigerators have the same number of dimples: many have none, some have one, some two, and so on. You ask for the average number of imperfections on a refrigerator. How many total imperfections of both kinds (on the average) are there on a refrigerator? Easy: If the average number of dimples is 0.7 and the average number of sags is 1.4, then counting both gives an average of .7 + 1.4 = 2.1 flaws.

Putting this example into more formal language, the number of dimples on a refrigerator is a random variable X that takes values 0, 1, 2, and so on. X varies as we inspect one refrigerator after another. Only the mean number of dimples $\mu_X = .7$ was reported to you. The number of paint sags is a second random variable Y having mean $\mu_Y = 1.4$. (You see how the subscripts keep straight which variable we are talking about.) The total number of both dimples and sags is the sum $X + Y$. That sum is another random variable that varies from refrigerator to refrigerator. Its mean μ_{X+Y} is the average number of dimples and sags together and is just the sum of the individual means μ_X and μ_Y. That is an important rule for how means of random variables behave.

Another important rule says how the mean of a random variable changes when we multiply every outcome by the same fixed number or add the same fixed number to every outcome. Suppose X is the length in centimeters of a cockroach randomly chosen from those living in a dormitory and that the mean length is $\mu_X = 2.2$ centimeters. If we decide to measure in millimeters, we multiply every value of X by 10 because there are 10 millimeters in a centimeter. The length in millimeters is the new random variable 10X formed by multiplying X by 10 to convert centimeters to millimeters. What is the mean μ_{10X} of this new variable? If we multiply every value of X by 10, surely we also multiply the mean (the average value) by 10. That's right: the mean length is $2.2 \times 10 = 22$ millimeters. In more formal language, the mean μ_{10X} of 10X is $10\mu_X$. Similarly, if we add the same fixed number to every value of a random variable X, we add that same number to the mean. The first rule in the box below combines the results of multiplying by a fixed number and adding a fixed number.

RULES FOR MEANS

Rule 1. If X is a random variable and a and b are fixed numbers, then

$$\mu_{a+bX} = a + b\mu_X$$

Rule 2. If X and Y are random variables, then

$$\mu_{X+Y} = \mu_X + \mu_Y$$

Means are a special kind of average. Rules 1 and 2 just say that means behave like averages. Here is an example that applies these rules.

EXAMPLE 7.9

Gain Communications sells aircraft communications units to both the military and civilian markets. Next year's sales depend on market conditions that cannot be predicted exactly. Gain follows the modern practice of using probability estimates of sales. The military division estimates its sales as follows:

Units sold	1000	3000	5000	10,000
Probability	.1	.3	.4	.2

These are personal probabilities that express the informed opinion of Gain's executives. The corresponding sales estimates for the civilian division are

Units sold	300	500	750
Probability	.4	.5	.1

Take X to be the number of military units sold and Y the number of civilian units. From the probability distributions we compute that

$$\mu_X = (1000)(.1) + (3000)(.3) + (5000)(.4) + (10,000)(.2)$$
$$= 100 + 900 + 2000 + 2000$$
$$= 5000 \text{ units}$$

$$\mu_Y = (300)(.4) + (500)(.5) + (750)(.1)$$
$$= 120 + 250 + 75$$
$$= 445 \text{ units}$$

Gain makes a profit of $2000 on each military unit sold and $3500 on each civilian unit. Next year's profit from military sales will be 2000X, $2000 times the number X of units sold. By the first rule for means, the mean military profit is

$$\mu_{2000X} = 2000\mu_X = (2000)(5000) = \$10,000,000$$

Similarly, the civilian profit is 3500Y and the mean profit from civilian sales is

$$\mu_{3500Y} = 3500\mu_Y = (3500)(445) = \$1,557,500$$

The total profit is the sum of the military and civilian profit:

$$Z = 2000X + 3500Y$$

The second rule for means says that the mean of this sum of two variables is the sum of the two individual means:

$$\mu_Z = \mu_{2000X} + \mu_{3500Y}$$
$$= 10,000,000 + 1,557,500$$
$$= \$11,557,500$$

This mean is the company's best estimate of next year's profit, combining the probability estimates of the two divisions.

Once you have gained some experience in applying the rules for means of random variables, the calculation of mean total profit in Example 7.9 can be done more quickly by applying both rules at once as follows:

$$\mu_Z = \mu_{2000X+3500Y}$$
$$= 2000\mu_X + 3500\mu_Y$$
$$= (2000)(5000) + (3500)(445) = \$11,557,500$$

EXERCISES

7.24   Laboratory data show that the time required to complete two chemical reactions in a production process varies. The first reaction has a mean time of 40 minutes and a standard deviation of 2 minutes; the second has a mean time of 25 minutes and a standard deviation of 1 minute. The two reactions are run in sequence during production. There is a fixed period of 5 minutes between them as the product of the first reaction is pumped into the vessel where the second reaction will take place. What is the mean time required for the entire process?

The variance of a random variable

The mean is a measure of the center of a distribution. Even the most basic numerical description requires in addition a measure of the spread or variability of the distribution. The variance and the standard deviation are the measures of spread that accompany the choice of the mean to measure center. Just as for the mean, we need a distinct symbol to distinguish the variance of a random variable from the variance s^2 of a data set. We write the variance of a random variable X as σ_X^2. Once again the subscript reminds us which variable we have in mind. The definition of the variance σ_X^2 of a random variable is similar to the definition of the sample variance s^2 given in Chapter 1. That is, the variance is an average of the squared deviation $(X - \mu_X)^2$ of the variable X from its mean μ_X. As for the mean, the average we use is a weighted average in which each outcome is weighted by its probability in order to take account of outcomes that are not equally likely. Calculating this weighted average is straightforward for discrete random variables but requires advanced mathematics in the continuous case. Here is the resulting definition.

VARIANCE OF A DISCRETE RANDOM VARIABLE

Suppose that X is a discrete random variable whose distribution is

Value of X	x_1	x_2	x_3	$\cdots$	x_k
Probability	p_1	p_2	p_3	$\cdots$	p_k

and that μ is the mean of X. The **variance** of X is

$$\sigma_X^2 = (x_1 - \mu_X)^2 p_1 + (x_2 - \mu_X)^2 p_2 + \ldots + (x_k - \mu_X)^2 p_k$$
$$= \Sigma(x_i - \mu_X)^2 p_i$$

The **standard deviation** of σ_X of X is the square root of the variance.

EXAMPLE 7.10

In Example 7.9 we saw that the number X of communications units sold by the Gain Communications military division has distribution

Units sold	1000	3000	5000	10,000
Probability	.1	.3	.4	.2

We can find the mean and variance of X by arranging the calculation in the form of a table. Both μ_X and σ_X^2 are sums of columns in this table.

x_i	p_i	$x_i p_i$	$(x_i - \mu_x)^2 p_i$	
1,000	.1	100	$(1,000 - 5,000)^2(.1)$	$=$ 1,600,000
3,000	.3	900	$(3,000 - 5,000)^2(.3)$	$=$ 1,200,000
5,000	.4	2,000	$(5,000 - 5,000)^2(.4)$	$=$ 0
10,000	.2	2,000	$(10,000 - 5,000)^2(.2)$	$=$ 5,000,000

$$\mu_x = 5,000 \qquad\qquad \sigma_x^2 = 7,800,000$$

We see that $\sigma_X^2 = 7,800,000$. The standard deviation of X is $\sigma_X = \sqrt{7,800,000} = 2792.8$. The standard deviation is a measure of how variable the number of units sold is. As in the case of distributions for data, the standard deviation of a probability distribution is easiest to understand for normal distributions.

Rules for variances

Rules for variances similar to Rules 1 and 2 for means are often helpful for understanding statistical questions. Although the mean of a sum of random variables is always the sum of their means, this addition rule is not always true for variances. To understand why, take X to be the percent of a family's after-tax income that is spent and Y the percent that is saved. When X increases, Y decreases by the same amount. Though X and Y may vary widely from year to year, their sum $X + Y$ is always 100% and does not vary at all. It is the association between the variables X and Y that prevents their variances from adding. If random variables are independent, this kind of association between their values is ruled out and their *independence* variances do add. Two random variables X and Y are **independent** if any event involving X alone is independent of any event involving Y alone. Probability models often assume independence when the random variables describe outcomes that appear unrelated to each other. You should ask in each instance whether the assumption of independence seems reasonable.

RULES FOR VARIANCES

Rule 1. If X is a random variable and a and b are fixed numbers, then

$$\sigma^2_{a+bX} = b^2\sigma^2_X$$

Rule 2. If X and Y are independent random variables, then

$$\sigma^2_{X+Y} = \sigma^2_X + \sigma^2_Y$$

$$\sigma^2_{X-Y} = \sigma^2_X + \sigma^2_Y$$

Notice that because a variance is the average of *squared* deviations from the mean, multiplying X by a constant b multiplies σ^2_X by the *square* of the constant. Adding a constant a to a random variable changes its mean but does not change its variability. The variance of $X + a$ is therefore the same as the variance of X. Because the square of -1 is 1, the addition rule says that the variance of a difference is the *sum* of the variances. The difference $X - Y$ is more variable than either X or Y alone because variations in both X and Y contribute to variation in their difference.

As with data, we prefer the standard deviation to the variance as a measure of variability. The addition rule for variances implies that standard deviations do *not* generally add. Standard deviations are most easily combined by using the rules for variances rather than by giving separate rules for standard deviations. For example, the standard deviations of $2X$ and $-2X$ are both equal to $2\sigma_X$ because this is the square root of the variance $4\sigma^2_X$.

EXAMPLE 7.11

The payoff X of a \$1 ticket in the Tri-State Pick 3 game is \$500 with probability 1/1000 and \$0 the rest of the time. Here is the combined calculation of mean and variance:

x_i	p_i	x_ip_i	$(x_i - \mu_X)^2 p_i$	
0	.999	0	$(0 - .5)^2(.999)$ =	.24975
500	.001	.5	$(500 - .5)^2(.001)$ =	249.50025
	$\mu_X = .5$		σ^2_X =	249.75

The standard deviation is $\sigma_X = \sqrt{249.75} = \$15.80$. It is usual for games of chance to have large standard deviations, because large variability makes gambling exciting.

If you buy a Pick 3 ticket, your winnings are $W = X - 1$ because the dollar you paid for the ticket must be subtracted from the payoff. By the rules for means, the mean amount you win is

$$\mu_W = \mu_X - 1 = -\$.50$$

That is, you lose an average of 50 cents on a ticket. The rules for variances remind us that the variance and standard deviation of the winnings $W = X - 1$ are the same as those of X. Subtracting a fixed number changes the mean but not the variance.

Suppose now that you buy a $1 ticket on each of two different days. The payoffs X and Y on the two tickets are independent because separate drawings are held each day. Your total payoff $X + Y$ has mean

$$\mu_{X+Y} = \mu_X + \mu_Y = \$.50 + \$.50 = \$1.00$$

Because X and Y are independent, the variance of $X + Y$ is

$$\sigma_{X+Y}^2 = \sigma_X^2 + \sigma_Y^2 = 249.75 + 249.75 = 499.5$$

The standard deviation of the total payoff is

$$\sigma_{X+Y} = \sqrt{499.5} = \$22.35$$

This is not the same as the sum of the individual standard deviations, which is $\$15.80 + \$15.80 = \$31.60$. Variances of independent random variables add; standard deviations do not.

If you buy a ticket every day (365 tickets in a year), your mean payoff is the sum of 365 daily payoffs. That's 365 times 50 cents, or $182.50. Of course, it costs $365 to play, so the state's mean take from a daily Pick 3 player is $182.50. Results for individual players will vary, but the law of large numbers assures the state its profit.

EXAMPLE 7.12

A college uses SAT scores as one criterion for admission. Experience has shown that the distribution of SAT scores among its entire population of applicants is such that

SAT math score X	$\mu_X = 625$	$\sigma_X = 90$
SAT verbal score Y	$\mu_Y = 590$	$\sigma_Y = 100$

What are the mean and standard deviation of the total score $X + Y$ among students applying to this college?

The mean overall SAT score is

$$\mu_{X+Y} = \mu_X + \mu_Y = 625 + 590 = 1215$$

The variance and standard deviation of the total *cannot be computed* from the information given. SAT verbal and math scores are not independent, because students who score high on one exam will tend to score high on the other also. Therefore, the addition rule for variances does not apply.

EXAMPLE 7.13

Tom and George play golf at the same club. Tom's score X varies from round to round but has

$$\mu_X = 110 \text{ and } \sigma_X = 10$$

George's score Y also varies, with

$$\mu_Y = 100 \text{ and } \sigma_Y = 8$$

Tom and George are playing the first round of the club tournament. Because they are not playing together, we will assume that their scores vary independently of each other. The difference between their scores on this first round has mean

$$\mu_{X-Y} = \mu_X - \mu_Y = 110 - 100 = 10$$

The variance of the difference between the scores is

$$\sigma^2_{X-Y} = \sigma^2_X + \sigma^2_Y = 10^2 + 8^2 = 164$$

The standard deviation is found from the variance

$$\sigma_{X-Y} = \sqrt{164} = 12.8$$

The variation in the difference between the scores is greater than the variation in the score of either player because the difference contains two independent sources of variation. It also makes the tournament interesting, because George will not always finish ahead of Tom even though he has the lower mean score. Computations of the variance of sums and differences of statistics play an important role in statistical inference.

EXERCISES

7.25   For each of the following situations, would you expect the random variables X and Y to be independent? Explain your answers.

(a)   X is the rainfall (in inches) on November 6 of this year, and Y is the rainfall at the same location on November 6 of next year.

(b)   X is the amount of rainfall today, and Y is the rainfall at the same location tomorrow.

(c)   X is today's rainfall at the Orlando, Florida airport, and Y is today's rainfall at Disney World just outside Orlando.

7.26   A time and motion study measures the time required for an assembly line worker to perform a repetitive task. The data show that the time required to bring a part from a bin to its position on an automobile chassis varies from car to car with mean 11 seconds and standard deviation 2 seconds. The time required to attach the part to the chassis varies with mean 20 seconds and standard deviation 4 seconds.

(a)   What is the mean time required for the entire operation of positioning and attaching the part?

(b)   If the variation in the worker's performance is reduced by better training, the standard deviations will decrease. Will this decrease change the mean you found in (a) if the mean times for the two steps remain as before?

(c)   The study finds that the times required for the two steps are independent. A part that takes a long time to position, for example, does not take more or less time to attach than other parts. Would your answer in (a) change if the two variables were dependent?

7.27   Find the standard deviation σ_X of the distribution of grades in Exercise 7.17.

7.28   In an experiment on the behavior of young children, each subject is placed in an area with five toys. The response of interest is the number of toys that the child plays with. Past experiments with many subjects have shown that the probability distribution of the number X of toys played with is as follows:

Number of toys x_i	0	1	2	3	4	5
Probability p_i	.03	.16	.30	.23	.17	.11

(a)   Calculate the mean μ_X and the standard deviation σ_X.

(b)   Describe the details of a simulation you could carry out to approximate the mean number of toys μ_X and the standard deviation σ_X. Then carry out your simulation. Are the mean and standard deviation produced from your simulation close to the values you calculated in (a)?

SUMMARY

The probability distribution of a random variable X, like a distribution of data, has a **mean μ_X** and a **standard deviation σ_X**.

The **law of large numbers** says that the average of the values of X observed in many trials must approach μ.

The **mean** μ is the balance point of the probability histogram or density curve. If X is discrete with possible values x_i having probabilities p_i, the mean is the average of the values of X, each weighted by its probability:

$$\mu_X = x_1 p_1 + x_2 p_2 + \cdots + x_k p_k$$

The **variance** σ_X^2 is the average squared deviation of the values of the variable from their mean. For a discrete random variable,

$$\sigma_X^2 = (x_1 - \mu)^2 p_1 + (x_2 - \mu)^2 p_2 + \cdots + (x_k - \mu)^2 p_k$$

The **standard deviation** σ_X is the square root of the variance. The standard deviation measures the variability of the distribution about the mean. It is easiest to interpret for normal distributions.

The mean and variance of a continuous random variable can be computed from the density curve, but to do so requires more advanced mathematics.

The means and variances of random variables obey the following rules. If a and b are fixed numbers, then

$$\mu_{a+bX} = a + b\mu_X$$

$$\sigma_{a+bX}^2 = b^2 \sigma_X^2$$

If X and Y are any two random variables, then

$$\mu_{X+Y} = \mu_X + \mu_Y$$

and if X and Y are independent, then

$$\sigma_{X+Y}^2 = \sigma_X^2 + \sigma_Y^2$$

$$\sigma_{X-Y}^2 = \sigma_X^2 + \sigma_Y^2$$

SECTION 7.2 EXERCISES

7.29   Find the standard deviation of the time required for the two-step assembly operation studied in Exercise 7.26.

7.30   The times for the two reactions in the chemical production process described in Exercise 7.24 are independent. Find the standard deviation of the time required to complete the process.

7.31 The academic motivation and study habits of female students as a group are better than those of males. The Survey of Study Habits and Attitudes (SSHA) is a psychological test that measures these factors. The distribution of SSHA scores among the women at a college has mean 120 and standard deviation 28, and the distribution of scores among men students has mean 105 and standard deviation 35. You select a single male student and a single female student at random and give them the SSHA test.

(a) Explain why it is reasonable to assume that the scores of the two students are independent.

(b) What are the mean and standard deviation of the difference (female minus male) between their scores?

(c) From the information given, can you find the probability that the woman chosen scores higher than the man? If so, find this probability. If not, explain why you cannot.

7.32 In a process for manufacturing glassware, glass stems are sealed by heating them in a flame. The temperature of the flame varies a bit. Here is the distribution of the temperature X measured in degrees Celsius:

Temperature	540°	545°	550°	555°	560°
Probability	.1	.25	.3	.25	.1

(a) Find the mean temperature μ_X and the standard deviation σ_X.

(b) The target temperature is 550° C. What are the mean and standard deviation of the number of degrees off target $X - 550$?

(c) A manager asks for results in degrees Fahrenheit. The conversion of X into degrees Fahrenheit is given by

$$Y = \frac{9}{5}X + 32$$

What are the mean μ_Y and the standard deviation σ_Y of the temperature of the flame in the Fahrenheit scale?

7.33 In continuation of the previous exercise, describe the details of a simulation you could carry out to approximate the mean temperature and the standard deviation in degrees Celsius. Then carry out your simulation. Are the mean and standard deviation produced from your simulation close to the values you calculated in (a)?

CHAPTER REVIEW

A random variable defines what is counted or measured in a statistics application. If the random variable X is a count, such as the number of heads
in four tosses of a coin, then X is discrete, and its distribution can be pictured as a histogram. If X is measured, as in the number of inches of rainfall in Richmond in April, then X is continuous, and its distribution is
pictured as a density curve. Among the continuous random variables, the
normal random variable is the most important. First introduced in Chapter 2, the normal distribution is revisited, with emphasis this time on it as a
probability distribution. The mean and variance of a random variable are
calculated, and rules for the sum or difference of two random variables are
developed. Here is a checklist of the major skills you should have acquired
by studying this chapter.

A. RANDOM VARIABLES

1. Recognize and define a discrete random variable, and construct a
   probability distribution table and a probability histogram for the
   random variable.
2. Recognize and define a continuous random variable, and determine
   probabilities of events as areas under density curves.
3. Given a normal random variable, use the standard normal table or
   a TI-83 to find probabilities of events as areas under the standard
   normal distribution curve.

B. MEANS AND VARIANCES OF RANDOM VARIABLES

1. Calculate the mean and variance of a discrete random variable. Find
   the expected payout in a raffle or similar game of chance.
2. Use simulation methods and the law of large numbers to approximate
   the mean of a distribution.
3. Use rules for means and rules for variances to solve problems
   involving sums and differences of random variables.

CHAPTER 7 EXERCISES

7.34   A life insurance company sells a term insurance policy to a 21-year-old
        male that pays $100,000 if the insured dies within the next 5 years. The
        probability that a randomly chosen male will die each year can be found

in mortality tables. The company collects a premium of $250 each year as payment for the insurance. The amount X that the company earns on this policy is $250 per year, less the $100,000 that it must pay if the insured dies. Here is the distribution of X. Fill in the missing probability in the table and calculate the mean profit μ_X.

Age at death	21	22	23	24	25	≥ 26
Profit	$-\$99,750$	$-\$99,500$	$-\$99,250$	$-\$99,000$	$-\$98,750$	$\$1250$
Probability	.00183	.00186	.00186	.00191	.00193	

7.35 It would be quite risky for you to insure the life of a 21-year-old friend under the terms of the previous exercise. There is a high probability that your friend would live and you would gain $1250 in premiums. But if he were to die, you would lose almost $100,000. Explain carefully why selling insurance is not risky for an insurance company that insures many thousands of 21-year-old men.

7.36 In which of the following games of chance would you be willing to assume independence of X and Y in making a probability model? Explain your answer in each case.

(a) In blackjack, you are dealt two cards and examine the total points X on the cards (face cards count 10 points). You can choose to be dealt another card and compete based on the total points Y on all three cards.

(b) In craps, the betting is based on successive rolls of two dice. X is the sum of the faces on the first roll, and Y the sum of the faces on the next roll.

7.37 Amarillo Slim is back and he's got another deal for you. We have a fair coin (heads and tails each have probability 1/2). Toss it twice. If two heads come up, you win. If you get any other result, you get another chance: toss the coin twice more, and if you get two heads, you win. If you fail to get two heads on the second try, you lose. You pay a dollar to play. If you win, you get your dollar back plus another dollar.

(a) Explain how to simulate one play of this game using Table B. How could you simulate one play using a TI-83? Simulate two tosses of a fair coin.

(b) Simulate 50 plays, using Table B or your TI-83. Use your simulation to estimate the expected value of the game.

(c) There are two outcomes in this game: win or lose. Let the random variable X be the (monetary) outcome. What are the two values X can take? Calculate the actual probabilities of each value of X. Then calculate μ_X. How does this compare with your estimate from the simulation in (b)?

7.38 A couple plans to have children until they have a girl or until they have four children, whichever comes first. Example 5.21 estimated the probability that they will have a girl among their children. Now we ask a different question: How many children, on the average, will couples who follow this plan have?

(a) To answer this question, construct a simulation similar to that in Example 5.21 but this time keep track of the number of children in each repetition. Carry out 25 repetitions and then average the results to estimate the expected value.

(b) Construct the probability distribution table for the random variable X = number of children.

(c) Use the table from (b) to calculate the expected value of X. Compare this number with the result from your simulation in (a).

7.39 In government data, a household consists of all occupants of a dwelling unit, while a family consists of two or more persons who live together and are related by blood or marriage. Here are the distributions of household size and of family size in the United States:

Number of persons	1	2	3	4	5	6	7
Household probability	.25	.32	.17	.15	.07	.03	.01
Family probability	0	.42	.23	.21	.09	.03	.02

Figure 7.9(b) (page 388) shows the mean and probability histogram for household size. Compare the two distributions using probability histograms, means, and standard deviations. Then write a brief comparison, using your calculations to back up your statements.

7.40 Here is a simple way to create a random variable X that has mean μ and standard deviation σ: X takes only the two values $\mu - \sigma$ and $\mu + \sigma$, each with probability 0.5. Use the definition of the mean and variance for discrete random variables to show that X does have mean μ and standard deviation σ.

7.41 Examples 7.9 (page 396) and 7.10 (page 399) concern a probabilistic projection of sales and profits by an electronics firm, Gain Communications.

(a) Find the variance and standard deviation of the estimated sales X of Gain's civilian unit, using the distribution and mean from Example 7.9.

(b) Because the military budget and the civilian economy are not closely linked, Gain is willing to assume that its military and civilian sales vary independently. Combine your result from (a) with the results for the military unit from Example 7.10 to obtain the standard deviation of the total sales X + Y.

(c) Find the standard deviation of the estimated profit, $Z = 2000X + 3500Y$.

7.42 You have two scales for measuring weights in a chemistry lab. Both scales give answers that vary a bit in repeated weighings of the same item. If the true weight of a compound is 2.00 grams (g), the first scale produces readings X that have mean 2.000 g and standard deviation 0.002 g. The second scale's readings Y have mean 2.001 g and standard deviation 0.001 g.

(a) What are the mean and standard deviation of the difference $Y - X$ between the readings? (The readings X and Y are independent.)

(b) You measure once with each scale and average the readings. Your result is $Z = (X + Y)/2$. What are μ_Z and σ_Z? Is the average Z more or less variable than the reading Y of the less variable scale?

7.43 The risk of an investment is often measured by the standard deviation of the return on the investment. The more variable the return is (the larger σ is), the riskier the investment. We can measure the great risk of insuring a single person's life in Exercise 7.34 by computing the standard deviation of the income X that the insurer will receive. Find σ_X, using the distribution and mean found in Exercise 7.34.

7.44 The risk of insuring one person's life is reduced if we insure many people. Use the result of the previous exercise and rules for means and variances to answer the following questions.

(a) Suppose that we insure two 21-year-old males, and that their ages at death are independent. If X and Y are the insurer's income from the two insurance policies, the total income is $T = X + Y$. The mean and variance of each of X and Y are as you computed them in the previous exercise. What are the mean μ_T and the standard deviation σ_T of the insurer's total income T?

(b) The insurer's average income on the two policies is

$$Z = \frac{X + Y}{2} = \frac{1}{2}X + \frac{1}{2}Y$$

Find the mean and standard deviation of Z. You see that the mean income is the same as for a single policy but the standard deviation is less.

(c) If four 21-year-old men are insured, the insurer's average income is

$$Z = \frac{1}{4}(X_1 + X_2 + X_3 + X_4)$$

where X_i is the income from insuring one man. The X_i are independent and each has the same distribution as before. Find the mean and standard deviation of Z. Compare your results with the results of (b).

NOTES AND DATA SOURCES

1. We use $\bar{x}$ both for the random variable, which takes different values in repeated sampling, and for the numerical value of the random variable in a particular sample. Similarly, s and $\hat{p}$ stand both for random variables and for specific values. This notation is mathematically imprecise but statistically convenient.

2. In most applications X takes a finite number of possible values. The same ideas, implemented with more advanced mathematics, apply to random variables with an infinite but still countable collection of values. An example is a geometric random variable, considered in Section 8.2.

3. The mean of a continuous random variable X with density function $f(x)$ can be found by integration:

$$\mu_X = \int xf(x)dx$$

This integral is a kind of weighted average, analogous to the discrete-case mean

$$\mu_X = \Sigma xP(X = x)$$

The variance of a continuous random variable X is the average squared deviation of the values of X from their mean, found by the integral

$$\sigma_X^2 = \int (x - \mu)^2 f(x)dx$$

4. See A. Tversky and D. Kahneman, "Belief in the law of small numbers," *Psychological Bulletin*, 76 (1971), pp. 105–110, and other writings of these authors for a full account of our misperception of randomness.

5. Probabilities involving runs can be quite difficult to compute. That the probability of a run of three or more heads in 10 independent tosses of a fair coin is $(1/2) + (1/128)$ = .508 can be found by clever counting, as can the other results given in the text. A general treatment using advanced methods appears in Section XIII.7 of William Feller, *An Introduction to Probability Theory and Its Applications*, vol. 1, 3rd ed., Wiley, New York, 1968.

6. R. Vallone and A. Tversky, "The hot hand in basketball: on the misperception of random sequences," *Cognitive Psychology*, 17 (1985), pp. 295–314. A later series of articles that debate the independence question is A. Tversky and T. Gilovich, "The cold facts about the 'hot hand' in basketball," *Chance*, 2, no. 1 (1989), pp. 16–21; P. D. Larkey, R. A. Smith, and J. B. Kadane, "It's OK to believe in the 'hot hand,' " *Chance*, 2, no. 4 (1989), pp. 22–30; and A. Tversky and T. Gilovich, "The 'hot hand': statistical reality or cognitive illusion?" *Chance*, 2, no. 4 (1989), pp. 31–34.

The Binomial and Geometric Distributions

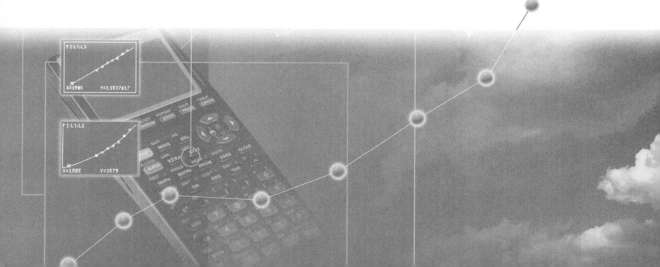

INTRODUCTION

In practice, we frequently encounter experimental situations where there are two outcomes of interest. Some examples are:

- We use a coin toss to see which of the two football teams gets the choice of kicking off or receiving to begin the game.
- A basketball player shoots a free throw; the outcomes of interest are {she makes the shot; she misses}.
- A young couple prepares for their first child; the possible outcomes are {boy; girl}.
- A quality control inspector selects a widget coming off the assembly line; he is interested in whether or not the widget meets production requirements.

In this chapter we will explore two important classes of distributions—the binomial distributions and the geometric distributions—and learn some of their properties. We will use what we have learned about probability and random variables from previous chapters, with the view toward completing the necessary foundation to study inference.

| ACTIVITY 8 | A Gaggle of Girls |

The Ferrells have 3 children: Jennifer, Jessica, and Jaclyn. If we assume that a couple is equally likely to have a girl or a boy, then how unusual is it for a family like the Ferrells to have 3 children who are all girls? We have encountered problems like this in an earlier chapter. But this time we're going to use the method of simulation. If success = girl, and failure = boy, then p(success) = .5. We will define the random variable X as the number of girls. Then we want to simulate families with 3 children. Our goal is to determine the long-term relative frequency of a family with 3 girls, that is, $P(X = 3)$.

1. Using a random number table, let even digits represent "girl" and odd digits represent "boy." Select a row, and beginning at that row, read off numbers 3 digits at a time. Each 3 digits will constitute one trial. Use tally marks in a table like this one to record the results:

(continued)

3 girls	
Not 3 girls	

Do at least 40 trials. Then combine your results with those of other students in the class to obtain at least 200 trials. Calculate the relative frequency of the event {3 girls}.

2. For variety, do the same thing as before, but this time using the TI-83. Using the codes 1 = girl and 0 = boy, enter the command `randInt(0,1,3)`. This command instructs the calculator to randomly pick a whole number from the set {0, 1} and to do this 3 times. The outcome {0, 0, 1}, using our codes, means {*boy, boy, girl*}, in that order. Continue to press the ENTER key and count until you have 40 trials. Use a tally mark to record each time you observe a {1, 1, 1} result. Calculate the relative frequency for the event {3 girls}.

3. *Extra for programming experts:* Write a TI-83 program to carry out the process described above. Allow the user to specify the number of trials, and have the calculator report the relative frequency of {3 girls} as a decimal number.

4. Determine the total number of outcomes for this experiment. List the outcomes in the sample space. Then complete the probability distribution table for the random variable X = number of girls.

X	0	1	2	3
P(X)				

Do the results of your simulations come close to the theoretical value for $P(X = 3)$?

8.1 THE BINOMIAL DISTRIBUTIONS

In Activity 8, we simulated families with 3 children to discover how often the children would be all girls. Flipping a fair coin 3 times and letting heads represent having a girl and tails represent having a boy would produce exactly the same results. The characterizing features of this experiment are as follows: A *trial* consists of flipping the coin once. There are two outcomes: heads = girl (success), and tails = boy (failure). We will flip the coin 3 times. The coin flips are independent in the sense that the outcome of one coin flip has no influence on the outcome of the next flip. And last, the probability of success (girl) is the same for each coin flip (trial). A situation where these four conditions are satisfied is said to be a *binomial setting*.

binomial setting

THE BINOMIAL SETTING

1. Each observation falls into one of just two categories, which for convenience we call "success" or "failure."
2. There is a fixed number n of observations.
3. The n observations are all **independent**. That is, knowing the result of one observation tells you nothing about the other observations.
4. The probability of success, call it p, is the same for each observation.

If you are presented with an experimental setting, it is important to be able to recognize it as a binomial setting or a geometric setting (covered in the next section) or neither. If you can verify that each of these four conditions is satisfied, you will be able to make use of known properties of binomial situations to gain more insights.

If data are produced in a binomial setting, then the random variable X = number of successes is called a *binomial random variable*, and the probability distribution of X is called a *binomial distribution*.

binomial random variable

BINOMIAL DISTRIBUTION

The distribution of the count X of successes in the binomial setting is the **binomial distribution** with parameters n and p. The parameter n is the number of observations, and p is the probability of a success on any one observation. The possible values of X are the whole numbers from 0 to n. As an abbreviation, we say that X is $B(n, p)$.

The binomial distributions are an important class of discrete probability distributions. Pay attention to the binomial setting, because not all counts have binomial distributions.

EXAMPLE 8.1

Blood type is inherited. If both parents carry genes for the O and A blood types, each child has probability 0.25 of getting two O genes and so of having blood type O. Different children inherit independently of each other. The number of O blood types among 5 children of these parents is the count X of successes in 5 independent observations with probability 0.25 of a success on each observation. So X has the binomial distribution with $n = 5$ and $p = .25$. We say that X is $B(5, 0.25)$.

EXAMPLE 8.2

Deal 10 cards from a shuffled deck and count the number X of red cards. There are 10 observations, and each gives either a red or a black card. A "success" is a red card. But the observations are *not* independent. If the first card is black, the second is more likely to be red because there are more red cards than black cards left in the deck. The count X does *not* have a binomial distribution.

EXAMPLE 8.3

An engineer chooses an SRS of 10 switches from a shipment of 10,000 switches. Suppose that (unknown to the engineer) 10% of the switches in the shipment are bad. The engineer counts the number X of bad switches in the sample.

This is not quite a binomial setting. Just as removing one card in Example 8.2 changed the makeup of the deck, removing one switch changes the proportion of bad switches remaining in the shipment. So the state of the second switch chosen is not independent of the first. But removing one switch from a shipment of 10,000 changes the makeup of the remaining 9999 switches very little. In practice, the distribution of X is very close to the binomial distribution with $n = 10$ and $p = .1$.

Example 8.3 shows how we can use the binomial distributions in the statistical setting of selecting an SRS. When the population is much larger than the sample, a count of successes in an SRS of size n has approximately the binomial distribution with n equal to the sample size and p equal to the proportion of successes in the population.

EXAMPLE 8.4

Engineers define reliability as the probability that an item will perform its function under specific conditions for a specific period of time. If an aircraft engine turbine has probability 0.999 of performing properly for an hour of flight, the number of turbines in a fleet of 350 engines that fly for an hour without failure has the $B(350, 0.999)$ distribution. This binomial distribution is obtained by assuming, as seems reasonable, that the turbines fail independently of each other. A common cause of failure, such as sabotage, would destroy the independence and make the binomial model inappropriate.

EXERCISES

In each of Exercises 8.1 to 8.4, X is a count. Does X have a binomial distribution? Give your reasons in each case.

8.1   You observe the sex of the next 20 children born at a local hospital; X is the number of girls among them.

8.2   A couple decides to continue to have children until their first girl is born; X is the total number of children the couple has.

8.3   A student studies binomial distributions using computer-assisted instruction. After the lesson, the computer presents 10 problems. The student solves each problem and enters her answer. The computer gives additional instruction between problems if the answer is wrong. The count X is the number of problems that the student gets right.

8.4   Joe buys a state lottery ticket every week. The count X is the number of times in a year that he wins a prize.

Finding binomial probabilities

We will give a formula later for the probability that a binomial random variable takes any of its values. In practice, you will rarely have to use this formula for calculations. The TI-83 and most statistical software packages calculate binomial probabilities.

EXAMPLE 8.5

A quality engineer selects an SRS of 10 switches from a large shipment for detailed inspection. Unknown to the engineer, 10% of the switches in the shipment fail to meet the specifications. What is the probability that no more than 1 of the 10 switches in the sample fail inspection?

The count X of bad switches in the sample has approximately the $B(10, 0.1)$ distribution. Figure 8.1 is a probability histogram for this distribution. The distribution is strongly skewed. Although X can take any whole-number

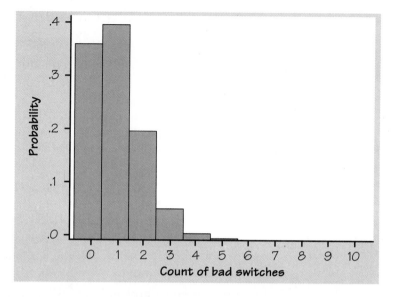

FIGURE 8.1   Probability histogram for the binomial distribution with $n = 10$ and $p = .1$.

value from 0 to 10, the probabilities of values larger than 5 are so small that they do not appear in the histogram. We want to calculate

$$P(X \leq 1) = P(X = 0) + P(X = 1)$$

when X is $B(10, 0.1)$. The TI-83 command binompdf(n,p,X) calculates the binomial probability of the value X. The suffix pdf stands for "probability distribution function." The binompdf command is found under 2nd / DISTR / 0:binompdf.

P.D.F.

Given a discrete random variable X, the **probability distribution function** assigns a probability to each value of X. The probabilities must satisfy the rules for probabilities given in Chapter 6.

The command binompdf(10,.1,0) calculates the binomial probability that $X = 0$ to be 0.3486784401, and the command binompdf(10,.1,1) returns probability 0.387420489. Thus,

$$P(X \le 1) = P(X = 0) + P(X = 1)$$
$$= .3487 + .3874 = .7361$$

About 74% of all samples will contain no more than 1 bad switch. A sample of size 10 cannot be trusted to alert the engineer to the presence of unacceptable items in the shipment.

To have the TI-83 calculate the probability distribution table and plot a histogram for the distribution of defective switches in a sample of 10 switches, proceed as follows:

1. Enter the values of X into list L_1 either directly through the STAT / EDIT mode or by entering the command $\texttt{seq(X,X,0,10,1)} \rightarrow L_1$. (The $\texttt{seq}$ command is under 2nd / LIST / OPS / 5:seq. The syntax is: the first X is the function, the second X is the counting variable, the next two numbers define the starting and ending values, and the last number is the increment.)

2. Enter the binomial probabilities into list L_2 by means of the command $\texttt{binompdf(10,.1,}L_1\texttt{)} \rightarrow L_2$. Note that the largest probability listed is about 0.3874. This will help us define our viewing window.

```
L1      | L2      | L3     2
 0      |.34868   |------
 1      |.38742   |
 2      |.19371   |
 3      |.0574    |
 4      |.01116   |
 5      |.00149   |
 6      |1.4E-4   |
L2(1)=.3486784401...
```

3. Disable or delete any active defined functions in the Y= window.

4. Turn the axes off: 2nd / FORMAT / AxesOff.

5. Enter STAT PLOT and specify a histogram with Xlist: L_1 and Freq: L_2.

6. Set the viewing Window: $X[-.5, 10.5]_1$ and $Y[-.1, .45]_1$.

7. Press TRACE and use the left and right cursor keys to inspect heights of various bars in the histogram.

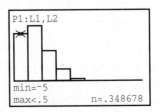

Outcomes larger than 6 do not have probability exactly 0, but their probabilities are so small that the rounded values are 0.0000. Use the command 1-Var Stats L_2 to verify that the sum of the probabilities given is 1, as it should be.

EXAMPLE 8.6

Corinne is a basketball player who makes 75% of her free throws over the course of a season. In a key game, Corinne shoots 12 free throws and makes only 7 of them. The fans think that she failed because she was nervous. Is it unusual for Corinne to perform this poorly? To answer this question, assume that free throws are independent with probability 0.75 of a success on each shot. (Studies of long sequences of free throws have found no evidence that they are dependent, so this is a reasonable assumption.) The number X of baskets (successes) in 12 attempts has the $B(12, 0.75)$ distribution.

We want the probability of making a basket on at most 7 free throws. This is

$$P(X \leq 7) = P(X = 0) + P(X = 1) + P(X = 2) + \ldots + P(X = 7)$$
$$= .0000 + .0000 + .0000 + .0004 + .0024 + .0115 + .0401 + .1032$$
$$= .1576$$

Corinne will make at most 7 of her 12 free throws about 16% of the time, or roughly in one of every six games. While below her average level, this performance is well within the range of the usual chance variation in her shooting.

EXAMPLE 8.7

In Activity 8 we wanted to determine the probability that all 3 children in a family are girls. In this case, the random variable of interest, X = the number of girls, has the $B(3, 0.5)$ distribution. We want to find the probability that the number of girls is 3, that is, $P(X = 3)$. The TI-83 command binompdf(3,.5,3) returns the probability 0.125. If you do these additional keystrokes: MATH / 1:▶FRAC, the answer will be converted to the fraction 1/8.

In applications we frequently want to find the probability that a random variable takes a range of values. The *cumulative* binomial probability is useful in these cases.

C.D.F.

Given a random variable X, the **cumulative distribution function** of X calculates the sum of the probabilities for 0, 1, 2, ..., up to the value X. That is, it calculates the probability of obtaining at most X successes in n trials.

For the count X of defective switches in Example 8.5, the command `binomcdf(10,.1,1)` outputs 0.736098903 for the cumulative probability $P(X \le 1)$.

EXAMPLE 8.8

In Example 8.6. Corinne shoots $n = 12$ free throws and makes only 7 of them. Since she is a 75% free throw shooter ($p = .75$), we wanted to know if it was unusual for Corinne to perform this poorly. If $X =$ number of baskets made on free throws, then X has the $B(12, 0.75)$ distribution, and we need to find the probability that she makes at most 7 of her free throws, that is, $P(X \le 7)$. The TI-83 command `binomcdf(12,.75,7)` calculates the cumulative probability $P(X \le 7)$ to be 0.1576436761. We round the answer to four decimal places and report that the probability that Corinne makes *at most* 7 of her 12 free throws is 0.1576.

The p.d.f. table for Corinne's shots looks like this:

X	0	1	2	3	4	5	6	7	8	9	10	11	12
P(X)	.000	.000	.000	.000	.002	.011	.040	.103	.194	.258	.232	.127	.032

If we denote the cumulative distribution function by $F(X)$, then we can record the cumulative sum of the probabilities in a third row of the table:

X	0	1	2	3	4	5	6	7	8	9	10	11	12
P(X)	.000	.000	.000	.000	.002	.011	.040	.103	.194	.258	.232	.127	.032
F(X)	.000	.000	.000	.000	.003	.014	.054	.158	.351	.609	.842	.968	1

Notice that terms sometimes don't appear to add up as they should. The cumulative function $F(4)$, for example, should equal $P(0) + P(1) + P(2) + P(3) + P(4)$. Of course, the culprit is roundoff error. Enter the integers 0 to 12 into list L_1, the corresponding binomial probabilities into list L_2, and use the command `binomcdf(12,.75,L₁)→ L₃` to enter the cumulative probabilities into list L_3.

In addition to being helpful in answering questions involving wording like "find the probability that it takes at most 6 trials," the c.d.f. is also particularly useful for calculating the probability that it takes *more* than a certain number of trials to see the first success. This calculation uses the complement rule:

$$P(X > n) = 1 - P(X \le n) \quad n = 2, 3, 4, \ldots$$

Returning to Corinne and her free throws, the probability distribution histogram peaks at $X = 9$ and is strongly left-skewed, because $p = .75$ is closer to 1 than it is to 0. Consequently, the cumulative distribution histogram stays fairly flat from $X = 0$ to $X = 5$ and then increases rapidly to 1.

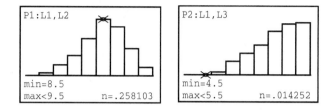

The rightmost bar of the cumulative distribution histogram for *every* discrete random variable will have height 1. Do you see why?

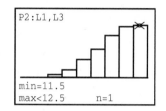

EXERCISES

Use the TI-83's `binompdf` or `binomcdf` commands to find the following probabilities.

8.5 (a) Each child born to a particular set of parents has probability 0.25 of having blood type O. If these parents have 5 children, what is the probability that exactly 2 of them have type O blood? Let X = number of children who have type O blood. Then X has the $B(5, 0.25)$ distribution. Find $P(X = 2)$.

(b) Make a table for the p.d.f. of the random variable X. Then use the calculator to find the probabilities of all possible values of X, and complete the table.

(c) Verify that the sum of the probabilities is 1.

(d) Construct a histogram of the p.d.f. on your calculator.

(e) Use the calculator to find the cumulative probabilities and add these values to your p.d.f. table. Then construct a cumulative distribution histogram. How is this histogram different from the histogram for Corinne's free throws?

8.6   Use lists L_1 and L_2 on your calculator to construct a p.d.f. for Corinne's
       free throw probabilities. (Refer to Examples 8.6 and 8.8.) Use the ran-
       dom variable X = number of baskets made on free throws. Then execute
       the command `cumSum(L₂)→L₃`. (`cumSum` is found under 2nd / LIST /
       OPS / 6:cumSum. What do you think this command does?) Then use the
       `binomcdf` command to enter the cumulative probabilities into list L_4.
       Compare L_3 and L_4. Are they the same?

8.7   This exercise is an extension of Activity 8. There's a movie classic entitled
       Seven Brides for Seven Brothers. Even if these brothers had a few sisters,
       this many brothers is unusual. We will assume that there are no sisters.

       (a)   Let X = number of boys in a family of 7 children. Assume that sons and
             daughters are equally likely outcomes. Do you think the distribution
             of X will be skewed left, symmetric, or skewed right? The answer to
             this question depends on what fact?

       (b)   Use the `binompdf` command to construct a p.d.f. table for X. Then
             construct a probability distribution histogram and a cumulative dis-
             tribution histogram for X. Keep a written record of your numerical
             results as they are produced by your calculator, as well as sketches of
             the histograms.

       (c)   What is the probability that all of the 7 children are boys?

Binomial formulas

We can find a formula for the probability that a binomial random variable
takes any value by adding probabilities for the different ways of getting
exactly that many successes in n observations. Here is the example we will
use to show the idea.

EXAMPLE 8.9

Each child born to a particular set of parents has probability 0.25 of having
blood type O. If these parents have 5 children, what is the probability that
exactly 2 of them have type O blood?
        The count of children with type O blood is a binomial random variable
X with $n = 5$ tries and probability $p = .25$ of a success on each try. We want
$P(X = 2)$.

        Because the method doesn't depend on the specific example, let's use
"S" for success and "F" for failure for short. Do the work in two steps.

Step 1. Find the probability that a specific 2 of the 5 tries give successes,
say the first and the third. This is the outcome SFSFF. Here's how to find
the probability of this outcome:

- The probability that the first try is a success is 0.25. That is, in many repetitions, we succeed on the first try 25% of the time.
- Out of all the repetitions with a success on the first try, 75% have a failure on the second try. So the proportion of repetitions on which the first two tries are SF is (.25)(.75). We can multiply here because the tries are *independent.* That is, the first try has no influence on the second.
- Keep going: Of these repetitions, the proportion 0.25 have S on the third try. So the probability of SFS is (.25)(.75)(.25). After two more tries, the probability of SFSFF is the product of the try-by-try probabilities:

$$(.25)(.75)(.25)(.75)(.75) = (.25)^2(.75)^3$$

Step 2. Observe that the probability of *any one* arrangement of 2 S's and 3 F's has this same probability. That's true because we multiply together 0.25 twice and 0.75 three times whenever we have 2 S's and 3 F's. The probability that $X = 2$ is the probability of getting 2 S's and 3 F's in any arrangement whatsoever. Here are all the possible arrangements:

$$\begin{array}{ccccc} \text{SSFFF} & \text{SFSFF} & \text{SFFSF} & \text{SFFFS} & \text{FSSFF} \\ \text{FSFSF} & \text{FSFFS} & \text{FFSSF} & \text{FFSFS} & \text{FFFSS} \end{array}$$

There are 10 of them, all with the same probability. The overall probability of 2 successes is therefore

$$P(X = 2) = 10(.25)^2(.75)^3 = .2637$$

The pattern of this calculation works for any binomial probability. To use it, we need to be able to count the number of arrangements of k successes in n observations without actually listing them. We use the following fact to do the counting.

BINOMIAL COEFFICIENT

The number of ways of arranging k successes among n observations is given by the **binomial coefficient**

$$\binom{n}{k} = \frac{n!}{k!\,(n-k)!}$$

for $k = 0, 1, 2, \ldots, n$.

factorial

The formula for binomial coefficients uses the **factorial** notation. For any positive whole number n, its factorial $n!$ is

$$n! = n \times (n-1) \times (n-2) \times \cdots \times 3 \times 2 \times 1$$

Also, $0! = 1$.

Notice that the larger of the two factorials in the denominator of a binomial coefficient will cancel much of the $n!$ in the numerator. For example, the binomial coefficient we need for Example 8.9 is

$$\binom{5}{2} = \frac{5!}{2!\,3!}$$

$$= \frac{(5)(4)(3)(2)(1)}{(2)(1) \times (3)(2)(1)}$$

$$= \frac{(5)(4)}{(2)(1)} = \frac{20}{2} = 10$$

The notation $\binom{n}{k}$ is *not* related to the fraction $\frac{n}{k}$. A helpful way to remember its meaning is to read it as "binomial coefficient n choose k." Binomial coefficients have many uses in mathematics, but we are interested in them only as an aid to finding binomial probabilities. The binomial coefficient $\binom{n}{k}$ counts the number of ways in which k successes can be distributed among n observations. The binomial probability $P(X = k)$ is this count multiplied by the probability of any specific arrangement of the k successes. Here is the formula we seek.

BINOMIAL PROBABILITY

If X has the binomial distribution with n observations and probability p of success on each observation, the possible values of X are $0, 1, 2, \ldots, n$. If k is any one of these values,

$$P(X = k) = \binom{n}{k} p^k (1-p)^{n-k}$$

EXAMPLE 8.10

The number X of switches that fail inspection in Example 8.3 has approximately the binomial distribution with $n = 10$ and $p = .1$. The probability that no more than 1 switch fails is

$$P(X \le 1) = P(X = 1) + P(X = 0)$$

$$= \binom{10}{1}(.1)^1(.9)^9 + \binom{10}{0}(.1)^0(.9)^{10}$$

$$= \frac{10!}{1!9!}(.1)(.3874) + \frac{10!}{0!10!}(1)(.3487)$$

$$= (10)(.1)(.3874) + (1)(1)(.3487)$$

$$= .3874 + .3487$$

$$= .7361$$

Notice that the calculation uses the facts that $0! = 1$ and that $a^0 = 1$ for any number a other than 0.

EXERCISES

8.8   A factory employs several thousand workers, of whom 30% are Hispanic. If the 15 members of the union executive committee were chosen from the workers at random, the number of Hispanics on the committee would have the binomial distribution with $n = 15$ and $p = .3$.

   (a) What is the probability that exactly 3 members of the committee are Hispanic?

   (b) What is the probability that 3 or fewer members of the committee are Hispanic?

8.9   A university claims that 80% of its basketball players get degrees. An investigation examines the fate of all 20 players who entered the program over a period of several years that ended six years ago. Of these players, 10 graduated and the remaining 10 are no longer in school. If the university's claim is true, the number of players among the 20 who graduate should have the binomial distribution with $n = 20$ and $p = .8$. What is the probability that exactly 10 out of 20 players graduate?

8.10   Among employed women, 25% have never been married. Select 10 employed women at random.

   (a) The number in your sample who have never been married has a binomial distribution. What are n and p?

   (b) What is the probability that exactly 2 of the 10 women in your sample have never been married?

   (c) What is the probability that 2 or fewer have never been married?

SIMULATING BINOMIAL EXPERIMENTS

In order to simulate a binomial experiment, you need to know how the random variable X and "success" are defined, the probability of success, and the number of trials. But if you know these things, you can apply the rules learned in this section to calculate the probabilities of events exactly. So perhaps simulation methods are not as important in a binomial setting as they are in other settings. On the other hand, being able to simulate a binomial experiment can give credence to results obtained by applying formulas and rules when the results may be less than convincing to someone who knows no statistics.

EXAMPLE 8.11

Recall that Corinne's free throw percentage was 75% (see Example 8.6). In a particular game, she had 12 attempts and she made only 7. The question was, "How unusual was it for Corinne to make at most 7 shots out of 12 attempts?" In Example 8.6, we calculated this binomial probability to be $P(X \leq 7) = .1576$. Now we will use the TI-83 to simulate 12 attempted shots and we will count the number of hits (baskets). Let X = number of hits in 12 free throw attempts. Note that the probability of "success" is 0.75. To set up the simulation, we will assign the digit 0 to a miss, and a 1 to a hit. The command randBin(1,.75,12) simulates 12 free throw attempts. In the long run, this random function will select the number 1 75% of the time and the number 0 25% of the time.

Here are the results of one simulated game: the first shot was a miss, the next two were hits, the fourth shot was a miss, and so forth. If we repeated this many times and counted the proportion of times Corinne had 7 or fewer hits, then that would give an estimate of the probability $P(X \leq 7)$ that Corinne made at most 7 of her 12 attempts. One way to automate this more is to assign these results to list L_1, then add the entries in L_1, and enter the sum as the first entry in list L_2. Enter the commands randBin(1,.75,12)→L₁:sum L₁→L₂(1) and press ENTER.

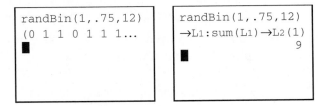

Continue pressing the ENTER key until you have 5 numbers. Record these numbers, press CLEAR, and then press ENTER five more times. Record these numbers.

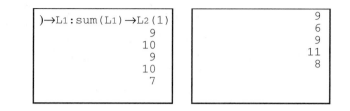

This makes 10 repetitions (i.e., simulates 10 games). So far Corinne has made 7 or fewer shots in 2 out of 10 games, for a relative frequency of 0.20. Compare this with the binomial probability of 0.1576 for this event. Continue to press ENTER to simulate 10 more games. Calculate the relative frequency for 20 simulated games. Then calculate the relative frequency for 30 games, and so on. According to the law of large numbers, these relative frequencies should get closer to 0.1576 as the number of simulated games increases. Continue in this fashion until you have simulated 50 games. Are your cumulative results close to 0.1576?

EXERCISES

8.11   Refer to Exercise 8.8 (page 427). Construct a simulation to estimate the probability that in a committee of 15 members, 3 or fewer members are Hispanic. Describe the design of your experiment, including the correspondence between digits and outcomes in the experiment, and report the relative frequency for 30 repetitions.

8.12   Refer to Exercise 8.9 (page 427). Construct a simulation to estimate the probability that at most 10 of the 20 basketball players graduated. Describe the design of your experiment, including the correspondence between digits and outcomes in the experiment and the number of repetitions you carried out. Report your results.

8.13   Refer to Exercise 8.10 (page 427). Construct a simulation to estimate the probability that 2 or fewer of a random sample of 10 employed women have never been married. Describe the design of your experiment, including the correspondence between digits and outcomes in the experiment and the number of repetitions you carried out. Report your results.

Binomial mean and standard deviation

If a count X has the binomial distribution based on n observations with probability p of success, what is its mean μ? We can guess the answer. If a basketball player makes 75% of her free throws, the mean number made in 12 tries should be 75% of 12, or 9. In general, the mean of a binomial distribution should be $\mu = np$. With some hard work, we could use the definitions of the mean and standard deviation of a discrete random variable in Section 7.2 to verify that this is true and also find a formula for the standard deviation. Here are the facts.

MEAN AND STANDARD DEVIATION OF A BINOMIAL RANDOM VARIABLE

If a count X has the binomial distribution with number of observations n and probability of success p, the **mean** and **standard deviation** of X are

$$\mu = np$$
$$\sigma = \sqrt{np(1-p)}$$

These short formulas are good only for binomial distributions. They can't be used for other discrete random variables.

EXAMPLE 8.12

Continuing Example 8.10, the count X of bad switches is binomial with $n = 10$ and $p = .1$. This is the sampling distribution the engineer would see if she drew all possible SRSs of 10 switches from the shipment and recorded the value of X for each sample.

The mean and standard deviation of the binomial distribution are

$$\mu = np$$
$$= (10)(.1) = 1$$
$$\sigma = \sqrt{np(1-p)}$$
$$= \sqrt{(10)(.1)(.9)} = \sqrt{.9} = .9487$$

The mean is marked on the probability histogram in Figure 8.2.

EXERCISES

8.14   What are the mean and standard deviation of the number of children with type O blood in Exercise 8.5? Mark the location of the mean on the probability histogram you made in that exercise.

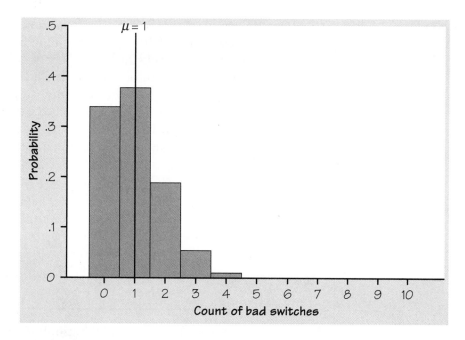

FIGURE 8.2 Probability histogram for the binomial distribution with $n = 10$ and $p = .1$.

8.15 (a) What is the mean number of Hispanics on randomly chosen committees of 15 workers in Exercise 8.8?

(b) What is the standard deviation σ of the count X of Hispanic members?

(c) Suppose that 10% of the factory workers were Hispanic. Then $p = .1$. What is σ in this case? What is σ if $p = .01$? What does your work show about the behavior of the standard deviation of a binomial distribution as the probability of a success gets closer to 0?

8.16 (a) Find the mean number of graduates out of 20 players in the setting of Exercise 8.9.

(b) Find the standard deviation σ of the count X.

(c) Suppose that the 20 players came from a population of which $p = .9$ graduated. What is the standard deviation σ of the count of graduates? If $p = .99$, what is σ? What does your work show about the behavior of the standard deviation of a binomial distribution as the probability p of success gets closer to 1?

8.17 You choose 10 employed women at random, as in Exercise 8.10. What is the mean number of women in such a sample who have never been married? What is the standard deviation?

8.18   (a) Refer to Example 8.10. Use the TI-83's `randBin` function to simulate the random selection of 10 switches from the $B(10, 0.1)$ distribution, and assign these 10 results to list L_1. Then use the `1-Var Stats` function to find the mean number of defective switches among the 10. Compare this result with the known mean $\mu = 1$. Repeat these steps to find the mean of 25 randomly selected switches, and then the mean of 50 randomly selected switches. What effect, if any, does the number of switches sampled have on the mean number of defective switches?

        (b) Do the same as in (a) for the distribution $B(12, 0.75)$ of Corinne's free throws made in 12 attempts (see Examples 8.6 and 8.8). How do your results for samples of size 10, 25, and 50 compare with the true mean number of successes?

SUMMARY

A count X of successes has a binomial distribution in the **binomial setting**: there are n observations; the observations are **independent** of each other; each observation results in a success or a failure; and each observation has the same probability p of a success.

If X has the binomial distribution with parameters n and p, the possible values of X are the whole numbers $0, 1, 2, \ldots, n$. The **binomial probability** that X takes any value is

$$P(X = k) = \binom{n}{k} p^k (1 - p)^{n-k}$$

The **binomial coefficient**

$$\binom{n}{k} = \frac{n!}{k!(n - k)!}$$

counts the number of ways k successes can be arranged among n observations. Here the **factorial** $n!$ is

$$n! = n \times (n - 1) \times (n - 2) \times \cdots \times 3 \times 2 \times 1$$

for positive whole numbers n, and $0! = 1$.

Given a random variable X, the **probability distribution function** (p.d.f.) assigns a probability to each value of X. For each value of X, the **cumulative distribution function** (c.d.f.) assigns the sum of the probabilities for values less than or equal to X.

The **mean** and **standard deviation** of a binomial count X are

$$\mu = np$$
$$\sigma = \sqrt{np(1 - p)}$$

SECTION 8.1 EXERCISES

8.19   In a test for ESP (extrasensory perception), a subject is told that cards the experimenter can see but he cannot contain either a star, a circle, a wave, or a square. As the experimenter looks at each of 20 cards in turn, the subject names the shape on the card. A subject who is just guessing has probability 0.25 of guessing correctly on each card.

   (a) The count of correct guesses in 20 cards has a binomial distribution. What are n and p?
   (b) What is the mean number of correct guesses?
   (c) What is the probability of exactly 5 correct guesses?

8.20   A believer in the "random walk" theory of stock markets thinks that an index of stock prices has probability 0.65 of increasing in any year. Moreover, the change in the index in any given year is not influenced by whether it rose or fell in earlier years. Let X be the number of years among the next 6 years in which the index rises.

   (a) X has a binomial distribution. What are n and p?
   (b) What are the possible values that X can take?
   (c) Find the probability of each value X. Draw a probability histogram for the distribution of X.
   (d) What are the mean and standard deviation of this distribution? Mark the location of the mean on the histogram.

8.21   A federal report finds that lie detector tests given to truthful persons have probability about 0.2 of suggesting that the person is deceptive.

   (a) A company asks 12 job applicants about thefts from previous employers, using a lie detector to assess their truthfulness. Suppose that all 12 answer truthfully. What is the probability that the lie detector says all 12 are truthful? What is the probability that the lie detector says at least one is deceptive?
   (b) What is the mean number among 12 truthful persons who will be classified as deceptive? What is the standard deviation of this number?

8.22   A test for the presence of antibodies to the AIDS virus in blood has proba-
         bility 0.99 of detecting the antibodies when they are present. Suppose that
         during a year 20 units of blood with AIDS antibodies pass through a blood
         bank.

         (a) Take X to be the number of these 20 units that the test detects. What
             is the distribution of X?
         (b) What is the probability that the test detects all 20 contaminated units?
             What is the probability that at least one unit is not detected?
         (c) What is the mean number of units among the 20 that will be detected?
             What is the standard deviation of the number detected?

8.23   A shipment contains 10,000 switches. Of these, 1000 are bad. An inspector
         draws switches at random, so that each switch has the same chance to be
         drawn.

         (a) Draw one switch. What is the probability that the switch you draw is
             bad? What is the probability that it is not bad?
         (b) Suppose the first switch drawn is bad. How many switches remain?
             How many of them are bad? Draw a second switch at random. What
             is the probability that this switch is bad?
         (c) Answer the questions in (b) again, but now suppose that the first switch
             drawn is not bad.

         Comment: Knowing the result of the first trial changes the probabilities
         for the second trial. But because the shipment is large, the probabilities
         change very little. The trials are almost independent.

8.2   THE GEOMETRIC DISTRIBUTIONS

geometric
distribution

In the case of a binomial random variable, the number of trials is fixed
beforehand, and the binomial variable X counts the number of successes
in that fixed number of trials. If there are n trials then the possible values of X are 0, 1 ,2, ..., n. By way of comparison, there are situations in
which the goal is to obtain a fixed number of successes. In particular, if
the goal is to obtain one success, a random variable X can be defined that
counts the number of trials needed to obtain that first success. A random
variable that satisfies the above description is called *geometric*, and the distribution produced by this random variable is called a **geometric distribution**.
The possible values of a geometric random variable are 1, 2, 3, ..., that is,
an infinite set, because it is theoretically possible to proceed indefinitely
without ever obtaining a success. Consider the following situations:

- Flip a coin until you get a <u>head</u>.
- Roll a die until you get a <u>3</u>.
- In basketball, attempt a three-point shot until you make a <u>basket</u>.

Notice that all of these situations involve counting the number of trials until an event of interest happens. We are now ready to characterize the geometric setting.

A random variable X is geometric provided that the following conditions are met.

THE GEOMETRIC SETTING

1. Each observation falls into one of just two categories, which for convenience we call "success" or "failure."
2. The probability of a success, call it p, is the same for each observation.
3. The observations are all **independent**.
4. The variable of interest is the number of trials required to obtain the first success.

EXAMPLE 8.13

An experiment consists of rolling a single die. The event of interest is rolling a 3; this event is called a success. The random variable is defined as X = the number of trials until a 3 occurs. To verify that this is a geometric setting, note that rolling a 3 will represent a success, and rolling any other number will represent a failure. The probability of rolling a 3 on each roll is the same: 1/6. The observations are independent. A trial consists of rolling the die once. We roll the die until a 3 appears. Since all of the requirements are satisfied, this experiment describes a geometric setting.

EXAMPLE 8.14

Suppose you repeatedly draw cards without replacement from a deck of 52 cards until you draw an ace. There are two categories of interest: ace = success; not ace = failure. But is the probability of success the same for each trial? No. The probability of an ace on the first card is 4/52. If you don't draw an ace on the first card, then the probability of an ace on the second card is 4/51. Since the result of the first draw affects probabilities on the second draw (and on all successive draws required), the trials are not independent. So this is not a geometric setting.

Using the setting of Example 8.13, let's calculate some probabilities.

$X = 1$: $P(X = 1) = P(\text{success of first roll}) = 1/6$

$X = 2$: $P(X = 2) = P(\text{success of second roll})$
$= P(\text{failure on first roll and success on second roll})$
$= P(\text{failure on first roll}) \times P(\text{success on second roll})$
$= (5/6) \times (1/6)$

(since trials are independent).

$X = 3$: $P(X = 3) = P(\text{failure on first roll}) \times P(\text{failure on second roll})$
$\times P(\text{success on third roll})$
$= (5/6) \times (5/6) \times (1/6)$

Continue the process. The pattern suggests that a general formula for the variable X is

$$P(X = n) = (5/6)^{n-1}(1/6)$$

Now we can state the following principle:

RULE FOR CALCULATING GEOMETRIC PROBABILITIES

If X has a geometric distribution with probability p of success and $(1 - p)$ of failure on each observation, the possible values of X are $1, 2, 3, \ldots$. If n is any one of these values, then the probability that the first success occurs on the nth trial is

$$P(X = n) = (1 - p)^{n-1}p$$

Although the setting for the geometric distribution is very similar to the binomial setting, there are some striking differences. In rolling a die, for example, it is possible that you will have to roll the die many times before you roll a 3. In fact, it is theoretically possible to roll the die forever without rolling a 3 (although the probability gets closer and closer to 0 the longer you roll the die without getting a 3). The probability of observing the first 3 on the fiftieth roll of the die is $P(X = 50) = .0000$.

A probability distribution table for the geometric random variable is strange indeed because it never ends; that is, the number of table entries is infinite. The rule for calculating geometric probabilities shown above can be used to construct the table:

X	1	2	3	4	5	6	7	
P(X)	p	$(1-p)p$	$(1-p)^2p$	$(1-p)^3p$	$(1-p)^4p$	$(1-p)^5p$	$(1-p)^6p$	...

The probabilities (i.e., the entries in the second row) are the terms of a *geometric sequence* (hence the name for this random variable). You may recall from your study of algebra that the general form for a geometric sequence is

$$a, \ ar, \ ar^2, \ ar^3, \ \dots, \ ar^{n-1}, \ \dots$$

where a is the first term, r is the ratio of one term in the sequence to the next, and the nth term is ar^{n-1}. You may also recall that even though the sequence continues forever, and even though you could never finish adding the terms, the sequence does have a sum (one of the implausible truths of the infinite!). This sum is

$$\frac{a}{1-r}$$

In order for the geometric random variable to have a valid p.d.f., the probabilities in the second row of the table must add to 1. Using the formula for the sum of a geometric sequence, we have

$$\sum_{i=1}^{\infty} P(x_i) = p + (1-p)p + (1-p)^2p + \dots$$

$$= \frac{p}{1-(1-p)} = \frac{p}{p} = 1$$

EXAMPLE 8.15

The rule for calculating geometric probabilities can be used to construct a probability distribution table for X = number of rolls of a die until a 3 occurs:

X	1	2	3	4	5	6	7	
P(X)	.1667	.1389	.1157	.0965	.0804	.0670	.0558	...

Here's one way to find these probabilities with the TI-83:

1. Enter the probability of success, 1/6. Press ENTER.
2. Enter *(5/6) and press ENTER.
3. Continue to press ENTER repeatedly.

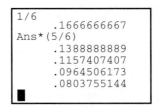

If you'd like to see the probabilities as fractions, modify step 2: enter *(5/6)▶FRAC and press ENTER. Verify that the entries in the second row are as shown:

X	1	2	3	4	...
P(X)	1/6	5/36	25/216	125/1296	...

Figure 8.3 is a graph of the distribution of X. As you might expect, the probability distribution histogram is strongly skewed to the right with a peak at the leftmost value, 1. It is easy to see why this must be so, since the height of each bar after the first is the height of the previous bar times the probability of failure $1 - p$. Since you're multiplying the height of each bar by a number less than 1, each new bar will be shorter than the previous bar, and hence the histogram will be right-skewed. Always.

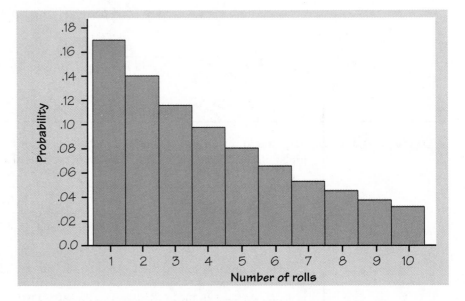

FIGURE 8.3   Probability histogram for the geometric distribution in Example 8.15.

EXERCISES

8.24   For each of the following, determine if the experiment describes a geometric distribution. If it does, describe the two events of interest (success
and failure), what constitutes a trial, and the probability of success on one
trial. If the random variable is not geometric, identify a condition of the
geometric setting that is not satisfied.

    (a) Flip a coin until you observe a tail.

    (b) Record the number of times a player makes both shots in a one-and-
one foul-shooting situation. (In this situation, you get to attempt a
second shot only if you make your first shot.)

    (c) Draw a card from a deck, observe the card, and replace the card within
the deck. Count the number of times you draw a card in this manner
until you observe a jack.

    (d) Buy a "match 6" lottery ticket every day until you win the lottery. (In
a "match 6" lottery, a player chooses 6 different numbers from the set
$\{1, 2, 3, \ldots, 44\}$. A lottery representative draws 6 different numbers
from this set. To win, the player must match all 6 numbers, in any
order.)

    (e) There are 10 red marbles and 5 blue marbles in a jar. You reach in
and, without looking, select a marble. You want to know how many
marbles you will have to draw (without replacement), on average, in
order to be sure that you have 3 red marbles.

8.25   An experiment consists of rolling a die until a prime number (2, 3, or 5) is
observed. Let X = number of rolls required to get the first prime number.

    (a) Verify that X has a geometric distribution.

    (b) Construct a probability distribution table to include at least 5 entries
for the probabilities of X. Record probabilities to four decimal places.

    (c) Construct a graph of the p.d.f. of X.

    (d) Compute the c.d.f. of X and plot its histogram.

    (e) Use the formula for the sum of a geometric sequence to show that the
probabilities in the p.d.f. table of X add to 1.

8.26   Suppose we have data that suggest that 3% of a company's hard disk drives
are defective. You have been asked to determine the probability that the
first defective hard drive is the fifth unit tested.

(a) Verify that this is a geometric setting. Identify the random variable; that is, write X = number of _____ and fill in the blank. What constitutes a success in this situation?

(b) Answer the original question: What is the probability that the first defective hard drive is the fifth unit tested?

(c) Find the first four entries in the table of the p.d.f. for the random variable X.

Exploring geometric distributions with the TI-83

The TI-83 command geometpdf (under 2nd / DISTR) takes two arguments: the probability p of success and the number of the trial on which the first success occurs. Consider the roll of a die of Example 8.13. The probability of rolling a 3 (success) is 1/6. So it should come as no surprise that geometpdf(1/6,1) gives the answer 0.1666666667, or 1/6. The next entry in the table is geometpdf(1/6,2), which returns 0.1388888889, or 5/36. The second argument can also be a list, such as geometpdf(.5,{1,2,3,4,5}) or, if you have values of X entered into list L_1, geometpdf(.5,L_1).

Here is an efficient way to quickly construct a p.d.f. table and plot the result as a histogram. From the Home screen, enter the value of X into L_1 with the command seq(X,X,1,10,1)→L_1. (We can't list all of the terms; we arbitrarily stop at 10.) Next, enter the probabilities into list L_2 with the command geometpdf(1/6,L_1)→L_2.

```
L1      L2       L3    2
1      .16667   ------
2      .13889
3      .11574
4      .09645
5      .08038
6      .06698
7      .05582
L2(1)=.1666666666...
```

Before you plot the probability histogram, you will want to specify the dimensions of an appropriate viewing window. Scanning the list of values gives you insight into reasonable dimensions for the window. The following appear to be good choices: X[-1, 11]$_1$ and Y[-.05, .2]$_{.1}$.

```
WINDOW
 Xmin=-1
 Xmax=11
 Xscl=1
 Ymin=-.05
 Ymax=.2
 Yscl=.1
 Xres=1
```

When you specify a histogram for the STAT / PLOT, specify list L_1 as Xlist, and specify list L_2 for the frequency. The resulting plot shows that the distribution is strongly right-skewed.

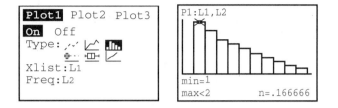

Suppose we are interested in finding the probability that it would take at most 6 rolls of the die to produce a 3. The c.d.f. can be used to answer questions like this. Recall that if $F(X)$ is the c.d.f. for the die experiment and X_0 is a positive integer, then $F(X_0)$ is defined as the sum of the probabilities of all positive integers less than or equal to X_0. The TI-83 command geometcdf(1/6,6) calculates the cumulative probability $F(6)$ for the first 6 values of X and reports the result as 0.6651020233.

The expected value and other noteworthy properties of the geometric random variable

If you're flipping a fair coin, how many times would you expect to have to flip the coin in order to observe the first head? If you're rolling a die, how many times would you expect to have to roll the die in order to observe the first 3? If you said 2 coin tosses and 6 rolls of the die, then your intuition is serving you well. Here is the principle.

THE MEAN OF A GEOMETRIC RANDOM VARIABLE

If X is a geometric random variable with probability of success p on each trial, then the **mean**, or **expected value**, of the random variable, that is, the expected number of trials required to get the first success, is $\mu = 1/p$.

The demonstration of the preceding fact proceeds as follows. The notation will be simplified if we let p = probability of success and let q = probability of failure. Then $q = 1 - p$ and the probability distribution table looks like this:

X	1	2	3	4	...
P(X)	p	pq	pq^2	pq^3	...

The mean (expected value) of X is calculated as follows:

$$\begin{aligned}
\mu &= 1(p) + 2(pq) + 3(pq^2) + 4(pq^3) + \cdots \\
&= p(1 + 2q + 3q^2 + 4q^3 + \cdots) \\
&= p\left(\frac{1}{1 - 2q + q^2}\right) \\
&= p\left[\frac{1}{(1 - q)^2}\right] \\
&= p\left(\frac{1}{p^2}\right) \\
&= \frac{1}{p}
\end{aligned}$$

There is another interesting result that relates to the probability that it takes *more* than a certain number of trials to achieve success. Here are the steps:

$$\begin{aligned}
P(X > n) &= 1 - P(X \leq n) \\
&= 1 - (p + qp + q^2 p + \cdots + q^{n-1}p) \\
&= 1 - p(1 + q + q^2 + \cdots + q^{n-1}) \\
&= 1 - p\left(\frac{1 - q^n}{1 - q}\right) \\
&= 1 - p\left(\frac{1 - q^n}{p}\right) \\
&= 1 - (1 - q^n) \\
&= q^n = (1 - p)^n
\end{aligned}$$

We summarize as follows:

P(X > n)

The probability that it takes *more* than n trials to see the first success is

$$P(X > n) = (1 - p)^n$$

Before we had the `geometcdf` function on the TI-83, we would habitually use this result to answer questions of the form $P(X > n)$. Although the importance of this result is somewhat diminished in an age of ready access to computers and graphing calculators, it is still quite useful.

EXERCISES

8.27 Consider the following experiment: flip a coin until a head appears.

(a) Use the TI-83's `geometpdf(` command to construct the p.d.f. table for this experiment. Then have the calculator plot the probability histogram.

(b) Use the techniques described in this section for plotting the p.d.f. to compute the c.d.f. and plot its histogram.

8.28 (a) Plot the cumulative distribution histogram for the die-rolling experiment described in Example 8.13 with the p.d.f. table in Example 8.15.

(b) Find the probability that it takes more than 6 rolls to observe a 3.

(c) Find the smallest positive integer k for which $P(X \leq k) > .99$.

8.29 A basketball player makes 80% of her free throws. We put her on the free throw line and ask her to shoot free throws until she misses one. Let $X =$ the number of free throws the player takes until she misses.

(a) What assumption do you need to make in order for the geometric model to apply? With this assumption, verify that X has a geometric distribution. What action constitutes "success" in this context?

(b) What is the probability that the player will make 5 shots before she misses?

(c) What is the probability that she will make at most 5 shots before she misses?

SUMMARY

A count X of successes has a **geometric distribution** in the geometric setting if the following are satisfied: each observation results in a success or a failure; each observation has the same probability p of success; observations are independent; and X counts the number of trials required to obtain the first success. A geometric random variable differs from a binomial variable because in the geometric setting the number of trials varies and the desired number of defined successes (1) is fixed in advance.

If X has the geometric distribution with probability of success p, the possible values of X are the positive integers 1, 2, 3, ... The **geometric probability** that X takes any value is

$$P(X = n) = (1 - p)^{n-1} p$$

The **mean** (expected value) of a geometric count X is $1/p$.

The probability that it takes *more* than n trials to see the first success is

$$P(X > n) = (1 - p)^n$$

SECTION 8.2 EXERCISES

8.30   There are 20 red marbles, 10 blue marbles, and 5 white marbles in a jar. An experiment consists of selecting a marble without looking, noting the color, and then replacing the marble in the jar. We're interested in the number of marbles you would have to draw in order to be sure you have a red marble.

(a) Is this a binomial or a geometric setting? Explain your choice and write a description of the random variable X.

(b) Calculate the probability of drawing a red marble on the second draw. Calculate the probability of drawing a red marble by the second draw. Calculate the probability that it would take more than 2 draws to get a red marble.

(c) What single TI-83 command will install the first 20 values of X into list L_1? What single command will install the corresponding probabilities into list L_2? What single command will install the cumulative probabilities into list L_3? Enter these commands in the Home screen. Copy this information from your calculator onto your paper to make an expanded probability distribution table (with the c.d.f. as the third row).

(d) Construct a probability distribution histogram as STAT PLOT1, and then construct a cumulative distribution histogram as STAT PLOT2.

8.31   This is a continuation of Exercise 8.30. Given the jar containing red, white, and blue marbles, Joey thinks a more interesting problem would be to find the number of marbles you would have to draw, without replacing them in the jar, to be sure that you have 2 red marbles. Does this experiment describe a geometric setting? Why or why not?

8.32 Carla makes random guesses on a multiple choice test that has five choices for each question. We want to know how many questions Carla answers until she gets one correct.

(a) Define a success in this context, and define the random variable X of interest. What is the probability of success?

(b) What is the probability that Carla's first correct answer occurs on problem 5?

(c) What is the probability that it takes more than 4 questions before Carla answers one correctly?

(d) Construct a probability distribution table for X.

(e) If Carla took a test like this test many times and randomly guessed at each question, what would be the average number of questions she would have to answer before she answered one correctly?

8.33 In some cultures, it is considered very important to have a son to carry on the family name. Suppose that a couple in one of these cultures plans to have children until they have exactly one son.

(a) Find the average number of children per family in such a culture.

(b) What is the expected number of girls in this family?

(c) Describe a simulation that could be used to find approximate answers to the questions in (a) and (b).

8.34 Example 5.21 used simulation techniques to explore the following situation: A couple plans to have children until they have a girl or until they have four children, whichever comes first.

(a) List the outcomes in the sample space for this "experiment." What event represents a success?

(b) Let X = the number of boys in this family. What values can X take? Use an appropriate probability rule to calculate the probability for each value of X, and make a probability distribution table for X. Then show that the sum of the probabilities is 1.

(c) Let Y = the number of children produced in this family until a girl is produced. Show that Y starts out as a geometric distribution but then is stopped abruptly. Make a probability distribution table for Y.

(d) What is the expected number of children for this couple?

(e) What is the probability that this couple will have more than the expected number of children?

(f) At the end of Example 5.21, it states that the probability of having a girl in this situation is 0.938. How can you prove this?

8.35   This is a continuation of Exercise 8.34. A couple plans to have children until they have a girl or until they have four children, whichever comes first. Use the random number table (Table B), beginning on line 130, to simulate 25 repetitions of this childbearing strategy. As in Example 5.21, since a girl and boy are equally likely, let the digits 0 to 4 represent a girl, and let digits 5 to 9 represent a boy. Write the digits in a string until you observe a girl, write B or G under each digit, and write the number of children noted at the bottom. The first two repetitions would be recorded as

$$
\begin{array}{ccc ccc}
6 & 9 & 0 & \quad & 5 & 1 \\
B & B & G & \quad & B & G \\
  & 3 &   & \quad &   & 2
\end{array}
$$

Then find the mean of the 25 repetitions. How do your results compare with the theoretical expected value of 1.8 children?

8.36   This is a continuation of Exercises 8.34 and 8.35. Devise a simulation procedure for the TI-83 to approximate the expected number of children. List the steps and commands you use as well as the number of repetitions and the results. Alternatively, incorporate these steps into a TI-83 program similar to the programs SPIN123, POP, or FREETHRO.

CHAPTER REVIEW

The previous chapter introduced discrete and continuous random variables and described methods for finding means and variances, as well as rules for means and variances. This chapter focused on two important classes of discrete random variables, each of which involves two outcomes or events of interest. Both require independent trials and the same probability of success on each trial. The **binomial** random variable requires a fixed number of trials; the **geometric** random variable has the property that the number of trials varies. Both the binomial and the geometric settings occur sufficiently often in applications that they deserve special attention. Here is a checklist of the major skills you should have acquired by studying this chapter.

A. BINOMIAL

1. Identify a random variable as binomial by verifying four conditions: two outcomes (success and failure); fixed number of trials; independent trials; and the same probability of success for each trial.

2. Use TI-83 or the formula to determine binomial probabilities and construct probability distribution tables and histograms.

3. Calculate cumulative distribution functions for binomial random variables and construct cumulative distribution tables and histograms.

4. Calculate means (expected values) and standard deviations of binomial random variables.

B. GEOMETRIC

1. Identify a random variable as geometric by verifying four conditions: two outcomes (success and failure); the same probability of success for each trial; independent trials; and the count of interest is the number of trials required to get the first success.

2. Use formulas or a TI-83 to determine geometric probabilities and construct probability distribution tables and histograms.

3. Calculate cumulative distribution functions for geometric random variables and construct cumulative distribution tables and histograms.

4. Calculate expected values of geometric random variables.

CHAPTER REVIEW EXERCISES

8.37   In 1996 there were 869 road fatalities in Virginia, according to the Virginia Department of Motor Vehicles. Of these, 346 were alcohol-related. A DMV analyst wants to randomly select several groups of 25 road fatalities for further study. Find the mean and standard deviation for the number of alcohol-related road fatalities in such groups of 25. What is the probability that such a group will have no more than 5 alcohol-related road fatalities?

8.38   You are planning a sample survey of small businesses in your area. You will choose an SRS of businesses in the telephone book's Yellow Pages. Experience shows that only about half of the businesses you contact will respond.

(a) If you contact 150 businesses, is it reasonable to use the $B(150, 0.5)$ distribution of the number X who respond? Explain why.

(b) What is the expected number of businesses that will respond?

(c) What is the probability that 70 or fewer businesses will respond?

(d) How large a sample must you take to increase the mean number of respondents to 100?

8.39   You operate a restaurant. You read that a sample survey by the National Restaurant Association shows that 40% of adults are committed to eating nutritious food when eating away from home. To help you plan your menu, you decide to conduct a sample survey in your own area. You will use random digit dialing to contact an SRS of 200 households by telephone.

(a) If the national result holds in your area, it is reasonable to use the $B(200, 0.4)$ distribution to describe the count X of adult respondents who seek nutritious food when eating out. Explain why.

(b) What is the mean number of nutrition-conscious people in your sample if $p = .4$ is true?

(c) What is the probability that X lies between 75 and 85?

(d) You find 100 of your 200 respondents concerned about nutrition. Is this reason to believe that the percent in your area is higher than the national 40%? To answer this question, find the probability that X is 100 or larger if $p = .4$ is true. If this probability is very small, that is reason to think that p is actually greater than 0.4 in your area.

8.40   The *Statistical Abstract of the United States, 1997,* reports that 77% of secondary schools in the United States have Internet access. Suppose that 8 U.S. secondary schools are selected at random. Let $X =$ the number of secondary schools chosen who have Internet access. This exercise uses the statistical software Minitab to answer several questions. To calculate binomial probabilites with Minitab, using pull-down menus, begin by entering the integers 0 to 8 in column 1 and naming this column VALUES. Then select **Calc > Probability Distributions > Binomial.** Then select **Probability** to indicate that you want individual probabilities. Specify the **Number of trials** and the **Probability of success.** Select **Input column** and specify VALUES in column C1 to tell Minitab to calculate binomial probabilities for each of the values in that column. Then click on **OK.** The following results are produced:

```
MTB > PDF 'VALUES';
SUBC>    Binomial 8 .77.
         K              P( X = K)
      0.00                 0.0000
      1.00                 0.0002
      2.00                 0.0025
      3.00                 0.0165
      4.00                 0.0689
      5.00                 0.1844
      6.00                 0.3087
      7.00                 0.2953
      8.00                 0.1236
MTB >
```

To calculate the cumulative distribution, make the same menu choices, except this time select **Cumulative probability** instead of **Probability**. The following output is produced:

```
MTB > CDF 'VALUES';
SUBC>    Binomial 8 .77.
         K   P( X LESS OR = K)
      0.00              0.0000
      1.00              0.0002
      2.00              0.0027
      3.00              0.0191
      4.00              0.0880
      5.00              0.2724
      6.00              0.5811
      7.00              0.8764
      8.00              1.0000
MTB >
```

From the printouts, calculate the probability that out of the 8 schools chosen, the number having Internet access is

(a) exactly 6

(b) at most 7

(c) less than 7

(d) at least 5 but no more than 7

(e) either less than 3 or more than 6

8.41   Three friends each toss a coin. The odd man wins; that is, if one coin comes up different from the other two, that person wins that round. If the coins all match, then no one wins and they toss again. We're interested in the number of times the players will have to toss the coins until someone wins.

(a) What is the probability that no one will win on a given coin toss?

(b) Define a success as "someone wins on a given coin toss." What is the probability of a success?

(c) Define the random variable of interest: X = number of _____. Is X binomial? Geometric? Justify your answer.

(d) Construct a probability distribution table for X. Then extend your table by the addition of cumulative probabilities in a third row.

(e) What is the probability that it takes no more than 2 rounds for someone to win?

(f) What is the probability that it takes more than 4 rounds for someone to win?

(g) What is the expected number of tosses needed for someone to win?

(h) Use the `randInt` function on your TI-83 to simulate 25 rounds of play. Then calculate the relative frequencies for X = 1, 2, 3, ... Compare the results of your simulation with the theoretical probabilities you calculated in (d).

8.42   This exercise provides visual reinforcement of the relationship between the probability of success and the mean (expected value) of a geometric random variable.

(a) Begin by completing the table below, where X = probability of success and Y = expected value.

X	.10	.20	.30	.40	.50	.60	.70	.80	.90
Y									

(b) Make a scatterplot of the points (X, Y).

(c) Enter the data into your TI-83 and perform power regression (STAT / CALC / A:PwrReg) on the data. Notice the r-value, and remember that the calculator transforms the data into a linear association and finds the correlation between the *transformed* values.

(d) Draw the power function curve on your scatterplot. Write the equation of the power function.

(e) Briefly explain the connection between this curve and what you have learned about the expected value of a geometric random variable.

8.43 Look at the probability histogram in Figure 8.3, where the probability p of success = P(roll a 3 with a fair die) = 1/6. Then look at the probability histogram from Exercise 8.41, where success was defined as one of the three coins was different, and p was much larger (you supply the p-value). Compare the two histograms and the two values of p. Which histogram has the longer tail: the histogram for the smaller value of p or the histogram for the larger value of p? Will this relationship always hold? Write a convincing argument that the higher the probability of success, the shorter the tail on the p.d.f. histogram.

8.44 Suppose that Roberto, a well-known major league baseball player, finished last season with a .325 batting average. He wants to calculate the probability that he will get his first hit of this new season in his first at-bat. You define a success as getting a hit and define the random variable X = number of at-bats until Roberto gets his first hit.

(a) What is the probability that Roberto will get a hit on his first at-bat (i.e., that $X = 1$)?

(b) What is the probability that it will take him at most 3 at-bats to get his first hit?

(c) What is the probability that it will take him more than 4 at-bats to get his first hit?

(d) Roberto wants to know the expected number of at-bats until he gets a hit. What would you tell him?

(e) Enter the first 10 values of X into list L_1, the corresponding geometric probabilities into list L_2, and the cumulative probabilities into list L_3.

(f) Construct a probability distribution histogram as STAT PLOT1, and then construct a cumulative distribution histogram as STAT PLOT2.

You show this analysis to Roberto and he is so impressed he gives you two free tickets to his first game.

8.45 Suppose we toss a penny repeatedly until we get a head. We want to determine the probability that the first head comes up in an *odd* number of tosses (1, 3, 5, and so on).

(a)   Toss a penny until the first head occurs, and repeat the procedure 50 times. Keep a record of the results of the first toss and of the number of tosses needed to get a head on each of your 50 repetitions.

(b)   Based on the result of your first toss in the 50 repetitions, estimate the probability of getting a head on the first toss.

(c)   Use your 50 repetitions to estimate the probability that the first head appears on an odd-numbered toss.

Sampling Distributions

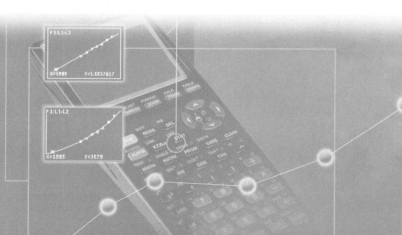

INTRODUCTION

The reasoning of statistical inference rests on asking, "How often would this method give a correct answer if I used it very many times?" If it doesn't make sense to imagine repeatedly producing your data in the same circumstances, statistical inference is not possible.[1] Exploratory data analysis makes sense for any data, but formal inference does not. Even experts can disagree about how widely statistical inference should be used. But all agree that inference is most secure when we produce data by random sampling or randomized comparative experiments. The reason is that when we use chance to choose respondents or assign subjects, the laws of probability answer the question "What would happen if we did this many times?" The purpose of this chapter is to prepare for the study of statistical inference by looking at the probability distributions of some very common statistics: sample proportions, sample counts, and sample means.

| **ACTIVITY 9A** | The Distribution of Height |

Materials: Several 3″×3″ or 3″×5″ Post-it Notes.

The height of young women varies approximately according to the $N(64.5, 2.5)$ distribution. That is to say, the population of young women is normally distributed with mean $\mu = 64.5$ inches and standard deviation $\sigma = 2.5$ inches. The random variable measured is $X =$ the height of a randomly selected young woman. In this activity you will use the TI-83 to sample from this distribution and then use Post-it Notes to construct several distributions.

1. If we choose one woman at random, the heights we get in repeated choices follow the $N(64.5, 2.5)$ distribution. On your TI-83, clear list L_1 and then simulate the heights of 100 randomly selected women. Store these heights in list L_1 with this command: `randNorm(64.5,2.5,100)→L₁` .

2. Plot a histogram of the 100 heights as follows. Deselect active functions in the Y= window, and turn off all STAT PLOTS. Set WINDOW dimensions to be $X[57, 72]_{2.5}$ and $Y[-10, 45]_5$ to extend three standard deviations to either side of the mean, 64.5. Define PLOT 1 to be a histogram using the heights in L_1. Press

(continued)

GRAPH to plot the histogram. Describe the approximate shape of your histogram. Is it fairly symmetric or clearly skewed?

3. Approximately how many heights should there be within 3σ of the mean (i.e., between 57 and 72)? Use TRACE to count the number of heights within 3σ. How many heights should there be within 1σ of the mean? Within 2σ of the mean? Again use TRACE to find these counts, and compare them with the numbers you would expect.

4. Use 1-Var Stats to find the mean, median, and standard deviation for your data. Compare $\bar{x}$ with the population mean $\mu = 64.5$. Compare the sample standard deviation s with $\sigma = 2.5$. How do the mean and median for your 100 heights compare? Recall that the closer the mean and the median are, the more symmetric the distribution. Define PLOT 2 to be a boxplot using L_1, and then GRAPH again. The boxplot will be plotted at the top of the screen. Does the boxplot appear symmetric? How close is the median in the boxplot to the mean of the histogram? Based on the appearance of the histogram and the boxplot, and a comparison of the mean and median, would you say that the distribution is nonsymmetric, moderately symmetric, or very symmetric?

5. Each student should repeat the process at least 2 or 3 more times, plot the histogram and boxplot, and calculate one-variable statistics. Each time, record the mean $\bar{x}$, median, and standard deviation s. (*Note:* While this is going on, the teacher should draw a baseline at the bottom of a clean blackboard and mark a scale from 63 to 66 with tick marks at 0.25 intervals. The tick marks should be spaced about an inch wider apart than the width of the Post-it Notes. Each tick mark will represent the center of a bar in a histogram.)

6. Each student should write the mean $\bar{x}$ and standard deviation s for each sample on a different Post-it Note. Next, the students will build a "Post-it Note histogram" of the distribution of the sample means $\bar{x}$. Each student should go to the blackboard and stick each of his or her notes above the tick mark that is closest to the mean written on the note. When the Post-it Note histogram is complete, answer the following questions:

   (a) What is the approximate shape of the distribution of $\bar{x}$?

   (b) Where is the center of the distribution of $\bar{x}$? How does this center compare with the mean of heights of the population of *all* young women?

(*continued*)

I Smell

(c) Roughly, how does the spread of the distribution of $\bar{x}$ compare with the spread of the original distribution ($\sigma = 2.5$)?

7. While someone calls out the values of $\bar{x}$ from the Post-it Notes, enter these values into a new list on the TI-83. Turn off PLOT 1 and define PLOT 3 to be a boxplot of the $\bar{x}$ data. How do these distributions of x and $\bar{x}$ compare visually? Use 1-Var Stats to calculate the standard deviation $s_{\bar{x}}$ for the distribution of $\bar{x}$. Compare this value with $\sigma/\sqrt{100}$.

8. Fill in the blanks in the following statement with a function of μ or σ: "The distribution of $\bar{x}$ is approximately normal with mean $\mu(\bar{x}) = $ _____ and standard deviation $\sigma(\bar{x}) = $ _____."

9.1   SAMPLING DISTRIBUTIONS

"Are you afraid to go outside at night within a mile of your home because of crime?" When the Gallup Poll asked this question, 45% of the people in the sample said "Yes." That 45% describes the sample, but we use it as an estimate for the entire population. We must now take care to keep straight whether a number describes a sample or a population. Here is the vocabulary we use.

PARAMETER, STATISTIC

A **parameter** is a number that describes the population. In statistical practice, the value of a parameter is not known.

A **statistic** is a number that can be computed from the sample data without making use of any unknown parameters. In practice, we often use a statistic to estimate an unknown parameter.

EXAMPLE 9.1

Are attitudes toward shopping changing? Sample surveys show that fewer people enjoy shopping than in the past. A recent survey asked a nationwide random sample of 2500 adults if they agreed or disagreed that "I like buying new clothes, but shopping is often frustrating and time-consuming." Of the respondents, 1650, or 66%, said they agreed.[2] The number 66% is a *statistic*. The population that the poll wants to draw conclusions about is all U.S. residents age 18 and over. The *parameter* of interest is the percent of all adult U.S. residents who would have said "Agree" if asked the same question. We don't know the value of this parameter.

State whether each boldface number in Exercises 9.1 to 9.4 is a *parameter* or a *statistic*.

9.1 The Bureau of Labor Statistics last month interviewed 60,000 members of the U.S. labor force, of whom **7.2%** were unemployed.

9.2 A carload lot of ball bearings has mean diameter **2.5003** centimeters (cm). This is within the specifications for acceptance of the lot by the purchaser. By chance, an inspector chooses 100 bearings from the lot that have mean diameter **2.5009** cm. Because this is outside the specified limits, the lot is mistakenly rejected.

9.3 A telemarketing firm in Los Angeles uses a device that dials residential telephone numbers in that city at random. Of the first 100 numbers dialed, **48%** are unlisted. This is not surprising because **52%** of all Los Angeles residential phones are unlisted.

9.4 A researcher carries out a randomized comparative experiment with young rats to investigate the effects of a toxic compound in food. She feeds the control group a normal diet. The experimental group receives a diet with 2500 parts per million of the toxic material. After 8 weeks, the mean weight gain is **335** grams for the control group and **289** grams for the experimental group.

Sampling variability

The sample survey of Example 9.1 produced information on attitudes about shopping. Let's examine it in more detail. We want to estimate the proportion of the population that find clothes shopping frustrating. Call this population proportion p. It is a parameter. The poll found that 1650 out of 2500 randomly selected adults agreed with the statement that shopping is often frustrating. The proportion of the sample who agreed was

$$\hat{p} = \frac{1650}{2500} = .66$$

The sample proportion $\hat{p}$ is a statistic. We use $\hat{p}$ to estimate the unknown parameter p.

How can $\hat{p}$, based on only 2500 of the more than 180 million American adults, be an accurate estimate of p? After all, a second random sample taken at the same time would choose different people and no doubt produce a different value of $\hat{p}$. This basic fact is called *sampling variability*: the value of a statistic varies in repeated random sampling.

sampling variability

To understand why sampling variability is not fatal, we ask, "What would happen if we took many samples?" Here's how to answer that question:

- Take a large number of samples from the same population.
- Calculate the sample proportion $\hat{p}$ for each sample.
- Make a histogram of the values of $\hat{p}$.
- Examine the distribution displayed in the histogram for overall pattern, center and spread, and outliers or other deviations.

In practice it is too expensive to take many samples from a population like all adult U.S. residents. But we can imitate many samples by using random digits. Recall that using random digits from a table or computer software to imitate chance behavior is called *simulation*.

EXAMPLE 9.2

We will simulate drawing simple random samples (SRSs) of size 100 from the population of all adult U.S. residents. Suppose that in fact 60% of the population find clothes shopping time-consuming and frustrating. Then the true value of the parameter we want to estimate is $p = .6$. (Of course, we would not sample in practice if we already knew that $p = .6$. We are sampling here to find out how sampling behaves.)

We can imitate the population by a huge table of random digits, with each entry standing for a person. Six of the ten digits (say 0 to 5) stand for people who find shopping frustrating. The remaining four digits, 6 to 9, stand for those who do not. Because all digits in a random number table are equally likely, this assignment produces a population proportion of frustrated shoppers equal to $p = .6$. We then imitate an SRS of 100 people from the population by taking 100 consecutive digits from Table B. The statistic $\hat{p}$ is the proportion of 0s to 5s in the sample.

For example, the first 100 entries in Table B contain 63 digits between 0 and 5, so $\hat{p} = 63/100 = .63$. A second SRS based on the second 100 entries in Table B gives a different result, $\hat{p} = .56$. The two sample results are different, and neither is equal to the true population value $p = .6$. That's sampling variability.

Simulation is a powerful tool for studying chance. It is much faster to use Table B than to actually draw repeated SRSs, and much faster yet to use a computer programmed to produce random digits. Figure 9.1 is the histogram of values of $\hat{p}$ from 1000 separate SRSs of size 100 drawn from a population with $p = .6$. This histogram shows what would happen if we drew many samples. It displays the *sampling distribution* of $\hat{p}$.

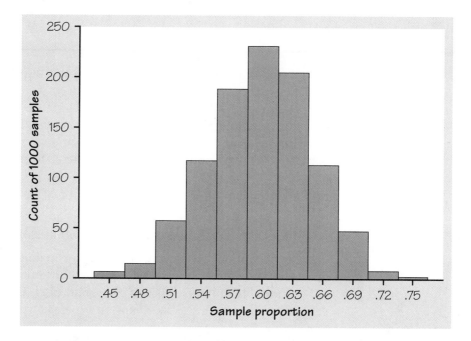

FIGURE 9.1 The sampling distribution of the sample proportion $\hat{p}$ from SRSs of size 100 drawn from a population with population proportions $p = .6$. The histogram shows the results of drawing 1000 SRSs.

SAMPLING DISTRIBUTION

The **sampling distribution** of a statistic is the distribution of values taken by the statistic in all possible samples of the same size from the same population.

Strictly speaking, the sampling distribution is the ideal pattern that would emerge if we looked at all possible samples of size 100 from our population. A distribution obtained from a fixed number of trials, like the 1000 trials in Figure 9.1, is only an approximation to the sampling distribution. One of the uses of probability theory in statistics is to obtain exact sampling distributions without simulation. The interpretation of a sampling distribution is the same, however, whether we obtain it by simulation or by the mathematics of probability.

EXERCISES

9.5   Coin tossing can illustrate the idea of a sampling distribution. The population is all outcomes (heads or tails) we would get if we tossed a coin forever. The parameter p is the proportion of heads in this population. We suspect that p is close to 0.5. That is, we think the coin will show about one-half heads in the long run. The sample is the outcomes of 20 tosses, and the statistic $\hat{p}$ is the proportion of heads in these 20 tosses.

(a) Toss a coin 20 times and record the value of $\hat{p}$.

(b) Repeat this sampling process 10 times. Make a histogram of the 10 values of $\hat{p}$. Is the center of this distribution close to 0.5?

(c) Ten repetitions give a very crude approximation to the sampling distribution. Pool your work with that of other students to obtain several hundred repetitions. Make a histogram of all the values of $\hat{p}$. Is the center close to 0.5? Is the shape approximately normal?

9.6   Use the TI-83 to replicate Exercise 9.5 as follows. The command `randBin(20,.5)` simulates tossing a coin 20 times. The output is the number of heads in 20 tosses. The command `randBin(20,.5,10)/20→` L_1 simulates 10 repetitions of tossing a coin 20 times, finding the proportions of heads, and storing these 10 values of $\hat{p}$ in list L_1.

(a) Execute this command and then plot a histogram of the 10 values of $\hat{p}$. Set WINDOW parameters to $X[-.05, 1.05]_{.1}$ and $Y[-2, 6]_1$ and then TRACE. Is the center of the histogram close to 0.5? Do this several times to see if you get similar results each time.

(b) Increase the number of repetitions to 100. Recall the command and then edit it to read `randBin(20,.5,100)/20→`L_1. Execute the command (it will take approximately 2 minutes to run) and then plot a histogram using these 100 values. Don't change the XMIN and XMAX values, but do adjust the Y-values to $Y[-20, 80]_{10}$ to accommodate the taller bars. Is the center close to 0.5? Is the shape approximately normal?

(c) Define PLOT 2 to be a boxplot using list L_1, and TRACE again. How close is the median (in the boxplot) to the center (mean) of the histogram?

(d) Note that we didn't increase the sample size, only the number of repetitions. Did the spread of the distribution change? What would you change to decrease the spread of the distribution?

9.7 Let us illustrate the idea of a sampling distribution in the case of a very small sample from a very small population. The population is the scores of 10 students on an exam:

Student	0	1	2	3	4	5	6	7	8	9
Score	82	62	80	58	72	73	65	66	74	62

The parameter of interest is the mean score in this population, which is 69.4. The sample is an SRS of size $n = 4$ drawn from the population. Because the students are labeled 0 to 9, a single random digit from Table B chooses one student for the sample.

(a) Use Table B to draw an SRS of size 4 from this population. Write the four scores in your sample and calculate the mean $\bar{x}$ of the sample scores. This statistic is an estimate of the population parameter.

(b) Repeat this process 10 times. Make a histogram of the 10 values of $\bar{x}$. You are constructing the sampling distribution of $\bar{x}$. Is the center of your histogram close to 69.4?

(c) Ten repetitions give a very crude approximation to the sampling distribution. Pool your work with that of other students—using different parts of Table B—to obtain several hundred repetitions. Make a histogram of all the values of $\bar{x}$. Is the center close to 69.4? Is the shape approximately normal? This histogram is a better approximation to the sampling distribution.

Describing sampling distributions

We can use the tools of data analysis to describe any distribution. Let's apply those tools to Figure 9.1. Remember that the figure shows the values of the sample proportion $\hat{p}$ in a large number of samples from the same population.

- The *overall shape* of the distribution is symmetric and approximately normal.
- There are no *outliers* or other important deviations from the overall pattern.
- The *center* of the distribution is very close to the true value $p = .6$ for the population from which the samples were drawn. In fact, the mean of the 1000 $\hat{p}$s is 0.598 and their median is exactly 0.6.
- The values of $\hat{p}$ have a large *spread*. They range from 0.45 to 0.75. Because the distribution is close to normal, we can use the standard deviation to describe its spread. The standard deviation is about 0.05.

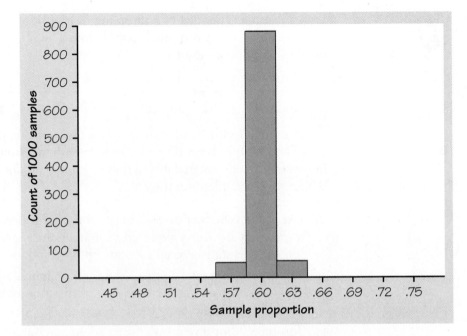

FIGURE 9.2   The sampling distribution of the sample proportion $\hat{p}$ from SRSs of size 2500 drawn from a population with population proportion p = .6. The histogram shows the results of drawing 1000 SRSs. The scale is the same as in Figure 9.1.

Figure 9.1 shows that a sample of 100 people often gives a $\hat{p}$ quite far from the population parameter p = .6. That is, a sample of 100 people does not produce a trustworthy estimate of the population proportion. That is why the sample survey of attitudes toward shopping interviewed, not 100, but 2500 people. Let's repeat our simulation, this time taking 1000 SRSs of size 2500 from a population with proportion p = .6 who find shopping frustrating.

Figure 9.2 displays the sampling distribution of the 1000 values of $\hat{p}$ from these new samples. Figure 9.2 uses the same horizontal scale as Figure 9.1 to make comparison easy. Here's what we see:

- The *center* of the distribution is again close to 0.6. In fact, the mean is 0.6002 and the median is exactly 0.6.
- The *spread* of Figure 9.2 is much less than that of Figure 9.1. The range of the values of $\hat{p}$ from 1000 samples is only 0.5728 to 0.6296. The standard deviation is about 0.01. Almost all samples of 2500 people give a $\hat{p}$ close to the population parameter p = .6.

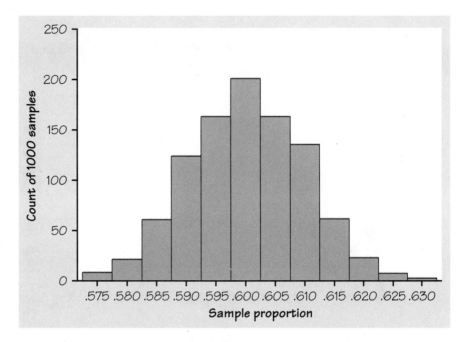

FIGURE 9.3 The sampling distribution from Figure 9.2, for samples of size 2500, redrawn on a different scale to better display the shape.

- Because the values of $\hat{p}$ cluster so tightly about 0.6, it is hard to see the *shape* of the distribution in Figure 9.2. Figure 9.3 displays the same 1000 values of $\hat{p}$ on an expanded scale that makes the shape clearer. The distribution is again approximately normal in shape.

The appearance of the sampling distributions in Figures 9.1 to 9.3 is a consequence of random sampling. Haphazard sampling does not give such regular and predictable results. When randomization is used in a design for producing data, statistics computed from the data have a definite pattern of behavior over many repetitions, even though the result of a single repetition is uncertain.

The bias of a statistic

The fact that statistics from random samples have definite sampling distributions allows a more careful answer to the question of how trustworthy a statistic is as an estimate of a parameter. Figure 9.4 shows the two sampling distributions of $\hat{p}$, for samples of 100 people and samples of 2500 people, side by side and drawn to the same scale. Both distributions are approximately normal, so we have also drawn normal curves for both. How

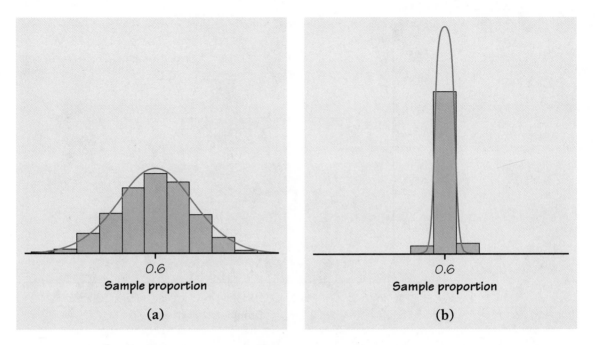

FIGURE 9.4   The sampling distributions for sample proportions $\hat{p}$ for SRSs of two sizes drawn from a population having population proportion $p = .6$. (*a*) Sample size 100. (*b*) Sample size 2500. Both statistics are unbiased because the means of their distributions equal the true population value $p = .6$. The statistic from the larger sample is less variable.

trustworthy is the sample proportion $\hat{p}$ as an estimator of the population proportion p in each case?

bias

First, we can describe **bias** more exactly by speaking of the bias of a statistic rather than bias in a sampling method. Bias concerns the center of the sampling distribution. The centers of the sampling distributions in Figure 9.4 are very close to the true value of the population parameter. Those distributions show the results of 1000 samples. In fact, the mean of the sampling distribution (think of taking all possible samples, not just 1000 samples) is *exactly* equal to 0.6, the parameter in the population.

UNBIASED STATISTIC

A statistic used to estimate a parameter is **unbiased** if the mean of its sampling distribution is equal to the true value of the parameter being estimated.

An unbiased statistic will sometimes fall above the true value of the parameter and sometimes below if we take many samples. Because its sampling distribution is centered at the true value, however, there is no systematic tendency to overestimate or underestimate the parameter. This makes the idea of lack of bias in the sense of "no favoritism" more precise. The sample proportion $\hat{p}$ from an SRS is an unbiased estimator of the population proportion p. If we draw an SRS from a population in which 60% find shopping frustrating, the mean of the sampling distribution of $\hat{p}$ is 0.6. If we draw an SRS from a population with 50% frustrated shoppers, the mean of $\hat{p}$ is then 0.5.

The variability of a statistic

The goal of archery is to shoot arrows as close as possible to the bull's-eye. To achieve this the archer needs small bias, but also small variability so that the arrows fall close together. As Figure 9.5 reminds us, bias and variability are separate properties. Both large and small bias can appear in combination with either large or small variability. Properly chosen statistics computed from random samples have little or no bias. But Figure 9.4(a) shows that estimates from an SRS can have quite a bit of variability.

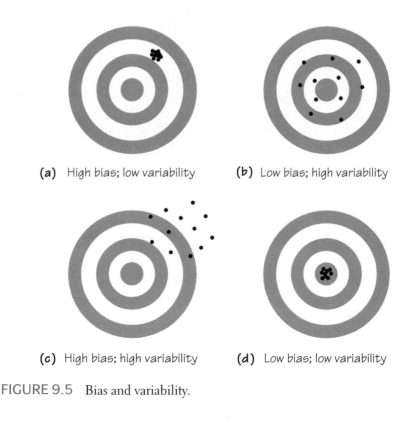

(**a**) High bias; low variability   (**b**) Low bias; high variability

(**c**) High bias; high variability   (**d**) Low bias; low variability

FIGURE 9.5   Bias and variability.

The statistics whose sampling distributions appear in Figure 9.4 are both unbiased. That is, both distributions are centered at 0.6, the true population parameter. The sample proportion $\hat{p}$ from a random sample of any size is an unbiased estimate of the parameter p. Larger samples have a clear advantage, however. They are much more likely to produce an estimate close to the true value of the parameter because there is much less variability among large samples than among small samples.

EXAMPLE 9.3

The sampling distribution of $\hat{p}$ for samples of size 100, shown in Figure 9.4(a), is close to the normal distribution with mean 0.6 and standard deviation 0.05. Recall the 68–95–99.7 rule for normal distributions. It says that 95% of values of $\hat{p}$ fall within two standard deviations of the mean of the distribution. So 95% of all samples give an estimate $\hat{p}$ between

$$\text{mean} \pm (2 \times \text{standard deviation}) = .6 \pm (2 \times .05) = .6 \pm .1$$

If in fact 60% of the population find clothes shopping frustrating, the estimates from repeated SRSs of size 100 will usually fall between 50% and 70%. That's not very satisfactory.

For samples of size 2500, Figure 9.4(b) shows that the standard deviation is only about 0.01. So 95% of these samples will give an estimate within about 0.02 of the mean, that is, between 0.58 and 0.62. An SRS of size 2500 can be trusted to give sample estimates that are very close to the truth about the entire population.

In Section 9.2 we will give the standard deviation of $\hat{p}$ for any size sample. We will then see Example 9.3 as part of a general rule that shows exactly how the variability of sample results decreases for larger samples. One important and surprising fact is that the spread of the sampling distribution does *not* depend very much on the size of the *population*.

VARIABILITY OF A STATISTIC

The variability of a statistic is described by the spread of its sampling distribution. This spread is determined by the sampling design and the size of the sample. Larger samples give smaller spread.

As long as the population is much larger than the sample (say, at least 10 times as large), the spread of the sampling distribution is approximately the same for any population size.

Why does the size of the population have little influence on the behavior of statistics from random samples? To see that this is plausible, imagine sampling harvested corn by thrusting a scoop into a lot of corn kernels. The scoop doesn't know whether it is surrounded by a bag of corn or by an entire truckload. As long as the corn is well mixed (so that the scoop selects a random sample), the variability of the result depends only on the size of the scoop.

The fact that the variability of sample results is controlled by the size of the sample has important consequences for sampling design. A statistic from an SRS of size 2500 from the more than 250,000,000 residents of the United States is just as precise as an SRS of size 2500 from the 740,000 inhabitants of San Francisco. This is good news for designers of national samples but bad news for those who want accurate information about the citizens of San Francisco. If both use an SRS, both must use the same size sample to obtain equally trustworthy results.

EXERCISES

9.8 The table below contains the results of simulating on a computer 100 repetitions of the drawing of an SRS of size 200 from a large lot of bearings. Ten percent of the bearings in the lot do not conform to the specifications. That is, $p = .10$ for this population. The numbers in the table are the counts of nonconforming bearings in each sample of 200.

17	23	18	27	15	17	18	13	16	18	20	15	18	16	21
17	18	19	16	23	20	18	18	17	19	13	27	22	23	26
17	13	16	14	24	22	16	21	24	21	30	24	17	14	16
16	17	24	21	16	17	23	18	23	22	24	23	23	20	19
20	18	20	25	16	24	24	24	15	22	22	16	28	15	22
9	19	16	19	19	25	24	20	15	21	25	24	19	19	20
28	18	17	17	25	17	17	18	19	18					

(a) Make a table that shows how often each count occurs. For each count in your table, give the corresponding value of the sample proportion

$$\hat{p} = \frac{\text{count}}{200}$$

Then draw a histogram for the values of the statistic $\hat{p}$.

(b) Is the shape of the distribution approximately normal?

(c) Find the mean of the 100 observations of $\hat{p}$. Mark the mean on your histogram to show its center. Does the statistic $\hat{p}$ appear to have large or small bias as an estimate of the population proportion p?

(d) The sampling distribution of $\hat{p}$ is the distribution of the values of $\hat{p}$ from all possible samples of size 200 from this population. What is the mean of this distribution?

(e) If we repeatedly selected SRSs of size 1000 instead of 200 from this same population, what would be the mean of the sampling distribution of the sample proportion $\hat{p}$? Would the spread be larger, smaller, or about the same when compared with the spread of your histogram in (a)?

9.9   Figure 9.6 shows histograms of four sampling distributions of statistics intended to estimate the same parameter. Label each distribution relative to the others as large or small bias and as large or small variability.

9.10   The Internal Revenue Service plans to examine an SRS of individual federal income tax returns from each state. One variable of interest is the proportion of returns claiming itemized deductions. The total number of

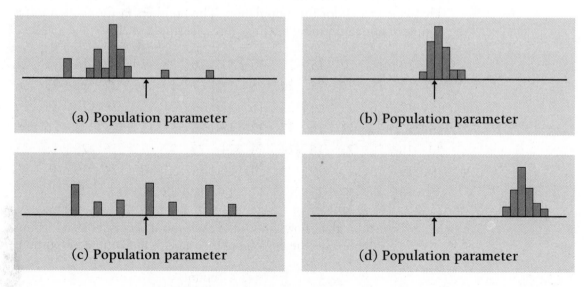

(a) Population parameter   (b) Population parameter

(c) Population parameter   (d) Population parameter

FIGURE 9.6   Which of these sampling distributions displays large or small bias and large or small variability? (Exercise 9.9)

tax returns in a state varies from almost 14 million in California to fewer than 210,000 in Wyoming.

(a) Will the sampling variability of the sample proportion change from state to state if an SRS of 2000 tax returns is selected in each state? Explain your answer.

(b) Will the sampling variability of the sample proportion change from state to state if an SRS of 1% of all tax returns is selected in each state? Explain your answer.

SUMMARY

A number that describes a population is called a **parameter**. A number that can be computed from the sample data is called a **statistic**. The purpose of sampling or experimentation is usually to use statistics to make statements about unknown parameters.

A statistic from a probability sample or randomized experiment has a **sampling distribution** that describes how the statistic varies in repeated data production. The sampling distribution answers the question, "What would happen if we repeated the sample or experiment many times?" Formal statistical inference is based on the sampling distributions of statistics.

A statistic as an estimator of a parameter may suffer from **bias** or from high **variability**. Bias means that the center of the sampling distribution is not equal to the true value of the parameter. The variability of the statistic is described by the spread of its sampling distribution.

Properly chosen statistics from randomized data production designs have no bias resulting from the way the sample is selected or the way the experimental units are assigned to treatments. The variability of the statistic is determined by the size of the sample or by the size of the experimental groups. Statistics from larger samples have less variability.

SECTION 9.1 EXERCISES

9.11   Suppose the true proportion of shoppers who would say that they find shopping frustrating and time-consuming is 67% (see Example 9.1). Here is a short TI-83 program that simulates sampling from this population:

```
PROGRAM:SHOPPER
:ClrHome
:ClrList L₁,L₂,L₃
:Disp "HOW MANY
TRIALS"
:Prompt N
:randInt(1,100,N
)→L₁
:(L₁≥1 and L₁≤67
)→L₂
:cumSum(L₂)→L₃
:Disp "SAMPLE PE
RCENT="
:Disp L₃(N)*100/
N
```

Enter this program or link it from your teacher or classmate.

(a) In the program, what digits are assigned to shoppers? What digits are assigned to shoppers who say that they find shopping frustrating and time-consuming? Does the program output a count of shoppers who answer "Yes" or a percent?

(b) Execute the program and specify 5 trials (sample size = 5). Do this 10 times, and record the 10 numbers

(c) Execute the program 10 more times, specifying a sample size of 25. Record the 10 results for sample size = 25.

(d) Execute the program 10 more times, specifying a sample size of 100. Record the 10 results.

(e) Enter the 10 outputs for sample size = 5 in list L_1, the 10 results for sample size = 25 in list L_2, and the 10 results for sample size = 25 in list L_3. Then do 1-Var Stats for L_1, L_2, and L_3 in order, and record the means and sample standard deviations s_x for each sample size. Complete the sentence "As the sample size increases, the variability _____."

9.12   An entomologist samples a field for egg masses of a harmful insect by placing a yard-square frame at random locations and examining the ground within the frame carefully. He wants to estimate the proportion of square yards in which egg masses are present. Suppose that in a large field egg masses are present in 20% of all possible yard-square areas. That is, $p = .2$ in this population.

(a) Use Table B to simulate the presence or absence of egg masses in each square yard of an SRS of 10 square yards from the field. Be sure to explain clearly which digits you used to represent the presence and the absence of egg masses. What proportion of your 10 sample areas had egg masses? This is the statistic $\hat{p}$.

(b) Repeat (a) with different lines from Table B, until you have simulated the results of 20 SRSs of size 10. What proportion of the square yards in each of your 20 samples had egg masses? Make a stemplot from these 20 values to display the distribution of your 20 observations on $\hat{p}$. What is the mean of this distribution? What is its shape?

(c) If you looked at all possible SRSs of size 10, rather than just 20 SRSs, what would be the mean of the values of $\hat{p}$? This is the mean of the sampling distribution of $\hat{p}$.

(d) In another field, 40% of all square-yard areas contain egg masses. What is the mean of the sampling distribution of $\hat{p}$ in samples from this field?

9.13 An opinion poll asks, "Are you afraid to go outside at night within a mile of your home because of crime?" Suppose that the proportion of all adults who would say "Yes" to this question is $p = .4$.

(a) Use Table B to simulate the result of an SRS of 20 adults. Be sure to explain clearly which digits you used to represent each of "Yes" and "No." What proportion of your 20 responses were "Yes"?

(b) Repeat (a) using different lines in Table B until you have simulated the results of 10 SRSs of size 20 from the same population. Compute the proportion of "Yes" responses in each sample. These are the values of the statistic $\hat{p}$ in 10 samples. Find the mean of your 10 values of $\hat{p}$. Is it close to p?

(c) The sampling distribution of $\hat{p}$ is the distribution of $\hat{p}$ from all possible SRSs of size 20 from this population. What would be the mean of this distribution?

(d) If the population proportion changed to $p = .5$, what would then be the mean of the sampling distribution of $\hat{p}$?

9.14 A national opinion poll recently estimated that 44% ($\hat{p} = .44$) of all adults agree that parents of school-age children should be given vouchers good for education at any public or private school of their choice. The polling organization used a probability sampling method for which the sample proportion $\hat{p}$ has a normal distribution with standard deviation about 0.015. If a sample were drawn by the same method from the state of New Jersey (population 7.8 million) instead of from the entire United States (population 250 million), would this standard deviation be larger, about the same, or smaller? Explain your answer.

9.2   SAMPLE PROPORTIONS

population
proportion p

sample
proportion $\hat{p}$

Retailers would like to know what proportion of all adults find clothes shopping frustrating and time-consuming. This unknown *population proportion* is a parameter p. A random sample of 2500 people found 1650 frustrated by clothes shopping. The *sample proportion*

$$\hat{p} = \frac{1650}{2500} = .66$$

is a statistic that we use to gain information about the parameter p. In everyday language we often express proportions as percents. We may say, "66% of the sample found clothes shopping frustrating." Statistical recipes work with proportions as decimal fractions, so 66% becomes 0.66.

The sampling distribution of $\hat{p}$

How good is the statistic $\hat{p}$ as an estimate of the parameter p? To find out, we ask, "What would happen if we took many samples?" The sampling distribution of $\hat{p}$ answers this question. In the simulation examples in Section 9.1, we found:

- The sampling distribution of the sample proportion $\hat{p}$ has a shape that is close to normal.
- Its mean is close to the population proportion p.
- Its standard deviation gets smaller as the size of the sample gets larger.

We looked at the results of 1000 samples. The mathematics of probability describes the actual sampling distribution from all possible samples. Here are the facts.

SAMPLING DISTRIBUTION OF A SAMPLE PROPORTION

Choose an SRS of size n from a large population with population proportion p having some characteristic of interest. Let $\hat{p}$ be the proportion of the sample having that characteristic. Then:

- The sampling distribution of $\hat{p}$ is **approximately normal** and is closer to a normal distribution when the sample size n is large.
- The **mean** of the sampling distribution is exactly p.
- The **standard deviation** of the sampling distribution is

$$\sqrt{\frac{p(1-p)}{n}}$$

Because the mean of the sampling distribution of $\hat{p}$ is always equal to the parameter p, the sample proportion $\hat{p}$ is an unbiased estimator of p. The standard deviation of $\hat{p}$ gets smaller as the sample size n increases because n appears in the denominator of the formula for the standard deviation. That is, $\hat{p}$ is less variable in larger samples. What is more, the formula shows just how quickly the standard deviation decreases as n increases. The sample size n is under the square root sign, so to cut the standard deviation in half, we must take a sample four times as large, not just twice as large.

The formula for the standard deviation of $\hat{p}$ doesn't apply when the sample is a large part of the population. You can't use this recipe if you choose an SRS of 50 of the 100 people in a class, for example. In practice, we usually take a sample only when the population is large. Otherwise, we could examine the entire population. Here is a practical guide.[3]

RULE OF THUMB 1

Use the recipe for the standard deviation of $\hat{p}$ only when the population is at least 10 times as large as the sample.

EXAMPLE 9.4

You ask an SRS of 1500 first-year college students whether they applied for admission to any other college. There are over 1.7 million first-year college students, so the rule of thumb is easily satisfied. In fact, 35% of all first-year students applied to colleges besides the one they are attending. What is the probability that your sample will give a result within 2 percentage points of this true value?

We have an SRS of size $n = 1500$ drawn from a population in which the proportion $p = .35$ applied to other colleges. The sample proportion $\hat{p}$ has mean 0.35 and standard deviation

$$\sqrt{\frac{p(1 - p)}{n}} = \sqrt{\frac{(.35)(.65)}{1500}} = \sqrt{.00015167} = .0123$$

We want the probability that $\hat{p}$ falls between 0.33 and 0.37 (within 2 percentage points, or 0.02, of 0.35). This is a normal distribution calculation.

Standardize $\hat{p}$ by subtracting its mean 0.35 and dividing by its standard deviation 0.0123. That produces a new statistic that has the standard normal distribution. It is usual to call such a statistic Z:

$$Z = \frac{\hat{p} - .35}{.0123}$$

Then draw a picture of the areas under the standard normal curve (Figure 9.7), and use Table A to find them. Here is the calculation.

$$P(.33 \leq \hat{p} \leq .37) = P\left(\frac{.33 - .35}{.0123} \leq \frac{\hat{p} - .35}{.0123} \leq \frac{.37 - .35}{.0123}\right)$$

$$= P(-1.63 \leq Z \leq 1.63)$$

$$= .9484 - .0516 = .8968$$

We see that almost 90% of all samples will give a result within 2 percentage points of the truth about the population.

The outline of the calculation in Example 9.4 is familiar from Chapter 2, but the language of probability is new. The sampling distribution of $\hat{p}$ gives probabilities for its values, so the entries in Table A are now probabilities. We used a brief notation that is common in statistics. The capital P in

$$P(.33 \leq \hat{p} \leq .37)$$

stands for "probability." The expression inside the parentheses tells us what event we are finding the probability of. This entire expression is a short way of writing "the probability that $\hat{p}$ lies between 0.33 and 0.37."

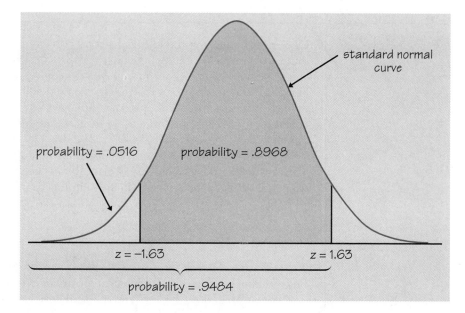

probability = .0516

probability = .8968

standard normal curve

$z = -1.63$ $z = 1.63$

probability = .9484

FIGURE 9.7   Probabilities in Example 9.4 as areas under the standard normal curve.

Using the normal approximation for $\hat{p}$

The sampling distribution of $\hat{p}$ is only approximately normal. For example, if we sample 100 individuals, the only possible values of $\hat{p}$ are 0, 1/100, 2/100, and so on. The statistic has only 101 possible values, so its distribution cannot be exactly normal. The accuracy of the normal approximation improves as the sample size n increases. For a fixed sample size n, the normal approximation is most accurate when p is close to 1/2, and least accurate when p is near 0 or 1. If $p = 1$, for example, then $\hat{p} = 1$ in every sample because every individual in the population has the characteristic we are counting. The normal approximation is no good at all when $p = 1$ or $p = 0$. Here is a rule of thumb that ensures that normal calculations are accurate enough for most statistical purposes. Unlike the first rule of thumb, this one rules out some settings of practical interest.

RULE OF THUMB 2

We will use the normal approximation to the sampling distribution of $\hat{p}$ for values of n and p that satisfy

$$np \geq 10 \quad \text{and} \quad n(1 - p) \geq 10$$

EXAMPLE 9.5

One way of checking the effect of undercoverage, nonresponse, and other sources of error in a sample survey is to compare the sample with known facts about the population. About 11% of American adults are black. The proportion $\hat{p}$ of blacks in an SRS of 1500 adults should therefore be close to 11%. It is unlikely to be exactly 11% because of sampling variability. If a national sample contains only 9.2% blacks, should we suspect that the sampling procedure is somehow underrepresenting blacks?

We will find the probability that a sample contains no more than 9.2% blacks when the population is 11% black. First, check our rule of thumb for using the normal approximation to the sampling distribution of $\hat{p}$:

$$np = (1500)(.11) = 165$$

and

$$n(1 - p) = (1500)(.89) = 1335$$

Both are much larger than 10, so the approximation will be quite accurate. The mean of $\hat{p}$ is $p = .11$. The standard deviation is

$$\sqrt{\frac{p(1 - p)}{n}} = \sqrt{\frac{(.11)(.89)}{1500}}$$

$$= \sqrt{.00006527} = .00808$$

Now do the normal probability calculation, illustrated by Figure 9.8.

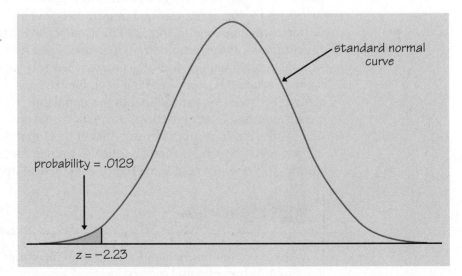

FIGURE 9.8   The probability in Example 9.5 as an area under the standard normal curve.

$$P(\hat{p} \leq .092) = P\left(\frac{\hat{p} - .11}{.00808} \leq \frac{.092 - .11}{.00808}\right)$$

$$= P(Z \leq -2.23)$$

$$= .0129$$

Only 1.29% of all samples would have so few blacks. Because it is unlikely that a sample would include so few blacks, we have good reason to suspect that the sampling procedure underrepresents blacks.

EXERCISES

9.15 The Gallup Poll once asked a random sample of 1540 adults, "Do you happen to jog?" Suppose that in fact 15% of all adults jog.

(a) Find the mean and standard deviation of the proportion $\hat{p}$ of the sample who jog. (Assume the sample is an SRS.)

(b) Explain why you can use the formula for the standard deviation of $\hat{p}$ in this setting (rule of thumb 1).

(c) Check that you can use the normal approximation for the distribution of $\hat{p}$ (rule of thumb 2).

(d) Find the probability that between 13% and 17% of the sample jog.

(e) What sample size would be required to reduce the standard deviation of the sample proportion to one-half the value you found in (a)?

9.16 The Gallup Poll asked a probability sample of 1785 adults whether they attended church or synagogue during the past week. Suppose that 40% of the adult population did attend. We would like to know the probability that an SRS of size 1785 would come within plus or minus 3 percentage points of this true value.

(a) If $\hat{p}$ is the proportion of the sample who did attend church or synagogue, what is the mean of $\hat{p}$? What is its standard deviation?

(b) Explain why you can use the formula for the standard deviation of $\hat{p}$ in this setting (rule of thumb 1).

(c) Check that you can use the normal approximation for the distribution of $\hat{p}$ (rule of thumb 2).

(d) Find the probability that $\hat{p}$ takes a value between 0.37 and 0.43. Will an SRS of size 1785 usually give a result $\hat{p}$ within plus or minus 3 percentage points of the true population proportion?

9.17   Suppose that 15% of all adults jog. Exercise 9.15 asks the probability that the sample proportion $\hat{p}$ from an SRS estimates $p = .15$ within ± 2 percentage points. Find this probability for SRSs of sizes 200, 800, and 3200. What general conclusion can you draw from your calculations?

9.18   Suppose that 40% of the adult population attended church or synagogue last week. Exercise 9.16 asks the probability that $\hat{p}$ from an SRS estimates $p = .4$ within ± 3 percentage points. Find this probability for SRSs of sizes 300, 1200, and 4800. What general fact do your results illustrate?

9.19   According to a market research firm, 52% of all residential telephone numbers in Los Angeles are unlisted. A telephone sales firm uses random digit dialing equipment that dials residential numbers at random, whether or not they are listed in the telephone directory. The firm calls 500 numbers in Los Angeles.

   (a) What are the mean and standard deviation of the proportion of unlisted numbers in the sample?
   (b) What is the probability that at least half the numbers dialed are unlisted? (Remember to check that you can use the normal approximation.)

9.20   Your mail-order company advertises that it ships 90% of its orders within three working days. You select an SRS of 100 of the 5000 orders received in the past week for an audit. The audit reveals that 86 of these orders were shipped on time.

   (a) What is the sample proportion of orders shipped on time?
   (b) If the company really ships 90% of its orders on time, what is the probability that the proportion in an SRS of 100 orders is as small as the proportion in your sample or smaller?
   (c) A critic says, "Aha! You claim 90%, but in your sample the on-time percentage is lower than that. So the 90% claim is wrong." Explain in simple language why your probability calculation in (b) shows that the result of the sample does not refute the 90% claim.

9.21   Exercise 9.19 asks for a probability calculation about telephone calls to a random sample of Los Angeles telephone numbers. Exercise 9.20 asks for a similar calculation about a random sample of mail orders. For which calculation does the normal approximation to the sampling distribution of $\hat{p}$ give a more accurate answer? Why? (You need not actually do either calculation.)

| | SUMMARY |

When we want information about the **population proportion** p of individuals with some special characteristic, we often take an SRS and use the **sample proportion** $\hat{p}$ to estimate the unknown parameter p.

The **sampling distribution** of $\hat{p}$ describes how the statistic varies in all possible samples from the population.

The **mean** of the sampling distribution is equal to the population proportion p. That is, $\hat{p}$ is an unbiased estimator of p.

The **standard deviation** of the sampling distribution is $\sqrt{p(1-p)/n}$ for an SRS of size n. This recipe can be used in practice if the population is at least 10 times as large as the sample.

The standard deviation of $\hat{p}$ gets smaller as the sample size n gets larger. Because of the square root, a sample four times larger is needed to cut the standard deviation in half.

When the sample size n is large, the sampling distribution of $\hat{p}$ is close to a normal distribution with mean p and standard deviation $\sqrt{p(1-p)/n}$. In practice, use this **normal approximation** when both $np \geq 10$ and $n(1-p) \geq 10$.

SECTION 9.2 EXERCISES

9.22 According to government data, 22% of American children under the age of 6 live in households with incomes less than the official poverty level. A study of learning in early childhood chooses an SRS of 300 children.

  (a) What is the probability that more than 20% of the sample are from poverty households? (Remember to check that you can use the normal approximation.)

  (b) What is the probability that more than 30% of the sample are from poverty households?

9.23 Here is a simple probability model for multiple-choice tests. Suppose that a student has probability p of correctly answering a question chosen at random from a universe of possible questions. (A strong student has a higher p than a weak student.) The correctness of an answer to any specific question doesn't depend on other questions. A test contains n questions. Then the proportion of correct answers that a student gives is a sample proportion $\hat{p}$ from an SRS of size n drawn from a population with population proportion p.

(a) Julie is a good student for whom $p = .75$. Find the probability that Julie scores 70% or lower on a 100-question test.

(b) If the test contains 250 questions, what is the probability that Julie will score 70% or lower?

(c) How many questions must the test contain in order to reduce the standard deviation of Julie's proportion of correct answers to half its value for a 100-item test?

(d) Laura is a weaker student for whom $p = .6$. Does the answer you gave in (c) for the standard deviation of Julie's score apply to Laura's standard deviation also?

9.24   The Helsinki Heart Study asks whether the anticholesterol drug gemfibrozil will reduce heart attacks. In planning such an experiment, the researchers must be confident that the sample sizes are large enough to enable them to observe enough heart attacks. The Helsinki study plans to give gemfibrozil to 2000 men and a placebo to another 2000. The probability of a heart attack during the 5-year period of the study for men this age is about 0.04. We can think of the study participants as an SRS from a large population, of which the proportion $p = .04$ will have heart attacks.

(a) What is the mean number of heart attacks that the study will find in one group of 2000 men if the treatment does not change the probability 0.04?

(b) What is the probability that the group will suffer at least 75 heart attacks?

9.25   Explain why you *cannot* use the methods of this section to find the following probabilities.

(a) A factory employs 3000 unionized workers, of whom 30% are Hispanic. The 15-member union executive committee contains 3 Hispanics. What would be the probability of 3 or fewer Hispanics if the executive committee were chosen at random from all the workers?

(b) A university is concerned about the academic standing of its intercollegiate athletes. A study committee chooses an SRS of 50 of the 316 athletes to interview in detail. Suppose that in fact 40% of the athletes have been told by coaches to neglect their studies on at least one occasion. What is the probability that at least 15 in the sample are among this group?

9.3 SAMPLE MEANS

Sample proportions arise most often when we are interested in categorical variables. We then ask questions like "What proportion of the shoppers sampled feel that clothes shopping is frustrating?" or "What percent of the sample respondents are black?" When we record quantitative variables—the income of a household, the diameter of a bearing, the blood pressure of a patient—we are interested in other statistics, such as the median or mean or standard deviation of the variable. Because sample means are just averages of observations, they are among the most common statistics. This section describes the sampling distribution of the mean of the responses in an SRS.

EXAMPLE 9.6

A basic principle of investment is that diversification reduces risk. That is, buying several securities rather than just one reduces the variability of the return on an investment. Figure 9.9 illustrates this principle in the case of common stocks listed on the New York Stock Exchange. Figure 9.9(a) shows the distribution of returns for all 1815 stocks listed on the Exchange for the entire year 1987.[4] This was a year of extreme swings in stock prices, including a record loss of over 20% in a single day. The mean return for all 1815 stocks was -3.5% and the distribution shows a very wide spread.

Figure 9.9(b) shows the distribution of returns for all possible portfolios that invested equal amounts in each of 5 stocks. A portfolio is just a sample of 5 stocks and its return is the average return for the 5 stocks chosen. The mean return for all portfolios is still -3.5%, but the variation among portfolios is much less than the variation among individual stocks. For example, 11% of all individual stocks had a loss of more than 40%, but only 1% of the portfolios had a loss that large.

The histograms in Figure 9.9 emphasize a principle that we will make precise in this section:

• Averages are less variable than individual observations.

More detailed examination of the distributions would point to a second principle:

• Averages are more normal than individual observations.

These two facts contribute to the popularity of sample means in statistical inference.

For means just as for proportions, we must distinguish carefully between a population parameter and a sample statistic. We write μ (the

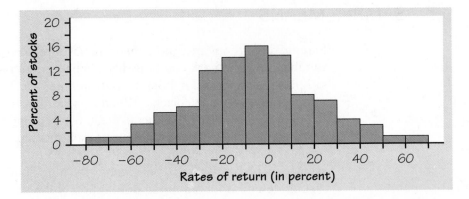

FIGURE 9.9(a)   The distribution of returns for New York Stock Exchange common stocks in 1987.

Greek letter mu) for the mean of a population. This is a fixed *parameter* that is unknown when we use a sample for inference. The mean of the sample is the familiar $\bar{x}$, the average of the observations in the sample. This is a *statistic* that would almost certainly take a different value if we chose another sample from the same population. The sample mean $\bar{x}$ from a sample or an experiment is an estimate of the mean μ of the underlying population, just as a sample proportion $\hat{p}$ is an estimate of a

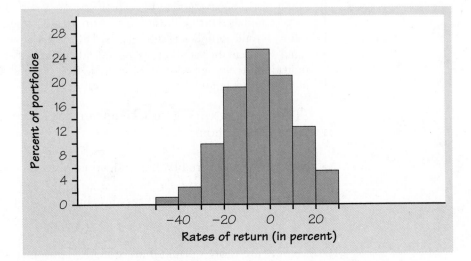

FIGURE 9.9(b)   The distribution of returns for portfolios of five stocks in 1987. Figure 9.9 is taken with permission from John K. Ford, "A method for grading 1987 stock recommendations," *American Association of Individual Investors Journal*, March 1988, pp. 16–17.

population proportion p. In the same way, we write σ (the Greek letter sigma) for the standard deviation of a population and the familiar s for the standard deviation of a sample. As long as we were just doing data analysis, the distinction between population and sample was not important. Now, however, it is essential.

PARAMETERS AND STATISTICS

The mean and standard deviation of a population are **parameters**. We use Greek letters to write these parameters: μ for the mean and σ for the standard deviation.

The mean and standard deviation calculated from sample data are **statistics**. We write the sample mean as $\bar{x}$ and the sample standard deviation as s.

The mean and the standard deviation of $\bar{x}$

The sampling distribution of $\bar{x}$ is the distribution of the values of $\bar{x}$ in all possible samples of the same size from the population. Figure 9.9(a) shows the distribution of a population, with mean $\mu = -3.5\%$. Figure 9.9(b) is the sampling distribution of the sample mean $\bar{x}$ from all samples of size $n = 5$ from this population. The mean of all the values of $\bar{x}$ is again -3.5%, but the values of $\bar{x}$ are less spread out than the individual values in the population. This is an example of a general fact.

MEAN AND STANDARD DEVIATION OF A SAMPLE MEAN

Suppose that $\bar{x}$ is the mean of an SRS of size n drawn from a large population with mean μ and standard deviation σ. Then the **mean** of the sampling distribution of $\bar{x}$ is μ and its **standard deviation** is $\sigma/\sqrt{n}$.

The behavior of $\bar{x}$ in repeated samples is much like that of the sample proportion $\hat{p}$:

- The sample mean $\bar{x}$ is an unbiased estimator of the population mean μ.

- The values of $\bar{x}$ are less spread out for larger samples. Their standard deviation decreases at the rate $\sqrt{n}$, so you must take a sample four times as large to cut the standard deviation of $\bar{x}$ in half.

- You should only use the recipe $\sigma/\sqrt{n}$ for the standard deviation of $\bar{x}$ when the population is at least 10 times as large as the sample. This is almost always the case in practice.

Notice that these facts about the mean and standard deviation of $\bar{x}$ are true no matter what the shape of the population distribution is.

EXAMPLE 9.7

The height of young women varies approximately according to the $N(64.5, 2.5)$ distribution. This is a population distribution with $\mu = 64.5$ and $\sigma = 2.5$. If we choose one young woman at random, the heights we get in repeated choices follow this distribution. That is, the distribution of the population is also the distribution of one observation chosen at random. So we can think of the population distribution as a distribution of probabilities, just like a sampling distribution.

Now measure the height of an SRS of 10 young women. The sampling distribution of their sample mean height $\bar{x}$ will have mean $\mu = 64.5$ inches and standard deviation

$$\frac{\sigma}{\sqrt{n}} = \frac{2.5}{\sqrt{10}} = .79 \text{ inch}$$

The heights of individual women vary widely about the population mean, but the average height of a sample of 10 women is less variable. Figure 9.10 compares the two distributions.

In Activity 9A, you plotted the distribution of $\bar{x}$ for samples of size $n = 100$, so the standard deviation of $\bar{x}$ is $\sigma/\sqrt{100} = 2.5/10 = .25$. How close did your class come to this number?

The fact that averages of several observations are less variable than individual observations is important in many settings. For example, it is common practice to repeat a careful measurement several times and report the average of the results. Think of the results of n repeated measurements as an SRS from the population of outcomes we would get if we repeated the measurement forever. The average of the n results (the sample mean $\bar{x}$) is less variable than a single measurement.

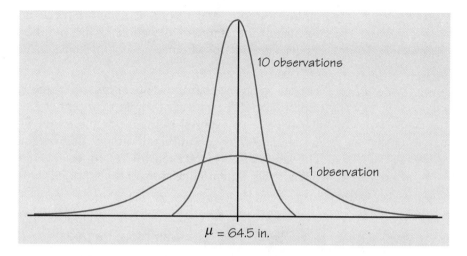

FIGURE 9.10   The sampling distribution of the mean height $\bar{x}$ for samples of 10 young women compared with the distribution of the height of a single woman chosen at random.

EXERCISES

9.26   Investors remember 1987 as the year stocks lost 20% of their value in a single day. For 1987 as a whole, the mean return of all common stocks on the New York Stock Exchange was $\mu = -3.5\%$. (That is, these stocks lost an average of 3.5% of their value in 1987.) The standard deviation of the returns was about $\sigma = 26\%$. Figure 9.9(a) shows the distribution of returns. Figure 9.9(b) is the sampling distribution of the mean returns $\bar{x}$ for all possible samples of 5 stocks. What are the mean and the standard deviation of the distribution in Figure 9.9(b)?

9.27   The scores of individual students on the American College Testing (ACT) composite college entrance examination have a normal distribution with mean 18.6 and standard deviation 5.9. At Northside High, 76 seniors take the test. If the scores at this school have the same distribution as national scores, what are the mean and standard deviation of the average (sample mean) score for the 76 students? Do your results depend on the fact that individual scores have a normal distribution?

9.28   An automatic grinding machine in an auto parts plant prepares axles with a target diameter $\mu = 40.125$ millimeters (mm). The machine has some variability, so the standard deviation of the diameters is $\sigma = .002$ mm. The machine operator inspects a sample of 4 axles each hour for quality control purposes and records the sample mean diameter. What will be the mean and standard deviation of the numbers recorded? Do your results depend on whether or not the axle diameters have a normal distribution?

9.29   The law requires coal mine operators to test the amount of dust in the atmosphere of the mine. A laboratory carries out the test by weighing filters that have been exposed to the air in the mine. The test has a standard deviation of $\sigma = .08$ milligram in repeated weighings of the same filter. The laboratory weighs each filter 3 times and reports the mean result. What is the standard deviation of the reported results?

ACTIVITY 9B | Sampling Pennies

Materials: For a week or so prior to this experiment, each student should collect 25 pennies from current circulation. Each student also should bring to class 2 nickels, 2 dimes, and 1 quarter and a small container such as a Styrofoam coffee cup or margarine tub.

1. This activity[5] begins by plotting the distribution of ages (in years) of the pennies students have brought to class. Sketch a density curve that you think will capture the shape of the distribution of ages of the pennies.

2. Make a table of years, beginning with the current year and counting backward. Make the second column the age of the penny. For the age subtract the date on the penny from the current year. Make a third column the frequency and use tally marks to record the number of pennies of each age. For example, if it is 1999:

Year	Age	Frequency
1999	0	|||
1998	1	|
1997	2	||||
etc.		

3. Put your 25 pennies in a cup, and randomly select 5 pennies. Find the average age of the 5 pennies in your sample, and record the mean age as $\bar{x}(5)$. If you are in a small class (fewer than about 15), each student should repeat this step. Replace the pennies in the cup, stir so they are randomly distributed, and then repeat the process.

(continued)

4. Repeat step 3, except this time randomly select 10 pennies. Calculate the average age of the sample of 10 pennies, and record this as $\bar{x}(10)$. If your class is small, do this twice to obtain two means.

5. Repeat step 3 but take all 25 pennies. Record the mean age as $\bar{x}(25)$.

6. Select a clear space on the floor, and use masking tape to make a number line (axis). Mark ages from 0 to about 30 on the axis. Make the intervals a little more than the width of a penny. Students should place their pennies on the axis according to age. When you are finished, look at the shape of the histogram. Are you surprised? How would you explain the shape?

7. Make a second axis on the floor, and label it 0, 0.5, 1, 1.5, etc. This time, nickels are used to plot the means for samples of size 5. What is the shape of this histogram for the distribution of $\bar{x}(5)$?

8. Make a third histogram for the means of samples of size 10. Use dimes to make this histogram.

9. Finally, use the quarters to make a histogram of the distribution of means for samples of size 25. Describe the shape of this histogram. Are you surprised?

10. Write a short description of what you have discovered by doing this activity.

The central limit theorem

We have described the mean and standard deviation of the sampling distribution of a sample mean $\bar{x}$, but not the distribution itself. The shape of the distribution of $\bar{x}$ depends on the shape of the population distribution. In particular, if the population distribution is normal, then so is the distribution of the sample mean.

SAMPLING DISTRIBUTION OF A SAMPLE MEAN

Draw an SRS of size n from a population that has the normal distribution with mean μ and standard deviation σ. Then the sample mean $\bar{x}$ has the normal distribution $N(\mu, \sigma/\sqrt{n})$ with mean μ and standard deviation $\sigma/\sqrt{n}$.

We already knew the mean and standard deviation of the sampling distribution. All that we have added now is the normal shape. In Activity 9A, we began with a normal distribution, $N(64.5, 2.5)$. The center (mean) of the sampling distribution of $\bar{x}$ should have been very close to the mean of the population: 64.5 inches. Was it? The spread of the distribution of $\bar{x}$ should have been very close to $\sigma/\sqrt{n}$. Was it? The reason that you don't observe exact agreement is sampling variability.

Although many populations have roughly normal distributions, very few indeed are exactly normal. What happens to $\bar{x}$ when the population distribution is not normal? In Activity 9B, the distribution of ages of pennies should have been right-skewed, but as the sample size increased from 1 to 5 to 10 and then to 25, the distribution should have gotten closer and closer to a normal distribution. This is true no matter what shape the population distribution has, as long as the population has a finite standard deviation σ. This famous fact of probability is called the *central limit theorem*. It is much more useful than the fact that the distribution of $\bar{x}$ is exactly normal if the population is exactly normal.

CENTRAL LIMIT THEOREM

Draw an SRS of size n from any population whatsoever with mean μ and finite standard deviation σ. When n is large, the sampling distribution of the sample mean $\bar{x}$ is close to the normal distribution $N(\mu, \sigma/\sqrt{n})$ with mean μ and standard deviation $\sigma/\sqrt{n}$.

How large a sample size n is needed for $\bar{x}$ to be close to normal depends on the population distribution. More observations are required if the shape of the population distribution is far from normal.

EXAMPLE 9.8

Figure 9.11 shows the central limit theorem in action for a very nonnormal population. Figure 9.11(a) displays the density curve for the distribution of the population. The distribution is strongly right-skewed, and the most probable outcomes are near 0 at one end of the range of possible values. The mean μ of this distribution is 1 and its standard deviation σ is also 1. This particular distribution is called an *exponential distribution* from the shape of its density curve. Exponential distributions are used to describe the lifetime in service of electronic components and the time required to serve a customer or repair a machine.

Figures 9.11(b), (c), and (d) are the density curves of the sample means of 2, 10, and 25 observations from this population. As n increases, the shape becomes more normal. The mean remains at $\mu = 1$ and the standard deviation decreases, taking the value $1/\sqrt{n}$. The density curve for 10 observations is still somewhat skewed to the right but already resembles a normal curve with $\mu = 1$ and $\sigma = 1/\sqrt{10} = .32$. The density curve for $n = 25$ is yet more normal. The contrast between the shape of the population distribution and the distribution of the mean of 10 or 25 observations is striking.

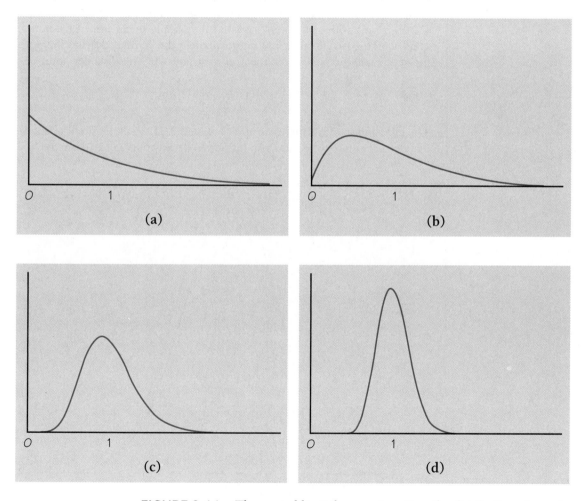

FIGURE 9.11   The central limit theorem in action: the distribution of sample means $\bar{x}$ from a strongly nonnormal population becomes more normal as the sample size increases. (a) The distribution of 1 observation. (b) The distribution of $\bar{x}$ for 2 observations. (c) The distribution of $\bar{x}$ for 10 observations. (d) The distribution of $\bar{x}$ for 25 observations.

The central limit theorem allows us to use normal probability calculations to answer questions about sample means from many observations even when the population distribution is not normal.

EXAMPLE 9.9

The time that a technician requires to perform preventive maintenance on an air-conditioning unit is governed by the exponential distribution whose density curve appears in Figure 9.11(a). The mean time is $\mu = 1$ hour and the standard deviation is $\sigma = 1$ hour. Your company operates 70 of these units. What is the probability that their average maintenance time exceeds 50 minutes?

The central limit theorem says that the sample mean time $\bar{x}$ (in hours) spent working on 70 units has approximately the normal distribution with mean equal to the population mean $\mu = 1$ hour and standard deviation

$$\frac{\sigma}{\sqrt{70}} = \frac{1}{\sqrt{70}} = .12 \text{ hour}$$

The distribution of $\bar{x}$ is therefore approximately $N(1, 0.12)$. Figure 9.12 shows this normal curve (solid) and also the actual density curve of $\bar{x}$ (dashed).

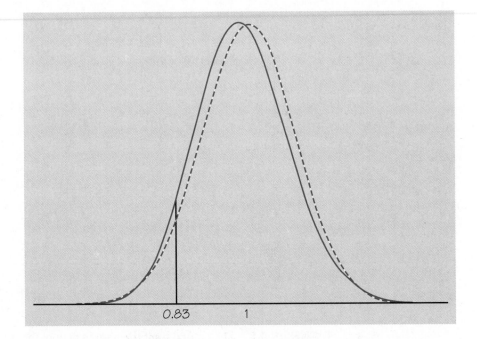

FIGURE 9.12   The exact distribution (*dashed*) and the normal approximation from the central limit theorem (*solid*) for the average time needed to maintain an air conditioner, for Example 9.9.

Because 50 minutes is 50/60 of an hour, or 0.83 hour, the probability we want is

$$P(\bar{x} > .83) = P\left(\frac{\bar{x} - 1}{.12} > \frac{.83 - 1}{.12}\right)$$

$$= P(Z > -1.42) = .9222$$

This is the area to the right of 0.83 under the solid normal curve in Figure 9.12. The exactly correct probability is the area under the dashed density curve in the figure. It is 0.9294. The central limit theorem normal approximation is off by only about 0.007.

EXERCISES

9.30 The scores of students on the ACT college entrance examination in a recent year had the normal distribution with mean $\mu = 18.6$ and standard deviation $\sigma = 5.9$.

(a) What is the probability that a single student randomly chosen from all those taking the test scores 21 or higher?

(b) Now take an SRS of 50 students who took the test. What is the probability that the mean score $\bar{x}$ of these students is 21 or higher?

9.31 A bottling company uses a filling machine to fill plastic bottles with cola. The bottles are supposed to contain 300 milliliters (ml). In fact, the contents vary according to a normal distribution with mean $\mu = 298$ ml and standard deviation $\sigma = 3$ ml.

(a) What is the probability that an individual bottle contains less than 295 ml?

(b) What is the probability that the mean contents of the bottles in a six-pack is less than 295 ml?

9.32 A laboratory weighs filters from a coal mine to measure the amount of dust in the mine atmosphere. Repeated measurements of the weight of dust on the same filter vary normally with standard deviation $\sigma = .08$ milligrams (mg) because the weighing is not perfectly precise. The dust on a particular filter actually weighs 123 mg. Repeated weighings will then have the normal distribution with mean 123 mg and standard deviation 0.08 mg.

(a) The laboratory reports the mean of 3 weighings. What is the distribution of this mean?

(b) What is the probability that the laboratory reports a weight of 124 mg or higher for this filter?

9.33   The number of flaws per square yard in a type of carpet material varies with mean 1.6 flaws per square yard and standard deviation 1.2 flaws per square yard. The population distribution cannot be normal, because a count takes only whole-number values. An inspector studies 200 square yards of the material, records the number of flaws found in each square yard, and calculates $\bar{x}$, the mean number of flaws per square yard inspected. Use the central limit theorem to find the approximate probability that the mean number of flaws exceeds 2 per square yard.

9.34   The distribution of annual returns on common stocks is roughly symmetric, but extreme observations are more frequent than in a normal distribution. Because the distribution is not strongly nonnormal, the mean return over even a moderate number of years is close to normal. In the long run, annual real returns on common stocks have varied with mean about 9% and standard deviation about 28%. Andrew plans to retire in 45 years and is considering investing in stocks. What is the probability (assuming that the past pattern of variation continues) that the mean annual return on common stocks over the next 45 years will exceed 15%? What is the probability that the mean return will be less than 5%?

The law of large numbers revisited

The central limit theorem says that in large samples the sample mean $\bar{x}$ must be close to the population mean μ. This is true because the values of $\bar{x}$ in all possible samples closely follow a normal distribution with mean μ and standard deviation that gets smaller and smaller as the sample size n increases. We can turn this fact around to help us better understand the population mean μ.

For any population, μ is the average value of the variable we are measuring for all the individuals in the population. If we use a density curve to describe the distribution of the population, the mean μ is the "balance point" of the curve. (See Chapter 2.) Now we can add that μ is the value approached by the mean $\bar{x}$ of observations drawn at random from the population as we draw more and more observations. This is true for any population whatever with a finite mean μ, even though the full central limit theorem requires that the population also have a finite variance.

Draw observations at random from any population with finite mean μ. As the number of observations drawn increases, the mean $\bar{x}$ of the observed values gets closer and closer to μ.

The law of large numbers can be proved mathematically starting from the basic laws of probability. The behavior of $\bar{x}$ is similar to the idea of probability. In the long run, the proportion of outcomes taking any value gets close to the probability of that value, and the average outcome gets close to the population mean.

EXERCISE

9.35 It would be quite risky for you to insure the life of a 21-year-old friend for $100,000. There is a high probability that your friend would live and you would gain the few dollars you charged him in insurance premiums. But if he were to die, you would lose almost $100,000. Explain carefully why selling insurance is not risky for an insurance company that insures many thousands of 21-year-old men.

SUMMARY

When we want information about the **population mean** μ for some variable, we often take an SRS and use the **sample mean** $\bar{x}$ to estimate the unknown parameter μ.

The **sampling distribution of** $\bar{x}$ describes how the statistic $\bar{x}$ varies in all possible samples from the population.

The **mean** of the sampling distribution is μ, so that $\bar{x}$ is an unbiased estimator of μ.

The **standard deviation** of the sampling distribution of $\bar{x}$ is $\sigma/\sqrt{n}$ for an SRS of size n if the population has standard deviation σ. This recipe can be used in practice if the population is at least 10 times as large as the sample.

If the population has a normal distribution, so does $\bar{x}$.

The **central limit theorem** states that for large n the sampling distribution of $\bar{x}$ is approximately normal for any population with finite standard deviation σ. The mean and standard deviation of the normal distribution are the mean μ and standard deviation $\sigma/\sqrt{n}$ of $\bar{x}$ itself.

The **law of large numbers** states that the actually observed mean outcome $\bar{x}$ of a large number of observations must approach the mean μ of the population.

SECTION 9.3 EXERCISES

9.36   A company that owns and services a fleet of cars for its sales force has found that the service lifetime of disc brake pads varies from car to car according to a normal distribution with mean $\mu = 55{,}000$ miles and standard deviation $\sigma = 4500$ miles. The company installs a new brand of brake pads on 8 cars.

   (a) If the new brand has the same lifetime distribution as the previous type, what is the distribution of the sample mean lifetime for the 8 cars?

   (b) The average life of the pads on these 8 cars turns out to be $\bar{x} = 51{,}800$ miles. What is the probability that the sample mean lifetime is 51,800 miles or less if the lifetime distribution is unchanged? The company takes this probability as evidence that the average lifetime of the new brand of pads is less than 55,000 miles.

9.37   The number of traffic accidents per week at an intersection varies with mean 2.2 and standard deviation 1.4. The number of accidents in a week must be a whole number, so the population distribution is not normal.

   (a) Let $\bar{x}$ be the mean number of accidents per week at the intersection during a year (52 weeks). What is the approximate distribution of $\bar{x}$ according to the central limit theorem?

   (b) What is the approximate probability that $\bar{x}$ is less than 2?

   (c) What is the approximate probability that there are fewer than 100 accidents at the intersection in a year? (*Hint:* Restate this event in terms of $\bar{x}$.)

9.38   Judy's doctor is concerned that she may suffer from hypokalemia (low potassium in the blood). There is variation both in the actual potassium level and in the blood test that measures the level. Judy's measured potassium level varies according to the normal distribution with $\mu = 3.8$ and

$\sigma = .2$. A patient is classified as hypokalemic if the potassium level is below 3.5.

(a) If a single potassium measurement is made, what is the probability that Judy is diagnosed as hypokalemic?

(b) If measurements are made instead on 4 separate days and the mean result is compared with the criterion 3.5, what is the probability that Judy is diagnosed as hypokalemic?

9.39 The level of nitrogen oxide (NOX) in the exhaust of a particular car model varies with mean 1.4 grams per mile (g/mi) and standard deviation 0.3 g/mi. A company has 125 cars of this model in its fleet. If $\bar{x}$ is the mean NOX emission level for these cars, what is the level L such that the probability that $\bar{x}$ is greater than L is only 0.01? (*Hint:* This requires a backward normal calculation. See Chapter 2 if you need to review.)

9.40 Children in kindergarten are sometimes given the Ravin Progressive Matrices Test (RPMT) to assess their readiness for learning. Experience at Southwark Elementary School suggests that the RPMT scores for its kindergarten pupils have mean 13.6 and standard deviation 3.1. The distribution is close to normal. Mr. Lavin has 22 children in his kindergarten class this year. He suspects that their RPMT scores will be unusually low because the test was interrupted by a fire drill. To check this suspicion, he wants to find the level L such that there is probability only 0.05 that the mean score of 22 children falls below L when the usual Southwark distribution remains true. What is the value of L? (*Hint:* This requires a backward normal calculation. See Chapter 2 if you need to review.)

CHAPTER REVIEW

This chapter lays the foundations for the study of statistical inference. Statistical inference uses data to draw conclusions about the population or process from which the data come. What is special about inference is that the conclusions include a statement, in the language of probability, about how reliable they are. The statement gives a probability that answers the question "What would happen if I used this method very many times?"

This chapter introduced sampling distributions of statistics. A sampling distribution describes the values a statistic would take in very many repetitions of a sample or an experiment under the same conditions.

Understanding that idea is the key to understanding statistical inference. The chapter gave details about the sampling distributions of two important statistics: a sample proportion $\hat{p}$ and a sample mean $\bar{x}$. These statistics behave much the same. In particular, their sampling distributions are approximately normal if the sample is large. This is a main reason why normal distributions are so important in statistics. We can use everything we know about normal distributions to study the sampling distributions of proportions and means.

   Here is a review list of the most important things you should be able to do after studying this chapter.

A. SAMPLING DISTRIBUTIONS

1. Identify parameters and statistics in a sample or experiment.
2. Recognize the fact of sampling variability: a statistic will take different values when you repeat a sample or experiment.
3. Interpret a sampling distribution as describing the values taken by a statistic in all possible repetitions of a sample or experiment under the same conditions.
4. Describe the bias and variability of a statistic in terms of the mean and spread of its sampling distribution.
5. Understand that the variability of a statistic is controlled by the size of the sample. Statistics from larger samples are less variable.

B. SAMPLE PROPORTIONS

1. Recognize when a problem involves a sample proportion $\hat{p}$.
2. Find the mean and standard deviation of a sample proportion $\hat{p}$ for an SRS of size n from a population having population proportion p.
3. Know that the standard deviation (spread) of the sampling distribution of $\hat{p}$ gets smaller at the rate $\sqrt{n}$ as the sample size n gets larger.
4. Recognize when you can use the normal approximation to the sampling distribution of $\hat{p}$. Use the normal approximation to calculate probabilities that concern $\hat{p}$.

C. SAMPLE MEANS

1. Recognize when a problem involves the mean $\bar{x}$ of a sample.
2. Find the mean and standard deviation of a sample mean $\bar{x}$ from an SRS of size n when the mean μ and standard deviation σ of the population are known.

3. Know that the standard deviation (spread) of the sampling distribution of $\bar{x}$ gets smaller at the rate $\sqrt{n}$ as the sample size n gets larger.

4. Understand that $\bar{x}$ has approximately a normal distribution when the sample is large (central limit theorem). Use this normal distribution to calculate probabilities that concern $\bar{x}$.

5. Use the law of large numbers to interpret the population mean μ as the average of an indefinitely large number of observations drawn from the population.

CHAPTER 9 REVIEW EXERCISES

9.41 An opinion poll asks a sample of 500 adults whether they favor giving parents of school-age children vouchers that can be exchanged for education at any public or private school of their choice. Each school would be paid by the government on the basis of how many vouchers it collected. Suppose that in fact 45% of the population favor this idea. What is the probability that more than half of the sample are in favor? (Assume that the sample is an SRS.)

9.42 High school dropouts make up 14.1% of all Americans aged 18 to 24. A vocational school that wants to attract dropouts mails an advertising flyer to 25,000 persons between the ages of 18 and 24.

(a) If the mailing list can be considered a random sample of the population, what is the mean number of high school dropouts who will receive the flyer?

(b) What is the probability that at least 3500 dropouts will receive the flyer?

9.43 The Wechsler Adult Intelligence Scale (WAIS) is a common "IQ test" for adults. The distribution of WAIS scores for persons over 16 years of age is approximately normal with mean 100 and standard deviation 15.

(a) What is the probability that a randomly chosen individual has a WAIS score of 105 or higher?

(b) What are the mean and standard deviation of the average WAIS score $\bar{x}$ for an SRS of 60 people?

(c) What is the probability that the average WAIS score of an SRS of 60 people is 105 or higher?

(d) Would your answers to any of (a), (b), or (c) be affected if the distribution of WAIS scores in the adult population were distinctly nonnormal?

9.44 The weight of the eggs produced by a certain breed of hen is normally distributed with mean 65 grams (g) and standard deviation 5 g. Think of cartons of such eggs as SRSs of size 12 from the population of all eggs. What is the probability that the weight of a carton falls between 750 g and 825 g?

9.45 A study of rush-hour traffic in San Francisco counts the number of people in each car entering a freeway at a suburban interchange. Suppose that this count has mean 1.5 and standard deviation 0.75 in the population of all cars that enter at this interchange during rush hours.

(a) Could the exact distribution of the count be normal? Why or why not?

(b) Traffic engineers estimate that the capacity of the interchange is 700 cars per hour. According to the central limit theorem, what is the approximate distribution of the mean number of persons $\bar{x}$ in 700 randomly selected cars at this interchange?

(c) What is the probability that 700 cars will carry more than 1075 people? (*Hint*: Restate this event in terms of the mean number of people $\bar{x}$ per car.)

9.46 Power companies trim trees growing near their lines to avoid power failures due to falling limbs in storms. Applying a chemical to slow the growth of the trees is cheaper than trimming, but the chemical kills some of the trees. Suppose that one such chemical would kill 20% of sycamore trees. The power company tests this chemical on 250 sycamores. Consider these as an SRS from the population of all sycamore trees.

(a) What are the mean and standard deviation of the number of trees in the sample that are killed?

(b) What is the probability that at least 60 trees (24% of the sample) are killed?

9.47 While he was a prisoner of the Germans during World War II, the British mathematician John Kerrich tossed a coin 10,000 times. He got 5067 heads. Take Kerrich's tosses to be an SRS from the population of all possible tosses of his coin. If the coin is perfectly balanced, $p = .5$. Is there reason to think that Kerrich's coin gave too many heads to be balanced? To answer this question, find the probability that a balanced coin would give 5067 or more heads in 10,000 tosses. What do you conclude?

NOTES AND DATA SOURCES

1. In this book we discuss only the most widely used kind of statistical inference. This is sometimes called *frequentist* inference because it is based on answering the question "What would happen in many repetitions?" Another approach to inference, called *Bayesian*, can be used even for one-time situations. Bayesian inference is important but is conceptually complex and much less widely used in practice.

2. The survey question and result are reported in Trish Hall, "Shop? Many say 'Only if I must,'" *New York Times*, November 28, 1990.

3. Strictly speaking, the recipes we give for the standard deviations of $\hat{p}$ and $\bar{x}$ assume that an SRS of size n is drawn from an *infinite* population. If the population has finite size N, the standard deviations in the recipes are multiplied by $\sqrt{1 - n/N}$. This "finite population correction" approaches 1 as N increases. When $n/N \leq .1$, it is $\geq .948$.

4. From John K. Ford, "A method for grading 1987 stock recommendations," *American Association of Individual Investors Journal*, March 1988, pp. 16–17.

5. This activity is suggested in Richard L. Scheaffer, Ann Watkins, Mrudulla Gnanadesikan, and Jeffrey A. Witmer, *Activity-Based Statistics*, Springer, New York, 1996.

Inference: Conclusions with Confidence

Introduction to Inference

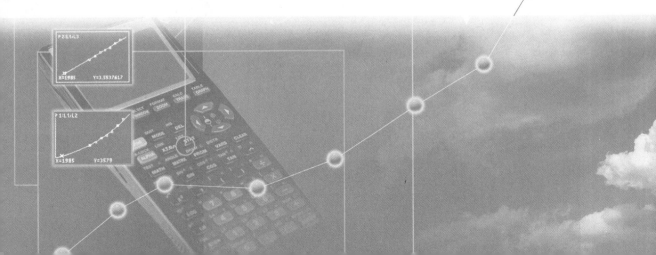

ACTIVITY 10	Pick a Card*

Materials: A deck of playing cards

Your teacher has what appears to be a standard deck of playing cards. Your task is to try to determine the distribution of red and black cards in the deck.

1. As a group, based on any experience you have with playing cards but without inspecting the cards, make your collective best guess as to the proportion of red cards in the deck. Write your conjecture (we will call it a "hypothesis") on the blackboard in the form "$p = $ ☐" with the hypothesized value in the box. The rules are that you may not look at the number or face sides of the cards except when you have drawn one at the direction of your teacher. In this activity, we will build on what we learned in the previous chapter about sampling.

2. The teacher will have one student draw a card, note the color, and replace the card in the deck. Write RED and BLACK on the blackboard, and use a tally mark to record the result of the draw. A second student should draw a card and again tally the result. Continue until there seems to be a consensus among the class members about the true proportion of red cards in the deck. Then write on the board the revised class guess about the proportion p of red cards, across from the original hypothesis.

3. The teacher will introduce a second deck and pose the same question about the proportion of red cards in this deck. Students, in turn, will draw one card and record the results on the board with tally marks. As soon as the class is ready to make a conjecture about the true proportion of red cards in the deck, write this down as Deck 2: $p = $ _____.

4. The teacher will introduce a third deck and pose the same question about the proportion of red cards. But this time, because someone remembers that larger sample sizes give better results, your teacher will have one student draw 2 cards at the same time, look at them, and report the proportion of red cards (either $\hat{p} = 0$,

(continued)

---

*Teachers: Refer to Teacher Commentary for instructions on setting up this activity.

$\hat{p} = .5$, or $\hat{p} = 1$). Replace the cards randomly in the deck (shuffle if necessary) and proceed to the next student. The second student does the same: randomly selects 2 cards, announces the proportion of red cards, and so on. We use a "hat" over the letter p to signify that this is a number calculated from the sample of 2 cards. The values of $\hat{p}$ should be written on the blackboard each time they are announced. After five or six students have drawn their samples and recorded $\hat{p}$, ask if the data suggest that the true proportion of red cards is different from 0.5. If not, then continue until the class feels that enough data have been collected to either make an "alternative hypothesis" or abandon the experiment. If you are ready to make an alternative hypothesis, choose from the following:

$p < .5$ if you really think there are *fewer red cards* than black cards

$p > .5$ if you really think there are *more red cards* than black cards

$p \neq .5$ if you can't decide between the first two choices but you still think that there are more cards of one color than of the other color

5. If you agreed on an alternative hypothesis, your task now is to try to gather evidence against the 0.5 proportion statement. We will continue to draw samples of cards, but knowing that larger samples reduce variability, we will draw larger samples. And to simplify the arithmetic, this time each student will draw 10 cards and calculate the proportion of red cards. After at least ten students have each drawn 10 cards and recorded the proportion of red cards, calculate the mean of these estimates of p. Will this produce a reasonable conjecture for the proportion of red cards in Deck 3?

Summary of the Process

A. We began by stating a hypothesis about the proportion of red cards in the deck.

B. We sampled from the population of all cards in the deck in order to gather data related to our hypothesis.

C. We asked, "If our original hypothesis were true, how likely is it that we would observe the data we actually collected?"

D. Based on the answer to (C), we formulated an "alternative" hypothesis.

(continued)

Question and Topics for Discussion

- Did some students make a decision before others? Is it reasonable that different people require different amounts of convincing?
- More generally, how much evidence do you need to gather before you are able to make a reasonable conjecture?
- Would it be helpful to be able to quantify one's conjecture about a population parameter?
- In typical statistical applications like this experiment, we usually have no way of determining the true population parameters. So if this experiment is supposed to model real-world inference procedures, the true proportion of red cards in the three decks should not be revealed; that will be left to the discretion of your teacher.

INTRODUCTION

To infer means to draw a conclusion. Statistical inference provides us with methods for drawing conclusions from data. We have, of course, been drawing conclusions from data all along. What is new in formal inference is that we use probability to express the strength of our conclusions. Probability allows us to take chance variation into account and so to correct our judgment by calculation. Here are two examples of how probability can correct our judgment.

| EXAMPLE 10.1 | In the Vietnam War years, a lottery determined the order in which men were drafted for army service. The lottery assigned draft numbers by choosing birth dates in random order. We expect a correlation near zero between birth dates and draft numbers if the draft numbers come from random choice. The actual correlation between birth date and draft number in the first draft lottery was $r = -.226$. That is, men born later in the year tended to get lower draft numbers. Is this small correlation evidence that the lottery was biased? Our unaided judgment can't tell, because any two variables will have some association in practice, just by chance. So we calculate that a correlation this far from zero has probability less than 0.001 in a truly random lottery. Because a correlation as strong as that observed would almost never occur in a random lottery, there is strong evidence that the lottery was unfair.

Probability calculations can also protect us from jumping to a conclusion when only chance variation is at work. Investigate 20 new businesses started by women and another 20 started by men. After two years, 12 of those headed by women have failed, but only 8 of the businesses started by men have failed. Can we conclude from this sample that failures are more frequent among all new businesses started by women? A difference this large or larger between the results for two groups of 20 businesses would occur about one time in five simply because of chance variation. An effect that could so easily be just chance is not convincing.

In this chapter we will meet the two most common types of formal statistical inference. Section 10.1 concerns *confidence intervals* for estimating the value of a population parameter. Section 10.2 presents *tests of significance*, which assess the evidence for a claim about a population. Both types of inference are based on the sampling distributions of statistics. That is, both report probabilities that state *what would happen if we used the inference method many times*. This kind of probability statement is characteristic of statistical inference. Users of statistics must understand the meaning of the probability statements that appear, for example, on computer output for statistical procedures.

The methods of formal inference require the long-run regular behavior that probability describes. Inference is most reliable when the data are produced by a properly randomized design. *When you use statistical inference you are acting as if the data are a random sample or come from a randomized experiment.* If this is not true, your conclusions may be open to challenge. Do not be overly impressed by the complex details of formal inference. This elaborate machinery cannot remedy basic flaws in producing the data such as voluntary response samples and uncontrolled experiments. Use the common sense developed in your study of the first five chapters of this book, and proceed to formal inference only when you are satisfied that the data deserve such analysis.

The purpose of this chapter is to describe the reasoning used in statistical inference. We will illustrate the reasoning by a few specific inference techniques, but these are oversimplified so that they are not very useful in practice. Later chapters will first show how to modify these techniques to make them practically useful and will then introduce inference methods for use in most of the settings we met in learning to explore data. There are libraries—both of books and of computer software—full of more elaborate statistical techniques. Informed use of any of these methods requires an understanding of the underlying reasoning. A computer will do the arithmetic, but you must still exercise judgment based on understanding.

10.1 ESTIMATING WITH CONFIDENCE

Young people have a better chance of full-time employment and good wages if they are good with numbers. How strong are the quantitative skills of young Americans of working age? One source of data is the National Assessment of Educational Progress (NAEP) Young Adult Literacy Assessment Survey, which is based on a nationwide probability sample of households.

| EXAMPLE 10.2 | The NAEP survey includes a short test of quantitative skills, covering mainly basic arithmetic and the ability to apply it to realistic problems. Scores on the test range from 0 to 500. For example, a person who scores 233 can add the amounts of two checks appearing on a bank deposit slip; someone scoring 325 can determine the price of a meal from a menu; a person scoring 375 can transform a price in cents per ounce into dollars per pound.

In a recent year, 840 men 21 to 25 years of age were in the NAEP sample. Their mean quantitative score was $\bar{x} = 272$. These 840 men are an SRS from the population of all young men. On the basis of this sample, what can we say about the mean score μ in the population of all 9.5 million young men of these ages?[1]

The sample mean $\bar{x}$ is an unbiased estimator of the unknown population mean μ. Because $\bar{x} = 272$, we guess that μ is "somewhere around 272." To make "somewhere around 272" more precise, we ask: *How would the sample mean $\bar{x}$ vary if we took many samples of 840 young men from this same population?* Recall the essential facts about the sampling distribution of $\bar{x}$:

- $\bar{x}$ has a normal distribution. (The central limit theorem tells us that the average of 840 scores has a distribution that is very close to normal.)
- The mean of this normal sampling distribution is the same as the unknown population mean μ.
- The standard deviation of $\bar{x}$ for an SRS of 840 men is $\sigma/\sqrt{840}$, where σ is the standard deviation of individual NAEP scores among all young men.

Let us suppose that we know from long experience that the standard deviation of scores in the population of all young men is $\sigma = 60$. The standard deviation of $\bar{x}$ is then

$$\frac{\sigma}{\sqrt{n}} = \frac{60}{\sqrt{840}} \doteq 2.1$$

(It is not realistic to assume we know σ. We will see in the next chapter how to proceed when σ is not known. For now, we are more interested in statistical reasoning than in details of realistic methods.)

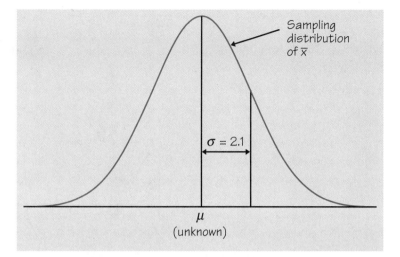

FIGURE 10.1 The sampling distribution of the mean score $\bar{x}$ of an SRS of 840 young men on the NAEP quantitative test.

In many repeated samples of size 840, the sample mean score $\bar{x}$ would vary according to the normal distribution with mean equal to the unknown μ and standard deviation 2.1. Inference about the unknown μ starts from this sampling distribution. Figure 10.1 displays the distribution. The different values of $\bar{x}$ appear along the axis in the figure, and the normal curve shows how probable these values are.

Statistical confidence

Figure 10.2 is another picture of the same sampling distribution. It illustrates the following line of thought:

- The 68–95–99.7 rule says that in about 95% of all samples, $\bar{x}$ will be within two standard deviations of the population mean score μ. That is, $\bar{x}$ will be within 4.2 points of μ in 95% of all samples.
- Whenever $\bar{x}$ is within 4.2 points of the unknown μ, then of course μ is within 4.2 points of the observed $\bar{x}$. This happens in 95% of all samples.
- So in 95% of all samples the unknown μ lies between $\bar{x} - 4.2$ and $\bar{x} + 4.2$.

This conclusion just restates a fact about the sampling distribution of $\bar{x}$. The language of statistical inference uses this fact about what would

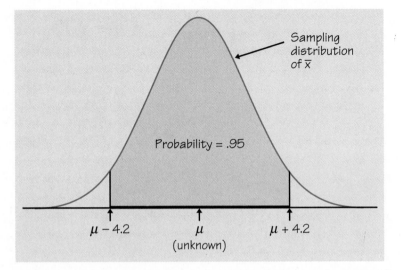

FIGURE 10.2   In 95% of all samples, $\bar{x}$ lies within ±4.2 of the unknown population mean μ. So μ also lies within ±4.2 of $\bar{x}$ in those samples.

happen in the long run to express our confidence in the results of any one sample. Our sample gave $\bar{x} = 272$. We say that we are 95% *confident* that the unknown mean NAEP quantitative score for all young men lies between

$$\bar{x} - 4.2 = 272 - 4.2 = 267.8$$

and

$$\bar{x} + 4.2 = 272 + 4.2 = 276.2$$

Be sure you understand the grounds for our confidence. There are only two possibilities:

1. The interval between 267.8 and 276.2 contains the true μ.
2. Our SRS was one of the few samples for which $\bar{x}$ is not within 4.2 points of the true μ. Only 5% of all samples give such inaccurate results.

We cannot know whether our sample is one of the 95% for which the interval $\bar{x} \pm 4.2$ catches μ, or one of the unlucky 5%. The statement that we are 95% confident that the unknown μ lies between 267.8 and 276.2 is shorthand for saying, "We got these numbers by a method that gives correct results 95% of the time."

The interval of numbers between the values $\bar{x} \pm 4.2$ is called a **95% confidence interval** for μ. Like most confidence intervals we will meet, this one has the form

confidence interval

$$\text{estimate} \pm \text{margin of error}$$

margin of error

The estimate ($\bar{x}$ in this case) is our guess for the value of the unknown parameter. The **margin of error** ± 4.2 shows how accurate we believe our guess is, based on the variability of the estimate. This is a 95% confidence interval because it catches the unknown μ in 95% of all possible samples.

Figure 10.3 illustrates the behavior of 95% confidence intervals in repeated sampling. The center of each interval is at $\bar{x}$ and therefore varies from sample to sample. The sampling distribution of $\bar{x}$ appears at the top

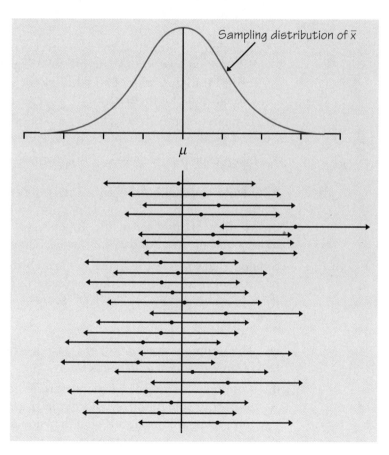

FIGURE 10.3 Twenty-five samples from the same population gave these 95% confidence intervals. In the long run, 95% of all samples give an interval that contains the population mean μ.

of the figure to show the long-term pattern of this variation. The 95% confidence intervals $\bar{x} \pm 4.2$ from 25 SRSs appear below. The center $\bar{x}$ of each interval is marked by a dot. The arrows on either side of the dot span the confidence interval. All except one of these 25 intervals cover the true value of μ. In a very large number of samples, 95% of the confidence intervals would contain μ.

EXERCISES

10.1   A *New York Times* poll on women's issues interviewed 1025 women randomly selected from the United States, excluding Alaska and Hawaii. The poll found that 47% of the women said they do not get enough time for themselves.

   (a) The poll announced a margin of error of ± 3 percentage points for 95% confidence in its conclusions. What is the 95% confidence interval for the percent of all adult women who think they do not get enough time for themselves?

   (b) Explain to someone who knows no statistics why we can't just say that 47% of all adult women do not get enough time for themselves.

   (c) Then explain clearly what "95% confidence" means.

10.2   A student reads that a 95% confidence interval for the mean NAEP quantitative score for men of ages 21 to 25 is 267.8 to 276.2. Asked to explain the meaning of this interval, the student says, "95% of all young men have scores between 267.8 and 276.2." Is the student right? Justify your answer.

10.3   Suppose that you give the NAEP test to an SRS of 1000 people from a large population in which the scores have mean 280 and standard deviation $\sigma = 60$. The mean $\bar{x}$ of the 1000 scores will vary if you take repeated samples.

   (a) The sampling distribution of $\bar{x}$ is approximately normal. What are its mean and standard deviation?

   (b) Sketch the normal curve that describes how $\bar{x}$ varies in many samples from this population. Mark its mean and the values one, two, and three standard deviations on either side of the mean.

   (c) According to the 68–95–99.7 rule, about 95% of all the values of $\bar{x}$ fall within _____ of the mean of this curve. What is the missing number? Call it m for "margin of error." Shade the region from the mean minus m to the mean plus m on the axis of your sketch, as in Figure 10.2.

(d) Whenever $\bar{x}$ falls in the region you shaded, the true value of the population mean, $\mu = 280$, lies in the confidence interval between $\bar{x} - m$ and $\bar{x} + m$. Draw the confidence interval below your sketch for one value of $\bar{x}$ inside the shaded region and one value of $\bar{x}$ outside the shaded region. (Use Figure 10.3 as a model for the drawing.)

(e) In what percent of all samples will the true mean $\mu = 280$ be covered by the confidence interval $\bar{x} \pm m$?

10.4 Oxides of nitrogen (called NOX for short) emitted by cars and trucks are important contributors to air pollution. The amount of NOX emitted by a particular model varies from vehicle to vehicle. For one light truck model, NOX emissions vary with mean μ that is unknown and standard deviation $\sigma = .4$ grams per mile. You test an SRS of 50 of these trucks. The sample mean NOX level $\bar{x}$ estimates the unknown μ. You will get different values of $\bar{x}$ if you repeat your sampling.

(a) The sampling distribution of $\bar{x}$ is approximately normal. What are its mean and standard deviation?

(b) Sketch the normal curve for the sampling distribution of $\bar{x}$. Mark its mean and the values one, two, and three standard deviations on either side of the mean.

(c) According to the 68–95–99.7 rule, about 95% of all values of $\bar{x}$ lie within a distance m of the mean of the sampling distribution. What is m? Shade the region on the axis of your sketch that is within m of the mean, as in Figure 10.2.

(d) Whenever $\bar{x}$ falls in the region you shaded, the unknown population mean μ lies in the confidence interval $\bar{x} \pm m$. For what percent of all possible samples does this happen?

(e) Following the style of Figure 10.3, draw the confidence intervals below your sketch for two values of $\bar{x}$, one that falls within the shaded region and one that falls outside it.

Confidence intervals

confidence level

Statisticians have constructed confidence intervals for many different parameters based on a variety of designs for producing data. We will meet a number of these in later chapters. Any confidence interval has two parts: an *interval* computed from the data and a **confidence level** giving the probability that the method produces an interval that covers the parameter. Users can choose the confidence level, most often 90% or higher because we most often want to be quite sure of our conclusions. We will use C to stand for the confidence level in decimal form. For example, a 95%

confidence level corresponds to $C = .95$. Here is the general definition of a confidence interval for an unknown parameter. In our examples, the parameter is the mean μ of the population, but it might be a population proportion p, the population standard deviation σ, or any other parameter.

CONFIDENCE INTERVAL

A **level** C **confidence interval** for a parameter is an interval computed from sample data by a method that has probability C of producing an interval containing the true value of the parameter.

We can now give the recipe for a level C confidence interval for the mean μ of a population when the data are an SRS of size n. The interval is based on the fact that the sampling distribution of the sample mean $\bar{x}$ is at least approximately normal. To get confidence level C we want to catch the central probability C under a normal curve. To do that, we must go out z^* standard deviations on either side of the mean. The number z^* is the same for any normal distribution, so we use the standard normal table. Here is an example of how to find z^*.

EXAMPLE 10.3

To find an 80% confidence interval, we must catch the central 80% of the normal sampling distribution of $\bar{x}$. In catching the central 80% we leave out 20%, or 10% in each tail. So z^* is the point with area 0.1 to its right (and 0.9 to its left) under the standard normal curve. Search the body of Table A to find the point with area 0.9 to its left. The closest entry is $z^* = 1.28$. There is area 0.8 under the standard normal curve between -1.28 and 1.28. Figure 10.4 shows how z^* is related to areas under the curve.

Figure 10.5 shows the general situation for any confidence level C. If we catch the central area C, the leftover tail area is $1 - C$, or $(1 - C)/2$ on each side. You can find z^* for any C by searching Table A. Here are the results for the most common confidence levels:

Confidence level	Tail area	z^*
90%	.05	1.645
95%	.025	1.960
99%	.005	2.576

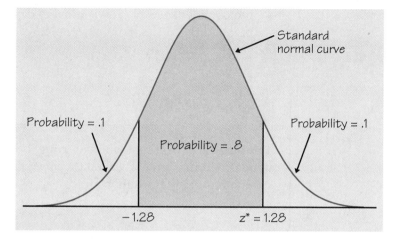

FIGURE 10.4   The central probability 0.8 under a standard normal curve lies between −1.28 and 1.28. That is, there is area 0.1 to the right of 1.28 under the curve.

Notice that for 95% confidence we use $z^* = 1.960$. This is more exact than the approximate value $z^* = 2$ given by the 68–95–99.7 rule. The bottom row in Table C gives the values of z^* for many confidence levels C. This row is labeled z^*. (You can find Table C in the back of the book and on the inside rear cover. We will use the other rows of the table in the next chapter.) Although we can find z^* from Table C by simply looking

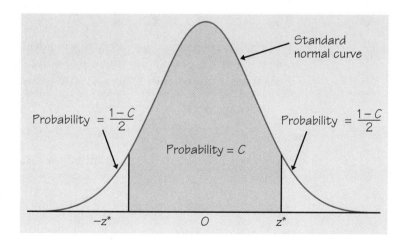

FIGURE 10.5   In general, the central probability C under a standard normal curve lies between $-z^*$ and z^*. Because z^* has area $(1 - C)/2$ to its right under the curve, we call it the upper $(1 - C)/2$ critical value.

above the confidence level C, it is usual to describe the point z^* in terms of the probability to its right. For example, we call 1.960 the upper 0.025 *critical value* of the standard normal distribution.

CRITICAL VALUES

The number z^* with probability p lying to its right under the standard normal curve is called the **upper p critical value** of the standard normal distribution.

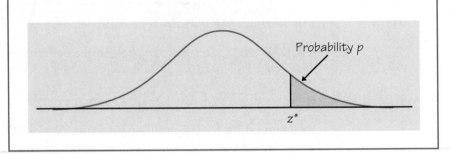

Probability p

z^*

Here's how to find the level C confidence interval:

- Any normal curve has probability C between the point z^* standard deviations below its mean and the point z^* standard deviations above its mean.
- The standard deviation of the sampling distribution of $\bar{x}$ is $\sigma/\sqrt{n}$, and its mean is the population mean μ. So there is probability C that the observed sample mean $\bar{x}$ takes a value between

$$\mu - z^*\frac{\sigma}{\sqrt{n}} \quad \text{and} \quad \mu + z^*\frac{\sigma}{\sqrt{n}}$$

- Whenever this happens, the population mean μ is contained between

$$\bar{x} - z^*\frac{\sigma}{\sqrt{n}} \quad \text{and} \quad \bar{x} + z^*\frac{\sigma}{\sqrt{n}}$$

That is our confidence interval. The estimate of the unknown μ is $\bar{x}$, and the margin of error is $z^*\sigma/\sqrt{n}$.

CONFIDENCE INTERVAL FOR A POPULATION MEAN

Draw an SRS of size n from a population having unknown mean μ and known standard deviation σ. A level C confidence interval for μ is

$$\bar{x} \pm z^* \frac{\sigma}{\sqrt{n}}$$

Here z^* is the upper $(1 - C)/2$ critical value for the standard normal distribution, found in Table C. This interval is exact when the population distribution is normal and is approximately correct for large n in other cases.

EXAMPLE 10.4

A manufacturer of pharmaceutical products analyzes a specimen from each batch of a product to verify the concentration of the active ingredient. The chemical analysis is not perfectly precise. Repeated measurements on the same specimen give slightly different results. The results of repeated measurements follow a normal distribution quite closely. The analysis procedure has no bias, so the mean μ of the population of all measurements is the true concentration in the specimen. The standard deviation of this distribution is known to be $\sigma = .0068$ grams per liter. The laboratory analyzes each specimen three times and reports the mean result.

Three analyses of one specimen give concentrations

$$0.8403 \quad 0.8363 \quad 0.8447$$

We want a 99% confidence interval for the true concentration μ.

The sample mean of the three readings is

$$\bar{x} = \frac{.8403 + .8363 + .8447}{3} = .8404$$

For 99% confidence, we see from Table C that $z^* = 2.576$. A 99% confidence interval for μ is therefore

$$\bar{x} \pm z^* \frac{\sigma}{\sqrt{n}} = .8404 \pm 2.576 \frac{.0068}{\sqrt{3}}$$

$$= .8404 \pm .0101$$

$$= (.8303, .8505)$$

We are 99% confident that the true concentration lies between 0.8303 and 0.8505 grams per liter.

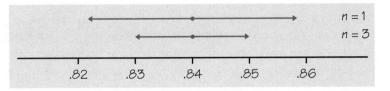

FIGURE 10.6   Confidence intervals for $n = 3$ and $n = 1$ for Example 10.4. Larger samples give shorter intervals.

Suppose that a single measurement gave $x = .8404$, the same value that the sample mean took in Example 10.4. Repeating the calculation with $n = 1$ shows that the 99% confidence interval based on a single measurement is

$$\bar{x} \pm z^* \frac{\sigma}{\sqrt{1}} = .8404 \pm (2.576)(.0068)$$

$$= .8404 \pm .0175$$

$$= (.8229, .8579)$$

The mean of three measurements gives a smaller margin of error and therefore a shorter interval than a single measurement. Figure 10.6 illustrates the gain from using three observations.

The form of confidence intervals for the population mean μ rests on the fact that the statistic $\bar{x}$ used to estimate μ has a normal distribution. Because many sample statistics have normal distributions (at least approximately), it is useful to notice that the confidence interval has the form

$$\text{estimate} \pm z^* \sigma_{\text{estimate}}$$

The estimate based on the sample is the center of the confidence interval. The margin of error is $z^* \sigma_{\text{estimate}}$. The desired confidence level determines z^* from Table C. The standard deviation of the estimate, σ_{estimate}, depends on the particular estimate we use. When the estimate is $\bar{x}$ from an SRS, the standard deviation of the estimate is $\sigma/\sqrt{n}$.

EXERCISES

10.5   A study of the career paths of hotel general managers sent questionnaires to an SRS of 160 hotels belonging to major U.S. hotel chains. There were

114 responses. The average time these 114 general managers had spent with their current company was 11.78 years. Give a 99% confidence interval for the mean number of years general managers of major-chain hotels have spent with their current company. (Take it as known that the standard deviation of time with the company for all general managers is 3.2 years.)

10.6 The Degree of Reading Power (DRP) is a test of the reading ability of children. Here are DRP scores for a sample of 44 third-grade students in a suburban school district:[2]

40	26	39	14	42	18	25	43	46	27	19
47	19	26	35	34	15	44	40	38	31	46
52	25	35	35	33	29	34	41	49	28	52
47	35	48	22	33	41	51	27	14	54	45

(a) We expect the distribution of DRP scores to be close to normal. Make a stemplot or histogram of the distribution of these 44 scores and describe its shape.

(b) Suppose that the standard deviation of the population of DRP scores is known to be $\sigma = 11$. Give a 99% confidence interval for the mean score in the school district.

(c) Would you trust your conclusion from (b) if these scores came from a single class in one school in the district? Why?

10.7 Here are measurements (in millimeters) of a critical dimension on a sample of auto engine crankshafts:

224.120	224.001	224.017	223.982	223.989	223.961
223.960	224.089	223.987	223.976	223.902	223.980
224.098	224.057	223.913	223.999		

The data come from a production process that is known to have standard deviation $\sigma = .060$ mm. The process mean is supposed to be $\mu = 224$ mm but can drift away from this target during production.

(a) We expect the distribution of the dimension to be close to normal. Make a stemplot or histogram of these data and describe the shape of the distribution.

(b) Give a 95% confidence interval for the process mean at the time these crankshafts were produced.

10.8   A test for the level of potassium in the blood is not perfectly precise. More-
         over, the actual level of potassium in a person's blood varies slightly from
         day to day. Suppose that repeated measurements for the same person on
         different days vary normally with $\sigma = .2$.

(a) Julie's potassium level is measured once. The result is $x = 3.2$. Give
     a 90% confidence interval for her mean potassium level.

(b) If three measurements were taken on different days and the mean
     result is $\bar{x} = 3.2$, what is a 90% confidence interval for Julie's mean
     blood potassium level?

How confidence intervals behave

The confidence interval $\bar{x} \pm z^*\sigma/\sqrt{n}$ for the mean of a normal population illustrates several important properties that are shared by all confidence intervals in common use. The user chooses the confidence level,
and the margin of error follows from this choice. We would like high
confidence and also a small margin of error. High confidence says that
our method almost always gives correct answers. A small margin of error
says that we have pinned down the parameter quite precisely. The margin
of error is

$$\text{margin of error} = z^* \frac{\sigma}{\sqrt{n}}$$

This expression has z^* and σ in the numerator and $\sqrt{n}$ in the denominator. So the margin of error gets smaller when

- z^* gets smaller. Smaller z^* is the same as smaller confidence level C
  (look at Figure 10.5 again). There is a trade-off between the
  confidence level and the margin of error. To obtain a smaller margin
  of error from the same data, you must be willing to accept lower
  confidence.

- σ gets smaller. The standard deviation σ measures the variation in the
  population. You can think of the variation among individuals in the
  population as noise that obscures the average value μ. It is easier to
  pin down μ when σ is small.

- n gets larger. Increasing the sample size n reduces the margin of error
  for any fixed confidence level. Because n appears under a square root
  sign, we must take four times as many observations in order to cut the
  margin of error in half.

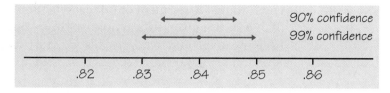

FIGURE 10.7   90% and 99% confidence intervals for Example 10.5. Higher confidence requires a wider interval.

EXAMPLE 10.5

Suppose that the pharmaceutical manufacturer in Example 10.4 is content with 90% confidence rather than 99%. Table C gives the critical value for 90% confidence as $z^* = 1.645$. The 90% confidence interval for μ based on three repeated measurements with mean $\bar{x} = .8404$ is

$$\bar{x} \pm z^* \frac{\sigma}{\sqrt{n}} = .8404 \pm 1.645 \frac{.0068}{\sqrt{3}}$$

$$= .8404 \pm .0065$$

$$= (.8339, .8469)$$

Settling for 90%, rather than 99%, confidence has reduced the margin of error from ± 0.0101 to ± 0.0065. Figure 10.7 compares these two intervals. Increasing the number of measurements from 3 to 12 will also reduce the width of the 99% confidence interval in Example 10.4. Check that replacing $\sqrt{3}$ by $\sqrt{12}$ cuts the ± 0.0101 margin of error in half, because we now have four times as many observations.

EXERCISES

10.9   Examples 10.4 and 10.5 give confidence intervals for the concentration μ based on 3 measurements with $\bar{x} = .8404$ and $\sigma = .0068$. The 99% confidence interval is 0.8303 to 0.8505 and the 90% confidence interval is 0.8339 to 0.8469.

(a) Find the 80% confidence interval for μ.

(b) Find the 99.9% confidence interval for μ.

(c) Make a sketch like Figure 10.7 to compare all four intervals. How does increasing the confidence level affect the length of the confidence interval?

10.10   Find the margin of error for 99% confidence in Example 10.4 if the laboratory measures the concentration of each specimen 12 times. Check that your result is half as large as the margin of error based on 3 measurements in Example 10.4.

10.11   The National Assessment of Educational Progress (NAEP) test (Example 10.2) was also given to a sample of 1077 women of ages 21 to 25 years. Their mean quantitative score was 275. Take it as known that the standard deviation of all individual scores is $\sigma = 60$.

   (a) Give a 95% confidence interval for the mean score μ in the population of all young women.

   (b) Give the 90% and 99% confidence intervals for μ.

   (c) What are the margins of error for 90%, 95%, and 99% confidence? How does increasing the confidence level affect the margin of error of a confidence interval?

10.12   The NAEP sample of 1077 young women had mean quantitative score $\bar{x} = 275$. Take it as known that the standard deviation of all individual scores is $\sigma = 60$.

   (a) Give a 95% confidence interval for the mean score μ in the population of all young women.

   (b) Suppose that the same result, $\bar{x} = 275$, had come from a sample of 250 women. Give the 95% confidence interval for the population mean μ in this case.

   (c) Then suppose that a sample of 4000 women had produced the sample mean $\bar{x} = 275$, and again give the 95% confidence interval for μ.

   (d) What are the margins of error for samples of size 250, 1077, and 4000? How does increasing the sample size affect the margin of error of a confidence interval?

Choosing the sample size

A wise user of statistics never plans data collection without at the same time planning the inference. You can arrange to have both high confidence and a small margin of error by taking enough observations. The margin of error of the confidence interval for the mean of a normally distributed population is $m = z^* \sigma / \sqrt{n}$. To obtain a desired margin of error m, substitute the value of z^* for your desired confidence level, set the expression for m less than or equal to the specified margin of error, and solve the inequality for n. The procedure is best illustrated with an example.

EXAMPLE 10.6   Management asks the laboratory of Example 10.4 to produce results accurate to within ±0.005 with 95% confidence. How many measurements must be averaged to comply with this request?

For 95% confidence, Table C gives $z^* = 1.960$. We know that $\sigma = .0068$. Set the margin of error to be at most 0.005:

$$m \leq .005$$

$$z^* \frac{\sigma}{\sqrt{n}} \leq .005$$

$$\frac{1.960 \times .0068}{.005} \leq \sqrt{n}$$

$$\sqrt{n} \geq 2.6656$$

$$n \geq 7.1, \text{ so take } n = 8$$

Because n is a whole number, the lab must take 8 measurements on each specimen to meet management's demand. On learning the cost of this many measurements, management may reconsider its request.

Here is the principle.

SAMPLE SIZE FOR DESIRED MARGIN OF ERROR

To determine the sample size n that will yield a confidence interval for a population mean with a specified margin of error m, set the expression for the margin of error to be less than or equal to m and solve for n:

$$z^* \frac{\sigma}{\sqrt{n}} \leq m$$

In practice, taking observations costs time and money. The required sample size may be impossibly expensive. Do notice once again that it is the size of the *sample* that determines the margin of error. The size of the *population* (as long as the population is much larger than the sample) does not influence the sample size we need.

EXERCISES

10.13   To assess the accuracy of a laboratory scale, a standard weight known to weigh 10 grams is weighed repeatedly. The scale readings are normally distributed with unknown mean (this mean is 10 grams if the scale has no bias). The standard deviation of the scale readings is known to be 0.0002 gram.

(a) The weight is weighed five times. The mean result is 10.0023 grams. Give a 98% confidence interval for the mean of repeated measurements of the weight.

(b) How many measurements must be averaged to get a margin of error of ±0.0001 with 98% confidence?

10.14   How large a sample of the hotel managers in Exercise 10.5 would be needed to estimate the mean μ within ±1 year with 99% confidence?

10.15   How large a sample of the crankshafts in Exercise 10.7 would be needed to estimate the mean μ within ±0.020 mm with 95% confidence?

Some cautions

Any formula for inference is correct only in specific circumstances. If statistical procedures carried warning labels like those on drugs, most inference methods would have long labels indeed. Our handy formula $\bar{x} \pm z^* \sigma/\sqrt{n}$ for estimating a normal mean comes with the following list of warnings for the user.

- The data must be an SRS from the population. We are completely safe if we actually carried out the random selection of an SRS. We are not in great danger if the data can plausibly be thought of as observations taken at random from a population. That is the case in Examples 10.4 to 10.6, where we have in mind the population resulting from a very large number of repeated analyses of the same specimen.

- The formula is not correct for probability sampling designs more complex than an SRS. Correct methods for other designs are available. We will not discuss confidence intervals based on multistage or stratified samples. If you plan such samples, be sure that you (or your statistical consultant) know how to carry out the inference you desire.

- There is no correct method for inference from data haphazardly collected with bias of unknown size. Fancy formulas cannot rescue badly produced data.

- Because $\bar{x}$ is strongly influenced by a few extreme observations, outliers can have a large effect on the confidence interval. You should search for outliers and try to correct them or justify their removal before computing the interval. If the outliers cannot be removed, ask your statistical consultant about procedures that are not sensitive to outliers.

- If the sample size is small and the population is not normal, the true confidence level will be different from the value C used in computing the interval. Examine your data carefully for skewness and other signs of nonnormality. The interval relies only on the distribution of $\bar{x}$, which even for quite small sample sizes is much closer to normal than the individual observations. When $n \geq 15$, the confidence level is not greatly disturbed by nonnormal populations unless extreme outliers or quite strong skewness are present. We will discuss this issue in more detail in the next chapter.

- You must know the standard deviation σ of the population. This unrealistic requirement renders the interval $\bar{x} \pm z^* \sigma/\sqrt{n}$ of little use in statistical practice. We will learn in the next chapter what to do when σ is unknown. However, if the sample is large, the sample standard deviation s will be close to the unknown σ. Then $\bar{x} \pm z^* s/\sqrt{n}$ is an approximate confidence interval for μ.

The most important caution concerning confidence intervals is a consequence of the first of these warnings. *The margin of error in a confidence interval covers only random sampling errors.* The margin of error is obtained from the sampling distribution and indicates how much error can be expected because of chance variation in randomized data production. Practical difficulties such as undercoverage and nonresponse in a sample survey can cause additional errors that may be larger than the random sampling error. Remember this unpleasant fact when reading the results of an opinion poll or other sample survey. The practical conduct of the survey influences the trustworthiness of its results in ways that are not included in the announced margin of error.

Every inference procedure that we will meet has its own list of warnings. Because many of the warnings are similar to those above, we will not print the full warning label each time. It is easy to state (from the mathematics of probability) conditions under which a method of inference is exactly correct. These conditions are *never* fully met in practice. For example, no population is exactly normal. Deciding when a statistical procedure should be used in practice often requires judgment assisted by exploratory analysis of the data.

Finally, you should understand what statistical confidence does not say. We are 95% confident that the mean NAEP quantitative score for all men aged 21 to 25 years lies between 267.8 and 276.2. That is, these numbers were calculated by a method that gives correct results in 95% of all possible samples. We *cannot* say that the probability is 95% that the true mean falls between 267.8 and 276.2. No randomness remains after we draw one particular sample and get from it one particular interval. The true mean either is or is not between 267.8 and 276.2. The probability calculations of standard statistical inference describe how often the *method* gives correct answers.

EXERCISES

10.16   A radio talk show invites listeners to enter a dispute about a proposed pay increase for city council members. "What yearly pay do you think council members should get? Call us with your number." In all, 958 people call. The mean pay they suggest is $\bar{x} = \$8740$ per year, and the standard deviation of the responses is $s = \$1125$. For a large sample such as this, s is very close to the unknown population σ. The station calculates the 95% confidence interval for the mean pay μ that all citizens would propose for council members to be $8669 to $8811.

(a)   Is the station's calculation correct?

(b)   Does their conclusion describe the population of all the city's citizens? Explain your answer.

10.17   A closely contested presidential election pitted Jimmy Carter against Gerald Ford in 1976. A poll taken immediately before the 1976 election showed that 51% of the sample intended to vote for Carter. The polling organization announced that they were 95% confident that the sample result was within ± 2 points of the true percent of all voters who favored Carter.

(a)   Explain in plain language to someone who knows no statistics what "95% confident" means in this announcement.

(b)   The poll showed Carter leading. Yet the polling organization said the election was too close to call. Explain why.

(c)   On hearing of the poll, a nervous politician asked, "What is the probability that over half the voters prefer Carter?" A statistician replied that this question can't be answered from the poll results, and that it doesn't even make sense to talk about such a probability. Explain why.

10.18   The New York Times/CBS News Poll asked the question, "Do you favor an amendment to the Constitution that would permit organized prayer in public schools?" Sixty-six percent of the sample answered "Yes." The article describing the poll says that it "is based on telephone interviews conducted from Sept. 13 to Sept. 18 with 1,664 adults around the United States, excluding Alaska and Hawaii....the telephone numbers were formed by random digits, thus permitting access to both listed and unlisted residential numbers."

(a)   The article gives the margin of error as 3 percentage points. Make a confidence statement about the percent of all adults who favor a school prayer amendment.

(b) The news article goes on to say: "The theoretical errors do not take into account a margin of additional error resulting from the various practical difficulties in taking any survey of public opinion." List some of the "practical difficulties" that may cause errors in addition to the ±3% margin of error. Pay particular attention to the news article's description of the sampling method.

SUMMARY

The purpose of a **confidence interval** is to estimate an unknown parameter with an indication of how accurate the estimate is and of how confident we are that the result is correct.

Any confidence interval has two parts: an interval computed from the data and a confidence level. The **interval** often has the form

estimate ± margin of error

The **confidence level** states the probability that the method will give a correct answer. That is, if you use 95% confidence intervals often, in the long run 95% of your intervals will contain the true parameter value. You do not know whether a 95% confidence interval calculated from a particular set of data contains the true parameter value.

A level C **confidence interval for the mean** μ of a normal population with known standard deviation σ, based on an SRS of size n, is given by

$$\bar{x} \pm z^* \frac{\sigma}{\sqrt{n}}$$

Here z^* is chosen so that the standard normal curve has area C between $-z^*$ and z^*. Because of the central limit theorem, this interval is approximately correct for large samples when the population is not normal.

The number z^* is called the **upper p critical value** of the standard normal distribution for $p = (1 - C)/2$. Critical values for many confidence levels appear in Table C.

Other things being equal, the **margin of error** of a confidence interval gets smaller as

- the confidence level C decreases,
- the population standard deviation σ decreases, and
- the sample size n increases.

The sample size required to obtain a confidence interval with specified margin of error m for a normal mean is found by setting

$$z^* \frac{\sigma}{\sqrt{n}} \leq m$$

and solving for n, where z^* is the critical value for the desired level of confidence. Always round n up when you use this procedure.

A specific confidence interval recipe is correct only under specific conditions. The most important conditions concern the method used to produce the data. Other factors such as the form of the population distribution may also be important.

SECTION 10.1 EXERCISES

10.19   The Acculturation Rating Scale for Mexican Americans (ARSMA) is a psychological test that measures the degree to which Mexican Americans have adopted Mexican/Spanish versus Anglo/English culture. The distribution of ARSMA scores in a population used to develop the test was approximately normal, with mean 3.0 and standard deviation 0.8. A further study gave ARSMA to 42 first-generation Mexican Americans. The mean of their scores was $\bar{x} = 2.13$. Assuming the standard deviation for the first-generation population is also $\sigma = .8$, give a 95% confidence interval for the mean ARSMA score for first-generation Mexican Americans.

10.20   How satisfied are hotel managers with the computer systems their hotels use? A survey was sent to 560 managers in hotels of size 200 to 500 rooms in Chicago and Detroit.[3] In all, 135 managers returned the survey. Two questions concerned their degree of satisfaction with the ease of use of their computer systems and with the level of computer training they had received. The managers responded using a seven-point scale, with 1 meaning "not satisfied," 4 meaning "moderately satisfied," and 7 meaning "very satisfied."

(a)   What do you think is the population for this study? There are some major shortcomings in the data production. What are they? These shortcomings reduce the value of the formal inference you are about to do.

(b)   The mean response for satisfaction with ease of use was $\bar{x} = 5.396$. Give a 95% confidence interval for the mean in the entire population. (Assume that the population standard deviation is $\sigma = 1.75$.)

(c) For satisfaction with training, the mean response was $\bar{x} = 4.398$. Taking $\sigma = 1.75$, give a 99% confidence interval for the population mean.

(d) The measurements of satisfaction are certainly not normally distributed, because they take only whole-number values from 1 to 7. Nonetheless, the use of confidence intervals based on the normal distribution is justified for this study. Why?

10.21 Consumers can purchase nonprescription medications at food stores, mass merchandise stores such as Kmart and Wal-Mart, or pharmacies. About 45% of consumers make such purchases at pharmacies. What accounts for the popularity of pharmacies, which often charge higher prices?

A study examined consumers' perceptions of overall performance of the three types of stores, using a long questionnaire that asked about such things as "neat and attractive store," "knowledgeable staff," and "assistance in choosing among various types of nonprescription medication." A performance score was based on 27 such questions. The subjects were 201 people chosen at random from the Indianapolis telephone directory. Here are the means and standard deviations of the performance scores for the sample:[4]

Store type	$\bar{x}$	s
Food stores	18.67	24.95
Mass merchandisers	32.38	33.37
Pharmacies	48.60	35.62

We do not know the population standard deviations, but a sample standard deviation s from so large a sample is usually close to σ. Use s in place of the unknown σ in this exercise.

(a) What population do you think the authors of the study want to draw conclusions about? What population are you certain they can draw conclusions about?

(b) Give 95% confidence intervals for the mean performance for each type of store in the population.

(c) Based on these confidence intervals, are you convinced that consumers think that pharmacies offer higher performance than the other types of stores?

10.22   The 1990 census "long form" was sent to a random sample of 17% of the nation's households. One question asked the total 1989 income of the householder. (The householder is the person in whose name the dwelling unit is owned or rented.) Suppose that the households that returned the long form are an SRS of the population of all households in each district. In Middletown, a city of 40,000 persons, 2621 householders reported their income. The mean of the responses was $\bar{x} = \$23,453$, and the standard deviation was $s = \$8721$. The sample standard deviation for so large a sample will be very close to the population standard deviation σ. Use these facts to give an approximate 99% confidence interval for the 1989 mean income of Middletown householders.

10.23   How large a sample of households would enable you to estimate the mean income of Middletown householders (see the previous exercise) within a margin of error of $1000 with 99% confidence?

10.24   A *New York Times* poll on women's issues interviewed 1025 women and 472 men randomly selected from the United States, excluding Alaska and Hawaii. The poll announced a margin of error of ± 3 percentage points for 95% confidence in conclusions about women. The margin of error for results concerning men was ± 4 percentage points. Why is this larger than the margin of error for women?

10.25   When the statistic that estimates an unknown parameter has a normal distribution, a confidence interval for the parameter has the form

$$\text{estimate} \pm z^* \sigma_{\text{estimate}}$$

In a complex sample survey design, the estimate of the population mean and the standard deviation of this estimate require elaborate computations. But when we are given the estimate and its standard deviation, we can calculate a confidence interval for μ without knowing the formulas that led to the numbers given.

A report based on the Current Population Survey estimates the median weekly earnings of families of wage or salary workers as $664 and also estimates that the standard deviation of this estimate is $3.50. The Current Population Survey uses an elaborate multistage sampling design to select a sample of about 60,000 households. The sampling distribution of the estimated median income is approximately normal. Give a 95% confidence interval for the median weekly earnings of all families of wage and salary workers.

10.26   (a) The TI-83 can be used to construct confidence intervals, using either data stored in a list or summary statistics. In Example 10.4, for example, we would enter the three specimen concentrations into list L_1 and use STAT / TESTS / 7:ZInterval to access the z confidence screen, as shown. Verify that the 99% confidence interval for the mean concentration of the active ingredient is $(0.8303, 0.8506)$.

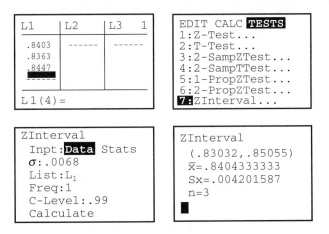

(b) Use your TI-83 to find the confidence interval for the mean reading score in Exercise 10.6(b).

(c) If you have summary statistics but not raw data, you would select "Stats" in the ZInterval screen and then provide the sample mean $\bar{x}$ and the number n of observations. Use your TI-83 in this way to find the confidence interval for the mean number of years hotel managers have spent with their current company in Exercise 10.5.

```
ZInterval
  Inpt:Data Stats
  σ:.0068
  x̄:.8404
  n:3
  C-Level:.99
  Calculate
```

10.2   TESTS OF SIGNIFICANCE

Confidence intervals are one of the two most common types of formal statistical inference. Use them when your goal is to estimate a population parameter. The second common type of inference has a different goal: to assess the evidence provided by the data in favor of some claim about the population. The reasoning of statistical tests, like that of confidence intervals, is based on asking what would happen if we repeated the sample or experiment many times. Here is the first example we will explore.

EXAMPLE 10.7

Diet colas use artificial sweeteners to avoid sugar. Colas with artificial sweeteners gradually lose their sweetness over time. Manufacturers therefore test new colas for loss of sweetness before marketing them. Trained tasters sip the cola along with drinks of standard sweetness and score the cola on a "sweetness score" of 1 to 10. The cola is then stored for a month at high temperature to imitate the effect of four months' storage at room temperature. After a month, each taster scores the stored cola. This is a matched pairs experiment. Our data are the differences (score before storage minus score after storage) in the tasters' scores. The bigger these differences, the bigger the loss of sweetness.

Here are the sweetness losses for a new cola, as measured by 10 trained tasters:

$$2.0 \quad 0.4 \quad 0.7 \quad 2.0 \quad -0.4 \quad 2.2 \quad -1.3 \quad 1.2 \quad 1.1 \quad 2.3$$

Most are positive. That is, most tasters found a loss of sweetness. But the losses are small, and two tasters (the negative scores) thought the cola gained sweetness. *Are these data good evidence that the cola lost sweetness in storage?*

The reasoning of a significance test

The average sweetness loss for our cola is given by the sample mean,

$$\bar{x} = \frac{2.0 + 0.4 + \cdots + 2.3}{10} = 1.02$$

significance test

That's not a large loss. Ten different tasters would almost surely give a different result. Maybe it's just chance that produced this result. A *test of significance* asks:

Does the sample result $\bar{x} = 1.02$ reflect a real loss of sweetness?

OR

Could we easily get the outcome $\bar{x} = 1.02$ just by chance?

The significance test starts with a careful statement of these alternatives. First, we always draw conclusions about some parameter of the population, so we must identify this parameter. In this case, it's the population mean μ. The mean μ is the average loss in sweetness that a very large number of tasters would detect in the cola. Our 10 tasters are a sample from this population.

null hypothesis

Next, state the **null hypothesis**. The null hypothesis says that there is no effect or no change in the population. If the null hypothesis is true, the sample result is just chance at work. Here, the null hypothesis says that the cola does not lose sweetness (no change). We can write that in terms of the mean sweetness loss μ in the population as

$$H_0: \mu = 0$$

We write H_0, read as "H-nought," to indicate the null hypothesis.

alternative hypothesis

The effect we suspect is true, the alternative to "no effect" or "no change," is described by the **alternative hypothesis**. We suspect that the cola does lose sweetness. In terms of the mean sweetness loss μ, the alternative hypothesis is

$$H_a: \mu > 0$$

The reasoning of a significance test goes like this.

- Suppose for the sake of argument that the null hypothesis is true, that on the average there is no loss of sweetness.
- *Is the sample outcome $\bar{x} = 1.02$ surprisingly large under that supposition?* If it is, that's evidence against H_0 and in favor of H_a.

To answer the question, we use our knowledge of how the sample mean $\bar{x}$ would vary in repeated samples if H_0 really were true. That's the sampling distribution of $\bar{x}$ once again.

From long experience we know that individual tasters' scores vary according to a normal distribution. The mean of this distribution is the parameter μ. We're asking what would happen if there is really no change in sweetness on the average, so μ is 0. That's just what the null hypothesis says. From long experience we also know that the standard deviation for all individual tasters is $\sigma = 1$. (It is not realistic to suppose that we know the population standard deviation σ. We will eliminate this assumption in the next chapter.) The sampling distribution of $\bar{x}$ from 10 tasters is then normal with mean $\mu = 0$ and standard deviation

$$\frac{\sigma}{\sqrt{n}} = \frac{1}{\sqrt{10}} = .316$$

We can judge whether any observed $\bar{x}$ is surprising by locating it on this distribution. Figure 10.8 shows the sampling distribution with the observed values of $\bar{x}$ for two types of cola.

- One cola had $\bar{x} = .3$ for a sample of 10 tasters. It is clear from Figure 10.8 that an $\bar{x}$ this large could easily occur just by chance when the population mean is $\mu = 0$. That 10 tasters find $\bar{x} = .3$ is not evidence of a sweetness loss.

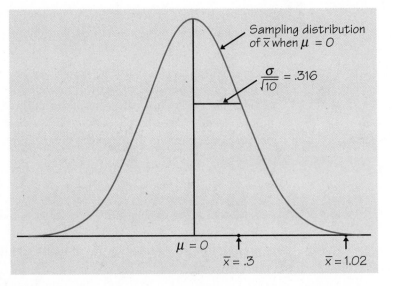

Sampling distribution of $\bar{x}$ when $\mu = 0$

$\dfrac{\sigma}{\sqrt{10}} = .316$

$\mu = 0$ $\bar{x} = .3$ $\bar{x} = 1.02$

FIGURE 10.8   If a cola does not lose sweetness in storage, the mean score $\bar{x}$ for 10 tasters will have this sampling distribution. The actual result for one cola was $\bar{x} = .3$. That could easily happen just by chance. Another cola had $\bar{x} = 1.02$. That's so far out on the normal curve that it is good evidence that this cola did lose sweetness.

- The taste test for our cola produced $\bar{x} = 1.02$. That's way out on the normal curve in Figure 10.8, so far out that an observed value this large would almost never occur just by chance if the true μ were 0. This observed value is good evidence that in fact the true μ is greater than 0, that is, that the cola lost sweetness. The manufacturer must reformulate the cola and try again.

A significance test works by asking how unlikely the observed outcome would be if the null hypothesis were really true. The final step in our test is to assign a number to measure how unlikely our observed $\bar{x}$ is if H_0 is true. The less likely this outcome is, the stronger is the evidence against H_0.

Look again at Figure 10.8. If the alternative hypothesis is true, there is a sweetness loss and we expect the mean loss $\bar{x}$ found by the tasters to be positive. The farther out $\bar{x}$ is in the positive direction, the more convinced we are that the population mean μ is not zero but positive. We measure the strength of the evidence against H_0 by the probability under the normal curve in Figure 10.8 to the right of the observed $\bar{x}$. This probability is called the **P-value.** It is the probability of a result at least as far out as the result we actually got. The lower this probability, the

P-value

more surprising our result, and the stronger the evidence against the null hypothesis.

- For one new cola, our 10 tasters gave $\bar{x} = .3$. Figure 10.9 shows the P-value for this outcome. It is the probability to the right of 0.3. This probability is about 0.17. That is, 17% of all samples would give a mean score as large or larger than 0.3 just by chance when the true population mean is 0. An outcome this likely to occur just by chance is not good evidence against the null hypothesis.

- Our cola showed a larger sweetness loss, $\bar{x} = 1.02$. The probability of a result this large or larger is only 0.0006. This probability is the P-value. Ten tasters would have an average score as large as 1.02 only 6 times in 10,000 tries if the true mean sweetness change were 0. An outcome this unlikely convinces us that the true mean is really greater than 0.

Small P-values are evidence against H_0, because they say that the observed result is unlikely to occur just by chance. Large P-values fail to give evidence against H_0. How small must a P-value be in order to persuade us? There's no fixed rule. But the level 0.05 (a result that would occur no more than once in 20 tries just by chance) is a common rule of thumb. *statistically significant* A result with a small P-value, say less than 0.05, is called **statistically significant**. That's just a way of saying that chance alone would rarely produce so extreme a result.

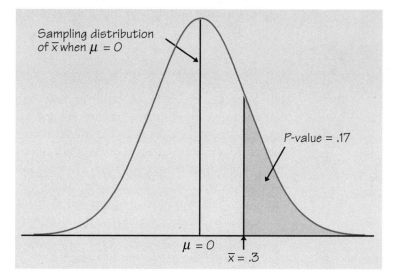

FIGURE 10.9   The P-value for the result $\bar{x} = .3$ in the cola taste test. The P-value is the probability (when H_0 is true) that $\bar{x}$ takes a value as large or larger than the actually observed value.

Outline of a test

Here is the reasoning of a significance test in outline form:

1. Describe the effect you are searching for in terms of a population parameter like the mean μ. (Never state a hypothesis in terms of a sample statistic like $\bar{x}$.)

2. The null hypothesis is the statement that this effect is *not* present in the population.

3. From the data, calculate a statistic like $\bar{x}$ that estimates the parameter. Is the value of this statistic far from the parameter value stated by the null hypothesis? If so, the data give evidence that the null hypothesis is false and that the effect you are looking for is really there.

4. The P-value says how unlikely a result at least as extreme as the one we observed would be if the null hypothesis were true. Results with small P-values would rarely occur if the null hypothesis were true. We call such results statistically significant.

This outline overlooks lots of detail and many fine points. But it is important to have it firmly in mind before we go on. Recipes for significance tests hide the underlying reasoning. In fact, statistical software often just gives a P-value. Look again at Figure 10.8, with its $\bar{x}$-values from taste tests of two colas. It should be clear that one result is not surprising if the true mean score in the population is 0, and that the other is surprising. A significance test simply says that more precisely.

EXERCISES

10.27   The Survey of Study Habits and Attitudes (SSHA) is a psychological test that measures the attitude toward school and study habits of students. Scores range from 0 to 200. The mean score for U.S. college students is about 115, and the standard deviation is about 30. A teacher suspects that older students have better attitudes toward school. She gives the SSHA to 25 students who are at least 30 years of age. Assume that scores in the population of older students are normally distributed with standard deviation $\sigma = 30$. The teacher wants to test the hypotheses

$$H_0: \mu = 115$$

$$H_a: \mu > 115$$

(a) What is the sampling distribution of the mean score $\bar{x}$ of a sample of 25 older students if the null hypothesis is true? Sketch the density curve of this distribution. (*Hint:* Sketch a normal curve first, then mark the axis using what you know about locating μ and σ on a normal curve.)

(b) Suppose that the sample data give $\bar{x} = 118.6$. Mark this point on the axis of your sketch. In fact, the result was $\bar{x} = 125.7$. Mark this point on your sketch. Using your sketch, explain in simple language why one result is good evidence that the mean score of all older students is greater than 115 and why the other outcome is not.

(c) Shade the area under the curve that is the *P*-value for the sample result $\bar{x} = 118.6$.

10.28 The Census Bureau reports that households spend an average of 31% of their total spending on housing. A homebuilders association in Cleveland believes that this average is lower in their area. They interview a sample of 40 households in the Cleveland metropolitan area to learn what percent of their spending goes toward housing. Take μ to be the mean percent of spending devoted to housing among all Cleveland households. We want to test the hypotheses

$$H_0: \mu = 31\%$$
$$H_a: \mu < 31\%$$

The population standard deviation is $\sigma = 9.6\%$.

(a) What is the sampling distribution of the mean percent $\bar{x}$ that the sample spends on housing if the null hypothesis is true? Sketch the density curve of the sampling distribution. (*Hint:* Sketch a normal curve first, then mark the axis using what you know about locating μ and σ on a normal curve.)

(b) Suppose that the study finds $\bar{x} = 30.2\%$ for the 40 households in the sample. Mark this point on the axis in your sketch. Then suppose that the study result is $\bar{x} = 27.6\%$. Mark this point on your sketch. Referring to your sketch, explain in simple language why one result is good evidence that average Cleveland spending on housing is less than 31% and the other result is not.

(c) Shade the area under the curve that gives the *P*-value for the result $\bar{x} = 30.2\%$. (Note that we are looking for evidence that spending is *less* than the null hypothesis states.)

More detail: stating hypotheses

We will now look in more detail at some aspects of significance tests. The first step in a test of significance is to state a claim that we will try to find evidence *against*.

NULL HYPOTHESIS H_0

> The statement being tested in a test of significance is called the **null hypothesis.** The test of significance is designed to assess the strength of the evidence against the null hypothesis. Usually the null hypothesis is a statement of "no effect" or "no difference."

The alternative hypothesis H_a is the claim about the population that we are trying to find evidence *for*. In Example 10.7, we were seeking evidence of a loss in sweetness. The null hypothesis says "no loss" on the average in a large population of tasters. The alternative hypothesis says "there is a loss." So the hypotheses are

$$H_0: \mu = 0$$
$$H_a: \mu > 0$$

one-sided alternative

This alternative hypothesis is **one-sided** because we are interested only in deviations from the null hypothesis in one direction.

EXAMPLE 10.8

Does the job satisfaction of assembly workers differ when their work is machine-paced rather than self-paced? One study chose 28 subjects at random from a group of women who worked at assembling electronic devices. Half of the subjects were assigned at random to each of two groups. Both groups did similar assembly work, but one work setup allowed workers to pace themselves and the other featured an assembly line that moved at fixed time intervals so that the workers were paced by machine. After two weeks, all subjects took the Job Diagnosis Survey (JDS), a test of job satisfaction. Then they switched work setups, and took the JDS again after two more weeks. This is another matched pairs design. The response variable is the difference in JDS scores, self-paced minus machine-paced.[5]

The parameter of interest is the mean μ of the differences in JDS scores in the population of all female assembly workers. The null hypothesis says that there is no difference between self-paced and machine-paced work, that is,

$$H_0: \mu = 0$$

The authors of the study wanted to know if the two work conditions have different levels of job satisfaction. They did not specify the direction of the difference. The alternative hypothesis is therefore **two-sided,**

two-sided alternative

$$H_a: \mu \neq 0$$

Hypotheses always refer to some population, not to a particular outcome. For this reason, always state H_0 and H_a in terms of population parameters. Because H_a expresses the effect that we hope to find evidence *for*, it is often easier to begin by stating H_a and then set up H_0 as the statement that the hoped-for effect is not present.

It is not always easy to decide whether H_a should be one-sided or two-sided. In Example 10.8, the alternative H_a: $\mu \neq 0$ is two-sided. That is, it simply says there is a difference in job satisfaction without specifying the direction of the difference. The alternative H_a: $\mu > 0$ in the taste test example is one-sided. Because colas can only lose sweetness in storage, we are interested only in detecting an upward shift in μ. The alternative hypothesis should express the hopes or suspicions we bring to the data. It is cheating to first look at the data and then frame H_a to fit what the data show. Thus the fact that the workers in the study of Example 10.8 were more satisfied with self-paced work should not influence our choice of H_a. If you do not have a specific direction firmly in mind in advance, use a two-sided alternative.

The choice of the hypotheses in Example 10.7 as

$$H_0: \mu = 0$$

$$H_a: \mu > 0$$

deserves a final comment. The cola maker is not concerned with the possibility that the tasters may detect a gain in sweetness, indicated by a negative mean loss μ. However, we can allow for the possibility that μ is less than zero by including this case in the null hypothesis. Then we would write

$$H_0: \mu \leq 0$$

$$H_a: \mu > 0$$

This statement is logically satisfying because the hypotheses account for all possible values of μ. However, only the parameter value in H_0 that is closest to H_a influences the form of the test in all common significance testing situations. We will therefore take H_0 to be the simpler statement that the parameter has a specific value, in this case H_0: $\mu = 0$.

EXERCISES

Each of the following situations calls for a significance test for a population mean μ. State the null hypothesis H_0 and the alternative hypothesis H_a in each case.

10.29 The diameter of a spindle in a small motor is supposed to be 5 mm. If the spindle is either too small or too large, the motor will not work properly. The manufacturer measures the diameter in a sample of motors to determine whether the mean diameter has moved away from the target.

10.30   Census Bureau data show that the mean household income in the area served by a shopping mall is $42,500 per year. A market research firm questions shoppers at the mall. The researchers suspect the mean household income of mall shoppers is higher than that of the general population.

10.31   The examinations in a large accounting class are scaled after grading so that the mean score is 50. The professor thinks that one teaching assistant is a poor teacher and suspects that his students have a lower mean score than the class as a whole. The TA's students this semester can be considered a sample from the population of all students in the course, so the professor compares their mean score with 50.

10.32   Last year, your company's service technicians took an average of 2.6 hours to respond to trouble calls from business customers who had purchased service contracts. Do this year's data show a different average response time?

More detail: *P*-values and statistical significance

test statistic

A significance test uses data in the form of a **test statistic.** The test statistic is usually based on a statistic that estimates the parameter that appears in the hypotheses. In our examples, the parameter is μ and the test statistic is the sample mean $\bar{x}$.

A test of significance assesses the evidence against the null hypothesis in terms of probability. If the test statistic falls far from the value suggested by the null hypothesis in the direction specified by the alternative hypothesis, it is good evidence against H_0 and in favor of H_a. To describe how strong the evidence is, find the probability of getting an outcome *as extreme or more extreme than the actually observed outcome.* "Extreme" means "far from what we would expect if H_0 were true." The direction or directions that count as "far from what we would expect" are determined by the alternative hypothesis H_a.

P-VALUE

The probability, computed assuming that H_0 is true, that the test statistic would take a value as extreme or more extreme than that actually observed is called the **P-value** of the test. The smaller the P-value is, the stronger is the evidence against H_0 provided by the data.

Computer software that carries out tests of significance usually calculates the P-value for us. In some cases we can find P-values from our knowledge of sampling distributions.

EXAMPLE 10.9

In Example 10.7 the observations are an SRS of size $n = 10$ from a normal population with $\sigma = 1$. The observed mean sweetness loss for one cola was $\bar{x} = .3$. The P-value for testing

$$H_0\colon \mu = 0$$

$$H_a\colon \mu > 0$$

is therefore

$$P(\bar{x} \geq .3)$$

calculated assuming that H_0 is true. When H_0 is true, $\bar{x}$ has the normal distribution with mean 0 and standard deviation

$$\frac{\sigma}{\sqrt{n}} = \frac{1}{\sqrt{10}} = .316$$

Find the P-value by a normal probability calculation. Start by drawing a picture that shows the P-value as an area under a normal curve. Figure 10.10 is the picture for this example. Then standardize $\bar{x}$ to get a standard normal Z and use Table A,

$$P(\bar{x} \geq .3) = P\left(\frac{\bar{x} - 0}{.316} \geq \frac{.3 - 0}{.316}\right)$$

$$= P(Z \geq .95)$$

$$= 1 - .8289 = .1711$$

This is the value that was reported in Example 10.7.

---

significance level

We sometimes take one final step to assess the evidence against H_0. We can compare the P-value with a fixed value that we regard as decisive. This amounts to announcing in advance how much evidence against H_0 we will insist on. The decisive value of P is called the **significance level**. We write it as α, the Greek letter alpha. If we choose $\alpha = .05$, we are requiring that the data give evidence against H_0 so strong that it would happen no more than 5% of the time (1 time in 20) when H_0 is true. If we choose $\alpha = .01$, we are insisting on stronger evidence against H_0, evidence so strong that it would appear only 1% of the time (1 time in 100) if H_0 is in fact true.

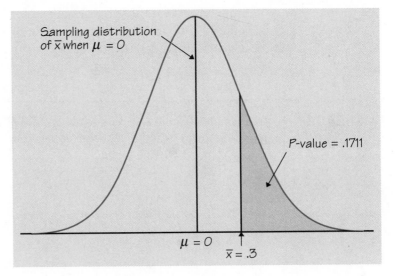

Sampling distribution
of $\bar{x}$ when $\mu = 0$

P-value = .1711

$\mu = 0$

$\bar{x} = .3$

FIGURE 10.10   The P-value for the one-sided test in Example 10.9.

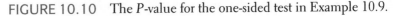

STATISTICAL SIGNIFICANCE

If the P-value is as small or smaller than α, we say that the data are statistically significant at level α.

"Significant" in the statistical sense does not mean "important." It means simply "not likely to happen just by chance." The significance level α makes "not likely" more exact. Significance at level 0.01 is often expressed by the statement "The results were significant ($P < .01$)." Here P stands for the P-value. The P-value is more informative than a statement of significance, because we can then assess significance at any level we choose. For example, a result with $P = .03$ is significant at the $\alpha = .05$ level, but not significant at the $\alpha = .01$ level.

EXERCISES

10.33   Return to Exercise 10.27. Starting from the picture you drew there, calculate the P-values for both $\bar{x} = 118.6$ and $\bar{x} = 125.7$. The two P-values express in numbers the comparison you made informally in Exercise 10.27.

10.34 Return to Exercise 10.28. Starting from the picture you drew there, calculate the P-values for both $\bar{x} = 30.2\%$ and $\bar{x} = 27.6\%$. The two P-values express in numbers the comparison you made informally in Exercise 10.28.

10.35 Weekly sales of regular ground coffee at a supermarket have in the recent past varied according to a normal distribution with mean $\mu = 354$ units per week and standard deviation $\sigma = 33$ units. The store reduces the price by 5%. Sales in the next three weeks are 405, 378, and 411 units. Is this good evidence that average sales are now higher? The hypotheses are

$$H_0: \mu = 354$$
$$H_a: \mu > 354$$

Assume that the standard deviation of the population of weekly sales remains $\sigma = 33$.

(a) Find the value of the test statistic $\bar{x}$.

(b) Sketch the normal curve for the sampling distribution of $\bar{x}$ when H_0 is true. Shade the area that represents the P-value for the observed outcome.

(c) Calculate the P-value.

(d) Is the result statistically significant at the $\alpha = .05$ level? Is it significant at the $\alpha = .01$ level? Do you think there is convincing evidence that mean sales are higher?

10.36 A study of the pay of corporate chief executive officers (CEOs) examined the increase in cash compensation of the CEOs of 104 companies, adjusted for inflation, in a recent year. The mean increase in real compensation was $\bar{x} = 6.9\%$ and the standard deviation of the increases was $s = 55\%$. Is this good evidence that the mean real compensation μ of all CEOs increased that year? The hypotheses are

$$H_0: \mu = 0 \quad \text{(no increase)}$$
$$H_a: \mu > 0 \quad \text{(an increase)}$$

Because the sample size is large, the sample s is close to the population σ, so take $\sigma = 55\%$.

(a) Sketch the normal curve for the sampling distribution of $\bar{x}$ when H_0 is true. Shade the area that represents the P-value for the observed outcome $\bar{x} = 6.9\%$.

(b) Calculate the P-value.

(c) Is the result significant at the $\alpha = .05$ level? Do you think the study gives strong evidence that the mean compensation of all CEOs went up?

10.37   A social psychologist reports that "in our sample, ethnocentrism was significantly higher $(P < .05)$ among church attenders than among non-attenders." Explain what this means in language understandable to someone who knows no statistics. Do not use the word "significance" in your answer.

10.38   The financial aid office of a university asks a sample of students about their employment and earnings. The report says that "for academic year earnings, a significant difference $(P = .038)$ was found between the sexes, with men earning more on the average. No difference $(P = .476)$ was found between the earnings of black and white students." Explain both of these conclusions, for the effects of sex and of race on mean earnings, in language understandable to someone who knows no statistics.[6]

Tests for a population mean

Although the reasoning of significance testing isn't simple, carrying out a test is. There are three steps:

1. State the hypotheses.
2. Calculate the test statistic.
3. Find the P-value.

Once you have stated your hypotheses and identified the proper test, you or your computer can do Steps 2 and 3 by following a recipe. We now develop the recipe for one significance test, the one we have used in our examples.

We have an SRS of size n drawn from a normal population with unknown mean μ. We want to test the hypothesis that μ has a specified value. Call the specified value μ_0. The null hypothesis is

$$H_0: \mu = \mu_0$$

The test is based on the sample mean $\bar{x}$. Because normal calculations require standardized variables, we will use as our test statistic the *standardized* sample mean

$$z = \frac{\bar{x} - \mu_0}{\sigma/\sqrt{n}}$$

z test statistic This **z test statistic** has the standard normal distribution when H_0 is true. If the alternative is one-sided on the high side

$$H_a: \mu > \mu_0$$

then the P-value is the probability that a standard normal variable Z takes a value at least as large as the observed z. That is,

$$P = P(Z \geq z)$$

Example 10.9 calculates this P-value for the cola taste test. There, $\mu_0 = 0$, the standardized sample mean was $z = .95$, and the P-value was $P(Z \geq .95) = .1711$. Similar reasoning applies when the alternative hypothesis states that the true μ lies below the hypothesized μ_0 (one-sided).

When H_a states that μ is simply unequal to μ_0 (two-sided), values of z away from zero in either direction count against the null hypothesis. The P-value is the probability that a standard normal Z is at least as far from zero *in either direction* as the observed z.

EXAMPLE 10.10 Suppose that the z test statistic for a two-sided test is $z = 1.7$. The two-sided P-value is the probability that $Z \leq -1.7$ or $Z \geq 1.7$. Figure 10.11 shows this probability as areas under the standard normal curve. Because the standard normal distribution is symmetric, we can calculate this probability by finding $P(Z \geq 1.7)$ and *doubling* it.

$$P(Z \leq -1.7 \text{ or } Z \geq 1.7) = 2P(Z \geq 1.7) = 2(1 - .9554) = .0892$$

We would make exactly the same calculation if we observed $z = -1.7$. It is the absolute value $|z|$ that matters, not whether z is positive or negative.

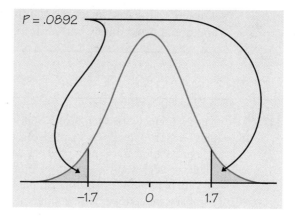

FIGURE 10.11   The P-value for the two-sided test in Example 10.10 is the sum of the area above 1.7 and below -1.7.

z TEST FOR A POPULATION MEAN

To test the hypothesis H_0: $\mu = \mu_0$ based on an SRS of size n from a population with unknown mean μ and known standard deviation σ, compute the z **test statistic**

$$z = \frac{\bar{x} - \mu_0}{\sigma/\sqrt{n}}$$

In terms of a variable Z having the standard normal distribution, the P-value for a test of H_0 against

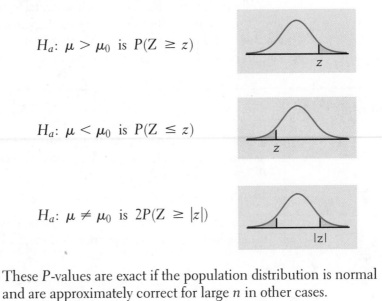

H_a: $\mu > \mu_0$ is $P(Z \geq z)$

H_a: $\mu < \mu_0$ is $P(Z \leq z)$

H_a: $\mu \neq \mu_0$ is $2P(Z \geq |z|)$

These P-values are exact if the population distribution is normal and are approximately correct for large n in other cases.

EXAMPLE 10.11

The National Center for Health Statistics reports that the mean systolic blood pressure for males 35 to 44 years of age is 128 and the standard deviation in this population is 15. The medical director of a large company looks at the medical records of 72 executives in this age group and finds that the mean systolic blood pressure in this sample is $\bar{x} = 126.07$. Is this evidence that the company's executives have a different mean blood pressure from the general population? As usual in this chapter, we make the unrealistic assumption that we know the population standard deviation. Assume that executives have the same $\sigma = 15$ as the general population of middle-aged males.

Step 1: Hypotheses. The null hypothesis is "no difference" from the national mean $\mu_0 = 128$. The alternative is two-sided, because the medical director did not have a particular direction in mind before examining the data. So the hypotheses about the unknown mean μ of the executive population are

$$H_0:\ \mu = 128$$
$$H_a:\ \mu \neq 128$$

Step 2: Test statistic. The z test statistic is

$$z = \frac{\bar{x} - \mu_0}{\sigma/\sqrt{n}}$$
$$= \frac{126.07 - 128}{15/\sqrt{72}}$$
$$= -1.09$$

Step 3: P-value. You should still draw a picture to help find the P-value, but now you can sketch the standard normal curve with the observed value of z. Figure 10.12 shows that the P-value is the probability that a standard normal variable Z takes a value at least 1.09 away from zero. From Table A we find that this probability is

$$P = 2P(Z \geq 1.09) = 2(1 - .8621) = .2758$$

Conclusion: More than 27% of the time, an SRS of size 72 from the general male population would have a mean blood pressure at least as far from 128 as that of the executive sample. The observed $\bar{x} = 126.07$ is therefore not good evidence that executives differ from other men.

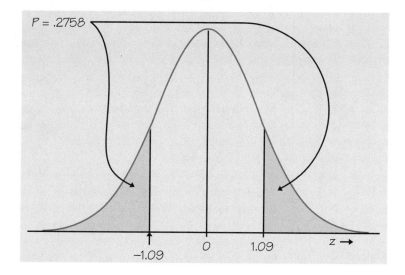

FIGURE 10.12   The P-value for the two-sided test in Example 10.11.

The z test assumes that the 72 executives in the sample are an SRS from the population of all middle-aged male executives in the company. We should check this assumption by asking how the data were produced. If medical records are available only for executives with recent medical problems, for example, the data are of little value for our purpose. It turns out that all executives are given a free annual medical exam, and that the medical director selected 72 exam results at random.

The data in Example 10.11 do *not* establish that the mean blood pressure μ for this company's executives is 128. We sought evidence that μ differed from 128 and failed to find convincing evidence. That is all we can say. No doubt the mean blood pressure of the entire executive population is not exactly equal to 128. A large enough sample would give evidence of the difference, even if it is very small. Tests of significance assess the evidence *against* H_0. If the evidence is strong, we can confidently reject H_0 in favor of the alternative. Failing to find evidence against H_0 means only that the data are consistent with H_0, not that we have clear evidence that H_0 is true.

EXAMPLE 10.12

In a discussion of the education level of the American workforce, someone says, "The average young person can't even balance a checkbook." The National Assessment of Educational Progress (NAEP) says that a score of 275 or higher on its quantitative test (see Example 10.2) reflects the skill needed to balance a checkbook. The NAEP random sample of 840 young men had a mean score of $\bar{x} = 272$, a bit below the checkbook-balancing level. Is this sample result good evidence that the mean for *all* young men is less than 275? As in Example 10.2, assume that $\sigma = 60$.

Step 1: Hypotheses. The hypotheses are

$$H_0: \mu = 275$$
$$H_a: \mu < 275$$

Step 2: Test statistic. The z statistic is

$$z = \frac{\bar{x} - \mu_0}{\sigma/\sqrt{n}} = \frac{272 - 275}{60/\sqrt{840}}$$
$$= -1.45$$

Step 3: P-value. Because H_a is one-sided on the low side, small values of z count against H_0. Figure 10.13 illustrates the P-value. Using Table A, we find that

$$P = P(Z \le -1.45)$$
$$= .0735$$

Conclusion: A mean score as low as 272 would occur about 7 times in 100 samples if the population mean were 275. This is modest evidence that the mean NAEP score for all young men is less than 275.

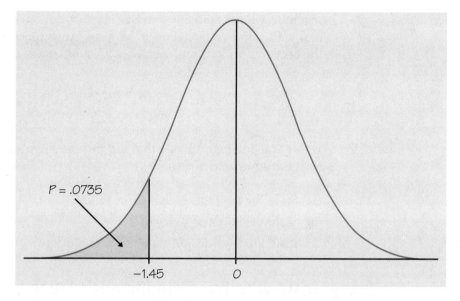

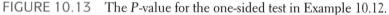

FIGURE 10.13   The *P*-value for the one-sided test in Example 10.12.

EXERCISES

10.39   Here are measurements (in millimeters) of a critical dimension on a sample of automobile engine crankshafts:

224.120	224.001	224.017	223.982
223.960	224.089	223.987	223.976
224.098	224.057	223.913	223.999
223.989	223.902	223.961	223.980

The manufacturing process is known to vary normally with standard deviation $\sigma = .060$ mm. The process mean is supposed to be 224 mm. Do these data give evidence that the process mean is not equal to the target value 224 mm?

(a)   State the H_0 and H_a that you will test.

(b)   Calculate the test statistic z.

(c)   Give the *P*-value of the test. Are you convinced that the process mean is not 224 mm?

10.40   Bottles of a popular cola are supposed to contain 300 milliliters (ml) of cola. There is some variation from bottle to bottle because the filling machinery is not perfectly precise. The distribution of the contents is normal with standard deviation $\sigma = 3$ ml. An inspector who suspects that the bottler is underfilling measures the contents of six bottles. The results are

$$299.4 \quad 297.7 \quad 301.0$$
$$298.9 \quad 300.2 \quad 297.0$$

Is this convincing evidence that the mean contents of cola bottles is less than the advertised 300 ml?

(a) State the hypotheses that you will test.

(b) Calculate the test statistic.

(c) Find the *P*-value and state your conclusion.

Tests with fixed significance level

Sometimes we demand a specific degree of evidence in order to reject the null hypothesis. A level of significance α says how much evidence we require. In terms of the *P*-value, the outcome of a test is significant at level α if $P \leq \alpha$. Significance at any level is easy to assess once you have the *P*-value. When you do not use statistical software, the *P*-value can be difficult to calculate. Fortunately, you can decide whether a result is statistically significant without calculating P. The following example illustrates how to assess significance at a fixed level α by using a table of critical values, the same table used to obtain confidence intervals.

EXAMPLE 10.13	In Example 10.12, we examined whether the mean NAEP quantitative score of young men is less than 275. The hypotheses are

$$H_0: \mu = 275$$
$$H_a: \mu < 275$$

The z statistic takes the value $z = -1.45$. Is the evidence against H_0 statistically significant at the 5% level?

To determine significance, we need only compare the observed $z = -1.45$ with the 5% critical value $z^* = 1.645$ from Table C. Because $z = -1.45$ is *not* farther from 0 than -1.645, it is *not* significant at level $\alpha = .05$.

Here is why. The *P*-value is the area to the left of -1.45 under the standard normal curve, shown in Figure 10.13. The result $z = -1.45$ is significant at the 5% level exactly when this area is no more than 5%. The area to the left of the critical value -1.645 is exactly 5%. So -1.645 separates values of z that are significant from those that are not. Figure 10.14 illustrates the procedure.

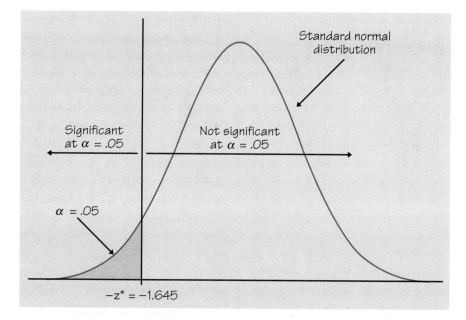

FIGURE 10.14   Deciding whether a z statistic is significant at the $\alpha = .05$
level in the one-sided test of Example 10.13.

FIXED SIGNIFICANCE LEVEL z TESTS FOR A POPULATION MEAN

To test the hypothesis $H_0: \mu = \mu_0$ based on an SRS of size n
from a population with unknown mean μ and known standard
deviation σ, compute the z test statistic

$$z = \frac{\bar{x} - \mu_0}{\sigma/\sqrt{n}}$$

Reject H_0 at significance level α against a one-sided alternative

$$H_a: \mu > \mu_0 \text{ if } z \geq z^*$$

$$H_a: \mu < \mu_0 \text{ if } z \leq -z^*$$

where z^* is the upper α critical value from Table C. Reject H_0 at
significance level α against a two-sided alternative

$$H_a: \mu \neq \mu_0 \text{ if } |z| \geq z^*$$

where z^* is the upper $\alpha/2$ critical value from Table C.

EXAMPLE 10.14

The analytical laboratory of Example 10.4 is asked to evaluate the claim that the concentration of the active ingredient in a specimen is 0.86%. The lab makes 3 repeated analyses of the specimen. The mean result is $\bar{x} = .8404$. The true concentration is the mean μ of the population of all analyses of the specimen. The standard deviation of the analysis process is known to be $\sigma = .0068$. Is there significant evidence at the 1% level that $\mu \neq .86$?

Step 1: Hypotheses. The hypotheses are

$$H_0: \mu = .86$$
$$H_a: \mu \neq .86$$

Step 2: Test statistic. The z statistic is

$$z = \frac{.8404 - .86}{.0068/\sqrt{3}}$$
$$= -4.99$$

Step 3: Significance. Because the alternative is two-sided, we compare $|z| = 4.99$ with the $\alpha/2 = .005$ critical value from Table C. This critical value is $z^* = 2.576$. Figure 10.15 illustrates the values of z that are statistically significant. Because $|z| > 2.576$, we reject the null hypothesis and conclude (at the 1% significance level) that the concentration is not as claimed.

The observed result in Example 10.14 was $z = -4.99$. The conclusion that this result is significant at the 1% level does not tell the whole

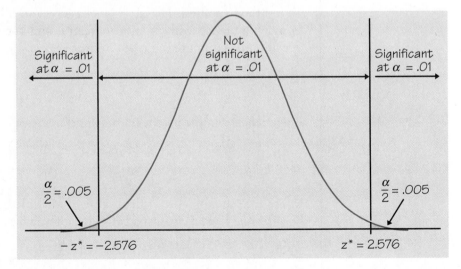

$\frac{\alpha}{2} = .005$

Significant at $\alpha = .01$ Not significant at $\alpha = .01$ Significant at $\alpha = .01$

$\frac{\alpha}{2} = .005$

$-z^* = -2.576$ $z^* = 2.576$

FIGURE 10.15   Deciding whether a z statistic is significant at the $\alpha = .01$ level in the two-sided test of Example 10.14.

story. The observed z is far beyond the 1% critical value, and the evidence against H_0 is far stronger than 1% significance suggests. The P-value

$$2P(Z \geq 4.99) = .0000006$$

gives a better sense of how strong the evidence is. *The P-value is the smallest level α at which the data are significant.* Knowing the P-value allows us to assess significance at any level.

Tables of critical values such as Table C allow us to estimate P-values without a probability calculation. In Example 10.14, compare the observed $z = -4.99$ with *all* of the normal critical values in the bottom row of the table. It is beyond even 3.291, the critical value for $P = .0005$. So we know that for the two-sided test, $P < .001$. In Example 10.13, $z = -1.45$ lies between the 0.05 and 0.10 entries in the table. So the P-value for the one-sided test lies between 0.05 and 0.10. This approximation is accurate enough for most purposes.

Because the practice of statistics almost always employs computer software that calculates P-values automatically, the use of tables of critical values is becoming outdated. The usual tables of critical values (such as Table C) appear in this book for learning purposes and to rescue students without good computing facilities.

EXERCISES

10.41 A computer has a random number generator designed to produce random numbers that are uniformly distributed on the interval from 0 to 1. If this is true, the numbers generated come from a population with $\mu = .5$ and $\sigma = .2887$. A command to generate 100 random numbers gives outcomes with mean $\bar{x} = .4365$. Assume that the population σ remains fixed. We want to test

$$H_0: \mu = .5$$

$$H_a: \mu \neq .5$$

(a) Calculate the value z of the z test statistic.

(b) Is the result significant at the 5% level ($\alpha = .05$)?

(c) Is the result significant at the 1% level ($\alpha = .01$)?

10.42 The `rand` function on the TI-83 (MATH / PRB / 1:rand) generates a pseudorandom real number in the interval $[0, 1)$—that is, in the interval $0 \leq X < 1$. The command `rand(100)` generates 100 random real numbers in the interval $[0, 1)$. Describe how you would use the TI-83 to carry out the experiment described in the previous exercise, using the sample mean calculated from your 100 calculator values. (*Hint:* store the 100 values in a list.) As in Exercise 10.41, take $\sigma = .2887$. Then carry out your plan and answer the questions in Exercise 10.41.

10.43   To determine whether the mean nicotine content of a brand of cigarettes is greater than the advertised value of 1.4 milligrams, a health advocacy group tests

$$H_0:\ \mu = 1.4$$

$$H_a:\ \mu > 1.4$$

The calculated value of the test statistic is $z = 2.42$.

(a) Is the result significant at the 5% level?

(b) Is the result significant at the 1% level?

Tests from confidence intervals

The calculation in Example 10.14 for a 1% significance test is very similar to that in Example 10.4 for a 99% confidence interval. In fact, a two-sided test at significance level α can be carried out directly from a confidence interval with confidence level $C = 1 - \alpha$.

CONFIDENCE INTERVALS AND TWO-SIDED TESTS

A level α two-sided significance test rejects a hypothesis $H_0:\ \mu = \mu_0$ exactly when the value μ_0 falls outside a level $1 - \alpha$ confidence interval for μ.

EXAMPLE 10.15   The 99% confidence interval for μ in Example 10.4 is

$$\bar{x} \pm z^* \frac{\sigma}{\sqrt{n}} = .8404 \pm .0101$$

$$= (.8303, .8505)$$

The hypothesized value $\mu_0 = 0.86$ in Example 10.14 falls outside this confidence interval, so we reject

$$H_0:\ \mu = .86$$

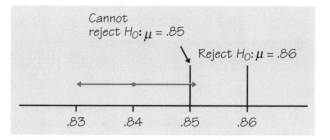

FIGURE 10.16   Values of μ falling outside a 99% confidence interval can be rejected at the 1% significance level. Values falling inside the interval cannot be rejected.

at the 1% significance level. On the other hand, we cannot reject

$$H_0: \mu = .85$$

at the 1% level in favor of the two-sided alternative $H_a: \mu \neq .85$, because 0.85 lies inside the 99% confidence interval for μ. Figure 10.16 illustrates both cases.

EXERCISES

10.44   Radon is a colorless, odorless gas that is naturally released by rocks and soils and may concentrate in tightly closed houses. Because radon is slightly radioactive, there is some concern that it may be a health hazard. Radon detectors are sold to homeowners worried about this risk, but the detectors may be inaccurate. University researchers placed 12 detectors in a chamber where they were exposed to 105 picocuries per liter of radon over 3 days. Here are the readings given by the detectors:[7]

$$\begin{array}{cccccc}
91.9 & 97.8 & 111.4 & 122.3 & 105.4 & 95.0 \\
103.8 & 99.6 & 96.6 & 119.3 & 104.8 & 101.7
\end{array}$$

Assume (unrealistically) that you know that the standard deviation of readings for all detectors of this type is $\sigma = 9$.

(a)   Give a 90% confidence interval for the mean reading μ for this type of detector.

(b)   Is there significant evidence at the 10% level that the mean reading differs from the true value 105? State hypotheses and base a test on your confidence interval from (a).

SUMMARY

A **test of significance** is intended to assess the evidence provided by data against a **null hypothesis** H_0 in favor of an **alternative hypothesis** H_a.

The hypotheses are stated in terms of population parameters. Usually H_0 is a statement that no effect is present, and H_a says that a parameter differs from its null value in a specific direction (**one-sided alternative**) or in either direction (**two-sided alternative**).

The essential reasoning of a significance test is as follows. Suppose for the sake of argument that the null hypothesis is true. If we repeated our data production many times, would we often get data as inconsistent with H_0 as the data we actually have? If the data are unlikely when H_0 is true, they provide evidence against H_0.

A test is based on a **test statistic**. The **P-value** is the probability, computed supposing H_0 to be true, that the test statistic will take a value at least as extreme as that actually observed. Small P-values indicate strong evidence against H_0. Calculating P-values requires knowledge of the sampling distribution of the test statistic when H_0 is true.

If the P-value is as small or smaller than a specified value α, the data are **statistically significant** at significance level α.

Significance tests for the hypothesis H_0: $\mu = \mu_0$ concerning the unknown mean μ of a population are based on the z **statistic**

$$z = \frac{\bar{x} - \mu_0}{\sigma/\sqrt{n}}$$

The z test assumes an SRS of size n, known population standard deviation σ, and either a normal population or a large sample. P-values are computed from the normal distribution (Table A). Fixed α tests use the table of standard normal **critical values** (bottom row of Table C).

SECTION 10.2 EXERCISES

10.45   The job satisfaction study of Example 10.8 measured the JDS job satisfaction score of 28 female assemblers doing both self-paced and machine-paced work. The parameter μ is the mean amount by which the self-paced score exceeds the machine-paced score in the population of all such workers. Scores are normally distributed. The population standard deviation is $\sigma = .60$. The hypotheses are

$$H_0: \mu = 0$$
$$H_a: \mu \neq 0$$

(a) What is the sampling distribution of the mean JDS score $\bar{x}$ for 28 workers if the null hypothesis is true? Sketch the density curve of this distribution. (*Hint:* Sketch a normal curve first, then mark the axis using what you know about locating μ and σ on a normal curve.)

(b) Suppose that the study had found $\bar{x} = .09$. Mark this point on the axis in your sketch. In fact, the study found $\bar{x} = .27$ for these 28 workers. Mark this point on your sketch. Referring to your sketch, explain in simple language why one result is good evidence that H_0 is not true, and why the other is not.

(c) Make another copy of your sketch. Shade the area under the curve that gives the P-value for the result $\bar{x} = .09$. Then calculate this P-value. (Note that H_a is two-sided.)

(d) Calculate the P-value for the result $\bar{x} = .27$ also. The two P-values express your explanation in (b) in numbers.

10.46 The mean area of the several thousand apartments in a new development is advertised to be 1250 square feet. A tenant group thinks that the apartments are smaller than advertised. They hire an engineer to measure a sample of apartments to test their suspicion. What are the null hypothesis H_0 and alternative hypothesis H_a?

10.47 Experiments on learning in animals sometimes measure how long it takes mice to find their way through a maze. The mean time is 18 seconds for one particular maze. A researcher thinks that a loud noise will cause the mice to complete the maze faster. She measures how long each of 10 mice takes with a noise as stimulus. What are the null hypothesis H_0 and alternative hypothesis H_a?

10.48 Cobra Cheese Company buys milk from several suppliers. Cobra suspects that some producers are adding water to their milk to increase their profits. Excess water can be detected by measuring the freezing point of the milk. The freezing temperature of natural milk varies normally, with mean $\mu = -.545°$ Celsius (C) and standard deviation $\sigma = .008°$ C. Added water raises the freezing temperature toward $0°$ C, the freezing point of water. Cobra's laboratory manager measures the freezing temperature of five consecutive lots of milk from one producer. The mean measurement is $\bar{x} = -.538°$ C. Is this good evidence that the producer is adding water to the milk? State hypotheses, carry out the test, give the P-value, and state your conclusion.

10.49 There are other z statistics that we have not yet studied. You can use Table C to assess the significance of any z statistic. A study compares

American-Japanese joint ventures in which the U.S. company is larger than its Japanese partner with joint ventures in which the U.S. company is smaller. One variable measured is the excess returns earned by shareholders in the American company. The null hypothesis is "no difference" between the means for the two populations. The alternative hypothesis is two-sided. The value of the test statistic is $z = -1.37$.

(a) Is this result significant at the 5% level?

(b) Is the result significant at the 10% level?

10.50   Use Table C to find the approximate P-value for the test in Exercise 10.41 without doing a probability calculation. That is, find from the table two numbers that contain the P-value between them.

10.51   Use Table C to find the approximate P-value for the test in Exercise 10.43. That is, between what two numbers obtained from the table must the P-value lie?

10.52   Between what values from Table C does the P-value for the outcome $z = -1.37$ in Exercise 10.49 lie? (Remember that H_a is two-sided.) Calculate the P-value using Table A, and verify that it lies between the values you found from Table C.

10.53   Market pioneers, companies that are among the first to develop a new product or service, tend to have higher market shares than latecomers to the market. What accounts for this advantage? Here is an excerpt from the conclusions of a study of a sample of 1209 manufacturers of industrial goods:

> Can patent protection explain pioneer share advantages? Only 21% of the pioneers claim a significant benefit from either a product patent or a trade secret. Though their average share is two points higher than that of pioneers without this benefit, the increase is not statistically significant ($z = 1.13$). Thus, at least in mature industrial markets, product patents and trade secrets have little connection to pioneer share advantages.[8]

Find the P-value for the given z. Then explain to someone who knows no statistics what "not statistically significant" in the study's conclusion means. Why does the author conclude that patents and trade secrets don't help, even though they contributed 2 percentage points to average market share?

10.54   The cigarette industry has adopted a voluntary code requiring that models appearing in its advertising must appear to be at least 25 years old. Studies have shown, however, that consumers think many of the models are younger. Here is a quote from a study that asked whether different brands of cigarettes use models that appear to be of different ages.[9]

The ANCOVA revealed that the brand variable is highly significant (P < .001), indicating that the average perceived age of the models is not equal across the 12 brands. As discussed previously, certain brands such as Lucky Strike Lights, Kool Milds, and Virginia Slims tended to have younger models . . .

ANCOVA is an advanced statistical technique, but significance and P-values have their usual meaning. Explain to someone who knows no statistics what "highly significant ($P < .001$)" means and why this is good evidence of differences among all advertisements of these brands even though the subjects saw only a sample of ads.

10.55 Explain in plain language why a significance test that is significant at the 1% level must always be significant at the 5% level.

10.56 Asked to explain the meaning of "statistically significant at the $\alpha = .05$ level," a student says: "This means that the probability that the null hypothesis is true is less than .05." Is this explanation correct? Why or why not?

10.57 The TI-83 can be used to conduct z tests of inference, using either data stored in a list or summary statistics. First, press STAT / TESTS / 1: Z-Test . . . to access the Z-Test screen, as shown. In the NAEP testing of Example 10.12, for example, you would enter 275 for the null hypothesized mean μ_0 Next enter 60 for σ, 272 for $\bar{x}$, and 840 for n. Select $<\mu_0$ for the alternative hypothesis, and choose "Calculate." The next screen shows the results of the z test: $z = -1.449$ and the P-value is 0.0736. If you select "Draw" instead of "Calculate," the left critical area under the normal curve will be shaded and the z-value and P-value will be displayed.

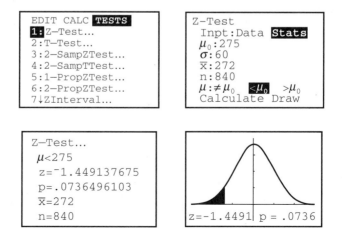

(a) Use your TI-83 in this way to find the z-value and P-value for the true value of the mean μ in Example 10.14. Check your calculator findings with the numbers given in Example 10.14.

If you have data, enter the data into any list and then specify a z test. In Example 10.4, for example, you would enter the three specimen concentrations into list L_1. Then from the Z-Test screen, specify "Data" and then enter the values shown in the screen to the right. The last step is to choose "Calculate" or "Draw." If you choose "Calculate," the z-test results screen is displayed, showing the z test statistic and the P-value, along with some other key numbers. If you select "Draw," the standard normal curve is displayed with the critical areas in both tails shaded (because the alternative hypothesis is two-sided). Notice here that since the test statistic $z = -4.98$ is so extreme, the shaded areas are not visible, and the P-value (which we know to be 0.0000006 from Example 10.14) is simply reported as $P = 0$.

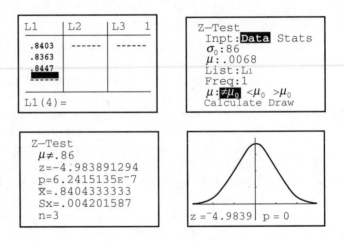

(b) Use your TI-83 to perform the z test of significance described in Exercise 10.39.

(c) Use your TI-83 to perform the z test of significance described in Exercise 10.40.

10.3 USING SIGNIFICANCE TESTS

Carrying out a test of significance is often quite simple, especially if you use a fixed significance level α or get the P-value effortlessly from a computer. Using tests wisely is not so simple. Some hesitation about the unthinking use of significance tests is a sign of statistical maturity.

Significance tests are widely used in reporting the results of research in many fields of applied science and in industry. Some products (such as pharmaceuticals) require significant evidence of effectiveness and safety. Courts inquire about statistical significance in hearing class action discrimination cases and in other legal proceedings. Marketers want to know whether a new ad campaign significantly outperforms the old one, and medical researchers want to know whether a new therapy performs significantly better. In all these uses, statistical significance is valued because it points to an effect that is unlikely to occur simply by chance. Here are some points to keep in mind when using or interpreting significance tests.

Choosing a level of significance

The purpose of a test of significance is to give a clear statement of the degree of evidence provided by the sample against the null hypothesis. The P-value does this. But sometimes you will make some decision or take some action if your evidence reaches a certain standard. A level of significance α sets such a standard. Perhaps you will publish a research finding if the effect is significant at the $\alpha = .01$ level. Or perhaps your company will lose a lawsuit alleging racial discrimination if the percent of blacks hired is significantly below the percent of blacks in the pool of potential employees at the $\alpha = .05$ level. Courts have in fact tended to accept this standard in discrimination cases.[10]

Making a decision is different in spirit from testing significance, though the two are often mixed in practice. Choosing a level α in advance makes sense if you must make a decision, but not if you wish only to describe the strength of your evidence. Using tests with fixed α for decision making is discussed at greater length in Section 10.4.

If you do use a fixed α significance test to make a decision, choose α by asking how much evidence is required to reject H_0. This depends mainly on two circumstances:

- *How plausible is H_0?* If H_0 represents an assumption that the people you must convince have believed for years, strong evidence (small α) will be needed to persuade them.
- *What are the consequences of rejecting H_0?* If rejecting H_0 in favor of H_a means making an expensive changeover from one type of product packaging to another, you need strong evidence that the new packaging will boost sales.

Both the plausibility of H_0 and H_a and the consequences of any action that rejection may lead to are somewhat subjective. Different people may want to use different levels of significance. It is better to report the P-value, which allows each of us to decide individually if the evidence is sufficiently strong.

Users of statistics have often emphasized certain standard levels of significance, such as 10%, 5%, and 1%. This emphasis reflects the time when tables of critical values rather than computer programs dominated statistical practice. The 5% level ($\alpha = .05$) is particularly common. *There is no sharp border between "significant" and "insignificant," only increasingly strong evidence as the P-value decreases.* There is no practical distinction between the P-values 0.049 and 0.051. It makes no sense to treat $\alpha = .05$ as a universal rule for what is significant.

EXERCISE

10.58   Suppose that in the absence of special preparation Scholastic Assessment Test mathematics (SATM) scores vary normally with mean $\mu = 475$ and $\sigma = 100$. One hundred students go through a rigorous training program designed to raise their SATM scores by improving their mathematics skills. Carry out a test of

$$H_0: \mu = 475$$
$$H_a: \mu > 475$$

in each of the following situations:

(a) The students' average score is $\bar{x} = 491.4$. Is this result significant at the 5% level?

(b) The average score is $\bar{x} = 491.5$. Is this result significant at the 5% level?

The difference between the two outcomes in (a) and (b) is of no importance. Beware attempts to treat $\alpha = .05$ as sacred.

Statistical significance and practical significance

When a null hypothesis ("no effect" or "no difference") can be rejected at the usual levels, $\alpha = .05$ or $\alpha = .01$, there is good evidence that an effect is present. But that effect may be very small. When large samples are available, even tiny deviations from the null hypothesis will be significant.

EXAMPLE 10.16

We are testing the hypothesis of no correlation between two variables. With 1000 observations, an observed correlation of only $r = .08$ is significant evidence at the $\alpha = .01$ level that the correlation in the population is not zero but positive. The low significance level does not mean there is a strong association, only that there is strong evidence of some association. The true population correlation is probably quite close to the observed sample value, $r = .08$. We might well conclude that for practical purposes we can ignore the association between these variables, even though we are confident (at the 1% level) that the correlation is positive.

Remember the wise saying: *Statistical significance is not the same thing as practical significance.* Exercise 10.59 demonstrates in detail the effect on P of increasing the sample size.

The remedy for attaching too much importance to statistical significance is to pay attention to the actual data as well as to the P-value. Plot your data and examine them carefully. Are there outliers or other deviations from a consistent pattern? A few outlying observations can produce highly significant results if you blindly apply common tests of significance. Outliers can also destroy the significance of otherwise convincing data. The foolish user of statistics who feeds the data to a computer without exploratory analysis will often be embarrassed. Is the effect you are seeking visible in your plots? If not, ask yourself if the effect is large enough to be practically important. It is usually wise to give a confidence interval for the parameter in which you are interested. A confidence interval actually estimates the size of an effect, rather than simply asking if it is too large to reasonably occur by chance alone. Confidence intervals are not used as often as they should be, while tests of significance are perhaps overused.

EXERCISES

10.59 Let us suppose that SATM scores in the absence of coaching vary normally with mean $\mu = 475$ and $\sigma = 100$. Suppose also that coaching may change μ but does not change σ. An increase in the SATM score from 475 to 478 is of no importance in seeking admission to college, but this unimportant change can be statistically very significant. To see this, calculate the P-value for the test of

$$H_0: \mu = 475$$
$$H_a: \mu > 475$$

in each of the following situations:

(a) A coaching service coaches 100 students. Their SATM scores average $\bar{x} = 478$.

(b) By the next year, the service has coached 1000 students. Their SATM scores average $\bar{x} = 478$.

(c) An advertising campaign brings the number of students coached to 10,000. Their average score is still $\bar{x} = 478$.

10.60   Give a 99% confidence interval for the mean SATM score μ after coaching in each part of the previous exercise. For large samples, the confidence interval tells us, "Yes, the mean score is higher than 475 after coaching, but only by a small amount."

Statistical inference is not valid for all sets of data

We emphasize again that badly designed surveys or experiments often produce invalid results. Formal statistical inference cannot correct basic flaws in the design. Each test is valid only in certain circumstances, with properly produced data being particularly important. The z test, for example, should bear the same warning label that we attached on page 524 to the corresponding confidence interval. Similar warnings accompany the other tests that we will learn.

EXAMPLE 10.17	You wonder whether background music would improve the productivity of the staff who process mail orders in your business. After discussing the idea with the workers, you add music and find a significant increase. You should not be impressed. In fact, almost any change in the work environment together with knowledge that a study is under way will produce a short-term productivity increase. This is the **Hawthorne effect**, named after the Western Electric manufacturing plant where it was first noted.
Hawthorne effect	

The significance test correctly informs you that an increase has occurred that is larger than would often arise by chance alone. It does not tell you *what* other than chance caused the increase. The most plausible explanation is that workers change their behavior when they know they are being studied. Your experiment was uncontrolled, so the significant result cannot be interpreted. A randomized comparative experiment would isolate the actual effect of background music and so make significance meaningful.

Tests of significance and confidence intervals are based on the laws of probability. Randomization in sampling or experimentation ensures that these laws apply. Yet we must often analyze data that do not arise from randomized samples or experiments. To apply statistical inference to such data, we must have confidence in the use of probability to describe the data. The diameters of successive holes bored in auto engine blocks during production, for example, may behave like a random sample from a normal distribution. We can check this probability model by examining the data. If the model appears correct, we can apply the recipes of this chapter to do inference about the process mean diameter μ. Always ask how the data were produced, and don't be too impressed by P-values on a printout until you are confident that the data deserve a formal analysis.

EXERCISES

10.61   A local television station announces a question for a call-in opinion poll on the six o'clock news, then gives the response on the eleven o'clock news. Today's question concerns a proposed gun-control ordinance. Of the 2372 calls received, 1921 oppose the new law. The station, following standard statistical practice, makes a confidence statement: "81% of the Channel 13 Pulse Poll sample oppose gun control. We can be 95% confident that the proportion of all viewers who oppose the law is within 1.6% of the sample result." Is the station's conclusion justified? Explain your answer.

Beware of multiple analyses

Statistical significance is a commodity much sought after. It ought to mean that you have found an effect that you were looking for. The reasoning behind statistical significance works well if you decide what effect you are seeking, design a study to search for it, and use a test of significance to weigh the evidence you get. In other settings, significance may have little meaning.

EXAMPLE 10.18   You want to learn what distinguishes managerial trainees who eventually become executives from those who, after expensive training, don't succeed and leave the company. You have abundant data on past trainees—data on their personalities and goals, their college preparation and performance, even their family backgrounds and their hobbies. Statistical software makes it easy to perform dozens of significance tests on these dozens of variables to see which ones best predict later success. Aha! You find that future executives are significantly more likely than washouts to have an urban or suburban upbringing and an undergraduate degree in a technical field.

Before basing future recruiting on these findings, pause for a moment of reflection. When you make dozens of tests at the 5% level, you expect a few of them to be significant by chance alone. After all, results significant at the 5% level do occur 5 times in 100 in the long run even when H_0 is true. Running one test and reaching the $\alpha = .05$ level is reasonably good evidence that you have found something. Running several dozen tests and reaching that level once or twice is not.

There are methods for testing many hypotheses simultaneously while controlling the risk of false findings of significance. But if you carry out many individual tests without these special methods, finding a few small P-values is only suggestive, not conclusive. The same is true of less formal analyses. Searching the trainee data for the variable with the biggest difference between future washouts and future executives, then testing whether that difference is significant, is bad statistics. The P-value assumes you had that specific difference in mind before you looked at the data. It is very misleading when applied to the largest of many differences.

Searching data for suggestive patterns is certainly legitimate. Exploratory data analysis is an important aspect of statistics. But the reasoning of formal inference does not apply when your search for a striking effect in the data is successful. The remedy is clear. Once you have a hypothesis, design a study to search specifically for the effect you now think is there. If the result of this study is statistically significant, you have real evidence.

EXERCISES

10.62   A researcher looking for evidence of extrasensory perception (ESP) tests 500 subjects. Four of these subjects do significantly better $(P < .01)$ than random guessing.

(a) Is it proper to conclude that these four people have ESP? Explain your answer.

(b) What should the researcher now do to test whether any of these four subjects have ESP?

SUMMARY

P-values are more informative than the reject-or-not result of a fixed level α test. Beware of placing too much weight on traditional values of α, such as $\alpha = .05$.

Very small effects can be highly significant (small P), especially when a test is based on a large sample. A statistically significant effect need not be practically important. Plot the data to display the effect you are seeking, and use confidence intervals to estimate the actual value of parameters.

On the other hand, lack of significance does not imply that H_0 is true, especially when the test is based on just a few observations.

Significance tests are not always valid. Faulty data collection, outliers in the data, and testing a hypothesis on the same data that suggested the hypothesis can invalidate a test.

Many tests run at once will probably produce some significant results by chance alone, even if all the null hypotheses are true.

SECTION 10.3 EXERCISES

10.63 Which of the following questions does a test of significance answer?

(a) Is the sample or experiment properly designed?

(b) Is the observed effect due to chance?

(c) Is the observed effect important?

10.64 A company compares two package designs for a laundry detergent by placing bottles with both designs on the shelves of several markets. Checkout scanner data on more than 5000 bottles bought show that more shoppers bought Design A than Design B. The difference is statistically significant ($P = .02$). Can we conclude that consumers strongly prefer Design A? Explain your answer.

10.65 A group of psychologists once measured 77 variables on a sample of schizophrenic people and a sample of people who were not schizophrenic. They compared the two samples using 77 separate significance tests. Two of these tests were significant at the 5% level. Suppose that there is in fact no difference on any of the 77 variables between people who are and people who are not schizophrenic in the adult population. Then all 77 null hypotheses are true.

(a) What is the probability that one specific test shows a difference significant at the 5% level?

(b) Why is it not surprising that 2 of the 77 tests were significant at the 5% level?

10.4   INFERENCE AS DECISION

Tests of significance assess the strength of evidence against the null hypothesis. We measure evidence by the P-value, which is a probability computed under the assumption that H_0 is true. The alternative hypothesis (the statement we seek evidence for) enters the test only to help us see what outcomes count against the null hypothesis.

Using significance tests with fixed level α, however, suggests another way of thinking. A level of significance α chosen in advance points to the outcome of the test as a *decision*. If our result is significant at level α, we reject H_0 in favor of H_a. Otherwise, we fail to reject H_0. The transition from measuring the strength of evidence to making a decision is not a small step. Many statisticians feel that making decisions should be left to the user rather than built into the statistical test. A test result is only one among many factors that influence a decision.

acceptance sampling

Yet there are circumstances that call for a decision or action as the end result of inference. **Acceptance sampling** is one such circumstance. A producer of bearings and the consumer of the bearings agree that each carload lot must meet certain quality standards. When a carload arrives, the consumer inspects a sample of the bearings. On the basis of the sample outcome, the consumer either accepts or rejects the carload. We will use acceptance sampling to show how a different concept—inference as decision—changes the reasoning used in tests of significance.

Type I and Type II errors

Tests of significance concentrate on H_0, the null hypothesis. If a decision is called for, however, there is no reason to single out H_0. There are simply two hypotheses, and we must accept one and reject the other. It is convenient to continue to call the two hypotheses H_0 and H_a, but H_0 no longer has the special status (the statement we try to find evidence against) that it had in tests of significance. In the acceptance sampling problem, we must decide between

H_0: the lot of bearings meets standards
H_a: the lot does not meet standards

on the basis of a sample of bearings.

We hope that our decision will be correct, but sometimes it will be wrong. There are two types of incorrect decisions. We can accept a bad lot of bearings, or we can reject a good lot. Accepting a bad lot injures the consumer, while rejecting a good lot hurts the producer. To distinguish these two types of error, we give them specific names.

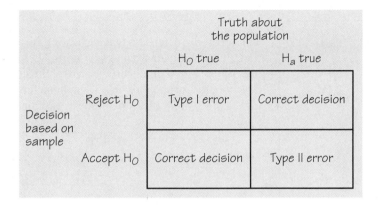

FIGURE 10.17   The two types of error in testing hypotheses.

TYPE I AND TYPE II ERRORS

If we reject H_0 (accept H_a) when in fact H_0 is true, this is a **Type I** error.

If we accept H_0 (reject H_a) when in fact H_a is true, this is a **Type II** error.

The possibilities are summed up in Figure 10.17. If H_0 is true, our decision is either correct (if we accept H_0) or is a Type I error. If H_a is true, our decision is either correct or is a Type II error. Only one error is possible at one time.

Error probabilities

We assess any rule for making decisions by looking at the probabilities of the two types of error. This is in keeping with the idea that statistical inference is based on asking, "What would happen if I used this procedure many times?"

Significance tests with fixed level α give a rule for making decisions, because the test either rejects H_0 or fails to reject it. If we adopt the decision-making way of thought, failing to reject H_0 means deciding that H_0 is true. We can then describe the performance of a test by the probabilities of Type I and Type II errors.

EXAMPLE 10.19

The mean diameter of a type of bearing is supposed to be 2.000 centimeters (cm). The bearing diameters vary normally with standard deviation $\sigma = .010$ cm. When a lot of the bearings arrives, the consumer takes an SRS of 5 bearings from the lot and measures their diameters. The consumer rejects the bearings if the sample mean diameter is significantly different from 2 at the 5% significance level.

This is a test of the hypotheses

$$H_0: \mu = 2$$
$$H_a: \mu \neq 2$$

To carry out the test, the consumer computes the z statistic

$$z = \frac{\bar{x} - 2}{.01/\sqrt{5}}$$

and rejects H_0 if $z < -1.96$ or $z > 1.96$. A Type I error is to reject H_0 when in fact $\mu = 2$.

What about Type II errors? Because there are many values of μ in H_a, we will concentrate on one value. The producer and the consumer agree that a lot of bearings with mean diameter 2.015 cm should be rejected. So a particular Type II error is to accept H_0 when in fact $\mu = 2.015$.

Figure 10.18 shows how the two probabilities of error are obtained from the *two* sampling distributions of $\bar{x}$, for $\mu = 2$ and for $\mu = 2.015$. When $\mu = 2$, H_0 is true and to reject H_0 is a Type I error. When $\mu = 2.015$, H_a is true and to accept H_0 is a Type II error. We will now calculate these error probabilities.

The probability of a Type I error is the probability of rejecting H_0 when it is really true. This is the probability that $|z| \geq 1.96$ when $\mu = 2$. But this is exactly the significance level of the test. The critical value 1.96 was chosen to make this probability 0.05, so we do not have to compute it again. The definition of "significance level 0.05" is that values of z this extreme will occur with probability 0.05 when H_0 is true.

SIGNIFICANCE AND TYPE I ERROR

The significance level α of any fixed level test is the probability of a Type I error. That is, α is the probability that the test will reject the null hypothesis H_0 when H_0 is in fact true.

The probability of a Type II error for the particular alternative $\mu = 2.015$ in Example 10.19 is the probability that the test will accept H_0 when

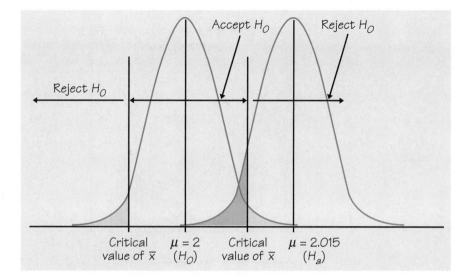

FIGURE 10.18   The two error probabilities for Example 10.19. The probability of a Type I error (*light shaded area*) is the probability of rejecting H_0: $\mu = 2$ when in fact $\mu = 2$. The probability of a Type II error (*dark shaded area*) is the probability of accepting H_0 when in fact $\mu = 2.015$.

μ has this alternative value. This is the probability that the test statistic z falls between -1.96 and 1.96, calculated assuming that $\mu = 2.015$. This probability is *not* $1 - .05$, because the probability 0.05 was found assuming that $\mu = 2$. Here is the calculation for Type II error.

EXAMPLE 10.20	To calculate the probability of a Type II error:

Step 1. *Write the rule for accepting H_0 in terms of $\bar{x}$.* The test accepts H_0 when

$$-1.96 \leq \frac{\bar{x} - 2}{.01/\sqrt{5}} \leq 1.96$$

This is the same as

$$2 - 1.96\left(\frac{.01}{\sqrt{5}}\right) \leq \bar{x} \leq 2 + 1.96\left(\frac{.01}{\sqrt{5}}\right)$$

or, doing the arithmetic,

$$1.9912 \leq \bar{x} \leq 2.0088$$

This step does not involve the particular alternative $\mu = 2.015$.

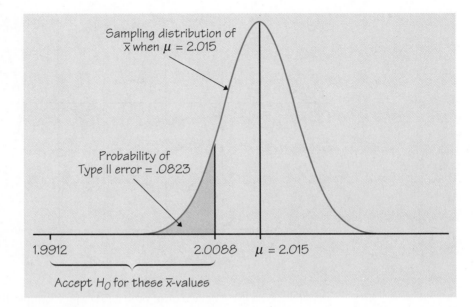

Sampling distribution of
$\bar{x}$ when $\mu = 2.015$

Probability of
Type II error = .0823

1.9912   2.0088 $\mu = 2.015$

Accept H_0 for these $\bar{x}$-values

FIGURE 10.19   The probability of a Type II error for Example 10.20. This is the probability that the test accepts H_0 when the alternative hypothesis is true.

Step 2. *Find the probability of accepting H_0 assuming that the alternative is true.* Take $\mu = 2.015$ and standardize to find the probability.

$$P(\text{Type II error}) = P(1.9912 \leq \bar{x} \leq 2.0088)$$

$$= P\left(\frac{1.9912 - 2.015}{.01/\sqrt{5}} \leq \frac{\bar{x} - 2.015}{.01/\sqrt{5}} \leq \frac{2.0088 - 2.015}{.01/\sqrt{5}}\right)$$

$$= P(-5.32 \leq Z \leq -1.39)$$

$$= .0823$$

Figure 10.19 illustrates this error probability in terms of the sampling distribution of $\bar{x}$ when $\mu = 2.015$. The test will wrongly accept the hypothesis that $\mu = 2$ in about 8% of all samples when in fact $\mu = 2.015$.

This test will reject 5% of all good lots of bearings (for which $\mu = 2$). It will accept 8% of lots so bad that $\mu = 2.015$. Calculations of error probabilities help the producer and consumer decide whether the test is satisfactory.

EXERCISES

10.66   Your company markets a computerized medical diagnostic program. The program scans the results of routine medical tests (pulse rate, blood tests, etc.) and either clears the patient or refers the case to a doctor. The program is used to screen thousands of people who do not have specific medical complaints. The program makes a decision about each person.

(a) What are the two hypotheses and the two types of error that the program can make? Describe the two types of error in terms of "false positive" and "false negative" test results.

(b) The program can be adjusted to decrease one error probability, at the cost of an increase in the other error probability. Which error probability would you choose to make smaller, and why? (This is a matter of judgment. There is no single correct answer.)

10.67   You have the NAEP quantitative scores for an SRS of 840 young men. You plan to test hypotheses about the population mean score,

$$H_0: \mu = 275$$
$$H_a: \mu < 275$$

at the 1% level of significance. The population standard deviation is known to be $\sigma = 60$. The z test statistic is

$$z = \frac{\bar{x} - 275}{60/\sqrt{840}}$$

(a) What is the rule for rejecting H_0 in terms of z?

(b) What is the probability of a Type I error?

(c) You want to know whether this test will usually reject H_0 when the true population mean is 270, 5 points lower than the null hypothesis claims. Answer this question by calculating the probability of a Type II error when $\mu = 270$.

10.68   You have an SRS of size $n = 9$ from a normal distribution with $\sigma = 1$. You wish to test

$$H_0: \mu = 0$$
$$H_a: \mu > 0$$

You decide to reject H_0 if $\bar{x} > 0$ and to accept H_0 otherwise.

(a) Find the probability of a Type I error. That is, find the probability that the test rejects H_0 when in fact $\mu = 0$.

(b) Find the probability of a Type II error when $\mu = .3$. This is the probability that the test accepts H_0 when in fact $\mu = .3$.

(c) Find the probability of a Type II error when $\mu = 1$.

Power

A test makes a Type II error when it fails to reject a null hypothesis that really is false. A high probability of a Type II error for a particular alternative means that the test is not sensitive enough to usually detect that alternative. Calculations of the probability of Type II errors are therefore useful even if you don't think that a statistical test should be viewed as making a decision. The language used is a bit different in the significance test setting. It is usual to report the probability that a test *does* reject H_0 when an alternative is true. The higher this probability is, the more sensitive the test is.

POWER

The probability that a fixed level α significance test will reject H_0 when a particular alternative value of the parameter is true is called the **power** of the test against that alternative.

The power of a test against any alternative is 1 minus the probability of a Type II error for that alternative.

Calculations of power are essentially the same as calculations of the probability of Type II error. In Example 10.20, the power is the probability of *rejecting* H_0 in Step 2 of the calculation. It is $1 - 0.0823$, or 0.9177.

Calculations of P-values and calculations of power both say what would happen if we repeated the test many times. A P-value describes what would happen supposing that the null hypothesis is true. Power describes what would happen supposing that a particular alternative is true.

In planning an investigation that will include a test of significance, a careful user of statistics decides what alternatives the test should detect and checks that the power is adequate. The power depends on which particular parameter value in H_a we are interested in. Values of the mean μ that are in H_a but lie close to the hypothesized value μ_0 are harder to detect (lower power) than values of μ that are far from μ_0. If the power is too low, a larger sample size will increase the power for the same significance level α. In order to calculate power, we must fix an α so that there is a fixed rule for rejecting H_0. We prefer to report P-values rather than to use a fixed significance level. The usual practice is to calculate the power at a common significance level such as $\alpha = .05$ even though you intend to report a P-value.

10.69   The cola maker of Example 10.7 determines that a sweetness loss is too large to accept if the mean response for all tasters is $\mu = 1.1$. Will a 5% significance test of the hypotheses

$$H_0: \mu = 0$$

$$H_a: \mu > 0$$

based on a sample of 10 tasters usually detect a change this great?
   We want the power of the test against the alternative $\mu = 1.1$. This is the probability that the test rejects H_0 when $\mu = 1.1$ is true. The calculation method is similar to that for Type II error.

(a) **Step 1.** *Write the rule for rejecting H_0 in terms of $\bar{x}$.* We know that $\sigma = 1$, so the test rejects H_0 at the $\alpha = .05$ level when

$$z = \frac{\bar{x} - 0}{1/\sqrt{10}} \geq 1.645$$

Restate this in terms of $\bar{x}$.

(b) **Step 2.** *The power is the probability of this event supposing that the alternative is true.* Standardize using $\mu = 1.1$ to find the probability that $\bar{x}$ takes a value that leads to rejection of H_0.

10.70   Exercise 10.40 concerns a test about the mean contents of cola bottles. The hypotheses are

$$H_0: \mu = 300$$

$$H_a: \mu < 300$$

The sample size is $n = 6$, and the population is assumed to have a normal distribution with $\sigma = 3$. A 5% significance test rejects H_0 if $z \leq -1.645$, where the test statistic z is

$$z = \frac{\bar{x} - 300}{3/\sqrt{6}}$$

Power calculations help us see how large a shortfall in the bottle contents the test can be expected to detect.

(a) Find the power of this test against the alternative $\mu = 299$.

(b) Find the power against the alternative $\mu = 295$.

(c) Is the power against $\mu = 290$ higher or lower than the value you found in (b)? (Don't actually calculate that power.) Explain your answer.

10.71   Increasing the sample size increases the power of a test when the level α is unchanged. Suppose that in the previous exercise a sample of n bottles had been measured. In that exercise, $n = 6$. The 5% significance test still rejects H_0 when $z \leq -1.645$, but the z statistic is now

$$z = \frac{\bar{x} - 300}{3/\sqrt{n}}$$

(a) Find the power of this test against the alternative $\mu = 299$ when $n = 25$.

(b) Find the power against $\mu = 299$ when $n = 100$.

Different views of statistical tests

The distinction between tests of significance and tests as rules for deciding between two hypotheses does not lie in the calculations but in the reasoning that motivates the calculations. In a test of significance we focus on a single hypothesis (H_0) and a single probability (the P-value). The goal is to measure the strength of the sample evidence against H_0. Calculations of power are done to check the sensitivity of the test. If we cannot reject H_0, we conclude only that there is not sufficient evidence against H_0, not that H_0 is actually true. If the same inference problem is thought of as a decision problem, we focus on two hypotheses and give a rule for deciding between them based on the sample evidence. We therefore must focus equally on two probabilities, the probabilities of the two types of error. We must choose one or the other hypothesis and cannot abstain on grounds of insufficient evidence.

There are clear distinctions between the two ways of thinking about statistical tests. But sometimes the two approaches merge. Jerzy Neyman advocated an approach called *testing hypotheses* that mixes the reasoning of significance tests and decision rules as follows:

testing hypotheses

1. State H_0 and H_a just as in a test of significance. In particular, we are seeking evidence against H_0.

2. Think of the problem as a decision problem, so that the probabilities of Type I and Type II errors are relevant.

3. Because of Step 1, Type I errors are more serious. So choose an α (significance level) and consider only tests with probability of Type I error no greater than α.

4. Among these tests, select one that makes the probability of a Type II error as small as possible (that is, power as large as possible). If this probability is too large, you will have to take a larger sample to reduce the chance of an error.

Hypothesis testing is often emphasized in mathematical presentations of statistics because Neyman developed an impressive mathematical theory. In simple settings, this theory shows how to find the test that has the smallest possible probability of a Type II error among all tests with a given probability (like 0.05) of a Type I error. In part because such pleasing results aren't available for many practical settings, the significance test way of thinking prevails in statistical practice.

SUMMARY

An alternative to significance testing regards H_0 and H_a as two statements of equal status that we must decide between. This **decision analysis** point of view regards statistical inference in general as giving rules for making decisions in the presence of uncertainty.

In the case of testing H_0 versus H_a, decision analysis chooses a decision rule on the basis of the probabilities of two types of error. A **Type I error** occurs if we reject H_0 when it is in fact true. A **Type II error** occurs if we accept H_0 when in fact H_a is true.

The **power** of a significance test measures its ability to detect an alternative hypothesis. The power against a specific alternative is the probability that the test will reject H_0 when the alternative is true.

In a fixed level α significance test, the significance level α is the probability of a Type I error, and the power against a specific alternative is 1 minus the probability of a Type II error for that alternative.

Increasing the size of the sample increases the power (reduces the probability of a Type II error) when the significance level remains fixed.

SECTION 10.4 EXERCISES

10.72 Power calculations for two-sided tests follow the same outline as for one-sided tests. Example 10.14 presents a test of

$$H_0: \mu = .86$$

$$H_a: \mu \ne .86$$

at the 1% level of significance. The sample size is $n = 3$ and $\sigma = .0068$. We will find the power of this test against the alternative $\mu = .845$.

(a) The test in Example 10.14 rejects H_0 when $|z| \ge 2.576$. The test statistic z is

$$z = \frac{\bar{x} - .86}{.0068/\sqrt{3}}$$

Write the rule for rejecting H_0 in terms of the values of $\bar{x}$. (Because the test is two-sided, it rejects when $\bar{x}$ is either too large or too small.)

(b) Now find the probability that $\bar{x}$ takes values that lead to rejecting H_0 if the true mean is $\mu = .845$. This probability is the power.

(c) What is the probability that this test makes a Type II error when $\mu = .845$?

10.73   In Example 10.11, a company medical director failed to find significant evidence that the mean blood pressure of a population of executives differed from the national mean $\mu = 128$. The medical director now wonders if the test used would detect an important difference if one were present. For the SRS of size 72 from a population with standard deviation $\sigma = 15$, the z statistic is

$$z = \frac{\bar{x} - 128}{15/\sqrt{72}}$$

The two-sided test rejects $H_0: \mu = 128$ at the 5% level of significance when $|z| \ge 1.96$.

(a) Find the power of the test against the alternative $\mu = 134$.

(b) Find the power of the test against $\mu = 122$. Can the test be relied on to detect a mean that differs from 128 by 6?

(c) If the alternative were farther from H_0, say $\mu = 136$, would the power be higher or lower than the values calculated in (a) and (b)?

10.74   In Exercise 10.70 you found the power of a test against the alternative $\mu = 295$. Use the result of that exercise to find the probabilities of Type I and Type II errors for that test and that alternative.

10.75   In Exercise 10.67 you found the probabilities of the two types of error for a test of $H_0: \mu = 275$, with the specific alternative that $\mu = 270$. Use the result of that exercise to give the power of the test against the alternative $\mu = 270$.

10.76   You are reading an article in a business journal that discusses the "efficient market hypothesis" for the behavior of securities prices. The author admits that most tests of this hypothesis have failed to find significant evidence against it. But he says this failure is a result of the fact that the tests used have low power. "The widespread impression that there is strong evidence for market efficiency may be due just to a lack of appreciation of the low power of many statistical tests."[11]

Explain in simple language why tests having low power often fail to give evidence against a hypothesis even when the hypothesis is really false.

CHAPTER REVIEW

Statistical inference draws conclusions about a population on the basis of sample data and uses probability to indicate how reliable the conclusions are. A confidence interval estimates an unknown parameter. A significance test shows how strong the evidence is for some claim about a parameter.

The probabilities in both confidence intervals and tests tell us what would happen if we used the recipe for the interval or test very many times. A confidence level is the probability that the recipe for a confidence interval actually produces an interval that contains the unknown parameter. A 99% confidence interval gives a correct result 99% of the time when we use it repeatedly. A P-value is the probability that the test would produce a result at least as extreme as the observed result if the null hypothesis really were true. That is, a P-value tells us how surprising the observed outcome is. Very surprising outcomes (small P-values) are good evidence that the null hypothesis is not true.

The rest of this book presents confidence intervals and tests for use in many specific settings. We will have a good deal to say about practical aspects of using statistical methods. But in every case, the basic reasoning of confidence intervals and significance tests remains the same. Sections 10.1 to 10.3 in this chapter are the foundation for your understanding of statistical inference. Here are the most important things you should be able to do after studying this chapter.

A. CONFIDENCE INTERVALS

1. State in nontechnical language what is meant by "95% confidence" or other statements of confidence in statistical reports.

2. Calculate a confidence interval for the mean μ of a normal population with known standard deviation σ, using the recipe

$$\bar{x} \pm z^* \sigma/\sqrt{n}$$

3. Recognize when you can safely use this confidence interval recipe and when the sample design or a small sample from a skewed population makes it inaccurate.

4. Understand how the margin of error of a confidence interval changes with the sample size and the level of confidence C.

5. Find the sample size required to obtain a confidence interval of specified margin of error m when the confidence level and other information are given.

B. SIGNIFICANCE TESTS

1. State the null and alternative hypotheses in a testing situation when the parameter in question is a population mean μ.

2. Explain in nontechnical language the meaning of the P-value when you are given the numerical value of P for a test.

3. Calculate the z statistic and the P-value for both one-sided and two-sided tests about the mean μ of a normal population.

4. Assess statistical significance at standard levels α, either by comparing P to α or by comparing z to standard normal critical values.

5. Recognize that significance testing does not measure the size or importance of an effect.

6. Recognize when you can use the z test and when the data collection design or a small sample from a skewed population makes it inappropriate.

CHAPTER 10 REVIEW EXERCISES

10.77   Sulfur compounds cause "off-odors" in wine, so winemakers want to know the odor threshold, the lowest concentration of a compound that the human nose can detect. The odor threshold for dimethyl sulfide (DMS) in trained wine tasters is about 25 micrograms per liter of wine (μg/l). The untrained noses of consumers may be less sensitive, however. Here are the DMS odor thresholds for 10 untrained students:

31   31   43   36   23   34   32   30   20   24

Assume that the standard deviation of the odor threshold for untrained noses is known to be $\sigma = 7\ \mu g/l$.

(a) Make a stemplot to verify that the distribution is roughly symmetric with no outliers. (More data confirm that there are no systematic departures from normality.)

(b) Give a 95% confidence interval for the mean DMS odor threshold among all students.

(c) Are you convinced that the mean odor threshold for students is higher than the published threshold, 25 $\mu g/l$? Carry out a significance test to justify your answer.

10.78   An agronomist examines the cellulose content of a variety of alfalfa hay. Suppose that the cellulose content in the population has standard deviation $\sigma = 8$ mg/g. A sample of 15 cuttings has mean cellulose content $\bar{x} = 145$ mg/g.

(a) Give a 90% confidence interval for the mean cellulose content in the population.

(b) A previous study claimed that the mean cellulose content was $\mu = 140$ mg/g, but the agronomist believes that the mean is higher than that figure. State H_0 and H_a and carry out a significance test to see if the new data support this belief.

(c) The statistical procedures used in (a) and (b) are valid when several assumptions are met. What are these assumptions?

10.79   Researchers studying iron deficiency in infants examined infants who were following different feeding patterns. One group of 26 infants was being breast-fed. At 6 months of age, these children had mean hemoglobin level $\bar{x} = 12.9$ grams per 100 milliliters of blood. Assume that the population standard deviation is $\sigma = 1.6$. Give a 95% confidence interval for the mean hemoglobin level of breast-fed infants. What assumptions (other than the unrealistic assumption that we know σ) does the method you used to get the confidence interval require?

10.80   Here are the Degree of Reading Power (DRP) scores for an SRS of 44 third-grade students from a suburban school district:

40	26	39	14	42	18	25	43	46	27	19
47	19	26	35	34	15	44	40	38	31	46
52	25	35	35	33	29	34	41	49	28	52
47	35	48	22	33	41	51	27	14	54	45

DRP scores are approximately normal. Suppose that the standard deviation of scores in this school district is known to be $\sigma = 11$. The researcher believes that the mean score μ of all third graders in this district is higher than the national mean, which is 32.

(a) State H_0 and H_a to test this suspicion.

(b) Carry out the test. Give the P-value, and then interpret the result in plain language.

10.81 A government report gives a 99% confidence interval for the 1994 median family income as $32,264 \pm \$397$. This result was calculated by advanced methods from the Current Population Survey, a multistage random sample of about 60,000 households.

(a) Would a 95% confidence interval be wider or narrower? Explain your answer.

(b) Would the null hypothesis that the 1994 median family income was $33,000 be rejected at the 1% significance level in favor of the two-sided alternative?

10.82 Statisticians prefer large samples. Describe briefly the effect of increasing the size of a sample (or the number of subjects in an experiment) on each of the following:

(a) The margin of error of a 95% confidence interval.

(b) The P-value of a test, when H_0 is false and all facts about the population remain unchanged as n increases.

(c) The power of a fixed level α test, when α, the alternative hypothesis, and all facts about the population remain unchanged.

10.83 A roulette wheel has 18 red slots among its 38 slots. You observe many spins and record the number of times that red occurs. Now you want to use these data to test whether the probability p of a red has the value that is correct for a fair roulette wheel. State the hypotheses H_0 and H_a that you will test. (We will describe the test for this situation in Chapter 12.)

10.84 When asked to explain the meaning of "the P-value was $P = .03$," a student says, "This means there is only probability 0.03 that the null hypothesis is true." Is this an essentially correct explanation? Explain your answer.

10.85 Another student, when asked why statistical significance appears so often in research reports, says, "Because saying that results are significant tells us that they cannot easily be explained by chance variation alone." Do you think that this statement is essentially correct? Explain your answer.

10.86   A study compares two groups of mothers with young children who were on welfare two years ago. One group attended a voluntary training program offered free of charge at a local vocational school and advertised in the local news media. The other group did not choose to attend the training program. The study finds a significant difference $(P < .01)$ between the proportions of the mothers in the two groups who are still on welfare. The difference is not only significant but quite large. The report says that with 95% confidence the percent of the nonattending group still on welfare is 21% ± 4% higher than that of the group who attended the program. You are on the staff of a member of Congress who is interested in the plight of welfare mothers, and who asks you about the report.

(a)   Explain in simple language what "a significant difference $(P < .01)$" means.

(b)   Explain clearly and briefly what "95% confidence" means.

(c)   Is this study good evidence that requiring job training of all welfare mothers would greatly reduce the percent who remain on welfare for several years?

NOTES AND DATA SOURCES

1.   Information from Francisco L. Rivera-Batiz, "Quantitative literacy and the likelihood of employment among young adults," *Journal of Human Resources*, 27 (1992), pp. 313–328.

2.   Data provided by Maribeth Cassidy Schmitt, from her Ph.D. dissertation, *The Effects of an Elaborated Directed Reading Activity on the Metacomprehension Skills of Third Graders*, Purdue University, 1987.

3.   Data provided by John Rousselle and Huei-Ru Shieh, Department of Restaurant, Hotel, and Institutional Management, Purdue University.

4.   Data provided by Mugdha Gore and Joseph Thomas, Purdue University School of Pharmacy.

5.   Based on G. Salvendy, G. P. McCabe, S. G. Sanders, J. L. Knight, and E. J. McCormick, "Impact of personality and intelligence on job satisfaction of assembly line and bench work—an industrial study," *Applied Ergonomics*, 13 (1982), pp. 293–299.

6.   From a study by M. R. Schlatter et al., Division of Financial Aid, Purdue University.

7.   Data provided by Diana Schellenberg, Purdue University School of Health Sciences.

8.   William T. Robinson, "Sources of market pioneer advantages: the case of industrial goods industries," *Journal of Marketing Research*, 25 (February 1988), pp. 87–94.

9.   Michael B. Maziz et al., "Perceived age and attractiveness of models in cigarette advertisements," *Journal of Marketing*, 56 (January 1992), pp. 22–37.

10. For a discussion of statistical significance in the legal setting, see D. H. Kaye, "Is proof of statistical significance relevant?" *Washington Law Review*, 61 (1986), pp. 1333–1365. Kaye argues: "Presenting the P-value without characterizing the evidence by a significance test is a step in the right direction. Interval estimation, in turn, is an improvement over P-values."

11. Robert J. Schiller, "The volatility of stock market prices," *Science*, 235 (1987), pp. 33–36.

Inference for Distributions

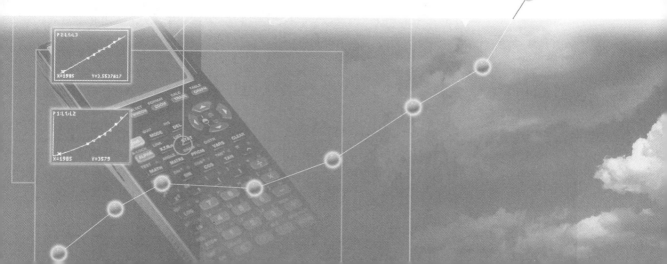

| ACTIVITY 11 | Paper Airplane Experiment |

Materials: Two paper airplane pattern sheets (in Teacher Resource Binder), scissors, masking tape, 25-meter steel tape measure, TI-83 graphing calculator

The Experiment   The purpose of this activity is to see which of two prototype paper airplane models has superior flight characteristics. Specifically, the object is to determine the average distance flown for each prototype plane and to compare these average distances flown. The null hypothesis will be that there is no difference between the average distance flown by Prototype A and the average distance flown by Prototype B. So H_0: $\mu_A = \mu_B$. Equivalently, we could write H_0: $\mu_A - \mu_B = 0$. What form should the alternative hypotheses take? (Remember that the alternative hypothesis should be stated *before* you conduct the experiment.)

The Task   Your task, as a class, is to design an experiment to determine which of the two prototype paper airplanes has the superior flight characteristics (i.e., flies the farthest). Then you will carry out your plan and gather the necessary data. You may want to explore the data both numerically and graphically prior to conducting formal inference. Unfortunately, we don't know the population standard deviation σ for either prototype, so we can't apply the methods of Chapter 10 to conduct significance tests. We will develop methods in this chapter that will enable us to calculate a test statistic so that we can answer the question about which prototype paper airplane flies the farthest. We will also calculate confidence intervals for the true population difference in flight distances. Keep your data at hand so that you can perform this analysis later. *Note:* These data will also be used in Activity 15 (in Chapter 15).

INTRODUCTION

With the principles in hand, we proceed to practice. This chapter describes confidence intervals and significance tests for the mean of a single population and for comparing the means of two populations. Later chapters present procedures for inference about population proportions, comparing the means of more than two populations, and studying relationships among variables.

11.1   INFERENCE FOR THE MEAN OF A POPULATION

Confidence intervals and tests of significance for the mean μ of a normal population are based on the sample mean $\bar{x}$. The sampling distribution of $\bar{x}$ has μ as its mean. (That is, $\bar{x}$ is an unbiased estimator of the unknown μ.) Its spread depends on the sample size and also on the population standard deviation σ. In the previous chapter we made the unrealistic assumption that we knew the value of σ. In practice, σ is unknown. We must then estimate σ from the data even though we are primarily interested in μ. The need to estimate σ changes some details of tests and confidence intervals for μ, but not their interpretation.

Here are the assumptions we make in order to do inference about a population mean:

ASSUMPTIONS FOR INFERENCE ABOUT A MEAN

- Our data are a **simple random sample** (SRS) of size n from the population.
- Observations from the population have a **normal distribution** with mean μ and standard deviation σ. Both μ and σ are unknown parameters.

In this setting, the sample mean $\bar{x}$ has the normal distribution with mean μ and standard deviation $\sigma/\sqrt{n}$. Because we don't know σ, we estimate it by the sample standard deviation s. We then estimate the standard deviation of $\bar{x}$ by $s/\sqrt{n}$. This quantity is called the *standard error* of the sample mean $\bar{x}$.

STANDARD ERROR

When the standard deviation of a statistic is estimated from the data, the result is called the **standard error** of the statistic. The standard error of the sample mean $\bar{x}$ is $s/\sqrt{n}$.

The t distributions

When we know the value of σ, we base confidence intervals and tests for μ on the standardized sample mean

$$z = \frac{\bar{x} - \mu}{\sigma/\sqrt{n}}$$

This z statistic has the standard normal distribution $N(0, 1)$. When we do not know σ, we substitute the standard error $s/\sqrt{n}$ of $\bar{x}$ for its standard deviation $\sigma/\sqrt{n}$. The statistic that results does *not* have a normal distribution. It has a distribution that is new to us, called a t *distribution*.

THE ONE-SAMPLE t STATISTIC AND THE t DISTRIBUTIONS

Draw an SRS of size n from a population that has the normal distribution with mean μ and standard deviation σ.

The **one-sample t statistic**

$$t = \frac{\bar{x} - \mu}{s/\sqrt{n}}$$

has the **t distribution** with $n - 1$ degrees of freedom.

---

degrees of freedom

The t statistic has the same interpretation as any standardized statistic: it says how far $\bar{x}$ is from its mean μ in standard deviation units. There is a different t distribution for each sample size. We specify a particular t distribution by giving its *degrees of freedom*. The degrees of freedom for the one-sample t statistic come from the sample standard deviation s in the denominator of t. We saw in Chapter 1 that s has $n - 1$ degrees of freedom. There are other t statistics with different degrees of freedom, some of which we will meet later in this chapter. We will write the t distribution with k degrees of freedom as $t(k)$ for short.

Figure 11.1 compares the density curves of the standard normal distribution and the t distributions with 2 and 9 degrees of freedom. The figure illustrates these facts about the t distributions:

- The density curves of the t distributions are similar in shape to the standard normal curve. They are symmetric about zero and are bell-shaped.

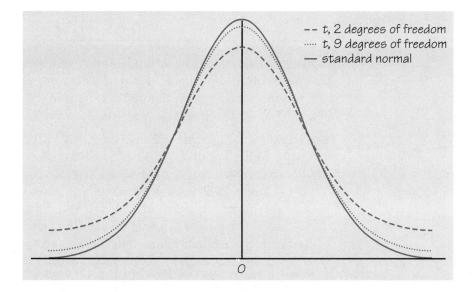

FIGURE 11.1   Density curves for the *t* distributions with 2 and 9 degrees of freedom and the standard normal distribution. All are symmetric with center 0. The *t* distributions have more probability in the tails than does the standard normal distribution.

- The spread of the *t* distributions is a bit greater than that of the standard normal distribution. The *t* distributions in Figure 11.1 have more probability in the tails and less in the center than does the standard normal. This is true because substituting the estimate *s* for the fixed parameter σ introduces more variation into the statistic.
- As the degrees of freedom *k* increase, the *t*(*k*) density curve approaches the N(0, 1) curve ever more closely. This happens because *s* estimates σ more accurately as the sample size increases. So using *s* in place of σ causes little extra variation when the sample is large.

Table C in the back of the book gives critical values for the *t* distributions. Table C also appears inside the rear cover. Each row in the table contains critical values for one of the *t* distributions; the degrees of freedom appear at the left of the row. For convenience, we label the table entries both by *p*, the upper tail probability needed for significance tests, and by the confidence level *C* (in percent) required for confidence intervals. You have already used the standard normal critical values in the bottom row of Table C. By looking down any column, you can check that the *t* critical values approach the normal values as the degrees of freedom increase. As in the case of the normal table, computer software often makes Table C unnecessary.

EXERCISES

11.1   This exercise uses the TI-83 to duplicate the comparison between the standard normal distribution and the two t distribution curves shown in Figure 11.1. Begin by clearing any defined functions in the Y= window, turning off any STAT PLOTS, and clearing the Graphics screen (ClrDraw).

(a)   Define Y_1 = normalpdf(X), press ENTER, and then move the cursor to the left of Y_1. Press ENTER again to change the plotting style to a thick line. Next define Y_2 = tpdf(X,2). Note that tpdf is found under DISTR. Note also that the second parameter, 2, specifies the degrees of freedom. Set the WINDOW to $X[-3, 3]_1$ and $Y[-.1, .4]_1$ and then GRAPH. Sketch the graphs of the two curves, and write a brief description of the differences.

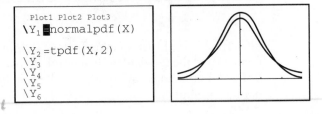

(b)   Change the plotting style for Y_1 to be the dotted line. Deselect Y_2 and define Y_3 = tpdf(X,9). GRAPH these two functions. How do these two curves differ?

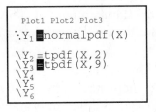

(c)   Deselect Y_3 and define Y_4 = tpdf(X,30). Graph these two functions. How do these two curves differ? What appears to be happening to the shape of the t distribution curve as the number of degrees of freedom increases?

11.2   This exercise uses the TI-83 to compare areas in the tails of the standard normal and several representative t distributions. Begin by clearing the Graphics screen (ClrDraw) and setting the WINDOW to $X[-3, 3]_1$ and $Y[-.1, .4]_1$.

(a)   Enter ShadeNorm(2,100) to shade the area under the standard normal curve to the right of $Z = 2$. Record this area, rounded off to four decimal places. Then make a table like the one following.

df	$P(t > 2)$	Absolute difference
2		
10		
30		
50		
100		

Clear the Graphics screen (ClrDraw) and enter `Shade_t(2,100,2)`. Note that the syntax is Shade_t(leftendpoint,rightendpoint,df). Round off the P-value shown to four decimal places and enter it in the table. Calculate the difference between the tail areas for the normal curve and the $t(2)$ curve, and enter this difference in the table.

(b) Calculate the areas to the right of $t = 2$ for the $t(10)$, $t(30)$, $t(50)$, and $t(100)$ curves, and enter the P-values (rounded to four decimal places) in the table.

(c) Describe what happens to the area to the right of 2 under the $t(k)$ distribution as the degrees of freedom increase.

11.3   The scores of four roommates on the Law School Aptitude Test have mean $\bar{x} = 589$ and standard deviation $s = 37$. What is the standard error of the mean?

11.4   What critical value t^* from Table C satisfies each of the following conditions:

(a) The t distribution with 5 degrees of freedom has probability 0.05 to the right of t^*.

(b) The t distribution with 21 degrees of freedom has probability 0.99 to the left of t^*.

11.5   What critical value t^* from Table C satisfies each of the following conditions:

(a) The one-sample t statistic from a sample of 15 observations has probability 0.025 to the right of t^*.

(b) The one-sample t statistic from an SRS of 20 observations has probability 0.75 to the left of t^*.

The *t* confidence intervals and tests

To analyze samples from normal populations with unknown σ, just re-place the standard deviation $\sigma/\sqrt{n}$ of $\bar{x}$ by its standard error $s/\sqrt{n}$ in the z procedures of Chapter 10. The z procedures then become *one-sample t procedures*. Use P-values or critical values from the t distribution with $n-1$ degrees of freedom in place of the normal values. The one-sample t procedures are similar in both reasoning and computational detail to the z procedures of Chapter 10. So we will now pay more attention to questions about using these methods in practice.

THE ONE-SAMPLE *t* PROCEDURES

Draw an SRS of size n from a population having unknown mean μ. A level C confidence interval for μ is

$$\bar{x} \pm t^* \frac{s}{\sqrt{n}}$$

where t^* is the upper $(1-C)/2$ critical value for the $t(n-1)$ distribution. This interval is exact when the population distribution is normal and is approximately correct for large n in other cases.

To test the hypothesis H_0: $\mu = \mu_0$ based on an SRS of size n, compute the one-sample t statistic

$$t = \frac{\bar{x} - \mu_0}{s/\sqrt{n}}$$

In terms of a variable T having the $t(n-1)$ distribution, the P-value for a test of H_0 against

H_a: $\mu > \mu_0$ is $P(T \geq t)$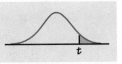

H_a: $\mu < \mu_0$ is $P(T \leq t)$

$$H_a: \mu \neq \mu_0 \text{ is } 2P(T \geq |t|)$$

These P-values are exact if the population distribution is normal and are approximately correct for large n in other cases.

EXAMPLE 11.1

To study the metabolism of insects, researchers fed cockroaches measured amounts of a sugar solution. After 2, 5, and 10 hours, they dissected some of the cockroaches and measured the amount of sugar in various tissues.[1] Five roaches fed the sugar D-glucose and dissected after 10 hours had the following amounts (in micrograms) of D-glucose in their hindguts:

$$55.95 \quad 68.24 \quad 52.73 \quad 21.50 \quad 23.78$$

The researchers gave a 95% confidence interval for the mean amount of D-glucose in cockroach hindguts under these conditions.
   First calculate that

$$\bar{x} = 44.44 \quad \text{and} \quad s = 20.741$$

The degrees of freedom are $n - 1 = 4$. From Table C we find that for 95% confidence $t^* = 2.776$. The confidence interval is

$$\bar{x} \pm t^* \frac{s}{\sqrt{n}} = 44.44 \pm 2.776 \frac{20.741}{\sqrt{5}}$$

$$= 44.44 \pm 25.75$$

$$= (18.69, \ 70.19)$$

Comparing this estimate with those for other body tissues and different times before dissection led to new insight into cockroach metabolism and to new ways of eliminating roaches from homes and restaurants. The large margin of error is due to the small sample size and the rather large variation among the cockroaches, reflected in the large value of s.

The one-sample t confidence interval has the form

$$\text{estimate} \pm t^* \, \text{SE}_{\text{estimate}}$$

where "SE" stands for "standard error." We will meet a number of confidence intervals that have this common form. Like the confidence interval, t tests are close in form to the z tests we met earlier. Here is an example. In

Chapter 10 we used the z test on these data. That required the unrealistic assumption that we knew the population standard deviation σ. Now we can do a realistic analysis.

EXAMPLE 11.2

Cola makers test new recipes for loss of sweetness during storage. Trained tasters rate the sweetness before and after storage. Here are the sweetness losses (sweetness before storage minus sweetness after storage) found by 10 tasters for one new cola recipe.

$$2.0 \quad 0.4 \quad 0.7 \quad 2.0 \quad -0.4 \quad 2.2 \quad -1.3 \quad 1.2 \quad 1.1 \quad 2.3$$

Are these data good evidence that the cola lost sweetness?

Step 1: Hypotheses. Tasters vary in their perception of sweetness loss. So we ask the question in terms of the mean loss μ for a large population of tasters. The null hypothesis is "no loss," and the alternative hypothesis says "there is a loss."

$$H_0: \mu = 0$$
$$H_a: \mu > 0$$

Step 2: Test statistic. The basic statistics are

$$\bar{x} = 1.02 \quad \text{and} \quad s = 1.196$$

The one-sample t test statistic is

$$t = \frac{\bar{x} - \mu_0}{s/\sqrt{n}} = \frac{1.02 - 0}{1.196/\sqrt{10}}$$
$$= 2.70$$

df = 9

p	.02	.01
t^*	2.398	2.821

Step 3: P-value. The P-value for $t = 2.70$ is the area to the right of 2.70 under the t distribution curve with degrees of freedom $n - 1 = 9$. Figure 11.2 shows this area. We can't find the exact value of P without software. But we can pin P between two values by using Table C. Search the df = 9 row of Table C for entries that bracket $t = 2.70$. Because the observed t lies between the critical values for 0.02 and 0.01, the P-value lies between 0.01 and 0.02. Computer software gives the more exact result $P = .012$. There is quite strong evidence for a loss of sweetness.

The t confidence interval in Example 11.1 and the t test in Example 11.2 rest on assumptions that are reasonable but are not all easy to check. Both examples are experiments. In both cases the researchers took pains to avoid bias. The cockroaches were assigned at random to different sugars and different times before dissection and treated identically in every other way. The taste testers worked in isolation booths to avoid influence by other tasters. So we trust the results for these particular cockroaches and tasters.

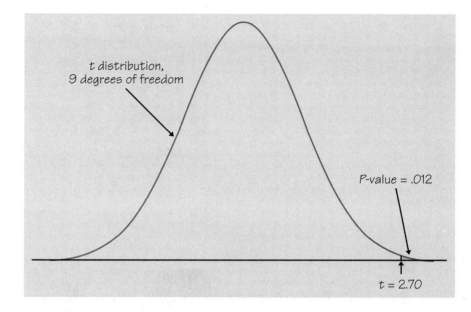

FIGURE 11.2   The *P*-value for the one-sided *t* test in Example 11.2.

The statistical analysis rests on two assumptions: random sampling and normal population distributions. We must be willing to treat the cockroaches and the tasters as SRSs from larger populations if we want to draw conclusions about cockroaches in general or tasters in general. The roaches were chosen at random from a population grown in the laboratory for research purposes. The tasters all have the same training. Even though we don't actually have SRSs from the populations we are interested in, we are willing to act as if we did. This is a matter of judgment.

The assumption that the population distribution is normal cannot be effectively checked with only 5 or 10 observations. In part, the researchers rely on experience with similar variables. They also look at the data. Stemplots of both distributions appear in Figure 11.3 (we rounded the cockroach data). The distribution of the 10 taste scores doesn't have a regular shape, but there are no gaps or outliers or other signs of nonnormal behavior. The cockroach data, on the other hand, have a wide gap between the two smallest and the three largest observations. In observational data, this might suggest two different species of roaches. In this case we know that all five cockroaches came from a single population grown in the laboratory. The gap is just chance variation in a very small sample.

Because the *t* procedures are so common, all statistical software systems will do the calculations for you. Figure 11.4 shows the output from three statistical packages: Data Desk, Minitab, and S-PLUS. In each case,

```
2 | 2 4          −1 | 3
3 |             −0 | 4
4 |              0 | 4 7
5 | 3 6          1 | 1 2
6 | 8            2 | 0 0 2 3
7 |              3 |
8 |
     (a)            (b)
```

FIGURE 11.3   Stemplots of the data in (a) Example 11.1 and (b) Example 11.2.

we entered the 10 sweetness losses as values of a variable called "cola" and asked for the one-sample t test of H_0: $\mu = 0$ against H_a: $\mu > 0$. The three outputs report slightly different information, but all include the basic facts: $\bar{x} = 1.02$, $t = 2.70$, $P = .012$. These are the results we found in Example 11.2.

```
                        Data Desk
cola:
Test Ho: mu (cola) = 0 vs Ha: mu (cola) > 0
Sample Mean = 1.02000 t-Statistic = 2.697 w/9 df
Reject Ho at Alpha = 0.0500
p = 0.0123
                        Minitab

TEST OF MU = 0.000 VS   MU   G. T.   0.000
           N    MEAN    STDEV    SE MEA    T    P   VALUE
cola      10    1.02    1.196    0.3      2.7     0.012


                        S-PLUS
data:      cola
t = 2.6967,   df = 9,   p-value = 0.0123
alternative hypothesis: true mean is greater than 0
sample estimates:
     mean of x
             1.02
```

FIGURE 11.4   Output for the one-sample t test of Example 11.2 from three statistical software packages. You can easily locate the basic results in output from any statistical software.

EXERCISES

11.6 What critical value t^* from Table C would you use for a confidence interval for the mean of the population in each of the following situations?

(a) A 95% confidence interval based on $n = 10$ observations.
(b) A 99% confidence interval from an SRS of 20 observations.
(c) An 80% confidence interval from a sample of size 7.

11.7 The one-sample t statistic for testing

$$H_0: \mu = 0$$
$$H_a: \mu > 0$$

from a sample of $n = 15$ observations has the value $t = 1.82$.

(a) What are the degrees of freedom for this statistic?
(b) Give the two critical values t^* from Table C that bracket t. What are the right-tail probabilities p for these two entries?
(c) Between what two values does the P-value of the test fall?
(d) Is the value $t = 1.82$ significant at the 5% level? Is it significant at the 1% level?

11.8 The one-sample t statistic from a sample of $n = 25$ observations for the two-sided test of

$$H_0: \mu = 64$$
$$H_a: \mu \neq 64$$

has the value $t = 1.12$.

(a) What are the degrees of freedom for t?
(b) Locate the two critical values t^* from Table C that bracket t. What are the right-tail probabilities p for these two values?
(c) Between what two values does the P-value of the test fall? (Note that H_a is two-sided.)
(d) Is the value $t = 1.12$ statistically significant at the 10% level? At the 5% level?

11.9 Poisoning by the pesticide DDT causes tremors and convulsions. In a study of DDT poisoning, researchers fed several rats a measured amount of DDT. They then made measurements on the rats' nervous systems that might explain how DDT poisoning causes tremors. One important variable was the "absolutely refractory period," the time required for a nerve to recover after a stimulus. This period varies normally. Measurements on four rats gave the data below (in milliseconds).[2]

$$1.6 \quad 1.7 \quad 1.8 \quad 1.9$$

(a) Find the mean refractory period $\bar{x}$ and the standard error of the mean.

(b) Give a 90% confidence interval for the mean absolutely refractory period for all rats of this strain when subjected to the same treatment.

11.10 The level of various substances in the blood of kidney dialysis patients is of concern because kidney failure and dialysis can lead to nutritional problems. A researcher did blood tests on several dialysis patients on six consecutive clinic visits. One variable she measured was the level of phosphate in the blood. An individual's phosphate levels tend to vary normally over time. The data on one patient, in milligrams of phosphate per deciliter of blood, are[3]

$$5.6 \quad 5.1 \quad 4.6 \quad 4.8 \quad 5.7 \quad 6.4$$

(a) Calculate the sample mean $\bar{x}$ and also its standard error.

(b) Use the t procedures to give a 90% confidence interval for this patient's mean phosphate level.

11.11 Suppose that the mean absolutely refractory period for unpoisoned rats is known to be 1.3 milliseconds. DDT poisoning should slow nerve recovery and so increase this period. Do the data in Exercise 11.9 give good evidence for this supposition? State H_0 and H_a and do a t test. Between what levels from Table C does the P-value lie? What do you conclude from the test?

Matched pairs t procedures

The cockroach study in Example 11.1 estimated the mean amount of sugar in the hindgut, but the researchers then compared results for several body tissues and several times before dissection to get a bigger picture. The taste test in Example 11.2 was a matched pairs study in which the same 10 tasters rated before-and-after sweetness. Comparative studies are more convincing than single-sample investigations. For that reason, one-sample inference is less common than comparative inference. One

matched pairs
design

common design to compare two treatments makes use of one-sample procedures. In a ***matched pairs design***, subjects are matched in pairs and each treatment is given to one subject in each pair. The experimenter can toss a coin to assign two treatments to the two subjects in each pair. Another situation calling for matched pairs is before-and-after observations on the same subjects, as in the taste test of Example 11.2.

MATCHED PAIRS t PROCEDURES

To compare the responses to the two treatments in a matched pairs design, apply the one-sample t procedures to the observed differences.

The parameter μ in a matched pairs t procedure is the mean difference in the responses to the two treatments within matched pairs of subjects in the entire population.

EXAMPLE 11.3

The National Endowment for the Humanities sponsors summer institutes to improve the skills of high school language teachers. One institute hosted 20 French teachers for four weeks. At the beginning of the period, the teachers took the Modern Language Association's listening test of understanding of spoken French. After four weeks of immersion in French in and out of class, they took the listening test again. (The actual spoken French in the two tests was different, so that simply taking the first test should not improve the score on the second test.) Table 11.1 gives the pretest and posttest scores. The maximum possible score on the test is 36.[4]

To analyze these data, subtract the pretest score from the posttest score to obtain the improvement for each teacher. These 20 differences form a single sample.

They appear in the "Gain" column in Table 11.1. The first teacher, for example, improved from 32 to 34, so the gain is $34 - 32 = 2$.

Step 1: Hypotheses. To assess whether the institute significantly improved the teachers' comprehension of spoken French, we test

$$H_0: \mu = 0$$
$$H_a: \mu > 0$$

Here μ is the mean improvement that would be achieved if the entire population of French teachers attended a summer institute. The null hypothesis says that no improvement occurs, and H_a says that posttest scores are higher on the average.

Step 2: Test statistic. The 20 differences have

$$\bar{x} = 2.5 \quad \text{and} \quad s = 2.893$$

TABLE 11.1	MLA listening scores for 20 French teachers						
Teacher	Pretest	Posttest	Gain	Teacher	Pretest	Posttest	Gain
1	32	34	2	11	30	36	6
2	31	31	0	12	20	26	6
3	29	35	6	13	24	27	3
4	10	16	6	14	24	24	0
5	30	33	3	15	31	32	1
6	33	36	3	16	30	31	1
7	22	24	2	17	15	15	0
8	25	28	3	18	32	34	2
9	32	26	−6	19	23	26	3
10	20	26	6	20	23	26	3

The one-sample t statistic is therefore

$$t = \frac{\bar{x} - 0}{s/\sqrt{n}} = \frac{2.5 - 0}{2.893/\sqrt{20}} = 3.86$$

df = 19

p	.001	.0005
t^*	3.579	3.883

Step 3: *P*-value. Find the *P*-value from the $t(19)$ distribution. (Remember that the degrees of freedom are 1 less than the sample size.) Table C shows that 3.86 lies between the upper 0.001 and 0.0005 critical values of the $t(19)$ distribution. The *P*-value therefore lies between these values. A computer statistical package gives the value $P = .00053$. The improvement in listening scores is very unlikely to be due to chance alone. We have strong evidence that the institute was effective in raising scores. In scholarly publications, the details of routine statistical procedures are usually omitted. This test would be reported in the form "The improvement in scores was significant ($t = 3.86$, df = 19, $P = .00053$)."

A 90% confidence interval for the mean improvement in the entire population requires the critical value $t^* = 1.729$ from Table C. The confidence interval is

$$\bar{x} \pm t^* \frac{s}{\sqrt{n}} = 2.5 \pm 1.729 \frac{2.893}{\sqrt{20}} = 2.5 \pm 1.12$$

$$= (1.38, \ 3.62)$$

The estimated average improvement is 2.5 points, with margin of error 1.12 for 90% confidence. Though statistically significant, the effect of attending the institute was rather small.

```
-6 │ 0
-5 │
-4 │
-3 │
-2 │
-1 │
-0 │
 0 │ 000
 1 │ 00
 2 │ 000
 3 │ 000000
 4 │
 5 │
 6 │ 00000
```

FIGURE 11.5   Stemplot of the gains in French comprehension for the 20 teacher in Example 11.3. All the leaves are 0 because all scores are whole numbers that are used as stems.

Example 11.3 illustrates how to restate matched pairs data as single-sample data by taking differences within each pair. We are in fact making inferences about a single population, the population of all differences within matched pairs. It is incorrect to ignore the pairs and analyze the data as if we had two samples, one from teachers who attended an institute and a second from teachers who did not. Inference procedures for comparing two samples assume that the samples are selected independently of each other. This assumption does not hold when the same subjects are measured twice. The proper analysis depends on the design used to produce the data.

What about the assumptions of simple random sampling and normality? The use of the t procedures in Example 11.3 is a bit questionable. First, the teachers are not an SRS from the population of high school French teachers. There is some selection bias in favor of energetic, committed teachers who are willing to give up four weeks of their summer vacation. It is therefore not clear to exactly what population the results apply. This vagueness is common when we don't actually take an SRS from a population.

Second, a look at the data shows that several of the teachers had pretest scores close to the maximum of 36. They could not improve their scores very much even if their mastery of French increased substantially. This is a weakness in the listening test that is the measuring instrument in this study. The differences in scores may not adequately indicate the effectiveness of the institute. This is one reason why the average increase was small.

A final difficulty facing the t procedures in Example 11.3 is that the data show departures from normality. In a matched pairs analysis, the population of *differences* must have a normal distribution because the t procedures are applied to the differences. Figure 11.5 is a stemplot of the 20 differences. One teacher actually lost 6 points between the pretest and the posttest. This one subject lowered the sample mean from 2.95 for the other 19 subjects to 2.5 for all 20. The stemplot shows this outlier, and a gap between 3 and 6 that may be due to chance. The fact that all the leaves in the stemplot are 0 reminds us that only whole-number scores are possible. The distribution is not normal even if we remove the outlier. Does this nonnormality forbid use of the t test? The behavior of the t procedures when the population does not have a normal distribution is one of their most important properties. We will discuss this issue later.

TI-83 techniques

Confidence intervals and t tests of significance can be performed on the TI-83, thus avoiding table look-ups. Here is a brief summary of the techniques. Return to the cola sweetness tasting of Example 11.2. The 10 sweetness losses were given as

$$2.0 \quad 0.4 \quad 0.7 \quad 2.0 \quad -0.4 \quad 2.2 \quad -1.3 \quad 1.2 \quad 1.1 \quad 2.3$$

Confidence Intervals   The first step is to enter the data into a list, such as L_1. All of the inference routines are found under STAT / TESTS. To determine a confidence interval, select choice 8:TInterval. In the TInterval screen, specify "Data" (vs. Stats), list L_1, and confidence level .95. Highlight "Calculate" and press ENTER. The results tell us that the 95% confidence interval for the true mean population sweetness loss is between 0.1644 and 1.8756. Note that if the company wanted to keep the 95% confidence level but wanted a shorter, more precise confidence interval, they would need to hire more taste testers, that is, increase the sample size n.

L1	L2	L3	2
2	▀▀▀▀	------	
.4			
.7			
2			
-.4			
2.2			
-1.3			
L2(1)=			

```
EDIT CALC TESTS
2↑T-Test…
3:2-SampZTest…
4:2-SampTTest…
5:1-PropZTest…
6:2-PropZTest…
7:ZInterval…
8↓TInterval…
```

```
TInterval
 Inpt:Data Stats
 List:L1
 Freq:1
 C-Level:.95
 Calculate
```

```
TInterval
 (.16436,1.8756)
 x̄=1.02
 Sx=1.196104789
 n=10
```

Test of Significance   Recall that the null hypothesis was "no sweetness loss," while the alternative hypothesis said that "there is some loss of sweetness":

$$H_0: \mu = 0$$

$$H_a: \mu > 0$$

With the data stored in list L_1, select STAT / TESTS / 2:T-Test. In the T-Test screen, specify the requested values: "Data," $\mu_0 = 0$, data in list L_1, and alternative hypothesis $\mu > 0$. Then you have a choice: Calculate or Draw.

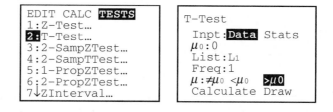

If you select "Calculate," the following screen appears:

```
T-Test
 μ>0
 t=2.69668949
 p=.0122631561
 x̄=1.02
 Sx=1.196104789
 n=10
```

The test statistic is $t = 2.69$ and the P-value is 0.0123.

If you specify "Draw," you see a $t(9)$ distribution curve with the upper critical area shaded. In either case, the P-value is 0.0123, the same probability that was calculated earlier using Table C.

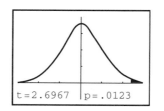

Remember that if you conduct a test of significance, it's always good practice to follow this up by determining a confidence interval for the true mean. The significance test simply tells us that in a month of storage under normal temperature conditions, there is some loss of sweetness. The confidence interval gives us some insight into how much sweetness was lost, on average.

If you are given the mean $\bar{x}$ instead of the original data, you would select the option "Stats" instead of "Data" in either the TInterval or the T-Test screen and then provide the values requested. The next exercise will follow this route.

EXERCISES

Many exercises from this point on ask you to give the P-value of a t test. If you have a suitable calculator or computer software, give the exact P-value. Otherwise, use Table C to give two values between which P lies.

11.12   An agricultural field trial compares the yield of two varieties of tomatoes for commercial use. The researchers divide in half each of 10 small plots of land in different locations and plant each tomato variety on one half of each plot. After harvest, they compare the yields in pounds per plant at each location. The 10 differences (Variety A − Variety B) give $\bar{x} = .34$ and $s = .83$. Is there convincing evidence that Variety A has the higher mean yield?

(a)   Describe in words what the parameter μ is in this setting.

(b)   State H_0 and H_a.

(c)   Find the t statistic and give a P-value. What do you conclude?

11.13   The design of controls and instruments affects how easily people can use them. A student project investigated this effect by asking 25 right-handed students to turn a knob (with their right hands) that moved an indicator by screw action. There were two identical instruments, one with a right-hand thread (the knob turns clockwise) and the other with a left-hand thread (the knob must be turned counterclockwise). The following table gives the times in seconds each subject took to move the indicator a fixed distance:[5]

Subject	Right thread	Left thread	Subject	Right thread	Left thread
1	113	137	14	107	87
2	105	105	15	118	166
3	130	133	16	103	146
4	101	108	17	111	123
5	138	115	18	104	135
6	118	170	19	111	112
7	87	103	20	89	93
8	116	145	21	78	76
9	75	78	22	100	116
10	96	107	23	89	78
11	122	84	24	85	101
12	103	148	25	88	123
13	116	147			

(a) Each of the 25 students used both instruments. Discuss briefly how you would use randomization in arranging the experiment.

(b) The project hoped to show that right-handed people find right-hand threads easier to use. What is the parameter μ for a matched pairs t test? State H_0 and H_a in terms of μ.

(c) Carry out a test of your hypotheses. Give the P-value and report your conclusions.

11.14   Give a 90% confidence interval for the mean time advantage of right-hand over left-hand threads in the setting of Exercise 11.13. Do you think that the time saved would be of practical importance if the task were performed many times—for example, by an assembly line worker? To help answer this question, find the mean time for right-hand threads as a percent of the mean time for left-hand threads.

Robustness of t procedures

The one-sample t procedures are exactly correct only when the population is normal. Real populations are never exactly normal. The usefulness of the t procedures in practice therefore depends on how strongly they are affected by lack of normality.

ROBUST PROCEDURES

A confidence interval or significance test is called **robust** if the confidence level or P-value does not change very much when the assumptions of the procedure are violated.

Because the tails of normal curves drop off quickly, samples from normal distributions will have very few outliers. Outliers suggest that your data are not a sample from a normal population. Like $\bar{x}$ and s, *the t procedures are strongly influenced by outliers.* If we dropped the single outlier in Example 11.3, the test statistic would change from $t = 3.86$ to $t = 5.98$ and the *P*-value would be much smaller. In this case, the outlier makes the test result *less* significant and the margin of error of the confidence interval *larger* than they would otherwise be. The results of the *t* procedures in Example 11.3 are conservative in the sense that the conclusions show a smaller effect than would be the case if the outlier were not present.

Fortunately, the *t* procedures are quite robust against nonnormality of the population when there are no outliers, especially when the distribution is roughly symmetric. Larger samples improve the accuracy of *P*-values and critical values from the *t* distributions when the population is not normal. The main reason for this is the central limit theorem. The *t* statistic uses the sample mean $\bar{x}$, which becomes more nearly normal as the sample size gets larger even when the population does not have a normal distribution.

Always make a plot to check for skewness and outliers before you use the *t* procedures for small samples. For most purposes, you can safely use the one-sample *t* procedures when $n \geq 15$ unless an outlier or quite strong skewness is present. If we can justify removing the outlier in Example 11.3 (perhaps that teacher was ill when she took the posttest), we can use the *t* procedures on the 19 remaining observations. Here are practical guidelines for inference on a single mean.[6]

USING THE *t* PROCEDURES

- Except in the case of small samples, the assumption that the data are an SRS from the population of interest is more important than the assumption that the population distribution is normal.
- *Sample size less than 15.* Use *t* procedures if the data are close to normal. If the data are clearly nonnormal or if outliers are present, do not use *t*.
- *Sample size at least 15.* The *t* procedures can be used except in the presence of outliers or strong skewness.
- *Large samples.* The *t* procedures can be used even for clearly skewed distributions when the sample is large, roughly $n \geq 40$.

EXAMPLE 11.4   Consider several of the data sets we graphed in Chapter 1. Figure 11.6 shows the histograms.

- Figure 11.6(a) is a histogram of the percent of each state's residents who are over 65 years of age. *We have data on the entire population of 50 states, so formal inference makes no sense.* We can calculate the exact mean for the population. There is no uncertainty due to having only a sample from the population, and no need for a confidence interval or test.

- Figure 11.6(b) shows the time of the first lightning flash each day in a mountain region in Colorado. The data contain more than 70 observations

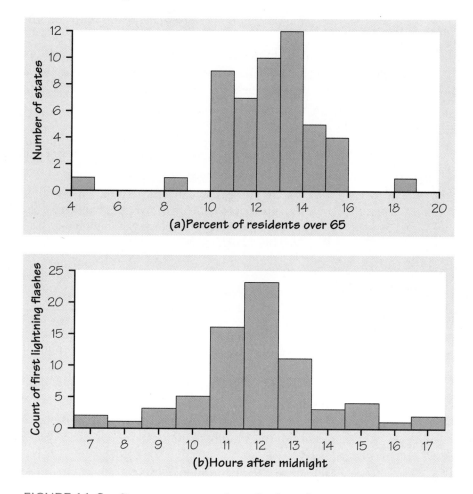

FIGURE 11.6   Can we use *t* procedures for these data? (a) Percent of residents over 65 years of age in the states. No: this is an entire population, not a sample. (b) Time of first lightning flash each day at a site in Colorado. Yes: there are over 70 observations with a symmetric distribution.

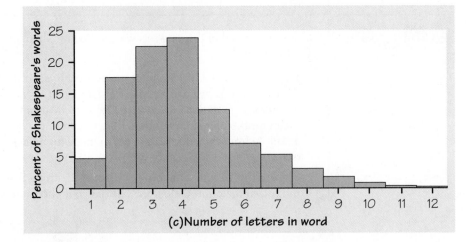

FIGURE 11.6   (*continued*) Can we use *t* procedures for these data? (c) Word lengths in Shakespeare's plays. *Yes, if the sample is large enough* to overcome the right-skewness.

that have a symmetric distribution. You can use the *t* procedures to draw conclusions about the mean time of a day's first lightning flash with complete confidence.

- Figure 11.6(c) shows that the distribution of word lengths in Shakespeare's plays is skewed to the right. We aren't told how large the sample is. You can use the *t* procedures for a distribution like this if the sample size is roughly 40 or larger.

EXERCISES

11.15   The Acculturation Rating Scale for Mexican Americans (ARSMA) measures the extent to which Mexican Americans have adopted Anglo/English culture. During the development of ARSMA, the test was given to a group of 17 Mexicans. Their scores, from a possible range of 1.00 to 5.00, had a symmetric distribution with $\bar{x} = 1.67$ and $s = .25$. Because low scores should indicate a Mexican cultural orientation, these results helped to establish the validity of the test.[7]

(a)   Give a 95% confidence interval for the mean ARSMA score of Mexicans.

(b)   What assumptions does your confidence interval require? Which of these assumptions is most important in this case?

11.16 A bank wonders whether omitting the annual credit card fee for customers who charge at least $2400 in a year would increase the amount charged on its credit cards. The bank makes this offer to an SRS of 200 of its credit card customers. It then compares how much these customers charge this year with the amount that they charged last year. The mean increase is $332, and the standard deviation is $108.

(a) Is there significant evidence at the 1% level that the mean amount charged increases under the no-fee offer? State H_0 and H_a and carry out a t test.

(b) Give a 99% confidence interval for the mean amount of the increase.

(c) The distribution of the amount charged is skewed to the right, but outliers are prevented by the credit limit that the bank enforces on each card. Use of the t procedures is justified in this case even though the population distribution is not normal. Explain why.

(d) A critic points out that the customers would probably have charged more this year than last even without the new offer, because the economy is more prosperous and interest rates are lower. Briefly describe the design of an experiment to study the effect of the no-fee offer that would avoid this criticism.

11.17 Here are measurements (in millimeters) of a critical dimension for 16 auto engine crankshafts.

224.120	224.001	224.017	223.982	223.989	223.961
223.960	224.089	223.987	223.976	223.902	223.980
224.098	224.057	223.913	223.999		

The dimension is supposed to be 224 mm and the variability of the manufacturing process is unknown. Is there evidence that the mean dimension is not 224 mm?

(a) Check the data graphically for outliers or strong skewness that might threaten the validity of the t procedures. What do you conclude?

(b) State H_0 and H_a and carry out a t test. Give the P-value (from Table C or software). What do you conclude?

11.18 Many homeowners buy detectors to check for the invisible gas radon in their homes. How accurate are these detectors? To answer this question, university researchers placed 12 radon detectors in a chamber that exposed them to 105 picocuries per liter of radon. The detector readings were as follows.[8]

$$91.9 \quad 97.8 \quad 111.4 \quad 122.3 \quad 105.4 \quad 95.0$$
$$103.8 \quad 99.6 \quad 96.6 \quad 119.3 \quad 104.8 \quad 101.7$$

(a) Make a stemplot of the data. The distribution is somewhat skewed to the right, but not strongly enough to forbid use of the t procedures.

(b) Is there convincing evidence that the mean reading of all detectors of this type differs from the true value 105? Carry out a test in detail, then write a brief conclusion.

The power of the t test

The power of a statistical test measures its ability to detect deviations from the null hypothesis. In practice we carry out the test in the hope of showing that the null hypothesis is false, so high power is important. The power of the one-sample t test against a specific alternative value of the population mean μ is the probability that the test will reject the null hypothesis when the mean has this alternative value. To calculate the power, we assume a fixed level of significance, usually $\alpha = .05$.

Calculation of the exact power of the t test takes into account the estimation of σ by s and is a bit complex. But an approximate calculation that acts as if σ were known is usually adequate for planning a study. This calculation is very much like that for the power of the z test, presented on pages 545–546.

EXAMPLE 11.5

It is the winter before the summer language institute of Example 11.3. The director, thinking ahead to the report he must write, hopes that enrolling 20 teachers will enable him to be quite certain of detecting an average improvement of 2 points in the mean listening score. Is this realistic?
We wish to compute the power of the t test for

$$H_0\text{: } \mu = 0$$
$$H_a\text{: } \mu > 0$$

against the alternative $\mu = 2$ when $n = 20$. We must have a rough guess of the size of σ in order to compute the power. People planning a large study often run a small pilot study for this and other purposes. In this case, listening-score improvements in past summer language institutes have had sample standard deviations of about 3. We therefore take both $\sigma = 3$ and $s = 3$ in our approximate calculation.

Step 1. *Write the rule for rejecting H_0 in terms of $\bar{x}$.* The t test with 20 observations rejects H_0 at the 5% significance level if the t statistic

$$t = \frac{\bar{x} - 0}{s/\sqrt{20}}$$

exceeds the upper 5% point of $t(19)$, which is 1.729. Taking $s = 3$, the test rejects H_0 when

$$t = \frac{\bar{x}}{3/\sqrt{20}} \geq 1.729$$

$$\bar{x} \geq 1.729\frac{3}{\sqrt{20}}$$

$$\bar{x} \geq 1.160$$

Step 2. *The power is the probability of rejecting H_0 assuming that the alternative is true.* We want the probability that $\bar{x} \geq 1.160$ when $\mu = 2$. Taking $\sigma = 3$, standardize $\bar{x}$ to find this probability:

$$P(\bar{x} \geq 1.160) = P\left(\frac{\bar{x} - 2}{3/\sqrt{20}} \geq \frac{1.160 - 2}{3/\sqrt{20}}\right)$$

$$= P(Z \geq -1.252)$$

$$= 1 - .1056 = .8944$$

A true difference of 2 points in the population mean scores will produce significance at the 5% level in 89% of all possible samples. The director can be reasonably confident of detecting a difference this large.

EXERCISES

11.19   The bank in Exercise 11.16 tested a new idea on a sample of 200 customers. The bank wants to be quite certain of detecting a mean increase of $\mu = \$100$ in the amount charged, at the $\alpha = .01$ significance level. Perhaps a sample of only $n = 50$ customers would accomplish this. Find the approximate power of the test with $n = 50$ against the alternative $\mu = \$100$ as follows:

(a) What is the critical value t^* for the one-sided test with $\alpha = .01$ and $n = 50$?

(b) Write the rule for rejecting H_0: $\mu = 0$ in terms of the t statistic. Then take $s = 108$ (an estimate based on the data in Exercise 11.16) and state the rejection rule in terms of $\bar{x}$.

(c) Assume that $\mu = 100$ (the given alternative) and that $\sigma = 108$ (an estimate from the data in Exercise 11.16). The approximate power is the probability of the event you found in (b), calculated under these assumptions. Find the power. Would you recommend that the bank do a test on 50 customers, or should more customers be included?

11.20   The tomato experts who carried out the field trial described in Exercise 11.12 suspect that the large P-value there is due to low power. They would like to be able to detect a mean difference in yields of 0.5 pound per plant at the 0.05 significance level. Based on the previous study, use 0.83 as an estimate of both the population σ and the value of s in future samples.

(a) What is the power of the test from Exercise 11.12 with $n = 10$ against the alternative $\mu = .5$?

(b) If the sample size is increased to $n = 25$ plots of land, what will be the power against the same alternative?

SUMMARY

Tests and confidence intervals for the mean μ of a normal population are based on the sample mean $\bar{x}$ of an SRS. Because of the central limit theorem, the resulting procedures are approximately correct for other population distributions when the sample is large.

The standardized sample mean is the **one-sample z statistic,**

$$z = \frac{\bar{x} - \mu}{\sigma/\sqrt{n}}$$

When we know σ, we use the z statistic and the standard normal distribution.

In practice, we do not know σ. Replace the standard deviation $\sigma/\sqrt{n}$ of $\bar{x}$ by the **standard error** $s/\sqrt{n}$ to get the **one-sample t statistic**

$$t = \frac{\bar{x} - \mu}{s/\sqrt{n}}$$

The t statistic has the **t distribution** with $n - 1$ degrees of freedom.

There is a t distribution for every positive **degrees of freedom** k. All are symmetric distributions similar in shape to the standard normal distribution. The $t(k)$ distribution approaches the $N(0, 1)$ distribution as k increases.

An exact level C **confidence interval** for the mean μ of a normal population is

$$\bar{x} \pm t^* \frac{s}{\sqrt{n}}$$

where t^* is the upper $(1 - C)/2$ critical value of the $t(n - 1)$ distribution.

Significance tests for H_0: $\mu = \mu_0$ are based on the t statistic. Use P-values or fixed significance levels from the $t(n - 1)$ distribution.

Use these one-sample procedures to analyze **matched pairs** data by first taking the difference within each matched pair to produce a single sample.

The t procedures are relatively **robust** when the population is nonnormal, especially for larger sample sizes. The t procedures are useful for nonnormal data when $n \geq 15$ unless the data show outliers or strong skewness.

SECTION 11.1 EXERCISES

When an exercise asks for a P-value, give P exactly if you have a suitable calculator or computer software. Otherwise, use Table C to give two values between which P lies.

11.21 The one-sample t statistic for a test of

$$H_0\text{: } \mu = 10$$
$$H_a\text{: } \mu < 10$$

based on $n = 10$ observations has the value $t = -2.25$.

(a) What are the degrees of freedom for this statistic?

(b) Between what two probabilities p from Table C does the P-value of the test fall?

11.22 A manufacturer of small appliances employs a market research firm to estimate retail sales of its products by gathering information from a sample of retail stores. This month an SRS of 75 stores in the Midwest sales region finds that these stores sold an average of 24 of the manufacturer's hand mixers, with standard deviation 11.

(a) Give a 95% confidence interval for the mean number of mixers sold by all stores in the region.

(b) The distribution of sales is strongly right skewed, because there are many smaller stores and a few very large stores. The use of t in (a) is reasonably safe despite this violation of the normality assumption. Why?

11.23   In a randomized comparative experiment on the effect of calcium in the diet on blood pressure, researchers divided 54 healthy white males at random into two groups. One group received calcium; the other, a placebo. At the beginning of the study, the researchers measured many variables on the subjects. The paper reporting the study gives $\bar{x} = 114.9$ and $s = 9.3$ for the seated systolic blood pressure of the 27 members of the placebo group.

(a) Give a 95% confidence interval for the mean blood pressure in the population from which the subjects were recruited.

(b) What assumptions about the population and the study design are required by the procedure you used in (a)? Which of these assumptions are important for the validity of the procedure in this case?

11.24   Gas chromatography is a sensitive technique used to measure small amounts of compounds. The response of a gas chromatograph is calibrated by repeatedly testing specimens containing a known amount of the compound to be measured. A calibration study for a specimen containing 1 nanogram (that's 10^{-9} gram) of a compound gave the following response readings.[9]

$$21.6 \quad 20.0 \quad 25.0 \quad 21.9$$

The response is known from experience to vary according to a normal distribution unless an outlier indicates an error in the analysis. Estimate the mean response to 1 nanogram of this substance, and give the margin of error for your choice of confidence level. Then explain to a chemist who knows no statistics what your margin of error means.

11.25   The embryos of brine shrimp can enter a dormant phase in which metabolic activity drops to a low level. Researchers studying this dormant phase measured the level of several compounds important to normal metabolism. They reported their results in a table, with the note, "Values are means ± SEM for three independent samples." The table entry for the compound ATP was $0.84 \pm .01$. Biologists reading the article must be able to decipher this.[10]

(a) What does the abbreviation SEM stand for?

(b) The researchers made three measurements of ATP, which had $\bar{x} =$ .84. What was the sample standard deviation s for these measurements?

(c) Give a 90% confidence interval for the mean ATP level in dormant brine shrimp embryos.

11.26 The table below gives the pretest and posttest scores on the Modern Language Association's listening test in Spanish for 20 high school Spanish teachers who attended an intensive summer course in Spanish. The setting is identical to the French institute described in Example 11.3.[11]

Subject	Pretest	Posttest	Subject	Pretest	Posttest
1	30	29	11	30	32
2	28	30	12	29	28
3	31	32	13	31	34
4	26	30	14	29	32
5	20	16	15	34	32
6	30	25	16	20	27
7	34	31	17	26	28
8	15	18	18	25	29
9	28	33	19	31	32
10	20	25	20	29	32

(a) We hope to show that attending the institute improves listening skills. State an appropriate H_0 and H_a. Be sure to identify the parameter appearing in the hypotheses.

(b) Make a graphical check for outliers or strong skewness in the data that you will use in your statistical test, and report your conclusions on the validity of the test.

(c) Carry out a test. Can you reject H_0 at the 5% significance level? At the 1% significance level?

(d) Give a 90% confidence interval for the mean increase in listening score due to attending the summer institute.

11.27 The ARSMA test (Exercise 11.15) was compared with a similar test, the Bicultural Inventory (BI), by administering both tests to 22 Mexican Americans. Both tests have the same range of scores (1.00 to 5.00) and are scaled to have similar means for the groups used to develop them. There was a high correlation between the two scores, giving evidence that both are measuring the same characteristics. The researchers wanted to know whether the population mean scores for the two tests are the same. The differences in scores (ARSMA − BI) for the 22 subjects had $\bar{x} = .2519$ and $s = .2767$.

(a) Describe briefly how to arrange the administration of the two tests to the subjects, including randomization.

(b) Carry out a significance test for the hypothesis that the two tests have the same population mean. Give the P-value and state your conclusion.

(c) Give a 95% confidence interval for the difference between the two population mean scores.

11.28 A study of the pay of corporate CEOs (chief executive officers) examined the cash compensation, adjusted for inflation, of the CEOs of 104 corporations over the period 1977 to 1988. Among the data are the average annual pay increases for each of the 104 CEOs. The mean percent increase in pay was 6.9%. The data showed great variation, with a standard deviation of 17.4%. The distribution was strongly skewed to the right.[12]

(a) Despite the skewness of the distribution, there were no extreme outliers. Explain why we can use t procedures for these data.

(b) What are the degrees of freedom? When the exact degrees of freedom do not appear in Table C, use the next lower degrees of freedom in the table.

(c) Give a 99% confidence interval for the mean increase in pay for all corporate CEOs. What essential condition must the data satisfy if we are to trust your result?

11.29 Table 1.3 (page 31) gives the ages of U.S. presidents when they took office. It does not make sense to use the t procedures (or any other statistical procedures) to give a 95% confidence interval for the mean age of the presidents. Explain why not.

11.30 Exercise 11.27 reports a small study comparing ARSMA and BI, two tests of the acculturation of Mexican Americans. Would this study usually detect a difference in mean scores of 0.2? To answer this question, calculate the approximate power of the test (with $n = 22$ subjects and $\alpha = .05$) of

$$H_0: \mu = 0$$
$$H_a: \mu \neq 0$$

against the alternative $\mu = .2$. Note that this is a two-sided test.

(a) From Table C, what is the critical value for $\alpha = .05$?

(b) Write the rule for rejecting H_0 at the $\alpha = .05$ level. Then take $s = .3$, the approximate value observed in Exercise 11.27, and restate the rejection criterion in terms of $\bar{x}$.

(c) Find the probability of this event when $\mu = .2$ (the alternative given) and $\sigma = .3$ (estimated from the data in Exercise 11.27) by a normal probability calculation. This is the approximate power.

11.2 COMPARING TWO MEANS

Comparing two populations or two treatments is one of the most common situations encountered in statistical practice. We call such situations *two-sample problems*.

TWO-SAMPLE PROBLEMS

- The goal of inference is to compare the responses to two treatments or to compare the characteristics of two populations.
- We have a separate sample from each treatment or each population.

Two-sample problems

A two-sample problem can arise from a randomized comparative experiment that randomly divides subjects into two groups and exposes each group to a different treatment. Comparing random samples separately selected from two populations is also a two-sample problem. Unlike the matched pairs designs studied earlier, there is no matching of the units in the two samples and the two samples can be of different sizes. Inference procedures for two-sample data differ from those for matched pairs. Here are some typical two-sample problems.

EXAMPLE 11.6

(a) A medical researcher is interested in the effect on blood pressure of added calcium in our diet. She conducts a randomized comparative experiment in which one group of subjects receives a calcium supplement and a control group gets a placebo.

(b) A psychologist develops a test that measures social insight. He compares the social insight of male college students with that of female college students by giving the test to a large group of students of each sex.

(c) A bank wants to know which of two incentive plans will most increase the use of its credit cards. It offers each incentive to a random sample of credit card customers and compares the amount charged during the following six months.

EXERCISES

11.31   The following situations require inference about a mean or means. Identify each as (1) single sample, (2) matched pairs, or (3) two samples. The procedures of Section 11.1 apply to cases (1) and (2). We are about to learn procedures for (3).

(a) An education researcher wants to learn whether it is more effective to put questions before or after introducing a new concept in an elementary school mathematics text. He prepares two text segments that teach the concept, one with motivating questions before and the other with review questions after. He uses each text segment to teach a separate group of children. The researcher compares the scores of the groups on a test over the material.

(b) Another researcher approaches the same issue differently. She prepares text segments on two unrelated topics. Each segment comes in two versions, one with questions before and the other with questions after. The subjects are a single group of children. Each child studies both topics, one (chosen at random) with questions before and the other with questions after. The researcher compares test scores for each child on the two topics to see which topic he or she learned better.

11.32 The following situations require inference about a mean or means. Identify each as (1) single sample, (2) matched pairs, or (3) two samples. The procedures of Section 11.1 apply to cases (1) and (2). We are about to learn procedures for (3).

(a) To check a new analytical method, a chemist obtains a reference specimen of known concentration from the National Institute of Standards and Technology. She then makes 20 measurements of the concentration of this specimen with the new method and checks for bias by comparing the mean result with the known concentration.

(b) Another chemist is checking the same new method. He has no reference specimen, but a familiar analytic method is available. He wants to know if the new and old methods agree. He takes a specimen of unknown concentration and measures the concentration 10 times with the new method and 10 times with the old method.

Comparing two population means

We can examine two-sample data graphically by comparing stemplots (for small samples) or histograms or boxplots (for larger samples). Now we will apply the ideas of formal inference in this setting. When both population distributions are symmetric, and especially when they are at least approximately normal, a comparison of the mean responses in the two populations is the most common goal of inference. Here are the assumptions we will make.

ASSUMPTIONS FOR COMPARING TWO MEANS

- We have **two SRSs**, from two distinct populations. The samples are **independent**. That is, one sample has no influence on the other. Matching violates independence, for example. We measure the same variable for both samples.
- Both populations are **normally distributed**. The means and standard deviations of the populations are unknown.

Call the variable we measure x_1 in the first population and x_2 in the second because the variable may have different distributions in the two populations. Here is the notation we will use to describe the two populations:

Population	Variable	Mean	Standard deviation
1	x_1	μ_1	σ_1
2	x_2	μ_2	σ_2

There are four unknown parameters, the two means and the two standard deviations. The subscripts remind us which population a parameter describes. We want to compare the two population means, either by giving a confidence interval for their difference $\mu_1 - \mu_2$ or by testing the hypothesis of no difference, H_0: $\mu_1 = \mu_2$.

We use the sample means and standard deviations to estimate the unknown parameters. Again, subscripts remind us which sample a statistic comes from. Here is the notation that describes the samples:

Population	Sample size	Sample mean	Sample standard deviation
1	n_1	$\bar{x}_1$	s_1
2	n_2	$\bar{x}_2$	s_2

To do inference about the difference $\mu_1 - \mu_2$ between the means of the two populations, we start from the difference $\bar{x}_1 - \bar{x}_2$ between the means of the two samples.

EXAMPLE 11.7

Does increasing the amount of calcium in our diet reduce blood pressure? Examination of a large sample of people revealed a relationship between calcium intake and blood pressure. The relationship was strongest for black men. Such observational studies do not establish causation. Researchers therefore designed a randomized comparative experiment.

The subjects in part of the experiment were 21 healthy black men. A randomly chosen group of 10 of the men received a calcium supplement for 12 weeks. The control group of 11 men received a placebo pill that looked identical. The experiment was double-blind. The response variable is the decrease in systolic (heart contracted) blood pressure for a subject after 12 weeks, in millimeters of mercury. An increase appears as a negative response.[13]

Take Group 1 to be the calcium group and Group 2 the placebo group. Here are the data for the 10 men in Group 1 (calcium),

$$7 \quad -4 \quad 18 \quad 17 \quad -3 \quad -5 \quad 1 \quad 10 \quad 11 \quad -2$$

and for the 11 men in Group 2 (placebo),

$$-1 \quad 12 \quad -1 \quad -3 \quad 3 \quad -5 \quad 5 \quad 2 \quad -11 \quad -1 \quad -3$$

From the data, calculate the summary statistics:

Group	Treatment	n	$\bar{x}$	s
1	Calcium	10	5.000	8.743
2	Placebo	11	-.273	5.901

The calcium group shows a drop in blood pressure, $\bar{x}_1 = 5.000$, while the placebo group had almost no change, $\bar{x}_2 = -.273$. Is this outcome good evidence that calcium decreases blood pressure in the entire population of healthy black men more than a placebo does?

Example 11.7 fits the two-sample setting. We write hypotheses in terms of the mean decreases we would see in the entire population, μ_1 for men taking calcium for 12 weeks and μ_2 for men taking a placebo. The hypotheses are

$$H_0: \mu_1 = \mu_2$$
$$H_a: \mu_1 > \mu_2$$

We want to test these hypotheses and also estimate the size of calcium's advantage, $\mu_1 - \mu_2$.

Are the assumptions satisfied? Because of the randomization, we are willing to regard the calcium and placebo groups as two independent SRSs. Although the samples are small, we check for serious nonnormality by examining the data. Here is a back-to-back stemplot of the responses. (We have split the stems. Notice that negative responses require -0 and 0 to be separate stems, and that the ordering of leaves out from the stems recognizes that -3 is smaller than -1.)

```
      Calcium          Placebo
                  −1 | 1
             5    −0 | 5
           234    −0 | 33111
             1     0 | 23
             7     0 | 5
            10     1 | 2
            87     1 |
```

The placebo responses appear roughly normal. The calcium group has an irregular distribution, which is not unusual when we have only a few observations. There are no outliers, and no departures from normality that prevent use of t procedures.

The natural estimator of the difference $\mu_1 - \mu_2$ is the difference between the sample means:

$$\bar{x}_1 - \bar{x}_2 = 5.000 - (-.273) = 5.273$$

This statistic measures the average advantage of calcium over a placebo. In order to use it for inference, we must know its sampling distribution.

The sampling distribution of $\bar{x}_1 - \bar{x}_2$

Here are the facts about the sampling distribution of the difference $\bar{x}_1 - \bar{x}_2$ between the sample means of two independent SRSs. These facts can be derived using the mathematics of probability or made plausible by simulation.

- The mean of $\bar{x}_1 - \bar{x}_2$ is $\mu_1 - \mu_2$. That is, the difference of sample means is an unbiased estimator of the difference of population means.

- The variance of the difference is the *sum* of the variances of $\bar{x}_1$ and $\bar{x}_2$, which is

$$\frac{\sigma_1^2}{n_1} + \frac{\sigma_2^2}{n_2}$$

Note that the *variances* add. The standard deviations do not.

- If the two population distributions are both normal, then the distribution of $\bar{x}_1 - \bar{x}_2$ is also normal.

Because the statistic $\bar{x}_1 - \bar{x}_2$ has a normal distribution, we can standardize it to obtain a standard normal z statistic. Subtract its mean, then

two-sample   divide by its standard deviation to get the **two-sample z statistic:**
z statistic

$$z = \frac{(\bar{x}_1 - \bar{x}_2) - (\mu_1 - \mu_2)}{\sqrt{\dfrac{\sigma_1^2}{n_1} + \dfrac{\sigma_2^2}{n_2}}}$$

Two-sample t procedures

two-sample t statistic

We don't know the population standard deviations σ_1 and σ_2. Following the pattern of the one-sample case, substitute the standard errors $s_i/\sqrt{n_i}$ for the standard deviations $\sigma_i/\sqrt{n_i}$ in the two-sample z statistic. The result is the **two-sample t statistic**:

$$t = \frac{(\bar{x}_1 - \bar{x}_2) - (\mu_1 - \mu_2)}{\sqrt{\dfrac{s_1^2}{n_1} + \dfrac{s_2^2}{n_2}}}$$

The statistic t has the same interpretation as any z or t statistic: it says how far $\bar{x}_1 - \bar{x}_2$ is from its mean in standard deviation units. Unfortunately, the two-sample t statistic does *not* have a t distribution. A t distribution replaces a $N(0, 1)$ distribution when we replace just one standard deviation in a z statistic by a standard error. In this case, we replaced two standard deviations by the corresponding standard errors. This does not produce a statistic having a t distribution.

Nonetheless, the two-sample t statistic is used with t critical values in inference for two-sample problems. There are two ways to do this.

Option 1: Use procedures based on the statistic t with critical values from a t distribution with degrees of freedom computed from the data. The degrees of freedom are generally not a whole number. This is a very accurate approximation to the distribution of t.

Option 2: Use procedures based on the statistic t with critical values from the t distribution with degrees of freedom equal to the smaller of $n_1 - 1$ and $n_2 - 1$. These procedures are always conservative for any two normal populations.

Most statistical software systems and the TI-83 use the two-sample t statistic with Option 1 for two-sample problems unless the user requests another method. Using this option without software is a bit complicated. We will therefore present the second, simpler, option first. We recommend that you use Option 2 when doing calculations without a TI-83 or computer. If you use a computer package, it should automatically do the calculations for Option 1. Here is a statement of the Option 2 procedures that includes a statement of just how they are "conservative."

THE TWO-SAMPLE t PROCEDURES

Draw an SRS of size n_1 from a normal population with unknown mean μ_1, and draw an independent SRS of size n_2 from another normal population with unknown mean μ_2. The confidence interval for $\mu_1 - \mu_2$ given by

$$(\bar{x}_1 - \bar{x}_2) \pm t^* \sqrt{\frac{s_1^2}{n_1} + \frac{s_2^2}{n_2}}$$

has confidence level *at least* C no matter what the population standard deviations may be. Here t^* is the upper $(1 - C)/2$ critical value for the $t(k)$ distribution with k the smaller of $n_1 - 1$ and $n_2 - 1$.

To test the hypothesis H_0: $\mu_1 = \mu_2$, compute the two-sample t statistic

$$t = \frac{\bar{x}_1 - \bar{x}_2}{\sqrt{\dfrac{s_1^2}{n_1} + \dfrac{s_2^2}{n_2}}}$$

and use P-values or critical values for the $t(k)$ distribution. The true P-value or fixed significance level will always be *equal to or less than* the value calculated from $t(k)$ no matter what values the unknown population standard deviations have.

These two-sample t procedures always err on the safe side, reporting *higher* P-values and *lower* confidence than are actually true. The gap between what is reported and the truth is quite small unless the sample sizes are both small and unequal. As the sample sizes increase, probability values based on t with degrees of freedom equal to the smaller of $n_1 - 1$ and $n_2 - 1$ become more accurate.[14] The following examples illustrate the two-sample t procedures.

EXAMPLE 11.8

The medical researchers in Example 11.7 can use the two-sample t procedures to compare calcium with a placebo. The test statistic for the null hypothesis H_0: $\mu_1 = \mu_2$ is

$$t = \frac{\bar{x}_1 - \bar{x}_2}{\sqrt{\dfrac{s_1^2}{n_1} + \dfrac{s_2^2}{n_2}}}$$

$$= \frac{5.000 - (-.273)}{\sqrt{\dfrac{8.743^2}{10} + \dfrac{5.901^2}{11}}}$$

$$= \frac{5.273}{3.2878} = 1.604$$

df = 9

p	.10	.05
t^*	1.383	1.833

There are 9 degrees of freedom, the smaller of $n_1 - 1 = 9$ and $n_2 - 1 = 10$. Because H_a is one-sided on the high side, the P-value is the area to the right of $t = 1.604$ under the $t(9)$ curve. Figure 11.7 illustrates this P-value. Table C shows that it lies between 0.05 and 0.10. The experiment found evidence that calcium reduces blood pressure, but the evidence falls a bit short of the traditional 5% and 1% levels.

For a 90% confidence interval, Table C shows that the $t(9)$ critical value is $t^* = 1.833$. We are 90% confident that the mean advantage of calcium over a placebo, $\mu_1 - \mu_2$, lies in the interval

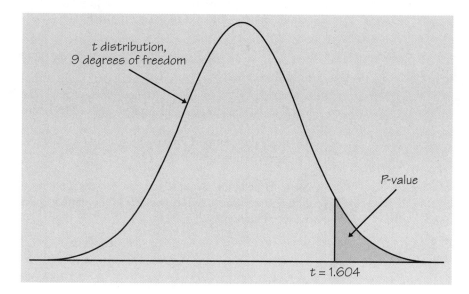

FIGURE 11.7 The P-value in Example 11.8. This example uses the conservative method, which leads to the t distribution with 9 degrees of freedom.

$$(\bar{x}_1 - \bar{x}_2) \pm t^* \sqrt{\frac{s_1^2}{n_1} + \frac{s_2^2}{n_2}} = [5.000 - (-.273)] \pm 1.833 \sqrt{\frac{8.743^2}{10} + \frac{5.901^2}{11}}$$

$$= 5.273 \pm 6.026$$

$$= (-.753, 11.299)$$

That the 90% confidence interval covers 0 tells us that we cannot reject H_0: $\mu_1 = \mu_2$ against the two-sided alternative at the $\alpha = .10$ level of significance.

Sample size strongly influences the P-value of a test. An effect that fails to be significant at a specified level α in a small sample will be significant in a larger sample. In the case of the rather small samples in Example 11.8, we suspect that more data might show that calcium has a significant effect. The published account of the study combined these results for blacks with results for whites and adjusted for pretest differences among the subjects. Using this more detailed analysis, the researchers were able to report the P-value $P = .008$.

EXAMPLE 11.9

The Chapin Social Insight Test is a psychological test designed to measure how accurately a person appraises other people. The possible scores on the test range from 0 to 41. During the development of the Chapin test, it was given to several different groups of people. Here are the results for male and female college students majoring in the liberal arts:[15]

Group	Sex	n	$\bar{x}$	s
1	Male	133	25.34	5.05
2	Female	162	24.94	5.44

Do these data support the contention that female and male students differ in average social insight?

Step 1: Hypotheses. Because we had no specific direction for the male/female difference in mind before looking at the data, we choose the two-sided alternative. The hypotheses are

$$H_0: \mu_1 = \mu_2$$
$$H_a: \mu_1 \neq \mu_2$$

Step 2: Test statistic. The two-sample t statistic is

$$t = \frac{\bar{x}_1 - \bar{x}_2}{\sqrt{\frac{s_1^2}{n_1} + \frac{s_2^2}{n_2}}} = \frac{25.34 - 24.94}{\sqrt{\frac{5.05^2}{133} + \frac{5.44^2}{162}}} = .654$$

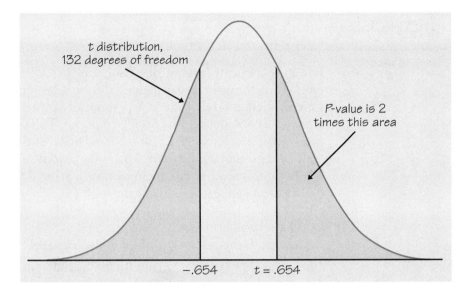

FIGURE 11.8   The P-value in Example 11.9. To find P, find the area above
$t = .654$ and double it because the alternative is two-sided.

df = 100		
p	.25	.20
t^*	0.677	0.845

Step 3: P-value. There are 132 degrees of freedom, the smaller of

$$n_1 - 1 = 133 - 1 = 132 \quad \text{and}$$
$$n_2 - 1 = 162 - 1 = 161$$

Figure 11.8 illustrates the P-value. Find it by comparing 0.654 to critical values for the $t(132)$ distribution and then doubling p because the alternative is two-sided. Degrees of freedom 132 do not appear in Table C, so we use the next smaller table value, degrees of freedom 100. Table C shows that 0.654 does not reach the 0.25 critical value, which is the largest upper tail probability in Table C. The P-value is therefore greater than 0.50. The data give no evidence of a male/female difference in mean social insight score ($t = .654$, df = 132, $P > .5$).

The researcher in Example 11.9 did not do an experiment but compared samples from two populations. The large samples imply that the assumption that the populations have normal distributions is of little importance. The sample means will be nearly normal in any case. The major question concerns the population to which the conclusions apply. The student subjects are certainly not an SRS of all liberal arts majors in the country. If they are volunteers from a single college, the sample results may not extend to a wider population.

EXERCISES

11.33   In a study of heart surgery, one issue was the effect of drugs called beta-blockers on the pulse rate of patients during surgery. The available subjects were divided at random into two groups of 30 patients each. One group received a beta-blocker; the other, a placebo. The surgical team recorded the pulse rate of each patient at a critical point during the operation. The treatment group had mean 65.2 beats per minute and standard deviation 7.8. For the control group, the mean was 70.3 and the standard deviation 8.3. The data appear roughly normal.

    (a) Do beta-blockers reduce the pulse rate? State the hypotheses and do a t test. Is the result significant at the 5% level? At the 1% level?

    (b) Give a 99% confidence interval for the difference in mean pulse rates.

11.34   In a study of cereal leaf beetle damage on oats, researchers measured the number of beetle larvae per stem in small plots of oats after randomly applying one of two treatments: no pesticide, or malathion at the rate of 0.25 pound per acre. The data appear roughly normal. Here are the summary statistics.[16]

Group	Treatment	n	$\bar{x}$	s
1	Control	13	3.47	1.21
2	Malathion	14	1.36	0.52

Is there significant evidence at the 1% level that malathion reduces the mean number of larvae per stem? Be sure to state H_0 and H_a.

11.35   A business school study compared a sample of Greek firms that went bankrupt with a sample of healthy Greek businesses. One measure of a firm's financial health is the ratio of current assets to current liabilities, called *CA/CL*. For the year before bankruptcy, the study found the mean *CA/CL* to be 1.72565 in the healthy group and 0.78640 in the group that failed. The paper reporting the study says that $t = 7.36$.[17]

    (a) You can draw a conclusion from this t without using a table and even without knowing the sizes of the samples (as long as the samples are not tiny). What is your conclusion? Why don't you need the sample size and a table?

(b) In fact, the study looked at 33 firms that failed and 68 healthy firms. What degrees of freedom would you use for the t test if you follow the conservative approach recommended for use without software?

Two-sample inference with the TI-83

Constructing confidence intervals and t tests of significance for two-sample models on the TI-83 is very similar to the one-sample case. To illustrate, we will use the data on calcium supplements to lower blood pressure from Examples 11.7 and 11.8. The data represent a decrease in systolic blood pressure after 12 weeks, in millimeters of mercury. The data for the 10 men in Group 1 (calcium) were

$$7 \quad -4 \quad 18 \quad 17 \quad -3 \quad -5 \quad 1 \quad 10 \quad 11 \quad -2$$

and for the 11 men in Group 2 (placebo),

$$-1 \quad 12 \quad -1 \quad -3 \quad 3 \quad -5 \quad 5 \quad 2 \quad -11 \quad -1 \quad -3$$

Test of Significance The first step is to enter the Group 1 data into a list, say L_1, and the Group 2 data into another list, say L_2. To specify the significance test, select STAT / TESTS / 4:2-SampTTest. In the 2-SampTTest screen, specify "Data" (vs. Stats), lists L_1 and L_2, $\mu_1 > \mu_2$ for the alternative hypothesis, and "No" for Pooled. Highlight "Calculate" and press ENTER. (The Pooled option will be discussed later.)

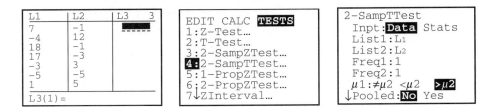

The results tell us that the t test statistic is $t = 1.6037$, and the P-value is $P = .0644$. This represents only modest evidence to reject H_0.

```
2-SampTTest
μ1>μ2
 t=1.603717288
 p=.0644196844
 df=15.59051297
 x̄1=5
↓x̄2=-.272727273
```

If you select "Draw" in the 2-SampTTest screen instead of "Calculate," the $t(k)$ distribution will be displayed, showing the t test statistic $t = 1.6037$ and the upper 0.0644 critical area shaded.

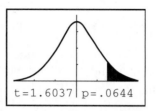

Confidence Interval With the data still stored in lists L_1 and L_2, selects STAT / TESTS / 0:2-SampTInt.

```
EDIT CALC TESTS         2-SampTInt
4↑2-SampTTest…            Inpt:Data  Stats
5:1-PropZTest…           List1:L1
6:2-PropZTest…           List2:L2
7:ZInterval…             Freq1:1
8:TInterval…             Freq2:1
9:2-SampZInt…            C-Level:.9
0↓2-SampTInt…           ↓Pooled:No  Yes
```

In the 2-SampTInt screen, specify the requested values: "Data," Group 1 data in list L_1 and Group 2 data in list L_2, C-level: .9(90% confidence level), and "No" for Pooled. Highlight "Calculate" and press ENTER. The following screen appears:

```
2-SampTInt
  (−.4767,11.022)
  df=15.59051297
  x̄1 =5
  x̄2=−.272727273
  Sx1=8.74325137
 ↓Sx2=5.90069333
```

The 90% confidence interval for the true difference of the means $\mu_1 - \mu_2$ is $(-.4767, 11.022)$.

If you are given the means $\bar{x}_1$ and $\bar{x}_2$ and sample standard deviations s_1 and s_2 instead of the original data, select the option "Stats" instead of "Data" in either the 2-SampTTest or the 2-SampTInt screen. Then provide the values requested.

Robustness again

The two-sample t procedures are more robust than the one-sample t methods, particularly when the distributions are not symmetric. When the sizes of the two samples are equal and the two populations being compared have distributions with similar shapes, probability values from the t table are quite accurate for a broad range of distributions when the sample sizes are as small as $n_1 = n_2 = 5$.[18] When the two population distributions have different shapes, larger samples are needed.

As a guide to practice, adapt the guidelines given on page 606 for the use of one-sample t procedures to two-sample procedures by replacing "sample size" with the "sum of the sample sizes," $n_1 + n_2$. These guidelines err on the side of safety, especially when the two samples are of equal size. In planning a two-sample study, you should usually choose equal sample sizes. The two-sample t procedures are most robust against nonnormality in this case, and the conservative probability values are most accurate.

EXERCISES

11.36   In Activity 11, you collected data on the distances flown by two prototype paper airplane models. Now perform a test of significance to test your hypotheses. Then construct an appropriate confidence interval for the true difference in distances flown, $\mu_A - \mu_B$. Write your conclusions in simple language.

11.37   College financial aid offices expect students to use summer earnings to help pay for college. But how large are these earnings? One college studied this question by asking a sample of students how much they earned. Omitting students who were not employed, there were 1296 responses. Here are the data in summary form:[19]

Group	n	$\bar{x}$	s
Males	675	$1884.52	$1368.37
Females	621	$1360.39	$1037.46

(a)   The distribution of earnings is strongly skewed to the right. Nevertheless, use of t procedures is justified. Why?

(b)   Give a 90% confidence interval for the difference between the mean summer earnings of male and female students.

(c) Once the sample size was decided, the sample was chosen by taking every 20th name from an alphabetical list of all undergraduates. Is it reasonable to consider the samples as SRSs chosen from the male and female undergraduate populations?

(d) What other information about the study would you request before accepting the results as describing all undergraduates?

11.38   Ordinary corn doesn't have as much of the amino acid lysine as animals need in their feed. Plant scientists have developed varieties of corn that have increased amounts of lysine. In a test of the quality of high-lysine corn as animal feed, an experimental group of 20 one-day-old male chicks ate a ration containing the new corn. A control group of another 20 chicks received a ration that was identical except that it contained normal corn. Here are the weight gains (in grams) after 21 days:[20]

Control				Experimental			
380	321	366	356	361	447	401	375
283	349	402	462	434	403	393	426
356	410	329	399	406	318	467	407
350	384	316	272	427	420	477	392
345	455	360	431	430	339	410	326

(a) Present the data graphically. Are there outliers or strong skewness that might prevent the use of t procedures?

(b) Is there good evidence that chicks fed high-lysine corn gain weight faster? Carry out a test and report your conclusions.

(c) Give a 95% confidence interval for the mean extra weight gain in chicks fed high-lysine corn.

11.39   The Survey of Study Habits and Attitudes (SSHA) is a psychological test that measures the motivation, attitude toward school, and study habits of students. Scores range from 0 to 200. A selective private college gives the SSHA to an SRS of both male and female first-year students. The data for the women are as follows:

154	109	137	115	152	140	154	178	101
103	126	126	137	165	165	129	200	148

Here are the scores of the men:

$$108 \quad 140 \quad 114 \quad 91 \quad 180 \quad 115 \quad 126 \quad 92 \quad 169 \quad 146$$
$$109 \quad 132 \quad 75 \quad 88 \quad 113 \quad 151 \quad 70 \quad 115 \quad 187 \quad 104$$

(a) Examine each sample graphically, with special attention to outliers and skewness. Is use of a t procedure acceptable for these data?

(b) Most studies have found that the mean SSHA score for men is lower than the mean score in a comparable group of women. Is this true for first-year students at this college? Carry out a test and give your conclusions.

(c) Give a 90% confidence interval for the mean difference between the SSHA scores of male and female first-year students at this college.

More accurate levels in the t procedures*

The two-sample t statistic does not have a t distribution. Moreover, the exact distribution changes as the unknown population standard deviations σ_1 and σ_2 change. However, an excellent approximation is available.

APPROXIMATE DISTRIBUTION OF THE TWO-SAMPLE t STATISTIC

The distribution of the two-sample t statistic is close to the t distribution with degrees of freedom df given by

$$df = \frac{\left(\dfrac{s_1^2}{n_1} + \dfrac{s_2^2}{n_2}\right)^2}{\dfrac{1}{n_1 - 1}\left(\dfrac{s_1^2}{n_1}\right)^2 + \dfrac{1}{n_2 - 1}\left(\dfrac{s_2^2}{n_2}\right)^2}$$

This approximation is quite accurate when both sample sizes n_1 and n_2 are 5 or larger.

The t procedures remain exactly as before except that we use the t distribution with df degrees of freedom to give critical values and P-values.

---

*This section can be omitted unless you are using statistical software and wish to understand what the software does.

EXAMPLE 11.10

In the calcium experiment of Examples 11.7 and 11.8 the data gave

Group	Treatment	n	$\bar{x}$	s
1	Calcium	10	5.000	8.743
2	Placebo	11	$-.273$	5.901

For improved accuracy, we can use critical points from the t distribution with degrees of freedom df given by

$$\text{df} = \frac{\left(\dfrac{8.743^2}{10} + \dfrac{5.901^2}{11}\right)^2}{\dfrac{1}{9}\left(\dfrac{8.743^2}{10}\right)^2 + \dfrac{1}{10}\left(\dfrac{5.901^2}{11}\right)^2}$$

$$= \frac{116.848}{7.494} = 15.59$$

Notice that the degrees of freedom df is not a whole number.

The conservative 90% confidence interval for $\mu_1 - \mu_2$ in Example 11.8 used the critical value $t^* = 1.833$ based on 9 degrees of freedom. A more exact confidence interval replaces this critical value with the critical value for df $= 15.59$ degrees of freedom. We cannot find this critical value exactly without using a computer package. For a close approximation, use the next smaller entry (15 degrees of freedom) in Table C. The critical value is $t^* = 1.753$. The 90% confidence interval is now

$$(\bar{x}_1 - \bar{x}_2) \pm t^* \sqrt{\frac{s_1^2}{n_1} + \frac{s_2^2}{n_2}}$$

$$= [5.000 - (-.273)] \pm 1.753\sqrt{\frac{8.743^2}{10} + \frac{5.901^2}{11}}$$

$$= 5.273 \pm 5.764$$

$$= (-.491, 11.037)$$

This confidence interval is a bit shorter (margin of error 5.764 rather than 6.026) than the conservative interval in Example 11.8.

As Example 11.10 illustrates, the two-sample t procedures are exactly as before, except that we use a t distribution with more degrees of freedom. The number df from the box on page 633 is always at least as large as the smaller of $n_1 - 1$ and $n_2 - 1$. On the other hand, df is never larger than the sum $n_1 + n_2 - 2$ of the two individual degrees of freedom. The number of degrees of freedom df is generally not a whole number. There is a t distribution for any positive degrees of freedom, even though Table C contains entries only for whole-number degrees of freedom. Some software packages find df and then use the t distribution with the next smaller whole-number degrees of freedom. Others take care to use $t(\text{df})$ even when df is not a whole number. We do not recommend regular use of this method unless a computer is doing the arithmetic. With a TI-83 or computer, the more accurate procedures are painless, as the following examples illustrate.

EXAMPLE 11.11

In the calcium study of Example 11.8, the pencil-and-paper-with-tables solution selected the following as the degrees of freedom: $\min(n_1 - 1, n_2 - 1) = \min(9, 10) = 9$. Using df = 9, Example 11.8 calculated the 90% confidence interval for $\mu_1 - \mu_2$ to be $(-0.753, 11.299)$. In Example 11.10, the more accurate fractional degrees of freedom were calculated by the formula to be 15.59, and using the smaller whole-number value df = 15 and Table C, the critical value was determined to be $t^* = 1.753$. With this t^*-value, the 90% confidence interval was calculated to be $(-0.491, 11.037)$. By comparison, the TI-83 calculates this same 90% confidence interval to be $(-0.4767, 11.022)$. The confidence interval that the calculator produces is shorter and more precise. The reason, of course, is that the calculator has been programmed to use the formula on page 633 to give the more accurate fractional degrees of freedom (15.59) and to calculate the t test statistic using this fractional df value.

```
2-SampTInt
 (-.4767,11.022)
 df=15.59051297
 x̄₁=5
 x̄₂=-.272727273
 Sx1=8.74325137
↓Sx2=5.90069333
```

EXAMPLE 11.12

Poisoning by the pesticide DDT causes convulsions in humans and other mammals. Researchers seek to understand how the convulsions are caused. In a randomized comparative experiment, they compared 6 white rats poisoned with DDT with a control group of 6 unpoisoned rats. Electrical measurements of nerve activity are the main clue to the nature of DDT poisoning. When a nerve is stimulated, its electrical response shows a sharp spike followed by a much smaller second spike. The experiment found that the second spike is larger in rats fed DDT than in normal rats. This finding helped biologists understand how DDT poisoning works.[21]

The researchers measured the height of the second spike as a percent of the first spike when a nerve in the rat's leg was stimulated. For the poisoned rats the results were

$$12.207 \quad 16.869 \quad 25.050 \quad 22.429 \quad 8.456 \quad 20.589$$

The control group data were

$$11.074 \quad 9.686 \quad 12.064 \quad 9.351 \quad 8.182 \quad 6.642$$

Both populations are reasonably normal, as far as can be judged from six observations. The DDT data are much more spread out than the control data. The difference in means is quite large, but in such small samples the sample mean is highly variable. A significance test can help confirm that we are seeing a real effect. Because the researchers did not conjecture in advance that the size of the second spike would increase in rats fed DDT, we use the two-sided alternative:

$$H_0 : \mu_1 = \mu_2$$
$$H_a : \mu_1 \neq \mu_2$$

Here is the output from the SAS statistical software system for these data.[22]

TTEST PROCEDURE

Variable: SPIKE

GROUP	N	Mean	Std Dev	Std Error
DDT	6	17.60000000	6.34014839	2.58835474
CONTROL	6	9.49983333	1.95005932	0.79610839

Variances	T	DF	Prob>|T|
Unequal	2.9912	5.9	0.0247
Equal	2.9912	10.0	0.0135

SAS reports the results of two t procedures: the general two-sample procedure ("Unequal" variances) and a special procedure that assumes that the two population variances are equal. We are interested in the first of these procedures. The two-sample t statistic has the value $t = 2.9912$, the degrees of freedom are df $= 5.9$, and the P-value from the $t(5.9)$ distribution is 0.0247. There is good evidence that the mean size of the secondary spike is larger in rats fed DDT.

Would the conservative test based on 5 degrees of freedom (both $n_1 - 1$ and $n_2 - 1$ are 5) have given a different result in Example 11.12? The statistic is exactly the same: $t = 2.9912$. The conservative P-value is $2P(T \geq 2.9912)$, where T has the $t(5)$ distribution. Table C shows that 2.9912 lies between the 0.02 and 0.01 upper critical values of the $t(5)$ distribution, so P for the two-sided test lies between 0.02 and 0.04. For practical purposes this is the same result as that given by the software. As this example and Example 11.10 suggest, the difference between the t procedures using the conservative and the approximately correct distributions is rarely of practical importance. That is why we recommend the simpler conservative procedure for inference without a computer.

EXERCISES

11.40   Example 11.11 demonstrates that if all other statistics stay the same, a higher number of degrees of freedom will produce a narrower (and hence more precise) confidence interval. Briefly explain why this is so.

11.41   Use your TI-83 and the two-sample procedures to replicate the results of the DDT–nerve stimulus experiment in Example 11.12. Verify that you get the same t text statistic and P-value.

11.42   Example 11.12 reports the analysis of data on the effects of DDT poisoning. The software uses the two-sample t test with degrees of freedom given in the box on page 633. Starting from the computer's results for $\bar{x}_i$ and s_i, verify the computer's values for the test statistic $t = 2.99$ and the degrees of freedom df $= 5.9$.

11.43   What aspects of rowing technique distinguish between novice and skilled competitive rowers? Researchers compared two groups of female competitive rowers: a group of skilled rowers and a group of novices. The researchers measured many mechanical aspects of rowing style as the subjects rowed on a Stanford Rowing Ergometer. One important variable is the angular velocity of the knee, which describes the rate at which the knee joint opens as the legs push the body back on the sliding seat. The data show no outliers or strong skewness. Here is the SAS computer output:[23]

TTEST PROCEDURE

Variable: KNEE

GROUP	N	Mean	Std Dev	Std Error
SKILLED	10	4.18283335	0.47905935	0.15149187
NOVICE	8	3.01000000	0.95894830	0.33903942

Variances	T	DF	Prob>|T|
Unequal	3.1583	9.8	0.0104
Equal	3.3918	16.0	0.0037

(a) The researchers believed that the knee velocity would be higher for skilled rowers. State H_0 and H_a.

(b) What is the value of the two-sample t statistic and its P-value? (Note that SAS provides two-sided P-values. If you need a one-sided P-value, divide the two-sided value by 2.) What do you conclude?

(c) Give a 90% confidence interval for the mean difference between the knee velocities of skilled and novice female rowers.

11.44   The researchers in the previous exercise also wondered whether skilled and novice rowers differ in weight or other physical characteristics. Here is the SAS computer output for weight in kilograms.

TTEST PROCEDURE

Variable: WEIGHT

GROUP	N	Mean	Std Dev	Std Error
SKILLED	10	70.3700000	6.10034898	1.92909973
NOVICE	8	68.4500000	9.03999930	3.19612240

Variances	T	DF	Prob>|T|
Unequal	0.5143	11.8	0.6165
Equal	0.5376	16.0	0.5982

Is there significant evidence of a difference in the mean weights of skilled and novice rowers? State H_0 and H_a, report the two-sample t statistic and its P-value, and state your conclusion. (Note that SAS provides two-sided P-values. If you need a one-sided P-value, divide the two-sided value by 2.)

The pooled two-sample t procedures*

In Example 11.12, the software offered a choice between two t tests. One is labeled for "unequal" variances, the other for "equal" variances. The "unequal" variance procedure is our two-sample t. This test is valid whether or not the population variances are equal. The other choice is a special version of the two-sample t statistic that assumes that the two populations have the same variance. This procedure averages (the statistical term is "pools") the two sample variances to estimate the common population variance. The resulting statistic is called the pooled two-sample t statistic.

---

*This is a special topic that is optional.

It is equal to our t statistic if the two sample sizes are the same, but not otherwise. We could choose to use the pooled t for both tests and confidence intervals.

The pooled t statistic has the advantage that it has exactly the t distribution with $n_1 + n_2 - 2$ degrees of freedom *if* the two population variances really are equal. Of course, the population variances are often not equal. Moreover, the assumption of equal variances is hard to check from the data. The pooled t was in common use before software made it easy to use the accurate approximation to the distribution of our two-sample t statistic. Now it is useful only in special situations. We cannot use the pooled t in Example 11.12, for example, because it is clear that the variance is much larger among rats fed DDT.

SUMMARY

The data in a **two-sample problem** are two independent SRSs, each drawn from a separate normally distributed population.

Tests and confidence intervals for the difference between the means μ_1 and μ_2 of the two populations start from the difference $\bar{x}_1 - \bar{x}_2$ of the two sample means. Because of the central limit theorem, the resulting procedures are approximately correct for other population distributions when the sample sizes are large.

Draw independent SRSs of sizes n_1 and n_2 from two normal populations with parameters μ_1, σ_1 and μ_2, σ_2. The **two-sample t statistic** is

$$t = \frac{(\bar{x}_1 - \bar{x}_2) - (\mu_1 - \mu_2)}{\sqrt{\dfrac{s_1^2}{n_1} + \dfrac{s_2^2}{n_2}}}$$

The statistic t does *not* have exactly a t distribution.

For conservative inference procedures to compare μ_1 and μ_2, use the two-sample t statistic with the $t(k)$ distribution. The degrees of freedom k is the smaller of $n_1 - 1$ and $n_2 - 1$. For more accurate probability values, use the $t(k)$ distribution with degrees of freedom k estimated from the data. This is the usual procedure in statistical software.

The **confidence interval** for $\mu_1 - \mu_2$ given by

$$(\bar{x}_1 - \bar{x}_2) \pm t^* \sqrt{\dfrac{s_1^2}{n_1} + \dfrac{s_2^2}{n_2}}$$

has confidence level at least C if t^* is the upper $(1 - C)/2$ critical value for $t(k)$ with k the smaller of $n_1 - 1$ and $n_2 - 1$.

Significance tests for $H_0 : \mu_1 = \mu_2$ based on

$$t = \frac{\bar{x}_1 - \bar{x}_2}{\sqrt{\dfrac{s_1^2}{n_1} + \dfrac{s_2^2}{n_2}}}$$

have a true P-value no higher than that calculated from $t(k)$.

The guidelines for practical use of two-sample t procedures are similar to those for one-sample t procedures. Equal sample sizes are recommended.

SECTION 11.2 EXERCISES

Many of these exercises ask you to think about issues of statistical practice as well as to carry out t procedures.

11.45 The Johns Hopkins Regional Talent Searches give the Scholastic Assessment Tests (intended for high school juniors and seniors) to 13-year-olds. In all, 19,883 males and 19,937 females took the tests between 1980 and 1982. The mean scores of males and females on the verbal test are nearly equal, but there is a clear difference between the sexes on the mathematics test. The reason for this difference is not understood. Here are the data:[24]

Group	$\bar{x}$	s
Males	416	87
Females	386	74

Give a 99% confidence interval for the difference between the mean score for males and the mean score for females in the population that Johns Hopkins searches. Must SAT scores have a normal distribution in order for your confidence interval to be valid? Why?

11.46 A study of iron deficiency in infants compared samples of infants whose mothers chose different ways of feeding them. One group contained breast-fed infants. The children in another group were fed a standard baby formula without any iron supplements. Here are summary results on blood hemoglobin levels at 12 months of age:[25]

Group	n	$\bar{x}$	s
Breast-fed	23	13.3	1.7
Formula	19	12.4	1.8

(a) Is there significant evidence that the mean hemoglobin level is different among breast-fed babies? State H_0 and H_a and carry out a t test. Give the P-value. What is your conclusion?

(b) Give a 95% confidence interval for the mean difference in hemoglobin level between the two populations of infants.

(c) State the assumptions that your procedures in (a) and (b) require in order to be valid.

(d) Is this study an experiment? Why? How does this affect the conclusions we can draw from the study?

11.47   Physical fitness is related to personality characteristics. In one study of this relationship, middle-aged college faculty who had volunteered for a fitness program were divided into low-fitness and high-fitness groups based on a physical examination. The subjects then took the Cattell Sixteen Personality Factor Questionnaire. Here are the data for the "ego strength" personality factor:[26]

Group	Fitness	n	$\bar{x}$	s
1	Low	14	4.64	0.69
2	High	14	6.43	0.43

(a) Is the difference in mean ego strength significant at the 5% level? At the 1% level? Be sure to state H_0 and H_a.

(b) You should be hesitant to generalize these results to the population of all middle-aged men. Explain why.

11.48   A market research firm supplies manufacturers with estimates of the retail sales of their products from samples of retail stores. Marketing managers are prone to look at the estimate and ignore sampling error. An SRS of 75 stores this month shows mean sales of 52 units of a small appliance, with standard deviation 13 units. During the same month last year, an SRS of 53 stores gave mean sales of 49 units, with standard deviation 11 units. An increase from 49 to 52 is a rise of 6%. The marketing manager is happy, because sales are up 6%.

(a) Use the two-sample t procedure to give a 95% confidence interval for the difference between this year and last year in the mean number of units sold at all retail stores.

(b) Explain in language that the manager can understand why he cannot be confident that sales rose by 6%, and that in fact sales may even have dropped.

11.49   A bank compares two proposals to increase the amount that its credit card customers charge on their cards. (The bank earns a percentage of the amount charged, paid by the stores that accept the card.) Proposal A offers to eliminate the annual fee for customers who charge $2400 or more during the year. Proposal B offers a small percent of the total amount charged as a cash rebate at the end of the year. The bank offers each proposal to an SRS of 150 of its credit card customers. At the end of the year, the total amount charged by each customer is recorded. Here are the summary statistics:

Group	n	$\bar{x}$	s
A	150	$1987	$392
B	150	$2056	$413

(a) Do the data show a significant difference between the mean amounts charged by customers offered the two plans? Give the null and alternative hypotheses, and calculate the two-sample t statistic. Obtain the P-value. State your practical conclusions.

(b) The distributions of amounts charged are skewed to the right, but outliers are prevented by the limits that the bank imposes on credit balances. Do you think that skewness threatens the validity of the test that you used in (a)? Explain your answer.

(c) Is the bank's study an experiment? Why? How does this affect the conclusions the bank can draw from the study?

11.50   An educator believes that new reading activities in the classroom will help elementary school pupils improve their reading ability. She arranges for a third-grade class of 21 students to follow these activities for an 8-week period. A control classroom of 23 third graders follows the same curriculum without the activities. At the end of the 8 weeks, all students are given the Degree of Reading Power (DRP) test, which measures the aspects of reading ability that the treatment is designed to improve. Here are the data:[27]

Treatment					Control				
24	43	58	71	43	42	43	55	26	62
49	61	44	67	49	37	33	41	19	54
53	56	59	52	62	20	85	46	10	17
54	57	33	46	43	60	53	42	37	42
57					55	28	48		

(a) Examine the data with a graph. Are there strong outliers or skewness that could prevent use of the t procedures?

(b) Is there good evidence that the new activities improve the mean DRP score? Carry out a test and report your conclusions.

(c) Although this study is an experiment, its design is not ideal because it had to be done in a school without disrupting classes. What aspect of good experimental design is missing?

11.51   Researchers studying the learning of speech often compare measurements made on the recorded speech of adults and children. One variable of interest is called the voice onset time (VOT). Here are the results for 6-year-old children and adults asked to pronounce the word "bees." The VOT is measured in milliseconds and can be either positive or negative.[28]

Group	n	$\bar{x}$	s
Children	10	−3.67	33.89
Adults	20	−23.17	50.74

(a) The researchers were investigating whether VOT distinguishes adults from children. State H_0 and H_a and carry out a two-sample t test. Give a P-value and report your conclusions.

(b) Give a 95% confidence interval for the difference in mean VOTs when pronouncing the word "bees." Explain why you knew from your result in (a) that this interval would contain 0 (no difference).

11.52   The researchers in the study discussed in Exercise 11.51 looked at VOTs for adults and children pronouncing many different words. Explain why they should not do a separate two-sample t test for each word and conclude that those words with a significant difference (say $P < .05$) distinguish children from adults. (The researchers did not make this mistake.)

The remaining exercises concern the power of the two-sample t test, an optional topic. If you have read Section 10.4 and the discussion of the power of the one-sample t test on pages 610–611, Exercise 11.53 guides you in finding the power of the two-sample t.

11.53   In Example 11.8, a small study of black men suggested that a calcium supplement can reduce blood pressure. Now we are planning a larger clinical trial of this effect. We plan to use 100 subjects in each of the two groups. Are these sample sizes large enough to make it very likely that the study will give strong evidence ($\alpha = .01$) of the effect of calcium if in fact calcium lowers blood pressure by 5 millimeters more than a placebo? To answer this question, we will compute the power of the two-sample t test of

$$H_0 : \mu_1 = \mu_2$$
$$H_a : \mu_1 > \mu_2$$

against the specific alternative $\mu_1 - \mu_2 = 5$. Based on the pilot study reported in Example 11.8, we take 8, the larger of the two observed s-values, as a rough estimate of both the population σ's and future sample s's.

(a) What is the approximate value of the $\alpha = .01$ critical value t^* for the two-sample t statistic when $n_1 = n_2 = 100$?

(b) **Step 1** *Write the rule for rejecting H_0 in terms of $\bar{x}_1 - \bar{x}_2$.* The test rejects H_0 when

$$\frac{\bar{x}_1 - \bar{x}_2}{\sqrt{\dfrac{s_1^2}{n_1} + \dfrac{s_2^2}{n_2}}} \geq t^*$$

Take both s_1 and s_2 to be 8, and n_1 and n_2 to be 100. Find the number c such that the test rejects H_0 when $\bar{x}_1 - \bar{x}_2 \geq c$.

(c) **Step 2** *The power is the probability of rejecting H_0 when the alternative is true.* Suppose that $\mu_1 - \mu_2 = 5$ and that both σ_1 and σ_2 are 8. The power we seek is the probability that $\bar{x}_1 - \bar{x}_2 \geq c$ under these assumptions. Calculate the power.

11.54   You are planning a larger study of VOTs, based on the pilot study reported in Exercise 11.51. Not all words distinguish children (Group 1) from adults (Group 2) as well as "bees," so you want high power against the alternative that $\mu_1 - \mu_2 = 10$ in a *two-sided t* test. From the pilot study, take 30 as an estimate of σ_1 and s_1, and 50 as an estimate of σ_2 and s_2.

(a) State H_0 and H_a, and write the formula for the test statistic.

(b) Give the $\alpha = .05$ critical value for the test when $n_1 = 100$ and $n_2 = 300$. (Although we recommend equal sample sizes to improve the robustness of t procedures against nonnormality, sample sizes that are proportional to the variances give higher power for the same total number of observations.)

(c) Find the approximate power against the alternative $\mu_1 - \mu_2 = 10$.

11.55   A bank asks you to compare two ways to increase the use of its credit cards. Plan A would offer customers a cash-back rebate based on their total amount charged. Plan B would reduce the interest rate charged on card balances. The response variable is the total amount a customer charges during the test period. You decide to offer each of Plan A and Plan B to a separate SRS of the bank's credit card customers. In the past, the mean amount charged in a six-month period has been about $1100, with a standard deviation of $400. Will a two-sample t test based on SRSs of 350 customers in each group detect a difference of $100 in the mean amounts charged under the two plans?

(a) State H_0 and H_a, and write the formula for the test statistic.

(b) Give the $\alpha = .05$ critical value for the test when $n_1 = n_2 = 350$.

(c) Calculate the power of the test with $\alpha = .05$, using $400 as a rough estimate of all standard deviations.

CHAPTER REVIEW

This chapter presented t tests and confidence intervals for inference about the mean of a single population and for comparing the means of two populations. The one-sample t procedures do inference about one mean and the two-sample t procedures compare two means. Matched pairs studies use one-sample procedures because you first create a single sample by taking the differences in the responses within each pair.

The t procedures require that the data be random samples and that the distribution of the population or populations be normal. One reason for the wide use of t procedures is that they are not very strongly affected by lack of normality. If you can't regard your data as a random sample, however, the results of inference may be of little value.

Chapter 10 concentrated on the reasoning of confidence intervals and tests. Understanding the reasoning is essential for wise use of the t and other inference methods. The discussion in this chapter paid more attention to practical aspects of using the methods. We saw that there are several versions of the two-sample t, for example. Which one you use depends largely on whether or not you use statistical software. Before you use any inference method, think about the design of the study and examine the data for outliers and other problems.

The chapter exercises are important in this and later chapters. You must now recognize problem settings and decide which of the methods presented in the chapter fits. In this chapter, you must recognize one-sample studies, matched pairs studies, and two-sample studies.

Here are the most important skills you should have after reading this chapter.

A. RECOGNITION

1. Recognize when a problem requires inference about a mean or comparing two means.
2. Recognize from the design of a study whether one-sample, matched pairs, or two-sample procedures are needed.

B. ONE-SAMPLE t PROCEDURES

1. Use the t procedure to obtain a confidence interval at a stated level of confidence for the mean μ of a population.
2. Carry out a t test for the hypothesis that a population mean μ has a specified value against either a one-sided or a two-sided alternative. Use Table C of t critical values to approximate the P-value or carry out a fixed α test.
3. Recognize when the t procedures are appropriate in practice, in particular that they are quite robust against lack of normality but are influenced by outliers.
4. Also recognize when the design of the study, outliers, or a small sample from a skewed distribution make the t procedures risky.
5. Recognize matched pairs data and use the t procedures to obtain confidence intervals and to perform tests of significance for such data.

C. TWO-SAMPLE t PROCEDURES

1. Give a confidence interval for the difference between two means. Use the two-sample t statistic with conservative degrees of freedom if you do not have statistical software. Use the TI-83 or software if you have it.

2. Test the hypothesis that two populations have equal means against either a one-sided or a two-sided alternative. Use the two-sample t test with conservative degrees of freedom if you do not have statistical software. Use the TI-83 or software if you have it.

3. Recognize when the two-sample t procedures are appropriate in practice.

CHAPTER 11 REVIEW EXERCISES

11.56   In a study of the effectiveness of a weight-loss program, 47 subjects who were at least 20% overweight took part in the program for 10 weeks. Private weighings determined each subject's weight at the beginning of the program and 6 months after the program's end. The matched pairs t test was used to assess the significance of the average weight loss. The paper reporting the study said, "The subjects lost a significant amount of weight over time, $t(46) = 4.68$, $p < .01$." It is common to report the results of statistical tests in this abbreviated style.[29]

(a) Why was the matched pairs t test appropriate?

(b) Explain to someone who knows no statistics but is interested in weight-loss programs what the practical conclusion is.

(c) The paper follows the tradition of reporting significance only at fixed levels such as $\alpha = .01$. In fact, the results are more significant than "$p < .01$" suggests. Use Table C to say more about the P-value of the t test.

11.57   Consumers who think a product's advertising is expensive often also think the product must be of high quality. Can other information undermine this effect? To find out, marketing researchers did an experiment. The subjects were 90 women from the clerical and administrative staff of a large organization. All subjects read an ad that described a fictional line of food products called "Five Chefs." The ad also described the major TV commercials that would soon be shown, an unusual expense for this type of product. The 45 women in the control group read nothing else. The 45 in the "undermine group" also read a news story headlined "No Link between Advertising Spending and New Product Quality."

All the subjects then rated the quality of Five Chefs products on a seven-point scale. The study report said, "The mean quality ratings were significantly lower in the undermine treatment ($\overline{X}_A = 4.56$) than in the control treatment ($\overline{X}_C = 5.05$; $t = 2.64$, $p < .01$)."[30]

(a) Is the matched pairs *t* test or the two-sample *t* test the right test in this setting? Why?

(b) What degrees of freedom would you use for the *t* statistic you chose in (a)?

(c) The distribution of individual responses is not normal, because there is only a seven-point scale. Why is it nonetheless proper to use a *t* test?

11.58   A major study of alternative welfare programs randomly assigned women on welfare to one of two programs, called "WIN" and "Options." WIN was the existing program. The new Options program gave more incentives to work. An important question was how much more (on the average) women in Options earned than those in WIN. Here is Minitab output for earnings in dollars over a three-year period:[31]

```
TWOSAMPLE T FOR 'OPT' VS 'WIN'

            N      MEAN     STDEV     SE MEAN
OPT     1362      7638       289      7.8309
WIN     1395      6595       247      6.6132

95 PCT CI FOR MU OPT - MU WIN: (1022.90, 1063.10)
```

(a) Give a 99% confidence interval for the amount by which the mean earnings of Options participants exceeded the mean earnings of WIN subjects. (Minitab will give a 99% confidence interval if you instruct it to do so. Here we have only the basic output, which includes the 95% confidence interval.)

(b) The distribution of incomes is strongly skewed to the right but includes no extreme outliers because all the subjects were on welfare. What fact about these data allows us to use *t* procedures despite the strong skewness?

11.59   The one-hole test is used to test the manipulative skill of job applicants. This test requires subjects to grasp a pin, move it to a hole, insert it, and return for another pin. The score on the test is the number of pins the subject inserts in a fixed time interval. A study compared the scores of male college students with those of experienced female industrial workers. Here are the data for the first minute of the test, summarized by the Data Desk statistical software. ("Cases" is the count of subjects.)[32]

```
Summary statistics for students        Summary statistics for workers

Mean 35.1242                           Mean 37.3234
Cases 750                              Cases 412
StdDev 4.3108                          StdDev 3.8317
```

(a) We expect that the experienced workers will outperform the students. State the hypotheses for a statistical test of this expectation and perform the test. Give a *P*-value and state your conclusions.

(b) The distribution of scores is slightly skewed to the left. Explain why the procedure you used in (a) is nonetheless acceptable.

(c) One purpose of the study was to develop performance norms for job applicants. Based on the data above, estimate the range that covers the middle 95% of experienced workers. (Be careful! This is not the same as a 95% confidence interval for the mean score of experienced workers.)

11.60 A college golf coach thinks that her players improve their scores in the second round of tournaments because they are less nervous after the first round. Here are the scores of the 12 team members in the two rounds of a tournament:

Golfer	1	2	3	4	5	6	7	8	9	10	11	12
Round 1	89	90	87	95	86	81	102	105	83	88	91	79
Round 2	94	85	89	89	81	76	107	89	87	91	88	80

The coach asks you whether the scores support her opinion. Remember that in golf lower scores are better.

(a) State the hypotheses you would like to test. What test procedure do you plan to use? Explain your choice.

(b) Graph the data in a way that matches your planned analysis. Unfortunately, the graph shows that the *t* procedures cannot be safely used to analyze these data. Why? (There are other procedures that can be used. You should seek expert advice.)

11.61 Nitrites are often added to meat products as preservatives. In a study of the effect of nitrites on bacteria, researchers measured the rate of uptake of an amino acid for 60 cultures of bacteria: 30 growing in a medium to which nitrites had been added and another 30 growing in a standard medium as a control group. Here are the data from this study:

Control			Nitrite		
6,450	8,709	9,361	8,303	8,252	6,594
9,011	9,036	8,195	8,534	10,227	6,642
7,821	9,996	8,202	7,688	6,811	8,766
6,579	10,333	7,859	8,568	7,708	9,893
8,066	7,408	7,885	8,100	6,281	7,689
6,679	8,621	7,688	8,040	9,489	7,360
9,032	7,128	5,593	5,589	9,460	8,874
7,061	8,128	7,150	6,529	6,201	7,605
8,368	8,516	8,100	8,106	4,972	7,259
7,238	8,830	9,145	7,901	8,226	8,552

Examine the data and briefly describe their distribution. Carry out a test of the research hypothesis that nitrites decrease amino acid uptake, and report your results.

11.62   The composition of the earth's atmosphere may have changed over time. To try to discover the nature of the atmosphere long ago, we can examine the gas in bubbles inside ancient amber. Amber is tree resin that has hardened and been trapped in rocks. The gas in bubbles within amber should be a sample of the atmosphere at the time the amber was formed. Measurements on specimens of amber from the late Cretaceous era (75 to 95 million years ago) give these percents of nitrogen:

63.4   65.0   64.4   63.3   54.8   64.5   60.8   49.1   51.0

These values are quite different from the present 78.1% of nitrogen in the atmosphere. Assume (this is not yet agreed on by experts) that these observations are an SRS from the late Cretaceous atmosphere.[33]

(a)   Graph the data, and comment on skewness and outliers.

(b)   The t procedures will be only approximate in this case. Give a 95% t confidence interval for the mean percent of nitrogen in ancient air.

11.63   A pharmaceutical manufacturer does a chemical analysis to check the potency of products. The standard release potency for cephalothin crystals is 910. An assay of 16 lots gives the following potency data:

897   914   913   906   916   918   905   921
918   906   895   893   908   906   907   901

(a)   Check the data for outliers or strong skewness that might threaten the validity of the t procedures.

(b) Give a 95% confidence interval for the mean potency.

(c) Is there significant evidence at the 5% level that the mean potency is not equal to the standard release potency?

11.64   Do various occupational groups differ in their diets? A British study compared the food and drink intake of 98 drivers and 83 conductors of London double-decker buses. The conductors' jobs require more physical activity. The article reporting the study gives the data as "Mean daily consumption (± s. e.)." Some of the study results appear below.[35]

	Drivers	Conductors
Total calories	2821 ± 44	2844 ± 48
Alcohol (grams)	.24 ± .06	.39 ± .11

(a) What does "s. e." stand for? Give $\bar{x}$ and s for each of the four sets of measurements.

(b) Is there significant evidence at the 5% level that conductors and drivers consume different numbers of calories per day?

(c) How significant is the observed difference in mean alcohol consumption?

(d) Give a 90% confidence interval for the mean daily alcohol consumption of London double-decker bus conductors.

(e) Give an 80% confidence interval for the difference in mean daily alcohol consumption between drivers and conductors.

11.65   You look up a census report that gives the populations of all 92 counties in the state of Indiana. Is it proper to apply the one-sample t method to these data to give a 95% confidence interval for the mean population of an Indiana county? Explain your answer.

11.66   The amount of lead in a type of soil, measured by a standard method, averages 86 parts per million (ppm). A new method is tried on 40 specimens of the soil, yielding a mean of 83 ppm lead and a standard deviation of 10 ppm.

(a) Is there significant evidence at the 1% level that the new method frees less lead from the soil?

(b) A critic argues that because of variations in the soil, the effectiveness of the new method is confounded with characteristics of the particular soil specimens used. Briefly describe a better data production design that avoids this criticism.

11.67   High levels of cholesterol in the blood are not healthy in either humans or dogs. Because a diet rich in saturated fats raises the cholesterol level, it is plausible that dogs owned as pets have higher cholesterol levels than dogs owned by a veterinary research clinic. "Normal" levels of cholesterol based on the clinic's dogs would then be misleading. A clinic compared healthy dogs it owned with healthy pets brought to the clinic to be neutered. The summary statistics for blood cholesterol levels (milligrams per deciliter of blood) appear below.[35]

Group	n	$\bar{x}$	s
Pets	26	193	68
Clinic	23	174	44

(a) Is there strong evidence that pets have higher mean cholesterol level than clinic dogs? State the H_0 and H_a and carry out an appropriate test. Give the P-value and state your conclusion.

(b) Give a 95% confidence interval for the difference in mean cholesterol levels between pets and clinic dogs.

(c) Give a 95% confidence interval for the mean cholesterol level in pets.

(d) What assumptions must be satisfied to justify the procedures you used in (a), (b), and (c)? Assuming that the cholesterol measurements have no outliers and are not strongly skewed, what is the chief threat to the validity of the results of this study?

11.68   Exercise 1.41 gives 29 measurements of the density of the earth, made in 1798 by Henry Cavendish. Display the data graphically to check for skewness and outliers. Then give an estimate for the density of the earth from Cavendish's data and a margin of error for your estimate.

11.69   Elite distance runners are thinner than the rest of us. Here are data on skinfold thickness, which indirectly measures body fat, for 20 elite runners and 95 ordinary men in the same age group. The data are in millimeters and are given in the form "mean (standard deviation)."[36]

	Runners	Others
Abdomen	7.1 (1.0)	20.6 (9.0)
Thigh	6.1 (1.8)	17.4 (6.6)

Use confidence intervals to describe the difference between runners and typical young men.

TABLE 11.2		Particulate levels (grams) in two nearby locations			
Day	Rural	City	Day	Rural	City
1	NA	39	19	43	42
2	67	68	20	39	38
3	42	42	21	NA	NA
4	33	34	22	52	57
5	46	48	23	48	50
6	NA	82	24	56	58
7	43	45	25	44	45
8	54	NA	26	51	69
9	NA	NA	27	21	23
10	NA	60	28	74	72
11	NA	57	29	48	49
12	NA	NA	30	84	86
13	38	39	31	51	51
14	88	NA	32	43	42
15	108	123	33	45	46
16	57	59	34	41	NA
17	70	71	35	47	44
18	42	41	36	35	42

The remaining exercises concern a study of air pollution. One component of air pollution is airborne particulate matter such as dust and smoke. To measure particulate pollution, a vacuum motor draws air through a filter for 24 hours. Weigh the filter at the beginning and end of the period. The weight gained is a measure of the concentration of particles in the air. A study of air pollution made measurements every 6 days with identical instruments in the center of a small city and at a rural location 10 miles southwest of the city. Because the prevailing winds blow from the west, we suspect that the rural readings will be generally lower than the city readings, but that the city readings can be predicted from the rural readings. Table 11.2 gives readings taken every 6 days over a 7-month period. The entry NA means that the reading for that date is not available, usually because of equipment failure.[37]

Missing data are common, especially in field studies like this one. We think that equipment failures are not related to pollution levels. If that is true, the missing data do not introduce bias. We can work with the data that are not missing as if they are a random sample of days. We can analyze these data in different ways to answer different questions. For each of the three exercises below, do a careful descriptive analysis with graphs and summary statistics and whatever formal inference is called for. Then present and interpret your findings.

11.70   We want to assess the level of particulate pollution in the city center. Describe the distribution of city pollution levels, and estimate the mean particulate level in the city center. (All estimates should include a statistically justified margin of error.)

11.71   We want to compare the mean level of particulates in the city with the rural level on the same day. We suspect that pollution is higher in the city, and we hope that a statistical test will show that there is significant evidence to confirm this suspicion. Make a graph to check for conditions that might prevent the use of the test you plan to employ. Your graph should reflect the type of procedure that you will use. Then carry out a significance test and report your conclusion. Also estimate the mean amount by which the city particulate level exceeds the rural level on the same day.

11.72   We hope to use the rural particulate level to predict the city level on the same day. Make a graph to examine the relationship. Does the graph suggest that using the least-squares regression line for prediction will give approximately correct results over the range of values appearing in the data? Calculate the least-squares line for predicting city pollution from rural pollution. What percent of the observed variation in city pollution levels does this straight-line relationship account for? On the fourteenth date in the series, the rural reading was 88 and the city reading was not available. What do you estimate the city reading to be for that date? (In Chapter 14, we will learn how to give a margin of error for the predictions we make from the regression line.)

NOTES AND DATA SOURCES

1. This example is based on information in D. L. Shankland et al., "The effect of 5-thio-D-glucose on insect development and its absorption by insects," *Journal of Insect Physiology*, 14 (1968), pp. 63–72.

2. Data from D. L. Shankland, "Involvement of spinal cord and peripheral nerves in DDT-poisoning syndrome in albino rats," *Toxicology and Applied Pharmacology*, 6 (1964), pp. 197–213.

3. The data are from Joan M. Susic, *Dietary Phosphorus Intakes, Urinary and Peritoneal Phosphate Excretion and Clearance in Continuous Ambulatory Peritoneal Dialysis Patients*, M.S. thesis, Purdue University, 1985.

4. Data provided by Joseph Wipf, Department of Foreign Languages and Literatures, Purdue University.

5. Data provided by Timothy Sturm.

6. These recommendations are based on extensive computer work. See, for example, Harry O. Posten, "The robustness of the one-sample t-test over the Pearson system," *Journal of Statistical Computation and Simulation*, 9 (1979), pp. 133–149, and E. S. Pearson and N. W. Please, "Relation between the shape of population distribution and the robustness of four simple test statistics," *Biometrika*, 62 (1975), pp. 223–241.

7. Based on I. Cuellar, L. C. Harris, and R. Jasso, "An acculturation scale for Mexican American normal and clinical populations," *Hispanic Journal of Behavioral Sciences*, 2 (1980), pp. 199–217.

8. Data provided by Diana Schellenberg, Purdue University School of Health Sciences.

9. Data from the appendix of D. A. Kurtz (ed.), *Trace Residue Analysis*, American Chemical Society Symposium Series, no. 284, 1985.

10. From S. C. Hand and E. Gnaiger, "Anaerobic dormancy quantified in *Artemia* embryos," *Science*, 239 (1988), pp. 1425–1427.

11. Data provided by Joseph Wipf, Department of Foreign Languages and Literatures, Purdue University.

12. Based on Charles W. L. Hill and Phillip Phan, "CEO tenure as a determinant of CEO pay," *The Academy of Management Journal*, 34 (1991), pp. 707–717.

13. This study is reported in Roseann M. Lyle et al., "Blood pressure and metabolic effects of calcium supplementation in normotensive white and black men," *Journal of the American Medical Association*, 257 (1987), pp. 1772–1776. The data were provided by Dr. Lyle.

14. Detailed information about the conservative t procedures can be found in Paul Leaverton and John J. Birch, "Small sample power curves for the two sample location problem," *Technometrics*, 11 (1969), pp. 299–307; in Henry Scheffé, "Practical solutions of the Behrens-Fisher problem," *Journal of the American Statistical Association*, 65 (1970), pp. 1501–1508; and in D. J. Best and J. C. W. Rayner, "Welch's approximate solution for the Behrens-Fisher problem," *Technometrics*, 29 (1987), pp. 205–210.

15. From H. G. Gough, *The Chapin Social Insight Test*, Consulting Psychologists Press, Palo Alto, Calif., 1968.

16. Based on M. C. Wilson et al., "Impact of cereal leaf beetle larvae on yields of oats," *Journal of Economic Entomology*, 62 (1969), pp. 699–702.

17. From Costas Papoulias and Panayiotis Theodossiou, "Analysis and modeling of recent business failures in Greece," *Managerial and Decision Economics*, 13 (1992), pp. 163–169.

18. See the extensive simulation studies in Harry O. Posten, "The robustness of the two-sample t-test over the Pearson system," *Journal of Statistical Computation and Simulation*, 6 (1978), pp. 295–311, and in Harry O. Posten, H. Yeh, and Donald B. Owen, "Robustness of the two-sample t-test under violations of the homogeneity assumption," *Communications in Statistics*, 11 (1982), pp. 109–126.

19. Data for 1982, provided by Marvin Schlatter, Division of Financial Aid, Purdue University.

20. Based on G. L. Cromwell et al., "A comparison of the nutritive value of *opaque-2, floury-2* and normal corn for the chick," *Poultry Science*, 47 (1968), pp. 840–847.

21. This example is loosely based on D. L. Shankland, "Involvement of spinal cord and peripheral nerves in DDT-poisoning syndrome in albino rats," *Toxicology and Applied Pharmacology*, 6 (1964), pp. 197–213.

22. We did not use Minitab or Data Desk in Example 11.12 because these packages shortcut the two-sample *t* procedure. They calculate the degrees of freedom df using the formula in the box on page 633 but then truncate to the next lower whole-number degrees of freedom to obtain the *P*-value. The result is slightly less accurate than the *P*-value from the t(df) distribution.

23. Based on W. N. Nelson and C. J. Widule, "Kinematic analysis and efficiency estimate of intercollegiate female rowers," unpublished manuscript, 1983.

24. From a news article in *Science*, 224 (1983), pp. 1029–1031.

25. From M. F. Picciano and R. H. Deering, "The influence of feeding regimens on iron status during infancy," *The American Journal of Clinical Nutrition*, 33 (1980), pp. 746–753.

26. From A. H. Ismail and R. J. Young, "The effect of chronic exercise on the personality of middle-aged men," *Journal of Human Ergology*, 2 (1973), pp. 47–57.

27. Adapted from Maribeth Cassidy Schmitt, *The Effects of an Elaborated Directed Reading Activity on the Metacomprehension Skills of Third Graders*, Ph.D. dissertation, Purdue University, 1987.

28. From M. A. Zlatin and R. A. Koenigsknecht, "Development of the voicing contrast: a comparison of voice onset time in stop perception and production," *Journal of Speech and Hearing Research*, 19 (1976), pp. 93–111.

29. Based loosely on D. R. Black et al., "Minimal interventions for weight control: a cost-effective alternative," *Addictive Behaviors*, 9 (1984), pp. 279–285.

30. Based on Amna Kirmani and Peter Wright, "Money talks: perceived advertising expense and expected product quality," *Journal of Consumer Research*, 16 (1989), pp. 344–353.

31. Based on D. Friedlander, *Supplemental Report on the Baltimore Options Program*, Manpower Demonstration Research Corporation, 1987.

32. Based on G. Salvendy, "Selection of industrial operators: the one-hole test," *International Journal of Production Research*, 13 (1973), pp. 303–321.

33. Data from R. A. Berner and G. P. Landis, "Gas bubbles in fossil amber as possible indicators of the major gas composition of ancient air," *Science*, 239 (1988), pp. 1406–1409.

34. From J. W. Marr and J. A. Heady, "Within- and between-person variation in dietary surveys: number of days needed to classify individuals," *Human Nutrition: Applied Nutrition*, 40A (1986), pp. 347–364.

35. From V. D. Bass, W. E. Hoffmann, and J. L. Dorner, "Normal canine lipid profiles and effects of experimentally induced pancreatitis and hepatic necrosis on lipids," *American Journal of Veterinary Research*, 37 (1976), pp. 1355–1357.

36. From M. L. Pollock et al., "Body composition of elite class distance runners," in P. Milvey (ed.), *The Marathon: Physiological, Medical, Epidemiological, and Psychological Studies*, New York Academy of Sciences, 1977, p. 366.

37. Data provided by Matthew Moore.

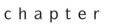

Inference for Proportions

- Introduction

- 12.1 Inference for A Population Proportion

- 12.2 Comparing Two Proportions

- Chapter Review

INTRODUCTION

We often want to answer questions about the proportion of some outcome in a population or to compare proportions in several populations. Here are some examples that call for inference about population proportions.

EXAMPLE 12.1

How common is behavior that puts people at risk of AIDS? The National AIDS Behavioral Surveys interviewed a random sample of 2673 adult heterosexuals. Of these, 170 had more than one sexual partner in the past year. That's 6.36% of the sample.[1] Based on these data, what can we say about the percent of all adult heterosexuals who have multiple partners? We want to *estimate a single population proportion.*

EXAMPLE 12.2

Do preschool programs for poor children make a difference in later life? A study looked at 62 children who were enrolled in a Michigan preschool in the late 1960s and at a control group of 61 similar children who were not enrolled. At 27 years of age, 61% of the preschool group and 80% of the control group had required the help of a social service agency (mainly welfare) in the previous ten years.[2] Is this significant evidence that preschool for poor children reduces later use of social services? We want to *compare two population proportions.*

EXAMPLE 12.3

What is the relationship between time spent in extracurricular activities and success in a tough course in college? North Carolina State University looked at the 123 students in an introductory chemical engineering course. Students needed a grade of C or better to advance to the next course. The passing rates were 55% for students who spent less than 2 hours per week in extracurricular activities, 75% for those who spent between 2 and 12 hours per week, and 38% for those who spent more than 12 hours per week.[3] Are the differences in passing rates statistically significant? We must *compare more than two population proportions.*

Our study of inference for proportions will follow the same pattern as these examples. Section 12.1 discusses inference for one population proportion, and Section 12.2 presents methods for comparing two proportions. Comparing more than two proportions raises new issues and requires more elaborate methods that also apply to some other inference problems. These methods are the topic of Chapter 13.

| ACTIVITY 12 | Is One Side of a Coin Heavier? |

Materials: 20 pennies for each student

Using a coin to randomly determine an outcome, most people would flip the coin. Is it equivalent to hold the coin vertically on a tabletop and spin the coin with a quick flick of your finger? In this activity, we will try a third variation. We will stand pennies on edge and then bang the table to make the pennies fall. We are interested in the proportion of times the pennies fall heads up. If the pennies are equally heavy on both sides of the coin, then it would be reasonable to expect the long-term proportion of heads to be about 0.5. We state the following hypotheses:

$$H_0 : p = .5$$
$$H_a : p \neq .5$$

Procedure

1. Each student should stand 20 pennies on edge on a horizontal tabletop. Take your time—this may take a steady hand and some patience.
2. Bang the table just hard enough to make all of the pennies fall.
3. Count the number of pennies that fall heads up.
4. Accumulate your results with those of other students in the class.

Questions

• Are the results about what you expected? Or are you surprised by the results?
• Do you think it is likely, by chance alone, to obtain results like the results you actually observed?

Keep these results handy. As soon as we develop the necessary theory, we will test to see if your results are significant, and we will construct a confidence interval for the true proportion of heads obtained in this manner.

12.1   INFERENCE FOR A POPULATION PROPORTION

We are interested in the unknown proportion p of a population that has some outcome. For convenience, call the outcome we are looking for a "success." In Example 12.1, the population is adult heterosexuals, and the parameter p is the proportion who have had more than one sexual partner in the past year. To estimate p, the National AIDS Behavioral Surveys used random dialing of telephone numbers to contact a sample of 2673 people. Of these, 170 said they had multiple sexual partners. The statistic that estimates the parameter p is the *sample proportion*

sample proportion

$$\hat{p} = \frac{\text{count of successes in the sample}}{\text{count of observations in the sample}}$$

$$= \frac{170}{2673} = .0636$$

Read the sample proportion $\hat{p}$ as "p-hat."

EXERCISES

In each of the following settings:

(a) Describe the population and explain in words what the parameter p is.

(b) Give the numerical value of the statistic $\hat{p}$ that estimates p.

12.1   Tonya wants to estimate what proportion of the students in her dormitory like the dorm food. She interviews an SRS of 50 of the 175 students living in the dormitory. She finds that 14 think the dorm food is good.

12.2   Glenn wonders what proportion of the students at his school think that tuition is too high. He interviews an SRS of 50 of the 2400 students at his college. Thirty-eight of those interviewed think tuition is too high.

12.3   A college president says, "99% of the alumni support my firing of Coach Boggs." You contact an SRS of 200 of the college's 15,000 living alumni and find that 76 of them support firing the coach.

Assumptions for inference

As always, inference is based on the sampling distribution of a statistic. We described the sampling distribution of a sample proportion $\hat{p}$ in Section 2 of Chapter 9. The mean is p. That is, the sample proportion $\hat{p}$ is an unbiased estimator of the population proportion p. The standard deviation of $\hat{p}$ is $\sqrt{p(1-p)/n}$. The normal approximation says that for large samples, the distribution of $\hat{p}$ is approximately normal. Figure 12.1 displays this sampling distribution.

Standardize $\hat{p}$ by subtracting its mean and dividing by its standard deviation. The result is a z statistic:

$$z = \frac{\hat{p} - p}{\sqrt{\dfrac{p(1-p)}{n}}}$$

The statistic z has approximately the standard normal distribution $N(0, 1)$. Inference about p uses this z statistic and standard normal critical values.

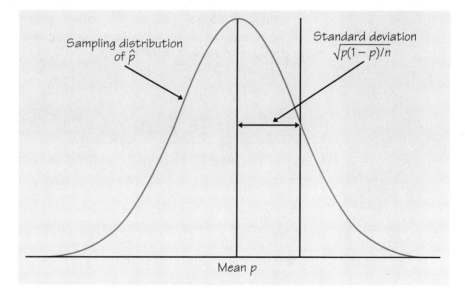

FIGURE 12.1   Select a large SRS from a population that contains proportion p of successes. The sampling distribution of the proportion $\hat{p}$ of successes in the sample is approximately normal. The mean is p and the standard deviation is $\sqrt{p(1-p)/n}$.

However, we need to deal with the fact that we don't know the standard deviation $\sqrt{p(1-p)/n}$ because we don't know p. Here's what we do:

- To test the null hypothesis $H_0 : p = p_0$ that the unknown p has a specific value p_0, just replace p by p_0 in the z statistic.

- In a confidence interval for p, we have no specific value to substitute. In large samples, $\hat{p}$ will be close to p, so replace the standard deviation by the **standard error of $\hat{p}$**.

*standard error
of $\hat{p}$*

$$\text{SE} = \sqrt{\frac{\hat{p}(1-\hat{p})}{n}}$$

The confidence interval has the form

$$\text{estimate} \pm z^{*}\text{SE}_{\text{estimate}}$$

The normal approximation is accurate if the sample is large and the population proportion p is not too close to 0 or 1. We don't know the value of p. For a test, the null hypothesis states that p has a specified value p_0. For confidence intervals, we use the statistic $\hat{p}$ to estimate the unknown p. So the requirements for using the z procedures for inference about a proportion are stated in terms of p_0 or $\hat{p}$.

ASSUMPTIONS FOR INFERENCE ABOUT A PROPORTION

- The data are an SRS from the population of interest.

- The population is at least 10 times as large as the sample.

- For a test of $H_0 : p = p_0$, the sample size n is so large that both np_0 and $n(1 - p_0)$ are 10 or more. For a confidence interval, n is so large that both the count of successes $n\hat{p}$ and the count of failures $n(1 - \hat{p})$ are 10 or more.

If you have a small sample or a sampling design more complex than an SRS, you can still do inference, but the details are more complicated. Get expert advice.

EXAMPLE 12.4

We want to use the National AIDS Behavioral Surveys data to give a confidence interval for the proportion of adult heterosexuals who have had multiple sexual partners. Does the sample meet the requirements for inference?

- The sampling design was in fact a complex stratified sample, and the survey used inference procedures for that design. The overall effect is close to an SRS, however.

- The number of adult heterosexuals (the population) is much larger than 10 times the sample size, $n = 2673$.

- The counts of "Yes" and "No" responses are much greater than 10:

$$n\hat{p} = (2673)(.0636) = 170$$
$$n(1 - \hat{p}) = (2673)(.9364) = 2503$$

The second and third requirements are easily met. The first requirement, that the sample be an SRS, is only approximately met.

As usual, the practical problems of a large sample survey pose a greater threat to the AIDS survey's conclusions. Only people in households with telephones could be reached. This is acceptable for surveys of the general population, because about 94% of American households have telephones. However, some groups at high risk for AIDS, like intravenous drug users, often don't live in settled households and are underrepresented in the sample. About 30% of the people reached refused to cooperate. A nonresponse rate of 30% is not unusual in large sample surveys, but it may cause some bias if those who refuse differ systematically from those who cooperate. The survey used statistical methods that adjust for unequal response rates in different groups. Finally, some respondents may not have told the truth when asked about their sexual behavior. The survey team tried hard to make respondents feel comfortable. For example, Hispanic women were interviewed only by Hispanic women, and Spanish speakers were interviewed by Spanish speakers with the same regional accent (Cuban, Mexican, or Puerto Rican). Nonetheless, the survey report says that some bias is probably present:

It is more likely that the present figures are underestimates; some respondents may underreport their numbers of sexual partners and intravenous drug use because of embarrassment and fear of reprisal, or they may forget or not know details of their own or of their partner's HIV risk and their antibody testing history.[4]

Reading the report of a large study like the National AIDS Behavioral Surveys reminds us that statistics in practice involves much more than recipes for inference.

EXERCISES

12.4   In which of the following situations can you safely use the methods of this section to get a confidence interval for the population proportion p? Explain your answers.

(a) Tonya wants to estimate what proportion of the students in her dormitory like the dorm food. She interviews an SRS of 50 of the 175 students living in the dormitory. She finds that 14 think the dorm food is good.

(b) Glenn wonders what proportion of the students at his school think that tuition is too high. He interviews an SRS of 50 of the 2400 students at his college. Thirty-eight of those interviewed think tuition is too high.

(c) In the National AIDS Behavioral Surveys sample of 2673 adult heterosexuals, 0.2% (that's 0.002 as a decimal fraction) had both received a blood transfusion and had a sexual partner from a group at high risk of AIDS. (We want to estimate the proportion p in the population who share these two risk factors.)

12.5   In which of the following situations can you safely use the methods of this section for a significance test? Explain your answers.

(a) You toss a coin 10 times in order to test the hypothesis $H_0 : p = .5$ that the coin is balanced.

(b) A college president says, "99% of the alumni support my firing of Coach Boggs." You contact an SRS of 200 of the college's 15,000 living alumni to test the hypothesis $H_0 : p = .99$.

(c) Do a majority of the 250 students in a statistics course agree that knowing statistics will help them in their future careers? You interview an SRS of 20 students to test $H_0 : p = .5$.

The *z* procedures

Here are the *z* procedures for inference about *p*.

LARGE-SAMPLE INFERENCE FOR A POPULATION PROPORTION

Draw an SRS of size *n* from a large population with unknown proportion *p* of successes. An approximate level *C* confidence interval for *p* is

$$\hat{p} \pm z^* \sqrt{\frac{\hat{p}(1 - \hat{p})}{n}}$$

where z^* is the upper $(1 - C)/2$ standard normal critical value.

To test the hypothesis $H_0 : p = p_0$, compute the *z* statistic

$$z = \frac{\hat{p} - p_0}{\sqrt{\dfrac{p_0(1 - p_0)}{n}}}$$

In terms of a variable Z having the standard normal distribution, the approximate *P*-value for a test of H_0 against

$H_a : p > p_0$ is $P(Z \geq z)$

$H_a : p < p_0$ is $P(Z \leq z)$

$H_a : p \neq p_0$ is $2P(Z \geq |z|)$

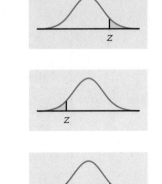

EXAMPLE 12.5

The National AIDS Behavioral Surveys found that 170 of a sample of 2673 adult heterosexuals had multiple partners. That is, $\hat{p} = .0636$. We will act as if the sample were an SRS.

A 99% confidence interval for the proportion p of all adult heterosexuals with multiple partners uses the standard normal critical value $z^* = 2.576$. (Look in the bottom row of Table C for standard normal critical values.) The confidence interval is

$$\hat{p} \pm z^* \sqrt{\frac{\hat{p}(1 - \hat{p})}{n}} = .0636 \pm 2.576 \sqrt{\frac{(.0636)(.9364)}{2673}}$$

$$= .0636 \pm .0122$$

$$= (.0514, .0758)$$

We are 99% confident that the percent of adult heterosexuals who had more than one sexual partner in the past year lies between about 5.1% and 7.6%.

EXAMPLE 12.6

A coin that is balanced should come up heads half the time in the long run. The population for coin tossing contains the results of tossing the coin forever. The parameter p is the probability of a head, which is the proportion of all tosses that give a head. The tosses we actually make are an SRS from this population.

The French naturalist Count Buffon (1707–1788) tossed a coin 4040 times. He got 2048 heads. The sample proportion of heads is

$$\hat{p} = \frac{2048}{4040} = .5069$$

That's a bit more than one-half. Is this evidence that Buffon's coin was not balanced? This is a job for a significance test.

Step 1: Hypotheses. The null hypothesis says that the coin is balanced ($p = .5$). The alternative hypothesis is two-sided, because we did not suspect before seeing the data that the coin favored either heads or tails. We therefore test the hypotheses

$$H_0 : p = .5$$
$$H_a : p \neq .5$$

The null hypothesis gives p the value $p_0 = .5$.

Step 2: Test statistic. The z test statistic is

$$z = \frac{\hat{p} - p_0}{\sqrt{\dfrac{p_0(1 - p_0)}{n}}}$$

$$= \frac{.5069 - .5}{\sqrt{\dfrac{(.5)(.5)}{4040}}} = .88$$

Step 3: P-value. Because the test is two-sided, the P-value is the area under the standard normal curve more than 0.88 away from 0 in either direction. Figure 12.2 shows this area. From Table A we find that the area below -0.88 is 0.1894. The P-value is twice this area:

$$P = 2(.1894) = .3788$$

Conclusion. A proportion of heads as far from one-half as Buffon's would happen 38% of the time when a balanced coin is tossed 4040 times. Buffon's result doesn't show that his coin is unbalanced.

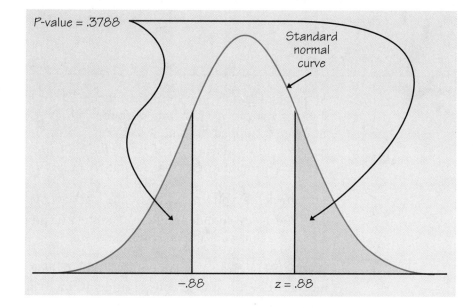

FIGURE 12.2   The P-value for the two-sided test of Example 12.6.

In Example 12.6, we failed to find good evidence against $H_0 : p = .5$. We *cannot* conclude that H_0 is true, that is, that the coin is perfectly balanced. No doubt p is not exactly 0.5. The test of significance only shows that the results of Buffon's 4040 tosses can't distinguish this coin from one that is perfectly balanced. To see what values of p are consistent with the sample results, use a confidence interval.

EXAMPLE 12.7

The 95% confidence interval for the probability p that Buffon's coin gives a head is

$$\hat{p} \pm z^* \sqrt{\frac{\hat{p}(1-\hat{p})}{n}} = .5069 \pm 1.960 \sqrt{\frac{(.5069)(.4931)}{4040}}$$

$$= .5069 \pm .0154$$

$$= (.4915, .5223)$$

We are 95% confident that the probability of a head is between 0.4915 and 0.5223.

The confidence interval is more informative than the test in Example 12.6. It tells us that any null hypothesis $H_0 : p = p_0$ for a p_0 between 0.4915 and 0.5223 would not be rejected at the $\alpha = .05$ level of significance. We would not be surprised if the true probability of a head for Buffon's coin were something like 0.51.

EXERCISES

12.6 As part of a quality improvement program, your mail-order company is studying the process of filling customer orders. Company standards say an order is shipped on time if it is sent within 3 working days after it is received. You audit an SRS of 100 of the 5000 orders received in the past month. The audit reveals that 86 of these orders were shipped on time.

(a) Check that you can safely use the methods in this section.

(b) Find a 95% confidence interval for the true proportion of the month's orders that were shipped on time.

12.7 The 1995 Harvard School of Public Health College Alcohol Study examined alcohol use among college students, including the practice called "binge drinking." Binge drinking for men was defined as consuming five or more drinks on at least one occasion during the two weeks prior to the survey (four drinks for women). Binge drinkers experience a higher percentage of alcohol-related problems such as disciplinary problems, violence, irresponsible sexual activity, personal injury, and poor academic performance. In a representative sample of 140 colleges and 17,592 students, 7741 students identified themselves as binge drinkers. Considering this an SRS of 17,592 from the population of all college students, does this constitute strong evidence that more than 40% of all college students engaged in binge drinking?

12.8 In a recent year, 73% of first-year college students responding to a national survey identified "being very well-off financially" as an important personal goal. A state university finds that 132 of an SRS of 200 of its first-year students say that this goal is important.

(a) Give a 95% confidence interval for the proportion of all first-year students at the university who would identify being well-off as an important personal goal.

(b) Is there good evidence that the proportion of first-year students at this university who think being very well-off is important differs from the national value, 73%? (Be sure to state hypotheses, give the P-value, and state your conclusion.)

(c) Check that you could safely use the methods of this section in both (a) and (b).

12.9 Around the year 1900, the English statistician Karl Pearson tossed a coin 24,000 times. He obtained 12,012 heads.

(a) Test the null hypothesis that Pearson's coin had probability 0.5 of coming up heads versus the two-sided alternative. Give the P-value. Do you reject H_0 at the 1% significance level?

(b) Find a 99% confidence interval for the probability of heads for Pearson's coin. This is the range of probabilities that cannot be rejected at the 1% significance level.

Choosing the sample size

In planning a study, we may want to choose a sample size that will allow us to estimate the parameter within a given margin of error. We saw earlier how to do this for a population mean. The method is similar for estimating a population proportion.

The margin of error in the approximate confidence interval for p is

$$m = z^* \sqrt{\frac{\hat{p}(1 - \hat{p})}{n}}$$

Here z^* is the standard normal critical value for the level of confidence we want. Because the margin of error involves the sample proportion of successes $\hat{p}$, we need to guess this value when choosing n. Call our guess p^*. Here are two ways to get p^*:

1. Use a guess p^* based on a pilot study or on past experience with similar studies. You should do several calculations that cover the range of $\hat{p}$-values you might get.
2. Use $p^* = .5$ as the guess. The margin of error m is largest when $\hat{p} = .5$, so this guess is conservative in the sense that if we get any other $\hat{p}$ when we do our study, we will get a margin of error smaller than planned.

Once you have a guess p^*, the recipe for the margin of error can be solved to give the sample size n needed. Here is the result.

SAMPLE SIZE FOR DESIRED MARGIN OF ERROR

To determine the sample size n that will yield a level C confidence interval for a population proportion p with a specified margin of error m, set the following expression for the margin of error to be less than or equal to m, and solve for n:

$$z^* \sqrt{\frac{p^*(1 - p^*)}{n}} \leq m$$

where p^* is a guessed value for the sample proportion. The margin of error will be less than or equal to m if you take the guess p^* to be 0.5.

Which method for finding the guess p^* should you use? The n you get doesn't change much when you change p^* as long as p^* is not too far from 0.5. So use the conservative guess $p^* = .5$ if you expect the true $\hat{p}$ to be roughly between 0.3 and 0.7. If the true $\hat{p}$ is close to 0 or 1, using $p^* = .5$ as your guess will give a sample much larger than you need. So try to use a better guess from a pilot study when you suspect that $\hat{p}$ will be less than 0.3 or greater than 0.7.

EXAMPLE 12.8

Gloria Chavez and Ronald Flynn are the candidates for mayor in a large city. You are planning a sample survey to determine what percent of the voters plan to vote for Chavez. This is a population proportion p. You will contact an SRS of registered voters in the city. You want to estimate p with 95% confidence and a margin of error no greater than 3%, or 0.03. How large a sample do you need?

The winner's share in all but the most lopsided elections is between 30% and 70% of the vote. So use the guess $p^* = .5$. Then you want

$$z^* \sqrt{\frac{p^*(1 - p^*)}{n}} \le .03$$

$$\frac{1.960 \sqrt{.5(.5)}}{\sqrt{n}} \le .03$$

$$\frac{1.960(.5)}{.03} \le \sqrt{n}$$

$$\sqrt{n} \ge 32.6\overline{6}$$

$$n \ge (32.66)^2 = 1067.1$$

Since the number of people in the sample must be a whole number, n must be 1068 to satisfy the inequality. If you want a 2.5% margin of error, you can show in similar fashion that $n = 1537$ is the required sample size. For a 2% margin of error, the sample size you need is 2401. (Work these out for practice!) As usual, smaller margins of error call for larger samples.

EXERCISES

12.10   A national opinion poll found that 44% of all American adults agree that parents should be given vouchers good for education at any public or private school of their choice. The result was based on a small sample. How large an SRS is required to obtain a margin of error of 0.03 (that is, ±3%) in a 95% confidence interval?

(a)   Answer this question using the previous poll's result as the guessed value p^*.

(b)   Do the problem again using the conservative guess $p^* = .5$. By how much do the two sample sizes differ?

12.11   PTC is a substance that has a strong bitter taste for some people and is tasteless for others. The ability to taste PTC is inherited. About 75% of Italians can taste PTC, for example. You want to estimate the proportion of Americans with at least one Italian grandparent who can taste PTC. Starting with the 75% estimate for Italians, how large a sample must you test in order to estimate the proportion of PTC tasters within ± 0.04 with 95% confidence?

TI-83 techniques

The TI-83 can be used to perform inference for a population proportion and to construct confidence intervals. Examples 12.9 and 12.10 revisit the Buffon coin-tossing experiments.

EXAMPLE 12.9

In $n = 4040$ coin tosses, Count Buffon observed $X = 2048$ heads. Recall that our hypotheses were

$$H_0 : p = .5$$
$$H_a : p \neq .5$$

To perform a test of the null hypothesis on the TI-83, select STAT / TESTS / 5:1−PropZTest.

```
EDIT CALC TESTS
1:Z−Test...
2:T−Test...
3:2−SampZTest...
4:2−SampTTest...
5:1−PropZTest...
6:2−PropZTest...
7↓ZInterval...
```

When the 1−PropZTest screen appears, enter the values shown: $p_0 = .5$, $x = 2048$, and $n = 4040$. Specify the alternative hypothesis as "prop $\neq p_0$."

```
1-PropZTest
p0:.5
x:2048
n:4040
prop≠p0  <p0  >p0
Calculate Draw
```

If you select the "Calculate" choice and press ENTER, you will see that the z-statistic is 0.88 and the P-value is 0.3783.

```
1-PropZTest
 prop≠.5
 z=.8810434857
 p=.3782942021
 p̂=.5069306931
 n=4040
```

If you select the "Draw" option, you will see the screen shown here. Compare these results with those in Example 12.6.

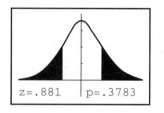

EXAMPLE 12.10

To use the TI-83 to construct a 95% confidence interval for the true proportion of heads, select STAT / TESTS / A:1−PropZInt.

```
EDIT CALC TESTS
5↑1−PropZTest...
6:2−PropZTest...
7:ZInterval...
8:TInterval...
9:2−SampZInt...
0:2−SampTInt...
A↓1−PropZInt...
```

When the 1−PropZInt screen appears, enter $x = 2048$, $n = 4040$, and confidence level = .95. Highlight "Calculate" and press ENTER.

```
1-PropZInt
 x:2048
 n:4040
 C-Level:.95
 Calculate
```

The 95% confidence interval for p is reported, along with the sample proportion $\hat{p}$ and the sample size, as shown here.

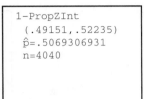

```
1-PropZInt
 (.49151,.52235)
 p̂=.5069306931
 n=4040
```

SUMMARY

Tests and confidence intervals for a population proportion p when the data are an SRS of size n are based on the **sample proportion $\hat{p}$**. When n is large, $\hat{p}$ has approximately the normal distribution with mean p and standard deviation $\sqrt{p(1-p)/n}$.

The level C **confidence interval** for p is

$$\hat{p} \pm z^* \sqrt{\frac{\hat{p}(1-\hat{p})}{n}}$$

where z^* is the upper $(1-C)/2$ standard normal critical value. **Tests** of $H_0 : p = p_0$ are based on the z **statistic**

$$z = \frac{\hat{p} - p_0}{\sqrt{\dfrac{p_0(1-p_0)}{n}}}$$

with P-values calculated from the standard normal distribution.

These inference procedures are approximately correct when the population is at least 10 times as large as the sample and the sample is large enough to satisfy $n\hat{p} \geq 10$ and $n(1-\hat{p}) \geq 10$ for a confidence interval or $np_0 \geq 10$ and $n(1-p_0) \geq 10$ for a test of $H_0 : p = p_0$.

The **sample size** needed to obtain a confidence interval with approximate margin of error m for a population proportion involves solving

$$z^* \sqrt{\frac{p^*(1-p^*)}{n}} \leq m$$

for n, where p^* is a guessed value for the sample proportion $\hat{p}$, and z^* is the standard normal critical point for the level of confidence you want. If you use $p^* = .5$ in this formula, the margin of error of the interval will be less than or equal to m no matter what the value of $\hat{p}$ is.

SECTION 12.1 EXERCISES

12.12   (a) Calculate the proportion of heads you obtained in Activity 12. Then use your TI-83 to test the null hypothesis: $H_0 : p = .5$ against the alternative hypothesis: $H_a : p \neq .5$. Report the P-value and state your conclusion.

(b) Find the proportion of heads for your entire class. What is n? Using the same null and alternative hypotheses as in (a), find the new P-value, and compare this with the value you obtained in (a).

(c) Using the data from your experiment, find the 95% confidence interval for the true proportion of heads obtained by the method in Activity 12.

(d) This part should be done as a class activity. Draw a horizontal line at the top of the blackboard, and mark a scale wide enough to accommodate each student confidence interval. Then below this scaled line, each student can draw his or her confidence interval. These intervals should vary somewhat. Looking at all of the confidence intervals, make a conjecture about the 95% confidence interval for the whole class.

(e) Use the cumulative data collected by the whole class to calculate the 95% confidence interval, and compare this interval with the interval conjectured in (d). Each student should also compare his or her confidence interval with the confidence interval for the whole class. Which confidence interval do you prefer, and why? What accounts for the difference in the width?

12.13   Tonya, Frank, and Sarah are investigating student attitudes toward college food for an assignment in their introductory statistics class. Based on comments overheard from other students, they believed that fewer than 1 in 3 students like college food. To test this hypothesis, each selected an SRS of students who regularly eat in the cafeteria, and asked them if they like college food. Fourteen in Tonya's SRS of 50 replied, "Yes," while 98 in Frank's sample of 350, and 140 in Sarah's sample of 500 said they like college food. Use your TI-83 to perform a test of significance on all three results and fill in a table like this:

X	n	$\hat{p}$	z	P-value
14	50			
98	350			
140	500			

Describe your findings in a short narrative.

12.14   How likely are patients who file complaints with a health maintenance organization (HMO) to leave the HMO? In one recent year, 639 of the more than 400,000 members of a large New England HMO filed complaints. Fifty-four of the complainers left the HMO voluntarily. (That is, they were not forced to leave by a move or a job change.)[5] Consider this year's complainers as an SRS of all patients who will complain in the future. Give a 90% confidence interval for the proportion of complainers who voluntarily leave the HMO.

12.15   An experiment on the side effects of pain relievers assigned arthritis patients to one of several over-the-counter pain medications. Of the 440 patients who took one brand of pain reliever, 23 suffered some "adverse symptom." Does the experiment provide strong evidence that fewer than 10% of patients who take this medication have adverse symptoms?

12.16   The Gallup Poll asked a sample of 1785 adults, "Did you, yourself, happen to attend church or synagogue in the last 7 days?" Of the respondents, 750 said "Yes." Suppose (it is not, in fact, true) that Gallup's sample was an SRS of all American adults.

   (a)   Give a 99% confidence interval for the proportion of all adults who attended church or synagogue during the week preceding the poll.

   (b)   Do the results provide good evidence that less than half of the population attended church or synagogue?

   (c)   How large a sample would be required to obtain a margin of error of 0.01 in a 99% confidence interval for the proportion who attend church or synagogue? (Use the conservative guess $p^* = .5$, and explain why this method is reasonable in this situation.)

12.17   A study of the survival of small businesses chose an SRS from the telephone directory's Yellow Pages listings of food-and-drink businesses in 12 counties in central Indiana. For various reasons, the study got no response from 45% of the businesses chosen. Interviews were completed with 148 businesses. Three years later, 22 of these businesses had failed.[6]

   (a)   Give a 95% confidence interval for the percent of all small businesses in this class that fail within three years.

   (b)   Based on the results of this study, how large a sample would you need to reduce the margin of error to 0.04?

   (c)   The authors hope that their findings describe the population of all small businesses. What about the study makes this unlikely? What population do you think the study findings describe?

12.18   One-sample procedures for proportions, like those for means, are used to analyze data from matched pairs designs. Here is an example.

Each of 50 subjects tastes two unmarked cups of coffee and says which he or she prefers. One cup in each pair contains instant coffee; the other, fresh-brewed coffee. Thirty-one of the subjects prefer the fresh-brewed coffee. Take p to be the proportion of the population who would prefer fresh-brewed coffee in a blind tasting.

(a)   Test the claim that a majority of people prefer the taste of fresh-brewed coffee. State hypotheses and report the z statistic and its P-value. Is your result significant at the 5% level? What is your practical conclusion?

(b)   Find a 90% confidence interval for p.

(c)   When you do an experiment like this, in what order should you present the two cups of coffee to the subjects?

12.19   An automobile manufacturer would like to know what proportion of its customers are not satisfied with the service provided by their local dealer. The customer relations department will survey a random sample of customers and compute a 99% confidence interval for the proportion who are not satisfied.

(a)   From past studies, they believe that this proportion will be about 0.2. Find the sample size needed if the margin of error of the confidence interval is to be about 0.015.

(b)   When the sample is actually contacted, 10% of the sample say they are not satisfied. What is the margin of error of the 99% confidence interval?

12.20   You are planning a survey of students at a large university to determine what proportion favor an increase in student fees to support an expansion of the student newspaper. Using records provided by the registrar you can select a random sample of students. You will ask each student in the sample whether he or she is in favor of the proposed increase. Your budget will allow a sample of 100 students.

(a)   For a sample of size 100, construct a table of the margins of error for 95% confidence intervals when $\hat{p}$ takes the values 0.1, 0.2, 0.3, 0.4, 0.5, 0.6, 0.7, 0.8, and 0.9.

(b)   A former editor of the student newspaper offers to provide funds for a sample of size 500. Repeat the margin of error calculations in (a) for the larger sample size. Then write a short thank-you note to the former editor describing how the larger sample size will improve the results of the survey.

12.2   COMPARING TWO PROPORTIONS

two-sample problem

In a **two-sample problem,** we want to compare two populations or the responses to two treatments based on two independent samples. When the comparison involves the mean of a quantitative variable, we use the two-sample t methods of Section 11.2. To compare the standard deviations of a variable in two groups, we use (under restrictive conditions) the F statistic, which will be described in optional Section 15.1. Now we turn to methods to compare the proportions of successes in two groups.

We will use notation similar to that used in our study of two-sample t statistics. The groups we want to compare are Population 1 and Population 2. We have a separate SRS from each population or responses from two treatments in a randomized comparative experiment. A subscript shows which group a parameter or statistic describes. Here is our notation:

Population	Population proportion	Sample size	Sample proportion
1	p_1	n_1	$\hat{p}_1$
2	p_2	n_2	$\hat{p}_2$

We compare the populations by doing inference about the difference $p_1 - p_2$ between the population proportions. The statistic that estimates this difference is the difference between the two sample proportions, $\hat{p}_1 - \hat{p}_2$.

EXAMPLE 12.11

To study the long-term effects of preschool programs for poor children, the High/Scope Educational Research Foundation has followed two groups of Michigan children since early childhood. One group of 62 attended preschool as 3- and 4-year-olds. This is a sample from Population 2, poor children who attend preschool. A control group of 61 children from the same area and similar backgrounds represents Population 1, poor children with no preschool. Thus the sample sizes are $n_1 = 61$ and $n_2 = 62$.

One response variable of interest is the need for social services as adults. In the past ten years, 38 of the preschool sample and 49 of the control sample have needed social services (mainly welfare). The sample proportions are

$$\hat{p}_1 = \frac{49}{61} = .803$$

$$\hat{p}_2 = \frac{38}{62} = .613$$

That is, about 80% of the control group uses social services, as opposed to about 61% of the preschool group.

To see if the study provides significant evidence that preschool reduces the later need for social services, we test the hypotheses

$$H_0 : p_1 - p_2 = 0 \qquad \text{or} \qquad H_0 : p_1 = p_2$$

$$H_a : p_1 - p_2 > 0 \qquad \text{or} \qquad H_a : p_1 > p_2$$

To estimate how large the reduction is, we give a confidence interval for the difference, $p_1 - p_2$. Both the test and the confidence interval start from the difference of sample proportions:

$$\hat{p}_1 - \hat{p}_2 = .803 - .613 = .190$$

The sampling distribution of $\hat{p}_1 - \hat{p}_2$

Here are the facts about the sampling distribution of $\hat{p}_1 - \hat{p}_2$ that we need for inference:

- The mean of $\hat{p}_1 - \hat{p}_2$ is $p_1 - p_2$. That is, the difference of sample proportions is an unbiased estimator of the difference of population proportions.

- The variance of the difference is the *sum* of the variances of $\hat{p}_1$ and $\hat{p}_2$, which is

$$\frac{p_1(1 - p_1)}{n_1} + \frac{p_2(1 - p_2)}{n_2}$$

Note that the *variances* add. The standard deviations do not.

- When the samples are large, the distribution of $\hat{p}_1 - \hat{p}_2$ is approximately normal.

Figure 12.3 displays the distribution of $\hat{p}_1 - \hat{p}_2$. The standard deviation of $\hat{p}_1 - \hat{p}_2$ involves the unknown parameters p_1 and p_2. Just as in the previous section, we must replace these by estimates in order to do inference. And just as in the previous section, we do this a bit differently for confidence intervals and for tests.

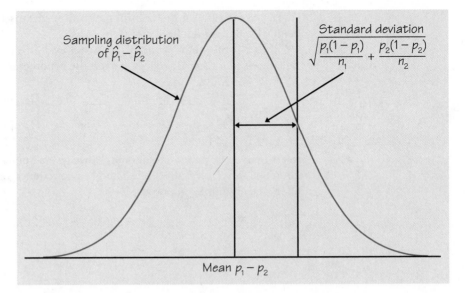

FIGURE 12.3   Select independent SRSs from two populations having proportions of successes p_1 and p_2. The proportions of successes in the two samples are $\hat{p}_1$ and $\hat{p}_2$. When the samples are large, the sampling distribution of the difference $\hat{p}_1 - \hat{p}_2$ is approximately normal.

Confidence intervals for $p_1 - p_2$

The standard deviation of $\hat{p}_1 - \hat{p}_2$ is the square root of the variance:

$$\sqrt{\frac{p_1(1 - p_1)}{n_1} + \frac{p_2(1 - p_2)}{n_2}}$$

To obtain a confidence interval, replace the population proportions p_1 and p_2 in this expression by the sample proportions. The result is the **standard error** of the statistic $\hat{p}_1 - \hat{p}_2$:

standard error

$$\text{SE} = \sqrt{\frac{\hat{p}_1(1 - \hat{p}_1)}{n_1} + \frac{\hat{p}_2(1 - \hat{p}_2)}{n_2}}$$

The confidence interval again has the form

$$\text{estimate} \pm z^*\text{SE}_{\text{estimate}}$$

> ## CONFIDENCE INTERVALS FOR COMPARING TWO PROPORTIONS
>
> Draw an SRS of size n_1 from a population having proportion p_1 of successes and draw an independent SRS of size n_2 from another population having proportion p_2 of successes. When n_1 and n_2 are large, an approximate level C confidence interval for $p_1 - p_2$ is
>
> $$(\hat{p}_1 - \hat{p}_2) \pm z^* \text{SE}$$
>
> In this formula the standard error SE of $\hat{p}_1 - \hat{p}_2$ is
>
> $$\text{SE} = \sqrt{\frac{\hat{p}_1(1 - \hat{p}_1)}{n_1} + \frac{\hat{p}_2(1 - \hat{p}_2)}{n_2}}$$
>
> and z^* is the upper $(1 - C)/2$ standard normal critical value. In practice, use this confidence interval when the populations are at least 10 times as large as the samples and $n_1\hat{p}_1$, $n_1(1 - \hat{p}_1)$, $n_2\hat{p}_2$, and $n_2(1 - \hat{p}_2)$ are all 5 or more.

EXAMPLE 12.12

Example 12.11 describes a study of the effect of preschool on later use of social services. The facts are:

Population	Population description	Sample size	Sample proportion
1	Control	$n_1 = 61$	$\hat{p}_1 = .803$
2	Preschool	$n_2 = 62$	$\hat{p}_2 = .613$

To check that our approximate confidence interval is safe, look at the counts of successes and failures in the two samples. The smallest of these four quantities is

$$n_1(1 - \hat{p}_1) = (61)(1 - .803) = 12$$

This is larger than 5, so the interval will be accurate.

The difference $p_1 - p_2$ measures the effect of preschool in reducing the proportion of people who later need social services. To compute a 95% confidence interval for $p_1 - p_2$, first find the standard error

$$SE = \sqrt{\frac{\hat{p}_1(1 - \hat{p}_1)}{n_1} + \frac{\hat{p}_2(1 - \hat{p}_2)}{n_2}}$$

$$= \sqrt{\frac{(.803)(.197)}{61} + \frac{(.613)(.387)}{62}}$$

$$= \sqrt{.00642} = .0801$$

The 95% confidence interval is

$$(\hat{p}_1 - \hat{p}_2) \pm z^*SE = (.803 - .613) \pm (1.960)(.0801)$$
$$= .190 \pm .157$$
$$= (.033, .347)$$

We are 95% confident that the percent needing social services is somewhere between 3.3% and 34.7% lower among people who attended preschool. The confidence interval is wide because the samples are quite small.

EXERCISES

12.21   The 1958 Detroit Area Study was an important investigation of the influence of religion on everyday life. The sample "was basically a simple random sample of the population of the metropolitan area" of Detroit, Michigan. Of the 656 respondents, 267 were white Protestants and 230 were white Catholics.

  The study took place at the height of the cold war. One question asked if the right of free speech included the right to make speeches in favor of communism. Of the 267 white Protestants, 104 said "Yes," while 75 of the 230 white Catholics said "Yes."[7]

(a)   Give a 95% confidence interval for the difference between the proportion of Protestants who agreed that communist speeches are protected and the proportion of Catholics who held this opinion.

(b)   Check that it is safe to use the z confidence interval.

12.22   Exercise 12.14 describes a study of whether patients who file complaints leave a health maintenance organization (HMO). We want to know whether complainers are more likely to leave than patients who do not file complaints. In the year of the study, 639 patients filed complaints, and 54 of these patients left the HMO voluntarily. For comparison, the HMO chose an SRS of 743 patients who had not filed complaints. Twenty-two of these patients left voluntarily.

(a) How much higher is the proportion of complainers who leave? Give a 90% confidence interval.

(b) The HMO has more than 400,000 members. Check that you can safely use the methods of this section.

Significance tests for $p_1 - p_2$

An observed difference between two sample proportions can reflect a difference in the populations, or it may just be due to chance variation in random sampling. Significance tests help us decide if the effect we see in the samples is really there in the populations. The null hypothesis says that there is no difference between the two populations:

$$H_0 : p_1 = p_2$$

The alternative hypothesis says what kind of difference we expect.

EXAMPLE 12.13

High levels of cholesterol in the blood are associated with higher risk of heart attacks. Will using a drug to lower blood cholesterol reduce heart attacks? The Helsinki Heart Study looked at this question. Middle-aged men were assigned at random to one of two treatments: 2051 men took the drug gemfibrozil to reduce their cholesterol levels, and a control group of 2030 men took a placebo. During the next five years, 56 men in the gemfibrozil group and 84 men in the placebo group had heart attacks.

The sample proportions who had heart attacks are

$$\hat{p}_1 = \frac{56}{2051} = .0273 \qquad \text{(gemfibrozil group)}$$

$$\hat{p}_2 = \frac{84}{2030} = .0414 \qquad \text{(placebo group)}$$

That is, about 4.1% of the men in the placebo group had heart attacks, against only about 2.7% of the men who took the drug. Is the apparent benefit of gemfibrozil statistically significant? We hope to show that gemfibrozil reduces heart attacks, so we have a one-sided alternative:

$$H_0 : p_1 = p_2$$
$$H_a : p_1 < p_2$$

To do a test, standardize $\hat{p}_1 - \hat{p}_2$ to get a z statistic. If H_0 is true, all the observations in both samples really come from a single population of men of whom a single unknown proportion p will have a heart attack in a five-year period. So instead of estimating p_1 and p_2 separately, we pool the

pooled sample proportion

two samples and use the overall sample proportion to estimate the single population parameter p. Call this the ***pooled sample proportion***. It is

$$\hat{p} = \frac{\text{count of successes in both samples combined}}{\text{count of observations in both samples combined}}$$

Use $\hat{p}$ in place of both $\hat{p}_1$ and $\hat{p}_2$ in the expression for the standard error SE of $\hat{p}_1 - \hat{p}_2$ to get a z statistic that has the standard normal distribution when H_0 is true. Here is the test.

SIGNIFICANCE TEST FOR COMPARING TWO PROPORTIONS

To test the hypothesis

$$H_0 : p_1 = p_2$$

first find the pooled proportion $\hat{p}$ of successes in both samples combined. Then compute the z statistic

$$z = \frac{\hat{p}_1 - \hat{p}_2}{\sqrt{\hat{p}(1 - \hat{p})\left(\dfrac{1}{n_1} + \dfrac{1}{n_2}\right)}}$$

In terms of a variable Z having the standard normal distribution, the P-value for a test of H_0 against

$H_a : p_1 > p_2$ is $P(Z \geq z)$

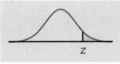

$H_a : p_1 < p_2$ is $P(Z \leq z)$

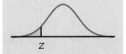

$H_a : p_1 \neq p_2$ is $2P(Z \geq |z|)$

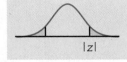

Use these tests in practice when the populations are at least 10 times as large as the samples and $n_1\hat{p}$, $n_1(1 - \hat{p})$, $n_2\hat{p}$, and $n_2(1 - \hat{p})$ are all 5 or more.

EXAMPLE 12.14   The pooled proportion of heart attacks for the two groups in the Helsinki Heart Study is

$$\hat{p} = \frac{\text{count of heart attacks in both samples combined}}{\text{count of subjects in both samples combined}}$$

$$= \frac{56 + 84}{2051 + 2030}$$

$$= \frac{140}{4081} = .0343$$

The z test statistic is

$$z = \frac{\hat{p}_1 - \hat{p}_2}{\sqrt{\hat{p}(1 - \hat{p})\left(\dfrac{1}{n_1} + \dfrac{1}{n_2}\right)}}$$

$$= \frac{.0273 - .0414}{\sqrt{(.0343)(.9657)\left(\dfrac{1}{2051} + \dfrac{1}{2030}\right)}}$$

$$= \frac{-.0141}{.005698} = -2.47$$

The one-sided *P*-value is the area under the standard normal curve to the left of -2.47. Figure 12.4 shows this area. Table A gives $P = .0068$. Because $P < .01$, the results are statistically significant at the $\alpha = .01$ level. There is strong evidence that gemfibrozil reduced the rate of heart attacks. The large samples in the Helsinki Heart Study helped the study get highly significant results.

TI-83 techniques

The TI-83 can be used to perform inference for comparing two population proportions and to construct confidence intervals. Examples 12.15 and 12.16 use the information from the Helsinki Heart Study of Examples 12.13 and 12.14.

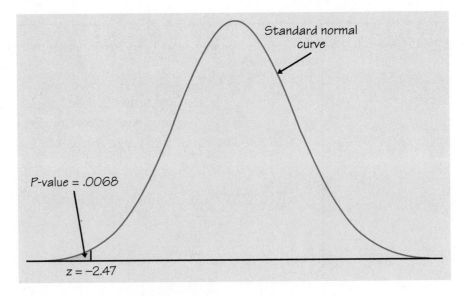

FIGURE 12.4   The *P*-value for the one-sided test of Example 12.14.

EXAMPLE 12.15

In the treatment (gemfibrozil) group of 2051 middle-aged men, 56 had heart attacks. In the control (placebo) group, 84 of the 2030 men had heart attacks. The hypotheses were

$$H_0 : p_1 = p_2$$

$$H_a : p_1 < p_2$$

The alternative hypothesis says that gemfibrozil reduces heart attacks. To perform a test of the null hypothesis on the TI-83, select STAT / TESTS / 6:2−PropZTest.

```
EDIT CALC TESTS
1:Z-Test...
2:T-Test...
3:2-SampZTest...
4:2-SampTTest...
5:1-PropZTest...
6:2-PropZTest...
7↓ZInterval...
```

When the 2−PropZTest screen appears, enter the following values: $x_1 = 56$, $n_1 = 2051$, $x_2 = 84$ and $n_2 = 2030$. Specify the alternative hypothesis $p_1 < p_2$.

```
2-PropZTest
 x1:56
 n1:2051
 x2:84
 n2:2030
 p1:≠p2  <p2   >p2
 Calculate Draw
```

If you select the "Calculate" choice and press ENTER, you are told that the z statistic is $z = -2.47$ and the P-value is 0.0068, as shown here. Do you see the pooled proportion of heart attacks? Does this agree with the calculated value in Example 12.14?

```
2-PropZTest
 p1<p2
 z=-2.470088266
 p=.0067539941
 p̂1=.0273037543
 p̂2=.0413793103
↓p̂=.0343053173
```

If you select the "Draw" option, you will see the screen shown here. Compare these results with those in Example 12.14.

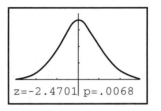

```
z=-2.4701 p=.0068
```

EXAMPLE 12.16

To use the TI-83 to construct a 95% confidence interval for the difference $p_1 - p_2$, select STAT / TESTS / B:2−PropZInt.

```
EDIT CALC TESTS
6↑2-PropZTest...
7:ZInterval...
8:TInterval...
9:2−SampZInt...
0:2−SampTInt...
A:1−PropZInt...
B↓2-PropZInt...
```

When the 2−PropZInt screen appears, verify the values $x_1 = 56$, $n_1 = 2051$, $x_2 = 84$, and $n_2 = 2030$, and specify the confidence level, 0.95. Highlight "Calculate" and press ENTER.

```
2-PropZInt
  x1:56
  n1:2051
  x2:84
  n2:2030
  C-Level:.95
  Calculate
```

The 95% confidence interval for $p_1 - p_2$ is reported, along with the two sample proportions and the two sample sizes, as shown here.

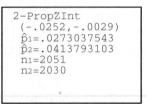

```
2-PropZInt
 (-.0252,-.0029)
  p̂1=.0273037543
  p̂2=.0413793103
  n1=2051
  n2=2030
```

EXERCISES

12.23   The 1958 Detroit Area Study was an important investigation of the influence of religion on everyday life. The sample "was basically a simple random sample of the population of the metropolitan area" of Detroit, Michigan. Of the 656 respondents, 267 were white Protestants and 230 were white Catholics.

One question asked whether the government was doing enough in areas such as housing, unemployment, and education; 161 of the Protestants and 136 of the Catholics said "No." Is there evidence that white Protestants and white Catholics differed on this issue?

(a)   State hypotheses and check that you can safely use the z test.

(b)   Carry out the test, find the P-value, and give your conclusion.

12.24   A study of "adverse symptoms" in users of over-the-counter pain relief medications assigned subjects at random to one of two common pain relievers: acetaminophen and ibuprofen. (Both of these pain relievers are sold under various brand names, sometimes combined with other ingredients.) In all, 650 subjects took acetaminophen, and 44 experienced some adverse symptom. Of the 347 subjects who took ibuprofen, 49 had an adverse symptom. How strong is the evidence that the two pain relievers differ in the proportion of people who experience an adverse symptom?

(a) State hypotheses and check that you can use the z test.

(b) Find the P-value of the test and give your conclusion.

SUMMARY

We want to **compare the proportions p_1 and p_2 of successes in two populations.** The comparison is based on the difference $\hat{p}_1 - \hat{p}_2$ between the sample proportions of successes. When the sample sizes n_1 and n_2 are large enough, we can use z procedures because the sampling distribution of $\hat{p}_1 - \hat{p}_2$ is close to normal.

An approximate level C **confidence interval** for $p_1 - p_2$ is

$$(\hat{p}_1 - \hat{p}_2) \pm z^* SE$$

where the **standard error** of $\hat{p}_1 - \hat{p}_2$ is

$$SE = \sqrt{\frac{\hat{p}_1(1 - \hat{p}_1)}{n_1} + \frac{\hat{p}_2(1 - \hat{p}_2)}{n_2}}$$

and z^* is a standard normal critical value.

Significance tests of $H_0 : p_1 = p_2$ use the **pooled sample proportion**

$$\hat{p} = \frac{\text{count of successes in both samples combined}}{\text{count of observations in both samples combined}}$$

and the **z statistic**

$$z = \frac{\hat{p}_1 - \hat{p}_2}{\sqrt{\hat{p}(1 - \hat{p})\left(\frac{1}{n_1} + \frac{1}{n_2}\right)}}$$

P-values come from the standard normal table.

SECTION 12.2 EXERCISES

12.25   The drug AZT was the first drug that seemed effective in delaying the onset of AIDS. Evidence for AZT's effectiveness came from a large randomized comparative experiment. The subjects were 1300 volunteers who were infected with HIV, the virus that causes AIDS, but did not yet have AIDS. The study assigned 435 of the subjects at random to take 500 milligrams of AZT each day, and another 435 to take a placebo. (The others were assigned to a third treatment, a higher dose of AZT. We will compare only two groups.) At the end of the study, 38 of the placebo subjects and 17 of the AZT subjects had developed AIDS. We want to test the claim that taking AZT lowers the proportion of infected people who will develop AIDS in a given period of time.

(a)   State hypotheses, and check that you can safely use the z procedures.

(b)   How significant is the evidence that AZT is effective?

(c)   The experiment was double-blind. Explain what this means.

(**Comment**: Medical experiments on treatments for AIDS and other fatal diseases raise hard ethical questions. Some people argue that because AIDS is always fatal, infected people should get any drug that has any hope of helping them. The counterargument is that we will then never find out which drugs really work. The placebo patients in this study were given AZT as soon as the results were in.)

12.26   North Carolina State University looked at the factors that affect the success of students in a required chemical engineering course. Students must get a C or better in the course in order to continue as chemical engineering majors. There were 65 students from urban or suburban backgrounds, and 52 of these students succeeded. Another 55 students were from rural or small-town backgrounds; 30 of these students succeeded in the course.[8]

(a)   Is there good evidence that the proportion of students who succeed is different for urban/suburban versus rural/small-town backgrounds? State hypotheses, give the P-value of a test, and state your conclusion.

(b)   Give a 90% confidence interval for the size of the difference.

(**Comment**: This study did not select two separate samples. A single sample of students was divided after the fact into two groups from different backgrounds. The two-sample z procedures for comparing proportions are valid in such situations. This is an important fact about these methods.)

12.27   The study of small-business failures described in Exercise 12.17 looked at 148 food-and-drink businesses in central Indiana. Of these, 106 were headed by men and 42 were headed by women. During a three-year period, 15 of the men's businesses and 7 of the women's businesses failed. Is there a significant difference between the rate at which businesses headed by men and women fail?

(**Comment**: This study did not select two separate samples. A single sample of businesses was divided after the fact into those headed by men and those headed by women. The two-sample z procedures for comparing proportions are valid in such situations. This is an important fact about these methods.)

12.28   The North Carolina State University study (see Exercise 12.26) also looked at possible differences in the proportions of female and male students who succeeded in the course. They found that 23 of the 34 women and 60 of the 89 men succeeded. Is there evidence of a difference between the proportions of women and men who succeed?

12.29   Different kinds of companies compensate their key employees in different ways. Established companies may pay higher salaries, while new companies may offer stock options that will be valuable if the company succeeds. Do high-tech companies tend to offer stock options more often than other companies? One study looked at a random sample of 200 companies. Of these, 91 were listed in the *Directory of Public High Technology Corporations* and 109 were not listed. Treat these two groups as SRSs of high-tech and non-high-tech companies. Seventy-three of the high-tech companies and 75 of the non-high-tech companies offered incentive stock options to key employees.[9]

(a)   Is there evidence that a higher proportion of high-tech companies offer stock options?

(b)   Give a 95% confidence interval for the difference in the proportions of the two types of companies that offer stock options.

12.30   Nonresponse to sample surveys may differ with the season of the year. In Italy, for example, many people leave town during the summer. The Italian National Statistical Institute called random samples of telephone numbers between 7 p.m. and 10 p.m. at several seasons of the year. Here are the results for two seasons:[10]

Dates	Number of calls	No answer	Total nonresponse
Jan. 1 to Apr. 13	1558	333	491
July 1 to Aug. 31	2075	861	1174

(a) How much higher is the proportion of "no answers" in July and August compared with the early part of the year? Give a 99% confidence interval.

(b) The difference between the proportions of "no answers" is so large that it is clearly statistically significant. How can you tell from your work in (a) that the difference is significant at the $\alpha = .01$ level?

(c) Use the information given to find the counts of calls which had nonresponse for some reason other than "no answer." Do the rates of nonresponse due to other causes also differ significantly for the two seasons?

12.31   The Physicians' Health Study examined the effects of taking an aspirin every other day. Earlier studies suggested that aspirin might reduce the risk of heart attacks. The subjects were 22,071 healthy male physicians at least 40 years old. The study assigned 11,037 of the subjects at random to take aspirin. The others took a placebo pill. The study was double-blind. Here are the counts for some of the outcomes of interest to the researchers:

	Aspirin group	Placebo group
Fatal heart attacks	10	26
Nonfatal heart attacks	129	213
Strokes	119	98

For which outcomes is the difference between the aspirin and placebo groups significant? (Use two-sided alternatives. Check that you can apply the z test. Write a brief summary of your conclusions.)

12.32   The National Assessment of Educational Progress (NAEP) Young Adult Literacy Assessment Survey interviewed a random sample of 1917 people 21 to 25 years old. The sample contained 840 men, of whom 775 were fully employed. There were 1077 women, and 680 of them were fully employed.[11]

(a) Use a 99% confidence interval to describe the difference between the proportions of young men and young women who are fully employed. Is the difference statistically significant at the 1% significance level?

(b) The mean and standard deviation of scores on the NAEP's test of quantitative skills were $\bar{x}_1 = 272.40$ and $s_1 = 59.2$ for the men in the sample. For the women, the results were $\bar{x}_2 = 274.73$ and $s_2 = 57.5$. Is the difference between the mean scores for men and women significant at the 1% level?

12.33   The Current Population Survey (CPS) is the monthly government sample survey of 60,000 households that provides data on employment in the United States. A study of child-care workers drew a sample from the CPS data tapes. We can consider this sample to be an SRS from the population of child-care workers.[12]

(a)   Out of 2455 child-care workers in private households, 7% were black. Of 1191 nonhousehold child-care workers, 14% were black. Give a 99% confidence interval for the difference in the percents of these groups of workers who are black. Is the difference statistically significant at the $\alpha = .01$ level?

(b)   The study also examined how many years of school child-care workers had. For household workers, the mean and standard deviation were $\bar{x}_1 = 11.6$ years and $s_1 = 2.2$ years. For nonhousehold workers, $\bar{x}_2 = 12.2$ years and $s_2 = 2.1$ years. Give a 99% confidence interval for the difference in mean years of education for the two groups. Is the difference significant at the $\alpha = .01$ level?

12.34   Never forget that even small effects can be statistically significant if the samples are large. To illustrate this fact, return to the study of 148 small businesses in Exercise 12.27.

(a)   Find the proportions of failures for businesses headed by women and businesses headed by men. These sample proportions are quite close to each other. Give the P-value for the z test of the hypothesis that the same proportion of women's and men's businesses fail. (Use the two-sided alternative.) The test is very far from being significant.

(b)   Now suppose that the same sample proportions came from a sample 30 times as large. That is, 210 out of 1260 businesses headed by women and 450 out of 3180 businesses headed by men fail. Verify that the proportions of failures are exactly the same as in (a). Repeat the z test for the new data, and show that it is now significant at the $\alpha = .05$ level.

(c)   It is wise to use a confidence interval to estimate the size of an effect, rather than just giving a P-value. Give 95% confidence intervals for the difference between the proportions of women's and men's businesses that fail for the settings of both (a) and (b). What is the effect of larger samples on the confidence interval?

CHAPTER REVIEW

Statistical inference always draws conclusions about one or more parameters of a population. When you think about doing inference, ask first what the population is and what parameter you are interested in. The t procedures of Chapter 11 allow us to give confidence intervals and carry out tests about population means. We use the z procedures of this chapter for inference about population proportions.

Inference about population proportions is based on sample proportions. We rely on the fact that a sample proportion has a distribution that is close to normal unless the sample is quite small. All the z procedures in this chapter work well when the samples are large enough. You must check this before using them. Here are the things you should now be able to do.

A. RECOGNITION

1. Recognize when a problem requires inference about a proportion or comparing two proportions.
2. Calculate from sample counts the sample proportion or proportions that estimate the parameters of interest.

B. INFERENCE ABOUT ONE PROPORTION

1. Use the z procedure to give a confidence interval for a population proportion p.
2. Use the z statistic to carry out a test of significance for the hypothesis $H_0 : p = p_0$ about a population proportion p against either a one-sided or a two-sided alternative.
3. Check that you can safely use these z procedures in a particular setting.

C. COMPARING TWO PROPORTIONS

1. Use the two-sample z procedure to give a confidence interval for the difference $p_1 - p_2$ between proportions in two populations based on independent samples from the populations.
2. Use a z statistic to test the hypothesis $H_0 : p_1 = p_2$ that proportions in two distinct populations are equal.
3. Check that you can safely use these z procedures in a particular setting.

CHAPTER 12 REVIEW EXERCISES

12.35   A television news program conducts a call-in poll about a proposed city ban on handgun ownership. Of the 2372 calls, 1921 oppose the ban. The station, following recommended practice, makes a confidence statement: "81% of the Channel 13 Pulse Poll sample opposed the ban. We can be 95% confident that the true proportion of citizens opposing a handgun ban is within 1.6% of the sample result." Is this conclusion justified?

12.36   Some people think that chemists are more likely than other parents to have female children. (Perhaps chemists are exposed to something in their laboratories that affects the sex of their children.) The Washington State Department of Health lists the parents' occupations on birth certificates. Between 1980 and 1990, 555 children were born to fathers who were chemists. Of these births, 273 were girls. During this period, 48.8% of all births in Washington State were girls.[13] Is there evidence that the proportion of girls born to chemists is higher than the state proportion?

12.37   The first law requiring child restraints in motor vehicles went into effect in 1978, in Tennessee. "Before enactment of child restraint laws, most children in the United States were unrestrained when traveling in motor vehicles." The writer of this statement cites a 1974 survey of drivers in Maryland and Virginia that found that "about 93% ($n = 8,933$) of children under the age of 10 were traveling unrestrained."[14] Assuming that the children observed are an SRS from the areas involved, give a 99% confidence interval for the proportion of unrestrained children in 1974.

12.38   Sickle-cell trait is a hereditary condition that is common among blacks and can cause medical problems. Some biologists suggest that sickle-cell trait protects against malaria. That would explain why it is found in people who originally came from Africa, where malaria is common. A study in Africa tested 543 children for the sickle-cell trait and also for malaria. In all, 136 of the children had the sickle-cell trait, and 36 of these had heavy malaria infections. The other 407 children lacked the sickle-cell trait, and 152 of them had heavy malaria infections.[15]

(a)   Give a 95% confidence interval for the difference in the proportion of all children in the population studied who have the sickle-cell trait.

(b)   Is there good evidence that the proportion of heavy malaria infections is lower among children with the sickle-cell trait?

12.39   The National Collegiate Athletic Association (NCAA) requires colleges to report the graduation rates of their athletes. Here are data from a Big Ten university's report.[16]

(a) Forty-five of the 74 athletes admitted in a specific year graduated within six years. Does the proportion of athletes who graduate differ significantly from the all-university proportion, which is 68%?

(b) The graduation rates that year were 21 of 28 female athletes and 24 of 46 male athletes. Is there evidence that a smaller proportion of male athletes than of female athletes graduate within six years?

(c) We are willing to regard athletes admitted in a specific year as an SRS from the large population of athletes the university will admit under its present standards. Explain why you can safely use the z procedures in parts (a) and (b). Then explain why you cannot use these procedures to test whether baseball players (3 out of 5 admitted that year graduated) differ from other athletes.

12.40   A study comparing American and Australian corporations examined a sample of 133 American and 63 Australian corporations. There are the usual practical difficulties involving nonresponse and the question of what population the samples represent. Ignore these issues and treat the samples as SRSs from the United States and Australia. The average percent of revenues from "highly regulated businesses" was 27% for the Australian companies and 41% for the American companies.[17]

(a) The data are given as percents. Explain carefully why comparing the percent of revenues from highly regulated businesses for U.S. and Australian corporations is *not* a comparison of two population proportions.

(b) What test would you use to make the comparison? (Don't try to carry out a test.)

NOTES AND DATA SOURCES

1. Data from Joseph H. Catania et al., "Prevalence of AIDS-related risk factors and condom use in the United States," *Science*, 258 (1992), pp. 1101–1106.

2. The study is reported in William Celis III, "Study suggests Head Start helps beyond school," *New York Times*, April 20, 1993.

3. Data from Richard M. Felder et al., "Who gets it and who doesn't: a study of student performance in an introductory chemical engineering course," *1992 ASEE Annual*

Conference Proceedings, American Society for Engineering Education, Washington, D.C., 1992, pp. 1516–1519.

4. The quotation is from page 1104 of the article cited in Note 1.

5. Sara J. Solnick and David Hemenway, "Complaints and disenrollment at a health maintenance organization," *The Journal of Consumer Affairs*, 26 (1992), pp. 90–103.

6. Arne L. Kalleberg and Kevin T. Leicht, "Gender and organizational performance: determinants of small business survival and success," *The Academy of Management Journal*, 34 (1991), pp. 136–161.

7. The Detroit Area Study is described in Gerhard Lenski, *The Religious Factor*, Doubleday, New York, 1961.

8. The data are from the source cited in Note 3.

9. Based on Greg Clinch, "Employee compensation and firms' research and development activity," *Journal of Accounting Research*, 29 (1991), pp. 59–78.

10. Giuliana Coccia, "An overview of non-response in Italian telephone surveys," *Proceedings of the 99th Session of the International Statistical Institute*, 1993, Book 3, pp. 271–272.

11. Francisco L. Rivera-Batiz, "Quantitative literacy and the likelihood of employment among young adults," *Journal of Human Resources*, 27 (1992), pp. 313–328.

12. David M. Blau, "The child care labor market," *Journal of Human Resources*, 27 (1992), pp. 9–39.

13. Eric Ossiander, letter to the editor, *Science*, 257 (1992), p. 1461.

14. William N. Evans and John D. Graham, "An estimate of the lifesaving benefit of child restraint use legislation," *Journal of Health Economics*, 9 (1990), pp. 121–132.

15. A. C. Allison and D. F. Clyde, "Malaria in African children with deficient erythrocyte dehydrogenase," *British Medical Journal*, 1 (1961), pp. 1346–1349.

16. Office of the Registrar, Purdue University, *Summary of the 1992–1993 NCAA Graduation-Rates Disclosure Form, 1985–1986 Academic Year Cohort*, West Lafayette, Ind., 1993.

17. From Noel Capon et al., "A comparative analysis of the strategy and structure of United States and Australian corporations," *Journal of International Business Studies*, 18 (1987), pp. 51–74.

Inference for Tables: Chi-Square Procedures

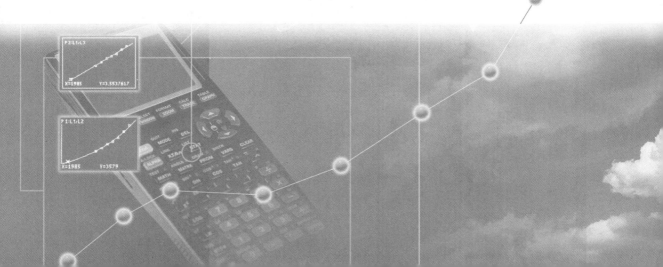

INTRODUCTION

In the previous chapter we discussed inference procedures for comparing population proportions. The *chi-square goodness of fit* test extends inference on proportions to more than two proportions by enabling us to determine if a particular population distribution has changed from a specified form.

The *chi-square test of independence* further extends this notion by allowing us to test whether or not the distribution of one variable has been influenced by another variable, based on the information provided in a two-way table.

| ACTIVITY 13 | "I Didn't Get Enough Blues!" |

Materials needed: One 1.69-ounce bag of plain M&M's per student.

The M&M/Mars Company, headquartered in Hackettstown, New Jersey, makes plain and peanut chocolate candies. In 1995, they decided to replace the tan-colored M&M's with a new color. After conducting an extensive national preference survey, they decided to replace the tan M&M's with blue M&M's. The company's Consumer Affairs Department announced:

> *On average, the new mix of colors of M&M's Plain Chocolate Candies will contain 30 percent browns, 20 percent each of yellows and reds and 10 percent each of oranges, greens, and blues.*

They explained:

> *While we mix the colors as thoroughly as possible, the above ratios may vary somewhat, especially in the smaller bags. This is because we combine the various colors in large quantities for the last production stage (printing). The bags are then filled on high-speed packaging machines by weight, not by count.*

In preparation for this activity, each student should have purchased one 1.69-ounce bag of plain M&M's. The purpose of this activity is to compare the color distribution of M&M's in the students' bags (considering the entire count of M&M's in the class as one large sample from the production process) with the advertised distribution. We will want

(continued)

to see if there is sufficient evidence to dispute the company's claim for their distribution. In order to use as random a sample as possible, it is best if the bags of M&M's are purchased at different stores and not obtained from one or a few sources of supply.

1. Each student should open a bag and carefully count the number of M&M's for each color—brown, yellow, red, orange, green, and blue—as well as count the total number of M&M's in the bag.

2. Accumulating the counts for all students, you will obtain a total count of M&M's of each color. First add up the number of M&M's in all of the bags to obtain a grand total. Fill in the counts, by color, in the "Observed" row in a table like this:

Color	Brown	Yellow	Red	Orange	Green	Blue
Observed						
Expected						
$(O - E)^2/E$						

3. For the expected counts, multiply the grand total number of M&M's by the company's stated percentages (expressed in decimal form) for each of the colors.

4. For each color, perform this calculation:

$$(\text{observed} - \text{expected})^2/\text{expected}$$

and enter the result in the last row of the table. Then add up all of these calculated values, and name the sum X^2. Keep this number handy—we will use it later in the chapter.

5. If your class sample reflects the distribution advertised by the M&M/Mars Company, then there should be very little difference between the observed counts and the expected counts. Hence the calculated values making up the sum X^2 should be very small. Are the entries in the last row all about the same, or do any of the quantities stand out because they are "significantly" larger? Did you get more of a particular color than you expected? Did you get fewer of a particular color than you expected?

13.1   TEST FOR GOODNESS OF FIT

Suppose you open a 1.69-ounce bag of plain M&M chocolate candies and discover that out of approximately 56 M&M's, there is only a single *blue* M&M. Knowing that 10% of all plain chocolate M&M's made by the M&M/Mars Company are blue, and that in your sample of size 56, the proportion of blue M&M's is only $1/56 = .018$, you might feel that you didn't get your fair share of blues. You could use the z test, as described in the last chapter, to test the hypotheses

$$H_0: p = .10$$
$$H_a: p < .10$$

chi-square (χ^2) test for goodness of fit

where p is the proportion of blue M&M's. You could then perform additional tests of significance for each of the remaining colors. But this would be inefficient. More important, it wouldn't tell us how likely it is that six sample proportions differ from the stated values as much as our sample does. There is a single test that can be applied to see if the observed sample distribution is different from the hypothesized population distribution. It is called the *chi-square (χ^2) test for goodness of fit*.

EXAMPLE 13.1

In recent years, the expression "the graying of America" has been used to refer to the belief that with better medicine and healthier lifestyles, people are living longer, and consequently a larger percentage of the population is of retirement age. We want to investigate whether this perception is accurate. The distribution of the U.S. population in 1980 is shown in Table 13.1. We want

TABLE 13.1   U.S. population by age group, 1980

Age group	Population (in thousands)	Percent
0 to 24	93,777	41.39
25 to 44	62,716	27.68
45 to 64	44,503	19.64
65 and older	25,550	11.28
Total	226,546	100.00

Source: *Statistical Abstract of the United States, 1997*, U.S. Department of Commerce, Bureau of the Census.

to determine if the distribution of age groups in the United States in 1996 has changed significantly from the 1980 distribution. We will perform the following hypothesis test:

H_0: the age group distribution in 1996 is *the same as* the 1980 distribution

H_a: the age group distribution in 1996 is *different from* the 1980 distribution

The idea of the test is this: We compare the observed counts for a sample from the 1996 population with the counts that would be expected if the 1996 distribution were the same as the 1980 distribution, that is, if H_0 were in fact true. The 1980 distribution is the *population*. The more the observed counts differ from the expected counts, the more evidence we have to reject H_0 and to conclude that the population distribution in 1996 is significantly different from that of 1980.

A random *sample* of 500 U.S. residents in 1996 is selected and the age of each subject is recorded. The results are shown in Table 13.2.

Before proceeding with a significance test, it's always a good idea to plot the data. In this case, a segmented bar chart allows the observer to compare segments from 1980 to 1996. To do this, you need to calculate marginal percents, as shown in Table 13.3. The segmented bar chart is shown in Figure 13.1.

The next step in the test is to calculate the expected counts for each age category. If the age group distribution seen in 1980 has not changed, then in a random sample of 500 U.S. residents in 1996, we would expect 41.39% of the 500 to be in the 0 to 24 age category, 27.68% to be in the 25 to 44 age category, and so forth. For each of the categories, we would expect the appropriate percentage (from Table 13.1) of the 500 to be in the corresponding age category. The expected counts are displayed in Table 13.4.

TABLE 13.2	Sample results for 500 randomly selected individuals

Age group	Count
0 to 24	177
25 to 44	158
45 to 64	101
65 and older	64
Total	500

TABLE 13.3	Sample results for 500 randomly selected individuals	
Age group	Count	Percent
0 to 24	177	35.4
25 to 44	158	31.6
45 to 64	101	20.2
65 and older	64	12.8
	500	100

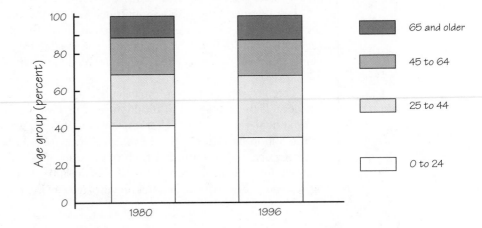

FIGURE 13.1   Segmented bar chart for the data in Table 13.3.

TABLE 13.4	Expected counts	
Age group	1980 population percents	Expected counts
0 to 24	41.39	500(.4139) = 207.0
25 to 44	27.68	500(.2768) = 138.4
45 to 64	19.64	500(.1964) = 98.2
65 and older	11.28	500(.1128) = 56.4
	100	500

TABLE 13.5	Calculating the goodness of fit		
Age group	Observed O	Expected E	$(O - E)^2/E$
0 to 24	177	207.0	4.3478
25 to 44	158	138.4	2.7757
45 to 64	101	98.2	.0798
65 and older	64	56.4	1.0241

In order to determine whether the distribution has changed since 1980, we need a way to measure how well the observed counts (O) from 1996 fit the expected counts (E) under H_0. The procedure is to calculate the quantity

$$\frac{(O - E)^2}{E}$$

*chi-square
statistic*

for each age category and then add up these terms. The sum is labeled X^2 and is called the ***chi-square statistic***. A summary of the calculations is shown in Table 13.5. The sum of the terms in the last column is $X^2 = 8.2275$. The larger the differences between the observed and expected values, the larger X^2 will be, and the more evidence there will be against H_0.

The χ^2 family of distribution curves is used to assess the evidence against H_0 represented in the value of X^2. The member of the family that is used is determined by the ***degrees of freedom***. Since we are working with percentages, three of the four percentages are free to vary, but the fourth is not, since all four have to add to 100. In this case, we say that there are $4 - 1 = 3$ degrees of freedom. In the back of the book, Table E, Chi-Square Distribution Critical Values, shows a typical chi-square curve with a right-tailed area shaded. The chi-square test statistic is a point on the horizontal axis, and the area to the right is the *P*-value of the test. This *P*-value is the probability of observing a value X^2 at least as extreme as the one actually observed. The larger the value of the chi-square statistic, the smaller the *P*-value, and the more evidence you have against the null hypothesis, H_0. In Table E, for a *P*-value of 0.05 and degrees of freedom $= 3$, we find that the critical value is 7.81. Since our $X^2 = 8.2275$ is more extreme (larger) than the critical value, we say that the probability of observing a result as extreme as the one we actually observed, by chance alone, is less than 5%. There is sufficient evidence to reject H_0 and conclude that the population distribution in 1996 is significantly different from the 1980 distribution, at the 5% significance level.

*degrees of
freedom*

Table 13.5 suggests a technique for easing the computations a bit. On your TI-83, clear lists L_1, L_2, and L_3, and enter the observed counts in list

L_1. Calculate the expected counts separately and enter them in list L_2. Then define list L_3 to be $(L_1 - L_2)^2/L_2$. The command sum(L_3) returns the test statistic X^2 (sum is found under 2nd / LIST / MATH / 5:sum).

L1	L2	L3 3
177	207	**4.3478**
158	138.4	2.7757
101	98.2	.07984
64	56.4	1.0241
---	----	-----

L3(1)=4.347826086...

```
sum(L3)
           8.227499173
```

Finally, using the χ^2cdf command from the distributions menu (choice 7), we ask for the area between $X^2 = 8.2275$ and a very large number (1E99) and specify the degrees of freedom, as shown. The P-value, 0.04, indicates that $X^2 = 8.277$ is a very unlikely result if H_0 is true; this represents strong evidence against H_0.

```
χ²cdf(8.227,1E99,3)
              .0415460217
```

Properties of the chi-square distributions

Software usually finds P-values for us. The P-value for a chi-square test comes from comparing the value of the chi-square statistic with critical values for a *chi-square distribution*.

THE CHI-SQUARE DISTRIBUTIONS

The **chi-square distributions** are a family of distributions that take only positive values and are skewed to the right. A specific chi-square distribution is specified by one parameter, called the **degrees of freedom**.

To see several representative members of the chi-square family of distributions, specify a WINDOW as shown: X[0, 14]$_1$ and Y[$-.05$, .3]$_{.1}$.

```
WINDOW
 Xmin=0
 Xmax=14
 Xscl=1
 Ymin=-.05
 Ymax=.3
 Yscl=.1
 Xres=1
```

Enter as Y_1: $\chi^2\text{pdf}(X,1)$ ($\chi^2\text{pdf}$ is found under 2nd / DISTR). Your chi-square distribution with df = 1 should look like this:

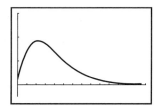

Deselect Y_1 and define $Y_2 = \chi^2\text{pdf}(X,4)$. Here is the chi-square distribution with df = 4:

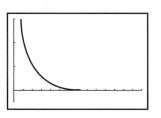

Deselect Y_2 and define Y_3 as $\chi^2\text{pdf}(X,8)$. This is the chi-square distribution with df = 8:

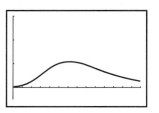

Finally, here are the chi-square density curves for df = 1, 4, and 8, all on the same axes:

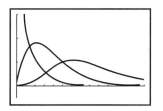

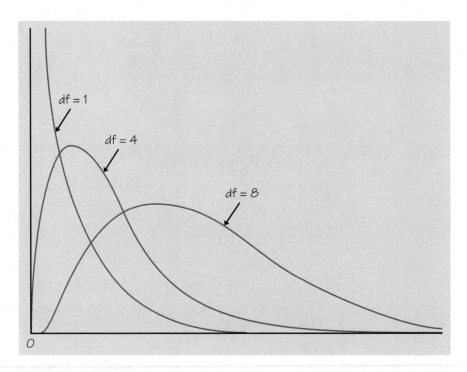

FIGURE 13.2   Density curves for the chi-square distributions with 1, 4, and 8 degrees of freedom. Chi-square distributions take only positive values.

Figure 13.2 shows the density curves for three members of the chi-square family of distributions. As the degrees of freedom increase, the density curves become less skewed and larger values become more probable. Table E in the back of the book gives critical values for chi-square distributions. You can use Table E if software does not give you *P*-values for a chi-square test.

The chi-square density curves have the following properties:

1. The total area under a chi-square curve is equal to 1.
2. Each chi-square curve begins at 0 on the horizontal axis, increases to a peak, and then approaches the horizontal axis asymptotically from above.
3. Each chi-square curve is skewed to the right. As the number of degrees of freedom increase, the curve becomes more and more symmetrical and looks more like a normal curve.

The *P*-value calculated in Example 13.1 for the population problem is about 0.04. This value can be visualized on the TI-83 as follows. First, define an appropriate viewing window. Then use the command $\text{Shade}\chi^2$ (8.227,100,3) (2nd / DISTR / DRAW / 3:$\text{Shade}\chi^2$) to show the critical area in the right tail of the χ^2 curve to the right of $X^2 = .8227$ with df = 3. Note that the $\text{Shade}\chi^2$ command requires a left endpoint and a right endpoint, so take a sufficiently large right endpoint to achieve four-decimal-place accuracy. The shaded area is about 0.04. This area is the P-value.

```
WINDOW
  Xmin=0
  Xmax=14
  Xscl=1
  Ymin=-.08
  Ymax=.3
  Yscl=.1
  Xres=1
```

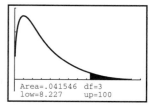

Let us summarize.

GOODNESS OF FIT TEST

A **goodness of fit test** is used to help determine whether a population has a certain hypothesized distribution, expressed as percents of population members falling into various outcome categories. Suppose that the hypothesized distribution has n outcome categories. To test the hypothesis

H_0: the actual population percents are *equal to* the hypothesized percentages

first calculate the chi-square test statistic

$$X^2 = \sum (O - E)^2 / E$$

Then X^2 has approximately a χ^2 distribution with $(n - 1)$ degrees of freedom.

For a test of H_0 against the alternative hypothesis

H_a: the actual population percentages are *different from* the hypothesized percentages

the P-value is $P(\chi^2 \geq X^2)$.

You may use this test with critical values from the chi-square distribution when all individual expected counts are at least 1 and no more than 20% of the expected counts are less than 5.

EXERCISES

13.1   (a)   Find the area to the right of $X^2 = 1.41$ under the chi-square curve with 2 degrees of freedom.

           (b)   Find the area to the right of $X^2 = 19.62$ under the chi-square curve with 9 degrees of freedom.

13.2   For the distribution of M&M's in Activity 13, complete the test of significance for the goodness of fit test. State the null and alternative hypotheses, the value of the statistic X^2, and the P-value. Then write your conclusion in nontechnical language.

13.3   According to the *Statistical Abstract of the United States, 1997*, the marital status distribution of the U.S. adult population in 1996 is as follows:

Marital status	Never married	Married	Widowed	Divorced
Percent	23.26	60.31	7.00	9.43

A random sample of 500 U.S. males, aged 25 to 29 years old, yielded the following frequency distribution:

Marital status	Never married	Married	Widowed	Divorced
Frequency	260	220	0	20

We want to determine if the marital status distribution of U.S. males 25 to 29 years old differs from that of the U.S. adult population.

   (a)   State the null and alternative hypotheses for a chi-square goodness of fit test.

   (b)   Calculate the expected counts.

   (c)   Find the statistic X^2. What are the degrees of freedom?

   (d)   Find the P-value and state your conclusions.

13.4   In recent years, a national effort has been made to enable more members of minority groups to have increased educational opportunities. You want to know if the policy of "affirmative action" and similar initiatives have had any effect in this regard. You obtain information on the ethnicity distribution of holders of the highest academic degree, the doctor of philosophy degree, for the year 1981:

Race/ethnicity	Percent
White, non-Hispanic	78.9
Black, non-Hispanic	3.9
Hispanic	1.4
Asian or Pacific Islander	2.7
American Indian/Alaskan Native	0.4
Nonresident alien	12.8

A random sample of 300 doctoral degree recipients in 1994 showed the following frequency distribution:

Race/ethnicity	Count
White, non-Hispanic	189
Black, non-Hispanic	10
Hispanic	6
Asian or Pacific Islander	14
American Indian/Alaskan Native	1
Nonresident alien	80

(a) Perform a goodness of fit test to determine if the distribution of doctoral degrees in 1994 is significantly different from the distribution in 1981. Don't forget to state your hypotheses, and don't forget to state you conclusion.

(b) In which categories have the greatest changes occurred, and in what direction?

Conducting inference by simulation

Let's return to the "graying of America" problem. Suppose that we didn't have the resources to select a representative sample from the current population of the United States or that there is some other reason why we can't gather the actual data we need. We still may be able to obtain an approximate solution by means of *simulation*.

| TABLE 13.6 | U.S. population distribution for 1996 |

Age group	Population percent
0 to 24	35.4
25 to 44	31.6
45 to 64	20.2
65 and older	12.8
	100

EXAMPLE 13.2

In the population study of Example 13.1, we can use recent census figures to obtain percents for the age categories for the year 1996. The relative frequency distribution in Table 13.6 is calculated from data in the *Statistical Abstract of the United States, 1997*.

Step 1. Establish a correspondence between random numbers and ages. To simplify matters we will round off the percents from Table 13.6 to whole numbers: 35%, 32%, 20%, and 13%, respectively. One possible scheme is as follows. Let a number from 1 to 100 represent a randomly selected person from the U.S. population.

Let the numbers

1 to 35 (35% of randomly generated numbers) represent persons 0 to 24 years of age.
36 to 67 (32% of randomly generated numbers) represent persons 25 to 44 years of age.
68 to 87 (20% of randomly generated numbers) represent persons 45 to 64 years of age.
88 to 100 (13% of randomly generated numbers) represent persons 65 years or older.

Step 2. Determine a sample size n (typically a number between 100 and 500). If the simulation is done on a TI-83, your sample size will be limited by the available memory in the calculator.

Step 3. Randomly generate n numbers in the range 1 to 100, and count the numbers that fall into each of the four age categories. The TI-83 program POP carries out this plan.

```
PROGRAM:POP
:ClrHome
:ClrList L₁,L₂,L₃
:Disp "HOW MANY TRIALS"
:Prompt N
:randInt(1,100,N)→L₁
:(L₁≥1 and L₁≤35)→L₂:sum(L₂)→L₃(1)
:(L₁≥36 and L₁≤67)→L₂:sum(L₂)→L₃(2)
:(L₁≥68 and L₁≤87)→L₂:sum(L₂)→L₃(3)
:(L₁≥88 and L₁≤100)→L₂:sum(L₂)→L₃(4)
:Disp ""
:Disp "OBSERVED COUNTS"
:Disp "ARE IN L3"
```

The calculator randomly generates a specified number of ages and stores them in list L_1. It then looks at each age in L_1 and records a 1 in list L_2 if the age is in the first age category and a 0 otherwise. Then it adds the 1s in L_2 and records the result in list L_3. It repeats the process for each of the other age categories. When the simulation is finished, the four numbers in list L_3 are the simulated counts of people in each age category.

```
HOW MANY TRIALS
N=?300

OBSERVED COUNTS
ARE IN L3
                      Done
■
```

Figure 13.3 shows the results for a sample run of the program for a specified sample size $n = 300$.

```
L1      L2      L3      3
1       0
45      0      103
77      0      85
65      0      67
86      0      45
35      0
57      0      -----

L3(1)=103
```

FIGURE 13.3   Simulation results for Example 13.2.

Step 4. Using the results of the simulated counts in each age category, perform a chi-square goodness of fit test. Copy the four simulated counts from L_3 to L_1, and enter the expected counts (300 times the 1980 decimals) in L_2. Define list L_3 as $(L_1 - L_2)^2/L_2$.

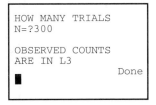

```
L1      L2       L3      3
103     124.17  3.6093
85      83.04   .04626
67      58.92   1.1081
45      33.84   3.6804
---     ----    -----

L3(1)=3.609317065...
```

To obtain the value of X^2, sum the four terms (observed − expected)^2/expected in list L$_3$. Then clear the graphics screen (2nd / DRAW / 1:ClrDraw), and ask for a sketch of the χ^2 curve with the area to the right of X^2 shaded.

```
sum(L3)
                  8.444056235
ClrDraw
                        Done
Shadeχ²(8.444,100,3)
```

The results show that for this particular simulation, with sample size 300, $X^2 = 8.444$, and df = 3, the P-value is 0.0377. Note that if you replicate this simulation on your calculator, especially with a smaller sample size (e.g., $n = 100$), you may get very different results. Remember that the X^2 statistic will show much greater variability with smaller sample sizes. For this reason, if you simulate a sampling problem like this with the TI-83, you should use as large a sample size as you can without getting a memory overflow error message.

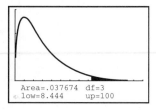

Area=.037674 df=3
low=8.444   up=100

Follow-up analysis

Do our results show the "graying of America" phenomenon? Only somewhat; the big story appears elsewhere. The sample in this example was calculated to accurately reflect the 1996 age group distribution, so a meaningful comparison with the 1980 distribution is possible. If you inspect the four terms, $(O − E)^2/E$, that are added together to give the test statistic X^2 in Example 13.1, the largest contribution to X^2 was from the first age category (see Table 13.5). The observed count for the 65 and over category was only slightly more than the expected count. So the greatest change in the distribution was in the 0 to 24 age category, where the observed (1996) population was *smaller* than the expected population size. The birthrate from 1972 until about 1987 was fairly stable at about 15.8 per 1000 residents. It increased from 1987 to 1990 and then began a steady decline. Here are the birthrates from 1987 through 1996:[1]

Year	1987	1988	1989	1990	1991	1992	1993	1994	1995	1996
Birthrate	15.7	16.0	16.4	16.6	16.3	15.9	15.5	15.2	14.8	14.5

Even though there is evidence that the distribution of ages has changed significantly from 1980 to 1996, one must look at the individual components of X^2 to see where the largest changes have occurred.

EXERCISES

13.5   This exercise is a continuation of Example 13.2. Downlink the POP program to your TI-83 and then execute the program. Specify a sample of size 100. While the calculator is working, write null and alternative hypotheses for a goodness of fit test. Then complete your analysis of the results of your simulation, and determine the P-value. How do the results of your simulation compare with the results in this section?

13.6   This exercise will test your TI-83's built-in random number generator. Use the calculator to generate 200 random digits. Then perform a goodness of fit test to see if the distribution of digits in your sample is different from the distribution that you would expect.

13.7   Simulate rolling a fair die 300 times on the TI-83 calculator. Plot a histogram of the results, and then perform a goodness of fit test of the hypothesis that the die is fair.

13.8   A statistics student suspected that his 1982 penny was not a fair coin, so he held it upright on a tabletop with a finger of one hand and spun the penny repeatedly with the thumb and index finger of the other hand. In 200 spins of the coin, the tail side of the coin came up 152 times. Perform a goodness of fit test to see if there is sufficient evidence to conclude that spinning the coin does not produce equally likely results.

SUMMARY

The **chi-square test for goodness of fit** tests the null hypothesis that a population distribution is the same as a reference distribution.

The **expected count** for any variable category is obtained by multiplying the percent of the distribution for each category times the sample size.

The **chi-square statistic** is $X^2 = \Sigma(O - E)^2/E$, where the sum is over n variable categories.

The chi-square test compares the value of the statistic X^2 with critical values from the **chi-square distribution** with $n - 1$ **degrees of freedom**.

For a test of H_0 against

H_a: the population percents are *different from* the hypothesized values the *P*-value is the area under the density curve to the right of X^2. Large values of X^2 are evidence against H_0.

The chi-square distribution is an approximation to the distribution of the statistic X^2. You can safely use this approximation when all expected counts are at least 1 and no more than 20% are less than 5. If the chi-square test finds a statistically significant *P*-value, do a follow-up analysis that compares the observed counts with the expected counts and that looks for the largest **components of chi-square**.

SECTION 13.1 EXERCISES

13.9   A die is tossed 200 times with the faces 1, 2, 3, 4, 5, and 6 turning up with frequencies 26, 36, 39, 30, 38, and 31, respectively. Is there reason to believe that the die is "loaded" (i.e., unfair)?

13.10 Trix cereal comes in five fruit flavors, and each flavor has a different shape. A curious student methodically sorted an entire box of the cereal and found the following distribution of flavors for the pieces of cereal in the box:

Flavor	Grape	Lemon	Lime	Orange	Strawberry
Frequency	530	470	420	610	585

Test the null hypothesis that the flavors are uniformly distributed versus the alternative that they are not.

13.11 The M&M/Mars Company reports the following distribution for other M&M varieties: for Peanut Chocolate Candies, the ratio is 20% each of browns, yellows, reds, and blues, and 10% each of greens and oranges. For Peanut Butter and Almond M&M's, the distribution is 20% each of browns, yellows, reds, greens, and blues. Buy a bag of one of these varieties of M&M's, perform a goodness of fit test of the company's reported distribution, and report your results. Better still, obtain a larger sample by using multiple bags and do this problem as another class activity.

13.12 A "wheel of fortune" at a carnival is divided into four equal parts:

            Part I: Win a doll
            Part II: Win a candy bar
            Part III: Win a free ride
            Part IV: Win nothing

You suspect that the wheel is unbalanced (i.e., not all parts of the wheel are equally likely to be landed upon when the wheel is spun). The results of 500 spins of the wheel are as follows:

Part	I	II	III	IV
Frequency	95	105	135	165

Perform a goodness of fit test. Is there evidence that the wheel is not in balance?

13.2 INFERENCE FOR TWO-WAY TABLES

The two-sample z procedures of Chapter 12 allow us to compare the proportions of success in two groups, either two populations or two treatment groups in an experiment. What if we want to compare more than two groups? We need a new statistical test. The new test starts by presenting the data in a new way, as a two-way table. Two-way tables have more general uses than comparing the proportions of successes in several groups. As we saw in Section 3 of Chapter 4, they describe relationships between any two categorical variables. The same test that compares several proportions tests whether the row and column variables are related in any two-way table. We will start with the problem of comparing several proportions.

EXAMPLE 13.3

Chronic users of cocaine need the drug to feel pleasure. Perhaps giving them a medication that fights depression will help them stay off cocaine. A three-year study compared an antidepressant called desipramine with lithium (a standard treatment for cocaine addiction) and a placebo. The subjects were 72 chronic users of cocaine who wanted to break their drug habit. Twenty-four of the subjects were randomly assigned to each treatment. Here are the counts and proportions of the subjects who avoided relapse into cocaine use during the study:[2]

Group	Treatment	Subjects	No relapse	Proportion
1	Desipramine	24	14	.583
2	Lithium	24	6	.250
3	Placebo	24	4	.167

The sample proportions of subjects who stayed off cocaine are quite different. The bar chart in Figure 13.4 compares the results visually. Are these data good evidence that the proportions of successes for the three treatments differ in the population of all cocaine users?

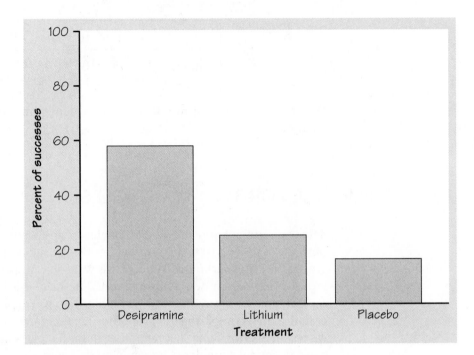

FIGURE 13.4   Bar chart comparing the success rates of three treatments for cocaine addiction (Example 13.3).

The problem of multiple comparisons

Call the population proportions of successes in the three groups p_1, p_2, and p_3. We again use a subscript to remind us which group a parameter or statistic describes. To compare these three population proportions, we might use the two-sample z procedures several times:

- Test $H_0: p_1 = p_2$ to see if the success rate of desipramine differs from that of lithium.
- Test $H_0: p_1 = p_3$ to see if desipramine differs from a placebo.
- Test $H_0: p_2 = p_3$ to see if lithium differs from a placebo.

The weakness of doing three tests is that we get three P-values, one for each test alone. That doesn't tell us how likely it is that *three* sample proportions are spread apart as far as these are. It may be that 0.167 and 0.583 are significantly different if we look at just two groups, but not significantly different if we know that they are the smallest and largest proportions in three groups. As we look at more groups, we expect the gap between the smallest and largest sample proportion to get larger. (Think of comparing the tallest and shortest person in larger and larger groups of people.)

We can't safely compare many parameters by doing tests or confidence intervals for two parameters at a time.

The problem of how to do many comparisons at once with some overall measure of confidence in all our conclusions is common in more advanced statistics. We are leaving the elementary parts of statistical inference now that we have met this problem. Statistical methods for dealing with many comparisons usually have two parts:

1. An *overall test* to see if there is good evidence of *any* differences among the parameters that we want to compare.
2. A detailed *follow-up analysis* to decide which of the parameters differ and to estimate how large the differences are.

The overall test is one with which we are familiar—the chi-square test—but in this new setting it will be used for comparing several population proportions. The follow-up analysis can be quite elaborate.

Two-way tables

two-way table

The first step in the overall test for comparing several proportions is to arrange the data in a **two-way table** that gives counts for both successes and failures. Here is two-way table of the cocaine addiction data:

	Relapse	
	No	Yes
Desipramine	14	10
Lithium	6	18
Placebo	4	20

r × c table

cell

We call this a 3×2 table because it has 3 rows and 2 columns. A table with r rows and c columns is an $r \times c$ **table**. The table shows the relationship between two categorical variables. The explanatory variable is the treatment (one of three drugs). The response variable is success (no relapse) or failure (relapse). The two-way table gives the counts for all 6 combinations of values of these variables. Each of the 6 counts occupies a **cell** of the table.

Expected counts

We want to test the null hypothesis that there are *no differences* among the proportions of successes for addicts given the three treatments:

$$H_0: p_1 = p_2 = p_3$$

The alternative hypothesis is that there *is* some difference, that not all three proportions are equal:

$$H_a: \text{ not all of } p_1, p_2, \text{ and } p_3 \text{ are equal}$$

The alternative hypothesis is no longer one-sided or two-sided. It is "many-sided," because it allows any relationship other than "all three equal." For example, H_a includes the situation in which $p_2 = p_3$ but p_1 has a different value.

To test H_0, we compare the observed counts in a two-way table with the *expected counts*, the counts we would expect—except for random variation—if H_0 were true. If the observed counts are far from the expected counts, that is evidence against H_0. It is easy to find the expected counts.

EXPECTED COUNTS

The **expected count** in any cell of a two-way table when H_0 is true is

$$\text{expected count} = \frac{\text{row total} \times \text{column total}}{\text{table total}}$$

To understand why this recipe works, think first about just one proportion.

EXAMPLE 13.4

Linda is a basketball player who makes 70% of her free throws. If she shoots 10 free throws in a game, we expect her to make 70% of them, or 7 of the 10. Of course, she won't make exactly 7 every time she shoots 10 free throws in a game. There is chance variation from game to game. But in the long run, 7 of 10 is what we expect. It is, in fact, the *mean* number of shots Linda makes when she shoots 10 times.

In more formal language, if we have n independent tries and the probability of a success on each try is p, we expect np successes. If we draw an SRS of n individuals from a population in which the proportion of successes is p, we expect np successes in the sample. That's the fact behind the formula for expected counts in a two-way table.

Let's apply this fact to the cocaine study. The two-way table with row and column totals is

	Relapse		
	No	Yes	Total
Desipramine	14	10	24
Lithium	6	18	24
Placebo	4	20	24
Total		24	48	72

We will find the expected count for the cell in row 1 (desipramine) and column 2 (relapse). The proportion of relapses among all 72 subjects is

$$\frac{\text{count of relapses}}{\text{table total}} = \frac{\text{column 2 total}}{\text{table total}} = \frac{48}{72} = \frac{2}{3}$$

Think of this as p, the overall proportion of relapses. If H_0 is true, we expect (except for random variation) this same proportion of relapses in all three groups. So the expected count of relapses among the 24 subjects who took desipramine is

$$np = (24)\left(\frac{2}{3}\right) = 16.00$$

This expected count has the form announced in the box:

$$\frac{\text{row 1 total} \times \text{column 2 total}}{\text{table total}} = \frac{(24)(48)}{72}$$

EXAMPLE 13.5

Here are the observed and expected counts side-by-side:

	Observed		Expected	
	No	Yes	No	Yes
Desipramine	14	10	8	16
Lithium	6	18	8	16
Placebo	4	20	8	16

Because 2/3 of all subjects relapsed, we expect 2/3 of the 24 subjects in each group to relapse if there are no differences among the treatments. In fact, desipramine has fewer relapses (10) and more successes (14) than expected. The placebo has fewer successes (4) and more relapses (20). That's another way of saying what the sample proportions in Example 13.3 say more directly: desipramine does much better than the placebo, with lithium in between.

EXERCISES

13.13   North Carolina State University studied student performance in a course required by its chemical engineering major. One question of interest is the relationship between time spent in extracurricular activities and whether a student earned a C or better in the course. Here are the data for the 119 students who answered a question about extracurricular activities:[3]

	Extracurricular activities (hours per week)		
	<2	2 to 12	>12
C or better	11	68	3
D or F	9	23	5

(a) This is an $r \times c$ table. What are the numbers r and c?

(b) Find the proportion of successful students (C or better) in each of the three extracurricular activity groups. What kind of relationship between extracurricular activities and succeeding in the course do these proportions seem to show?

(c) Make a bar chart to compare the three proportions of successes.

(d) The null hypothesis says that the proportions of successes are the same in all three groups if we look at the population of all students. Find the expected counts if this hypothesis is true, and display them in a two-way table.

(e) Compare the observed counts with the expected counts. Are there large deviations between them? These deviations are another way of describing the relationship you described in (b).

13.14   How are the smoking habits of students related to their parents' smoking? Here are data from a survey of students in eight Arizona high schools:[4]

	Student smokes	Student does not smoke
Both parents smoke	400	1380
One parent smokes	416	1823
Neither parent smokes	188	1168

(a) This is an $r \times c$ table. What are the numbers r and c?

(b) Calculate the proportion of students who smoke in each of the three parent groups. Then describe in words the association between parent smoking and student smoking.

(c) Make a graph to display the association.

(d) Explain in words what the null hypothesis $H_0 : p_1 = p_2 = p_3$ says about student smoking.

(e) Find the expected counts if H_0 is true, and display them in a two-way table similar to the table of observed counts.

(f) Compare the tables of observed and expected counts. Explain how the comparison expresses the same association you see in (b) and (c).

The chi-square test

Comparing the sample proportions of successes describes the differences among the three treatments for cocaine addiction. But the statistical test that tells us whether those differences are statistically significant doesn't use the sample proportions. It compares the observed and expected counts. The test statistic that makes the comparison is the *chi-square statistic*.

CHI-SQUARE STATISTIC

The **chi-square statistic** is a measure of how far the observed counts in a two-way table are from the expected counts. The formula for the statistic is

$$X^2 = \sum \frac{(\text{observed count} - \text{expected count})^2}{\text{expected count}}$$

The sum is over all $r \times c$ cells in the table.

The chi-square statistic is a sum of terms, one for each cell in the table. In the cocaine example, 14 of the desipramine group succeeded in avoiding a relapse. The expected count for this cell is 8. So the component of the chi-square statistic from this cell is

$$\frac{(\text{observed count} - \text{expected count})^2}{\text{expected count}} = \frac{(14 - 8)^2}{8} = \frac{36}{8} = 4.5$$

As in the test for goodness of fit, you should think of the chi-square statistic X^2 as a measure of the distance of the observed counts from the expected counts. Like any distance, it is always zero or positive, and it is zero only when the observed counts are exactly equal to the expected counts. Large values of X^2 are evidence against H_0 because they say that the observed counts are far from what we would expect if H_0 were true. Although the alternative hypothesis H_a is many-sided, the chi-square test is one-sided because any violation of H_0 tends to produce a large value of X^2. Small values of X^2 are not evidence against H_0.

DEGREES OF FREEDOM

> The chi-square test for a two-way table with r rows and c columns uses critical values from the chi-square distribution with $(r - 1)(c - 1)$ *degrees of freedom*. The P-value is the area to the right of X^2 under the chi-square density curve.

EXAMPLE 13.6

The two-way table of 3 treatments by 2 outcomes for the cocaine study has 3 rows and 2 columns. That is, $r = 3$ and $c = 2$. The chi-square statistic therefore has degrees of freedom

$$(r - 1)(c - 1) = (3 - 1)(2 - 1) = (2)(1) = 2$$

Calculating chi-square with technology

Calculating the expected counts and then the chi-square statistic by hand is a bit time-consuming. As usual, computer software saves time and always gets the arithmetic right. The TI-83, on a smaller scale, has also been programmed to conduct inference for two-way tables.

EXAMPLE 13.7

Enter the two-way table (the 6 counts) for the cocaine study into the Minitab software package and request the chi-square test. The output appears in Figure 13.5. Most statistical software packages produce chi-square output similar to this.

Minitab repeats the two-way table of observed counts and puts the expected count for each cell below the observed count. It numbers the rows (1, 2, and 3) and the columns (C1 and C2) and also puts in the row and column totals. Then the software calculates the chi-square statistic X^2. For these data,

```
            Expected counts are printed below observed counts

                    C1          C2      Total
            1        14          10        24
                    8.00       16.00

            2         6          18        24
                    8.00       16.00

            3         4          20        24
                    8.00       16.00

        Total        24          48        72

        ChiSq =   4.500  +    2.250  +
                  0.500  +    0.250  +
                  2.000  +    1.000  =  10.500
        df = 2
                                     D. F.
            Chisquare  2.
            10.5000       0.9948  -  Complement of
                                        P. value
```

FIGURE 13.5   Minitab output for that two-way table in the cocaine study. The output gives the observed counts, the expected counts, and the value 10.500 for the chi-square statistic. The last line gives 0.9948 as the probability of a value *less than* 10.500 if the null hypothesis is true. The P-value is therefore $1 - .9948$.

$X^2 = 10.500$. The statistic is a sum of 6 terms, one for each cell in the table. The ''ChiSq'' display in the output shows the individual terms, as well as their sum. The first term is 4.500, just as we calculated.

The P-value is the probability that X^2 would take a value as large as 10.500 if H_0 were really true. Many software systems give the P-value. Minitab requires us to ask for the probability of a value of 10.500 or smaller. This probability is 0.9948 (at the bottom of the output), so the P-value is $1 - .9948 = .0052$. The small P-value gives us good reason to conclude that there *are* differences among the effects of the three treatments.

EXAMPLE 13.8   To perform a chi-square test for the cocaine study on the TI-83, use a matrix, say matrix [A], to store the observed counts. Here are the keystrokes, along with several calculator screens for you to check your progress.

Step 1. Enter the observed counts in the matrix: MATRIX / EDIT / 1:[A]. Enter the number of rows (first number) and columns (second number), pressing ENTER after each number to record the number. The arrangement of the numbers in the matrix should look exactly like the observed counts in the two-way table.

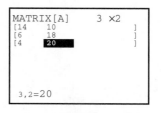

Step 2. Specify the chi-square test, the matrix where observed counts are found, and the matrix where the expected counts will be stored: STAT / TESTS / C:χ^2-Test...

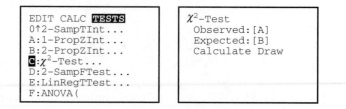

Step 3. When the next screen appears, simply choose "Calculate" or "Draw." If you specify "Calculate," you should get these results:

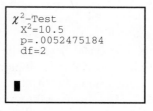

If you specify "Draw," the χ^2 curve with 2 degrees of freedom will be drawn, the critical area in the tail will be shaded, and the *P*-value will be displayed.

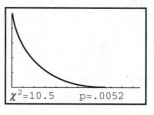

If you want to see the expected counts, then simply ask for a display of the matrix [B]: MATRIX / NAMES / 2:[B]

```
┌─────────────────────────────────┐
│                                 │
│  [B]                            │
│            [ [8  16]            │
│              [8  16]            │
│              [8  16]  ]         │
│ ▪                               │
└─────────────────────────────────┘
```

Verify that these TI-83 results agree with the Minitab results.

Using the chi-square table

df = 2

p	.01	.005
x^*	9.21	10.60

The computer output in Figure 13.5 for the cocaine study gives 2 as the degrees of freedom. The observed value of the chi-square statistic is $X^2 =$ 10.500. Look in the df = 2 row of Table E. The value $X^2 = 10.500$ falls between the 0.01 and 0.005 critical values of the chi-square distribution with 2 degrees of freedom. Remember that the chi-square test is always one-sided. So the P-value of $X^2 = 10.500$ is between 0.01 and 0.005. The differences among the three proportions of successes are significant at the $\alpha = .01$ level.

Follow-up analysis

The chi-square test is the overall test for comparing any number of population proportions. If the test allows us to reject the null hypothesis that all the proportions are equal, we then want to do a follow-up analysis that examines the differences in detail. We won't describe how to do a formal follow-up analysis, but you should look at the data to see what specific effects they suggest.

EXAMPLE 13.9

The cocaine study found significant differences among the proportions of successes for three treatments for cocaine addiction. We can see the specific differences in three ways.

Look first at the *sample proportions*:

$$\hat{p}_1 = .583 \qquad \hat{p}_2 = .250 \qquad \hat{p}_3 = .167$$

These suggest that the major difference between the proportions is that desipramine has a much higher success rate than either lithium or a placebo. That is the effect that the study hoped to find.

Next, *compare the observed and expected counts* in Figure 13.5. Treatment 1 (desipramine) has more successes and fewer failures than we would expect if all three treatments had the same success rate in the population. The other two treatments had fewer successes and more failures than expected.

*components of
chi-square*

Finally, Minitab prints under the table the 6 individual "distances" between the observed and expected counts that are added to get X^2. The arrangement of these **components of** X^2 is the same as the 3×2 arrangement of the table. The largest components show which cells contribute the most to the overall distance X^2. The largest component by far is for the top left cell in the table: desipramine has more successes than would be expected.

All three ways of examining the data point to the same conclusion: desipramine works better than the other treatments. This is an informal conclusion. More advanced methods provide tests and confidence intervals that make this follow-up analysis formal.

EXERCISES

13.15   In Exercise 13.13, you began to analyze data on the relationship between time spent on extracurricular activities and success in a tough course. Figure 13.6 gives Minitab output for the two-way table in Exercise 13.13.

```
Expected counts are printed below observed counts

              C1        C2        C3      Total
    1         11        68         3         82
           13.78     62.71     5.51
                   -.261    .686

    2          9        23         5         37
            6.22     28.29      2.49
             .44      -.18      1.608

Total         20        91         8        119

ChiSq =   0.561 +   0.447 +   1.145 +
          1.244 +   0.991 +   2.538 = 6.926
df = 2
1 cells with expected counts less than 5.0

      Chisquare 2.
      6.9260         0.9687
```

FIGURE 13.6   Minitab output for the study of extracurricular activity and success in a tough course (Exercise 13.15).

(a) Starting from the table of expected counts, find the 6 components of the chi-square statistic and then the statistic X^2 itself. Check your work against the computer output.

(b) What is the P-value for the test? Explain in simple language what it means to reject H_0 in this setting.

(c) Which term contributes the most to X^2? What specific relation between extracurricular activities and academic success does this term point to?

(d) Does the North Carolina State study convince you that spending more or less time on extracurricular activities *causes* changes in academic success? Explain your answer.

13.16   In Exercise 13.14, you began to analyze data on the relationship between smoking by parents and smoking by high school students. Figure 13.7 gives the Minitab output for the two-way table in Exercise 13.14.

(a) Starting from the table of expected counts, find the 6 components of the chi-square statistic and then the statistic X^2 itself. Check your work against the computer output.

(b) What is the P-value for the test? Explain in simple language what it means to reject H_0 in this setting.

(c) Which two terms contribute the most to X^2? What specific relation between parent smoking and student smoking do these terms point to?

(d) Does the study convince you that parent smoking *causes* student smoking? Explain your answer.

13.17   The computer output in Figure 13.6 gives the degrees of freedom for the table in Exercise 13.13 as 2.

(a) Verify that this is correct.

(b) The computer gives the value of the chi-square statistic as $X^2 = 6.926$. Between what two entries in Table E does this value lie? What does the table tell you about the P-value?

13.18   The computer output in Figure 13.7 gives the degrees of freedom for the table in Exercise 13.14 as 2.

(a) Verify that this is correct.

(b) The computer gives the value of the chi-square statistic as $X^2 = 37.568$. Where in Table E does this value lie? What does the table tell you about the P-value?

```
              Expected counts are printed below observed counts

                       C1         C2      Total
              1        400       1380      1780
                    332.49    1447.51

              2        416       1823      2239
                    418.22    1820.78

              3        188       1168      1356
                    253.29    1102.71

     Total          1004       4371      5375

     ChiSq = 13.709 +   3.149 +
              0.012 +   0.003 +
             16.829 +   3.866 = 37.566
     df = 2

        Chisquare 2.
        37.5680     1.0000
```

FIGURE 13.7   Minitab output for the study of parent smoking and student smoking (Exercise 13.16).

More uses of the chi-square test

Two-way tables can arise in several ways. The cocaine study is an experiment that assigned 24 addicts to each of three groups. Each group is a sample from a separate population corresponding to a separate treatment. The study design fixes the size of each sample in advance, and the data record which of two outcomes occurred for each subject. The null hypothesis of "no difference" among the treatments takes the form of "equal proportions of successes" in the three populations. The next example illustrates a different setting for a two-way table.

EXAMPLE 13.10   A study of the relationship between men's marital status and the level of their jobs used data on all 8235 male managers and professionals employed by a large manufacturing firm. Each man's job has a grade set by the company that reflects the value of that particular job to the company. The authors of the study grouped the many job grades into quarters. Grade 1 contains jobs in the lowest quarter of job grades, and grade 4 contains those in the highest quarter. Here are the data.[5]

		Single	Married	Divorced	Widowed
			Marital status		
	1	58	874	15	8
Job	2	222	3927	70	20
grade	3	50	2396	34	10
	4	7	533	7	4

Do these data show a statistically significant relationship between marital status and job grade?

In Example 13.10 we do not have four separate samples from the four marital statuses. We have a single group of 8235 men, each classified in two ways, by marital status and job grade. The number of men in each marital status is not fixed in advance but is known only after we have the data. Both marital status and job grade have four levels, so a careful statement of the null hypothesis

H_0 : there is no relationship between marital status and job grade

in terms of population parameters is complicated.

In fact, we should probably regard these 8235 men as an entire population rather than as a sample. The data include all the managers and professionals employed by this company. The company no doubt has many special features. Its employees are not necessarily a random sample from any larger population. Nevertheless, we would still like to decide if the relationship between marital status and job grade is statistically significant in the sense that it is too strong to happen just by chance if job grades were handed out at random to men of all marital statuses. "Not likely to happen just by chance if H_0 is true" is the usual meaning of statistical significance. In this example, that meaning makes sense even though we have data on an entire population.

The setting of Example 13.10 is very different from a comparison of several proportions. Nevertheless, we can apply the chi-square test. One of the most useful properties of chi-square is that it tests the hypothesis "the row and column variables are not related to each other" whenever this hypothesis makes sense for a two-way table.

USES OF THE CHI-SQUARE TEST

Use the chi-square test to test the null hypothesis

H_0: there is no relationship between two categorical variables

when you have a two-way table from any of these situations:

- Independent SRSs from each of several populations, with each individual classified according to one categorical variable. (The other variable says which sample the individual comes from.)
- A single SRS, with each individual classified according to both of two categorical variables.
- An entire population, with each individual classified according to both of two categorical variables.

```
Expected counts are printed below observed counts

              C1          C2         C3         C4      Total
     1        58         874         15          8        955
            39.08      896.44      14.61       4.87

     2       222        3927         70         20       4239
           173.47     3979.05      64.86      21.62

     3        50        2396         34         10       2490
           101.90     2337.30      38.10      12.70

     4         7         533          7          4        551
            22.55      517.21       8.43       2.81

Total        337        7330        126         42       8235

ChiSq =   9.158 +    0.562 +    0.010 +    2.011 +
         13.575 +    0.681 +    0.407 +    0.121 +
         26.432 +    1.474 +    0.441 +    0.574 +
         10.722 +    0.482 +    0.243 +    0.504 =  67.397
df = 9
2 cells with expected counts less than 5.0

     Chisquare 9.
     67.3970      1.0000
```

FIGURE 13.8   Minitab output for the 4 × 4 table in Example 13.11.

EXAMPLE 13.11

To analyze the job grade data in Example 13.10, first do the overall chi-square test. The Minitab chi-square output appears in Figure 13.8. The observed chi-square is very large, $X^2 = 67.397$. The probability of a value smaller than this when H_0 is true is 1.0000 (to four decimal places). That is, the P-value is close to 0. We have overwhelming evidence that job grade is related to marital status.

Table E gives a similar result. For a 4×4 table, the degrees of freedom are $(r-1)(c-1) = 9$. Look in the df = 9 row of Table E. The largest critical value is 29.67, corresponding to the P-value 0.0005. The observed $X^2 = 67.397$ is beyond that value, so $P < .0005$.

Next, do a follow-up analysis to describe the relationship. As in Section 3 of Chapter 4, we describe a relationship between two categorical variables by comparing percents. Here is a table of the percent of men in each marital status whose jobs have each grade. Each column of this table gives the conditional distribution of job grade among men with a specific marital status. Each column adds to 100% because it accounts for all the men in one marital status.

		Marital status			
		Single	Married	Divorced	Widowed
	1	17.2%	11.3%	11.9%	19.1%
Job	2	65.9%	50.8%	55.5%	47.7%
grade	3	14.9%	31.0%	26.9%	23.8%
	4	2.0%	6.9%	5.6%	9.6%
		100%	100%	100%	100%

The bar charts in Figure 13.9 help us compare these four conditional distributions. We see at once that smaller percents of single men have jobs in the higher grades 3 and 4. Not only married men but men who were once married and are now divorced or widowed are more likely to hold higher-grade jobs. Look at the 16 components of the chi-square sum in the computer output for confirmation. The four cells for single men have the four largest components of X^2. Minitab's table of counts shows that the observed counts for single men are higher than expected in grades 1 and 2 and lower than expected in grades 3 and 4.

Of course, this association between marital status and job grade does not show that being single *causes* lower-grade jobs. The explanation might be as simple as the fact that single men tend to be younger and so have not yet advanced to higher grades.

Cell counts required for the chi-square test

The computer output in Figure 13.8 has one more feature. It warns us that the expected counts in two of the 16 cells are less than 5. The chi-

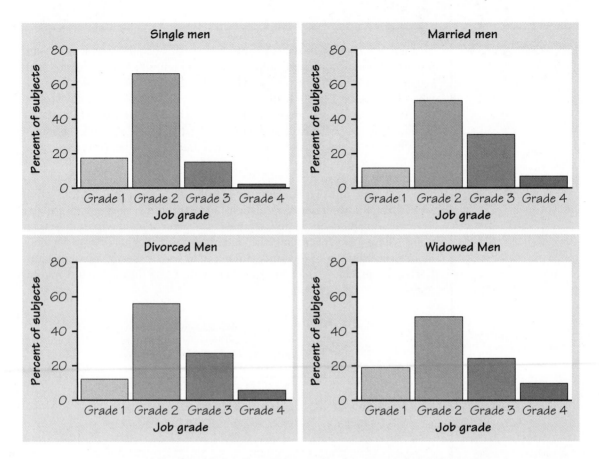

FIGURE 13.9   Bar charts for the data in Example 13.11. Each chart presents the percents of each job grade among men of one marital status.

square test, like the *z* procedures for comparing two proportions, is an approximate method that becomes more accurate as the counts in the cells of the table get larger. Fortunately, the approximation is accurate for quite modest counts. Here is a practical guideline.[6]

CELL COUNTS REQUIRED FOR THE CHI-SQUARE TEST

You can safely use the chi-square test with critical values from the chi-square distribution when no more than 20% of the expected counts are less than 5 and all individual expected counts are 1 or greater. In particular, all four expected counts in a 2 × 2 table should be 5 or greater.

Example 13.11 easily passes this test. All the expected counts are greater than 1, and only 2 out of 16 (12.5%) are less than 5.

The chi-square test and the z test

We can use the chi-square test to compare any number of proportions. If we are comparing r proportions and make the columns of the table "success" and "failure," the counts form an $r \times 2$ table. P-values come from the chi-square distribution with $r - 1$ degrees of freedom. If $r = 2$, we are comparing just two proportions. We have two ways to do this: the z test from Section 12.2 and the chi-square test with 1 degree of freedom for a 2×2 table. *These two tests always agree.* In fact, the chi-square statistic X^2 is just the square of the z statistic, and the P-value for X^2 is exactly the same as the two-sided P-value for z. We recommend using the z test to compare two proportions, because it gives you the choice of a one-sided test and is related to a confidence interval for $p_1 - p_2$.

EXERCISES

13.19   The North Carolina State study described in Exercise 13.13 also looked at the relationship between student goals and success in getting a C or better in the course. The study report says: "The probability of passing CHE 205 was different for students who would be satisfied with a grade of C or better (36% of 14 students), B or better (64% of 64), A (90% of 30), creative work beyond an A (82% of 11)."[7]

(a) Use the information given to make a 4×2 table of goal by success or failure for the 119 students.

(b) Find the expected counts. The expected counts don't pass our rule of safe practice. Why? Chi-square distribution P-values are nonetheless reasonably accurate.

(c) Confirm that the chi-square statistic for this table is $X^2 = 14.986$. Give the degrees of freedom. Use Table E to approximate the P-value for this statistic.

(d) Describe briefly the relationship between students' goals and their grades in the course.

13.20   Exercise 12.19 compared HMO members who filed complaints with an SRS of members who did not complain. The study actually broke the complainers into two subgroups: those who filed complaints about medical treatment and those who filed nonmedical complaints. Here are the data on the total number in each group and the number who voluntarily left the HMO:

	No complaint	Medical complaint	Nonmedical complaint
Total	743	199	440
Left	22	26	28

(a) Find the percent of each group who left.

(b) Make a two-way table of complaint status by left or not.

(c) Find the expected counts and check that you can safely use the chi-square test.

(d) Determine the chi-square statistic for this table. What null and alternative hypotheses does this statistic test? What are its degrees of freedom? Use Table E to approximate the P-value.

(e) What do you conclude from these data?

13.21   A large study of child care used samples from the data tapes of the Current Population Survey over a period of several years. The result is close to an SRS of child-care workers. The Current Population Survey has three classes of child-care workers: private household, nonhousehold, and preschool teacher. Here are data on the number of blacks among women workers in these three classes:[8]

	Total	Black
Household	2455	172
Nonhousehold	1191	167
Teachers	659	86

(a) What percent of each class of child-care workers is black?

(b) Make a two-way table of class of worker by race (black or other).

(c) Can we safely use the chi-square test? What null and alterative hypotheses does X^2 test?

(d) Calculate the chi-square statistic for this table. What are its degrees of freedom? Use Table E to approximate the P-value.

(e) What do you conclude from these data?

13.22   Gastric freezing was once a recommended treatment for ulcers in the up-
         per intestine. Use of gastric freezing stopped after experiments showed it
         had no effect. One randomized comparative experiment found that 28 of
         the 82 gastric freezing patients improved, while 30 of the 78 patients in the
         placebo group improved.[9] We can test the hypothesis of "no difference"
         between the two groups in two ways: using the two-sample z statistic or
         using the chi-square statistic.

   (a) State the null hypothesis with a two-sided alternative and carry out the
        z test. What is the P-value from Table A?
   (b) Present the data in a 2×2 table. Use the chi-square test to test the
        hypothesis from (a). Verify that the X^2 statistic is the square of the z
        statistic. Use Table E to verify that the chi-square P-value agrees with
        the z result up to the accuracy of the table.
   (c) What do you conclude about the effectiveness of gastric freezing as a
        treatment for ulcers?

SUMMARY

The **chi-square test** for a two-way table tests the null hypothesis that there
is no relationship between the row variable and the column variable.

One common use of the chi-square test is to **compare several population proportions**. The null hypothesis states that all of the population proportions
are equal. The alternative hypothesis states that they are not all equal but
allows any other relationship among the population proportions.

The **expected count** in any cell of a two-way table when H_0 is true is

$$\text{expected count} = \frac{\text{row total} \times \text{column total}}{\text{table total}}$$

The **chi-square statistic** is

$$X^2 = \sum \frac{(\text{observed count} - \text{expected count})^2}{\text{expected count}}$$

The chi-square test compares the value of the statistic X^2 with critical
values from the **chi-square distribution** with $(r - 1)(c - 1)$ **degrees of freedom**.
Large values of X^2 are evidence against H_0 so the P-value is the area under
the chi-square density curve to the right of X^2.

The chi-square distribution is an approximation to the distribution of the
statistic X^2. You can safely use this approximation when all expected cell
counts are at least 1 and no more than 20% are less than 5.

If the chi-square test finds a statistically significant relationship between the row and column variables in a two-way table, do a follow-up analysis to describe the nature of the relationship. An informal follow-up analysis compares well-chosen percents, compares the observed counts with the expected counts, and looks for the largest **components of chi-square**.

SECTION 13.2 EXERCISES

13.23   It seems that the attitude of cancer patients can influence the progress of their disease. We can't experiment with humans, but here is a rat experiment on this theme. Inject 60 rats with tumor cells and then divide them at random into two groups of 30. All the rats receive electric shocks, but rats in Group 1 can end the shock by pressing a lever. (Rats learn this sort of thing quickly.) The rats in Group 2 cannot control the shocks, which presumably makes them feel helpless and unhappy. We suspect that the rats in Group 1 will develop fewer tumors. The results: 11 of the Group 1 rats and 22 of the Group 2 rats developed tumors.[10]

(a) State the null and alternative hypotheses for this investigation. Explain why the z test rather than the chi-square test for a 2×2 table is the proper test.

(b) Carry out the test and report your conclusion.

13.24   Do unregulated providers of child care in their homes follow different health and safety practices in different cities? A study looked at people who regularly provided care for someone else's children in poor areas of three cities. The numbers who required medical releases from parents to allow medical care in an emergency were 42 of 73 providers in Newark, N.J., 29 of 101 in Camden, N.J., and 48 of 107 in South Chicago, Ill.[11]

(a) Use the chi-square test to see if there are significant differences among the proportions of child-care providers who require medical releases in the three cities.

(b) How should the data be produced in order for your test to be valid? (In fact, the samples came in part from asking parents who were subjects in another study who provided their child care. The author of the study wisely did not use a statistical test. He wrote: "Application of conventional statistical procedures appropriate for random samples may produce biased and misleading results.")

13.25   Sample surveys on sensitive issues can give different results depending on how the question is asked. A University of Wisconsin study divided 2400 respondents into 3 groups at random. All were asked if they had ever used cocaine. One group of 800 was interviewed by phone; 21% said they had used cocaine. Another 800 people were asked the question in a one-on-one personal interview; 25% said "Yes." The remaining 800 were allowed to make an anonymous written response; 28% said "Yes."[12] Are there statistically significant differences among these proportions? (State the hypotheses, convert the information given into a two-way table, give the test statistic and its *P*-value, and state your conclusions.)

13.26   In 1912 the luxury liner *Titanic*, on its first voyage across the Atlantic, struck an iceberg and sank. Some passengers got off the ship in lifeboats, but many died. Think of the *Titanic* disaster as an experiment in how the people of that time behaved when faced with death in a situation where only some can escape. The passengers are a sample from the population of their peers. Here is information about who lived and who died, by sex and economic status. (The data leave out a few passengers whose economic status is unknown.)[13]

| | Men | | | Women | |
Status	Died	Survived	Status	Died	Survived
Highest	111	61	Highest	6	126
Middle	150	22	Middle	13	90
Lowest	419	85	Lowest	107	101
Total	680	168	Total	126	317

(a) Compare the percents of men and of women who died. Is there strong evidence that a higher proportion of men die in such situations? Why do you think this happened?

(b) Look only at the women. Describe how the three economic classes differ in the percent of women who died. Are these differences statistically significant?

(c) Now look only at the men and answer the same questions.

13.27   How accurate are pre-election polls of voters' intentions? The 1992 presidential election featured the Republican incumbent, George Bush, the Democrat Bill Clinton, and the third-party candidate Ross Perot. Clinton was elected. Here are results for three polls taken a few days before the election, as well as the actual election result. The poll results don't add to 100% because some voters were undecided.[14]

	Sample size	Bush	Clinton	Perot
ABC News	912	38%	41%	18%
USA Today/CNN	1610	39%	42%	14%
New York Times/CBS	1912	34%	43%	15%
Actual vote		38%	43%	19%

The polls use complex multistage sample designs. For the purposes of this exercise, treat each poll as an SRS of registered voters.

(a) Did the three polls differ significantly in the percent of their samples who favored the winner, Bill Clinton?

(b) For each poll individually, test whether its percent for Ross Perot differs significantly at the $\alpha = .05$ level from the actual election result, which is 19%.

(c) A single test at the $\alpha = .05$ level will wrongly reject H_0 only 5% of the time when H_0 is actually true. In (b) you made three separate tests at the $\alpha = .05$ level. Explain why at least one of three tests will be wrong on more than 5% of the occasions you do three tests.

CHAPTER REVIEW

This chapter develops several settings where a variation of the chi-square test of significance is useful. In a goodness of fit test, the object is to determine if a population distribution has changed. The null hypothesis states that there is no difference between two distributions, while the alternative hypothesis states that there is a difference. The chi-square test tells whether there is sufficient reason to reject the null hypothesis, but further analysis is needed to determine how and where the changes have occurred.

A goodness of fit test begins by finding the expected counts for each category, if the assumed distribution has not changed. The chi-square statistic is a measure of how much the sample distribution diverges from the hypothesized distribution. For a given number of degrees of freedom, large chi-square statistic values provide evidence to reject the null hypothesis of no difference.

A chi-square procedure is also useful in testing the equality of proportions of successes in any number of populations. The alternative to this hypothesis is "many-sided," because it allows any relationship other than "all equal." The chi-square test is an overall test that tells us whether the data give good reason to reject the hypothesis that all the population proportions are equal. You should always accompany the chi-square test by data analysis to see what kind of inequality is present.

To do a chi-square test of independence, arrange the counts of successes and failures in a two-way table. Two-way tables can display counts for any two categorical variables, not just successes and failures, in r populations. The chi-square test is also more general than just a test for equal population proportions of successes. It tests the null hypothesis that there is "no relationship" between the row variable and the column variable in a two-way table. The chi-square test is actually an approximate test that becomes more accurate as the cell counts in the two-way table increase. Fortunately, chi-square P- values are quite accurate even for small counts.

After studying this chapter, you should be able to do the following.

A. GOODNESS OF FIT

1. Calculate expected counts for each category in a distribution, the chi-squared statistic, and the P-value.
2. State null and alternative hypotheses for a difference between two distributions.
3. If the test is significant, use components of the chi-square statistic to identify the most important deviations between observed and expected counts.

B. TWO-WAY TABLES

1. Arrange data on successes and failures in several groups into a two-way table of counts of successes and failures in all groups.
2. Use percents to describe the relationship between two categorical variables starting from the counts in a two-way table.

C. INTERPRETING CHI-SQUARE TESTS OF INDEPENDENCE

1. Locate expected cell counts, the chi-square statistic, and its P-value in output from your software or calculator.
2. Explain what null hypothesis the chi-square statistic tests in a specific two-way table.
3. If the test is significant, use percents, comparison of expected and observed counts, and the components of the chi-square statistic to see what deviations from the null hypothesis are most important.

D. DOING CHI-SQUARE TESTS

1. Calculate the expected count for any cell from the observed counts in a two-way table.
2. Calculate the component of the chi-square statistic for any cell, as well as the overall statistic.
3. Give the degrees of freedom of a chi-square statistic.
4. Use the chi-square critical values in Table E to approximate the P-value of a chi-square test.

CHAPTER 13 REVIEW EXERCISES

13.28   The Advanced Placement (AP) Statistics examination was first administered in May 1997. Students' papers are graded on a scale of 1 to 5, with 5 being the highest score. Over 7600 students took the exam in the first year, and the distribution of scores was as follows (not including exams that were scored late):

Score	5	4	3	2	1
Percent	15.3	22.0	24.8	19.8	18.1

A sample of students who took the exam had the following distribution of grades:

Score	5	4	3	2	1
Frequency	167	158	101	79	30

Calculate marginal percents and make a segmented bar chart of the population scores and the sample scores, so that the two distributions can be compared visually. Then perform a goodness of fit test to determine if the distribution of scores for this particular sample is significantly different from the distribution of scores for all students who took the inaugural exam.

13.29   In this exercise you will use the TI-83 to simulate sampling from the following uniform distribution:

X	0	1	2	3	4	5	6	7	8	9
$P(X)$	.1	.1	.1	.1	.1	.1	.1	.1	.1	.1

You will then perform a goodness of fit test to see if a randomly generated sample distribution comes from a population that is different from this distribution.

(a) State your null and alternative hypotheses for this test.

(b) Use the `randInt` function to randomly generate 200 digits from 0 to 9, and store these values in list L_4.

(c) Plot the data as a histogram with Window dimensions set as follows: $X[-.5, 9.5]_1$ and $Y[-5, 30]_5$. (You may have to increase the vertical scale.) Then TRACE to see the frequencies of each digit. Record these frequencies (observed values) in list L_1.

(d) Determine the expected counts for a sample of size 200, and store them in list L_2.

(e) Complete a goodness of fit test. Report your chi-square statistic, the P-value, and your conclusions with regard to the null and alternative hypotheses.

13.30   A study of the career plans of young women and men sent questionnaires to all 722 members of the senior class in the College of Business Administration at the University of Illinois. One question asked which major within the business program the student had chosen. Here are the data from the students who responded:[15]

	Female	Male
Accounting	68	56
Administration	91	40
Economics	5	6
Finance	61	59

(a) Test the null hypothesis that there is no relation between the sex of students and their choice of major. Give a P-value and state your conclusion.

(b) Describe the differences between the distributions of majors for women and men with percents, with a graph, and in words.

(c) Which two cells have the largest components of the chi-square statistic? How do the observed and expected counts differ in these cells? (This should strengthen your conclusions in (b).)

(d) Two of the observed cell counts are small. Do these data satisfy our guidelines for safe use of the chi-square test?

(e) What percent of the students did not respond to the questionnaire? The nonresponse weakens conclusions drawn from these data.

13.31   To study the export activity of manufacturing firms on Taiwan, researchers mailed questionnaires to an SRS of firms in each of five industries that export many of their products. The response rate was only 12.5%, because private companies don't like to fill out long questionnaires from academic researchers. Here are data on the planned sample sizes and the actual number of responses received from each industry:[16]

	Sample size	Responses
Metal products	185	17
Machinery	301	35
Electrical equipment	552	75
Transportation equipment	100	15
Precision instruments	90	12

If the response rates differ greatly, comparisons among the industries may be difficult. Is there good evidence of unequal response rates among the five industries? (Start by creating a two-way table of response or nonresponse by industry.)

13.32   Shopping at secondhand stores is becoming more popular and has even attracted the attention of business schools. A study of customers' attitudes toward secondhand stores interviewed samples of shoppers at two secondhand stores of the same chain in two cities. The breakdown of the respondents by sex is as follows:[17]

	City 1	City 2
Men	38	68
Women	203	150
Total	241	218

Is there a significant difference between the proportions of women customers in the two cities?

(a) State the null hypothesis, find the sample proportions of women in both cities, do a two-sided z test, and give a P-value using Table A.

(b) Calculate the chi-square statistic X^2 and show that it is the square of the z statistic. Show that the P-value from Table E agrees (up to the accuracy of the table) with your result from (a).

(c) Give a 95% confidence interval for the difference between the proportions of women customers in the two cities.

13.33   The study of shoppers in secondhand stores cited in the previous exercise also compared the income distributions of shoppers in the two stores. Here is a two-way table of counts:

Income	City 1	City 2
Under $10,000	70	62
$10,000 to $19,999	52	63
$20,000 to $24,999	69	50
$25,000 to $34,999	22	19
$35,000 or more	28	24

A statistical calculator gives the chi-square statistic for this table as $X^2 = 3.955$. Is there good evidence that customers at the two stores have different income distributions? (Give the degrees of freedom, the P-value, and your conclusion.)

13.34   The success of a sample survey can depend on the season of the year. The Italian National Statistical Institute kept records of nonresponse to one of its national telephone surveys. All calls were made between 7 p.m. and 10 p.m. Here is a table of the percents of responses and of three types of nonresponse at different seasons. The percents in each row add to 100% (up to roundoff error).[18]

Season	Calls made	Successful interviews	Nonresponse		
			No answer	Busy signal	Refusal
Jan. 1 to Apr. 13	1558	68.5%	21.4%	5.8%	4.3%
Apr. 21 to June 20	1589	52.4%	35.8%	6.4%	5.4%
July 1 to Aug. 31	2075	43.4%	41.5%	8.6%	6.5%
Sept. 1 to Dec. 15	2638	60.0%	30.0%	5.3%	4.7%

(a) What are the degrees of freedom for the chi-square test of the hypothesis that the distribution of responses varies with the season? (Don't do the test. The sample sizes are so large that the results are sure to be highly significant.)

(b) Consider just the proportion of successful interviews. Describe how this proportion varies with the seasons, and assess the statistical significance of the changes. What do you think explains the changes? (Look at the full table for ideas.)

(c) It is incorrect to apply the chi-square test to percents rather than to counts. If you enter the 4×4 table of percents above into statistical software and ask for a chi-square test, well-written software should give an error message. (Counts must be whole numbers, so the software should check that.) Try this using your software or calculator, and report the result.

13.35   Continue the analysis of the data in the previous exercise by considering just the proportion of people called who refused to participate. We might think that the refusal rate changes less with the season than, for example, the rate of "no answer." State the hypothesis that the refusal rate does not change with the season. Check that you can safely use the chi-square test. Carry out the test. What do you conclude?

13.36   Example 4.12 (page 223) presents artificial data that illustrate Simpson's paradox. The data concern the survival rates of surgery patients at two hospitals.

(a) Apply the chi-square test to the data for all patients combined and summarize the results.

(b) Do separate chi-square tests for the patients in good condition and for those in poor condition. Summarize these results.

(c) Are the effects that illustrate Simpson's paradox in this example statistically significant?

13.37   During the 1991 Persian Gulf War, the military's control of news reporting from the combat zone was a public issue. Princeton Survey Research Associates interviewed a random sample of 924 adults in January of 1991. One question asked was

> *Do you think the military should exert more control over how news organizations report about the war or do you think that most decisions about how to report about the war should be left to news organizations themselves?*

(a) Fifty-seven percent of the sample thought the military should have more control. Give a 90% confidence interval for the proportion of all adults who feel this way. (Treat the sample as an SRS.)

(b) Professor Ted Chang of the University of Virginia did an experiment on the effect of the wording of questions in a survey. He asked 174 students the question above; 55% said the military should have more control. Is the student response significantly different from the national response?

(c) Professor Chang asked another 199 students this question:

> *Do you feel the amount of control the military is exerting over news organizations reporting the war is about right?*

Now only 16% said that the military should exert more control. Is this significantly different from the student response to the first wording as given in (b)?

(d) All three samples were also asked the question

> *To the best of your knowledge are news reports from the Gulf being censored by the American military?*

The percents who said "Yes" were 76% in the national sample ($n = 924$), 67% in the first student sample ($n = 174$), and 65% in the second student sample ($n = 199$). Is there significant evidence that the three populations did not all have the same proportion who thought the news is censored? (The two student samples represent different populations because they were asked different questions about military control of the news, and this might influence their response to other questions.)

13.38 Substances suspected of causing cancer are often tested (in high doses) on mice or rats. How well do results from mice agree with the results of testing the same substances on rats? Here, in a 2×2 table, are the results of testing 249 chemicals on both mice and rats (+ indicates that the substance was found to cause cancer; −, that it was not):[19]

		Rats	
		+	−
	+	111	21
Mice			
	−	17	100

(a) Explain in words what hypothesis the chi-square test for this table examines.

(b) We are interested in how often results from rats and results from mice for the same chemical agree. Explain why the chi-square test does *not* help assess agreement. Beware the temptation to apply chi-square to every two-way table!

(c)   On what percent of the 249 chemicals do mice and rats agree? What percent of chemicals to which mice give a + also get a + from rats? What percent of chemicals that get a − from mice also get a − from rats? These percents help answer our question.

NOTES AND DATA SOURCES

1. *Statistical Abstract of the United States, 1997*, U.S. Department of Commerce, Bureau of the Census.

2. D. M. Barnes, "Breaking the cycle of addiction," *Science*, 241 (1988), pp. 1029–1030.

3. Richard M. Felder et al., "Who gets it and who doesn't: a study of student performance in an introductory chemical engineering course," *1992 ASEE Annual Conference Proceedings*, American Society for Engineering Education, Washington, D.C., 1992, pp. 1516–1519.

4. S. V. Zagona (ed.), *Studies and Issues in Smoking Behavior*, University of Arizona Press, Tucson, 1967, pp. 157–180.

5. Sanders Korenman and David Neumark, "Does marriage really make men more productive?" *Journal of Human Resources*, 26 (1991), pp. 282–307.

6. There are many computer studies of the accuracy of chi-square critical values for X^2. For a brief discussion and some references, see Section 3.2.5 of David S. Moore, "Tests of chi-squared type," in Ralph B. D'Agostino and Michael A. Stephens (eds.), *Goodness-of-Fit Techniques*, Marcel Dekker, New York, 1986, pp. 63–95. If the expected cell counts are roughly equal, the chi-square approximation is adequate when the average expected counts are as small as 1 or 2. The guideline given in the text protects against unequal expected counts. For a survey of inference for smaller samples, see Alan Agresti, "A survey of exact inference for contingency tables," *Statistical Science*, 7 (1992), pp. 131–177.

7. The quotation is from p. 1517 of the article cited in Note 3.

8. David M. Blau, "The child care labor market," *Journal of Human Resources*, 27 (1992), pp. 9–39.

9. Lillian Lin Miao, "Gastric freezing: an example of the evaluation of medical therapy by randomized clinical trials," in John P. Bunker, Benjamin A. Barnes, and Frederick Mosteller (eds.), *Costs, Risks, and Benefits of Surgery*, Oxford University Press, New York, 1977, pp. 198–211.

10. Adapted from M. A. Visintainer, J. R. Volpicelli, and M. E. P. Seligman, "Tumor rejection in rats after inescapable or escapable shock," *Science*, 216 (1982), pp. 437–439.

11. James R. Walker, "New evidence on the supply of child care," *Journal of Human Resources*, 27 (1991), pp. 40–69.

12. Modified from Felicity Barringer, "Measuring sexuality through polls can be shaky," *New York Times*, April 25, 1993.

13. Data provided by Don Bentley, Pomona College.

14. Poll results reported in the *New York Times*, November 1, 1992.

15. Francine D. Blau and Marianne A. Ferber, "Career plans and expectations of young women and men," *Journal of Human Resources*, 26 (1991), pp. 581–607.

16. Erdener Kaynak and Wellington Kang-yen Kuan, "Environment, strategy, structure, and performance in the context of export activity: an empirical study of Taiwanese manufacturing firms," *Journal of Business Research*, 27 (1993), pp. 33–49.

17. William D. Darley, "Store-choice behavior for pre-owned merchandise," *Journal of Business Research*, 27 (1993), pp. 17–31.

18. Giuliana Coccia, "An overview of non-response in Italian telephone surveys," *Proceedings of the 99th Session of the International Statistical Institute*, 1993, Book 3, pp. 271–272.

19. Cited by J. K. Haseman in his discussion of D. A. Freedman and H. Zeisel, "From mouse to man: the quantitative assessment of cancer risks," *Statistical Science*, 3 (1988), pp. 3–28. The citation is on p. 34.

chapter

Inference
for Regression

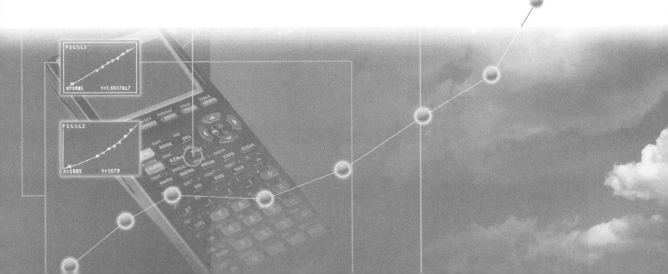

ACTIVITY 14 | The Legacy of Jet Skis

Jet skis have become one of the most popular types of recreational vehicle sold today. But critics say that they're noisy, dangerous, and damaging to the environment. An article in the August 1997 issue of the *Journal of the American Medical Association* reported on a survey that tracked emergency room visits at randomly selected hospitals nationwide. Here are data on the number of jet skis in use, the number of accidents, and the number of fatalities for the years 1987 to 1996:[1]

Year	Jet skis in use	Accidents	Fatalities
1987	92,756	376	5
1988	126,881	650	20
1989	178,510	844	20
1990	241,376	1,162	28
1991	305,915	1,513	26
1992	372,283	1,650	34
1993	454,545	2,236	35
1994	600,000	3,002	56
1995	760,000	4,028	68
1996	900,000	4,010	55

(a) In Exercises 3.5 and 3.7, we explored the association between the number of jet skis in use and the number of accidents. Now we want to examine the relationship between the number of jet skis in use and the number of *fatalities*. Make a scatterplot of these data. (Be sure to label the axes with the variable names, not just x and y.) What does the scatterplot show about the relationship between these variables?

(b) Perform least-squares regression on your TI-83 and report the regression equation, the correlation coefficient r, and the coefficient of determination r^2. Write a sentence that interprets the r-value. Do the same for the r^2-value.

(continued)

(c) Construct a residual plot. What does the residual plot tell you about the suitability of the least-squares line as a model for these data?

(d) Use the least-squares equation to predict the number of fatalities in 1996.

(e) There were 900,000 jet skis in use in 1996, with 4010 accidents and 55 fatalities. Compare the predicted number of fatalities in 1996 with the actual figure of 55. Does this newest point (900,000, 55) continue the pattern or does it depart somewhat from the pattern?

INTRODUCTION

When a scatterplot shows a linear relationship between a quantitative explanatory variable x and a quantitative response variable y, we can use the least-squares line fitted to the data to predict y for a given value of x. Now we want to do tests and confidence intervals in this setting.

EXAMPLE 14.1

The Sanchez household is about to install solar panels to reduce the cost of heating their house. In order to know how much the solar panels help, they record their consumption of natural gas before the panels are installed. Gas consumption is higher in cold weather, so the relationship between outside temperature and gas consumption is important.

Table 14.1 gives data for 16 months.[2] The response variable y is the average amount of natural gas consumed each day during the month, in hundreds of cubic feet. The explanatory variable x is the average number of heating degree-days each day during the month. (One heating degree-day is accumulated for each degree a day's average temperature falls below 65° F. An average temperature of 20° F, for example, corresponds to 45 degree-days.)

We met the Sanchez's natural gas consumption data in Chapter 3. Before attempting inference, examine the data as follows.

scatterplot

1. Make a **scatterplot.** Plot the explanatory variable x horizontally and the response variable y vertically. Fitting a line only makes sense if the overall pattern of the scatterplot is roughly linear (straight-line).

| | TABLE 14.1 | | Average degree-days and natural gas consumption for the Sanchez household | | |

Month	Degree-days	Gas (100 cu ft)	Month	Degree-days	Gas (100 cu ft)
Nov.	24	6.3	July	0	1.2
Dec.	51	10.9	Aug.	1	1.2
Jan.	43	8.9	Sept.	6	2.1
Feb.	33	7.5	Oct.	12	3.1
Mar.	26	5.3	Nov.	30	6.4
Apr.	13	4.0	Dec.	32	7.2
May	4	1.7	Jan.	52	11.0
June	0	1.2	Feb.	30	6.9

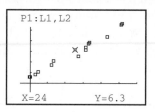

least-squares line

2. Use a calculator or computer to fit the **least-squares regression line** to the data. This line lies as close as possible to the points (in the sense of least squares) in the vertical (y) direction. We use it to predict y for a given x.

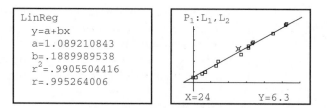

outliers
influential observations

3. Look for **outliers** and **influential observations.** Outliers are points that lie far from the overall linear pattern. Influential observations are points that move the fitted line, usually points that are far out in the x direction and isolated from other points. Inference is not safe if there are influential points, because the results will depend strongly on these few points.

correlation

4. Use a calculator or computer to calculate the **correlation** r and then its square r^2. The squared correlation r^2 describes how well the regression line fits the data. It is the proportion of the observed variation in y that is accounted for by the straight-line relationship of y with x.

EXAMPLE 14.2

Figure 14.1 is a scatterplot of the Sanchez gas consumption data. Degree-days, the explanatory variable, appears on the horizontal scale. There is a strong linear relationship.

Enter the data into a calculator or computer software. The least-squares regression line is

$$\hat{y} = a + bx$$
$$= 1.0892 + .1890x$$

We use the notation $\hat{y}$ to remind ourselves that this line gives *predictions* of y. The predictions usually won't agree exactly with the observed values of y. Drawing the least-squares line on the scatterplot helps us see that there are no strong outliers or influential observations.

The calculator or computer also tells us the correlation is $r = .9953$. So $r^2 = .9906$. The scatterplot shows that almost all the variation in gas consumption is explained by degree-days. In fact, $r^2 = .99$ says that 99% is explained. Prediction of gas consumption from degree-days should be quite accurate.

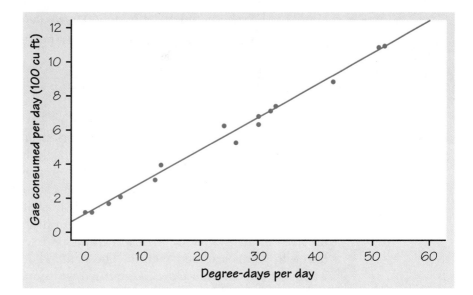

FIGURE 14.1 Scatterplot of natural gas consumption against heating degree-days for the Sanchez household (Example 14.2). The line is the least-squares regression line for predicting gas consumption from degree-days.

The regression model

model

The slope b and intercept a of the least-squares line are *statistics*. They are computed from the sample data and would no doubt be different if we monitored the Sanchez house for another 16 months. To do formal inference, we need to know what unknown *parameters* a and b estimate. The parameters appear in a description, often called a **model**, of the process that produces our data. Here are the assumptions that describe the regression model.

ASSUMPTIONS FOR REGRESSION INFERENCE

We have n observations on an explanatory variable x and a response variable y. Our goal is to study or predict the behavior of y for given values of x.

- For any fixed value of x, the response y varies according to a normal distribution. Repeated responses y are independent of each other.
- The mean response μ_y has a straight-line relationship with x:

$$\mu_y = \alpha + \beta x$$

  The slope β and intercept α are unknown parameters.
- The standard deviation of y (call it σ) is the same for all values of x. The value of σ is unknown.

---

true regression line

The heart of this model is that there is an "on the average" straight-line relationship between y and x. The ***true regression line*** $\mu_y = \alpha + \beta x$ says that the *mean* response μ_y moves along a straight line as the explanatory variable x changes. We can't observe the true regression line. The values of y that we do observe vary about their means according to a normal distribution. If we hold x fixed and take many observations on y, the normal pattern will eventually appear in a stemplot or histogram. In practice, we observe y for many different values of x, and so we see an overall linear pattern with points scattered about it. The standard deviation σ determines whether the points fall close to the true regression line (small σ) or are widely scattered (large σ).

Figure 14.2 shows the regression model in picture form. The line in the figure is the true regression line. The mean of the response y moves along this line as the explanatory variable x takes different values. The

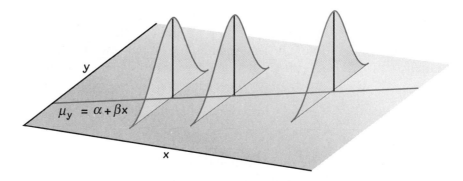

FIGURE 14.2   The regression model. The line is the true regression line, which shows how the mean response μ_y changes as the explanatory variable x changes. For any fixed value of x, the observed response y varies according to a normal distribution having mean μ_y.

normal curves show how y will vary when x is held fixed at different values. All of the curves have the same σ, so the variability of y is the same for all values of x. You should check the assumptions for inference when you do inference about regression. We will see later how to do that.

14.1   INFERENCE ABOUT THE MODEL

The first step in inference is to estimate the unknown parameters α, β, and σ. When the regression model describes our data and we calculate the least-squares line $\hat{y} = a + bx$:

- The slope b of the least-squares line is an unbiased estimator of the true slope β.
- The intercept a of the least-squares line is an unbiased estimator of the true intercept α.

EXAMPLE 14.3

The Sanchez household gas consumption data in Figure 14.1 fit the regression model of scatter about an invisible true regression line quite well. The least-squares line is $\hat{y} = 1.0892 + .1890x$. The slope is particularly important. A *slope is a rate of change*. The true slope β says how much the average gas consumption changes when the number of degree-days x increases by 1. Because $b = .1890$ estimates the unknown β, we estimate that on the average the Sanchez household will use 0.1890 hundred cubic feet more gas per day for each degree the outside temperature drops.

residuals

The remaining parameter of the model is the standard deviation σ, which describes the variability of the response y about the true regression line. The least-squares line estimates the true regression line. So the ***residuals*** estimate how much y varies about the true line. Recall that the residuals are the vertical deviations of the data points from the least-squares line:

$$\text{residual} = \text{observed } y - \text{predicted } y$$
$$= y - \hat{y}$$

There are n residuals, one for each data point. Because σ is the standard deviation of responses about the true regression line, we estimate it by a sample standard deviation of the residuals. We call this sample standard deviation a *standard error* to emphasize that it is estimated from data. The residuals from a least-squares line always have mean zero. That simplifies their standard error.

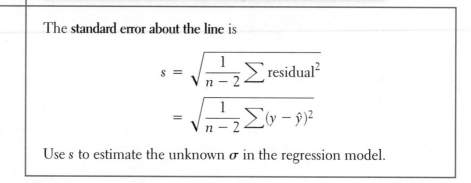

STANDARD ERROR ABOUT THE LEAST-SQUARES LINE

The **standard error about the line** is

$$s = \sqrt{\frac{1}{n-2} \sum \text{residual}^2}$$

$$= \sqrt{\frac{1}{n-2} \sum (y - \hat{y})^2}$$

Use s to estimate the unknown σ in the regression model.

Because we use the standard error about the line so often in regression inference, we just call it s. Notice that s^2 is an average of the squared deviations of the data points from the line, so it qualifies as a variance. We average the squared deviations by dividing by $n - 2$, the number of data points less 2. It turns out that if we know $n - 2$ of the n residuals, the other two are determined. So $n - 2$ is the ***degrees of freedom*** of s. We first met the idea of degrees of freedom in the case of the ordinary sample standard deviation of n observations, which has $n - 1$ degrees of freedom. Now we observe two variables rather than one, and the proper degrees of freedom is $n - 2$ rather than $n - 1$.

degrees of freedom

Calculating s begins with finding the predicted response for each value of x in your data set, then the residuals, and then s using the above formula. In practice, you will use technology that does this arithmetic instantly. The next example shows how to use the TI-83 to help calculate s.

EXAMPLE 14.4

The first observation in Table 14.1, for the month of November, has $x = 24$ degree-days per day and $y = 6.3$ hundred cubic feet of gas used per day. The predicted gas usage for $x = 24$ is

$$\hat{y} = 1.0892 + .1890x$$
$$= 1.0892 + (.1890)(24) = 5.6252$$

The residual for November's observation is

$$\text{residual} = y - \hat{y}$$
$$= 6.3 - 5.6252 = .6748$$

That is, the November data point on the scatterplot lies 0.6748 above the least-squares line.

Repeat this calculation 15 more times, once for each data point.

Or, with the TI-83, insert a new list after L_1, and L_2 and name it RES for residuals. Define LRES to be the *observed* minus the *predicted* values: $L_2 - Y_1(L_1)$. Verify that the residuals are

.6748	.1718	−.3162	.1738	−.7032	.4538	−.1452	.1108
.1108	−.0782	−.1232	−.2572	−.3592	.0628	.0828	.1408

If you need to arrange the three lists for easy viewing in the STATS/Edit window, enter STATS / SetUpEditor and then specify the lists you want to see in order.

```
SetUpEditor L1,L2,
LRES
                Done
1-Var Stats LRES
```

To verify that the sum of the residuals is approximately 0, specify 1-Var Stats LRES. Notice that the sum of the residuals Σx and the mean $\bar{x}$ are both 0 (up to round-off error). Notice also that the sum of the squares of the residuals $\Sigma x^2 = 1.6082$.

```
1-Var Stats
x̄=1.25E-14
Σx=2E-13
Σx²=1.60821443
Sx=.327435941
σx=.3170384867
↓n=16
```

The variance about the line is

$$s^2 = \frac{1}{n-2} \sum \text{residual}^2$$

$$= \frac{1}{16-2}(.6748^2 + .1718^2 + \cdots + .1408^2)$$

$$= \frac{1}{14}(1.6082) = .1149$$

Finally, the standard error about the line is

$$s = \sqrt{.1149} = .3390$$

We rounded each step to four decimal places. As a result, s is off by 0.0001. The correct value is $s = .3389$ to four places. If you do calculations by hand, carry at least one more decimal place than the accuracy you want in the final result.

   We will study several kinds of inference in the regression setting. The standard error s about the line is the key measure of the variability of the responses in regression. It is part of the standard error of all the statistics we will use for inference.

EXERCISES

14.1   *Archaeopteryx* is an extinct beast having feathers like a bird but teeth and a long bony tail like a reptile. Here are the lengths in centimeters of the femur (a leg bone) and the humerus (a bone in the upper arm) for the five fossil specimens that preserve both bones:[3]

Femur	38	56	59	64	74
Humerus	41	63	70	72	84

The strong linear relationship between the lengths of the two bones helped persuade scientists that all five specimens belong to the same species.

(a) Examine the data. Make a scatterplot with femur length as the explanatory variable. Use your calculator to obtain the correlation r and the equation of the least-squares regression line. Do you think that femur length will allow good prediction of humerus length?

(b) Explain in words what the slope β of the true regression line says about *Archaeopteryx*. What is the estimate of β from the data? What is your estimate of the intercept α of the true regression line?

(c) Calculate the residuals for the five data points. Check that their sum is 0 (up to roundoff error). Use the residuals to estimate the standard deviation σ in the regression model. You have now estimated all three parameters in the model.

14.2   Good runners take more steps per second as they speed up. Here are the average number of steps per second for a group of top female runners at different speeds. The speeds are in feet per second.[4]

Speed (ft/s)	15.86	16.88	17.50	18.62	19.97	21.06	22.11
Steps per second	3.05	3.12	3.17	3.25	3.36	3.46	3.55

(a) You want to predict steps per second from running speed. Make a scatterplot of the data with this goal in mind. Use your calculator to find the correlation r and the equation of the least-squares regression line. Describe the form and strength of the relationship. Do steps per second increase with running speed at a steady rate?

(b) Find the residuals for all 7 data points. Check that their sum is 0 (up to roundoff error).

(c) The model for regression inference has three parameters, which we call α, β, and σ. Estimate these parameters from the data.

14.3   Return to the jet ski data from Activity 14.

(a) Verify that the sum of the residuals is zero. Hence the mean of the residuals is zero. Note that you can't calculate one-variable statistics on the data in the list RESID because RESID is a reserved word. Instead, you can do the following: Define a new list named R, and copy the data into list R with the command LRESID→LR. Then you can calculate statistics on the residuals in list R.

(b) From the 1-Var Stats printout, what is the sum of the squares of the residuals? Use this sum to calculate the standard error s about the line.

(c) The model for regression inference has three parameters: α, β, and σ. Estimate these parameters from the data.

Confidence intervals for the regression slope

The slope β of the true regression line is usually the most important parameter in practical regression problems. The slope is the rate of change of the mean response as the explanatory variable increases. We often want to estimate β. The slope b of the least-squares line is an unbiased estimator of β. A confidence interval is more useful because it shows how accurate the estimate b is likely to be. The confidence interval for β has the familiar form

$$\text{estimate} \pm t^* \text{SE}_{\text{estimate}}$$

Because b is our estimate, the confidence interval becomes

$$b \pm t^* \text{SE}_b$$

Here are the details.

CONFIDENCE INTERVAL FOR REGRESSION SLOPE

A level C confidence interval for the slope β of the true regression line is

$$b \pm t^* \text{SE}_b$$

In this recipe, the standard error of the least-squares slope b is

$$\text{SE}_b = \frac{s}{\sqrt{\sum (x - \bar{x})^2}}$$

and t^* is the upper $(1 - C)/2$ critical value from the t distribution with $n - 2$ degrees of freedom.

As advertised, the standard error of b is a multiple of s. Although we give the recipe for this standard error, you should rarely have to calculate it by hand. Regression software gives the standard error SE_b along with b itself.

EXAMPLE 14.5 Figure 14.3 shows the basic output for the gas consumption data from the regression command in the Minitab software package. Most statistical software

```
The regression equation is
gas used = 1.09 + 0.189 ddays

Predictor          Coef        Stdev      t-ratio          P
Constant         1.0892       0.1389         7.84      0.000
ddays          0.188999     0.004934        38.31      0.000

s = 0.3389          R-sq = 99.1%       R-sq(adj) = 99.0%
```

FIGURE 14.3 Minitab regression output for the Sanchez household gas consumption data (Example 14.5).

provides similar output. (Minitab, like other software, produces more than this basic output. When you use software, just ignore the parts you don't need.)

The first line gives the equation of the least-squares regression line. The slope and intercept are rounded off there, so look in the "Coef" column of the table that follows for more accurate values. The intercept $a = 1.0892$ appears in the "Constant" row. The slope $b = .188999$ appears in the "ddays" row because we named the x variable "ddays" when we entered the data.

The next column of output, headed "Stdev," gives standard errors. In particular, $SE_b = .004934$. The standard error about the line, $s = .3389$, appears below the table.

There are 16 data points, so the degrees of freedom are $n - 2 = 14$. A 95% confidence interval for the true slope β uses the critical value $t^* = 2.145$ from the df = 14 row of Table C. The interval is

$$b \pm t^* SE_b = .188999 \pm (2.145)(.004934)$$
$$= .1890 \pm .0106$$
$$= .1784 \text{ to } .1996$$

We are 95% confident that each additional degree-day causes the Sanchez household to use between 0.1784 and 0.1996 additional hundred cubic feet of natural gas on the average.

You can find a confidence interval for the intercept α of the true regression line in the same way, using a and SE_a from the "Constant" line of the printout. It is not common to want to estimate α.

Testing the hypothesis of no linear relationship

We can also test hypotheses about the value of the slope β. The most common hypothesis is

$$H_0 : \beta = 0$$

A regression line with slope 0 is horizontal. That is, the mean of y does not change at all when x changes. So this H_0 says that there is *no true linear relationship* between x and y. Put another way, H_0 says that *straight-line dependence on x is of no value for predicting y*. Put yet another way, H_0 says that there is *no correlation* between x and y in the population from which we drew our data. You can use the test for zero slope to test the hypothesis of zero correlation between any two quantitative variables. That's a useful trick.

The test statistic is just the standardized version of the least-squares slope b. It is another t statistic. Here are the details.

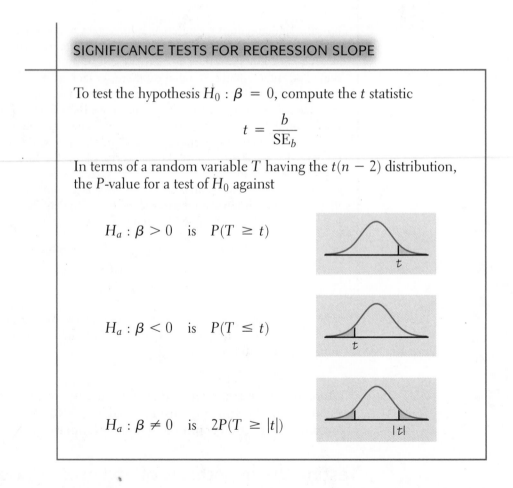

SIGNIFICANCE TESTS FOR REGRESSION SLOPE

To test the hypothesis $H_0 : \beta = 0$, compute the t statistic

$$t = \frac{b}{SE_b}$$

In terms of a random variable T having the $t(n-2)$ distribution, the P-value for a test of H_0 against

$H_a : \beta > 0$ is $P(T \geq t)$

$H_a : \beta < 0$ is $P(T \leq t)$

$H_a : \beta \neq 0$ is $2P(T \geq |t|)$

Regression output from statistical software usually gives t and its *two-sided* P-value. For a one-sided test, divide the P-value in the output by 2.

EXAMPLE 14.6

The computer output in Figure 14.3 gives $t = 38.31$ with two-sided P-value 0.000 for the hypothesis $H_0 : \beta = 0$ that there is no linear relationship between degree-days and natural gas used. (The P-value is rounded to three decimal places.) Because the scatterplot shows a very strong relationship, we are not surprised that the P-value is very small.

The TI-83 can produce most of these results as well. With the x-values (degree-days) in list L_1 and the y-values (natural gas used) in the list L_2, press STAT / EDIT / TESTS / E:LinRegTTest, as shown here.

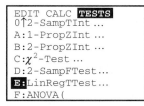

In the LinRegTTest screen, specify L_1 for Xlist, L_2 for Ylist, and $\neq$ 0 for the hypothesized slope β. You can leave the RegEQ space blank. Highlight the command "Calculate" and press ENTER.

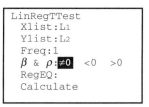

The linear regression t test results take more than one screen to present.

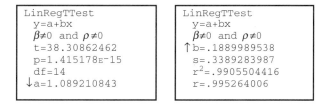

The first screen reports that the t statistic is 38.31 with df = 14. The P-value is 1.415×10^{-15}, a number so small it is essentially 0. Scrolling down, you find the LSRL constants a and b, as well as the correlation coefficient r, the coefficient of determination r^2, and the standard error about the line s.

EXAMPLE 14.7

How well do golfers' scores in the first round of a two-round tournament predict their scores in the second round? Here are data for the 12 members of a college's women's golf team in a recent tournament. (A golf score is the number of strokes required to complete the course, so low scores are better.)

Golfer	1	2	3	4	5	6	7	8	9	10	11	12
Round 1	89	90	87	95	86	81	102	105	83	88	91	79
Round 2	94	85	89	89	81	76	107	89	87	91	88	80

The scatterplot in Figure 14.4 shows a clear linear relationship. There are two unusual points. Golfer 7 had poor scores on both rounds (102 and 107). This point is close to the linear pattern of the other points but may be influential. Golfer 8 is an outlier. She scored 105 on the first round, then improved to 89

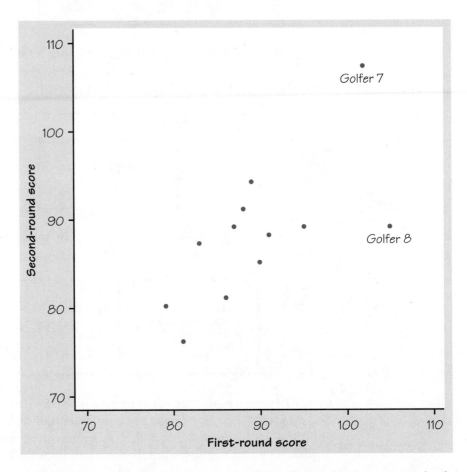

FIGURE 14.4   Scatterplot of the scores of 12 college golfers in two rounds of a tournament (Example 14.7).

```
Dependent variable is:  round2

R squared = 47.2%
s =  5.974  with  12 - 2 = 10  degrees of freedom

Variable    Coefficient    s.e. of Coeff    t-ratio    prob
Constant    26.3320        20.69            1.27       0.2320
round1      0.687747       0.2300           2.99       0.0136
```

FIGURE 14.5   Data Desk output for the regression of second-round score on first-round score (Example 14.7).

on the second round. We will verify later (Example 14.10) that the regression assumptions are quite well satisfied except for these unusual observations.

Figure 14.5 gives part of the output from the regression command in the Data Desk statistical software package. You should be able to interpret this output. The least-squares line for predicting second-round score y from first-round score x is

$$\hat{y} = 26.332 + .688x$$

The t statistic for testing

$$H_0 : \beta = 0$$
$$H_a : \beta \neq 0$$

is $t = 2.99$ with two-sided P-value $P = .0136$. Check that t is the slope $b = .688$ divided by its standard error, $SE_b = .230$.

We think that there is a positive association between the same golfer's scores in two rounds of play. So we want to test the one-sided alternative,

$$H_0 : \beta = 0$$
$$H_a : \beta > 0$$

The value of t is unchanged. The one-sided P is half the two-sided $P : P = .0136/2 = .0068$. Alternatively, we can use Table C. There are $n - 2 = 10$ degrees of freedom. Compare the observed $t = 2.99$ with the entries in the df $= 10$ row of the table. The one-sided P-value lies between 0.005 and 0.01. This agrees with the computer's more exact value. There is (as the plot suggests) strong evidence that second-round score increases with first-round score in a straight-line fashion.

df = 10

p	.01	.005
t^*	2.764	3.169

SECTION 14.1 EXERCISES

14.4   Figure 14.3 gives computer output for the regression of natural gas consumed by the Sanchez household on degree-days. Use the information in this output to give a 90% confidence interval for the slope β of the true regression line.

14.5   Example 14.7 finds strong evidence that golfers' scores in the second round of tournament play are linearly related to their scores in the first round. Use the computer output in Figure 14.5 to give a 95% confidence interval for the slope β of the true regression line. Then explain in plain language what β means in this setting.

14.6   Exercise 14.1 presents data on the lengths of two bones in five fossil specimens of the extinct beast *Archaeopteryx*. Here is part of the output from the S-PLUS statistical software when we regress the length y of the humerus on the length x of the femur:

```
                 coef   std.err   t.stat   p.value
Intercept   -3.6596    4.4590   -0.8207   0.4719
        X    1.1969    0.0751        *        *
```

(a)   What is the equation of the least-squares regression line?

(b)   We left out the t statistic for testing $H_0 : \beta = 0$ and its P-value. Use the output to find t.

(c)   How many degrees of freedom does t have? Use Table C to approximate the P-value of t against the one-sided alternative $H_a : \beta > 0$.

(d)   Confirm your results for parts (a) to (c) by entering the data from Exercise 14.1 into your TI-83 and performing a LinRegTTest.

14.7   Return to the jet ski data of Activity 14 once again. There appears to be clear evidence that more jet skis in use will mean more fatalities.

(a)   Formulate null and alternative hypotheses about the slope of the true regression line. State a one-sided alternative hypothesis.

(b)   Perform a LinRegTTest on the TI-83. Report the t statistic, the degrees of freedom, and the P-value.

(c)   Write your conclusion in plain language.

14.8   Exercise 14.2 presents data on the relationship between the speed of runners (x, in feet per second) and the number of steps y that they take in a second. Here is part of the Data Desk regression output for these data:

```
R squared =   99.8%
s =   0.0091   with 7 - 2 = 5 degrees of freedom

Variable   Coefficient   s.e. of Coeff   t-ratio   prob
Constant   1.76608       0.0307          57.6      <0.0001
Speed      0.080284      0.0016          49.7      <0.0001
```

(a) How can you tell from this output, even without the scatterplot, that there is a very strong straight-line relationship between running speed and steps per second?

(b) What parameter in the regression model gives the rate at which steps per second increase as running speed increases? Give a 99% confidence interval for this rate.

14.9   The Leaning Tower of Pisa leans more as time passes. Here are measurements of the lean of the tower for the years 1975 to 1987.[5] The lean is the distance between where a point on the tower would be if the tower were straight and where it actually is. The distances are tenths of a millimeter in excess of 2.9 meters. For example, the 1975 lean, which was 2.9642 meters, appears in the table as 642. We use only the last two digits of the year as our time variable.

Year	75	76	77	78	79	80	81	82	83	84	85	86	87
Lean	642	644	656	667	673	688	696	698	713	717	725	742	757

Here is part of the output from the Data Desk regression procedure with year as the explanatory variable and lean as the response variable:

```
Variable    Coefficient  s.e. of Coeff  t-ratio   prob
Constant    -61.1209     25.13                     -2.43     0.0333
year          9.31868     0.3099                    30.1    <0.0001
```

(a) Plot the data. Briefly describe the shape, strength, and direction of the relationship. The tower is tilting at a steady rate.

(b) The main purpose of the study is to estimate how fast the tower is tilting. What parameter in the regression model gives the rate at which the tilt is increasing, in tenths of a millimeter per year?

(c) We want a 95% confidence interval for this rate. How many degrees of freedom does t have? Find the critical value t^* and the confidence interval.

14.2   INFERENCE ABOUT PREDICTION

One of the most common reasons to fit a line to data is to predict the response to a particular value of the explanatory variable. The method is simple: just substitute the value of x into the equation of the line.

EXAMPLE 14.8

In Example 14.2 we found that the least-squares line for predicting gas consumption from degree-days for the Sanchez household is

$$\hat{y} = 1.0892 + .1890x$$

To predict gas consumption at 20 degree-days (that's average temperature 45° F), substitute $x = 20$:

$$\hat{y} = 1.0892 + (.1890)(20)$$
$$= 1.0892 + 3.78 = 4.869$$

We predict that the household will use 4.869 hundred cubic feet of gas per day.

We would like to give a confidence interval that describes how accurate this prediction is. To do that, you must answer these questions: Do you want to predict the *mean* gas consumption for *all* months with 20 degree-days per day? Or do you want to predict the gas consumption in *one specific* month that has 20 degree-days per day? Both of these predictions may be interesting, but they are two different problems. The actual prediction is the same, $\hat{y} = 4.869$ hundred cubic feet. But the margin of error is different for the two kinds of prediction. Individual months with 20 degree-days per day don't all have the same gas consumption. So we need a larger margin of error to pin down one month's result than to estimate the mean response for all months that have 20 degree-days per day.

Write the given value of the explanatory variable x as x^*. In the example, $x^* = 20$. The distinction between predicting a single outcome and predicting the mean of all outcomes when $x = x^*$ determines what margin of error is correct. To emphasize the distinction, we use different terms for the two intervals.

- To estimate the *mean* response, we use a *confidence interval*. It is an ordinary confidence interval for the parameter

$$\mu_y = \alpha + \beta x^*$$

The regression model says that μ_y is the mean of responses y when x has the value x^*. It is a fixed number whose value we don't know.

prediction interval
- To estimate an *individual* response y, we use a **prediction interval**. A prediction interval estimates a single random response y rather than a parameter like μ_y. The response y is not a fixed number. If we took more observations with $x = x^*$, we would get different responses.

Fortunately, the meaning of a prediction interval is very much like the meaning of a confidence interval. A 95% prediction interval, like a 95%

confidence interval, is right 95% of the time in repeated use. "Repeated use" now means that we take an observation on y for each of the n values of x in the original data, and then take one more observation y at the point $x = x^*$. Form the prediction interval from the n observations, then see if it covers the one more y. It will in 95% of all repetitions.

The interpretation of prediction intervals is a minor point. The main point is that it is harder to predict one response than to predict a mean response. Both intervals have the usual form

$$\hat{y} \pm t^* \text{SE}$$

but the prediction interval is wider than the confidence interval. Here are the details.

CONFIDENCE AND PREDICTION INTERVALS FOR REGRESSION RESPONSE

A level C **confidence interval for the mean response** μ_y when x takes the value x^* is

$$\hat{y} \pm t^* \text{SE}_{\hat{\mu}}$$

The standard error $\text{SE}_{\hat{\mu}}$ is

$$\text{SE}_{\hat{\mu}} = s \sqrt{\frac{1}{n} + \frac{(x^* - \bar{x})^2}{\sum (x - \bar{x})^2}}$$

The sum runs over all the observations on the explanatory variable x.

A level C **prediction interval for a single observation** on y when x takes the value x^* is

$$\hat{y} \pm t^* \text{SE}_{\hat{y}}$$

The standard error for prediction $\text{SE}_{\hat{y}}$ is[6]

$$\text{SE}_{\hat{y}} = s \sqrt{1 + \frac{1}{n} + \frac{(x^* - \bar{x})^2}{\sum (x - \bar{x})^2}}$$

In both recipes, t^* is the upper $(1 - C)/2$ critical value of the t distribution with $n - 2$ degrees of freedom.

There are two standard errors: $SE_{\hat{\mu}}$ for estimating the mean response μ_y and $SE_{\hat{y}}$ for predicting an individual response y. The only difference between the two standard errors is the extra 1 under the square root sign in the standard error for prediction. The extra 1 makes the prediction interval wider. Both standard errors are multiples of s. The degrees of freedom are again $n - 2$, the degrees of freedom of s. Calculating these standard errors by hand is a nuisance, which software spares us.

EXAMPLE 14.9

The Sanchez household installs solar panels. The following March, there are 20 degree-days per day. What would their gas usage have been without the solar panels? We want to predict gas consumption in a specific single month when x has the value $x^* = 20$.

Despite the importance of prediction, many statistical software systems neglect it and some omit it entirely. Minitab does prediction but offers only 95% intervals. Here is the output from the prediction option in the Minitab regression command for $x^* = 20$:

```
  Fit Stdev.Fit      95% C.I.           95% P.I.
4.8692    0.0855   (4.6858, 5.0526)   (4.1193, 5.6191)
```

The "Fit" entry gives the prediction: 4.8692 hundred cubic feet. This agrees with our result in Example 14.8. Minitab gives both 95% intervals. You must choose which one you want. We are predicting a single response, so the prediction interval "95% P.I." is the right choice. We are 95% confident that gas usage without the solar panels for this month would fall between 4.119 and 5.619 hundred cubic feet.

The 95% confidence interval for the mean gas used in all months with 20 degree-days per day, given as "95% C.I.," is narrower.

By the way, the standard deviation Minitab gives as "Stdev.Fit" is $SE_{\hat{\mu}}$, used for confidence intervals for the mean response. You can use that in the recipe

$$\hat{y} \pm t^* SE_{\hat{\mu}}$$

to estimate the mean response with different confidence levels.

SECTION 14.2 EXERCISES

14.10   Figure 14.3 gives computer output for the regression of natural gas consumed by the Sanchez household on degree-days. Use this output to answer the following questions.

(a) In the month of January after solar panels were installed, there were 40 degree-days per day. How much gas do you predict the Sanchez household would have used per day without the solar panels? They actually used 7.5 hundred cubic feet per day. How much gas per day did the solar panels save?

(b) Here is the output from the prediction option in Minitab for 40 degree-days per day. Give a 95% interval for the amount of gas that would have been used this January without the solar panels.

```
  Fit   Stdev.Fit          95% C.I.            95% P.I.
8.6492      0.1216   ( 8.3882, 8.9101)   ( 7.8767, 9.4217)
```

(c) Give a 95% interval for the mean gas consumption per day in months with 40 degree-days per day.

14.11   Manatees are large, gentle sea creatures that live along the Florida coast. Many manatees are killed or injured by powerboats. Here are data on powerboat registrations (in thousands) and the number of manatees killed by boats in Florida in the years 1977 to 1990:

Year	Powerboat registrations	Manatees killed	Year	Powerboat registrations	Manatees killed
1977	447	13	1984	559	34
1978	460	21	1985	585	33
1979	481	24	1986	614	33
1980	498	16	1987	645	39
1981	513	24	1988	675	43
1982	512	20	1989	711	50
1983	526	15	1990	719	47

(a) Make a scatterplot showing the relationship between powerboats registered and manatees killed. (Which is the explanatory variable?)

(b) Is the overall pattern roughly linear? Are there clear outliers or strongly influential data points?

(c) Here is part of the output from the Minitab regression command:

```
Predictor         Coef        Stdev       t-ratio        p
Constant       -41.430        7.412         -5.59    0.000
Boats          0.12486      0.01290          9.68    0.000

s = 4.276            R-sq = 88.6%
```

What does $r^2 = .886$ tell you about the relationship between boats and manatees killed?

(d) Explain what the slope β of the true regression line means in this setting. Then give a 90% confidence interval for β.

(e) If Florida decided to freeze powerboat registrations at 700,000, how many manatees do you predict would be killed by boats each year?

(f) Here is the result of asking Minitab to do prediction for $x^* = 700$:

```
    Fit   Stdev.Fit       95% C.I.              95% P.I.
   45.97      2.06     ( 41.49, 50.46)    (  35.63, 56.31)
```

Check that the prediction 45.97 agrees with your result in (e). Then give a 95% interval for the mean number of manatees that would be killed each year if Florida froze boat registrations at 700,000.

14.12   The 95% interval in part (f) of the previous exercise is quite wide. Changing to 90% confidence will give a smaller margin of error. Use the computer output in the previous exercise, along with Table C, to give a 90% interval for the mean number of manatees killed when there are 700,000 powerboats registered.

14.3   CHECKING THE REGRESSION ASSUMPTIONS

You can fit a least-squares line to any set of explanatory-response data when both variables are quantitative. If the scatterplot doesn't show a roughly linear pattern, the fitted line may be almost useless. But it is still the line that fits the data best in the least-squares sense. To use regression inference, however, the data must satisfy the regression model assumptions. Before we do inference, we must check these assumptions one by one.

The true relationship is linear. We can't observe the true regression line, so we will almost never see a perfect straight-line relationship in our data. Look at the scatterplot to check that the overall pattern is roughly linear. A plot of the residuals against x magnifies any unusual pattern. Draw a horizontal line at zero on the residual plot to orient your eye. Because the sum of the residuals is always zero, zero is also the mean of the residuals.

The standard deviation of the response about the true line is the same every-where. Look at the scatterplot again. The scatter of the data points about the line should be roughly the same over the entire range of the data. A plot of the residuals against x, with a horizontal line at zero, makes this easier to check. It is quite common to find that as the response y gets larger, so does the scatter of the points about the fitted line. Rather than remaining fixed, the standard deviation σ about the line is changing with x as the mean response changes with x. You cannot safely use our inference recipes when this happens. There is no fixed σ for s to estimate.

The response varies normally about the true regression line. We can't observe the true regression line. We can observe the least-squares line and the residuals, which show the variation of the response about the fitted line. The residuals estimate the deviations of the response from the true regression line, so they should follow a normal distribution. Make a histogram or stemplot of the residuals and check for clear skewness or other major departures from normality. Like other t procedures, inference for regression is (with one exception) not very sensitive to minor lack of normality, especially when we have many observations. Do beware of influential observations, which move the regression line and can greatly affect the results of inference.

The exception is the prediction interval for a single response y. This interval relies on normality of individual observations, not just on the approximate normality of statistics like the slope a and intercept b of the least-squares line. The statistics a and b become more normal as we take more observations. This contributes to the robustness of regression inference, but it isn't enough for the prediction interval. We will not study methods that carefully check normality of the residuals, so you should regard prediction intervals as rough approximations.

The assumptions for regression inference are a bit elaborate. Fortunately, it is not hard to check for gross violations. There are ways to deal with violations of any of the regression model assumptions. If your data don't fit the regression model, get expert advice. Checking assumptions uses the residuals. Most regression software will calculate and save the residuals for you.

EXAMPLE 14.10

Example 14.7 shows the regression of the scores of 12 college golfers in the second round of a tournament on their first-round scores. The Data Desk statistical software that did the regression calculations also calculates the 12 residuals. Here they are (Golfers 1 to 12 are in order reading across the rows):

6.45850	−3.22925	2.83399	−2.66798	−4.47826	−6.03953
10.5178	−9.54545	3.58498	4.14625	−.916996	−.664032

The computer has given us more decimal places than we need. The residual plot appears in Figure 14.6. The values of x are on the horizontal axis. The residuals are on the vertical axis, with a horizontal line at zero.

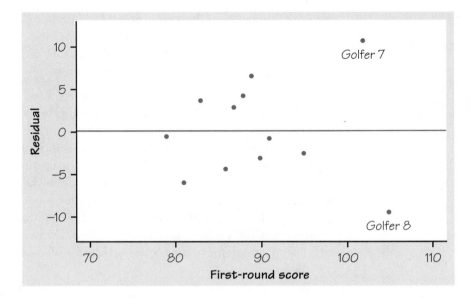

FIGURE 14.6   Plot of the regression residuals for the golf scores data against the explanatory variable, first-round score. The mean of the residuals is always 0.

Examine the residual plot to check that the relationship is roughly linear and that the scatter about the line is about the same from end to end. The two unusual data points produce large residuals, but overall there is no clear deviation from the even scatter about the line that should occur (except for chance variation) when the regression assumptions hold.

Now examine the distribution of the residuals for signs of strong nonnormality. Here is a stemplot of the residuals after rounding to the nearest whole number:

$$
\begin{array}{r|l}
-1 & 0 \\
-0 & 643311 \\
0 & 3446 \\
1 & 1
\end{array}
$$

The distribution is quite symmetric. Inference using the assumption that y varies normally will give approximately correct results.

As a reminder that regression is easily influenced by extreme data points, we recalculate the least-squares line leaving out first Golfer 7 and then Golfer 8. The slope is $b = .6877$ with all 12 golfers, $b = .4096$ if we drop Golfer 7, and $b = 1.0696$ if we drop Golfer 8. Study the scatterplot in Figure 14.4 to see why dropping Golfer 7 decreases the slope and dropping Golfer 8 increases it. The data are all correct, so we hesitate to throw out any observations. Because the regression results do depend strongly on two golfers with high first-round scores, we should gather more data before drawing any firm conclusions.

EXAMPLE 14.11   The residual plots in Figure 14.7 illustrate violations of the regression assumptions that require corrective action before using regression. Both plots come from a study of the salaries of major-league baseball players.[7] Salary is the response variable. There are several explanatory variables that measure the players' past performance. Regression with more than one explanatory variable is called *multiple regression*. Although interpreting the fitted model is more complex in multiple regression, we check assumptions by examining residuals as usual.

Figure 14.7(a) is a plot of the residuals against the predicted salary $\hat{y}$, produced by the SAS statistical software. When points on the plot overlap, SAS uses letters to show how many observations each point represents. A is one observation, B stands for two observations, and so on. The plot shows a clear violation of the assumption that the spread of responses about the model is everywhere the same. There is more variation among players with high salaries than among players with lower salaries.

Although we don't show a histogram, the distribution of salaries is strongly skewed to the right. Using the *logarithm* of the salary as the response variable gives a more normal distribution and also fixes the unequal-spread problem. It is common to work with some transformation of data in order to satisfy the regression assumptions. But all is not yet well. Figure 14.7(b) plots the new residuals against years in the major leagues. There is a clear curved pattern. The relationship between logarithm of salary and years in the majors is not linear but curved. The statistician must take more corrective action.

SECTION 14.3 EXERCISES

14.13   The residuals for the Sanchez household gas consumption example appear in Example 14.4.

(a) Display the distribution of the residuals in a plot. It is hard to assess the shape of a distribution from only 16 observations. Do the residuals appear roughly symmetric? Are there any outliers?

(b) Plot the residuals against the explanatory variable, degree-days. Draw a horizontal line at height 0 on the plot. Is there clear evidence of a nonlinear relationship? Does the variation about the line appear roughly the same as the number of degree-days changes?

14.14   In Exercise 14.9 we regressed the lean of the Leaning Tower of Pisa on year to estimate the rate at which the tower is tilting. Here are the residuals from that regression, in order by years across the rows:

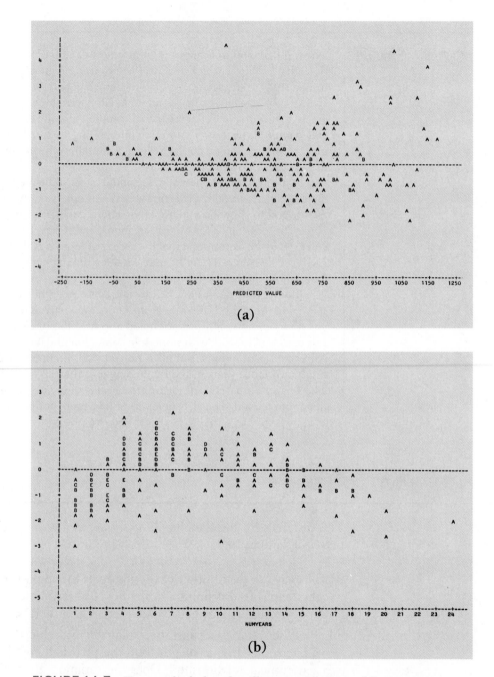

FIGURE 14.7   Two residual plots that illustrate violations of the regression assumptions. (a) The variation of the residuals is not constant. (b) There is a curved relationship between the response variable and the explanatory variable.

$$4.220 \quad -3.099 \quad -.418 \quad 1.264 \quad -2.055 \quad 3.626 \quad 2.308$$
$$-5.011 \quad .670 \quad -4.648 \quad -5.967 \quad 1.714 \quad 7.396$$

Use the residuals to check the regression assumptions, and describe your findings. Is the regression in Exercise 14.9 trustworthy?

SUMMARY

Least-squares regression fits a straight line to data in order to predict a response variable y from an explanatory variable x.

The **assumptions for regression inference** say that there is a **true regression line** $\mu_y = \alpha + \beta x$ that describes how the mean response varies as x changes. The observed response y for any x has a normal distribution with mean given by the true regression line and with the same standard deviation σ for any value of x. The parameters of the regression model are the intercept α, the slope β, and the standard deviation σ.

The slope a and intercept b of the least-squares line estimate the slope α and intercept β of the true regression line. To estimate σ, use the **standard error about the line s.**

The standard error s has $n - 2$ **degrees of freedom.** All t procedures in regression inference have $n - 2$ degrees of freedom.

Confidence intervals for the slope of the true regression line have the form $b \pm t^* \text{SE}_b$. In practice, use computer software to find the slope b of the least-squares line and its standard error SE_b.

To test the hypothesis that the true slope is zero, use the t statistic $t = b/\text{SE}_b$, also given by software and the TI-83. This null hypothesis says that straight-line dependence on x has no value for predicting y. It also says that the population correlation between x and y is zero.

Confidence intervals for the mean response when x has value x^* have the form $\hat{y} \pm t^* \text{SE}_{\hat{\mu}}$. **Prediction intervals** for an individual future response y have a similar form with a larger standard error, $\hat{y} \pm t^* \text{SE}_{\hat{y}}$. Computer software often gives these intervals.

CHAPTER REVIEW

When a scatterplot shows a straight-line relationship between an explanatory variable x and a response variable y, we often fit a least-squares regression line to describe the relationship. Use this line to predict y from x. Statistical inference in the regression setting, however, requires more than just an overall linear pattern on a scatterplot.

The regression model says that there is a true straight-line relationship between x and the mean response μ_y. We can't observe this true regression line. The responses y that we do observe vary if we take several observations at the same x. The regression model says that for any fixed x, the responses have a normal distribution. Moreover, the standard deviation σ of this distribution is the same for all values of x.

The standard deviation σ describes how much variation there is in responses y when x is held fixed. Estimating σ is the key to inference about regression. Use the standard error s (roughly, the sample standard deviation of the residuals) to estimate σ. We can then do these types of inference:

- Give confidence intervals for the slope of the true regression line.
- Test the null hypothesis that this slope is zero. This hypothesis says that a straight-line relation between x and y is of no value for predicting y. It is the same as saying that the correlation between x and y in the entire population is zero.
- Give confidence intervals for the mean response for any fixed value of x.
- Give prediction intervals for an individual response y for a fixed value of x.

Here are the skills you should develop from studying this chapter.

A. PRELIMINARIES

1. Make a scatterplot to show the relationship between an explanatory and a response variable.
2. Use a calculator or software to find the correlation and the least-squares regression line.

B. RECOGNITION

1. Recognize the regression setting: a straight-line relationship between an explanatory variable x and a response variable y.
2. Recognize which type of inference you need in a particular regression setting.
3. Inspect the data to recognize situations in which inference isn't safe: a nonlinear relationship, influential observations, strongly skewed

residuals in a small sample, or nonconstant variation of the data points about the regression line.

C. DOING INFERENCE USING SOFTWARE AND TI-83 OUTPUT

1. Explain in any specific regression setting the meaning of the slope β of the true regression line.

2. Understand computer output for regression. Find in the output the slope and intercept of the least-squares line, their standard errors, and the standard error about the line.

3. Use that information to carry out tests and calculate confidence intervals for β.

4. Explain the distinction between a confidence interval for the mean response and a prediction interval for an individual response.

5. If software gives output for prediction, use that output to give either confidence or prediction intervals.

CHAPTER 14 REVIEW EXERCISES

14.15 Investors ask about the relationship between returns on investments in the United States and investments overseas. Here are data on the total returns on U.S. and overseas common stocks over a 22-year period. (The total return is change in price plus any dividends paid, converted into U.S. dollars. Both returns are averages over many individual stocks.)[8]

Year	Overseas % return	U.S. % return	Year	Overseas % return	U.S. % return
1971	29.6	14.6	1982	−1.9	21.5
1972	36.3	18.9	1983	23.7	22.4
1973	−14.9	−14.8	1984	7.4	6.1
1974	−23.2	−26.4	1985	56.2	31.6
1975	35.4	37.2	1986	69.4	18.6
1976	2.5	23.6	1987	24.6	5.1
1977	18.1	−7.4	1988	28.5	16.8
1978	32.6	6.4	1989	10.6	31.5
1979	4.8	18.2	1990	−23.0	−3.1
1980	22.6	32.3	1991	12.8	30.4
1981	−2.3	−5.0	1992	−12.1	7.6

(a) Make a scatterplot suitable for predicting overseas returns from U.S. returns.

(b) Here is part of the output from the Minitab regression command:

```
Predictor     Coef       Stdev      t-ratio      p
Constant      4.777      5.477      0.87         0.393
US return     0.8130     0.2628     3.09         0.006

s = 20.08     R-sq = 32.4%
```

Find the correlation and r^2. Describe the relationship between U.S. and overseas returns in words, using r and r^2 to make your description more precise. Does the scatterplot show clear outliers or strongly influential observations?

(c) Is there good evidence that there is a linear relationship between U.S. and foreign returns? State hypotheses and give the test statistic and its P-value.

(d) What is the predicted return on overseas stocks in a year when U.S. stocks return 20%?

(e) Here is the output for prediction of overseas returns when U.S. stocks return 20%:

```
Fit      Stdev.Fit          95% C.I.          95% P.I.
21.04       4.66      ( 11.32, 30.76) ( -21.97, 64.04)
```

Check the "Fit" against your result from (d). You think U.S. stocks will return 20% next year. Give a 95% interval for the return on foreign stocks next year if you are right about U.S. stocks. Is the regression prediction useful in practice?

14.16   The computer output in Exercise 14.10 gives 95% intervals for prediction. You want 99% confidence. Use the computer's results for the least-squares line and $SE_{\hat{\mu}}$ to obtain a 99% confidence interval for the mean gas consumption in months that average 40 degree-days per day.

14.17   Exercise 14.15 presents a regression of overseas stock returns on U.S. stock returns based on 22 years' data. The residuals for this regression (in order by years across the rows) are

12.953	16.157	−7.645	−6.514	.379	−21.464	19.339	22.620
−14.774	−8.437	−3.012	−24.157	.712	−2.337	25.732	49.501
15.676	10.064	−19.787	−25.257	−16.692	−23.056		

(a) Plot the residuals against x, the U.S. return. The plot suggests a mild violation of one of the regression assumptions. Which one? The violation is not strong enough to forbid regression inference.

(b) Display the distribution of the residuals in a graph. There is one possible outlier. Circle that point on the residual plot in (a). This point is not an extreme outlier, and redoing the regression without it does not greatly change the results. We are willing to do regression inference for these data.

14.18 The Franklin National Bank failed in 1974. Franklin was one of the 20 largest banks in the nation, and the largest ever to fail. Could regression have detected Franklin's weakened condition in advance? The table below gives the total assets (in billions of dollars) and net income (in millions of dollars) for the 20 largest banks in 1973, the year before Franklin failed.[9] Franklin is bank number 19.

Bank	1	2	3	4	5	6	7	8	9	10
Assets	49.0	42.3	36.3	16.4	14.9	14.2	13.5	13.4	13.2	11.8
Income	218.8	265.6	170.9	85.9	88.1	63.6	96.9	60.9	144.2	53.6

Bank	11	12	13	14	15	16	17	18	19	20
Assets	11.6	9.5	9.4	7.5	7.2	6.7	6.0	4.6	3.8	3.4
Income	42.9	32.4	68.3	48.6	32.2	42.7	28.9	40.7	13.8	22.2

(a) We expect banks with more assets to earn higher income. Make a scatterplot of these data with assets as the explanatory variable. Mark Franklin (Bank 19) with a separate symbol.

(b) The scatterplot suggests that regression calculations will be dominated by the three banks with the highest incomes. Circle these points on your scatterplot. Here is output from a Minitab regression of income on assets for the remaining 17 banks only:

```
Predictor      Coef      Stdev    t-ratio      p
Constant      -0.59      14.48     -0.04      0.968
Assets         5.840     1.363      4.28      0.001

s = 22.59      R-sq = 55.0%
```

(c) A report on this regression analysis says, "There is highly significant evidence that banks' incomes increase as their assets increase ($t = ?$, $df = ?$, $P = ?$)." State the null and alternative hypotheses for this test. Then give the numbers that replace the question marks in the report's summary.

(d) Predict Franklin's income from its assets, and find the residual for Franklin.

(e) Here is output for predicting income for Franklin's assets ($x^* = 3.8$):

```
    Fit   Stdev.Fit        95% C.I.          95% P.I.
  21.60        9.88  ( 0.55, 42.66) ( -30.96, 74.17)
```

Give a 95% interval for Franklin's income. Would this prediction have been helpful in showing that Franklin's actual income was lower than we would expect from its assets?

14.19   (a) Use the least-squares line from Example 14.7 to predict the second-round score for Golfer 1, who shot an 89 on the first round. Her actual second-round score was 94. Find the residual for this data point and check that your work agrees with the first residual given in Example 14.10.

(b) Calculate the sum of the squares of the 12 residuals given in Example 14.10. Divide by $n - 2 = 10$, then take the square root to obtain the standard error s about the line. Check that your work agrees with $s = 5.974$, given by the software in Figure 14.5.

14.20   Call the correlation between first-round and second-round tournament scores in the population of college female competitive golfers ρ. We want to test the null hypothesis of "no correlation":

$$H_0 : \rho = 0$$
$$H_a : \rho \neq 0$$

Regard the 12 golfers in Example 14.7 as an SRS from this population. Then the regression output in Figure 14.5 allows you to find both the sample correlation r that estimates ρ and the P-value for a test of these hypothesis.

(a) What is r? Explain clearly how you found r from the output.

(b) The test of $H_0 : \rho = 0$ is exactly the same as the test of $H_0 : \beta = 0$. What is the P-value of this test against the two-sided alternative?

(c) We expect the scores of the same golfers in two rounds of play to have a positive correlation. What is the P-value for the test of $H_0 : \rho = 0$ against the one-sided alternative $H_a : \rho > 0$?

14.21   (Optional) Figure 14.3 gives Minitab output for the regression of household natural gas consumption on degree-days. Example 14.5 uses the output to find a 95% confidence interval for the slope β of the true regression line. The intercept α of the true regression line is the average amount of gas the household consumes when there are zero degree-days. (There are zero degree-days when the average temperature is 65° or above.) Confidence intervals for α have the form

$$a \pm t^* \mathrm{SE}_a$$

Use the intercept a of the least-squares line and its standard error, given in the computer output, to find a 95% confidence interval for α. (The degrees of freedom are $n - 2$ once again.)

14.22 **(Optional)** Here is a guide to calculating confidence intervals for the mean response from basic regression output. You can find prediction intervals for an individual response in the same way.

Figure 14.5 gives regression output for predicting second-round score y from first-round score x for college female golfers. We want to predict the mean second-round score for golfers who score 90 in the first round. The box on page 771 shows that we must calculate the standard error

$$\mathrm{SE}_{\hat{\mu}} = s \sqrt{\frac{1}{n} + \frac{(x^* - \bar{x})^2}{\sum (x - \bar{x})^2}}$$

(a) Use your calculator to find the mean $\bar{x}$ and the standard deviation s_x of the first-round scores.

(b) What is x^*? Find $(x^* - \bar{x})^2$.

(c) From the recipe for the sample variance, you know that

$$\sum (x - \bar{x})^2 = (n - 1)s_x^2$$

Use this fact to find $\sum (x - \bar{x})^2$.

(d) Now use s from the output with your results in (b) and (c) to find $\mathrm{SE}_{\hat{\mu}}$.

(e) Finally, find the predicted second-round score $\hat{y}$ and the 95% confidence interval $\hat{y} \pm t^* \mathrm{SE}_{\hat{\mu}}$ for the mean response.

14.23 Table 14.2 contains data on the size of perch caught in a lake in Finland.[10] Statistical software will help you analyze these data.

(a) We want to know how well we can predict the width of a perch from its length. Make a scatterplot of width against length. There is a strong linear pattern, as expected. The heaviest perch had six newly eaten fish in its stomach and is an outlier in weight. Find this fish on your scatterplot and circle the point. Is this fish an outlier in your plot of width against length?

(b) Find the least-squares regression line to predict width from length.

| TABLE 14.2 | Measurements on 56 perch | | | | |

Length (centimeters)	Width (centimeters)	Weight (grams)	Length (centimeters)	Width (centimeters)	Weight (grams)
8.4	2.02	16.0	13.7	3.29	13.6
15.0	3.58	15.2	16.2	4.33	15.3
17.4	4.32	15.9	18.0	4.90	17.3
18.7	5.01	16.1	19.0	5.30	15.1
19.6	4.84	14.6	20.0	4.84	13.2
21.0	5.31	15.8	21.0	5.52	14.7
21.0	5.31	16.3	21.3	5.96	15.5
22.0	5.72	14.5	22.0	5.28	15.0
22.0	5.72	15.0	22.0	5.50	15.0
22.0	5.17	17.0	22.5	5.49	15.1
22.5	6.37	15.1	22.7	5.58	15.0
23.0	4.90	14.8	23.5	5.90	14.9
24.0	6.86	14.6	24.0	6.00	15.0
24.6	6.32	15.9	25.0	6.08	13.9
25.6	6.22	15.7	26.5	6.78	14.8
27.3	7.92	17.9	27.5	6.82	15.0
27.5	6.71	15.0	27.5	6.93	15.8
28.0	7.45	14.3	28.7	7.23	15.4
30.0	7.23	15.1	32.8	9.68	17.7
34.5	9.69	17.5	35.0	10.78	20.9
36.5	10.18	17.6	36.0	9.97	17.6
37.0	10.18	15.9	37.0	9.95	16.2
39.0	10.49	18.1	39.0	10.49	14.5
39.0	11.74	17.8	40.0	11.28	16.8
40.0	11.04	17.0	40.0	11.68	17.6
40.0	10.48	15.6	42.0	12.05	15.4
43.0	11.35	16.1	43.0	11.82	16.3
43.5	11.92	17.7	44.0	11.79	16.3

(c) The length of a typical perch is about $x^* = 25$ centimeters. Predict the mean width of such fish and give a 95% confidence interval.

(d) Examine the residuals. Is there any reason to mistrust inference?

14.24   We can also use the data in Table 14.2 to study the prediction of the weight of a perch from its length.

(a) Make a scatterplot of weight versus length, with length as the explanatory variable. Describe the pattern of the data and any clear outliers.

(b) It is more reasonable to expect the one-third power of the weight to have a straight-line relationship with length than to expect weight itself to have a straight-line relationship with length. Explain why this is true. (**Hint:** What happens to weight if length, width, and height all double?)

(c) Use your software to create a new variable that is the one-third power of weight. Make a scatterplot of this new response variable against length. Describe the pattern and any clear outliers.

(d) Is the straight-line pattern in (c) stronger or weaker than that in (a)? Compare the plots and also the values of r^2.

(e) Find the least-squares regression line to predict the new weight variable from length. Predict the mean of the new variable for perch 25 centimeters long, and give a 95% confidence interval.

(f) Examine the residuals from your regressions. Does it appear that any of the regression assumptions are not met?

NOTES AND DATA SOURCES

1. Data from Personal Watercraft Industry Association and U.S. Coast Guard.

2. Data provided by Robert Dale, Purdue University.

3. The data are from M. A. Houck et al., "Allometric scaling in the earliest fossil bird, *Archaeopteryx lithographica*," *Science*, 247 (1990), pp. 195–198.

4. Data from R. C. Nelson, C. M. Brooks, and N. L. Pike, "Biomechanical comparison of male and female distance runners," in P. Milvy (ed.), *The Marathon: Physiological, Medical, Epidemiological, and Psychological Studies*, New York Academy of Sciences, 1977, pp. 793–807.

5. Data from G. Geri and B. Palla, "Considerazioni sulle più recenti osservazioni ottiche alla Torre Pendente di Pisa," *Estratto dal Bollettino della Società Italiana di Topografia e Fotogrammetria*, 2 (1988), pp. 121–135. Professor Julia Mortera of the University of Rome provided a translation.

6. Strictly speaking, this quantity is the estimated standard deviation of $\hat{y} - y$, where y is the additional observation taken at $x = x^*$.

7. The data are for 1987 salaries and measures of past performance. They were collected and distributed by the Statistical Graphics Section of the American Statistical Association for an annual data analysis contest. The analysis here was done by Crystal Richard of Purdue University.

8. The U.S. returns are for the Standard & Poor's 500 stock index. The overseas returns are for the Morgan Stanley Europe, Australia, Far East (EAFE) index.

9. Data from D. E. Booth, *Regression Methods and Problem Banks*, COMAP, Inc., 1986.

10. The data in Table 14.2 are part of a larger data set in the *Journal of Statistics Education* archive, accessible via Internet. The original source is Pekka Brofeldt, "Bidrag till kaennedom on fiskbestondet I vaara sjoear. Laengelmaevesi," in T. H. Jaervi, *Finlands Fiskeriet,* Band 4, *Meddelanden utgivna av fiskerifoereningen i Finland,* Helsinki, 1917. The data were contributed to the archive (with information in English) by Juha Puranen of the University of Helsinki.

Post-Exam Topic

■ 15. Analysis of Variance

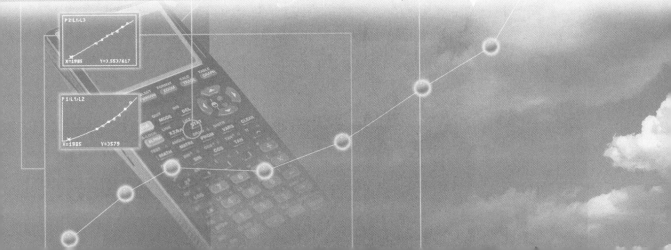

Analysis of Variance

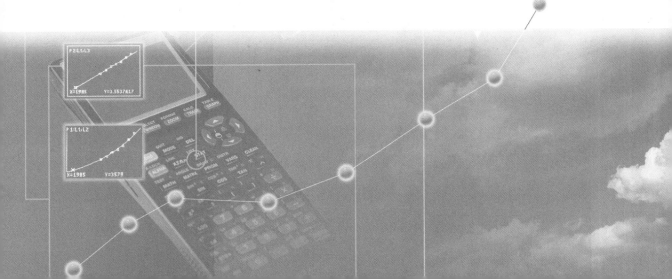

INTRODUCTION

In this chapter you are introduced to a very useful procedure that extends the idea of comparing two means. Curiously, it is called *analysis of variance*, for reasons to be explained in Section 15.2, and it is widely known by its acronym, ANOVA. In the first part of the chapter we shall briefly consider inference procedures for comparing two standard deviations, partly to make the point that inference can be performed on population parameters other than means and proportions. Although the objectives of Section 15.1 (Inference for Population Spread) and Section 15.2 (One-Way Analysis of Variance) are quite different, and the assumptions and characteristics of the procedures are also quite different, they both involve the same distribution, the *F distribution*.

ACTIVITY 15 | The Return of Paper Airplanes

Materials: Paper airplane pattern sheet (in Teacher Resource Binder), scissors, masking tape, 25-meter steel tape measure, TI-83 graphing calculator

In Activity 11, you determined which of two paper airplane models flew farther. In this activity, you will test a new (third) paper airplane model and record distances flown. Later in the chapter, after sufficient theory has been presented, you will conduct a one-way analysis of variance procedure using the data from all three paper airplane models.

Task 1. Extend your experiment from Activity 11 for producing data on the flight distances of the two original paper airplane models to include the third model. Then carry out your plan and gather the necessary data. (These data will be added to the data collected in Activity 11.)

Task 2. In preparation for performing inference on the mean distances of the three models, calculate the descriptive statistics for each of the three sets of distances, and assess the normality of the data collected. Compare the three mean distances graphically. Does it appear that all of the means are about the same, or is at least one mean different from the other means?

Note: Keep these data at hand for analysis later in the chapter.

15.1 INFERENCE FOR POPULATION SPREAD

The two most basic descriptive features of a distribution are its center and spread. In a normal population, we measure center and spread by the mean and the standard deviation. We use the t procedures for inference about population means for normal populations, and we know that t procedures are often useful for nonnormal populations as well. It is natural to turn next to inference about the standard deviations of normal populations. Our advice here is short and clear: Don't do it without expert advice.

Avoid inference about standard deviations

There are methods for inference about the standard deviations of normal populations. We will describe the most common such method, the F test for comparing the spread of two normal populations. Unlike the t procedures for means, the F test and other procedures for standard deviations are extremely sensitive to nonnormal distributions. This lack of robustness does not improve in large samples. It is difficult in practice to tell whether a significant F-value is evidence of unequal population spreads or simply a sign that the populations are not normal.

The deeper difficulty underlying the very poor robustness of normal population procedures for inference about spread already appeared in our work on describing data. The standard deviation is a natural measure of spread for normal distributions but not for distributions in general. In fact, because skewed distributions have unequally spread tails, no single numerical measure does a good job of describing the spread of a skewed distribution. In summary, the standard deviation is not always a useful parameter, and even when it is (for symmetric distributions), the results of inference are not trustworthy. Consequently, we do not recommend trying to do inference about population standard deviations in basic statistical practice.[1]

It was once common to test equality of standard deviations as a preliminary to performing the pooled two-sample t test for equality of two population means. It is better practice to check the distributions graphically, with special attention to skewness and outliers, and to use the version of the two-sample t featured in Section 11.2. This test does not require equal standard deviations.

The F test for comparing two standard deviations

Because of the limited usefulness of procedures for inference about the standard deviations of normal distributions, we will present only one such procedure. Suppose that we have independent SRSs from two normal populations, a sample of size n_1 from $N(\mu_1, \sigma_1)$ and a sample of size n_2

from $N(\mu_2, \sigma_2)$. The population means and standard deviations are all unknown. The two-sample t test examines whether the means are equal in this setting. To test the hypothesis of equal spread,

$$H_0: \sigma_1 = \sigma_2$$
$$H_a: \sigma_1 \neq \sigma_2$$

we use the ratio of sample variances. This is the F *statistic*.

THE F STATISTIC AND F DISTRIBUTIONS

When s_1^2 and s_2^2 are sample variances from independent SRSs of sizes n_1 and n_2 drawn from normal populations, the F **statistic**

$$F = \frac{s_1^2}{s_2^2}$$

has the F **distribution** with $n_1 - 1$ and $n_2 - 1$ degrees of freedom when $H_0: \sigma_1 = \sigma_2$ is true.

The F distributions are a family of distributions with two parameters. The parameters are the degrees of freedom of the sample variances in the numerator and denominator of the F statistic. The numerator degrees of freedom are always mentioned first. Interchanging the degrees of freedom changes the distribution, so the order is important. Our brief notation will be $F(j, k)$ for the F distribution with j degrees of freedom in the numerator and k in the denominator. The F distributions are not symmetric but are right-skewed. The density curve in Figure 15.1 illustrates the shape. Because sample variances cannot be negative, the F statistic takes only positive values, and the F distribution has no probability below 0. The peak of the F density curve is near 1. When the two populations have the same standard deviation, we expect the two sample variances to be close in size, so that F takes a value near 1. Values of F far from 1 in either direction provide evidence against the hypothesis of equal standard deviations.

Tables of F critical points are awkward, because we need a separate table for every pair of degrees of freedom j and k. Table D in the back of the book gives upper p critical points of the F distributions for $p = .10$, .05, .025, .01, and .001. For example, these critical points for the $F(9, 10)$ distribution shown in Figure 15.1 are

p	.10	.05	.025	.01	.001
F^*	2.35	3.02	3.78	4.94	8.96

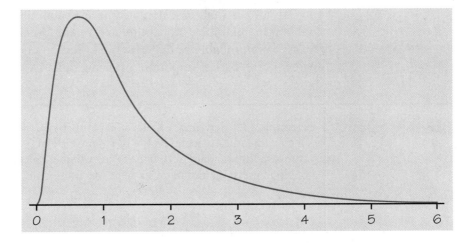

FIGURE 15.1 The density curve for the $F(9, 10)$ distribution. The F distributions are skewed to the right.

The skewness of the F distributions causes additional complications. In the symmetric normal and t distributions, the point with probability 0.05 below it is just the negative of the point with probability 0.05 above it. This is not true for F distributions. We therefore need either tables of both the upper and lower tails or some way to eliminate the need for lower tail critical values. The TI-83 graphing calculator and statistical software both do away with the need for tables. If you do not use the TI-83 or statistical software, arrange the two-sided F test as follows:

CARRYING OUT THE F TEST

Step 1 Take the test statistic to be

$$F = \frac{\text{larger } s^2}{\text{smaller } s^2}$$

This amounts to naming the populations so that Population 1 has the larger of the observed sample variances. The resulting F is always 1 or greater.

Step 2 Compare the value of F with critical values from Table D. Then *double* the significance levels from the table to obtain the significance level for the two-sided F test.

The idea is that we calculate the probability in the upper tail and double it to obtain the probability of all ratios on either side of 1 that are at least as improbable as that observed. Remember that the order of the degrees of freedom is important in using Table D.

EXAMPLE 15.1

Example 11.7 describes a medical experiment to compare the mean effects of calcium and a placebo on the blood pressure of black men. We might also compare the standard deviations to see whether calcium changes the spread of blood pressures among black men. We want to test

$$H_0: \sigma_1 = \sigma_2$$
$$H_a: \sigma_1 \neq \sigma_2$$

The larger of the two sample standard deviations is $s = 8.743$ from 10 observations. The other is $s = 5.901$ from 11 observations. The F test statistic is therefore

$$F = \frac{\text{larger } s^2}{\text{smaller } s^2} = \frac{8.743^2}{5.901^2} = 2.195$$

Compare the calculated value $F = 2.20$ with critical points for the $F(9, 10)$ distribution. Table D shows that 2.20 is less than the 0.10 critical value of the $F(9, 10)$ distribution, which is $F^* = 2.35$. Doubling 0.10, we know that the P-value for the two-sided test is greater than 0.20. The results are not significant at the 20% level (or any lower level). Statistical software shows that the exact upper tail probability is 0.118, and hence $P = .24$. If the populations were normal, the observed standard deviations would give little reason to suspect unequal population standard deviations. Because one of the populations shows some nonnormality, we cannot be fully confident of this conclusion.

Although the computations for a two-sample F test and the table lookup procedure are not complicated, performing an F test on the TI-83 has the advantage of not requiring a table of areas under the F distribution. The following example shows the steps.

EXAMPLE 15.2

Using the data on the effects of calcium on the blood pressure of black men from the previous example, first select STAT / TESTS / D:2-SampFTest.

```
EDIT CALC TESTS
0↑2-SampTInt...
A:1-PropZInt...
B:2-PropZInt...
C:χ²-Test...
D:2-SampFTest...
E:LinRegTTest...
F:ANOVA(
```

When the 2-SampFTest screen appears, select "Stats" and enter the following: $s_{x1} = 8.743, n_1 = 10, s_{x2} = 5.901, n_2 = 11$. Select the two-sided alternative hypothesis, $\sigma_1 \neq \sigma_2$.

```
2-SampFTest
 Inpt:Data Stats
 Sx1:8.743
 n1:10
 Sx2:5.901
 n2:11
 σ1:≠σ2   <σ2   >σ2
 Calculate Draw
```

If you select "Calculate," you will see the following results: $F = 2.195$ and $P = .237$. Compare these results with the results in Example 15.1.

```
2-SampFTest
 σ 1≠σ 2
 F=2.195177929
 p=.2365479014
 Sx1=8.743
 Sx2=5.901
↓n1=10
```

If you select "Draw," you will see the $F(9, 10)$ curve with the two tail areas shaded.

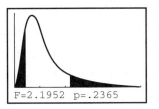

```
F=2.1952  p=.2365
```

SUMMARY

Inference procedures for comparing the **standard deviations** of two normal populations are based on the **F statistic**, which is the ratio of sample variances

$$F = \frac{s_1^2}{s_2^2}$$

If an SRS of size n_1 is drawn from Population 1 and an independent SRS of size n_2 is drawn from Population 2, the F statistic has the **F distribution**

$F(n_1 - 1, n_2 - 1)$ if the two population standard deviations σ_1 and σ_2 are in fact equal.

The **F distributions** are skewed to the right and take only values greater than 0. A specific F distribution $F(j, k)$ is fixed by the two **degrees of freedom** j and k.

The two-sided **F test** of H_0: $\sigma_1 = \sigma_2$ uses the statistic

$$F = \frac{\text{larger } s^2}{\text{smaller } s^2}$$

and doubles the upper tail probability to obtain the P-value.

The F tests and other procedures for inference on the spread of one or more normal distributions are so strongly affected by lack of normality that we do not recommend them for regular use.

SECTION 15.1 EXERCISES

15.1   This exercise investigates the effect of increasing both sample sizes on the shape of the F distribution curve.

(a) Using the TI-83, define $Y_1 = \text{Fpdf}(X,2,25)$. Note that Fpdf is found under 2nd / DISTR / 4:ShadeF(. Set the Window as follows: $X[0,5]_1$ and $Y[-.2,1.2]_1$. Briefly describe the shape of the curve.

(b) Define $Y_2 = \text{Fpdf}(X,5,50)$. How does this curve differ from Y_1?

(c) Define $Y_3 = \text{Fpdf}(X,12,100)$. How does this curve differ from Y_1 and Y_2?

(d) What values for df1 and df2 will yield a plot of $Y = \text{Fpdf}(X,\text{df1},\text{df2})$ that is approximately the shape of a normal distribution?

(e) Describe what happens to the shape of the F distribution curve as the sample sizes increase.

In all exercises calling for use of the F test, assume that both population distributions are very close to normal. The actual data are not always sufficiently normal to justify use of the F test.

15.2   The F statistic $F = s_1^2/s_2^2$ is calculated from samples of size $n_1 = 10$ and $n_2 = 8$. (Remember that n_1 is the numerator sample size.)

(a) What is the upper 5% critical value for this F?

(b) In a test of equality of standard deviations against the two-sided alternative, this statistic has the value $F = 3.45$. Is this value significant at the 10% level? Is it significant at the 5% level?

15.3 The F statistic for equality of standard deviations based on samples of sizes $n_1 = 21$ and $n_2 = 16$ takes the value $F = 2.78$.

(a) Is this significant evidence of unequal population standard deviations at the 5% level? At the 1% level?

(b) Between which two values obtained from Table D does the P-value of the test fall?

15.4 The sample variance for the treatment group in the DDT experiment of Example 11.12 (page 635) is more than 10 times as large as the sample variance for the control group. Calculate the F statistic. Can you reject the hypothesis of equal population standard deviations at the 5% significance level? At the 1% level?

15.5 Exercise 11.43 (page 637) records the results of comparing a measure of rowing style for skilled and novice female competitive rowers. Is there significant evidence of inequality between the standard deviations of the two populations?

(a) State H_0 and H_a.

(b) Calculate the F statistic. Between which two levels does the P-value lie?

15.6 Answer the same questions for the weights of the two groups, recorded in Exercise 11.44 (page 638).

15.7 The data for VOTs of children and adults in Exercise 11.51 (page 643) show quite different sample standard deviations. How statistically significant is the observed inequality?

15.8 Return to the SSHA data in Exercise 11.39 (page 632). We want to know if the spread of SSHA scores is different among women and among men at this college. Use the F test to obtain a conclusion.

15.2 ONE-WAY ANALYSIS OF VARIANCE

We use the two-sample t procedures of Chapter 11 to compare the means of two populations or the mean responses to two treatments in an experiment. Of course, studies don't always compare just two groups. We need a method for comparing any number of means.

EXAMPLE 15.3

Do smaller cars really have better gas mileage? Table 15.1 contains data on the city and highway gas mileage (in miles per gallon, as reported by the Environmental Protection Agency) for a sample of 59 compact, midsize, and large 1994 car models.[2] We want to compare the city gas mileages.

TABLE 15.1 City and highway gas mileage for 1994 car models

Model	Size	City MPG	Highway MPG
Acura Legend	Compact	19	24
Acura Vigor	Compact	20	26
Audi 100	Compact	18	24
Audi 90	Compact	18	26
BMW 525i	Compact	18	25
Buick Skylark	Compact	22	32
Chevrolet Beretta	Compact	25	31
Chrysler Lebaron	Compact	22	27
Dodge Shadow	Compact	23	30
Ford Tempo	Compact	22	27
Honda Accord	Compact	23	29
Jaguar XJ12	Compact	12	16
Mazda Protege	Compact	23	29
Mazda 323	Compact	26	33
Mercedes-Benz E320	Compact	19	25
Mercedes-Benz E420	Compact	18	24
Mitsubishi Diamante	Compact	18	24
Mitsubishi Galant	Compact	20	26
Mitsubishi Precis	Compact	27	35
Nissan Altima	Compact	21	29
Oldsmobile Achieva	Compact	22	32
Saturn SL	Compact	26	35
Subaru Legacy	Compact	22	29
Toyota Corolla	Compact	26	29
Volkswagen Golf	Compact	21	28
Volkswagen Jetta	Compact	21	27

| TABLE 15.1 | City and highway gas mileage for 1994 car models *(continued)* |

Model	Size	City MPG	Highway MPG
BMW 740i	Midsize	16	23
Buick Century	Midsize	25	31
Buick Regal	Midsize	19	29
Cadillac Eldorado	Midsize	16	25
Chevrolet Lumina	Midsize	19	29
Dodge Spirit	Midsize	22	27
Ford Taurus	Midsize	20	29
Ford Thunderbird	Midsize	19	26
Hyundai Sonata	Midsize	21	27
Infiniti Q45	Midsize	17	22
Lexus GS300	Midsize	17	23
Lexus LS400	Midsize	18	23
Lincoln-Mercury Mark VIII	Midsize	18	25
Mazda 626	Midsize	23	31
Mazda 929	Midsize	19	24
Nissan Maxima	Midsize	19	26
Rolls-Royce Silver Spur	Midsize	10	15
Saab 900	Midsize	19	26
Toyota Camry	Midsize	21	28
Volvo 850	Midsize	19	26
Buick LeSabre	Large	19	28
Buick Park Avenue	Large	19	27
Buick Roadmaster	Large	17	25
Cadillac DeVille	Large	16	25
Chevrolet Caprice	Large	18	26
Chrysler Concorde	Large	20	28
Chrysler New Yorker	Large	18	26
Ford LTD	Large	18	25
Lincoln-Mercury Continental	Large	18	26
Mercedes-Benz S320	Large	17	24
Mercedes-Benz S420	Large	15	20
Mercedes-Benz S500	Large	14	19
Saab 9000	Large	17	27

Compact		Midsize		Large	
10		10	0 ← Rolls-Royce	10	
11	Jaguar	11	Silver Spur	11	
12	0 XJ12	12		12	
13		13		13	
14		14		14	0
15		15		15	0
16		16	00	16	0
17		17	00	17	000
18	00000	18	00	18	0000
19	00	19	0000000	19	00
20	00	20	0	20	0
21	000	21	00	21	
22	00000	22	0	22	
23	000	23	0	23	
24		24		24	
25	0	25	0	25	
26	000	26		26	
27	0	27		27	

FIGURE 15.2   Side-by-side stemplots comparing the city gas mileages of compact, midsize, and large cars from Table 15.1.

Figure 15.2 shows side-by-side stemplots of the city mileages for the three types of cars. We used the same stems in all three for easier comparison. It does appear that gas mileage decreases as cars get larger. The plots show a low outlier in the compact group and another in the midsize group. The Jaguar XJ12 and the Rolls-Royce Silver Spur are responsible. These are rather exotic cars, so we will omit these outliers from all further analyses.

Here are the means, standard deviations, and five-number summaries for the three car types, from the Minitab statistical software:

```
            N    MEAN    MEDIAN  STDEV   MIN     MAX      Q1      Q3
Compact     25  21.600  22.000  2.814  18.000  27.000  19.000  23.000
Midsize     19  19.316  19.000  2.311  16.000  25.000  18.000  21.000
Large       13  17.385  18.000  1.660  14.000  20.000  16.500  18.500
```

We will use the mean to describe the center of the gas mileage distributions. As we expect, mean gas mileage goes down as we move from compact to midsize and then to large cars. The differences among the means are not large. Are they statistically significant?

The problem of multiple comparisons

Call the mean city gas mileage for the three populations of cars μ_1 for compact cars, μ_2 for midsize cars, and μ_3 for large cars. The subscript reminds us which group a parameter or statistic describes. To compare these three population means, we might use the two-sample t test several times:

- Test H_0: $\mu_1 = \mu_2$ to see if the mean miles per gallon for compact cars differs from the mean for midsize cars.

- Test H_0: $\mu_1 = \mu_3$ to see if compact cars differ from large cars.

- Test H_0: $\mu_2 = \mu_3$ to see if midsize cars differ from large cars.

The weakness of doing three tests is that we get three P-values, one for each test alone. That doesn't tell us how likely it is that *three* sample means are spread apart as far as these are. It may be that 17.385 and 21.600 are significantly different if we look at just two groups, but not significantly different if we know that they are the smallest and largest means in three groups. As we look at more groups, we expect the gap between the smallest and largest sample mean to get larger. (Think of comparing the tallest and shortest person in larger and larger groups of people.) We can't safely compare many parameters by doing tests or confidence intervals for two parameters at a time.

The problem of how to do many comparisons at once with some overall measure of confidence in all our conclusions is common in more advanced statistics. We are leaving the elementary parts of statistical inference now that we have met this problem. Statistical methods for dealing with many comparisons usually have two parts:

1. An *overall test* to see if there is good evidence of *any* differences among the parameters that we want to compare.

2. A detailed *follow-up analysis* to decide which of the parameters differ and to estimate how large the differences are.

The overall test is often reasonably straightforward, though still more complex than the tests we met in Chapters 11 and 12. The follow-up analysis can be quite elaborate. In our basic introduction to statistical practice, we will look only at some overall tests. Chapter 13 describes an overall test to compare several population proportions. In this chapter we present a test for comparing several population means.

The analysis of variance *F* test

We want to test the null hypothesis that there are *no differences* among the mean city gas mileages for the three car types:

$$H_0: \mu_1 = \mu_2 = \mu_3$$

The alternative hypothesis is that there *is* some difference, that not all three population means are equal:

$$H_a: \text{not all of } \mu_1, \mu_2, \text{ and } \mu_3 \text{ are equal}$$

*analysis of
variance F test*

The alternative hypothesis is no longer one-sided or two-sided. It is "many-sided," because it allows any relationship other than "all three equal." For example, H_a includes the case in which $\mu_2 = \mu_3$ but μ_1 has a different value. The test of H_0 against H_a is called the **analysis of variance F test**. Analysis of variance is usually abbreviated as ANOVA, (Don't confuse the ANOVA F, which compares several means, with the F statistic of Section 15.1, which compares two standard deviations and is *not* robust against nonnormality.)

EXAMPLE 15.4

Enter the city gas mileage data from Table 15.1 (omitting the two outliers) into the Minitab software package and request analysis of variance. The output appears in Figure 15.3. Most statistical software packages produce an ANOVA output similar to this one.

For now, look at just two parts of this output. First, check that the sample sizes, means, and standard deviations agree with those in Example 15.3. Then find the F test statistic, $F = 13.62$, and its P-value. The P-value is given as 0.000. This means that P is zero to three decimal places, or $P < .001$. There is extremely strong evidence that the three car sizes do not all have the same mean gas mileage.

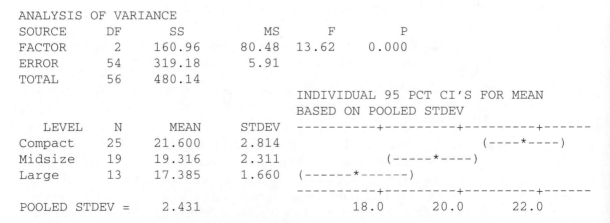

```
ANALYSIS OF VARIANCE
SOURCE       DF      SS         MS       F        P
FACTOR        2    160.96     80.48   13.62    0.000
ERROR        54    319.18      5.91
TOTAL        56    480.14
                                    INDIVIDUAL 95 PCT CI'S FOR MEAN
                                    BASED ON POOLED STDEV
     LEVEL    N     MEAN      STDEV  ----------+---------+---------+------
Compact      25   21.600      2.814                            (----*----)
Midsize      19   19.316      2.311                   (-----*----)
Large        13   17.385      1.660   (------*------)
                                    ----------+---------+---------+------
POOLED STDEV =     2.431                   18.0      20.0      22.0
```

FIGURE 15.3 Minitab output for analysis of variance of the city gas mileage data (Example 15.4).

> The F test does not say *which* of the three means are significantly different. It appears from our preliminary analysis of the data that midsize and large cars differ by less than do compact and midsize cars. The computer output includes confidence intervals for all three means that suggest the same conclusion. The midsize- and large-car intervals overlap, and the compact-car interval lies a bit above them on the mileage scale. These are 95% confidence intervals for each mean separately. We are not 95% confident that *all three* intervals cover the three means. There are follow-up procedures that provide 95% confidence that we have caught all three means at once, but we won't study them.
>
> Our conclusion: There is strong evidence ($P < .001$) that the means are not all equal, and the most important difference among the means is that compact cars have better gas mileage than midsize and large cars.

Example 15.4 illustrates our approach to comparing means. The ANOVA F test (often done by software) assesses the evidence for *some* difference among the population means. In most cases, we expect the F test to be significant. We would not undertake a study if we did not expect to find some effect. The formal test is nonetheless important to guard against being misled by chance variation. We will not do the formal follow-up analysis that is often the most useful part of an ANOVA study. Follow-up analysis would allow us to say which means differ and by how much, with (say) 95% confidence that all our conclusions are correct. We rely instead on a preliminary examination of the data to show what differences are present and whether they are large enough to be interesting. The gap of more than 4 miles per gallon between compact cars and large cars *is* large enough to be of practical interest.

EXERCISES

15.9   Table 15.2 gives the results of a study of fish caught in a lake in Finland.[3] We are willing to regard the fish caught by the researchers a random sample of fish in this lake. The weight of commercial fish is of particular interest. Is there evidence that the mean weights of all bream, perch, and roach found in the lake are different? Before doing ANOVA, we must examine the data.

    (a)   Display the distribution of weights for each species of fish with side-by-side stemplots. Are there outliers or strong skewness in any of the distributions?

    (b)   Find the five-number summary for each distribution. What do the data appear to show about the weights of these species?

15.10   Now we proceed to the ANOVA for the data in Table 15.2. The heaviest perch had six roach in its stomach when caught. This fish may be an out-

TABLE 15.2	Weights (grams) of three fish species

Bream			Perch				Roach	
13.4	13.9	14.1	16.0	14.5	15.7	17.6	14.0	13.6
13.8	15.0	14.9	13.6	15.0	14.8	17.6	13.9	15.4
15.1	13.8	15.5	15.2	15.0	17.9	15.9	13.7	14.0
13.3	13.5	14.3	15.3	15.0	14.6	16.2	14.3	15.4
15.1	13.3	14.3	15.9	17.0	15.0	18.1	16.1	15.6
14.2	13.7	14.9	17.3	16.3	15.0	14.5	14.7	15.3
15.3	14.8	14.7	16.1	15.1	15.8	17.8	14.7	
13.4	14.1		15.1	15.1	14.3	16.8	13.9	
13.8	13.7		14.6	15.0	15.4	17.0	15.2	
13.7	13.3		13.2	14.8	15.1	17.6	14.6	
14.1	15.1		15.8	14.9	17.7	15.6	15.1	
13.3	13.8		14.7	15.0	17.7	15.4	13.3	
12.0	14.8		16.3	15.9	17.5	16.1	15.2	
13.6	15.0		15.5	13.9	20.9	16.3	14.1	

lier and its condition is unusual, so remove this observation from the data.
Figure 15.4 gives the Minitab ANOVA output for the data with the outlier
removed.

(a) What null hypothesis does the ANOVA F statistic test? State this hy-
    pothesis both in words and in symbols.

```
ANALYSIS OF VARIANCE
SOURCE        DF           SS        MS         F         P
FACTOR         2        60.17     30.08     29.92     0.000
ERROR        107       107.60      1.01
TOTAL        109       167.77
                                    INDIVIDUAL 95 PCT CI'S FOR MEAN
                                    BASED ON POOLED STDEV
     LEVEL      N        MEAN     STDEV    ---+---------+---------+---------+---
bream          35      14.131     0.770    (----*----)
perch          55      15.747     1.186                              (---*---)
roach          20      14.605     0.780          (------*-----)
                                             ---+---------+---------+---------+---
POOLED STDEV =          1.003             14.00     14.70     15.40     16.10
```

FIGURE 15.4   Minitab output for the data in Table 15.2 of weights of three
species of fish (Exercise 15.10).

(b)   What is the value of the F statistic? What is its P-value?

(c)   Based on your work in this and the previous exercise, what do you conclude about the weights of these species of fish?

15.11   How much corn per acre should a farmer plant to obtain the highest yield? Too few plants will give a low yield. On the other hand, if there are too many plants, they will compete with each other for moisture and nutrients, and yields will fall. To find out, plant at different rates on several plots of ground and measure the harvest. (Be sure to treat all the plots the same except for the planting rate.) Here are data from such an experiment:[4]

Plants per acre	Yield (bushels per acre)			
12,000	150.1	113.0	118.4	142.6
16,000	166.9	120.7	135.2	149.8
20,000	165.3	130.1	139.6	149.9
24,000	134.7	138.4	156.1	
28,000	119.0	150.5		

(a)   Make side-by-side stemplots of yield for each number of plants per acre. Find the mean yield for each planting rate. What do the data appear to show about the influence of plants per acre on yield?

(b)   ANOVA will assess the statistical significance of the observed differences in yield. What are H_0 and H_a for the ANOVA F test in this situation?

(c)   The Minitab ANOVA output for these data appears in Figure 15.5. What does the ANOVA F test say about the significance of the effects you described in (a)?

(d)   The observed differences among the mean yields in the sample were quite large. Why are they not statistically significant?

The idea of analysis of variance

Here is the main idea for comparing means: what matters is not how far apart the sample means are but how far apart they are *relative to the variability of individual observations*. Look at the two sets of boxplots in Figure 15.6. For simplicity, these distributions are all symmetric, so that the mean and median are the same. The centerline in each boxplot is therefore the sample mean. Both figures compare three samples with the same three means. Like the three car types in Example 15.3, the means are dif-

```
ANALYSIS OF VARIANCE ON yield
SOURCE      DF       SS       MS     F     P
Rate         4      600      150   0.50  0.736
ERROR       12     3597      300
TOTAL       16     4197
                                   INDIVIDUAL 95 PCT CI'S FOR MEAN
                                   BASED ON POOLED STDEV
LEVEL      N      MEAN    STDEV   ---+---------+---------+---------+---
  12       4    131.03   18.09      (-----------*-----------)
  16       4    143.15   19.79          (----------*-----------)
  20       4    146.23   15.07           (---------*-----------)
  24       3    143.07   11.44       (-----------*------------)
  28       2    134.75   22.27    (--------------*----------------)
                                   ---+---------+---------+---------+---
POOLED STDEV =  17.31             112       128       144       160
```

FIGURE 15.5   Minitab output for yields of corn at five planning rates (Exercise 15.11).

ferent but not very different. Could differences this large easily arise just due to chance, or are they statistically significant?

- The boxplots in Figure 15.6(a) have tall boxes, which show lots of variation among the individuals in each group. With this much variation among individuals, we would not be surprised if another set of samples gave quite different sample means. The observed differences among the sample means could easily happen just by chance.

- The boxplots in Figure 15.6(b) have the same centers as those in Figure 15.6(a), but the boxes are much shorter. That is, there is much less variation among the individuals in each group. It is unlikely that any sample from the first group would have a mean as small as the mean of the second group. Because means as far apart as those observed would rarely arise just by chance in repeated sampling, they are good evidence of real differences among the means of the three populations we are sampling from.

This comparison of the two parts of Figure 15.6 is too simple in one way. It ignores the effect of the sample sizes, an effect that boxplots do not show. Small differences among sample means can be significant if the samples are very large. Large differences among sample means may fail to be significant if the samples are very small. All we can be sure of is that

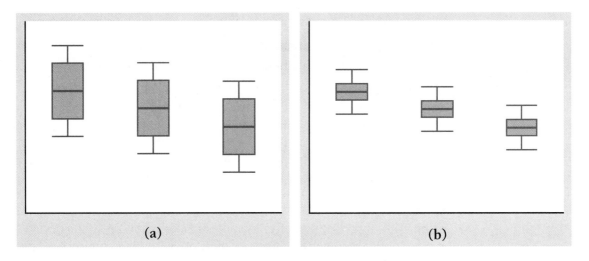

FIGURE 15.6   Boxplots for two sets of three samples each. The sample means are the same in (a) and (b). Analysis of variance will find a more significant difference among the means in (b) because there is less variation among the individuals within those samples.

for the same sample size, Figure 15.6(b) will give a much smaller *P*-value than Figure 15.6(a). Despite this qualification, the big idea remains: if sample means are far apart relative to the variation among individuals in the same group, that's evidence that something other than chance is at work.

THE ANALYSIS OF VARIANCE IDEA

Analysis of variance compares the variation due to specific sources with the variation among individuals who should be similar. In particular, ANOVA tests whether several populations have the same mean by comparing how far apart the sample means are with how much variation there is within the samples.

one-way ANOVA

It is one of the oddities of statistical language that methods for comparing means are named after the variance. The reason is that the test works by comparing two kinds of variation. Analysis of variance is a general method for studying sources of variation in responses. Comparing several means is the simplest form of ANOVA, called ***one-way*** *ANOVA*. One-way ANOVA is the only form of ANOVA that we will study.

The ANOVA F statistic for comparing several means has this form:

$$F = \frac{\text{variation among the sample means}}{\text{variation among individuals in the same sample}}$$

We give more detail later. Because ANOVA is in practice done by software, the idea is more important than the detail. The F statistic can only take values that are zero or positive. It is zero only when all the sample means are identical. Chance variation creates some differences among the sample means even when the population means are equal. In fact, when the null hypothesis is true, we expect F to take values near 1. As the sample means get farther apart, the value of F gets larger. Large values of F are evidence against the null hypothesis H_0 that all population means are the same. Although the alternative hypothesis H_a is many-sided, the ANOVA F test is one-sided because any violation of H_0 tends to produce a large value of F.

How large must F be to provide significant evidence against H_0? To answer questions of statistical significance, compare the F statistic with critical values from an **F** *distribution*. Recall from Section 15.1 that a specific F distribution is specified by two parameters: a numerator degrees of freedom and a denominator degrees of freedom. Table D in the back of the book contains critical values for F distributions with various degrees of freedom.

F distribution

EXAMPLE 15.5

Look again at the computer output for the city gas mileage data in Figure 15.3. The degrees of freedom for the F test appear in the first two rows of the "DF" column. There are 2 degrees of freedom in the numerator and 54 in the denominator.

In Table D, find the numerator degrees of freedom 2 at the top of the table. Then look for the denominator degrees of freedom 54 at the left of the table. There is no entry for 54, so we use the next smaller entry, 50 degrees of freedom. The critical values for 2 and 50 degrees of freedom are

df = 2, 50	
p	Critical value
.100	2.41
.050	3.18
.025	3.97
.010	5.06
.001	7.96

The observed $F = 13.62$ is larger than the upper 0.001 critical value, so $P <$ .001. Figure 15.7 shows the F density curve with 2 and 50 degrees of freedom. The observed $F = 13.62$ lies far to the right on this curve.

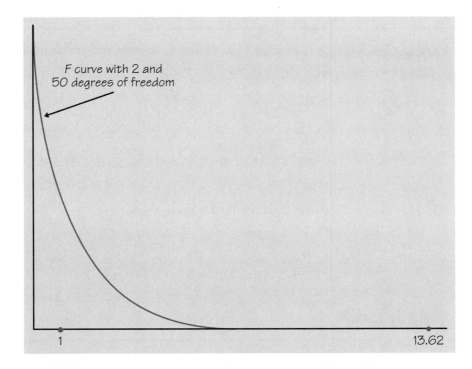

FIGURE 15.7 The density curve of the F distribution with 2 degrees of freedom in the numerator and 50 degrees of freedom in the denominator, with the observed value $F = 13.62$ from Example 15.5 marked. This observed F is highly significant.

The degrees of freedom of the F statistic depend on the number of means we are comparing and the number of observations in each sample. That is, the F test does take into account the number of observations. Here are the details.

DEGREES OF FREEDOM FOR THE F TEST

We want to compare the means of I populations. We have an SRS of size n_i from the ith population, so that the total number of observations in all samples combined is

$$N = n_1 + n_2 + \cdots + n_I$$

If the null hypothesis that all population means are equal is true, the ANOVA F statistic has the F distribution with $I - 1$ degrees of freedom in the numerator and $N - I$ degrees of freedom in the denominator.

| EXAMPLE 15.6 | In Examples 15.3 and 15.4, we compared the mean city gas mileage for three sizes of car, so $I = 3$. The three sample sizes are |

$$n_1 = 25 \qquad n_2 = 19 \qquad n_3 = 13$$

The total number of observations is therefore

$$N = 25 + 19 + 13 = 57$$

The ANOVA F test has numerator degrees of freedom

$$I - 1 = 3 - 1 = 2$$

and denominator degrees of freedom

$$N - I = 57 - 3 = 54$$

These degrees of freedom are given in the computer output as the first two entries in the "DF" column in Figure 15.3.

EXERCISES

15.12 Exercises 15.9 and 15.10 compare the weights of three species of fish.

(a) What are I, the n_i, and N for these data? (Remember that the heaviest perch was removed before doing ANOVA.) Identify these quantities in words, and give their numeric values.

(b) Find the degrees of freedom for the ANOVA F statistic. Check your work against the computer output in Figure 15.4.

(c) For these data, $F = 29.92$. What does Table D tell you about the P-value of this statistic?

15.13 Exercise 15.11 compares the yields for several planting rates for corn.

(a) What are I, the n_i, and N for these data? Identify these quantities in words, and give their numeric values.

(b) Find the degrees of freedom for the ANOVA F statistic. Check your work against the computer output in Figure 15.5.

(c) For these data, $F = .50$. What does Table D tell you about the P-value of this statistic?

15.14 In each of the following situations, we want to compare the mean response in several populations. For each setting, identify the populations and the response variable. Then give I, the n_i, and N. Finally, give the degrees of freedom of the ANOVA F statistic.

(a) Which of four tomato varieties has the highest mean yield? Grow ten plants of each variety and record the yield of each plant in pounds of tomatoes.

(b) A maker of detergents wants to know which of six package designs is most attractive to consumers. Each package is shown to 120 different consumers who rate the attractiveness of the design on a 1 to 10 scale.

(c) An experiment to compare the effectiveness of three weight-loss programs has 32 subjects who want to lose weight. Ten subjects are assigned at random to each of two programs, and the remaining 12 subjects follow the third program. After six months, each subject's change in weight is recorded.

Assumptions for ANOVA

Like all inference procedures, ANOVA is valid only in some circumstances. Here are the requirements for using ANOVA to compare population means.

ANOVA ASSUMPTIONS

- We have I **independent SRSs**, one from each of I populations.
- Each population has a **normal distribution** with unknown mean μ_i. The means may be different in the different populations. The ANOVA F statistic tests the null hypothesis that all of the populations have the same mean:

$$H_0: \mu_1 = \mu_2 = \cdots = \mu_I$$
$$H_a: \text{not all of the } \mu_i \text{ are equal}$$

- All of the populations have the **same standard deviation** σ, whose value is unknown.

The first two requirements are familiar from our study of the two-sample t procedures for comparing two means. As usual, the design of the data production is the most important foundation for inference. Biased sampling or confounding can make any inference meaningless. If we do not actually draw separate SRSs from each population or carry out a randomized comparative experiment, it is often unclear to what population the conclusions of inference apply. (This is the case in Example 15.3, for example.) ANOVA, like other inference procedures, is often used when random samples are not available. You must judge each use on its merits, a judgment that usually requires some knowledge of the subject of the study in addition to some knowledge of statistics.

robustness

Because no real population has exactly a normal distribution, the usefulness of inference procedures that assume normality depends on how sensitive they are to departures from normality. Fortunately, procedures for comparing means are not very sensitive to lack of normality. The ANOVA F test, like the t procedures, is **robust**. What matters is normality of the sample means, so ANOVA becomes safer as the sample sizes get larger, because of the central limit theorem effect. Remember to check for outliers that change the value of sample means and for extreme skewness. When there are no outliers and the distributions are roughly symmetric, you can safely use ANOVA for sample sizes as small as 4 or 5.

The third assumption is annoying: ANOVA assumes that the variability of observations, measured by the standard deviation, is the same in all populations. You may recall from Chapter 11 (page 638) that there is a special version of the two-sample t test that assumes equal standard deviations in both populations. The ANOVA F for comparing two means is exactly the square of this special t statistic. We prefer the t test that does not assume equal standard deviations, but for comparing more than two means there is no general alternative to the ANOVA F. It is not easy to check the assumption that the populations have equal standard deviations. Statistical tests for equality of standard deviations are very sensitive to lack of normality, so much so that they are of little practical value. You must either seek expert advice or rely on the robustness of ANOVA.

How serious are unequal standard deviations? ANOVA is not too sensitive to violations of the assumption, especially when all samples have the same or similar sizes and no sample is very small. When designing a study, try to take samples of the same size from all the groups you want to compare. The sample standard deviations estimate the population standard deviations, so check before doing ANOVA that the sample standard deviations are similar to each other. We expect some variation among them due to chance. Here is a rule of thumb that is safe in almost all situations.

CHECKING STANDARD DEVIATIONS IN ANOVA

The results of the ANOVA F test are approximately correct when the largest sample standard deviation is no more than twice as large as the smallest sample standard deviation.

EXAMPLE 15.7

In the gas mileage study, the sample standard deviations for compact, midsize, and large cars are

$$s_1 = 2.814 \qquad s_2 = 2.311 \qquad s_3 = 1.660$$

These standard deviations easily satisfy our rule of thumb. The three sample sizes are 25, 19, and 13, and we left out the two exotic cars that appeared as outliers in Figure 15.2. We can safely use ANOVA to compare the mean gas mileage for the three sizes of cars.

The report from which Table 15.1 was taken also contained data on 33 subcompact cars. The standard deviation of their city gas mileage is $s_4 = 5.888$ miles per gallon. This is more than twice the smallest standard deviation:

$$\frac{\text{largest } s}{\text{smallest } s} = \frac{5.888}{1.660} = 3.55$$

It would *not* be safe to use ANOVA to compare the mean gas mileage of all four car sizes.

A large standard deviation is often due to skewness in the distribution. Here is a stemplot of the subcompact-car gas mileages:

```
1 | 778888899
2 | 002233334
2 | 556666678899
3 | 3
3 | 6
4 | 4
```

The distribution is skewed to the right. Three models get 33, 36, and 44 miles per gallon in the city. The mean is $\bar{x} = 24.1$ for all 33 small cars, and $\bar{x} = 22.8$ when we omit these three. We are wise not to apply ANOVA to such data.

EXAMPLE 15.8

To detect the presence of harmful insects in farm fields, put up boards covered with a sticky material and examine the insects trapped on the boards. Which colors attract insects best? Experimenters placed six boards of each of four colors at random locations in a field of oats and measured the number of cereal leaf beetles trapped.[5]

Board color	Insects trapped					
Blue	16	11	20	21	14	7
Green	37	32	20	29	37	32
White	21	12	14	17	13	20
Yellow	45	59	48	46	38	47

	Blue		Green		White		Yellow
0	7	0		0		0	
1	146	1		1	2347	1	
2	01	2	09	2	01	2	
3		3	2277	3		3	8
4		4		4		4	5678
5		5		5		5	9

FIGURE 15.8   Side-by-side stemplots comparing the counts of insects attracted by six boards for each of four board colors (Example 15.8).

We would like to use ANOVA to compare the mean numbers of beetles that would be trapped by all boards of each color. Because the samples are small, we plot the data in side-by-side stemplots in Figure 15.8. Computer output for descriptive statistics and ANOVA appears in Figure 15.11. It appears that yellow boards attract by far the most insects ($\bar{x}_4 = 47.167$), with green next ($\bar{x}_2 = 31.167$) and blue and white far behind.

Check that we can safely use ANOVA to test equality of the four means. The largest of the four sample standard deviations is 6.795 and the smallest is 3.764. The ratio

$$\frac{\text{largest } s}{\text{smallest } s} = \frac{6.795}{3.764} = 1.8$$

is less than 2, so these data satisfy our rule of thumb for safe use of ANOVA. The shapes of the four distributions are irregular, as we expect with only 6 observations in each group, but there are no outliers. The ANOVA results will be approximately correct.

There are $I = 4$ groups and $N = 24$ observations overall, so the degrees of freedom for F are

$$\text{numerator: } I - 1 = 4 - 1 = 3$$
$$\text{denominator: } N - I = 24 - 4 = 20$$

This agrees with the computer results. The F statistic is $F = 42.84$, a very large F with P-value $P < .001$. Despite the small samples, the experiment gives very strong evidence of differences among the colors. Yellow boards appear best at attracting leaf beetles.

A one-way ANOVA test can be performed with the TI-83, as long as you are comparing no more than 6 means.

EXAMPLE 15.9

We will use the data from Example 15.8. Begin by entering the data into lists: Blue $\leftrightarrow$ L$_1$, Green $\leftrightarrow$ L$_2$, White $\leftrightarrow$ L$_3$, and Yellow $\leftrightarrow$ L$_4$. Then select STATS / TESTS / F:ANOVA and enter the lists that contain the data. Here, ANOVA(L$_1$,L$_2$,L$_3$,L$_4$). Press ENTER. See Figures 15.9(a) and (b).

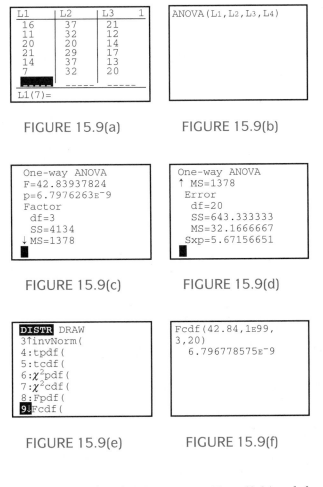

FIGURE 15.9(a) FIGURE 15.9(b)

FIGURE 15.9(c) FIGURE 15.9(d)

FIGURE 15.9(e) FIGURE 15.9(f)

The TI-83 reports that the F statistic is $F = 42.84$ and the P-value is 6.8×10^{-9}. The numerator degrees of freedom are $I - 1 = 3$, and by scrolling down, you see that the denominator degrees of freedom are $N - I = 24 - 4 = 20$. See Figures 15.9(c) and (d).

If you know the F statistic and the numerator and denominator degrees of freedom, then you can find the P-value with the command Fcdf, under the DISTR menu, as shown in Figure 15.9(e). The syntax is Fcdf(leftendpoint, rightendpoint, numerator df, denominator df). See Figure 15.9(f).

EXAMPLE 15.10

To draw an F distribution with the right critical area shaded, proceed as follows. First select 2nd / DISTR / DRAW / 4:ShadeF(. See Figure 15.10(a). Then supply the parameters. The syntax is ShadeF(leftendpoint, rightendpoint, df numerator, df denominator). See Figure 15.10(b). For illustration purposes, we'll change the F statistic to 3.12 so that the critical area will be

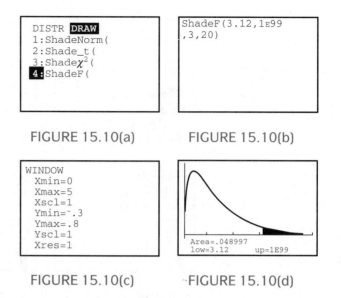

FIGURE 15.10(a)   FIGURE 15.10(b)

FIGURE 15.10(c)   FIGURE 15.10(d)

visible. Before you press ENTER, set the Window as follows: X[0,5]$_1$ and Y[-.3,.8]$_1$. See Figure 15.10(c).

Now press ENTER. The $F(3, 20)$ curve is drawn, with the area under the curve to the right of $F = 3.12$ shaded. See Figure 15.10(d).

EXERCISES

15.15   Do the sample standard deviations for the fish weights in Table 15.2 allow use of ANOVA to compare the mean weights? (Use the computer output in Figure 15.4.)

15.16   Do the standard deviations for the corn yields in Exercise 15.11 allow use of ANOVA to compare the mean yields? (Use the computer output in Figure 15.5.)

15.17   Use your TI-83 to analyze the paper airplane flight distance data from Activity 15.

(a)   First, enter the data into your calculator and explore the data graphically. Does it appear that at least one of the means is different from the other means?

(b)   Check the ANOVA assumptions. Can you state with some assurance that these assumptions are satisfied?

(c)   Apply the ANOVA standard deviation test. Is the quotient (largest standard deviation)/(smallest standard deviation) ≤ 2? If not, you probably should not proceed.

(d) If the data pass the tests in (b) and (c), then perform one-way ANOVA. Write your conclusions.

15.18   Married men tend to earn more than single men. An investigation of the relationship between marital status and income collected data on all 8235 men employed as managers or professionals by a large manufacturing firm in 1976. Suppose (this is risky) we regard these men as a random sample from the population of all men employed in managerial or professional positions in large companies. Here are descriptive statistics for the salaries of these men:[6]

	Single	Married	Divorced	Widowed
n_i	337	7,730	126	42
$\bar{x}_i$	$21,384	$26,873	$25,594	$26,936
s_i	$5,731	$7,159	$6,347	$8,119

(a) Briefly describe the relationship between marital status and salary.

(b) Do the sample standard deviations allow use of the ANOVA F test? (The distributions are skewed to the right. We expect right skewness in income distributions. The investigators actually applied ANOVA to the logarithms of the salaries, which are more symmetric.)

(c) What are the degrees of freedom of the ANOVA F test?

(d) The F test is a formality for these data, because we are sure that the P-value will be very small. Why are we sure?

(e) Single men earn less on the average than men who are or have been married. Do the highly significant differences in mean salary show that getting married raises men's mean income? Explain your answer.

15.19   What factors influence the success of students who plan to study computer science (CS)? Look at all 256 students who entered a major university planning a CS major in a specific year. We are willing to regard these students as a random sample of the students the university CS program will attract in subsequent years. After three semesters of study, some of these students were CS majors, some were majors in another field of science or engineering, and some had left science and engineering or left the university. The table below gives the sample means and standard deviations and the ANOVA F statistics for three variables that describe the students' high school performance. These are three separate ANOVA F tests.[7]

The first variable is a student's rank in the high school class, given as a percentile (so rank 50 is the middle of the class and rank 100 is the top). The next variable is the number of semester courses in mathematics the student took in high school. The third variable is the student's average grade in high school mathematics. The mean and standard deviation appear in a form common in published reports, with the standard deviation in parentheses following the mean.

Group	n	Mean (standard deviation)		
		High school class rank	Semesters of HS math	Average grade in HS math
CS majors	103	88.0 (10.5)	8.74 (1.28)	3.61 (.46)
Sci./Eng. majors	31	89.2 (10.8)	8.65 (1.31)	3.62 (.40)
Other	122	85.8 (10.8)	8.25 (1.17)	3.35 (.55)
F statistic		1.95	4.56	9.38

(a) What null and alternative hypotheses does F test for rank in the high school class? Express the hypotheses both in symbols and in words. The hypotheses are similar for the other two variables.

(b) What are the degrees of freedom for each F?

(c) Check that the standard deviations allow use of all three F tests. The shapes of the distributions also allow use of F. How significant is F for each of these variables?

(d) Write a brief summary of the differences among the three groups of students, taking into account both the significance of the F tests and the values of the means.

Some details of ANOVA

Now we will give the actual recipe for the ANOVA F statistic. We have SRSs from each of I populations. Subscripts from 1 to I tell us which sample a statistic refers to:

Population	Sample size	Sample mean	Sample std. dev.
1	n_1	$\bar{x}_1$	s_1
2	n_2	$\bar{x}_2$	s_2
⋮	⋮	⋮	⋮
I	n_I	$\bar{x}_I$	s_I

You can find the F statistic from just the sample sizes n_i, the sample means $\bar{x}_i$, and the sample standard deviations s_i. You don't need to go back to the individual observations.

The ANOVA F statistic has the form

$$F = \frac{\text{variation among the sample means}}{\text{variation among individuals}}$$

mean squares

The measures of variation in the numerator and denominator of F are called **mean squares**. A mean square is a more general form of a sample variance. An ordinary sample variance s^2 is an average (or mean) of the squared deviations of observations from their mean, so it qualifies as a "mean square."

The numerator of F is a mean square that measures variation among the I sample means $\bar{x}_1, \bar{x}_2, \ldots, \bar{x}_I$. Call the overall mean response, the mean of all N observations together, $\bar{x}$. You can find $\bar{x}$ from the I sample means by

$$\bar{x} = \frac{n_1\bar{x}_1 + n_2\bar{x}_2 + \cdots + n_I\bar{x}_I}{N}$$

The sum of each mean multiplied by the number of observations it represents is the sum of all the individual observations. Dividing this sum by N, the total number of observations, gives the overall mean $\bar{x}$. The numerator mean square in F is an average of the I squared deviations of the means of the samples from $\bar{x}$. It is called the **mean square for groups**, abbreviated as MSG:

MSG

$$\text{MSG} = \frac{n_1(\bar{x}_1 - \bar{x})^2 + n_2(\bar{x}_2 - \bar{x})^2 + \cdots + n_I(\bar{x}_I - \bar{x})^2}{I - 1}$$

Each squared deviation is weighted by n_i, the number of observations it represents.

The mean square in the denominator of F measures variation among individual observations in the same sample. For any one sample, the sample variance s_i^2 does this job. For all I samples together, we use an average of the individual sample variances. It is again a weighted average in which each s_i^2 is weighted by one fewer than the number of observations it represents, $n_i - 1$. Another way to put this is that each s_i^2 is weighted by its degrees of freedom $n_i - 1$. The resulting mean square is called the **mean square for error**, MSE:

MSE

$$\text{MSE} = \frac{(n_1 - 1)s_1^2 + (n_2 - 1)s_2^2 + \cdots + (n_I - 1)s_I^2}{N - I}$$

Here is a summary of the ANOVA test.

THE ANOVA F TEST

Draw an independent SRS from each of I populations. The ith population has the $N(\mu_i, \sigma)$ distribution, where σ is the common standard deviation in all the populations. The ith sample has size n_i, sample mean $\bar{x}_i$, and sample standard deviation s_i.

The **ANOVA F statistic** tests the null hypothesis that all I populations have the same mean:

$$H_0: \mu_1 = \mu_2 = \cdots = \mu_I$$
$$H_a: \text{not all of the } \mu_i \text{ are equal}$$

The statistic is

$$F = \frac{\text{MSG}}{\text{MSE}}$$

The **mean squares** that make up F are

$$\text{MSG} = \frac{n_1(\bar{x}_1 - \bar{x})^2 + n_2(\bar{x}_2 - \bar{x})^2 + \cdots + n_I(\bar{x}_I - \bar{x})^2}{I - 1}$$

and

$$\text{MSE} = \frac{(n_1 - 1)s_1^2 + (n_2 - 1)s_2^2 + \cdots + (n_I - 1)s_I^2}{N - I}$$

When H_0 is true, F has the F **distribution** with $I - 1$ and $N - I$ degrees of freedom.

The denominators in the recipes for MSG and MSE are the two degrees of freedom $I - 1$ and $N - I$ of the F test. The numerators are called *sums of squares*, from their algebraic form. It is usual to present the results of ANOVA in an **ANOVA** *table* like that in the Minitab output.

ANOVA table

Source	Degrees of freedom	Sum of squares	Mean square	F
Groups	$I - 1$	$\sum_{\text{groups}} n_i(\bar{x}_i - \bar{x})^2$	SSG/DFG	MSG/MSE
Error	$N - I$	$\sum_{\text{groups}}(n_i - 1)s_i^2$	SSE/DFE	
Total	$N - 1$	$\sum_{\text{obs}}(x_{ij} - \bar{x})^2$	SST/DFT	

The table has columns for degrees of freedom (DF), sums of squares (SS), and mean squares (MS). Check that each MS entry in Figure 15.11, for example, is the sum of squares SS divided by the degrees of freedom DF in the same row. The F statistic in the "F" column is MSG/MSE. The rows are labeled by sources of variation. In this output, variation among groups is labeled "FACTOR." Other statistical software calls this line "Treatments" or "Groups." Variation among observations in the same group is called "ERROR" by most software. This doesn't mean a mistake has been made. It's a traditional term for chance variation.

```
ANALYSIS OF VARIANCE
SOURCE    DF        SS         MS        F         P

FACTOR    3       4134.0     1378.0    42.84     0.000
ERROR     20       643.3       32.2

TOTAL     23      4777.3
```

FIGURE 15.11   ANOVA table from Minitab output for comparing the four board colors in Example 15.8.

Because MSE is an average of the individual sample variances, it is also called the *pooled sample variance*, written as s_p^2. When all I populations have the same population variance σ^2 (ANOVA assumes that they do), s_p^2 estimates the common variance σ^2. The square root of MSE is the ***pooled standard deviation*** s_p. It estimates the common standard deviation σ of observations in each group. Minitab, like most ANOVA programs, gives the value of s_p as well as MSE. It is the "POOLED STDEV" value in Figure 15.11.

pooled standard deviation

The pooled standard deviation s_p is a better estimator of the common σ than any individual sample standard deviation s_i because it combines (pools) the information in all I samples. We can get a confidence interval for any of the means μ_i from the usual form

$$\text{estimate} \pm t^* \text{SE}_\text{estimate}$$

using s_p to estimate σ. The confidence interval for μ_i is

$$\bar{x}_i \pm t^* \frac{s_p}{\sqrt{n_i}}$$

Use the critical value t^* from the t distribution with $N - I$ degrees of freedom, because s_p has $N - I$ degrees of freedom. These are the confidence intervals that appear in the Minitab ANOVA output.

EXAMPLE 15.11

We can do the ANOVA test comparing board colors in Example 15.8 using only the sample sizes, sample means, and sample standard deviations. Minitab gives these in Figure 15.11, but it is not hard to find them with a calculator.
The overall mean of the 24 counts is

$$\bar{x} = \frac{n_1\bar{x}_1 + n_2\bar{x}_2 + \cdots + n_I\bar{x}_I}{N}$$

$$= \frac{(6)(14.833) + (6)(31.167) + (6)(16.167) + (6)(47.167)}{24}$$

$$= \frac{656}{24} = 27.333$$

The mean square for groups is

$$\text{MSG} = \frac{n_1(\bar{x}_1 - \bar{x})^2 + n_2(\bar{x}_2 - \bar{x})^2 + \cdots + n_I(\bar{x}_I - \bar{x})^2}{I - 1}$$

$$= \frac{1}{4 - 1}[(6)(14.833 - 27.333)^2 + (6)(31.167 - 27.333)^2$$

$$+ (6)(16.167 - 27.333)^2 + (6)(47.167 - 27.333)^2]$$

$$= \frac{4134.100}{3} = 1378.033$$

The mean square for error is

$$\text{MSE} = \frac{(n_1 - 1)s_1^2 + (n_2 - 1)s_2^2 + \cdots + (n_I - 1)s_I^2}{N - I}$$

$$= \frac{(5)(5.345^2) + (5)(6.306^2) + (5)(3.764^2) + (5)(6.795^2)}{24 - 4}$$

$$= \frac{643.372}{20} = 32.169$$

Finally, the ANOVA test statistic is

$$F = \frac{\text{MSG}}{\text{MSE}} = \frac{1378.033}{32.169} = 42.84$$

Our work agrees with the computer output in Figure 15.11 and the TI-83 output in Example 15.9. We don't recommend doing these calculations, because tedium and roundoff errors cause frequent mistakes.

The pooled estimate of the standard deviation σ in any group is

$$s_p = \sqrt{\text{MSE}} = \sqrt{32.169} = 5.672$$

A 95% confidence interval for the mean count of insects trapped by yellow boards, using s_p and 20 degrees of freedom, is

$$\bar{x}_4 \pm t^* \frac{s_p}{\sqrt{n_4}} = 47.167 \pm 2.086 \frac{5.672}{\sqrt{6}}$$

$$= 47.167 \pm 4.830$$

$$= 42.34 \text{ to } 52.00$$

This confidence interval appears in the graph in the Minitab ANOVA output in Figure 15.11.

EXAMPLE 15.12

The results produced by the TI-83 in Example 15.9 comparing board colors included the sum of squares for groups, SSG = 4134, mean square for groups MSG = 1378, sum of squares for error, SSE = 643.33, mean square for error MSE = 32.17, and, of course, the F statistic, $F = $ MSG/MSE $= 1378/32.17 = 42.84$.

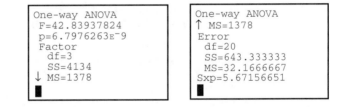

EXERCISES

15.20   Return to the study of fish weights in Table 15.2 and Figure 15.4.

(a) Starting from the sample standard deviations in Figure 15.4, calculate MSE and the pooled standard deviation s_p. Use the computer output to check your work.

(b) Give a 95% confidence interval for the mean weight of perch, using the pooled standard deviation s_p. A graph of this interval appears in the computer output.

15.21   Continue the ANOVA calculations for the fish data. Starting from the sample means in the computer output in Figure 15.4, find the overall mean weight $\bar{x}$. Then find MSG. Finally, combine MSG with MSE from the previous exercise to obtain F. Use the TI-83 command `Fcdf` to find the P-value.

15.22   Return to the data in Exercise 15.11 on corn yields for different planting rates. Starting from the sample means and standard deviations for the five groups (Figure 15.5), calculate MSE, the overall mean yield $\bar{x}$, and MSG. Then calculate the F-statistic and the P-value. Use the computer output in Figure 15.5 to check your work.

15.23   Give a 90% confidence interval for the mean yield of corn planted at 20,000 plants per acre under the growing conditions of the experiment described in Exercise 15.11. Use the pooled standard deviation s_p to estimate σ in the standard error.

SUMMARY

One-way analysis of variance (ANOVA) compares the means of several populations. The **ANOVA F test** tests the overall H_0 that all the populations have the same mean. If the F test shows significant differences, examine the data to see where the differences lie and whether they are large enough to be important.

ANOVA assumes that we have an **independent SRS** from each population; that each population has a **normal distribution**; and that all populations have the **same standard deviation.**

In practice, ANOVA is relatively **robust** when the populations are nonnormal, especially when the samples are large. Before doing the F test, check the observations in each sample for outliers or strong skewness. Also verify that the largest sample standard deviation is no more than twice as large as the smallest standard deviation.

When the null hypothesis is true, the **ANOVA F statistic** for comparing I means from a total of N observations in all samples combined has the F **distribution** with $I - 1$ and $N - I$ degrees of freedom.

ANOVA calculations are reported in an **ANOVA table** that gives sums of squares, mean squares, and degrees of freedom for variation among groups and for variation within groups. In practice, we use software to do the calculations.

CHAPTER REVIEW

The F test can be used to perform inference for comparing the spread of two normal populations, but there are serious cautions for basic users of statistics. The procedure is extremely sensitive to nonnormal populations, and interpreting significant F-values is therefore difficult. Consequently, inference about population standard deviations at this level is not recommended.

Advanced statistical inference often concerns relationships among several parameters. The second section in this chapter introduces the ANOVA F test for one such relationship: equality of the means of any number of populations. The alternative to this hypothesis is "many-sided," because it allows any relationship other than "all equal." The ANOVA F test is an overall test that tells us whether the data give good reason to reject the hypothesis that all the population means are equal. You should always accompany the test by data analysis to see what kind of inequality is present. Plotting the data in all groups side-by-side is particularly helpful.

ANOVA requires that the data satisfy conditions similar to those for t procedures. The most important assumption concerns the design of the data production. Inference is most trustworthy when we have random samples or responses to the treatments in a randomized comparative experiment. ANOVA assumes that the distribution of responses in each population has a normal distribution. Fortunately, the ANOVA F test shares the robustness of the t procedures, especially when the sample sizes are not very small. Do beware of outliers, which can greatly influence the mean responses and the F statistic.

ANOVA also requires a new assumption: the populations must all have the same standard deviation. Fortunately, ANOVA is not highly sensitive to unequal standard deviations, especially when the samples are similar in size. As a rule of thumb, you can use ANOVA when the largest sample standard deviation is no more than twice as large as the smallest.

After studying this chapter, you should be able to do the following.

A. RECOGNITION

1. Recognize when testing the equality of several means is helpful in understanding data.
2. Recognize that the statistical significance of differences among sample means depends on the sizes of the samples and on how much variation there is within the samples.

3. Recognize when you can safely use ANOVA to compare means. Check the data production, the presence of outliers, and the sample standard deviations for the groups you want to compare.

B. INTERPRETING ANOVA

1. Explain what null hypothesis F tests in a specific setting.

2. Locate the F statistic and its P-value on the output of a computer analysis of variance program.

3. Find the degrees of freedom for the F statistic from the number and sizes of the samples. Use Table D of the F distributions to approximate the P-value when software does not give it.

4. If the test is significant, use graphs and descriptive statistics to see what differences among the means are most important.

CHAPTER 15 REVIEW EXERCISES

15.24   Do the data in Example 11.9 (page 626) provide evidence of different standard deviations for Chapin test scores in the populations of female and male college liberal arts majors?

(a) State the hypotheses and carry out the test. Software can assess significance exactly, but inspection of the proper table is enough to draw a conclusion.

(b) Do the large sample sizes allow us to ignore the assumption that the population distributions are normal?

15.25   In each of the following situations, we want to compare the mean response in several populations. For each setting, identify the populations and the response variable. Then give I, the n_i, and N. Finally, give the degrees of freedom of the ANOVA F test.

(a) A study of the effects of smoking classifies subjects as nonsmokers, moderate smokers, or heavy smokers. The investigators interview a sample of 200 people in each group. Among the questions is "How many hours do you sleep on a typical night?"

(b) The strength of concrete depends on the mixture of sand, gravel, and cement used to prepare it. A study compares five different mixtures. Workers prepare six batches of each mixture and measure the strength of the concrete made from each batch.

(c) Which of four methods of teaching American Sign Language is most effective? Assign 10 of the 42 students in a class at random to each of three methods. Teach the remaining 12 students by the fourth method. Record the students' scores on a standard test of sign language after a semester's study.

15.26 How do nematodes (microscopic worms) affect plant growth? A botanist prepares 16 identical planting pots and then introduces different numbers of nematodes into the pots. He transplants a tomato seedling into each plot. Here are data on the increase in height of the seedlings (in centimeters) 16 days after planting:[8]

Nematodes	Seedling growth			
0	10.8	9.1	13.5	9.2
1,000	11.1	11.1	8.2	11.3
5,000	5.4	4.6	7.4	5.0
10,000	5.8	5.3	3.2	7.5

(a) Make a table of means and standard deviations for the four treatments. Make side-by-side stemplots to compare the treatments. What do the data appear to show about the effect of nematodes on growth?

(b) State H_0 and H_a for the ANOVA test for these data, and explain in words what ANOVA tests in this setting.

(c) Use your TI-83 or computer software to carry out the ANOVA. Report your overall conclusions about the effect of nematodes on plant growth.

15.27 Table 15.1 presents both the city and highway gas mileage for a number of cars of three sizes. We analyzed the city gas mileage data in Examples 15.3 and 15.4. There are significant differences ($P < .001$) among the mean miles per gallon for the three sizes, and the most important difference is that compact cars have better mileage than either of the other types. Now use computer software to analyze the highway gas mileage data and report your conclusions in detail. (Be sure to include graphs of the data and descriptive statistics as well as an ANOVA.)

15.28 We have two methods to compare the means of two groups: the two-sample t test of Section 11.2 and the ANOVA F test with $I = 2$. We prefer the t test because it allows one-sided alternatives and does not assume that both populations have the same standard deviation. Let us apply both tests to the same data.

There are two types of life insurance companies. "Stock" companies have shareholders, and "mutual" companies are owned by their policy-holders. Take an SRS of each type of company from those listed in a directory of the industry. Then ask the annual cost per $1000 of insurance for a $50,000 policy insuring the life of a 35-year-old man who does not smoke. Here are the data summaries:[9]

	Stock companies	Mutual companies
n_i	13	17
$\bar{x}_i$	$2.31	$2.37
s_i	$0.38	$0.58

(a) Calculate the two-sample t statistic for testing H_0: $\mu_1 = \mu_2$ against the two-sided alternative. Use the conservative method to find the P-value.

(b) Calculate MSG, MSE, and the ANOVA F statistic for the same hypotheses. What is the P-value of F?

(c) How close are the two P-values? (The square root of the F statistic is a t statistic with $N - I = n_1 + n_2 - 2$ degrees of freedom. This is the "pooled two-sample t" mentioned on page 638. So F for $I = 2$ is exactly equivalent to a t statistic, but it is a slightly different t from the one we use.)

15.29 Carry out the ANOVA calculations (MSG, MSE, and F) required for part of Exercise 15.26. Find the degrees of freedom for F and report its P-value as closely as Table D allows.

NOTES AND DATA SOURCES

1. The problem of comparing spreads is difficult even with advanced methods. Common distribution-free procedures do not offer a satisfactory alternative to the F test, because they are sensitive to unequal shapes when comparing two distributions. A good introduction to the available methods is W. J. Conover, M. E. Johnson, and M. M. Johnson, "A comparative study of tests for homogeneity of variances, with applications to outer continental shelf bidding data," *Technometrics*, 23 (1981), pp. 351–361. Modern resampling procedures often work well. See Dennis D. Boos and Colin Brownie, "Bootstrap methods for testing homogeneity of variances," *Technometrics*, 31 (1989), pp. 69–82.

2. The data in Table 15.1 are from the U.S. Department of Energy's *1994 Gas Mileage Guide*, October 1993. The table gives data for the basic engine/transmission combination for each model. Models that are essentially identical (such as the Ford Taurus and Mercury Sable) appear only once.

3. The data in Table 15.2 are part of a larger data set in the *Journal of Statistics Education* archive, accessible via Internet. The original source is Pekka Brofeldt,

"Bidrag till kaennedom on fiskbestondet I vaara sjoear. Laengelmaevesi," in T. H. Jaervi, *Finlands Fiskeriet*, Band 4, *Meddelanden utgivna av fiskerifoereningen I Finland*, Helsinki, 1917. The data were contributed to the archive (with information in English) by Juha Puranen of the University of Helsinki.

4. The data are from W. L. Colville and D. P. McGill, "Effect of rate and method of planting on several plant characters and yield of irrigated corn," *Agronomy Journal*, 54 (1962), pp. 235–238.

5. Modified from M. C. Wilson and R. E. Shade, "Relative attractiveness of various luminescent colors to the cereal leaf beetle and the meadow spittlebug," *Journal of Economic Entomology*, 60 (1967), pp. 578–580.

6. Sanders Korenman and David Neumark, "Does marriage really make men more productive?" *Journal of Human Resources*, 26 (1991), pp. 282–307.

7. Patricia F. Campbell and George P. McCabe, "Predicting the success of freshmen in a computer science major," *Communications of the ACM*, 27 (1984), pp. 1108–1113.

8. Data provided by Matthew Moore.

9. Mark Kroll, Peter Wright, and Pochera Theerathorn, "Whose interests do hired managers pursue? An examination of select mutual and stock life insurers," *Journal of Business Research*, 26 (1993), pp. 133–148.

Appendix

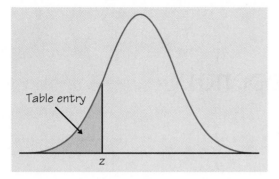

Table entry

Table entry for z is the area under the standard normal curve to the left of z.

Standard normal probabilities

z	.00	.01	.02	.03	.04	.05	.06	.07	.08	.09
−3.4	.0003	.0003	.0003	.0003	.0003	.0003	.0003	.0003	.0003	.0002
−3.3	.0005	.0005	.0005	.0004	.0004	.0004	.0004	.0004	.0004	.0003
−3.2	.0007	.0007	.0006	.0006	.0006	.0006	.0006	.0005	.0005	.0005
−3.1	.0010	.0009	.0009	.0009	.0008	.0008	.0008	.0008	.0007	.0007
−3.0	.0013	.0013	.0013	.0012	.0012	.0011	.0011	.0011	.0010	.0010
−2.9	.0019	.0018	.0018	.0017	.0016	.0016	.0015	.0015	.0014	.0014
−2.8	.0026	.0025	.0024	.0023	.0023	.0022	.0021	.0021	.0020	.0019
−2.7	.0035	.0034	.0033	.0032	.0031	.0030	.0029	.0028	.0027	.0026
−2.6	.0047	.0045	.0044	.0043	.0041	.0040	.0039	.0038	.0037	.0036
−2.5	.0062	.0060	.0059	.0057	.0055	.0054	.0052	.0051	.0049	.0048
−2.4	.0082	.0080	.0078	.0075	.0073	.0071	.0069	.0068	.0066	.0064
−2.3	.0107	.0104	.0102	.0099	.0096	.0094	.0091	.0089	.0087	.0084
−2.2	.0139	.0136	.0132	.0129	.0125	.0122	.0119	.0116	.0113	.0110
−2.1	.0179	.0174	.0170	.0166	.0162	.0158	.0154	.0150	.0146	.0143
−2.0	.0228	.0222	.0217	.0212	.0207	.0202	.0197	.0192	.0188	.0183
−1.9	.0287	.0281	.0274	.0268	.0262	.0256	.0250	.0244	.0239	.0233
−1.8	.0359	.0351	.0344	.0336	.0329	.0322	.0314	.0307	.0301	.0294
−1.7	.0446	.0436	.0427	.0418	.0409	.0401	.0392	.0384	.0375	.0367
−1.6	.0548	.0537	.0526	.0516	.0505	.0495	.0485	.0475	.0465	.0455
−1.5	.0668	.0655	.0643	.0630	.0618	.0606	.0594	.0582	.0571	.0559
−1.4	.0808	.0793	.0778	.0764	.0749	.0735	.0721	.0708	.0694	.0681
−1.3	.0968	.0951	.0934	.0918	.0901	.0885	.0869	.0853	.0838	.0823
−1.2	.1151	.1131	.1112	.1093	.1075	.1056	.1038	.1020	.1003	.0985
−1.1	.1357	.1335	.1314	.1292	.1271	.1251	.1230	.1210	.1190	.1170
−1.0	.1587	.1562	.1539	.1515	.1492	.1469	.1446	.1423	.1401	.1379
−0.9	.1841	.1814	.1788	.1762	.1736	.1711	.1685	.1660	.1635	.1611
−0.8	.2119	.2090	.2061	.2033	.2005	.1977	.1949	.1922	.1894	.1867
−0.7	.2420	.2389	.2358	.2327	.2296	.2266	.2236	.2206	.2177	.2148
−0.6	.2743	.2709	.2676	.2643	.2611	.2578	.2546	.2514	.2483	.2451
−0.5	.3085	.3050	.3015	.2981	.2946	.2912	.2877	.2843	.2810	.2776
−0.4	.3446	.3409	.3372	.3336	.3300	.3264	.3228	.3192	.3156	.3121
−0.3	.3821	.3783	.3745	.3707	.3669	.3632	.3594	.3557	.3520	.3483
−0.2	.4207	.4168	.4129	.4090	.4052	.4013	.3974	.3936	.3897	.3859
−0.1	.4602	.4562	.4522	.4483	.4443	.4404	.4364	.4325	.4286	.4247
−0.0	.5000	.4960	.4920	.4880	.4840	.4801	.4761	.4721	.4681	.4641

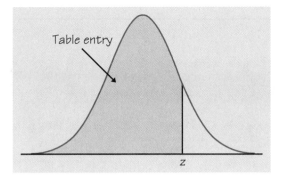

Table entry for z is the area under the standard normal curve to the left of z.

TABLE A			Standard normal probabilities (*continued*)							
z	.00	.01	.02	.03	.04	.05	.06	.07	.08	.09
0.0	.5000	.5040	.5080	.5120	.5160	.5199	.5239	.5279	.5319	.5359
0.1	.5398	.5438	.5478	.5517	.5557	.5596	.5636	.5675	.5714	.5753
0.2	.5793	.5832	.5871	.5910	.5948	.5987	.6026	.6064	.6103	.6141
0.3	.6179	.6217	.6255	.6293	.6331	.6368	.6406	.6443	.6480	.6517
0.4	.6554	.6591	.6628	.6664	.6700	.6736	.6772	.6808	.6844	.6879
0.5	.6915	.6950	.6985	.7019	.7054	.7088	.7123	.7157	.7190	.7224
0.6	.7257	.7291	.7324	.7357	.7389	.7422	.7454	.7486	.7517	.7549
0.7	.7580	.7611	.7642	.7673	.7704	.7734	.7764	.7794	.7823	.7852
0.8	.7881	.7910	.7939	.7967	.7995	.8023	.8051	.8078	.8106	.8133
0.9	.8159	.8186	.8212	.8238	.8264	.8289	.8315	.8340	.8365	.8389
1.0	.8413	.8438	.8461	.8485	.8508	.8531	.8554	.8577	.8599	.8621
1.1	.8643	.8665	.8686	.8708	.8729	.8749	.8770	.8790	.8810	.8830
1.2	.8849	.8869	.8888	.8907	.8925	.8944	.8962	.8980	.8997	.9015
1.3	.9032	.9049	.9066	.9082	.9099	.9115	.9131	.9147	.9162	.9177
1.4	.9192	.9207	.9222	.9236	.9251	.9265	.9279	.9292	.9306	.9319
1.5	.9332	.9345	.9357	.9370	.9382	.9394	.9406	.9418	.9429	.9441
1.6	.9452	.9463	.9474	.9484	.9495	.9505	.9515	.9525	.9535	.9545
1.7	.9554	.9564	.9573	.9582	.9591	.9599	.9608	.9616	.9625	.9633
1.8	.9641	.9649	.9656	.9664	.9671	.9678	.9686	.9693	.9699	.9706
1.9	.9713	.9719	.9726	.9732	.9738	.9744	.9750	.9756	.9761	.9767
2.0	.9772	.9778	.9783	.9788	.9793	.9798	.9803	.9808	.9812	.9817
2.1	.9821	.9826	.9830	.9834	.9838	.9842	.9846	.9850	.9854	.9857
2.2	.9861	.9864	.9868	.9871	.9875	.9878	.9881	.9884	.9887	.9890
2.3	.9893	.9896	.9898	.9901	.9904	.9906	.9909	.9911	.9913	.9916
2.4	.9918	.9920	.9922	.9925	.9927	.9929	.9931	.9932	.9934	.9936
2.5	.9938	.9940	.9941	.9943	.9945	.9946	.9948	.9949	.9951	.9952
2.6	.9953	.9955	.9956	.9957	.9959	.9960	.9961	.9962	.9963	.9964
2.7	.9965	.9966	.9967	.9968	.9969	.9970	.9971	.9972	.9973	.9974
2.8	.9974	.9975	.9976	.9977	.9977	.9978	.9979	.9979	.9980	.9981
2.9	.9981	.9982	.9982	.9983	.9984	.9984	.9985	.9985	.9986	.9986
3.0	.9987	.9987	.9987	.9988	.9988	.9989	.9989	.9989	.9990	.9990
3.1	.9990	.9991	.9991	.9991	.9992	.9992	.9992	.9992	.9993	.9993
3.2	.9993	.9993	.9994	.9994	.9994	.9994	.9994	.9995	.9995	.9995
3.3	.9995	.9995	.9995	.9996	.9996	.9996	.9996	.9996	.9996	.9997
3.4	.9997	.9997	.9997	.9997	.9997	.9997	.9997	.9997	.9997	.9998

TABLE B | Random digits

Line								
101	19223	95034	05756	28713	96409	12531	42544	82853
102	73676	47150	99400	01927	27754	42648	82425	36290
103	45467	71709	77558	00095	32863	29485	82226	90056
104	52711	38889	93074	60227	40011	85848	48767	52573
105	95592	94007	69971	91481	60779	53791	17297	59335
106	68417	35013	15529	72765	85089	57067	50211	47487
107	82739	57890	20807	47511	81676	55300	94383	14893
108	60940	72024	17868	24943	61790	90656	87964	18883
109	36009	19365	15412	39638	85453	46816	83485	41979
110	38448	48789	18338	24697	39364	42006	76688	08708
111	81486	69487	60513	09297	00412	71238	27649	39950
112	59636	88804	04634	71197	19352	73089	84898	45785
113	62568	70206	40325	03699	71080	22553	11486	11776
114	45149	32992	75730	66280	03819	56202	02938	70915
115	61041	77684	94322	24709	73698	14526	31893	32592
116	14459	26056	31424	80371	65103	62253	50490	61181
117	38167	98532	62183	70632	23417	26185	41448	75532
118	73190	32533	04470	29669	84407	90785	65956	86382
119	95857	07118	87664	92099	58806	66979	98624	84826
120	35476	55972	39421	65850	04266	35435	43742	11937
121	71487	09984	29077	14863	61683	47052	62224	51025
122	13873	81598	95052	90908	73592	75186	87136	95761
123	54580	81507	27102	56027	55892	33063	41842	81868
124	71035	09001	43367	49497	72719	96758	27611	91596
125	96746	12149	37823	71868	18442	35119	62103	39244
126	96927	19931	36809	74192	77567	88741	48409	41903
127	43909	99477	25330	64359	40085	16925	85117	36071
128	15689	14227	06565	14374	13352	49367	81982	87209
129	36759	58984	68288	22913	18638	54303	00795	08727
130	69051	64817	87174	09517	84534	06489	87201	97245
131	05007	16632	81194	14873	04197	85576	45195	96565
132	68732	55259	84292	08796	43165	93739	31685	97150
133	45740	41807	65561	33302	07051	93623	18132	09547
134	27816	78416	18329	21337	35213	37741	04312	68508
135	66925	55658	39100	78458	11206	19876	87151	31260
136	08421	44753	77377	28744	75592	08563	79140	92454
137	53645	66812	61421	47836	12609	15373	98481	14592
138	66831	68908	40772	21558	47781	33586	79177	06928
139	55588	99404	70708	41098	43563	56934	48394	51719
140	12975	13258	13048	45144	72321	81940	00360	02428
141	96767	35964	23822	96012	94591	65194	50842	53372
142	72829	50232	97892	63408	77919	44575	24870	04178
143	88565	42628	17797	49376	61762	16953	88604	12724
144	62964	88145	83083	69453	46109	59505	69680	00900
145	19687	12633	57857	95806	09931	02150	43163	58636
146	37609	59057	66967	83401	60705	02384	90597	93600
147	54973	86278	88737	74351	47500	84552	19909	67181
148	00694	05977	19664	65441	20903	62371	22725	53340
149	71546	05233	53946	68743	72460	27601	45403	88692
150	07511	88915	41267	16853	84569	79367	32337	03316

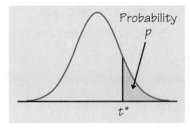

Table entry for p and C is the critical value t^* with probability p lying to its right and probability C lying between $-t^*$ and t^*.

TABLE C	t distribution critical values

df	\multicolumn{12}{c}{Upper tail probability p}
	.25	.20	.15	.10	.05	.025	.02	.01	.005	.0025	.001	.0005
1	1.000	1.376	1.963	3.078	6.314	12.71	15.89	31.82	63.66	127.3	318.3	636.6
2	0.816	1.061	1.386	1.886	2.920	4.303	4.849	6.965	9.925	14.09	22.33	31.60
3	0.765	0.978	1.250	1.638	2.353	3.182	3.482	4.541	5.841	7.453	10.21	12.92
4	0.741	0.941	1.190	1.533	2.132	2.776	2.999	3.747	4.604	5.598	7.173	8.610
5	0.727	0.920	1.156	1.476	2.015	2.571	2.757	3.365	4.032	4.773	5.893	6.869
6	0.718	0.906	1.134	1.440	1.943	2.447	2.612	3.143	3.707	4.317	5.208	5.959
7	0.711	0.896	1.119	1.415	1.895	2.365	2.517	2.998	3.499	4.029	4.785	5.408
8	0.706	0.889	1.108	1.397	1.860	2.306	2.449	2.896	3.355	3.833	4.501	5.041
9	0.703	0.883	1.100	1.383	1.833	2.262	2.398	2.821	3.250	3.690	4.297	4.781
10	0.700	0.879	1.093	1.372	1.812	2.228	2.359	2.764	3.169	3.581	4.144	4.587
11	0.697	0.876	1.088	1.363	1.796	2.201	2.328	2.718	3.106	3.497	4.025	4.437
12	0.695	0.873	1.083	1.356	1.782	2.179	2.303	2.681	3.055	3.428	3.930	4.318
13	0.694	0.870	1.079	1.350	1.771	2.160	2.282	2.650	3.012	3.372	3.852	4.221
14	0.692	0.868	1.076	1.345	1.761	2.145	2.264	2.624	2.977	3.326	3.787	4.140
15	0.691	0.866	1.074	1.341	1.753	2.131	2.249	2.602	2.947	3.286	3.733	4.073
16	0.690	0.865	1.071	1.337	1.746	2.120	2.235	2.583	2.921	3.252	3.686	4.015
17	0.689	0.863	1.069	1.333	1.740	2.110	2.224	2.567	2.898	3.222	3.646	3.965
18	0.688	0.862	1.067	1.330	1.734	2.101	2.214	2.552	2.878	3.197	3.611	3.922
19	0.688	0.861	1.066	1.328	1.729	2.093	2.205	2.539	2.861	3.174	3.579	3.883
20	0.687	0.860	1.064	1.325	1.725	2.086	2.197	2.528	2.845	3.153	3.552	3.850
21	0.686	0.859	1.063	1.323	1.721	2.080	2.189	2.518	2.831	3.135	3.527	3.819
22	0.686	0.858	1.061	1.321	1.717	2.074	2.183	2.508	2.819	3.119	3.505	3.792
23	0.685	0.858	1.060	1.319	1.714	2.069	2.177	2.500	2.807	3.104	3.485	3.768
24	0.685	0.857	1.059	1.318	1.711	2.064	2.172	2.492	2.797	3.091	3.467	3.745
25	0.684	0.856	1.058	1.316	1.708	2.060	2.167	2.485	2.787	3.078	3.450	3.725
26	0.684	0.856	1.058	1.315	1.706	2.056	2.162	2.479	2.779	3.067	3.435	3.707
27	0.684	0.855	1.057	1.314	1.703	2.052	2.158	2.473	2.771	3.057	3.421	3.690
28	0.683	0.855	1.056	1.313	1.701	2.048	2.154	2.467	2.763	3.047	3.408	3.674
29	0.683	0.854	1.055	1.311	1.699	2.045	2.150	2.462	2.756	3.038	3.396	3.659
30	0.683	0.854	1.055	1.310	1.697	2.042	2.147	2.457	2.750	3.030	3.385	3.646
40	0.681	0.851	1.050	1.303	1.684	2.021	2.123	2.423	2.704	2.971	3.307	3.551
50	0.679	0.849	1.047	1.299	1.676	2.009	2.109	2.403	2.678	2.937	3.261	3.496
60	0.679	0.848	1.045	1.296	1.671	2.000	2.099	2.390	2.660	2.915	3.232	3.460
80	0.678	0.846	1.043	1.292	1.664	1.990	2.088	2.374	2.639	2.887	3.195	3.416
100	0.677	0.845	1.042	1.290	1.660	1.984	2.081	2.364	2.626	2.871	3.174	3.390
1000	0.675	0.842	1.037	1.282	1.646	1.962	2.056	2.330	2.581	2.813	3.098	3.300
z^*	0.674	0.841	1.036	1.282	1.645	1.960	2.054	2.326	2.576	2.807	3.091	3.291
	50%	60%	70%	80%	90%	95%	96%	98%	99%	99.5%	99.8%	99.9%
	\multicolumn{12}{c}{Confidence level C}

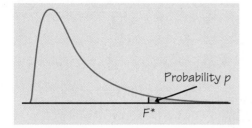

Probability p

F^*

Table entry for p is the critical value F^* with probability p lying to its right.

| *F* distribution critical values

		Degrees of freedom in the numerator							
	p	1	2	3	4	5	6	7	8
1	.100	39.86	49.50	53.59	55.83	57.24	58.20	58.91	59.44
	.050	161.45	199.50	215.71	224.58	230.16	233.99	236.77	238.88
	.025	647.79	799.50	864.16	899.58	921.85	937.11	948.22	956.66
	.010	4052.2	4999.5	5403.4	5624.6	5763.6	5859	5928.4	5981.1
	.001	405284	500000	540379	562500	576405	585937	592873	598144
2	.100	8.53	9.00	9.16	9.24	9.29	9.33	9.35	9.37
	.050	18.51	19.00	19.16	19.25	19.30	19.33	19.35	19.37
	.025	38.51	39.00	39.17	39.25	39.30	39.33	39.36	39.37
	.010	98.50	99.00	99.17	99.25	99.30	99.33	99.36	99.37
	.001	998.50	999.00	999.17	999.25	999.30	999.33	999.36	999.37
3	.100	5.54	5.46	5.39	5.34	5.31	5.28	5.27	5.25
	.050	10.13	9.55	9.28	9.12	9.01	8.94	8.89	8.85
	.025	17.44	16.04	15.44	15.10	14.88	14.73	14.62	14.54
	.010	34.12	30.82	29.46	28.71	28.24	27.91	27.67	27.49
	.001	167.03	148.50	141.11	137.10	134.58	132.85	131.58	130.62
4	.100	4.54	4.32	4.19	4.11	4.05	4.01	3.98	3.95
	.050	7.71	6.94	6.59	6.39	6.26	6.16	6.09	6.04
	.025	12.22	10.65	9.98	9.60	9.36	9.20	9.07	8.98
	.010	21.20	18.00	16.69	15.98	15.52	15.21	14.98	14.80
	.001	74.14	61.25	56.18	53.44	51.71	50.53	49.66	49.00
5	.100	4.06	3.78	3.62	3.52	3.45	3.40	3.37	3.34
	.050	6.61	5.79	5.41	5.19	5.05	4.95	4.88	4.82
	.025	10.01	8.43	7.76	7.39	7.15	6.98	6.85	6.76
	.010	16.26	13.27	12.06	11.39	10.97	10.67	10.46	10.29
	.001	47.18	37.12	33.20	31.09	29.75	28.83	28.16	27.65
6	.100	3.78	3.46	3.29	3.18	3.11	3.05	3.01	2.98
	.050	5.99	5.14	4.76	4.53	4.39	4.28	4.21	4.15
	.025	8.81	7.26	6.60	6.23	5.99	5.82	5.70	5.60
	.010	13.75	10.92	9.78	9.15	8.75	8.47	8.26	8.10
	.001	35.51	27.00	23.70	21.92	20.80	20.03	19.46	19.03
7	.100	3.59	3.26	3.07	2.96	2.88	2.83	2.78	2.75
	.050	5.59	4.74	4.35	4.12	3.97	3.87	3.79	3.73
	.025	8.07	6.54	5.89	5.52	5.29	5.12	4.99	4.90
	.010	12.25	9.55	8.45	7.85	7.46	7.19	6.99	6.84
	.001	29.25	21.69	18.77	17.20	16.21	15.52	15.02	14.63
8	.100	3.46	3.11	2.92	2.81	2.73	2.67	2.62	2.59
	.050	5.32	4.46	4.07	3.84	3.69	3.58	3.50	3.44
	.025	7.57	6.06	5.42	5.05	4.82	4.65	4.53	4.43
	.010	11.26	8.65	7.59	7.01	6.63	6.37	6.18	6.03
	.001	25.41	18.49	15.83	14.39	13.48	12.86	12.40	12.05

Degrees of freedom in the denominator

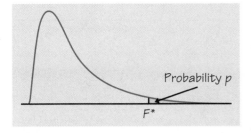

Table entry for p is the critical value F^* with probability p lying to its right.

TABLE D F distribution critical values (*continued*)

		\multicolumn{8}{c}{Degrees of freedom in the numerator}
	p	9	10	15	20	30	60	120	1000
1	.100	59.86	60.19	61.22	61.74	62.26	62.79	63.06	63.30
	.050	240.54	241.88	245.95	248.01	250.10	252.20	253.25	254.19
	.025	963.28	968.63	984.87	993.10	1001.4	1009.8	1014	1017.7
	.010	6022.5	6055.8	6157.3	6208.7	6260.6	6313	6339.4	6362.7
	.001	602284	605621	615764	620908	626099	631337	633972	636301
2	.100	9.38	9.39	9.42	9.44	9.46	9.47	9.48	9.49
	.050	19.38	19.40	19.43	19.45	19.46	19.48	19.49	19.49
	.025	39.39	39.40	39.43	39.45	39.46	39.48	39.49	39.50
	.010	99.39	99.40	99.43	99.45	99.47	99.48	99.49	99.50
	.001	999.39	999.40	999.43	999.45	999.47	999.48	999.49	999.50
3	.100	5.24	5.23	5.20	5.18	5.17	5.15	5.14	5.13
	.050	8.81	8.79	8.70	8.66	8.62	8.57	8.55	8.53
	.025	14.47	14.42	14.25	14.17	14.08	13.99	13.95	13.91
	.010	27.35	27.23	26.87	26.69	26.50	26.32	26.22	26.14
	.001	129.86	129.25	127.37	126.42	125.45	124.47	123.97	123.53
4	.100	3.94	3.92	3.87	3.84	3.82	3.79	3.78	3.76
	.050	6.00	5.96	5.86	5.80	5.75	5.69	5.66	5.63
	.025	8.90	8.84	8.66	8.56	8.46	8.36	8.31	8.26
	.010	14.66	14.55	14.20	14.02	13.84	13.65	13.56	13.47
	.001	48.47	48.05	46.76	46.10	45.43	44.75	44.40	44.09
5	.100	3.32	3.30	3.24	3.21	3.17	3.14	3.12	3.11
	.050	4.77	4.74	4.62	4.56	4.50	4.43	4.40	4.37
	.025	6.68	6.62	6.43	6.33	6.23	6.12	6.07	6.02
	.010	10.16	10.05	9.72	9.55	9.38	9.20	9.11	9.03
	.001	27.24	26.92	25.91	25.39	24.87	24.33	24.06	23.82
6	.100	2.96	2.94	2.87	2.84	2.80	2.76	2.74	2.72
	.050	4.10	4.06	3.94	3.87	3.81	3.74	3.70	3.67
	.025	5.52	5.46	5.27	5.17	5.07	4.96	4.90	4.86
	.010	7.98	7.87	7.56	7.40	7.23	7.06	6.97	6.89
	.001	18.69	18.41	17.56	17.12	16.67	16.21	15.98	15.77
7	.100	2.72	2.70	2.63	2.59	2.56	2.51	2.49	2.47
	.050	3.68	3.64	3.51	3.44	3.38	3.30	3.27	3.23
	.025	4.82	4.76	4.57	4.47	4.36	4.25	4.20	4.15
	.010	6.72	6.62	6.31	6.16	5.99	5.82	5.74	5.66
	.001	14.33	14.08	13.32	12.93	12.53	12.12	11.91	11.72
8	.100	2.56	2.54	2.46	2.42	2.38	2.34	2.32	2.30
	.050	3.39	3.35	3.22	3.15	3.08	3.01	2.97	2.93
	.025	4.36	4.30	4.10	4.00	3.89	3.78	3.73	3.68
	.010	5.91	5.81	5.52	5.36	5.20	5.03	4.95	4.87
	.001	11.77	11.54	10.84	10.48	10.11	9.73	9.53	9.36

Degrees of freedom in the denominator

F distribution critical values (*continued*)

		Degrees of freedom in the numerator							
	p	1	2	3	4	5	6	7	8
9	.100	3.36	3.01	2.81	2.69	2.61	2.55	2.51	2.47
	.050	5.12	4.26	3.86	3.63	3.48	3.37	3.29	3.23
	.025	7.21	5.71	5.08	4.72	4.48	4.32	4.20	4.10
	.010	10.56	8.02	6.99	6.42	6.06	5.80	5.61	5.47
	.001	22.86	16.39	13.90	12.56	11.71	11.13	10.70	10.37
10	.100	3.29	2.92	2.73	2.61	2.52	2.46	2.41	2.38
	.050	4.96	4.10	3.71	3.48	3.33	3.22	3.14	3.07
	.025	6.94	5.46	4.83	4.47	4.24	4.07	3.95	3.85
	.010	10.04	7.56	6.55	5.99	5.64	5.39	5.20	5.06
	.001	21.04	14.91	12.55	11.28	10.48	9.93	9.52	9.20
12	.100	3.18	2.81	2.61	2.48	2.39	2.33	2.28	2.24
	.050	4.75	3.89	3.49	3.26	3.11	3.00	2.91	2.85
	.025	6.55	5.10	4.47	4.12	3.89	3.73	3.61	3.51
	.010	9.33	6.93	5.95	5.41	5.06	4.82	4.64	4.50
	.001	18.64	12.97	10.80	9.63	8.89	8.38	8.00	7.71
15	.100	3.07	2.70	2.49	2.36	2.27	2.21	2.16	2.12
	.050	4.54	3.68	3.29	3.06	2.90	2.79	2.71	2.64
	.025	6.20	4.77	4.15	3.80	3.58	3.41	3.29	3.20
	.010	8.68	6.36	5.42	4.89	4.56	4.32	4.14	4.00
	.001	16.59	11.34	9.34	8.25	7.57	7.09	6.74	6.47
20	.100	2.97	2.59	2.38	2.25	2.16	2.09	2.04	2.00
	.050	4.35	3.49	3.10	2.87	2.71	2.60	2.51	2.45
	.025	5.87	4.46	3.86	3.51	3.29	3.13	3.01	2.91
	.010	8.10	5.85	4.94	4.43	4.10	3.87	3.70	3.56
	.001	14.82	9.95	8.10	7.10	6.46	6.02	5.69	5.44
25	.100	2.92	2.53	2.32	2.18	2.09	2.02	1.97	1.93
	.050	4.24	3.39	2.99	2.76	2.60	2.49	2.40	2.34
	.025	5.69	4.29	3.69	3.35	3.13	2.97	2.85	2.75
	.010	7.77	5.57	4.68	4.18	3.85	3.63	3.46	3.32
	.001	13.88	9.22	7.45	6.49	5.89	5.46	5.15	4.91
50	.100	2.81	2.41	2.20	2.06	1.97	1.90	1.84	1.80
	.050	4.03	3.18	2.79	2.56	2.40	2.29	2.20	2.13
	.025	5.34	3.97	3.39	3.05	2.83	2.67	2.55	2.46
	.010	7.17	5.06	4.20	3.72	3.41	3.19	3.02	2.89
	.001	12.22	7.96	6.34	5.46	4.90	4.51	4.22	4.00
100	.100	2.76	2.36	2.14	2.00	1.91	1.83	1.78	1.73
	.050	3.94	3.09	2.70	2.46	2.31	2.19	2.10	2.03
	.025	5.18	3.83	3.25	2.92	2.70	2.54	2.42	2.32
	.010	6.90	4.82	3.98	3.51	3.21	2.99	2.82	2.69
	.001	11.50	7.41	5.86	5.02	4.48	4.11	3.83	3.61
200	.100	2.73	2.33	2.11	1.97	1.88	1.80	1.75	1.70
	.050	3.89	3.04	2.65	2.42	2.26	2.14	2.06	1.98
	.025	5.10	3.76	3.18	2.85	2.63	2.47	2.35	2.26
	.010	6.76	4.71	3.88	3.41	3.11	2.89	2.73	2.60
	.001	11.15	7.15	5.63	4.81	4.29	3.92	3.65	3.43
1000	.100	2.71	2.31	2.09	1.95	1.85	1.78	1.72	1.68
	.050	3.85	3.00	2.61	2.38	2.22	2.11	2.02	1.95
	.025	5.04	3.70	3.13	2.80	2.58	2.42	2.30	2.20
	.010	6.66	4.63	3.80	3.34	3.04	2.82	2.66	2.53
	.001	10.89	6.96	5.46	4.65	4.14	3.78	3.51	3.30

Degrees of freedom in the denominator

		Degrees of freedom in the numerator							
	p	9	10	15	20	30	60	120	1000
9	.100	2.44	2.42	2.34	2.30	2.25	2.21	2.18	2.16
	.050	3.18	3.14	3.01	2.94	2.86	2.79	2.75	2.71
	.025	4.03	3.96	3.77	3.67	3.56	3.45	3.39	3.34
	.010	5.35	5.26	4.96	4.81	4.65	4.48	4.40	4.32
	.001	10.11	9.89	9.24	8.90	8.55	8.19	8.00	7.84
10	.100	2.35	2.32	2.24	2.20	2.16	2.11	2.08	2.06
	.050	3.02	2.98	2.85	2.77	2.70	2.62	2.58	2.54
	.025	3.78	3.72	3.52	3.42	3.31	3.20	3.14	3.09
	.010	4.94	4.85	4.56	4.41	4.25	4.08	4.00	3.92
	.001	8.96	8.75	8.13	7.80	7.47	7.12	6.94	6.78
12	.100	2.21	2.19	2.10	2.06	2.01	1.96	1.93	1.91
	.050	2.80	2.75	2.62	2.54	2.47	2.38	2.34	2.30
	.025	3.44	3.37	3.18	3.07	2.96	2.85	2.79	2.73
	.010	4.39	4.30	4.01	3.86	3.70	3.54	3.45	3.37
	.001	7.48	7.29	6.71	6.40	6.09	5.76	5.59	5.44
15	.100	2.09	2.06	1.97	1.92	1.87	1.82	1.79	1.76
	.050	2.59	2.54	2.40	2.33	2.25	2.16	2.11	2.07
	.025	3.12	3.06	2.86	2.76	2.64	2.52	2.46	2.40
	.010	3.89	3.80	3.52	3.37	3.21	3.05	2.96	2.88
	.001	6.26	6.08	5.54	5.25	4.95	4.64	4.47	4.33
20	.100	1.96	1.94	1.84	1.79	1.74	1.68	1.64	1.61
	.050	2.39	2.35	2.20	2.12	2.04	1.95	1.90	1.85
	.025	2.84	2.77	2.57	2.46	2.35	2.22	2.16	2.09
	.010	3.46	3.37	3.09	2.94	2.78	2.61	2.52	2.43
	.001	5.24	5.08	4.56	4.29	4.00	3.70	3.54	3.40
25	.100	1.89	1.87	1.77	1.72	1.66	1.59	1.56	1.52
	.050	2.28	2.2	2.09	2.01	1.92	1.82	1.77	1.72
	.025	2.68	2.61	2.41	2.30	2.18	2.05	1.98	1.91
	.010	3.22	3.13	2.85	2.70	2.54	2.36	2.27	2.18
	.001	4.71	4.56	4.06	3.79	3.52	3.22	3.06	2.91
50	.100	1.76	1.73	1.63	1.57	1.50	1.42	1.38	1.33
	.050	2.07	2.03	1.87	1.78	1.69	1.58	1.51	1.45
	.025	2.38	2.32	2.11	1.99	1.87	1.72	1.64	1.56
	.010	2.78	2.70	2.42	2.27	2.10	1.91	1.80	1.70
	.001	3.82	3.67	3.20	2.95	2.68	2.38	2.21	2.05
100	.100	1.69	1.66	1.56	1.49	1.42	1.34	1.28	1.22
	.050	1.97	1.93	1.77	1.68	1.57	1.45	1.38	1.30
	.025	2.24	2.18	1.97	1.85	1.71	1.56	1.46	1.36
	.010	2.59	2.50	2.22	2.07	1.89	1.69	1.57	1.45
	.001	3.44	3.30	2.84	2.59	2.32	2.01	1.83	1.64
200	.100	1.66	1.63	1.52	1.46	1.38	1.29	1.23	1.16
	.050	1.93	1.88	1.72	1.62	1.52	1.39	1.30	1.21
	.025	2.18	2.11	1.90	1.78	1.64	1.47	1.37	1.25
	.010	2.50	2.41	2.13	1.97	1.79	1.58	1.45	1.30
	.001	3.26	3.12	2.67	2.42	2.15	1.83	1.64	1.43
1000	.100	1.64	1.61	1.49	1.43	1.35	1.25	1.18	1.08
	.050	1.89	1.84	1.68	1.58	1.47	1.33	1.24	1.11
	.025	2.13	2.06	1.85	1.72	1.58	1.41	1.29	1.13
	.010	2.43	2.34	2.06	1.90	1.72	1.50	1.35	1.16
	.001	13.13	2.99	2.54	2.30	2.02	1.69	1.49	1.22

Degrees of freedom in the denominator

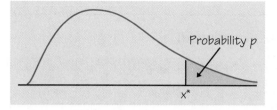

Table entry for p is the critical value x^* with probability p lying to its right.

TABLE E	Chi-square distribution critical values

df	.25	.20	.15	.10	.05	.025	.02	.01	.005	.0025	.001	.0005
1	1.32	1.64	2.07	2.71	3.84	5.02	5.41	6.63	7.88	9.14	10.83	12.12
2	2.77	3.22	3.79	4.61	5.99	7.38	7.82	9.21	10.60	11.98	13.82	15.20
3	4.11	4.64	5.32	6.25	7.81	9.35	9.84	11.34	12.84	14.32	16.27	17.73
4	5.39	5.99	6.74	7.78	9.49	11.14	11.67	13.28	14.86	16.42	18.47	20.00
5	6.63	7.29	8.12	9.24	11.07	12.83	13.39	15.09	16.75	18.39	20.51	22.11
6	7.84	8.56	9.45	10.64	12.59	14.45	15.03	16.81	18.55	20.25	22.46	24.10
7	9.04	9.80	10.75	12.02	14.07	16.01	16.62	18.48	20.28	22.04	24.32	26.02
8	10.22	11.03	12.03	13.36	15.51	17.53	18.17	20.09	21.95	23.77	26.12	27.87
9	11.39	12.24	13.29	14.68	16.92	19.02	19.68	21.67	23.59	25.46	27.88	29.67
10	12.55	13.44	14.53	15.99	18.31	20.48	21.16	23.21	25.19	27.11	29.59	31.42
11	13.70	14.63	15.77	17.28	19.68	21.92	22.62	24.72	26.76	28.73	31.26	33.14
12	14.85	15.81	16.99	18.55	21.03	23.34	24.05	26.22	28.30	30.32	32.91	34.82
13	15.98	16.98	18.20	19.81	22.36	24.74	25.47	27.69	29.82	31.88	34.53	36.48
14	17.12	18.15	19.41	21.06	23.68	26.12	26.87	29.14	31.32	33.43	36.12	38.11
15	18.25	19.31	20.60	22.31	25.00	27.49	28.26	30.58	32.80	34.95	37.70	39.72
16	19.37	20.47	21.79	23.54	26.30	28.85	29.63	32.00	34.27	36.46	39.25	41.31
17	20.49	21.61	22.98	24.77	27.59	30.19	31.00	33.41	35.72	37.95	40.79	42.88
18	21.60	22.76	24.16	25.99	28.87	31.53	32.35	34.81	37.16	39.42	42.31	44.43
19	22.72	23.90	25.33	27.20	30.14	32.85	33.69	36.19	38.58	40.88	43.82	45.97
20	23.83	25.04	26.50	28.41	31.41	34.17	35.02	37.57	40.00	42.34	45.31	47.50
21	24.93	26.17	27.66	29.62	32.67	35.48	36.34	38.93	41.40	43.78	46.80	49.01
22	26.04	27.30	28.82	30.81	33.92	36.78	37.66	40.29	42.80	45.20	48.27	50.51
23	27.14	28.43	29.98	32.01	35.17	38.08	38.97	41.64	44.18	46.62	49.73	52.00
24	28.24	29.55	31.13	33.20	36.42	39.36	40.27	42.98	45.56	48.03	51.18	53.48
25	29.34	30.68	32.28	34.38	37.65	40.65	41.57	44.31	46.93	49.44	52.62	54.95
26	30.43	31.79	33.43	35.56	38.89	41.92	42.86	45.64	48.29	50.83	54.05	56.41
27	31.53	32.91	34.57	36.74	40.11	43.19	44.14	46.96	49.64	52.22	55.48	57.86
28	32.62	34.03	35.71	37.92	41.34	44.46	45.42	48.28	50.99	53.59	56.89	59.30
29	33.71	35.14	36.85	39.09	42.56	45.72	46.69	49.59	52.34	54.97	58.30	60.73
30	34.80	36.25	37.99	40.26	43.77	46.98	47.96	50.89	53.67	56.33	59.70	62.16
40	45.62	47.27	49.24	51.81	55.76	59.34	60.44	63.69	66.77	69.70	73.40	76.09
50	56.33	58.16	60.35	63.17	67.50	71.42	72.61	76.15	79.49	82.66	86.66	89.56
60	66.98	68.97	71.34	74.40	79.08	83.30	84.58	88.38	91.95	95.34	99.61	102.7
80	88.13	90.41	93.11	96.58	101.9	106.6	108.1	112.3	116.3	120.1	124.8	128.3
100	109.1	111.7	114.7	118.5	124.3	129.6	131.1	135.8	140.2	144.3	149.4	153.2

Chapter 1

1.1 (a) Male and female members of the class. (b) Two. Pulse rates, gender. (c) Pulse rate: beats per minute. Gender: male, female (d) Pulse rate is quantitative; gender is categorical.

1.3 (a) Categorical; (b) quantitative; (c) categorical; (d) categorical; (e) quantitative; (f) quantitative.

1.5 (a) Roughly symmetric, though it might be viewed as SLIGHTLY skewed to the right. (b) About 15%. (c) Smallest: between -70% and -60%; largest: between 100% and 110%. (d) 23%.

1.7 Lightning: centered at noon (or "somewhere from 11:30 to 12:30"). Spread: 7 to 17 (or "6:30am to 5:30pm"). Shakespeare: centered at 4, spread from 1 to 12.

1.9 Outlier: 200. Center: between 137 and 140. Spread (ignoring outlier): 101 to 178.

1.11 Answers will vary, but your description should compare centers, spread, gaps, and outliers.

1.13 (a) Virginia Road Fatalities, 1986–96: The graph shows a general decline over the decade. Possible reasons include safer, newer, and better cars; more airbags and antilock brakes; increased use of seat belts (estimated at 70%, there is a mandatory seat belt law in Virginia); newer and better roads with safer design engineering; law enforcement and educational groups promoting safe driving habits. (b) Alcohol–related Fatalities, 1986–95: Again, there appears to be a gradual decline in recent years. Possible reasons include those in (a), as well as better education on the dangers of drunk driving; a growing national intolerance of drunk driving; and stiffer penalties.

1.15 Slightly skewed to the right, centered at 4.

1.17 (a) The distribution is skewed to the left, and centered at 46. 60 is *not* an outlier. (b) Maris's 61 is not an outlier. The center of Ruth's distribution is much higher than the center for Maris, and two-thirds of Ruth's numbers are higher than all but two of Maris's.

1.19 (a) Round to nearest integer before creating stemplot. (b) There is no particular observable shape (considering symmetry and skewness). (c) (Time plot). (d) The time plot shows an increasing trend—adjustments should be made to counteract the rising tensions.

1.21 There are two distinct groups of states—"less than 30%" and "more than 40%." There are no particular outliers.

1.23   Skewed to the right. New Jersey (at \$9159 per student) might be considered an outlier.

1.25   (a) $\bar{x} = 141.056$. (b) Without the outlier, $\bar{x}^* = 137.588$. The outlier makes the mean larger.

1.27   For Ruth: $M = 46$; for Maris: $M = 24.5$.

1.29 $\bar{x} = \$480,000 \div 8 = \$60,000$. Seven of the eight employees (everyone but the owner) earned less than the mean. $M = \$22,000$.

1.31   Yes. The middle 50% of the data are typically well insulated from the effects of a few extreme observations.

1.33   (a) There is no particular skewness, so M should be about the same as $\bar{x}$. (b) Five-number summary: 42, 51, 55, 58, 69; $\bar{x} = 54.833$ (c) Between $Q1$ and $Q3$: 51 to 58. (d) There is an outlier: Reagan (69). Harrison (68) is not an outlier.

1.35   (a) $\bar{x} = 5.4$. (b) $\Sigma(x_1 - \bar{x})^2 = 2.06$; $s^2 = 0.412$; $s = 0.6419$.

1.37   The stemplot reveals two peaks—one around 470 and one around 520—with a valley in between. The mean and median fall between these two peaks.

1.39   There seems to be little difference between beef and meat hot dogs, but poultry hot dogs are generally lower in calories than the other two.

1.41.   A stemplot (or histogram) shows the distribution to be fairly symmetrical, with a low outlier of $4.88 - \bar{x}$ and s should be reasonable in this setting. $\bar{x} = 5.4479$ and $s = 0.22095$; $\bar{x}$ is our best estimate of the earth's density.

1.43   The distribution is clearly skewed to the right, with the top two or three salaries as outliers. Use the five-number summary; 109, 158, 635, 2300, 6200.

1.45   (a) Mean. (b) Median.

1.47   (a) 1, 1, 1, 1. (Or 2, 2, 2, and 2, etc.) (b) 0, 0, 10, 10. (c) For (a), any four identical numbers will have $s = 0$. The answer for (b) is unique.

1.49   (a) Stemplot is symmetric with no *obvious* outliers (although 10.17 and 9.75 seem to be unusually high, and 6.75 is extraordinarily low). (b) Outliers show up more clearly in time plot. (c) $\bar{x} = 8.3628$ and $s = 0.4645$. (d) Within $\bar{x} \pm s$: 25 (64.1%); within $\bar{x} \pm 2s$: 37 (94.9%); 100% fall within $\bar{X} \pm 3s$.

1.51   DiMaggio: 12, 20.5, 30, 32, 46. Mantle: 13, 21) or 20.5), 28.5, 37 (or 37.75), 54. The first three numbers in both summaries are similar, but Mantle's Q_3 and maximum are higher—he apparently had higher "big seasons" than DiMaggio.

1.53   (a) Distribution is clearly skewed to the right (as expected). Main peak occurs from 50 to 150—the guinea pigs which lived over 500 days are apparent outliers. (b) The skewness makes the mean larger than the median.

(c) 43, 82.25, 102.5, 153.75, 598. The skewness shows up in the difference between Q_3 and the maximum; it is much larger than the other differences between successive numbers.

1.55   (a) After the first two years, the median return is above zero all but once. However, there is no particular evidence of a trend. (b) The spread of the boxplots is considerably smaller in recent years (with the exception of 1987). (c) Four of the five high outliers are visible: in 1973, 1975, 1979, and 1974. Most of the outliers occurred in the early years, and lately the variability has lessened considerably. The low in 1987 stands out as a "real" deviation from the pattern.

1.57   (b) The plot shows a decreasing trend—fewer disturbances overall in the later years—and more importantly, there is an apparent cyclic behavior. The spring and summer months (April through September) generally have the most disturbances because more people are outside during those periods.

1.59   Total value of stock is likely to be skewed to the right—there are a (relatively) few companies with high market values which increase the mean, but not the median.

Chapter 2

2.1   There are many correct drawings. Here are two possibilities:

(a)   (b)

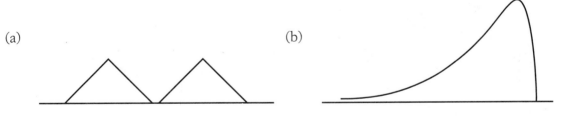

2.3   (a) 0.2. (b) 0.6. (c) 1. (d) 0.35.

2.5   The uniform distribution. Each of the 6 bars should have a height of 20.

2.7   (a) 2.5% (this is 2 standard deviations above the mean). (b) 69 ± 5; that is, 64 to 74 inches. (c) 16%. (d) 84%.

2.9   (a) 50. (b) 2.5. (c) 84. (d) 99.85.

2.11   Approximately 0.2 (for the tall one) and 0.5.

2.13   (a) 266 ± 32, or 234 to 298 days. (b) Less than 234 days. (c) More than 298 days.

2.15   No. Within 4 standard deviations is .999937 area. Going out to 5 standard deviations gives area $=$ .999999, which rounds to 1 for 4 decimal place accuracy.

2.17 (b) 68%: (58.3, 67.9). 95%: (53.5, 72.7). 99.7%: (48.7, 77.5).

2.19 (a) Outcomes around 25 are most likely. (d) The distribution should be roughly symmetric, with center at about 25, single peaked at the center, standard deviation about 3.5, and few or no outliers. The normal density curve should fit this histogram well.

2.21 The standard normal density function is defined by the formula $y = \frac{1}{\sigma\sqrt{2\pi}}e^{-\frac{1}{2}\left(\frac{x-\mu}{\sigma}\right)^2}$ when $\mu = 0$ and $\sigma = 1$.

2.23 (a) About $-0.675(-0.67449)$. (b) About 0.25 (0.253347).

2.25 (a) 65.5%. (b) 5.5%. (c) About 127 (or more).

2.27 Cobb: $z = 4.15$; Williams: $z = 4.26$; Brett: $z = 4.07$. Williams' z-score is highest.

2.29 (a) About 0.84. (b) About 0.385.

2.31 (a) About 5.2%. (b) About 55%. (c) Approximately 279 days or longer.

2.33 (a) At about ± 1.28. (b) 64.5 ± 3.2, or 61.3 and 67.7.

2.35 No answers required.

2.37 The normal probability plot shows a strong linear trend. The presidents' ages are approximately normally distributed.

2.39 (b) M = 0.5, $Q_1 \cong .3$, $Q_3 \cong .7$. (c) 25.2%. (d) 49.6%.

2.41 13.

2.43 Soldiers whose head circumference is outside the range 22.8 ± 1.81—approximately, less than 21 in or greater than 24.6 in.

2.45 (a) The distribution is slightly skewed left with no obvious outliers (a boxplot shows that 10.17 is an outlier). (b) The remaining data are approximately normally distributed; the mean is slightly less than the median.

Chapter 3

3.1 Age 6 height is explanatory, and age 16 height is the response. Both are quantitative.

3.3 "Treatment"—old or new—is the (categorical) explanatory variable. Survival time is the (quantitative) response variable.

3.5   (a) Number of jet skis in use. (b) There is a strong straight line relationship between number of jet skis in use and accidents.

3.7   (a) Positive association. (With more jet skis there are more accidents.) (b) The association is linear. (c) The association is very strong. For 1 million jet skis, there would be about 5,000 accidents.

3.9   (b) Positive association, linear, moderately strong. (c) The description of the male subjects' plot is much the same, though the scatter appears to be greater. The males typically have larger values for both variables.

3.11   (a) Lowest: about 107 calories (with about 145 mg of sodium); highest: about 195 calories, with about 510 mg of sodium. (b) There is a positive association; high-calorie hot dogs tend to be high in salt, and low-calorie hot dogs tend to have low sodium. (c) The lower left point is an outlier. Ignoring this point, the remaining points seem to fall roughly on a line. The relationship is moderately strong.

3.13   Either variable could go on the vertical axis. The plot shows a strong positive linear relationship, with no outliers; there appears to be only one species represented.

3.15   (a) Planting rate is explanatory. (b) See (d). (c) The pattern is curved—high in the middle, and lower on the ends. Not linear, and there is neither positive nor negative association. (d) 20,000 plants per acre seems to give the highest average yield.

3.17   (a) The means are (in the order given) 47.167, 15.667, 31.5, and 14.833. (b) Yellow seems to be the most attractive, and green is second. White and blue are poor attractors. (c) Positive or negative association make no sense here because color is a categorical variable (what is an "above-average" color?)

3.19   (a) With x as femur length and y as humerus length: $\bar{x} = 58.2$, $s_x = 13.20$; $\bar{y} = 66.0$, $s_y = 15.89$; $r = 0.994$.

3.21   Positive but not near 1 (there is positive association but a fair amount of scatter).

3.23   (a) The plot shows a strong positive linear relationship, with little scatter, so we expect that r is close to 1. (b) r would not change—it is computed from unitless standardized values.

3.25   (a) Both correlations should be positive; the men's may be slightly smaller since those points are more spread out. (b) Women: $r_w = 0.87645$; Men: $r_m = 0.59207$. (c) Women $\bar{x}_w = 43.03$; Men: $\bar{x}_m = 53.10$. The difference in means has no effect on the correlation. (d) There would be no change, since standardized measurements are dimensionless.

3.27 $r = 0.25310$ for both sets of data. The points have been transformed, but the distances between corresponding points and the strength of association have not changed.

3.29 $r = 0.17162$, close to zero because the relationship is a curve rather than a line.

3.31   (a) $a = 1.0892$ and $b = 0.1890$, as given. (b) $\bar{x} = 22.31$, $s_x = 17.74$; $\bar{y} = 5.306$, $s_y = 3.368$; $r = 0.99526$. Except for roundoff error, we again find $b = 0.1890$ and $a = 1.0892$.

3.33   (a) A negative association—the pH decreased (i.e., the acidity increased) over the 150 weeks. (b) The initial pH was 5.4247; the final pH was 4.6350. (c) The slope is -0.0053; the pH decreased by 0.0053 units per week (on average).

3.35 $(\bar{x}, \bar{y})$ satisfied the equation $a = \bar{y} - b\bar{x}$.

3.37   (a) Put speed on the horizontal axis. (b) There is a very strong positive linear relationship; $r = 0.9990$. (c) $\hat{y} = 1.76608 + 0.080284x$. (d) $r^2 = 0.998$, so nearly all (99.8%) the variation in steps per second is explained by the linear relationship. (e) The regression line would be different; the line in (c) minimizes the sum of the squared *vertical* distances on the graph, while this new regression would minimize the squared *horizontal* distances. r^2 would remain the same, however.

3.39   (b) The line is clearly *not* a good predictor of the actual data—it is too high in the middle and too low on each end. (c) The sum is -0.01.

3.41   (a) Without Child 19, $\hat{y}^* = 109.305 - 1.1933x$. Child 19 is not very influential, since removing this data point does not change the line substantially. (b) With all children, $r^2 = 0.410$; without Child 19, $r^2 = 0.572$. With Child 19's high Gesell score removed, there is less scatter around the regression line—more of the variation is explained by the regression.

3.43   (a) $y(\text{weight}) = 100 + 40x$ grams (c) When $x = 104$, $y = 4260$ grams, or about 9.4 pounds. The regression line is only reliable for "young" rats; rats do not grow at a constant rate throughout their entire life.

3.45   (b) $\hat{y} = 71.950 + 0.38333x$. (c) When $x = 40$ months, $\hat{y} = 87.2832$; when $x = 60$, $\hat{y} = 94.9498$. (d) Sarah's rate: 0.38 cm/month; normal rate: 0.5 cm/month ($0.5 = 6/60 - 48)$).

3.47   When $x = 480$, $\hat{y} = 255.95$ cm $= 100.77$ in $= 8.4$ feet!

3.49   (a) About 69.4% of the variation is explained ($r^2 = 0.694$). (b) the sum is zero. (c) the residuals change from negative to positive in 1996. In that year, the regression line changes from overestimating to underestimating.

3.51   The data show a downward trend. $\hat{y} = 950.5615 - .3647x$ and $r = .9831$. The LSRL appears to be a good model and the high r value shows a strong association. The first point ($x = 1868$) has the largest residual. About 0.44 seconds are lost each year on average. One might feel comfortable predicting the record in 2000, but it may be risky predicting the record in 2005.

3.53 (a) Start with points (1,1) and (2,2). Then add the influential point (0,4). (b) Consider {(1, 1), (1, 2), (2, 1.1), (2, 2)}. Then add the influential point (10,10).

3.55 (b) There is a moderately strong positive linear association. There are no really extreme observations, though Bank 9 did rather well. Franklin does not look out of place. (c) $\hat{y} = 7.573 + 4.9872x$. (d) Franklin's predicted income was $\hat{y} = 26.5$ million dollars — almost twice the actual income. The residual is -12.7.

3.57 The plot shows an apparent negative association between nematode count and seedling growth. The correlation supports this: $r = -0.78067$. About 61% of the variation in growth can be accounted for by a linear relationship with nematode count.

3.59 (a) No. See Exercise 3.53. (b) No. See Exercise 3.53.

Chapter 4

4.1 (e) The residual plot of the transformed data shows no clear pattern, so the line is a reasonable model for these points. (f) $\hat{y} = 10^{-1094.51} x 10^{.5558x}$. (g) The predicted number of acres defoliated in 1982 is 12,178,673.85 acres.

4.3 (a) The sum of the squares of the deviations is 7795.687. The quantity that was minimized was the sum of the squares of the deviation of the transformed points. (b) The sum of the squares of the residuals is .0143. (c) There's no reason to expect the answers to be the same.

4.5 (a) The predicted rate of a one-ounce letter in 2005 is \$.72. The cost should reach \$.50 in 1999, according to our model.

4.7 (a) All the ratios, except the first, are about 1.1. (b) $r = .998$. The residual plot shows a random scatter of points. (c) The predicted number of EFT transactions for the year 2000 is 17,622.7 million.

4.9 (a) The exponential equation $\hat{y} = 10^{-10.8137} x 10^{.0063891x}$. (b) The Louisiana Purchase in 1803 added a large territory but relatively few people, so density declined. A similar drop occurred between 1840 and 1850 due to the acquisition of Mexican territory, Texas and California.

4.11 (a) The last 3 points don't appear to fit the pattern. (b) Average ratio for first 8 points is about 1.05 (c) $r = .9864$. (d) $\hat{y} = 10^{-45.865} x 10^{.0247x}$ (e) The predicted number of violent crimes in 1986 is about 1424. The residual is 65.

4.13 (a) If price is proportional to surface *area*, then power regression ($\hat{y} = ax^b$) is an appropriate model: $\hat{y} = .121x^{1.5516}$ ($r = .9762$), even though a straight line and a logarithmic curve appear to fit these data better. Larger pizzas are priced to give better value. (b) The giant pizza (18″) is under-priced. Note that without the giant pizza, the power regression model becomes $\hat{y} = .0348x^{2.065}$. The power of x is very close to the 2 you would expect, and the r value is 0.9988. (c) A new 6″ pizza should cost $1.95, according to our power function. (d) A 24″ pizza would cost $16.76.

4.15   (b) The curve should be decreasing: as disk diameters increase, proportion of wins decreases. (d) y (0) should equal 1, and y should be 0 when disk diameter is about 229 mm (9 inches).

4.17   (a) The heavier the athlete, the more pounds he can lift. (b) The parabola $y = -.0163x^2 + 8.41x - 132.984$ appears to fit the data reasonably well. (c) The model predicts $\hat{y}(0) = -132.984$ and $\hat{y}(300) = 423.016$ both clearly wrong. The model is not appropriate outside the given weight limits. (d) The regression equation coefficients would change.

4.19   (a) $\hat{y} = 1166.93 - 0.58679x$. (b) The farm population decreased about 590 thousand (0.59 million) people per year. The regression line explains 97.7% of the variation. (c) $-782,100$—but a population must be greater than or equal to 0.

4.21   The correlation would be smaller because there is much more variation among the individual data points.

4.23   Age is the lurking variable: we would expect both quantities—shoe size and reading level—to increase as the child ages.

4.25   (a) $r^2 = 0.925$—more than 90% of the variation in one SAT score can be explained through a linear relationship with the other score. (b) The correlation would be much smaller, since individual students have much more variation between their scores. By averaging—or, as in this case, taking the median of—the scores of large groups of students, we muffle the effects of these individual variations.

4.27   The explanatory variable is whether or not a student has taken at least two years of foreign language, and the score on the test is the response. The lurking variable is the students' English skills *before* taking (or not taking) the foreign language: students who have a good command of English early in their high school career are more likely to choose (or be advised to choose) to take a foreign language.

4.29   Apparently drivers are typically larger and heavier men than conductors—and are therefore more predisposed to health problems such as heart disease.

4.31   24.9%, 43.9% and 31.3% (total is 100.1% due to rounding).

4.33   12.9%, 12.5% and 30.8%. The percentage of people who did not finish high school is about the same for the two younger groups, but more than double for those 55 and over.

4.35   Among 35 to 54 year-olds: 12.5% never finished high school, 33.0% finished high school, 27.3% had some college, and 27.2% completed college. This is more like the 25–34 age group than the 55 and over group.

4.37   Two possible answers: Row 1–30, 20; Row 2–30, 20; and Row 1–10, 40; Row 2–50, 0.

4.39   (a) Of students with two smoking parents, 22.5% smoke; with one smoking parent, 18.6% smoke; among students with no smoking parents, only 13.9% smoke. (c) It appears that children of smokers are more likely to smoke—even more so when both parents are smokers.

4.41   (a) White defendants: 19 got the death penalty; 141 did not. For black defendants, the numbers are 17 vs. 149.

4.43   (a) 11,374,000. (b) 51.2%. (c) 60.8%, 22.1%, 68.4%, and 11.0% (d) The 18–21 age group makes up more than 60% of full-time students, but comprises less than 20% of part-time students.

4.45   (a) Counts: 127, 5829, 3031, 1907, 480. Percents: 1.1%, 51.2%, 26.6%, 16.8%, 4.2%. (b) 0.2%, 22.1%, 33.4%, 34.1%, and 10.1%. (c) The biggest difference between the distributions in (a) and (b) is that among part-time students at 2-year colleges, there is a markedly lower percentage of 18–21 year-olds, and considerable increases in the higher age brackets—the last two age categories are more than twice as large in (b) as they were in (a).

4.47   The lurking variable is temperature or season. More flu cases occur in winter, and less ice cream is sold in winter. This is an example of common response.

4.49   (a) 59.0%. (b) Larger businesses were less likely to respond: only 37.5% of the small businesses did not respond, compared to 59.5% of medium-sized businesses and 80% of large businesses. (c) Bar chart below.

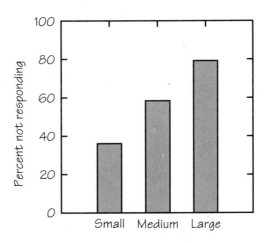

4.51 $2^{1 \times 4} = 16$, $2^{5 \times 4} = 1,048,576$.

4.53   (a) $y = 500(1.075)^{year}$. 537.50, 577.81, 621.15, 667.73, 717.81, 771.65, 829.52, 891.74, 958.62, 1030.52 (c) The logs are 2.73, 2.76, 2.79, 2.82, 2.86, 2.89, 2.92, 2.95, 2.98, 3.01.

4.55   (a) The pattern appears to be exponential. (b) Yes. (d) 5.617046, $414,043.00. (e) The point is somewhat below the line. The rate of growth appears to be slower.

4.57   (b) The log plot looks linear from 1790 to 1880 and also from 1880 to 1990. The slope is less in the second period. (c) Using only the data from 1900 to 1990, we obtain the least squares model $\log y = -8.8605 + .0056645x$, and the inverse transformation yields the exponential model $y = 10^{-8.8605} \times 10^{.0056645x}$. (d) When $x = 2000$, $y = 10^{2.4685} = 294.1$ million.

4.59   (a) There is a negative association. As year increases, infant mortality decreases. (b) The common ratio is about .935, so try an exponential model: $y = 10^{32.3563} \times 10^{-.0158x}$. (Note: $r = .98584$ for transformed data.) (c) For years 1994–99: 6.30, 6.07, 5.85, 5.64, 5.44, and 5.25. This (exponential decay) model should not be used to predict infant mortality out to the year 2005.

4.61   (a) and (b) The data follow a model of the form ax^b where a is a negative number. Power regression yields $y = .3x^{-2.013}$. (c) The intensity of the light source appears to vary inversely with the square of the distance from the bulb. (d) The formula for intensity as a function of distance is given by $y = 900/x^2$ where y is measured in candlepower, x in meters. With an appropriate units change to candelas, this formula appears to fit the data.

4.63   80% of all suicide victims were men—that in itself is a major difference. Firearms were most the common method for both sexes: 65.9% of male suicides used firearms, as did 42.1% of females. Poison was a close second for women at 35.6%, compared with only 13.0% for men. In the last two categories, the percentages were pretty close: Men chose hanging 14.9% of the time, and women chose it 12.2% of the time; 6.2% of men and 10.1% of women fell into the "other" group.

4.65   (a) 12,625; roundoff error. (b) 19.3%, 59.3%, 11.8%, and 9.6%. (c) 18–24: 71.3%, 26.5%, 0.06%, 2.0%. 40–64: 5.8%, 72.5%, 7.6%, 14.1%. Among the younger women, almost three-fourths have not yet married, and those that are married have had little time to become widowed or divorced. Most of the older group is or has been married—only about 6% are still single. (d) 48.6% of single women are 18–24, 35.9% are 25–39, 10.7% are 40–64, and 4.9% are 65 or older.

4.67   One very simple possibility is shown on the next page, using 10 smokers and 10 non-smokers. Lumped together, we find that there are 5 people in each classification (overweight/died early, etc.). There are, of course, in-finitely more examples.

Early Death?	Smoker Overweight? Yes	Smoker Overweight? No	Non-Smoker Overweight? Yes	Non-Smoker Overweight? No
Yes	1	4	3	1
No	0	2	6	3

Chapter 5

5.1 The (desired) population is employed adult women, the sample is the 48 club members who returned the survey.

5.3 Only persons with a strong opinion on the subject—strong enough that they are willing to spend the time, and at least 50 cents—will repond to this advertisement.

5.5 Starting with 01 and numbering down the columns, one chooses 04–Bonds, 10–Fleming, 17–Liao, 19–Naber, 12–Goel, and 13–Gomez.

5.7 Label from 001 to 440; select 400, 077, 172, 417, 350, 131, 211, 273, 208, and 074.

5.9 Label midsize accounts 001–500, and small accounts from 0001–4400. Select 417, 494, 322, 247, and 097 for the midsize group, and then 3698, 1452, 2605, 2480, and 3716 for the small group.

5.11 The higher no-answer was probably the second period—more families are likely to be gone for vacations, etc. Nonresponse of this type might underrepresent those who are more affluent (and are able to travel).

5.13 The increased sample size gives more accurate information about the population.

5.15 (a) An individual is a small business; the specific population is "eating and drinking establishments" in the large city. (b) An individual is an adult; the Congressman's constituents are the *desired* population. The letter-writers are a voluntary sample and do not represent that population well. (c) Individual: auto insurance claim; the population is all claims filed in a given month.

5.17 The call-in poll is faulty in part because it is a voluntary sample. Furthermore, even a small charge like 50 cents can discourage some people from calling in—especially poor people. Reagan's Republican policies appealed to upper-class voters, who would be less concerned about a 50-cent charge than lower-class voters who might favor Carter.

5.19   Number the bottles across the rows from 01 to 25, then select 12–B0986, 04–A1101, and 11–A2220. (If numbering is done down columns instead, the sample will be A1117, B1102, and A1098.)

5.21   (a) False—if it were true, then after looking at 39 digits, we would know whether or not the 40th digit was a 0, contrary to property 2. (b) True— there are 100 pairs of digits 00 through 99, and all are equally likely. (c) False—0000 is just as likely as any other string of four digits.

5.23   It is *not* an SRS, because some samples of size 250 have no chance of being selected (e.g., a sample containing 250 women).

5.25   A smaller sample gives less information about the population. "Men" constituted only about one-third of our sample, so we know less about that group than we know about all adults.

5.27   (a) The liners are the experimental units. (b) Factor: heat applied to the liners; levels: 250°F, 275°F, 300°F, and 325°F. (c) The force required to open the package is the response variable.

5.29   (a) The experimental units are the batches of the product; the yield of each batch is the response variable. (b) There are two factors: temperature (with 2 levels) and stirring rates (with 3 levels), for a total of 6 treatments. Assign 50°C and 60 rpm to treatment group 1, etc. (c) Since two experimental units will be used for each treatment, we need 12.

5.31   The diagram should show random allocation of 5 liner pairs to each of 4 groups, then applying the appropriate level of heat to each group, then measuring the force required.

5.33   Number the liners from 01 to 20. Group 1: 16, 04, 19, 07, and 10; Group 2: 13, 15, 05, 09, and 08; Group 3: 18, 03, 01, 06, and 11. The others are in group 4.

5.35   The possible differences between the two years would confound the effects of the treatments. For example, if this summer is warmer; the customers may run their air conditioners more often.

5.37   There almost certainly was *some* difference between the sexes and between blacks and whites; the difference between men and women was so large that it is unlikely to be due to chance. For black and white students, however, the difference was small enough that it could be attributed to random variation.

5.39   (a) If only the new drug is administered, and the subjects are then interviewed, their responses will not be useful, because there will be nothing to compare them to: How much "pain relief" does one expect to experience? (b) Randomly assign 20 patients to each of three groups: Group 1, the placebo group; Group 2, the aspirin group; and Group 3, which will receive the new medication. After treating the patients; ask them how much pain relief they feel, and then compare the average pain relief experienced by each group. (c) The subjects should certainly not know what drug they are getting—a patient told that she is receiving a placebo, for example, will probably not expect any pain relief. (d) Yes—presumably, the researchers would like to conclude that the new medication is better than aspirin. If it is not double-blind, the interviewers may subtly influence the subjects into giving reponses that support that conclusion.

5.41   (a) Split the subjects by gender. Randomly allocate the 24 women into 6 groups of 4 each; and allocate the men into 6 groups of 2. Each group receives the appropriate treatment (1–6), and then the subjects' attitudes about the product, etc., are measured. (b) Number the women from 01 to 24, and the men from 01 to 12. First we look for 20 women's numbers, and find Group 1: 12, 13, 04, 18; Group 2: 19, 24, 23, 16; Group 3: 02, 08, 17, 21; Group 4: 10, 05, 09, 06; Group 5: 01, 20, 22, and 07. The remaining women are group 6. For the men, we find Group 1: 05, 09; Group 2: 07, 02; Group 3: 01, 08; Group 4: 11, 06; Group 5: 12, 04; the rest go into Group 6.

5.43   The randomization will vary with the starting line in Table B. *Completely randomized design:* Randomly assign 10 students to "Group 1" (which has the trend-highlighting software) and the other 10 to "Group 2" (which does not). Compare the performance of Group 1 with that of Group 2. *Matched pairs design:* Each student does the activity twice, once with the software and once without. Randomly decide (for each student) whether they have the software the first or second time. Compare performance with the software and without it. *Alternate matched pairs design:* Again, all students do the activity twice. Randomly assign 10 students to Group 1 and 10 to Group 2. Group 1 uses the software the first time; Group 2 uses the software the second time.

5.45   (a) It is an experiment (albeit a poorly designed one), because a treatment (herbal tea) is imposed on the subjects. (b) No, it is a study—the scores on the English test are merely observed for the various subjects.

5.47   (a) Subjects: physicians; factor: medication (with two levels—aspirin and placebo); response: observing health, specifically whether the subjects have heart attacks or not. (b) The diagram should show random allocation into two groups of 11,000 each; one group receives aspirin, the other, the placebo. Then observe heart attacks.

5.49   (a) Randomly assign 20 men to each of two groups. Record each subject's blood pressure, then apply the treatments: a calcium supplement for Group 1, and a placebo for Group 2. After sufficient time has passed, measure blood pressure again and observe any change. (b) Number from 01 to 40 down the columns. Group 1 is 18–Howard, 20–Imrani,

26–Maldonado, 35–Tompkins, 39–Willis, 16–Guillen, 04–Bikalis, 21–James, 19–Hruska, 37–Tullock, 29–O'Brian, 07–Cranston, 34–Solomon, 22–Kaplan, 10–Durr, 25–Liang, 13–Fratianna, 38–Underwood, 15–Green, and 05–Chen.

5.51   Responding to a placebo does not imply that the complaint was not "real"—38% of the placebo group in the gastric freezing experiment improved, and those patient's really had ulcers. The placebo effect is a *psychological* response, but it may make an actual *physical* improvement in the patient's health.

5.53   (a)

Measure ↗
BP for all
subjects ↘

Black men → Random assignment ↗ Group 1 → Placebo ↘
                              ↘ Group 2 → Calcium ↗ Observe change in BP

White men → Random assignment ↗ Group 1 → Placebo ↘
                              ↘ Group 2 → Calcium ↗ Observe change in BP

(b) A larger group gives more information—when more subjects are involved, the random differences between individuals have less influence, and we can expect the average of our sample to be a better representation of the whole population.

5.55   For the choices made in the solution to Exercise 5.54 (a) D,R,R,R,R, R,R,D,R,D — 3 Democrats, 7 Republicans. (b) R,D,D,R,R,R,R,D,R,R — 3 Democrats, 7 Republicans. (c) R,U,R,D,R,U,U,U,D,R — 2 Democrats, 4 Republicans, 4 undecided. (d) R,R,R,D,D,D,D,D,D,R — 6 Democrats, 4 Republicans.

5.57   (a) A single random digit simulates one shot, with 0 to 6 a hit and 7,8, or 9 a miss. Then 5 consecutive digits simulate 5 independent shots. (b) It will be shown in Chapter 8 that the theoretical probability of missing 3 or more shots (i.e., making 2 or fewer shots) is 0.1631, or about one time in six.

5.59   Let 1 = girl and 0 = boy. The command randInt(0,1) produces a 0 to 1 with equal likelihood. Continue to press ENTER. In 50 repetitions, we get a girl 47 times, and all 4 boys three times. Our simulation produced a girl 94% of the time, vs. a theoretical probability of 0.938.

5.61   The command randInt(1,365,23) $\rightarrow$ L$_1$:SortA(L$_1$) randomly selects 23 birthdays and assigns them to L$_1$. Then it sorts the day in increasing order. Scroll through the list to see duplicate birthdays. Repeat many times. For a large number of repetitions, there should be duplicate birthdays about half the time. To simulate 41 people, change 23 to 41 in the command and repeat many times. We assume that there are 365 days for birthdays, and that all days are equally likely to be a birthday.

5.63 (a) Read two random digits at a time from Table B. Let 01 to 13 represent a Heart, let 14 to 52 represent another suit, and ignore other two-digit numbers. (b) You should beat Slim about 44% of the time

5.65 (a) One digit simulates system A's response: 0 to 8 shut down the reactor, and 9 fails to shut it down. (b) One digit simulates system B's response: 0 to 7 shut down the reactor, and 8 or 9 fail. (c) A pair of consecutive digits simulates the response of both systems, the first giving A's response as in (a) , and the second B's response as in (b). If a single digit were used to simulate both systems, the reactions of A and B would be dependent—for example, if A fails, then B must also fail. (d) The true probability that the reactor will shut down is $1 - (0.2)(0.1) = 0.98$.

5.67 (a) Explanatory: treatment method; response: survival times. (b) No treatment is actively imposed; the women (or their doctors) chose which treatment to use. (c) Doctors may make the decision of which treatment to recommend based in part on how advanced the case is—for example, some might be more likely to recommend the older treatment for advanced cases, in which case the chance of recovery is lower.

5.69 (a) Label the students from 0001 to 3478. (b) Taking four digits at a time gives 2940, 0769, 1481, 2975, and 1315.

5.71 (a) One possible population: all full-time undergraduate students in the fall term on a list provided by the Registrar. (b) A stratified sample with 125 students from each year is one possibility. (c) Mailed questionnaires might have high non-response rates. Telephone interviews exclude those without phones, and may mean repeated calling for those that are not home. Face-to-face interviews might be more costly than your funding will allow.

5.73 (a) Diagram should show random allocation into 6 groups (2 batches each). Each group is given the appropriate treatment (temperature/stirring rate combination), then the yield is observed. (b) The first 10 numbers (between 01 and 12) are 06, 09, 03, 05, 04, 07, 02, 08, 10 and 11. So the 6th and 9th batches will receive treatment 1; batches 3 and 5 will be processed with treatment 2, etc.

5.75 Each subject should taste both kinds of cheeseburger, in a randomly selected order, and then be asked about their preference. Both burgers should have the same "fixings" (ketchup, mustard, etc.). Since some subjects might be able to identify the cheeseburgers by appearance, one might need to take additional steps (such as blindfolding, or serving only the center part of the burger) in order to make this a true "blind" experiment.

5.77 It means that the correlation is large enough (presumably, though not necessarily, in the positive direction) that it is unlikely to have occurred by chance.

5.79 (a) Let 01 to 05 represent demand for 0 cheesecakes. Let 06 to 20 represent demand for 1 cheesecake. Let 21 to 45 represent demand for 2 cheesecakes. Let 46 to 70 represent demand for 3 cheesecakes. Let 71 to 90 represent demand for 4 cheesecakes and let 91 to 99 and 00 represent demand for 5 cheesecakes. (b) The baker should make 2 cheesecakes each day to maximize his profits.

5.81 (a) A single digit simulates one try, with 0 or 1 a pass and 2 to 9 a failure. Three independent tries are simulated by three successive random digits. (b) With the convention of (a), 50 tries beginning in line 120 gives 25 successes, so the probability of success is estimated as $25/50 = 1/2$. [In doing the simulation, remember that you can end a repetition after 1 or 2 tries if the student passes, so that some repetitions do not use three digits. Though this is a proper simulation of the student's behavior, the probability of at least one pass is the same if three digits are examined in every repetition. The true probability is $1 - (0.8)^3 = 0.488$, so this particular simulation was quite accurate.] (c) No—learning usually occurs in taking an exam, so the probability of passing probably increases on each trial.

Chapter 6

6.1 Long trials of this experiment often approach 40% heads.

6.3 Obviously, results will vary with the type of thumbtack used.

6.5 In the long run, of a large number of hands of five cards, about 2% (one out of 50) will contain three of a kind. [Note: This probability is actually $\frac{88}{4165} = 0.02113$.]

6.7 (a) The results have been stored in a list. (b) Shak hit 52% of his shots. (c) The longest sequence of misses was 6 and the longest sequence of hits was 9.

6.9 (a) S = {germinates, does not germinate}. (b) If measured in weeks, for example S = {0, 1, 2, . . .}. (c) S = {A, B, C, D, F}. (d) S = {misses both, makes one, makes both}, or S = {misses both, makes first/misses second, misses first/makes second, makes both}. (e) S = {1, 2, 3, 4, 5, 6, 7}.

6.11 S = {all numbers between___ and___}. The numbers in the blanks may vary. Table 1.8 has values from 86 to 195 cal; the range of values in S should include at *least* those numbers. Some students may play it safe and say "all numbers greater than 0."

6.13   (a) 10,000 (b) 5,040 (c) 11,110.

6.15   (a) 1,000,000 (b) 17,576,000 (c) 175,760,000.

6.17   (a) 0.7037.

6.19   (a) The sum of the given probabilities is 0.9, so P(blue) = 0.1. (b) The sum of the given probabilities is 0.7, so P(blue) = 0.3. (c) P (plain M&M is red, yellow, or orange) = 0.2 + 0.2 + 0.1 = 0.5. P (peanut M&M is red, yellow, or orange) = 0.1 + 0.2 + 0.1 = 0.4.

6.21 P (either CV disease or cancer) = 0.45 + 0.22 = 0.67; P (other cause) = 0.33.

6.23   (a) The sum is 1, as we expect since all possible outcomes are listed. (b) 1 − 0.41 = 0.59. (c) 0.41 + 0.23 = 0.64. (d) (0.41)(0.41) = 0.1681.

6.25 $(1 − 0.05)^{12} = (0.95)^{12} \doteq 0.540$.

6.27   (a) $P(A) = \frac{38,225}{166,438} \doteq 0.230$ since there are 38,225 (thousand) people who have completed 4+ years of college out of 166,438 (thousand). (b) $P(B) = \frac{52,022}{166,438} \doteq 0.313$. (c) $P(A \text{ and } B) = \frac{8,005}{166,438} \doteq 0.048$; A and B are not independent since $P(A \text{ and } B) \neq P(A)P(B)$.

6.29   (a) Legitimate. (b) Not legitimate, because probabilities sum to more than 1. (c) Not legitimate, because probabilities sum to less than 1.

6.31   (a) The probabilities sum to 1. (b) P (female) = 0.43. (c) 1 − 0.03 − 0.01 = 0.96. (d) 0.11 + 0.12 + 0.01 + 0.04 = 0.28. (e) 1 − 0.28 = 0.72.

6.33   Look at the first five rolls in each sequence. All have one G and four Rs, so those probabilities are the same. In the first sequence, you win regardless of the sixth roll; for the second you win if the sixth roll is G, and for the third sequence, you win if it is R. The respective probabilities are $\left(\frac{2}{6}\right)^4 \left(\frac{4}{6}\right) = \frac{2}{243} \doteq 0.00823$, $\left(\frac{2}{6}\right)^4 \left(\frac{4}{6}\right)^2 = \frac{4}{729} \doteq 0.00549$, and $\left(\frac{2}{6}\right)^5 \left(\frac{4}{6}\right) = \frac{2}{729} \doteq 0.00274$.

6.35   (a) $(0.65)^3 \doteq 0.275$ (under the random walk theory). (b) 0.35 (since performance in separate years is independent). (c) $(0.65)^2 = (0.35)^2 = 0.545$.

6.37 $P(A \text{ or } B) = P(A) = P(B) − P(A \text{ and } B) = 0.125 + 0.237 − 0.077 = 0.285$.

6.39   (a) {A and B}: household is both prosperous and educated. $P(A \text{ and } B) = 0.077$. (b) {$A$ and B^c}: household is prosperous but not educated. $P(A \text{ and } B^c) = 0.048$. (c) {$A^c$ and B}: household is not prosperous but is educated. $P(A^c \text{ and } B) = 0.160$. (d) {$A^c$ and B^c}: household is neither prosperous nor educated. $P(A^c \text{ and } B^c) = 0.715$.

6.41   (a) $\frac{18,262}{99,585} \doteq 0.1834$. (b) $\frac{7,767}{18,262} \doteq 0.4253$. (c) $\frac{7,767}{99,585} = 0.0780$. (d) P (over 65 and married) = P(over 65) P(married | over 65) = (0.1834)(0.4253).

6.43   (a) $\frac{3,046}{58,929} \doteq 0.0517$. (b) "...married...age 18 to 24." (c) "0.0517 is the proportion of women who are *age 18 to 24* among those women who are *married.*"

6.45   (a) $\frac{770}{1626} \doteq 0.4736$. (b) $\frac{529}{770} = 0.6870$. (d) Using multiplication rule: P(male and bachelor's degree) $= P$(male)P(bachelor's degree | male) $= (0.4736)(0.6870) = 0.3254$. (Answers will vary with how much previous answers had been rounded.) Directly: $\frac{529}{1626} \doteq 0.3253$. [Note that the difference between these answers is inconsequential, since the numbers in the table are rounded.]

6.47   (a) 15% drink only cola. (b) 20% drink none of these.

6.49   If F = {dollar falls} and R = {renegotiation demanded}, then $P(F$ and $R) = P(F)P(R \mid F) = (0.4)(0.8) = .032$.

6.51   If F = {dollar falls} and R = {renegotiation demanded}, then $P(R) = P(F$ and $R) + P(F^c$ and $R) = 0.32 + P(F^c)P(R \mid F^c) = 0.32 + (0.6)(0.2) = 0.44$.

6.53 P(correct) $= P$(knows answer)$+P$(doesn't know, but guesses correctly) $= 0.75 + (0.25)(0.20) = 0.8$.

6.55 P (knows the answer | gives the correct answer) $= \frac{0.75}{0.80} = \frac{15}{16} = 0.9375$.

6.57   (a) (points in quarter disk)/(points in square) $\cong (\pi(1)/4)/(1)$, so $\pi \cong$ proportion of points in quarter disk.

6.59   (a) There are 10 pairs. Just using initials: {(A,D), (A,J), (A,S), (A,R), (D,J), (D,S), (D,R), (J,S), (J,R), (S,R)}. (b) Each has probability 1/10 = 10%. (c) Julie is chosen in 4 of 10 possible outcomes: 4/10 = 40%. (d) There are 3 pairs with neither Sam nor Roberto, so the probability is 3/10.

6.61   (a) P(B or O) $= 0.13 + 0.44 = 0.57$. (b) P(wife has type B and husband has type A) $= (0.13)(0.37) = 0.0481$. (c) P (one has type A and other has type B) $= (0.13)(0.37) + (0.37)(0.13) = 0.0962$. (d) P(at least one has type O) $= 1-P$ (neither has type O) $= 1-(1-0.44)(1-0.44) = 0.6864$.

6.63   (a) $P(X \geq 50) = 0.14 + 0.05 = 0.19$. (b) $P(X \geq 100 | X \geq 50) = \frac{0.05}{0.19} = \frac{5}{19}$.

6.65   The response will be "no" with probability $0.35 = (0.5)(0.7)$. If the probability of plagiarism were 0.2, then P(student answers "no") $= 0.4 = (0.5)(0.8)$. If 39% of students surveyed answered "no," then we estimate that $2 \cdot 39\% = 78\%$ have *not* plagiarized, so about 22% have plagiarized.

Chapter 7

7.1 $\frac{2}{6} = \frac{1}{3}$.

7.3   (a) 1% (b) All probabilities are between 0 and 1; the probabilities add to 1. (c) $P(X \leq 3) = 0.48 + .038 + 0.08 = 1 - 0.01 - 0.05 = 0.94$.

(d) $P(X < 3) = 0.48 + 0.38 = 0.86$. (e) Write either $X \geq 4$ or $X > 3$. The probability is 0.06. (f) Read two random digits from Table B. Here is the correspondence: 01 to 48 $\leftrightarrow$ Class 1, 49 to 86 $\leftrightarrow$ Class 2, 87 to 94 $\leftrightarrow$ Class 3, 95 to 99 $\leftrightarrow$ Class 4, and 00 $\leftrightarrow$ Class 5. Repeatedly generate 2 digit random numbers. The proportion of numbers in the range 01 to 94 will be an estimate of the required probability.

7.5 (a) $P(X \leq 0.49) = 0.49$. (b) $P(X \geq 0.27) = 0.73$. (c) $P(0.27 \leq X \leq 1.27) = P(0.27 \leq X \leq 1) = 0.73$. (d) $P(0.1 \leq X \leq 0.2$ or $0.8 \leq X \leq 0.9) = 0.1 + 0.1 = 0.2$. (e) $P(\text{not } [0.3 \leq X \leq 0.8]) = 1 - 0.5 = 0.5$. (f) $P(X = 0.5) = 0$.

7.7 (a) All probabilities are between 0 and 1; the probabilities add to 1. (b) $P(X \geq 5) = 0.07 + 0.03 + 0.01 = 0.11$. (c) $P(X > 5) = 0.03 + 0.01 = 0.04$. (d) $P(2 < X \leq 4) = 0.17 + 0.15 = 0.32$. (e) $P(X \neq 1) = 1 - 0.25 = 0.75$. (f) Write either $X \geq 3$ or $X > 2$. The probability is $1 - (0.25 + 0.32) = 0.43$.

7.9 (a) $(0.6)(0.6)(0.4) = 0.144$. (b) The possible combinations are SSS, SSO, SOS, OSS, SOO, OSO, OOS, OOO (S = support, O = oppose). $P(\text{SSS}) = 0.6^3 = 0.216, P(\text{SSO}) = P(\text{SOS}) = P(\text{OSS}) = (0.6^2)(0.4) = 0.144$, $P(\text{SOO}) = P(\text{OSO}) = P(\text{OOS}) = (0.6)(0.4^2) = 0.096$, and $P(\text{OOO}) = 0.4^3 = 0.064$. (c) $P(X = 0) = 0.216$, $P(X = 1) = 0.432, P(X = 2) = 0.288, P(X = 3) = 0.064$. (d) Write either $X \geq 2$ or $X > 1$. The probability is $0.288 + 0.064 = 0.352$.

7.11 (a) The area of a triangle is $\frac{1}{2}bh = \frac{1}{2}(2)(1) = 1$. (b) $P(Y < 1) = 0.5$. (c) $P(Y < 0.5) = 0.125$.

7.13 (a) $P(\hat{p} \geq 0.5) = P(Z \geq \frac{0.5-0.3}{0.023}) \doteq P(Z \geq 8.7) \doteq 0$. (b) $P(\hat{p} < 0.25) = P(Z < -2.17) = 0.0150$. (c) $P(0.25 \leq \hat{p}0.35) \doteq P(-2.17 \leq Z \leq 2.17) = 0.070$.

7.15 (a) $P(\hat{p} \geq 0.16) = P(Z \geq 1.087) = 0.1379$. (b) $P(0.14 \leq \hat{p} \leq 0.16) = P(-1.087 \leq Z \leq 1.087) = 0.7242$.

7.17 $\mu = (0)(0.10) + (1)(0.15) + (2)(0.30) + (3)(0.30) + (4)(0.15) = 2.25$.

7.19 If your number is abc, then of the 1000 three-digit numbers, there are six — $abc, acb, bac, bca, cab, cba$ — for which you will win the box. Therefore, we win nothing with probability $\frac{994}{1000} = 0.994$ and win \$83.33 with probability $\frac{6}{1000} = 0.006$. The expected payoff on a \$1 bet is $\mu = (\$0)(0.994) + (\$83.33)(0.006) = \$0.50$.

7.21 Below is the probability distribution for L, the length of the longest run of heads or tails. $P(\text{You win}) = P(\text{run of 1 or 2}) = \frac{89}{512} \doteq 0.1738$, so the expected outcome is $\mu = (\$2)(0.1738) + (-\$1)(0.8262) \doteq \$0.4785$. On the average, you will lose about 48 cents each time you play. (Simulated results should be close to this exact results, how close depends on how many trials are used.)

Value of L	1	2	3	4	5	6	7	8	9	10
Probability	$\frac{1}{512}$	$\frac{88}{512}$	$\frac{185}{512}$	$\frac{127}{512}$	$\frac{63}{512}$	$\frac{28}{512}$	$\frac{12}{512}$	$\frac{5}{512}$	$\frac{2}{512}$	$\frac{1}{512}$

7.23 No: Assuming all "at-bats" are independent of each other, the 35% figure only applies to the "long run" of the season, not to "short runs."

7.25 (a) Independent: Weather conditions a year apart should be independent. (b) Not independent: Weather patterns tend to persist for several days; today's weather tells us something about tomorrow's. (c) Not independent: The two locations are very close together, and would likely have similar weather conditions.

7.27 In 4.51, we had $\mu = 2.25$, so $\sigma_x^2 = (0-2.25)^2(0.10)+(1-2.25)^2(0.15)+(2-2.25)^2(0.30)+(3-2.25)^2(0.30)+(4-2.25)^2(0.15) = 1.3875$, and $\sigma_x = \sqrt{1.3875} \doteq 1.178$.

7.29 Since the two times are independent, the total variance is $\sigma_{total}^2 = \sigma_{pos}^2 + \sigma_{all}^2 = 2^2 + 4^2 = 20$, so $\sigma_{total} = \sqrt{20} \doteq 4.472$.

7.31 (a) Randomly selected students would presumably be unrelated. (b) $\mu_{f-m} = \mu_f - \mu_m = 120 - 105 - 15$. $\sigma_{f-m}^2 = \sigma_m^2 + \sigma_m^2 = 28^2 + 35^2 = 2009$, so $\sigma_{f-m} = 44.82$. (c) Knowing only the mean and standard deviation, we cannot find that probability (unless we assume that the distribution is normal). Many different distributions can have the same mean and standard deviation.

7.33 Read two-digit random numbers. Establish the correspondence 01 to 10 $\leftrightarrow$ 540°, 11 to 35 $\leftrightarrow$ 545°, 36 to 65 $\leftrightarrow$ 550°, 66 to 90 $\leftrightarrow$ 555°, and 91 to 99, 00 $\leftrightarrow$ 560°. Repeat many times, and record the corresponding temperatures. Average the temperatures to approximate μ; find the standard deviations of the temperatures to approximate σ.

7.35 The mean μ of the company's "winnings" (premiums) and its "losses" (insurance claims) is positive. Even though the company will lose a large amount of money on a small number of policyholders who die, it will gain a small amount on the majority. The law of large numbers says that the average "winnings" minus "losses" should be close to μ, and overall the company will almost certainly show a profit.

7.37 (a) A single random digit simulates each toss, with (say) odd = heads and even = tails. The first round is two digits, with two odds a win; if you dont win, look at two more digits, again with two odds a win. (b) The probability of winning is $\frac{1}{4} + (\frac{3}{4})(\frac{1}{4}) = \frac{7}{16}$, so the expected value is $(\$1)(\frac{7}{16}) + (-\$1)(\frac{9}{16}) = -\frac{2}{16} = -\$0.125$.

7.39 Means: $\mu_H = 2.6$ and $\mu_F = 3.14$ persons. Variances: $\sigma_H^2 = 2.02$ and $\sigma_F^2 = 1.5604$. Standard deviations: $\sigma_H \doteq 1.421$ and $\sigma_F \doteq 1.249$ persons.

Since families must include at least two people, it is not too surprising that the average family is slightly larger (about 0.54 persons) than the average household. For large family/household sizes, the differences between the distributions are small.

7.41 (a) $\sigma_Y^2 = (-145)^2(0.4) = (55)(0.5) = (305)(0.1) = 19{,}225$ and $\sigma_Y \doteq 138.65$ units. (b) $\sigma_{X+Y}^2 = \sigma_X^2 + \sigma_Y^2 = 7{,}800{,}000 = 19{,}225 = 7{,}819{,}225$, so $\sigma_{X+Y} \doteq 2796.29$ units. (c) $\sigma_{2000X+3500Y}^2 = \sigma_{2000X}^2 + \sigma_{3500Y}^2 = (2000)^2\sigma_X^2 + (3500)^2\sigma_Y^2$, so $\sigma_{2000X+3500Y} \doteq \$5{,}606{,}738$.

7.43 $\sigma_X^2 = 94{,}236{,}826{,}64$, so that $\sigma_X \doteq 9707.57$.

Chapter 8

8.1 It may be binomial if we assume that there are no twins or other multiple births among the next 20 (these would not be independent), and that for all births, the probability that the baby is female is the same (requirement 4).

8.3 No—since she receives instruction after incorrect answers, her probability of success is likely to increase.

8.5 (a) .2637 (b) The binomial probabilities for $x = 0, \ldots, 5$ are: .2373, .3955, .2637, .0879, .0146, .0010. (e) The cumulative probabilities for $x = 0, \ldots, 5$ are: .2373, .6328, .8965, .9844, .9990, 1. Compare with Corinne's cdf histogram; the bars in this histogram get taller, sooner. Both peak at 1 on the extreme right.

8.7 (a) Symmetric; the shape depends on the value of the probability of success. (c) .0078125.

8.9 $P(X = 10) = \binom{20}{10}(0.8)^{10}(0.2)^{10} = 0.00203$.

8.11 Let $0,1,2 \leftrightarrow$ Hispanic and let 3 to $9 \leftrightarrow$ non-Hispanic. Use random digit table. Or, using the TI-83, repeat the command 30 times:

```
randBin(1,.3,15) → L₁:sum(L₁) → L₂(1)s
```

Our frequencies were:

1	2	4	5	11	3	4
0	1	2	3	4	5	6	7	8

The relative frequency of 3 or fewer Hispanics is $7/30 = .233$.

8.13 (a) Let $0 \leftrightarrow$ never married, let $1,2,3, \leftrightarrow$ married, and use Table B. Or, using the TI-83, repeat the command.

```
randBin(1,.25,10) → L₁:sum(L₁)
```

Our results for 30 repetitions were:

2	3	14	5	4	2
0	1	2	3	4	5

The relative frequency of 2 or fewer never married is $19/30 = .6\overline{3}$.

8.15   (a) $\mu = 4.5$. (b) $\sigma = \sqrt{3.15} = 1.77482$. (c) If $p = 0.1$, then $\sigma = \sqrt{1.35} = 1.16190$. If $p = 0.01$, then $\sigma = \sqrt{0.1485} = 0.38536$. As p gets close to 0, σ gets closer to 0.

8.17 $\mu = 2.5$, $\sigma = \sqrt{1.875} = 1.36931$.

8.19   (a) $n = 20$ and $p = 0.25$. (b) $\mu = 5$. (c) $\binom{20}{5}(0.25)^5(0.75)^{15} = 0.20233$.

8.21   (a) That all are assessed as truthful: $P = 0.06872$; the probability that at least one is reported to be a liar is $1 - 0.06872 = 0.93128$. (b) $\mu = 2.4$, $\sigma = \sqrt{1.92} = 1.38564$.

8.23   (a) $P(\text{switch is bad}) = \frac{1000}{10000} = .01$; $P(\text{switch is OK}) = \frac{9000}{10000} = 0.9$ (b) 9999 switches remain, of which 999 are bad. Under these conditions, $P(\text{switch is bad}) = \frac{999}{9999} = 0.09990999\ldots$, and $P(\text{switch is OK}) = \frac{9000}{9999} = 0.90009000\ldots$ (c) Again, 9999 switches remain; this time 1000 are bad, so that $P(\text{switch is bad}) = \frac{1000}{9999} = 0.10001000\ldots$, and $P(\text{switch is OK}) = \frac{8999}{9999} = 0.89998999\ldots$.

8.25   (a) The four conditions of a geometric setting hold, with probability of success $= 1/2$.
(b)

X	1	2	3	4	5	...
P(X)	.5	.25	.125	.0625	.03125	

(d) cdf   .5   .75   .875   .9375   .96875 (e) Sum $= \frac{a}{1-r} = \frac{.5}{1-r} = 1$

8.27

X	pdf	cdf
1	.5	.5
2	.25	.75
3	.125	.825
4	.0625	.9375
5	.03125	.96875
...		

8.29   (a) That the shots are independent, and that the probability of success is the same for each shot. (b) .0655 (c) .738.

8.31   No. The probability of success changes with each marble drawn.

8.33   (a) 2 (b) 1 (c) Let even digit = boy, and odd digit = girl. Read random digits until an even digit occurs. Count number of digits read. Repeat many times, and average the counts.

8.35   (simulation)

8.37 $\mu = 9.95$, $\sigma = 2.447$. $P(x \leq 5) = .0307$.

8.39   (a) The 4 requirements of a binomial setting are satisfied. (b) 80 (c) .5727 (d) .0026.

8.41   (a) .25 (b) .75 (c) X = number of coin tosses until someone wins. X is geometric.
(d)

X	1	2	3	4	5	...
$P(x)$	.75	.1875	.04688	.01172	.00293	
cdf	.75	.9375	.9844	.9961	.9990	

(e) .9375 (f) .0039 (g) $1.3\overline{3}$.

8.43   The larger the probability of success, the shorter the tail.

8.45   (simulation).

Chapter 9

9.1   7.2% is a statistic.

9.3   48% is a statistic; 52% is a parameter.

9.5   Answers will vary.

9.7   (a) The scores will vary depending on the starting row. Note that the smallest possible mean is 61.75 (from the sample 58, 62, 62, 65) and the largest is 77.25 (from 73,74, 80, 82). Answers to (b) and (c) will vary.

9.9   (a) Large bias and large variability. (b) Small bias and small variability. (c) Small bias, large variability. (d) Large bias, small variability.

9.11   (a) Shoppers: 00 to 100. Shoppers who find shopping frustrating: 01 to 67. A percent. (e) As the sample size increases, the variability decreases.

9.13   (a) Use 0–3 to represent persons who would answer "yes." Looking at the first 20 digits on line 136 gives YNNYY NNNNY NNYNN YNNNN— 6 yes and 14 no, so $\hat{p} = 0.3$. (b) Most answers should fall between 0.3 and 0.5 (c) 0.4 (d) 0.5.

9.15   (a) $\mu = p = 0.15$, $\sigma = \sqrt{(0.15)(0.85) \div 1540} = 0.0091$. (b) The population (U.S. adults) is considerably larger than 10 times the sample size (1540). (c) $np = 231$, $n(1 - p) = 1309$—both are much bigger than 10.

(d) $P(0.13 < \hat{p} < 0.17) = P(-2.198 < Z < 2.198) = 0.9722$. (e) To achieve $\sigma = 0.0045$, we need a sample four times as large: 6160.

9.17   For $n = 200$: $\sigma = 0.02525$, and the probability is $P = 0.5704$. For $n = 800$: $\sigma = 0.01262$ and $P = 0.8858$. For $n = 3200$: $\sigma = 0.00631$ and $P = 0.9984$. Larger sample sizes give more accurate results (the sample proportions are more likely to be close to the true proportion.)

9.19   (a) $\mu = 0.52$, $\sigma = 0.02234$, (b) np and $n(1 - p)$ are 260 and 240 respectively. $P(\hat{p} \geq 0.50) = P(Z \geq -0.8951) = 0.8159$.

9.21   Comparing the results of "Rule of Thumb 2," we see that it is clearly satisfied in the telephone number problem, and just barely satisfied in the mail-order problem—so the approximation is more accurate in the first of these.

9.23   (a) $P(\hat{p} \leq 0.70) \doteq P(Z \leq -1.155) = 0.1241$. (b) $P(\hat{p} \leq 0.70) \doteq P(Z \leq -1.826) = 0.0339$. (c) The test must contain 400 questions. (d) The answer is the same for Laura.

9.25   (a) $np = (15)(0.3) = 4.5$—this fails Rule of Thumb 2. (b) The population size (316) is not at least 10 times as large as the sample size (50)—this fails rule of Thumb 1.

9.27   Mean: 18.6 standard deviation: $5.9/\sqrt{76} = 0.67678$. The normality of individual scores is not necessary for this to be true.

9.29   Standard deviation: 0.04619.

9.31   (a) $P(X < 295) = P(Z < -1) = 0.1587$. (b) $P(\bar{x} < 295) = P(Z < -2.4495) = 0.0072$.

9.33 $\bar{x}$ has approximately a $N(1.6, 0.0849)$ distribution; the probability is $P(Z > 4.71)$—essentially 0.

9.35   The mean μ of the company's "winnings" (premiums) and their "losses" (insurance claims) is positive. Even though the company will lose a large amount of money on a small number of policyholders who die, it will gain a small amount on the majority. The law of large numbers guarantees that the average "winnings" minus "losses" will be close to μ, and overall the company will almost certainly show a profit.

9.37   (a) $N(2.2, 0.1941)$. (b) $P(Z < -1.0304) = 0.1515$. (c) $P(\bar{x} < \frac{100}{52} = P(Z < -1.4267) = 0.0768$.

9.39 $\mu + 2.33\sigma\sqrt{n} = 1.4625$.

9.41 $P(\hat{p} > 0.50) = P(Z > \frac{0.05 - 0.45}{0.02225} = P(Z > 2.247) = 0.01231$.

9.43   (a) $P(Z > \frac{105 - 100}{15}) = P(Z > \frac{1}{3}) = 0.36944$. (b) Mean: 100; standard deviation: 1.93649. (c) $P(Z > \frac{105 - 100}{1.93649}) = P(Z > 2.5820) = 0.00491$. (d) The answer to (a) could be quite different; (b) would be the same (it does not depend on normality at all). The answer we gave for (c) would still be fairly reliable because of the central limit theorem.

9.45 (a) Not normal—a count assumes only whole-number values. (b) $N(1.5,$ $0.02835)$. (c) $P(\bar{x} > \frac{1075}{700} = P(Z > 1.2599) = 0.10386$.

9.47 $P(5067 \text{ or more heads}) = P(\hat{p} \geq 0.5067) = P(Z \geq 1.34) = 0.0901$. If Kerrichs coin was "fair," we would see 5067 or more heads in about 9% of all repetitions of the experiment of flipping the coin 10,000 times, or about once every 11 attempts. This is *some* evidence against the coin being fair, but it is not by any means overwhelming.

Chapter 10

10.1 (a) 44% to 50%. (b) We do not have information about the whole population; we only know about a small sample. We expect our sample to give us a good estimate of the population value, but it will not be exactly correct. (c) The procedure used gives an estimate within 3 percentage points of the true value in 95% of all samples.

10.3 (a) Mean: 280; standard deviation: 1.89737. (c) 2 standard deviations— 3.8 points. (e) 95%.

10.5 11.78 ± 0.77, or 11.01 to 12.55 years.

10.7 (a) The distribution is slightly skewed to the right. (b) 224.002 ± 0.029, or 223.973 to 224.031.

10.9 (a) 0.8354 to 0.8454. (b) 0.8275 to 0.8533. (c) Increasing confidence makes the interval longer.

10.11 (a) 271.4 to 278.6. (b) 90%: 272.0 to 278.0; 99%: 270.3 to 279.7. (c) 90%: 3.0; 95%: 3.6; 99%: 4.7. Margin of error goes up with increasing confidence.

10.13 (a) 10.00209 to 10.00251. (b) 22 (21.64).

10.15 35 (34.57).

10.17 (a) The interval was based on a method that gives correct results 95% of the time. (b) With a margin of error of 2%, the true value of p could be as low as 49%, so the confidence interval contains some values of p which give the election to Ford. (c) The proportion in question is not random—discussing probabilities about this proportion has little meaning. The probability the politician asked about is either 1 or 0: either a majority favors Carter, or they don't.

10.19 1.888 to 2.372.

10.21 (a) The intended population is "the American public"; the population which was actually sampled was "citizens of Indianapolis (with listed phone numbers)." (b) Food stores: 15.22 to 22.12; Mass merchandisers: 27.77 to 36.99; Pharmacies: 43,68 to 53.52. (c) The confidence intervals do not overlap at all; in particular, the *lower* confidence limit of the rating for pharmacies is higher than the *upper* confidence limit for the other stores. This indicates that the pharmacies are really higher.

10.23   505 residents.

10.25 $657.14 to $670.86. (The sample size is not used; $\sigma_{estimate}$ is given.)

10.27   (a) N(115,6). (b) The actual result lies out toward the high tail of the curve, while 118.6 is fairly close to the middle. Assuming H_0 is true, observing a value like 118.6 would not be surprising, but 125.7 is less likely, and therefore provides evidence against H_0.

10.29 H_0: $\mu = 5$ mm; H_a: $\mu \neq 5$ mm.

10.31 H_0: $\mu = 50$; H_a: $\mu < 50$.

10.33   The P-values are 0.2743 and 0.0373, respectively. The P-values are 0.2991 and 0.0125, respectively.

10.35   (a) $\bar{x} = 398$. (b) A N(354, 19.053) density. (c) 0.0105. (d) It is significant at $\sigma = 0.05$, but not at $\sigma = 0.01$. This is pretty convincing evidence against H_0.

10.37   If church attenders are no more ethnocentric than nonattenders, then the outcomes observed for *this* sample would occur in less than 1 out of 20 instances. This is unlikely; we conclude that churchgoers are more ethnocentric.

10.39   (a) H_0: $\mu = 224$ vs. H_a: $\mu \neq 224$. (b) $z = 0.1292$. (c) $P = 0.8972$ — this is reasonable variation when the null hypothesis is true, so we do not reject, H_0.

10.41   (a) $z = -2.200$. (b) Yes, because $|z| > 1.960$. (c) No, because $|z| < 2.576$.

10.43   (a) Yes, because $|z| > 1.645$. (b) Yes, because $|z| > 2.326$.

10.45   (a) N(0, 0.11339). (b) $\bar{x} = 0.27$ lies out in the tail of the curve, while 0.09 is fairly close to the middle. Assuming H_0 is true, observing a value like 0.09 would not be surprising, but 0.27 is unlikely, and therefore provides evidence against H_0. (c) $P = 0.4274$. (d) $P = 0.0173$.

10.47 H_0: $\mu = 18$ vs. H_a: $\mu < 18$.

10.49   (a) No, because $|z| < 1.960$. (b) No, because $|z| < 1.645$.

10.51   P is between 0.005 and 0.01 (in fact, $P = 0.0078$).

10.53 $P = 0.1292$. Although this sample showed *some* difference in market share between pioneers with and without patents or trade secrets; the difference was small enough that it could have arisen merely by chance: it would occur in about 13% of all samples even if there is *no* difference between the two types of pioneer companies.

10.55   When a test is significant at the 1% level, it means that if the null hypothesis is true, outcomes similar to those seen are expected to occur less than once in 100 repetitions of the experiment or sampling. "Significant at the 5% level" means we have observed something which occurs in less than 5 out of 100 repetitions (when H_0 is true). Something that occurs "less than once in 100 repetitions" also occurs "less than 5 times in 100 repetitions," so significance at the 1% level implies significance at the 5% level (or any higher level).

10.57   (a) $Z = -4.99$, P-value: $p = 5.97 \times 10^{-7}$. (b) $Z = 0.129$, $p = .897$. Since the P-value is large, there is no evidence that the process mean μ is not equal to the target $\mu_0 = 224$. (c) $Z = -.7893$, $p = .2150$. Conclude that there is insufficient evidence that $\mu < 300$ ml.

10.59   (a) $P = 0.3821$. (b) $P = 0.1714$ (c) $P = 0.0014$.

10.61   No—the percentage was based on a voluntary response sample, and so cannot be assumed to be a fair representation of the population. Such a poll is likely to draw a higher-than-actual proportion of people with a strong opinion, esp. a strong negative opinion.

10.63   A test of significance answers question (b).

10.65   (a) 0.05. (b) Out of 77 tests, we can expect to see about 3 or 4 (3.85, to be precise) significant tests at the 5% level.

10.67   (a) Reject H_0 if $z < -2.326$. (b) 0.01 (the significance level). (c) We accept H_0 if $\bar{x} \geq 270.185$, so when $\mu = 270$, P(Type II error) $= P(\bar{x} \geq$

$270.185) = P\left(\dfrac{\bar{x}-270}{60/\sqrt{840}} \geq \dfrac{270.185-270}{60/\sqrt{840}}\right) = 0.4644$.

10.69   (a) Reject H_0 if $\bar{x} \geq 0.5202$. (b) 0.9666.

10.71   (a) 0.5086 (b) 0.9543.

10.73   (a) We reject H_0 if $\bar{x} \geq 131.46$ or $\bar{x} \leq 124.54$. Power: 0.9246. (b) Power: 0.9246 (same as (a)). Over 90% of the time, this test will detect a difference of 6 (in either the positive or negative direction). (c) The power would be higher—it is easier to detect greater differences than smaller ones.

10.75   Power $1 - 0.4644 = 0.5356$.

10.77   (a) The plot is reasonably symmetric for such a small sample. (b) 26.06 to 34.74. (c) H_0: $\mu = 25$ vs. H_a: $\mu > 25$; $z = 2.44$; P-value is .007. This is strong evidence against H_0.

10.79   12.285 to 13.515. This assumes that the babies are an SRS from the population. The population should not be too nonnormal (although a sample of size 26 will overcome quite a bit of skewness).

10.81   (a) Narrower; lowering confidence level decreases the interval size. (b) Yes. $33,000 falls outside the 99% confidence interval, indicating that $P < 0.01$.

10.83 H_0: $p = \frac{18}{38}$ vs. H_a: $p \neq \frac{18}{38}$.

10.85 Yes—significance tests allow us to discriminate between random differences ("chance variation") that might occur when the null hypothesis is true, and the differences that are unlikely to occur when H_0 is true.

Chapter 11

11.1 (a) The t_2 curve is a bit shorter at the peak and slightly higher in the tails. (b) The t_9 curve has moved toward coincidence with the standard normal curve. (c) The t_{30} curve cannot be distinguished from the standard normal curve. As the degrees of freedom increase, the $t(df)$ curve approaches the standard normal density graph.

11.3 $37/\sqrt{4} = 18.5$.

11.5 (a) 2.145. (b) 0.688.

11.7 (a) 14. (b) 1.82 is between 1.761 ($p = 0.05$) and 2.145 ($p = 0.025$). (c) The P-value is between 0.025 and 0.05 (in fact, $P = 0.0451$). (d) $t = 1.82$ is significant at $\alpha = 0.05$ but not at $\alpha = 0.01$.

11.9 (a) $\bar{x} = 1.75$ and $s = 0.1291$, so $SE(\bar{x}) = 0.06455$. (b) 1.598 to 1.902.

11.11 $H_0: \mu = 1.3$ vs. $H_a: \mu > 1.3$; $t = 6.9714$; P is between 0.005 and 0.0025 (in fact, $P = 0.003$). This is very strong evidence against H_0; reject it in favor of H_a.

11.13 (a) Randomly assign 12 (or 13) into a group which will use the right-hand knob first; the rest should use the left-hand knob first. Alternatively, for each student, randomly select which knob he or she should use first. (b) Let μ_R be the mean right-hand thread time for all right-handed people (or students), and μ_L be the mean left-hand thread time. Then $\mu = \mu_R - \mu_L$ is the difference between right-handed times and left-handed times; the hypotheses are $H_0: \mu = 0$ (no difference) and $H_a: \mu < 0$ (i.e., $\mu_R < \mu_L$). (c) $\bar{x} = -13.32$, $SE(\bar{x}) = 4.5872$, $t = -2.9037$, and $P = 0.0039$. Reject H_0 in favor of H_a.

11.15 (a) 1.54 to 1.80. (b) We are told the distribution is symmetric; because the scores range from 1 to 5, the possibility for skewness is limited. In this situation, the assumption that the 17 Mexicans are an SRS from the population is the most crucial.

11.17 (a) The distribution is slightly skewed, but there are no apparent outliers. (b) $H_0: \mu = 224$ vs. $H_a: \mu \neq 224$. $t = 0.12536$ and $P = 0.9019$, so we have very little evidence against H_0.

11.19 (a) Approximately 2.403 (from Table C), or 2.405 (using software). (b) Using $t^* = 2.403$: Reject H_0 if $t > 2.403$, which means $\bar{x} > 36.70$. (c) The power against $\mu = 100$ is 0.99998—basically 1. A sample of size 50 should be quite adequate.

11.21 (a) 9. (b) $P = 0.0255$; it lies between 0.05 and 0.025.

11.23 (a) 111.22 to 118.58. (b) We assume that the 27 members of the placebo group can be viewed as an SRS, and that the distribution of seated systolic BP in this population is normal, or at least not too nonnormal. Since the sample size is somewhat large, the procedure should be valid as long as the data show no outliers and no strong skewness.

11.25 (a) Standard error of the mean. (b) $s = 0.01\sqrt{3} = 0.01732$. (c) $0.84 \pm 0.0292 = 0.8108$ to 0.8692.

11.27 (a) For each subject, randomly choose which test to administer first. Or, randomly assign 11 subjects to the "ARSMA first" group, and the rest to the "BI first" group. (b) $t = 4.27$; $P < 0.001$, so we reject H_0: $\mu = 0$ in favor of H_a: $\mu \neq 0$. (c) 0.1292 to 0.3746.

11.29 We know the data for *all* presidents; we know about the whole population, not just a sample. (We might want to try to make statements about future presidents, but doing so from this data would be highly questionable; they can hardly be considered an SRS from the population.)

11.31 (a) (3)—two samples. (b) (2)—matched pairs.

11.33 (a) H_0: $\mu_1 = \mu_2$ vs. H_a: $\mu_1 < \mu_2$, where μ_1 is the beta-blocker population mean pulse rate and μ_2 is the placebo mean pulse rate. $t = -2.4525$; use a $t(29)$ distribution, which gives $P = 0.01022$. This makes the result significant at 5% but not at 1%. (b) -10.8311 to 0.6311 for $\mu_1 - \mu_2$.

11.35 (a) If k (the degrees of freedom) is reasonably large, the $t(k)$ distribution looks enough like the $N(0, 1)$ distribution that for $t = 7.36$, we can conclude that the P-value is tiny (based on the 68–95–99.7 rule), so the result is significant. (b) Use a $t(32)$ distribution.

11.37 (a) Because the sample sizes are so large (and the sample sizes are almost the same), deviation from the assumptions have little effect. (b) Using $t^* = 1.660$ from a $t(100)$ distribution, the interval is \$412.68 to \$635.58. Using $t^* = 1.6473$ from a $t(620)$ distribution (obtained with software), the interval is \$413.54 to \$634.72. (c) The sample is not *really* random, but there is no reason to expect that the method used should introduce any bias into the sample. (d) Students without employment were excluded, so the survey results can only (possibly) extend to *employed* undergraduates. Knowing the number of unreturned questionnaires would also be useful.

11.39 Both distributions are slightly skewed to the right, and have one or two moderate high outliers. A t procedure may be (cautiously) used nonetheless, since the sum of the sample sizes is almost 40. (b) H_0: $\mu_w = \mu_m$ vs. H_a: $\mu_w > \mu_m$. $\bar{x}_w = 141.056$, $s_w = 26.4363$, $\bar{x}_m = 121.250$, $s_m = 32.8519$, and $t = 2.0561$, so $P = 0.0277$ (for a $t(17)$ distribution). (c) For $\mu_m - \mu_w$: -36.57 to -3.05.

11.41 No answer required.

11.43　(a) H_0: $\mu_{skilled} = \mu_{novice}$ vs. H_a: $\mu_s > \mu_n$. (b) The t statistic we want is the "Unequal" value: $t = 3.1583$; its P-value is 0.0052. This is strong evidence against H_o. (c) Using $t^* - 1.833$ from a $t(9)$ distribution: 0.4922 to 1.8535. Using $t^* = 1.8162$ from a $t(9.8)$ distribution (from software): 0.4984 to 1.8473.

11.45　Conservative: 19882 d.f.; more exact method: 38786 d.f. Either way, the distribution we use is almost exactly a $N(0, 1)$ distribution. To three decimal places, $t^* = 2.576$, so the confidence interval is 27.91 to 32.09. Normality of SAT scores is not necessary since the sample sizes are so large.

11.47　(a) H_0: $\mu_1 = \mu_2$ vs. H_a: $\mu_1 \neq \mu_2$. $t = -8.2379$; using either a $t(13)$ or a $t(21.8)$ distribution, the P-value is smaller than 0.0001, so there is a significant difference. (b) The fact that all the subjects in this study are college professors may have some confounding effects on the results. Additionally, all the subjects volunteered for a fitness program, which could bring in some further confounding.

11.49　(a) H_0: $\mu_A = \mu_B$ vs. H_a: $\mu_A \neq \mu_B$; $t = -1.484$. Using $t(149)$ and $t(297.2)$ distributions, P equals 0.1399 and 0.1388, respectively; not significant in either case. The bank might choose to implement Proposal A even though the difference is not significant, since it may have a *slight* advantage over Proposal B. Otherwise, the bank should choose whichever option costs them less. (b) Because the sample sizes are equal and large, the t procedure is reliable in spite of the skewness. (c) This is an experiment—treatments are imposed by the bank. However, one other thing might be useful: statistics for a control group, to see if either plan increased spending.

11.51　(a) H_0: $\mu_c = \mu_a$ vs. H_a: $\mu_c \neq \mu_a$; $t = 1.249$. Using $t(9)$ and $t(25.4)$ distributions, P equals 0.2431 and 0.229, respectively; the difference is not significant. (b) -15.8 to 54.8 (using $t(9)$) or -12.6 to 51.6 (using $t(25.4)$). These intervals had to contain 0 because according to (a), the observed difference would occur in more than 22% of samples when the means are the same; thus 0 would appear in any confidence interval with a confidence level greater than 78%.

11.53　(a) $t^* \doteq 2.364$, the value for a $t(100)$ distribution (since values for a $t(99)$ distribution are not given). (b) Reject H_0 when $\bar{x}_1 - \bar{x}_2 \geq 2.6746$. (c) Power: $P(Z \geq -2.0554) = 0.9801$.

11.55　(a) H_0: $\mu_A = \mu_B$ vs. H_a: $\mu_A \neq \mu_B$; $t = (\bar{x}_A - \bar{x}_B)/\sqrt{\frac{s_A^2}{n_A} + \frac{s_B^2}{n_B}}$. (b) For a $t(349)$ distribution, $t^* = 1.967$; using a $t(100)$ distribution, take $t^* = 1.984$. (c) We reject H_0 when $|\bar{x}_a - \bar{x}_B| \geq 59.48$ (using $t^* = 1.967$). To find the power against $|\mu_A - \mu_B| = 100$, we choose *either* $\mu_A - \mu_B = 100$

or $\mu_A - \mu_B = -100$ (the probability is the same either way). Taking the former, we compute: $P[(\bar{x}_A - \bar{x}_B) \le -59.48 \text{ or } (\bar{x}_A - \bar{x}_B) \ge 59.48] = P(Z \le -5.274 \text{ or } Z \ge -1.340) = 0.9099$.

Repeating these computations with $t^* = 1.984$ gives power 0.9071.

11.57 (a) Two-sample t test—the two groups of women are (presumably) independent. (b) Use a $t(44)$ distribution. (c) The sample sizes are large enough that nonnormality has little effect on the reliability of the procedure.

11.59 (a) $H_0: \mu_1 = \mu_2$ vs. $H_a: \mu_1 < \mu_2$; $t = -8.947$; $P \doteq 0$ (however one chooses the degrees of freedom). Reject H_0 and conclude that the workers have higher output. (b) The t procedures are robust against skewness when the sample sizes are large. (c) Insertions for experienced workers have (approximately) a $N(\bar{x}_2, s_2)$ distribution; the 68–95–99.7 rule tells us that 95% of all workers can insert between $\bar{x}_2 - 2s_2 = 29.66$ and $\bar{x}_2 + 2s_2 = 44.99$ pins in the allotted time. (Or use $\bar{x}_2 \pm 1.96s_2$.)

11.61 Both stemplots are reasonably symmetrical, though the nitrite group may be slightly left-skewed. There are no *extreme* outliers. $H_0: \mu_c = \mu_n$ vs. $H_a: \mu_c > \mu_n$; $t = 0.8909$ and P equals 0.1902 (using a $t(29)$ distribution) or 0.1884 (with 56.8 d.f.). In either case, the difference is not significant.

11.63 (a) The stemplot is reasonably symmetrical given the small sample size. There are no outliers. (b) 903.23 to 912.27. (c) No, because 910 falls inside the 95% confidence interval.

11.65 No—you have information about all Indiana counties (not just a sample).

11.67 (a) $H_0: \mu_1 = \mu_2$ vs. $H_a: \mu_1 > \mu_2$; $t = 1.1738$, so $P = 0.1265$ (using $t(22)$) or 0.123453 (using $t(43.3)$). Not enough evidence to reject H_0. (b) -14.57 to 52.57 (using $t(22)$), or -13.64 to 51.64 (using $t(43.3)$). (c) 165.53 to 220.47. (d) We are assuming that we have two SRSs from each population, and that underlying distributions are normal. It is unlikely that we have random samples from either population, especially among pets.

11.69 Using 95% confidence—for abdomen skinfold: -15.5 to -11.5 (using $t(19)$), or -15.4 to -11.6 (using $t(103.6)$). For thigh skinfold: -12.95 to -9.65 (using $t(19)$), or -12.86 to -9.74 (using $t(106.4)$).

11.71 The distributions of differences (city minus rural) has 26 observations. There are two high outliers (15 and 18).

If we throw out these outliers, the data seem to be more suitable for a t procedure. Using the other 24 observations, $\bar{x} = 1.00$, $s = 2.1059$, $t = 2.326$, and $P = 0.0146$—we can reject H_0. A 95% confidence interval for $\mu_c - \mu_r$ is 0.111 to 1.889.

Chapter 12

12.1 (a) Population: the 175 dorm residents; p is the proportion who like the food. (b) $\hat{p} = 0.28$.

12.3 (a) The population is the 15,000 alumni, and p is the proportion who support the president's decision. (b) $\hat{p} = 0.38$.

12.5 (a) No—np_0 and $n(1 - p_0)$ are less than 10 (they both equal 5). (b) No—the expected number of failures is less than 10 ($n(1 - p_0) = 2$). (c) Yes—we have an SRS, the population is more than 10 times as large as the sample, and $np_0 = n(1 - p_0) = 10$.

12.7 $Z = 10.84$, P-value $= 1.17 = \times 10^{-27}$. This is strong evidence.

12.9 (a) $\hat{p} = 0.5005$, $z = 0.1549$, and $P = 0.8769$—do not reject H_0. (b) 0.4922 to 0.5088.

12.11 450.2—round up to 451.

12.13

X	n	$\hat{p}$	Z	P-value
14	50	.28	$-.752$	.226
98	350	.28	-1.989	.023
140	500	.28	-2.378	.009

12.15 $z = -3.337$ and $P < 0.0005$—very strong evidence against H_0 in favor of H_a: $p < 0.1$.

12.17 (a) 0.0913 to 0.2060. (b) 304. (c) The sample comes from a limited area in Indiana, focuses on only one kind of business, and leaves out any businesses not in the Yellow Pages (there might be a few of these; perhaps they are more likely to fail). It is more realistic to believe that this describes businesses that match the above profile; it *might* generalize to food-and-drink establishments elsewhere, but probably not to (e.g.) hardware stores and other types of business.

12.19 (a) 4719. (b) 0.01125.

12.21 (a) -0.0208 to 0.1476. (b) The population-to-sample ratio is certainly large enough, and the smallest count in any category is 75—much larger than 5.

12.23 (a) H_0: $p_1 = p_2$ vs. H_a: $p_1 \ne p_2$. The population-to-sample ratio is large enough, and the smallest count is 94 (Catholics answering "Yes"). (b) $\hat{p}_1 = 0.6030$, $\hat{p}_2 = 0.5913$, $\hat{p} = 0.5976$; $z = 0.2650$, and $P = 0.7910$—H_0 is quite plausible given this sample.

12.25 (a) H_0: $p_1 = p_2$ vs. H_a: $p_1 > p_2$; the populations are much larger than the samples, and 17 (the smallest count) is greater than 5. (b) $\hat{p} = 0.0632$, $z = 2.926$, and $P = 0.0017$—the difference is statistically significant. (c) Neither the subjects nor the researchers who had contact with them knew which subjects were getting which drug—if any-one had known, they might confound the outcome by letting their expectations or biases affect the results.

12.27 $H_0: p_1 = p_2$ vs. $H_a: p_1 \neq p_2$; $P = 0.6981$ —insufficient evidence to reject H_0.

12.29 (a) $H_0: p_1 = p_2$ vs. $H_a: p_1 > p_2$; $P = 0.0335$ —reject H_0 (at the 5% level). (b) -0.0053 to 0.2336.

12.31 The population-to-sample ratio is large enough, and the smallest count is 10. Fatal heart attacks: $z = -2.67$, $P = 0.0076$. Nonfatal heart attacks: $z = -4.58$, $P < 0.000005$. Strokes: $z = 1.43$, $P = 0.1525$. The proportions for both kinds of heart attacks were significantly different; the stroke proportions were not.

12.33 (a) -9.91% to -4.09%. 0 is not in this interval; reject $H_0: p_1 = p_2$ at the 1% level. (b) -0.7944 to -0.4056. 0 is not in this interval; reject $H_0: \mu_1 = \mu_2$ at the 1% level.

12.35 No—the data is not based on an SRS, and thus the z procedures are not reliable in this case. In particular, a voluntary response sample is typically biased.

12.37 92.3% to 93.7%.

12.39 (a) No—$P = 0.1849$. (b) Yes—$P = 0.0255$. (c) The population is (presumably) very large, so the ratio of population-to-sample is big enough. Also, all the counts—45 and 29 in (a); 21, 7, 24, and 22 in (b)—are bigger than 5. The counts for baseball players are too small.

Chapter 13

13.1 (a) .4941 (b) .0204.

13.3 (a) H_0: The marital-status distribution of 25-to-29 year old U.S. males (b) is the same as that of the population as a whole.

H_a: The marital-status distribution of 25-to-29 year old U.S. males is different from that of the population as a whole.

Expected counts: 116.3, 301.55, 35, and 47.15. (c) $X^2 = 250.24$, df $= 3$. (d) P-value $= 5.8 \times 10^{-54} = .0000$. Reject H_0. The two distributions are different.

13.5 (a) H_0: The age-group distribution in 1996 is the same as the 1980 distribution. H_a: The age-group distribution in 1996 is different from the 1980 distribution. One simulation produced observed counts: 35, 31, 21, 13. Using expected counts 41.39, 27.68, 19.64, 11.28, $X^2 = 1.74$, df $= 3$, P-value $= .628$. There is no evidence that the distributions are different.

13.7 You should be surprised if you get a significant P-value.

13.9 $X^2 = 3.831$, df $= 5$. P-value $= P(X^2 \quad 3.83) = .574$. There is no evidence that the die is "loaded."

13.11   No answer required.

13.13   (a) 2×3. (b) 55.0%, 74.7%, and 37.5%. Some (but not too much) time spent in extracurricular activities seems to be beneficial. (d) Expected counts—C or better: 13.78, 62.71, 5.51; D or F: 6.22, 28.29, 2.49. (e) The first and last columns have lower numbers than we expect in the "passing" row (and higher numbers in the "failing" row), while the middle column has this reversed—more passed than we would expect if all proportions were equal.

13.15   (b) $P = 0.0313$. Rejecting H_0 means that we conclude that there is a relationship between hours spent in extracurricular activities and performance in the course. (c) The highest contribution comes from ">12 hours of extracurricular activities, D or F in the course." Too much time spent on these activities seems to hurt academic performance. (d) No—this study demonstrates association, not causation. Certain types of students may tend to spend a moderate amount of time in extracurricular activities and also work hard on their classes—one does not necessarily cause the other.

13.17   (a) $(r - 1)(c - 1) = (2 - 1)(3 - 1) = 2$. (b) $X^2 = 6.926$ lies between 5.99 and 7.38; therefore, P is between 0.05 and 0.025.

13.19   (a) First row: 5, 9; Second: 41, 23; Third: 27, 3; Fourth: 9, 2. (b) First: 9.65, 4.35; Second: 44.10, 19.90; Third: 20.67, 9.33; Fourth: 7.58, 3.42. 25% (2 out of 8) of the expected counts are less than 5, which goes against our guidelines. (c) 3 d.f.; P-value is between 0.0025 and 0.001 (in fact, $P = 0.0018$). (d) Students with high goals show a higher proportion of passing grades than those who merely wanted to pass.

13.21   (a) 7.01%, 14.02%, and 13.05%. (b) 172, 2283; 167, 1024; 86, 573. (c) 242.36, 2212.64; 117.58, 1073.42; 65.06, 593.94. Expected counts are all much bigger than 5, so the chi-square test is safe. H_0: there is no relationship between worker class and race vs. H_a: there is some relationship. (d) 2 d.f.; $P < 0.0005$ (basically 0). (e) Black female child-care workers are more likely to work in non-household or preschool positions.

13.23   (a) H_0: $p_1 = p_2$ vs. H_a: $p_1 < p_2$. The z test must be used because the chi-square procedure will not work for a one-sided alternative. (b) $z = -2.8545$ and $P = 0.0022$. Reject H_0; there is strong evidence in favor of H_a.

13.25 H_0: all proportions are equal vs. H_a: some proportions are different. Phone interviews: 168 yes, 632 no; one-on-one: 200 and 600; Anonymous: 224 and 576. $X^2 = 10.619$ with 2 d.f., and $P = 0.0049$—good evidence against H_0, so we conclude that contact method makes a difference in response.

13.27   (a) No; $X^2 = 1.051$ with 2 d.f., which gives $P = 0.5913$. (b) ABC News: $z = -0.7698$; $P = 0.4414$ (not significant). USA Today/CNN: $z = -5.1140$; $P \doteq 0$ (significant). New York Times/CBS: $z = -4.4585$; $P \doteq 0$ (significant). (c) An individual test will be wrong for only 5% of all samples. Imagine doing three tests in a row: Assuming the first test comes out correct (which it does 95% of the time), there is still a 5% chance that the next test will come out wrong, etc. Altogether, all three will be correct only 85.7%($= 0.95^3$) of the time.

13.29   Using randInt (0,9,200) $\rightarrow$ L$_4$, we obtained these counts for digits 0 to 9: 19, 17, 23, 22, 19, 20, 25, 12, 27, 16. $X^2 = 8.9$, df = 9, P-value $= .447$. There is no evidence that the sample data were generated from a distribution that is different from the uniform distribution.

13.31 $X^2 = 3.277$ (4. d.f.); $P = 0.4874$, so we can accept H_0: all types of companies had the same response rate.

	Response	Nonresponse
Metal Products	17	168
Machinery	35	266
Electrical Equiment	75	477
Transportation Equipment	15	85
Precision Instruments	12	78

13.33   4 degrees of freedom; $P > 0.25$ (in fact, $P = 0.4121$). There is not enough evidence to reject H_0 at any reasonable level of significance; the difference in the two income distributions is not statistically significant.

13.35 H_0: all refusal proportions are equal. Actual counts: 67, 1491; 86, 1503; 135, 1940; 124, 2514. Expected counts: 81.67, 1476.33; 83.29, 1505.71; 108.77, 1966.23; 138.28, 2499.72. The smallest expected count is 81.67, so the chi-square test is safe. $X^2 = 11.106$ (3 d.f.) and $P = 0.0112$—pretty strong evidence against H_0. The two largest components of X^2 are the July–Aug refusals (which were higher than expected) and the Jan–April refusals (which were low). Perhaps more people had "other things to do" in the summer, and had less pressing business during the winter.

13.37   (a) 0.5432 to 0.5968. (b) No: $z = 0.4884$, $P = 0.6253$. Or, using the exact counts (527/924 national, 96/174 student), $z = 0.4548$, $P = 0.6492$—again, not at all significant. (c) Yes: $z = 7.9215$, P is essentially 0. (d) Actual counts—National group: 702 yes, 222 no; First student group: 117, 57; Second student group: 129, 70. Expected counts—

National group: 675.37, 248.63; First student group: 127.18, 46.82; Second student group: 145.45, 53.55. $X^2 = 13.847$ with 2 d.f.; $P < 0.001$. Both student groups were less likely to believe that the military was censoring the news.

Chapter 14

14.1 (a) $r = 0.99415$ and $\hat{y} = -3.660 + 1.19690x$. The scatterplot shows a strong linear relationship, which is confirmed by r. (b) β represents how much increase we can expect in humerus length when femur length increases by 1 cm. b(the estimate of β) is 1.1969; $a = -3.660$. (c) The residuals are $-0.82262, -0.36682, 3.04248, -0.94202$, and -0.91102; the sum is 0. $s = \sqrt{3.92843} = 1.9820$.

14.3 (a) Sum of residuals $= 9.61 \times 10^{-7} \cong 0$. (b) Sum of squares of residuals: $449.842737, s = 7.4987$. (c) 8.003 is estimate for α; .0000662 is estimate for β; and 7.4987 estimates σ.

14.5 Using a $t(10)$ distribution: $0.687747 \pm (2.228)(0.2300) = 0.1753$ to 1.2002. β is the increase in second-round score we expect based on an increase of one shot in round one.

14.7 (a) $H_0: \beta = 0$, $H_a: \beta > 0$ (positive slope). (b) $t = 7.26$, df $= 8$, P-value $= 4.38 \times 10^{-5} \cong .0000438$. (c) There is strong evidence to reject $H_0: \beta = 0$, and conclude that as the number of jet skis in use increases, the number of fatalities also increases.

14.9 (a) The plot shows a strong positive linear relationship. (b) β (the slope) is this rate; the estimate is listed as the coefficient of 'year' : 9.31868. (c) 11 d.f.; $t^* = 2.201$; $9.31868 \pm (2.201)(0.3099) = 8.6366$ to 10.0008.

14.11 (a) Powerboat registrations is explanatory. (b) The plot shows a moderately strong positive liner relationship; there are no clear outliers or strongly influential points. (c) $r^2 = 88.6\%$ indicates that much, but not all, of the variation in manatee deaths is explained by powerboat registrations. (d) β is the number of additional manatee deaths we can expect when there are 1000 additional powerboat registrations. Using a $t(12)$ distribution: $0.12486 + (1.782)(0.01290) = 0.1019$ to 0.1478. (e) $\hat{y} = 45.972$ (about 46 manatee deaths per year). (f) Use the confidence interval: 41.49 to 50.46.

14.13 (a) The stemplot does not show any *major* asymmetry, and has no particular outliers. (b) The plot does not suggest a nonlinear relationship. There is *some* indication that there may be less variation at the high and low ends of the plot, but nothing too strong—there are too few observations to make any judgments about that.

14.15 (b) $r^2 = 0.324$; $r = \sqrt{0.324} = 0.569$—use the *positive* square root. The regression of overseas returns on U.S. returns explains about 1/3 (32.4%) of the variation in overseas returns; the relationship is positive. There is one outlier (in 1986) and one potentially influential point (in 1974). (c) H_0: $\beta = 0$ vs. H_a:$\beta = 0$; $t = 3.09$, and $P = 0.006$ (so we reject H_0). (d) $\hat{y} = 21.037$ percent. (e) Use the prediction interval: -21.97 to 64.04 percent. In practice, this is of little value—the interval includes everything from a 20% loss to a 60% gain.

14.17 (a) It appears that the variation about the line is greater for larger values of x—on the left side of the plot, the residuals are less spread out. (b) Round residuals to whole numbers first.

14.19 (a) $\hat{y} = 26.3320 + (0.687747)(89) = 87.5415$, residual $= 6.4585$. (b) Σ residual$^2 = 356.885$; $s = \sqrt{35.6885} = 5.974$.

14.21 $t^* = 2.145$ from a $t(14)$ distribution: $1.0892 \pm (2.145)(0.1389) = 0.7913$ to 1.3871.

14.23 (a) The heavy fish does not appear to be out of place on the width vs. length plot. (b) $\hat{y} = -0.8831 + 2.297518x$. (c) Based on Minitab output: $\hat{y} = 6.5549$; confidence interval is $6.5549 \pm 0.1294 = 6.4255$ to 6.6842. There are no apparent gross violations, so inference should be fairly safe.

Chapter 15

15.1 (a) The $F(2, 25)$ curve begins at $(0, 1)$ and decreases, approaching the x-axis asymptotically. (b) The $F(5, 50)$ curve begins near the origin, rises to a peak and then decreases, approaching the x-axis asymptotically. (c) $F(12, 100)$ reaches a higher peak, which is slightly to the right of the peak for $F(5, 50)$. (e) As the sample sizes increase, the F-distribution curve becomes more symmetric and more like a normal distribution.

15.3 (a) Significant at 5%, but not at 1%. (b) Between 0.02 and 0.05 ($P = 0.0482$).

15.5 (a) H_0: $\sigma_{\text{skilled}} = \sigma_{\text{novice}}$ vs. H_a: $\sigma_s \neq \sigma_n$. (b) $F = 4.007$; P-value is between 0.05 and 0.10($P = 0.0574$).

15.7 $F = 2.242$; the $F(19, 9)$ distribution doesn't show up in the tables, but by comparing to the $F(20, 9)$ distribution we see that the P-value is greater than 0.20($P = 0.2152$). The difference is not significant.

15.9 (a) The distribution for perch is slightly higher than the other two, and has an extreme high outlier (20.9); the bream distribution has a mild low outlier (12.0). Otherwise there is no strong skewness (b) Bream: 12.0, 13.6, 14.1, 14.9, 15.5; Perch: 13.2, 15.0, 15.55, 16.675, 20.9, Roach: 13.3, 13.925, 14.65, 15.275, 16.1. The most important difference seems to be that perch are larger than the other two fish. It also appears that (typically) bream *may* be slightly smaller than roach.

15.11   (a) Mean yields: 131.03, 143.15, 146.23, 143.07, 134.8. The mean yields first increase with plant density, then decrease; the greatest yield occurs at or around 20,000 plants per acre. This is also reflected (to a lesser extent) in the stemplots. (b) $H_0: \mu_1 = \mu_2 = \mu_3 = \mu_4 = \mu_5$ (all plant densities give the same mean yield per acre) vs. H_a: not all means are the same. (c) $F = 0.50$ and $P = 0.736$. The differences are not significant. (d) The sample sizes were small, which means there is a lot of potential variation in the outcome.

15.13   (a) I, the number of populations, is 5; the sample sizes are $n_1 = 4$, $n_2 = 4$, $n_3 = 4$, $n_4 = 3$, and $n_5 = 2$; the total sample size is $N = 17$. (b) numerator ("factor"): $I - 1 = 4$, denominator ("error"): $N - I = 12$. (c) Since $F < 2.48$, the smallest critical value for an $F(4, 12)$ distribution in Table D, we conclude that $P > 0.100$.

15.15   Yes.

15.17   No answer required.

15.19   (a) $H_0: \mu_{r1} = \mu_{r2} = \mu_{r3}$ (all class rank means are same) vs. H_a: not all means are the same. (b) 2 and 253 (for all three tests). (c) Yes $\frac{10.8}{10.5} = 1.03$, $\frac{1.31}{1.17} = 1.12$, and $\frac{55}{40} = 1.375$. Comparing to $F(2, 200)$ critical values, we find $P_{\text{rank}} > 0.100$, P_{sem} is between 0.025 and 0.0100, and $P_{\text{grade}} < 0.001$. (d) Mean high school class rank varies little between the groups. Regarding the other two variables, there appears to be little difference between the CS and Sci/Eng majors. However, "semesters of HS math" and "average grade in HS math" both show a significant difference between CS/Sci/Eng majors and those in the "Other" category: on the average, the first two groups had about one half-semester more math, and had grades about 0.25 higher.

5.21 $\bar{x} = \frac{1}{110}[(35)(14.131) + (55)(15.747) + (20)(14.605)] = 15.025$. MSG $= \frac{1}{2}\left[(35)(14.131 - 15.025)^2 + (55)(15.747 - 15.025)^2 + (20)(14.605 - 15.025)^2\right] = 30.086$. $F = \frac{30.8086}{1.003} = 29.996$ — reasonably close to Minitab's output.

5.23   Use $t^* = 1.782$ from a $t(12)$ distribution: 130.81 to 161.05.

15.25   (a) Populations: nonsmokers, moderate smokers, heavy smokers; response variable: hours of sleep per night. $1 = 3$, $n_1 = n_1 = n_3 = 200$, $N = 600$; 2 and 597 d.f. (b) Populations: different concrete mixtures; response variable: strength. $I = 5$, $n_1 = \cdots = n_5 = 6$, and $N = 30$; 4 and 25 d.f. (c) Populations: teaching methods; response variable: test scores: $I = 5$, $n_1 = n_2 = n_3 = 10$, $n_4 = 12$, and $N = 42$; 3 and 38 d.f.

15.27 Side-by-side stemplots show that the Jaguar XJ12 and the Rolls Royce Silver Spur are again low outliers, so we omit them. The Mercedes-Benz S420 and S500 are also somewhat low among large cars. However, they are not as extreme as the other two, so one might decide to keep them in.

The mean and standard deviation for each size are (respectively) 28.240 and 3.345, 26.316 and 2.689, and 25.077 and 2.753 (or 26.091 and 1.300 if the two Mercedes are omitted).

Analysis of variance is significant at the 1% level (or the 5% level, when the Mercedes are omitted). The confidence intervals show some overlap, but suggest a conclusion similar to that for city mileage—the most important difference is that compact cars have better average mileage than the other two types.

15.29 $\bar{x} = \frac{128.5}{16} = 8.031$; MSG $= 33.55$; MSE $= 2.78$: $F = 12.08$. Table D places the P-value at less than 0.001; software gives $P = 0.0006$.

Page numbers in *italics* indicate figures.

TABLE B | Random digits

Line								
101	19223	95034	05756	28713	96409	12531	42544	82853
102	73676	47150	99400	01927	27754	42648	82425	36290
103	45467	71709	77558	00095	32863	29485	82226	90056
104	52711	38889	93074	60227	40011	85848	48767	52573
105	95592	94007	69971	91481	60779	53791	17297	59335
106	68417	35013	15529	72765	85089	57067	50211	47487
107	82739	57890	20807	47511	81676	55300	94383	14893
108	60940	72024	17868	24943	61790	90656	87964	18883
109	36009	19365	15412	39638	85453	46816	83485	41979
110	38448	48789	18338	24697	39364	42006	76688	08708
111	81486	69487	60513	09297	00412	71238	27649	39950
112	59636	88804	04634	71197	19352	73089	84898	45785
113	62568	70206	40325	03699	71080	22553	11486	11776
114	45149	32992	75730	66280	03819	56202	02938	70915
115	61041	77684	94322	24709	73698	14526	31893	32592
116	14459	26056	31424	80371	65103	62253	50490	61181
117	38167	98532	62183	70632	23417	26185	41448	75532
118	73190	32533	04470	29669	84407	90785	65956	86382
119	95857	07118	87664	92099	58806	66979	98624	84826
120	35476	55972	39421	65850	04266	35435	43742	11937
121	71487	09984	29077	14863	61683	47052	62224	51025
122	13873	81598	95052	90908	73592	75186	87136	95761
123	54580	81507	27102	56027	55892	33063	41842	81868
124	71035	09001	43367	49497	72719	96758	27611	91596
125	96746	12149	37823	71868	18442	35119	62103	39244
126	96927	19931	36809	74192	77567	88741	48409	41903
127	43909	99477	25330	64359	40085	16925	85117	36071
128	15689	14227	06565	14374	13352	49367	81982	87209
129	36759	58984	68288	22913	18638	54303	00795	08727
130	69051	64817	87174	09517	84534	06489	87201	97245
131	05007	16632	81194	14873	04197	85576	45195	96565
132	68732	55259	84292	08796	43165	93739	31685	97150
133	45740	41807	65561	33302	07051	93623	18132	09547
134	27816	78416	18329	21337	35213	37741	04312	68508
135	66925	55658	39100	78458	11206	19876	87151	31260
136	08421	44753	77377	28744	75592	08563	79140	92454
137	53645	66812	61421	47836	12609	15373	98481	14592
138	66831	68908	40772	21558	47781	33586	79177	06928
139	55588	99404	70708	41098	43563	56934	48394	51719
140	12975	13258	13048	45144	72321	81940	00360	02428
141	96767	35964	23822	96012	94591	65194	50842	53372
142	72829	50232	97892	63408	77919	44575	24870	04178
143	88565	42628	17797	49376	61762	16953	88604	12724
144	62964	88145	83083	69453	46109	59505	69680	00900
145	19687	12633	57857	95806	09931	02150	43163	58636
146	37609	59057	66967	83401	60705	02384	90597	93600
147	54973	86278	88737	74351	47500	84552	19909	67181
148	00694	05977	19664	65441	20903	62371	22725	53340
149	71546	05233	53946	68743	72460	27601	45403	88692
150	07511	88915	41267	16853	84569	79367	32337	03316

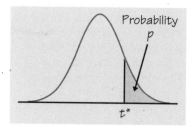

Probability
p

Table entry for p and C is the critical value t^* with probability p lying to its right and probability C lying between $-t^*$ and t^*.

t^*

TABLE C | t distribution critical values

df	\multicolumn{12}{c}{Upper tail probability p}
	.25	.20	.15	.10	.05	.025	.02	.01	.005	.0025	.001	.0005
1	1.000	1.376	1.963	3.078	6.314	12.71	15.89	31.82	63.66	127.3	318.3	636.6
2	0.816	1.061	1.386	1.886	2.920	4.303	4.849	6.965	9.925	14.09	22.33	31.60
3	0.765	0.978	1.250	1.638	2.353	3.182	3.482	4.541	5.841	7.453	10.21	12.92
4	0.741	0.941	1.190	1.533	2.132	2.776	2.999	3.747	4.604	5.598	7.173	8.610
5	0.727	0.920	1.156	1.476	2.015	2.571	2.757	3.365	4.032	4.773	5.893	6.869
6	0.718	0.906	1.134	1.440	1.943	2.447	2.612	3.143	3.707	4.317	5.208	5.959
7	0.711	0.896	1.119	1.415	1.895	2.365	2.517	2.998	3.499	4.029	4.785	5.408
8	0.706	0.889	1.108	1.397	1.860	2.306	2.449	2.896	3.355	3.833	4.501	5.041
9	0.703	0.883	1.100	1.383	1.833	2.262	2.398	2.821	3.250	3.690	4.297	4.781
10	0.700	0.879	1.093	1.372	1.812	2.228	2.359	2.764	3.169	3.581	4.144	4.587
11	0.697	0.876	1.088	1.363	1.796	2.201	2.328	2.718	3.106	3.497	4.025	4.437
12	0.695	0.873	1.083	1.356	1.782	2.179	2.303	2.681	3.055	3.428	3.930	4.318
13	0.694	0.870	1.079	1.350	1.771	2.160	2.282	2.650	3.012	3.372	3.852	4.221
14	0.692	0.868	1.076	1.345	1.761	2.145	2.264	2.624	2.977	3.326	3.787	4.140
15	0.691	0.866	1.074	1.341	1.753	2.131	2.249	2.602	2.947	3.286	3.733	4.073
16	0.690	0.865	1.071	1.337	1.746	2.120	2.235	2.583	2.921	3.252	3.686	4.015
17	0.689	0.863	1.069	1.333	1.740	2.110	2.224	2.567	2.898	3.222	3.646	3.965
18	0.688	0.862	1.067	1.330	1.734	2.101	2.214	2.552	2.878	3.197	3.611	3.922
19	0.688	0.861	1.066	1.328	1.729	2.093	2.205	2.539	2.861	3.174	3.579	3.883
20	0.687	0.860	1.064	1.325	1.725	2.086	2.197	2.528	2.845	3.153	3.552	3.850
21	0.686	0.859	1.063	1.323	1.721	2.080	2.189	2.518	2.831	3.135	3.527	3.819
22	0.686	0.858	1.061	1.321	1.717	2.074	2.183	2.508	2.819	3.119	3.505	3.792
23	0.685	0.858	1.060	1.319	1.714	2.069	2.177	2.500	2.807	3.104	3.485	3.768
24	0.685	0.857	1.059	1.318	1.711	2.064	2.172	2.492	2.797	3.091	3.467	3.745
25	0.684	0.856	1.058	1.316	1.708	2.060	2.167	2.485	2.787	3.078	3.450	3.725
26	0.684	0.856	1.058	1.315	1.706	2.056	2.162	2.479	2.779	3.067	3.435	3.707
27	0.684	0.855	1.057	1.314	1.703	2.052	2.158	2.473	2.771	3.057	3.421	3.690
28	0.683	0.855	1.056	1.313	1.701	2.048	2.154	2.467	2.763	3.047	3.408	3.674
29	0.683	0.854	1.055	1.311	1.699	2.045	2.150	2.462	2.756	3.038	3.396	3.659
30	0.683	0.854	1.055	1.310	1.697	2.042	2.147	2.457	2.750	3.030	3.385	3.646
40	0.681	0.851	1.050	1.303	1.684	2.021	2.123	2.423	2.704	2.971	3.307	3.551
50	0.679	0.849	1.047	1.299	1.676	2.009	2.109	2.403	2.678	2.937	3.261	3.496
60	0.679	0.848	1.045	1.296	1.671	2.000	2.099	2.390	2.660	2.915	3.232	3.460
80	0.678	0.846	1.043	1.292	1.664	1.990	2.088	2.374	2.639	2.887	3.195	3.416
100	0.677	0.845	1.042	1.290	1.660	1.984	2.081	2.364	2.626	2.871	3.174	3.390·
1000	0.675	0.842	1.037	1.282	1.646	1.962	2.056	2.330	2.581	2.813	3.098	3.300
z^*	0.674	0.841	1.036	1.282	1.645	1.960	2.054	2.326	2.576	2.807	3.091	3.291
	50%	60%	70%	80%	90%	95%	96%	98%	99%	99.5%	99.8%	99.9%
	\multicolumn{12}{c}{Confidence level C}